APPLICATION OF THERMODYNAMICS TO BIOLOGICAL AND MATERIALS SCIENCE

Edited by **Tadashi Mizutani**

Application of Thermodynamics to Biological and Materials Science
Edited by Tadashi Mizutani

Published by InTech
Janeza Trdine 9, 51000 Rijeka, Croatia

Publishing Process Manager Ana Nikolic

Technical Editor Teodora Smiljanic

Cover Designer InTech Design Team

Additional hard copies can be obtained from orders@intechopen.com

Application of Thermodynamics to Biological and Materials Science,
Edited by Tadashi Mizutani
p. cm.
ISBN 978-953-307-980-6

For EU product safety concerns:
IN TECH d.o.o., Prolaz Marije Krucifikse Kozulić 3, 51000 Rijeka, Croatia,
info@intechopen.com or visit our website at intechopen.com.

Meet the editor

Tadashi Mizutani is a professor of Biofunctional Chemistry, Faculty of Science and Engineering at Doshisha University, Japan. He was born in Osaka, Japan in 1957. He received a Doctor of Engineering degree from Kyoto University. He joined Toyota Central Research Laboratories, Inc., in 1986. After research and education in Tottori University and Kyoto University, he moved to Doshisha University in 2003. He has conducted experimental research on synthetic receptors, synthesis and supramolecular chemistry of tetrapyrrolic dyes, synthesis of clay minerals, nanoscience of organic-inorganic interfacial structures, and organic-inorganic hybrid materials. He has published over 100 papers in these fields including reviews and book contributions.

Contents

Preface

Studies on efficiency of thermal machines in the nineteenth century lead to the great discovery of entropy and the second law of thermodynamics. Classical and quantum statistical mechanics then emerged and thermodynamics played an important role in bridging between the properties of microscopic particles such as molecules and the properties of the macroscopic objects. Therefore, thermodynamics is a powerful tool for all scientists/engineers working in the field of biological science, chemistry, and materials science. Thermodynamics is more powerful when it covers irreversible processes and non-equilibrium systems, because important biological functions and materials functions arise from the non-equilibrium dynamic irreversible behaviour.

The first section of the book treats various applications of thermodynamics to biological studies. Recent progress in biology and molecular biology allowed us to visualise the structures of complex macromolecules. By using thermodynamic analysis, we can understand molecular mechanisms of a number of biological functions such as enzymatic catalysis, signal transduction, and gene duplication. In particular, the behaviours of solvents and electrolytes and their important contributions to the equilibria and kinetics are difficult to clarify by use of structural analysis, while the thermodynamic analysis is a powerful tool for quantitative evaluation. Protein structure, ligand binding to proteins, nucleic acid conformation/binding/reactions are described in detail. Cells and organs are also subjects of thermodynamic analysis, and cancer cell activity and the function of the heart are studied by use of thermodynamics. Structure and dynamics of interfaces, mesomechanics of biological membranes, and lyotropic liquid crystals of biological importance are discussed.

The topics of the second section are related to materials science and technology. Gas absorption and film formation on the solid surface are studied by a calorimetric equipment and thermodynamic analysis. Chemical equilibria and fluid phase equilibria are discussed. In the fields of ceramics and metallurgy, equilibria and phases of ceramics and metal alloys are described. Extended irreversible thermodynamics was applied to analyse the non-equilibrium behaviour of viscodielectric materials.

All these chapters demonstrate that thermodynamics is a useful tool to analyse biological functions, materials properties, and the process to fabricate materials. Similarities between biological functions and materials functions are obvious when viewed from the thermodynamic point. Readers can see how useful thermodynamics is in biological science, materials science and the interdisciplinary research.

Tadashi Mizutani
Doshisha University, Kyoto
Japan

Part 1

Application of Thermodynamics to Biology and Medicine

1

Thermodynamics of Protein Structure Formation and Function

Dan W. Urry
Bioelastics Inc., Vestavia Hills, Alabama
USA

1. Introduction

1.1 The birth of thermodynamics with the development of the steam-powered heat engine

Thermodynamics was born of the need to improve efficiency of the steam-powered heat engine in order that flooded salt mines of England could become more productive. The *water-to-vapor phase transition* provides the physical property whereby the steam-powered heat engine functions. Heat flows into the engine at the 100°C of the phase transition to effect a dramatic volume expansion. *For the steam-powered heat engine, heating causes expansion to perform mechanical work*. Principal contributors to the initial development of thermodynamics were Nicolas Léonard Sadi Carnot (1824), French physicist and military engineer who died of cholera in 1832 at the age of 36 and William Thomson (Lord Kelvin), a physicist and engineer of the University of Glasgow, whose contribution was in the period of 1840 to 1855 (Smith, 1977).

Looking back at this remarkable development, Prigogine and Stengers (1984a) state, under the section heading of "Heat, the Rival of Gravitation" that "Out of all this common knowledge, nineteenth-century science concentrated on the single fact that combustion produces heat and that heat may lead to an increase in volume; as a result, combustion produces work. Fire leads, therefore, to a new kind of machine, the heat engine, the technological innovation on which industrial society was founded." Heating water at 100°C converts water to steam, a phase transition, to an increase in disorder (in entropy). Perhaps Lord Kelvin's statement of the Second Law of Thermodynamics is most relevant to our concerns, which is "It is impossible to convert heat completely into work in a cyclic process." Greater efficiencies in the conversion of heat into work become possible when heat is poured into a system at the temperature of a transition. Biology utilizes a unique and unfailing two-component phase transition of protein-in-water, and biology does so with a particularly empowering twist made possible by the accuracy and diversity of its protein sequences.

1.2 The aqueous protein-based heat engine of biology

The heat engine of biology comprises a two-component system of protein-in-water. Heating the fully hydrated (soluble) protein effects a phase separation of hydrophobic association (an association of oil-like side chains) that results in contraction. As depicted in Figure 1A, a model protein of the repeating pentamer sequence, (glycyl-valyl-glycyl-valyl-prolyl)$_{251}$, in water (cross-linked by γ-irradiation to form a transparent elastic-contractile sheet) is swollen below the temperature of the transition and contracts on heating to raise the temperature from below to above the that of the phase transition. As seen in Fig. 1B, on heating the strip becomes transiently opaque, while contracting to lift a weight in the performance of mechanical work. *For the protein-in-water heat engine of biology, heating causes contraction to perform mechanical work*.

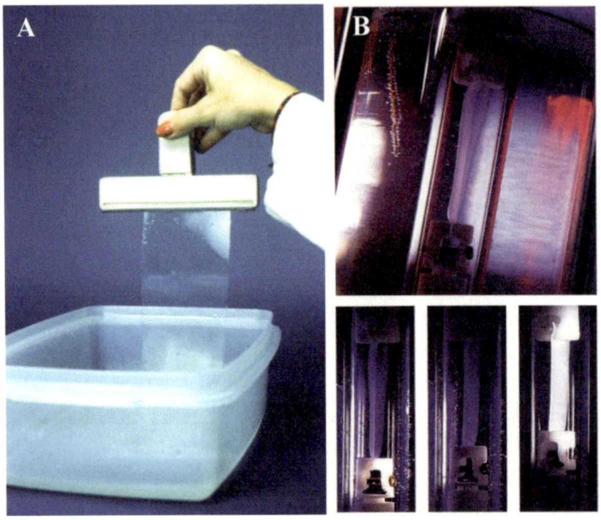

Fig. 1. An aqueous protein-based heat engine of biology, a water swollen sheet and a contracting strip of the cross-linked $(GVGVP)_{251}$, which is the basic elastic-contractile model protein of our study
A. Water-swollen transparent sheet below the temperature of the onset of the phase transition.
B. Upper: Aqueous chamber at a tilt containing a thermocouple and a strip, the heat engine, stretched by an attached weight. Lower: As the temperature is raised through that of the phase transition, the protein in water heat engine performs the work of lifting a weight by contraction. From Urry, 1995 with permission of Ann States Photography.

For warm-blooded animals, however, temperatures change very little. Importantly in these cases, the protein-in-water heat engine does not require heating to raise the temperature from below to above the temperature of the reversible phase separation of hydrophobic association in water to drive contraction. Instead contraction by hydrophobic association occurs by lowering the transition temperature from above to below body temperature, as attached biological functional groups are converted to their more hydrophobic states. The transition temperature is lowered by means of chemical or electrochemical energy inputs that convert a functional group from a more-polar to a more-hydrophobic state, such as occurs on charge neutralization or otherwise removal of charge. In mammals, when the temperature of the phase separation is lowered from above to below 37°C, contraction occurs as low entropy hydrophobic hydration becomes higher entropy bulk water (See section **9**: Summarizing Comments).

In your author's view, only when this increase in entropy (of pentagonal rings of hydrophobic hydration becoming less-ordered bulk water) is explicitly taken into consideration, can treatments of biological energy conversion involving changes in hydrophobic association in water be consistent with the Second Law of Thermodynamics.

That this performance of work, seen on charge neutralization, still represents an underlying protein-based heat engine is easily demonstrated. Here we note a family of model protein compositions that is considered in more detail below in Section **6**. At pH 7.5 in phosphate buffered saline, the glutamic acid (E, Glu) residue in **Model protein i**, (GVGVP GVGVP GEGVP GVGVP GVGVP GVGVP)$_{36}$GVGVP, is ionized as the carboxylate (-COO$^-$). This designed ECMP contracts when the temperature is raised from 55 to 70°C. For **Model protein i** lowering the pH to 3 forms the uncharged carboxyl (-COOH) and under this circumstance contraction occurs on raising the temperature from 15 to 30°C. Thus, at pH 7.5 **Model protein i** is a protein-in-water heat engine that contracts with a transition centered near 60°C, and at pH 3 **Model protein i** is a protein-in-water heat engine that contracts with a transition centered between 20 and 25°C. Thus, at pH 7.5 **Model protein i** *performs thermo-mechanical transduction* at elevated temperature, and at pH 3 **Model protein i** *performs thermo-mechanical transduction* below physiological temperature.

Also, **Model protein i**, at physiological temperature (37°C) and physiological pH, dissolves in water or occurs as a swollen cross-linked matrix. At 37°C, on lowering the pH to 3 the dissolved solution phase separates by hydrophobic association and the swollen cross-linked matrix contracts by hydrophobic association, with release of water, *to perform chemo-mechanical transduction.* Numerous functional groups of biology, attached to designed ECMP, drive contraction on conversion from their more polar state to their more hydrophobic state. Neutralization of charge results in formation of more hydrophobic hydration (See Figs. 10C and 12), with a negative $\delta\Delta H$ and a larger positive $\delta[-T\Delta S]$ (See Eqn. 4 of section **6.1.1** and associated discussion). This requires that the phase transition, where $\Delta H_t = T_t\Delta S_t$, occurs at a lower temperature. This ΔT_t-mechanism of energy conversion derives from input energies that shift the onset temperature, T_t, of phase transitions. The T_t-based Hydrophobicity Scale, of all amino acid residues in their different functional states (as applicable) and of additional functional groups, allows for the phenomenological design of ECMP capable of performing diverse free energy transductions (Urry, 2006a).

Experimental evaluations - 1) of the change in Gibbs free energy for hydrophobic association, ΔG_{HA}, to obtain a ΔG_{HA}-based Hydrophobicity Scale (Urry, 2004), 2) of an apolar-polar repulsive free energy of hydration, ΔG_{ap}, where charge disrupts hydrophobic hydration, and 3) of the mechanism of protein elasticity – allow insight into protein function, design of ECMP as transductional drug delivery/diseased cell targeting vehicles, and of many other medical and non-medical applications (Urry, 2006a; Urry et al., 2010).

1.3 Biology's inverse temperature transition, the rival of gravitation

Thus, for the biological world we note the Prigogine and Stengers (1984a) assertion that for the industrial world "Heat, the Rival of Gravitation" drives the phase transition of a more-ordered, condensed state of bulk water to the more-disordered, expanded gaseous state of steam to *achieve mechanical work by expansion.* And we extend it here to the biological world and argue that "Heat, the Rival of Gravitation" drives a phase transition to increased protein order by association of hydrophobic (oil-like) groups within and between protein chains to *achieve mechanical work by contraction,* (Urry, 1995; 1997; 2006a; Urry et al, 2010).

Central to understanding this phenomenon is that hydrophobic hydration is low entropy, structured water. Before the protein-in-water transition occurs, structured water arranges as pentagonal rings in association with hydrophobic groups (Stackelberg & Müller, 1951; 1954; Teeter, 1984), as may be seen in Fig. 2. During the phase transition of hydrophobic

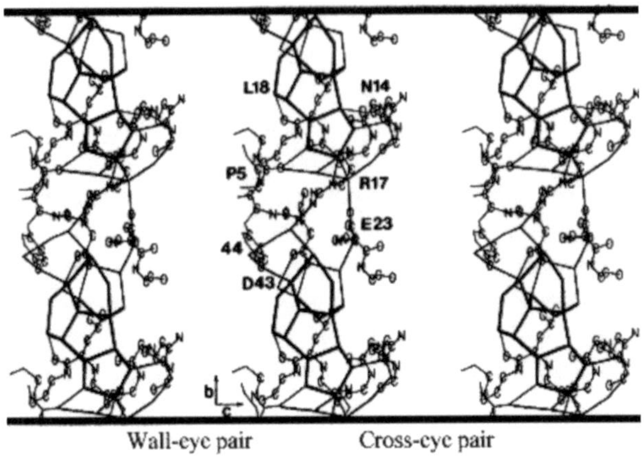

Fig. 2. Stereo views of residual pentagonal rings of hydrophobic hydration in association with hydrophobic moieties of L18 (leucine) and R17 (arginine) residues, after hydrophobic association of the small protein, crambin. From Teeter, 1984 with permission of M. M. Teeter.

association, the pentagonal rings of water of hydrophobic hydration become more-disordered as pentagonal rings of water become higher entropy bulk water (Urry et al., 1997). This decrease in order of water, i.e., increase in entropy, overwhelms in magnitude the increase in order on protein association, i.e., decrease in entropy, as hydrophobic groups of protein associate in the process of contraction (See section **6.1.3**). To emphasize this distinction, the ECMP-based phase transition to greater order of the model protein on raising the temperature is called an *inverse temperature transition*, (ITT). This is protein ordering on heating through the ITT of the ECMP, which ordering can be seen microscopically as the formation of twisted filaments that associate to form fibrils and fibers (Urry, 1992) and can even be seen with cyclic analogues of the model proteins as reversible crystallization on heating (Urry et al. 1978; Cook et al. 1980).

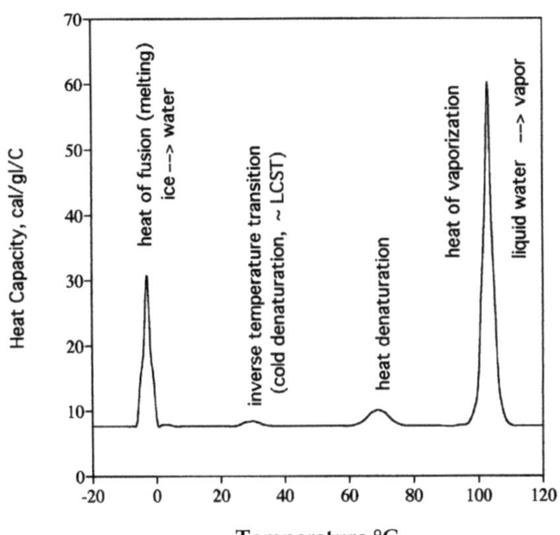

Fig. 3. Representation of endothermic phase transitions of $(GVGVP)_{251}$- in-water. From Figure 5.2 of Urry, 2006a.

Thus, without explicit consideration of water, which goes from being more-ordered to being less-ordered on raising the temperature from below to above the phase transition, the ITT of the protein-in-water heat engine of biology would seem to contradict the Second Law of Thermodynamics. But in fact, the heat driven increase in disorder (in entropy) as pentagonal rings of hydrophobic hydration become less-ordered bulk water is greater than the increase in order (decrease in entropy) as the model protein associates. Thus, in spite of the increase in order of the protein component, the ITT of ECMPs, is endothermic like those of the other transitions of water-to-vapor and ice-to-water, as water goes to a state of higher entropy of Fig. 3.

In summary, the water-to-vapor phase transition results in a dramatic increase in entropy of water and thereby enables the steam engine of the 19th Century Industrial Revolution to perform work by expansion. More profoundly, in your author's view, biology's inverse temperature transition results in a remarkable increase in entropy of water as pentagonal rings of hydrophobic hydration become higher entropy bulk water - whether driven by thermal energy input to raise the temperature through the phase transition or by chemical and other energy inputs that lower the temperature of the phase transition to hydrophobic association from above to below the operating temperature. This enables the diverse protein-based machines that sustain living organisms to perform work by contraction (Urry, 1995, 1997, 2006a; Urry et al., 2010).

1.4 Contrast between the arrow-of-time for the universe and the arrow-of-time for biology

Expressing his high esteem for the Second Law of Thermodynamics Eddington (1958) stated, "The law that entropy always increases - the second law of thermodynamics - holds, I think, the supreme position among the laws of Nature." With entropy measuring the increase in disorder, i.e., the increase in randomness, Eddington put forth the concept of "times arrow," (now commonly referred to as the arrow-of-time) using the argument, "Let us draw an arrow arbitrarily. If as we follow the arrow we find more and more of the random element in the state of the world, then the arrow is pointing towards the future; if the random element decreases the arrow points toward the past. That is the only distinction known to physics. I shall use the phrase 'times arrow' to express this one-way property of time which has no analogue in space. It is a singularly interesting property from a philosophical standpoint."

Considering the arrow-of-time, Toffler (1984), in the Forward to "Order Out of Chaos: Man's New Dialogue with Nature," (Prigogine & Stengers, 1984), addressed the dichotomy presented by biology with, "Imagine the problems introduced by Darwin and his followers! For evolution, far from pointing toward reduced organization and diversity, points in the opposite direction. Evolution proceeds from simple to complex, from 'lower' to 'higher' forms of Life, from undifferentiated to differentiated structures. And, from a human point of view, all is quite optimistic. The (biological) universe gets 'better' organized as it ages, continually advancing to a higher level as time sweeps by." The Toffler Forward set the stage for the Prigogine & Stengers thesis from the discipline of non-equilibrium thermodynamics, under which circumstances less-ordered systems may spontaneously give rise to complex more-ordered systems. Again quoting from Prigogine & Stengers, (1984b), "We can speak of a new coherence, of a mechanism of 'communication' among molecules. But this type of communication can arise only in far-from-equilibrium conditions. It is quite interesting that such communication seems to be the rule in the world of biology. It may in fact be taken as the very basis of the definition of a biological system."

Your author has previously argued (See the Epilogue of Urry, 2006a) that, while the energy required to produce the great macromolecules of biology is very large, the macromolecules themselves are not-so-far-from-equilibrium, due to discarding of 8 kcal/mol-residue with the addition of each residue. Yet repulsive free energies within complementary protein sequences can drive association between them. For further discussion of this issue see section **2**.

1.5 The components of this paper

Our perspective of the thermodynamics of protein structure formation and function unfolds below in seven parts: 1) Description of a key step in the biosynthesis of biomacromolecules, the nucleic acids and proteins, whereby biology achieves order out of chaos. The key step simply exemplifies an energy-fed reversal of biology's otherwise vaunted exception to the universal arrow-of-time. 2) Development of a model system of elastic-contractile model proteins (ECMPs) with which to establish the thermodynamics of hydration and of elasticity in protein function. 3) Phenomenological demonstration of a family of 15 pair-wise energy conversions achievable by designed ECMP capable of a thermally driven inverse temperature transition (ITT) to increased order by hydrophobic association. Thereby numerous inputs of intensive variables of the free energy - mechanical force, pressure, chemical potential, temperature, electrochemical potential, and electromagnetic radiation - act on different functional groups to change the temperature of the ITT. 4) Development of the thermodynamics of protein hydration ($\boldsymbol{\Delta G_{HA}}$ and ΔG_{ap}) and of elasticity (the internal energy, f_E, and entropy, f_S, components of force) as established by designed ECMP. 5) Noting how the Genetic Code (which is common to all characterized life on earth) facilitates protein-based machine evolution, new energy sources and improved machine efficiencies are, thereby, shown to be accessible at no increase in the energy required to produce new and/or more efficient protein machines. 6) The thermodynamics of protein hydration ($\boldsymbol{\Delta G_{HA}}$ and ΔG_{ap}) and of elasticity (f_E and f_S) are shown to be operative in biology's protein-based machines. 7) Application of the thermodynamics of Eyring's Absolute Rate Theory to the essential functions of trans-membrane transport processes of biology allows that the single image of the Gibbs free energy profile for ion passage from one side to the other of a cell membrane through a conduit of protein is sufficient to calculate trans-membrane ion currents as a function of ion activity and trans-membrane potential. This means of analysis, extrapolated to an array of essential biological trans-membrane transport processes, points to a future of a remarkable Eyring legacy, even to the trans-membrane transport processes of the energy factory of the living cell, the mitochondria of the animal kingdom and the chloroplasts of the plant kingdom.

2. How does biology reverse the universal arrow-of-time to achieve its order out of chaos?

In an early consideration relevant to biology's reversal of the universal arrow-of-time, Schrödinger (1944a) reasoned, "… we had to evade the tendency to disorder by 'inventing the molecule', in fact, an unusually large molecule which has to be a masterpiece of highly differentiated order…." Almost a decade later Sanger (Sanger, 1952; Sanger & Thompson, 1953a; 1953b) demonstrated that proteins have specified sequences. *The means whereby biology achieves specified sequences for large chain molecules and the Genetic Code (See section 5) provide the solution as to how biology reverses the universal arrow-of-time, given sufficient energy supply.* Anticipating construction of biological molecules different from anything as yet characterized by 1944, Schrödinger (1944b) further reasoned, "...from all that we have learnt about the structure of living matter, we must be prepared to find it working in a manner that cannot be reduced to the ordinary laws of physics." With remarkable foresight, he then went on to say, "... not on the grounds that there is any 'new force' or what not, directing the behaviour of the single atoms within a living organism, but because the construction is different from anything we have yet tested in the physical laboratory."

Indeed, a protein, in general, is in the words of Schrödinger (1944a) "an unusually large molecule" and always "a masterpiece of highly differentiated order." For a protein is a polymer, a polypeptide, in which each peptide unit may be formed of any one of 20 chemically and structurally diverse amino acid residues. So differentiated is the order that a 100 residue protein with the possibility of any one of twenty amino acid residues in each position gives the probability of a particular sequence as one in 10^{131}.

The key process in biology's reversal of the universal arrow-of-time resides within the synthesis of the magnificent macromolecules of biology, the nucleic acid and protein chain molecules of biology. These polymers exhibit precise sequences of subunits. The repeating units derive from four distinct nucleotides in each of the deoxyribonucleic acids (DNAs) and the ribonucleic acids (RNAs) and from 20 distinct amino acid residues of proteins. *Once these remarkably accurate sequences of diverse amino acids are obtained, three dimensional structure and function follow.* The primary structure, for example the accurate sequence of diverse amino acids of a protein, dictates protein folding and assembly, i.e., dictates three-dimensional structure (Anfinsen, 1973). Also, by the analysis reviewed here, the changes in structure that result in function, arise out of discrete energy inputs acting on biological functional groups attached to protein to bring about changes in hydrophobic association and often coupled with elastic deformation. Accordingly, an understanding, of how biology achieves order out of chaos and reverses the universal arrow-of-time, has as its basis an understanding of the thermodynamics whereby precise protein sequences are obtained, the Genetic Code, and the thermodynamics of protein function. In your author's view, central to understanding the energy conversions that constitute protein function are knowledge of the thermodynamics of hydrophobic hydration, elasticity, and Eyring Rate Theory.

2.1 A common key step whereby biology achieves order out of chaos in the biosynthesis for each of its great macromolecules – DNA, RNA, and protein

During construction of the nucleic acids and proteins of biology, the growing polymers are not-so-far-from-equilibrium. While protein and nucleic acid biosyntheses do require a very large amount of energy, the completed chain is *never*-very-far-from-equilibrium. The addition of each single amino acid residue for protein synthesis or of a triplet nucleotide codon of nucleic acid synthesis per amino acid, consumes ~24 kcal/mol of free energy. Discarding 24 kcal/mol to the environment, on adding each triplet codon to the growing nucleic acid and each amino acid residue to the growing protein chain, reproducibly produces accurate sequences. *A precise sequence dictates the three-dimensional structure of a protein in water for a given state of the functional groups of the sequence and of functional groups otherwise bound to the protein. And changes in state of the associated functional groups result in structural changes that give rise to function.*

In the biosynthesis of protein the activation of each amino acid (AA) and transfer to tRNA by aminoacyl-tRNA synthetase is given as follows: AA + ATP + tRNA = AA-tRNA + AMP +

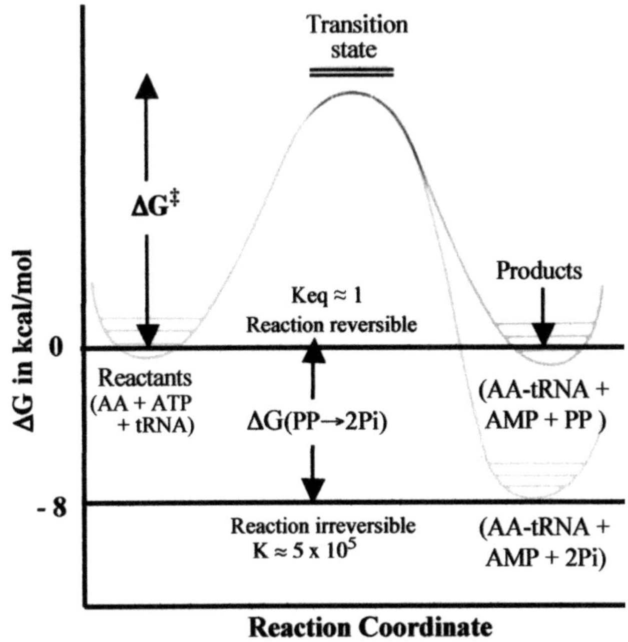

Fig. 4. Free energy profile for the reaction of amino acid (AA) plus ATP plus tRNA to produce the activated amino acid, i.e., AA-tRNA, ready for selective addition to the growing protein chain. The reaction may be seen in two steps: 1) The formation of AA-tRNA + AMP + PP, which is perfectly reversible with an equilibrium constant of one and the ratio of reactant to product of 1:1, 2) The enzymatic breakdown of pyrophosphate, PP → 2Pi + 8 kcal/mol, results in an irreversible overall reaction, i.e., $K \approx 5x10^5$. This very large cost of 800 kcal/mol-residue activation for production of a 100-residue-protein provides the free energy required for the peptide bond formation. There is yet another 1500 kcal-mol-(AA-tRNA) to bring the 100 AA-tRNA molecules out of disarray into alignment (see Eqns. 3b and 3c). Thus, some 2300 kcal/mol-residues added to take 100 amino acids (AA) out of chaos and to form a 100-residue protein of specified sequence.

PP(pyrophosphate), where AA stands for amino acid, ATP for adenosine triphosphate, tRNA for transfer-RNA, AA-tRNA for the activated amino acid as aminoacyl–tRNA energy-wise readied for addition to the growing protein chain, and PP for pyrophosphate. The equilibrium constant for this reaction required for attachment of each amino acid residue to tRNA is of the order of 1, i.e., K ≈ 1. The reactants and products occur at a ratio of approximately one. Due to the presence of an abundance of pyrophosphatase, catalytic breakdown of pyrophosphate immediately ensues, i.e., PP → 2Pi (inorganic phosphate) + 8kcal/mol. At each step of residue activation, a free energy of 8 kcal is released per mole of residue activated. As shown in **Figure 4**, this lowers the free energy of products by 8 kcal/mol. Based on this activation step alone, only one error would be made during the addition of some 500,000 residues. The free energy of pyrophosphate hydrolysis of 8 kcal/mol-residue-activated for addition to the growing chain immediately dissipates into the environment and is no longer associated with the process of chain growth. (For further discussion see Chapter 4 Likelihood of Life's Protein Machines: Extravagant in Construction Yet Efficient in Function of Urry, 2006a).

"Thus, (rather than employing far-from-equilibrium conditions) biology produces its macromolecules by means of an energetically extravagant, step-by-step, methodical march out of chaos" (See the Epilogue of Urry, 2006a).

2.1.1 Replication of DNA by G-C and A-T base pairings

Three steps lead to the biosynthesis of protein. These are: replication, wherein the strand of DNA that encodes protein sequence is duplicated for a daughter cell; transcription, the conversion of DNA into the equivalent sequence of RNA, and translation, the conversion of the ribonucleic acid sequence into the specified protein sequence. Beginning with replication of DNA, i.e.,

$$\text{parent DNA} \rightarrow \text{replication} \rightarrow \text{DNA of daughter cell} \qquad (1)$$

An overall expression for DNA replication may be written as,

$$\text{pATP} + \text{qGTP} + \text{rTTP} + \text{sCTP} = \text{DNA} + (\text{p} + \text{q} + \text{r} + \text{s})\text{PP} \qquad (1a)$$

where A (adenine), G (guanine), T (thymine) and C (cytosine) are the four bases, and the nucleotides - AMP (adenosine monophosphate), GMP (guanosine monophosphate), TMP (thymidine monophosphate), CMP (cytidine monophosphate) are the repeating units added one-by-one to form DNA. This applies to the synthesis of each strand of DNA to duplicate the DNA double helix. For biosynthesis of a 100-residue protein, the sum, $(p + q + r + s) = 300$.

A codon, which is a specific sequence of three bases, in general, encodes for one of the 20 amino acid residues, and there is a redundancy of codons for most amino acids. For example, there are four codons that encode for G (glycine, Gly) and a different four codons encode for V (valine, Val), and yet another set of four codons encode for P (proline, Pro), for A (alanine, Ala), and for L (leucine, Leu). On the other hand only one codon encodes for W (tryptophan, Trp) and six codons encode for R (arginine, Arg). The Genetic Code is a table that lists the codons that encode for each amino acid. As discussed in Section **5** below, the Genetic Code is arranged remarkably well for evolution of diverse and efficient protein-based machines that utilize modulation of inverse temperature transitions for function.

Again reaction (1a) occurs at near equilibrium for each nucleotide addition, but an abundant pyrophosphatase by way of reaction (1b) catalyzes the breakdown of pyrophosphate, PP, into 2 inorganic phosphates, 2Pi, and in the process releases 8 kcal/mol of energy to be dissipated into the environment, including heat that is no longer to be associated with the growing biomacromolecule.

$$(\text{p} + \text{q} + \text{r} + \text{s})\text{PP} \rightarrow \text{pyrophosphatase} \rightarrow 2(\text{p} + \text{q} + \text{r} + \text{s})\text{Pi} + (\text{p} + \text{q} + \text{r} + \text{s}) \times 8 \text{ kcal/mole} \qquad (1b)$$

Thus, when encoding for a 100-residue protein, which requires a sequence of 300 nucleotides, there would be a free energy of (300 x 8) kcal/mol-residue released into the environment, that is, 2400 kcal/mol-300 base daughter cell DNA, which by transcription gives a 300 base strand of RNA, see Eqns. (2), as required for production of a 100-residue protein.

2.1.2 Transcription of DNA to produce RNA by G-C and A-U base pairings

The four bases of RNA are – adenine (A), guanine (G), uracil (U), and cytosine (C) – and the added nucleotide residues are – adenosine monophosphate (AMP), guanosine monophosphate (GMP), uridine monophosphate (UMP), and cytidine monophosphate

(CMP). The reaction constitutes transcribing a strand of deoxyribonucleic acid, DNA, into a strand of RNA. The statement of which may be given as Eqn. (2), i.e.,

$$\text{DNA} \rightarrow \text{transcription} \rightarrow \text{RNA} \tag{2}$$

The stoichiometry of the reaction may be given as,

$$\text{pATP} + \text{qGTP} + \text{rUTP} + \text{sCTP} + \text{DNA} = \text{DNA} + \text{RNA} +$$

$$2(\text{p} + \text{q} + \text{r} + \text{s})\text{Pi} + (\text{p} + \text{q} + \text{r} + \text{s}) \times 8 \text{ kcal/mol} \tag{2a}$$

Again, to encode for a 100-residue protein would mean (300 x 8) kcal/mol, or again 2400 kcal/mol being released to the surrounding solution.

2.1.3 Translation of RNA to produce protein

The translation of an RNA sequence into protein of $\eta = 100$, i.e.,

$$\text{RNA} \rightarrow \text{translation} \rightarrow \text{protein} \tag{3}$$

stated in terms of four reactions: a) The activation of an amino acid residue, AA_i, to its specific $tRNA_i$, discussed above, wherein AA_i, ATP, and $tRNA_i$ react to give AA_i-$tRNA_i$, AMP, and 2Pi with release of 8 kcal/mol-residue, i.e.,

$$\eta\ AA_i + \eta\ tRNA_i + \eta\ ATP \rightarrow \eta\ AA_i\text{-}tRNA_i + \eta\ AMP + 2\eta\ Pi + (\eta \times 8) \text{ kcal/mol-residue.} \tag{3a}$$

Eqn. (3a) represents a selectivity step where the correct amino acid is attached to its appropriate tRNA that contains the correct triplet codon for the amino acid being attached in an activated state. The amino acid selectivity process continues in the following reactions.

$$\eta\ AA_i\text{-}tRNA_i + \eta \text{ ribosome(position 1)} + \eta\ GTP \rightarrow$$

$$\eta\ AA_i\text{-}tRNA_i\text{-ribosome(position 1)} + \eta GDP + \eta Pi + (\eta \times 7.5) \text{ kcal/mol-residue,} \tag{3b}$$

transfer to ribosome (position 2)

$$\eta\ AA_i\text{-}tRNA_i\text{-ribosome(position 1)} + \eta \text{ ribosome(position 2)} + \eta GTP \rightarrow$$

$$\eta \text{ ribosome(position 1)} + \eta\ AA_i\text{-}tRNA_i\text{-ribosome(position 2)} + \eta GDP$$

$$+ \eta Pi + (\eta \times 7.5) \text{ kcal/mol-residue} \tag{3c}$$

and finally the activated amino acid, AA_i-$tRNA_i$, bound at ribosome position 2, is added to the growing protein chain in its designated position in the sequence, i.e.,

$$\eta\ AA_i\text{-}tRNA_i\text{-ribosome(position 2)} \rightarrow \eta\ tRNA_i + \eta\ AA_i \text{ in protein} \tag{3d}$$

The cost in terms of Gibbs free energy to add a single amino acid to the growing protein chain is (8 + 2 x 7.5) kcal/mol-residue, and the cost of producing a 100-residue protein would be 2300 kcal/mol-100-residue protein.

As given above, the probability for a precise sequence of a 100-residue protein, with the possibility of one of 20 amino acid residues in each position, i.e., $(1/20)^{100} = 10^{-131}$. When the equilibrium constant is one, i.e., $K = 10^{-\Delta G/2.3RT} = 1$, ΔG is 0, and there is the probability of an equal number of reactants and products. When the probability of a product is one chance in 10^{-131} for the occurrence of the product, one may write that $K = 10^{-\Delta G/2.3RT} = 10^{-131}$, or $\Delta G/2.3RT = 131$. Solving for the Gibbs free energy, $\Delta G = 131 \times 2.3RT = 186$ kcal/mole-100-residue protein. Calculated in this manner the efficiency of the synthesis of the 100-residue protein becomes 186/2300 = 0.08, i.e., an efficiency of the order of some 8%.

As will be noted below, protein-based motors can function at very high efficiencies. The F_1-ATPase (the F_1-motor of ATP synthase acting in reverse) has been calculated as approaching 100% (Kinosita et al., 2000). This has led to the exclamation that Life's protein machines are ***extravagant in construction yet efficient in function*** (See Chapter 4 of Urry, 2006a). (Some of the 1500 kcal/mol pays for a repulsive free energy between hydrophobic and charged groups.)

2.2 Precise primary structure, i.e., sequence, dictates three dimensional structure and function!

As argued above, a high price in terms of Gibbs free energy is paid in order to obtain polymers of precise sequence. Consequences of this severe price for precise sequence are the beautiful functional structures of biology. The more diverse the "side chains" of the repeating sequence, the more diverse are the functional capabilities. This is why the nucleic acids with but four similar repeating nucleotides each with the capacity of base pairing, i.e., A-T and G-C of poly(deoxyribonucleic acid) DNA and U-T and G-C of poly(ribonucleic acid), RNA, are suitable for sequence replication and transcription as considered above in terms of free energy required to produce precise sequences in Eqns. (1) and (2).

At the root of the structuring that becomes a living organism is the primary structure of DNA, the poly(deoxyribonucleic acid). DNA provides the sequence of bases that ultimately specify the precise sequence of protein. Protein sequence utilizes 20 structurally diverse residues that may be broadly classified as aromatic and aliphatic hydrophobic residues, as negatively and positively charged residues, and as neutral residues with non-ionizable polar functional moieties, capable, for example, of hydrogen-bonding. Overlapping with the latter two groups is cysteine with its -SH functional group that is commonly used in disulfide, -S-S-, cross-linking on formation of cystine.

Again, the probability of a precise sequence, with the possibility of one specific residue out of 20 residues in each position of even a relatively small 100-residue protein, becomes $(1/20)^{100} = 10^{-131}$, that is, one out 10^{131} sequences (See Chapter 4 of Urry, 2006a). This truly enormous number of possible sequences allows for an extraordinary number of protein three-dimensional structures with which to perform the diverse work (functions) required to sustain a cell.

2.2.1 Protein performs the work of constructing and maintaining the cell

The precise sequence of a protein, under physiological conditions, dictates the three-dimensional structure of the protein itself and whether it associates with like subunits to form an oligomeric protein comprised of symmetrically related subunits and/or with unlike subunits to form more complex protein structures. A remarkable example is ATP synthase of more than 20 subunits (10a, 2b, 3α, 3β, γ, ε). This rotary protein motor combines ADP (adenosine diphosphate) and Pi (inorganic phosphate) to make 32 of the 36 ATP (adenosine triphosphate) molecules on complete oxidation of a single molecule of glucose to 6 CO_2 plus 6 H_2O. Recall, ATP is the biological energy currency utilized to perform the work that sustains and propagates the living cell.

Assemblies of subunits, such as those of the three-fold rotary F_1-motor of ATP synthase, are dominated by hydrophobic inter-subunit interactions (Privalov, 1990) under the control of temperature and biological functional groups that can occur in two or more functional states. The more polar (e.g., charged) state, disrupts hydrophobic association and the more apolar (the more hydrophobic) favors hydrophobic association, each in a cooperative manner.

2.2.2 Familiar insights into the changes in hydrophobic associations that give rise to function

Insight begins with the familiar adage, "Oil and water don't mix!" Of course, they simply phase separate. But if oil-like and polar (e.g., water-like) groups are constrained to coexist along a polymer chain, they can't phase separate. Instead, the oil-like groups, dispersed along the polymer chain, self-associate by chain folding and by association with other chain segments, and, thereby, separate from water. But once the most favorable, the lowest free energy state, is obtained at a given temperature and pressure, only substantial changes in solvent or such as phosphorylation can change the state.

A related and more interesting adage becomes, "Oil and vinegar don't mix!" The solute of vinegar is principally acetic acid, which can exist in two states, i.e., $CH_3\text{-}COOH \leftrightarrows CH_3\text{-}COO^- + H^+$, the very polar charged state, $CH_3\text{-}COO^-$, and the less-polar (more-hydrophobic) uncharged state, $CH_3\text{-}COOH$. Again as for oil and water, phase separation dominates the mixture of oil and vinegar. When hydrophobic and ionizable groups are forced by sequence to coexist as demonstrated with certain designed ECMP, it has been shown by means of substantial physical characterization of ECMP containing a glutamic acid (E, Glu) residue with the R-group of $-CH_2\text{-}CH_2\text{-}COOH$ that the formation of the more polar state of $-CH_2\text{-}CH_2\text{-}COO^-$ disrupts hydrophobic association (See for example Urry et al., 1997). It will be seen below, using the crystal structure of the closed conformation of the full-length KcsA potassium ion channel (Uysal et al., 2009) that the absence of carboxylate is seen associated with hydrophobic association that opens the channel, whereas the presence of carboxylate is seen associated with hydrophobic dissociation (Urry et al., 2010) and a closed channel. And the pH dependence of the conductance of the KcsA K^+-channel of Thompson et al. (2008) demonstrates conductance to turn off on the titration of glutamic acids to form charged glutamates.

Particularly, when the oil-like and charged groups are constrained to coexist by protein structure, they can be shown to reach out for hydration unperturbed by the other, that is, there is a competition for hydration between hydrophobic and charged residues (See for example Urry et al., 1997). This results in an *apolar-polar repulsive free energy of hydration*, ΔG_{ap}. (See Section **6.2.6** below and Urry, 1992; 1997).

2.2.3 Biological polymers of reproducible precise sequence add a new wrinkle to the *"laws of physics"*

Anticipating construction of biological molecules different from anything as yet characterized at the time, Schrödinger (1944b) further reasoned, "...from all that we have learnt about the structure of living matter, we must be prepared to find it (living matter) working in a manner that cannot be reduced to the ordinary laws of physics." With remarkable foresight, he then went on to say, "... not on the grounds that there is any 'new force' or what not, directing the behaviour of the single atoms within a living organism, but because the construction is different from anything we have yet tested in the physical laboratory." Different constructions arise due to the capacity of biology to synthesize long proteins of precise sequence. This is because near physiological temperature the fundamental activation reaction, essentially independent of amino acid structure, has an equilibrium constant, K, of $10^{5.7}$. This translates into the order of one error in a half a million residue additions. Again, assuming that the twenty different residues possible at each position had an equal probability of being added, there would be 10^{131} different sequences possible for a 100-residue protein. This results in protein constructions that were simply inconceivable prior to the elucidation protein sequences and protein biosynthesis.

Again as Schrödinger (1944a) stated, "... living matter, while not eluding the 'laws of physics' as established up to date, is likely to involve 'other laws of physics' hitherto unknown, which, however, once they have been revealed, will form just as integral a part of this science as the former." As indicated above, the *new wrinkle to the "laws of physics"* derives from *an apolar-polar repulsive free energy of hydration, ΔG_{ap},* that can be seen with the disparate side-chains (e.g., hydrophobic and charged) constrained to coexist along the precise sequence of which a protein chain is comprised.

It has been seen above that the reproducibly-achieved precise protein sequence (with an error as small as of one in one-half million residue additions) is achieved at an extraordinary cost in energy, and as such is consistent with the Second Law of Thermodynamics. It is not yet understood, however, just how the protein biosynthetic apparatus came into existence with which to achieve this protein construction so essential to the existence of life as we understand it.

2.3 Is the construction and maintenance of the biosynthetic apparatus for protein in accordance with the Second Law of Thermodynamics?

The biomacromolecular composition of the biosynthetic apparatus for production of protein requires RNA to specify protein sequence and protein catalysis to transcribe DNA into RNA, to produce tRNA, to attach amino acid (AA_i) to $tRNA_i$, i.e., to produce AA_i-$tRNA_i$, and to catalyze the steps in which the correct AA_i of an AA_i-$tRNA_i$ becomes attached to the correct position in the growing protein chain. The energy required for the latter, some 15 kcal/mol amino acid residue added of Eqns (3b) and (3c) in addition to precise protein sequence also pays for repulsive free energies that occur between disparate residues. DNA, RNA and protein chains of precise sequence are all simultaneously required in the first instance to achieve replication, transcription, translation to protein. How the initial biosynthetic apparatus came into existence is unknown. Once the ribosomal biosynthetic apparatus has been assembled with its accessory enzymes and nucleic acids all available, however, synthesis of protein does not contravene the Second Law of Thermodynamics.

3. A model protein system with which to establish thermodynamics of protein structure formation and function!

3.1 The composition of the basic model protein, $(GVGVP)_n$

Our model protein system, with which to establish thermodynamic elements of protein function, originates from the mammalian elastic protein, elastin, as a repeating pentapeptide sequence, $(GVGVP)_n$ with $n \leq 15$, depending on the species. A polypeptide chain may be represented as $[\text{-NH-CHR-CO-}]_n$ or as $[\text{-CO-NH-CHR-}]_n$ where the side chain (the R-group) of G (Gly, glycine) is the hydrogen atom, -H, the R-group of V (Val, valine) is $-CH(CH_3)_2$, and the R-group of P (Pro, proline) is N_i-CH_2-CH_2-CH_2-C_i^{α}, i.e., three CH_2 groups spanning from the nitrogen atom, N, to the α-carbon of the same residue, i. Therefore, all side chains in $(GVGVP)_n$ are either the hydrophobic aliphatic hydrocarbons or the near neutral hydrogen atom and the only polar group is the recurring dipolar peptide moiety, -CO-NH-. Chemically and biologically synthesized $(GVGVP)_n$, with n ranging up to 200 or more, may be modified with sparse substitution of V by one or more functional groups, such as the carboxyls of glutamic and aspartic acids and the amino function of lysine (K, Lys), and additional biological functional groups such as redox functions, other prosthetic groups, phosphate, etc. Also, V residues may be replaced by the more hydrophobic F (Phe or phenylalanine) systematically to raise the hydrophobicity with the result of increased positive cooperativity giving an increased efficiency of energy conversion. These modified $(GVGVP)_n$ are called *designed* elastic-contractile model proteins (ECMP).

3.2 The molecular structure of the basic model protein, $(GVGVP)_n$

Figure 5A schematically represents the molecular structure of the basic model protein, $(GVGVP)_n$, as a series of VPGV β-turns with G spacers. Extending the spacer G residue to the adjacent V residue α-carbons, the VGV segment allows dynamic torsional oscillations of the intervening two peptide moieties. The damping of the amplitude of these peptide torsional oscillations gives rise to the librational entropy mechanism of protein elasticity (Urry et al., 1982d).

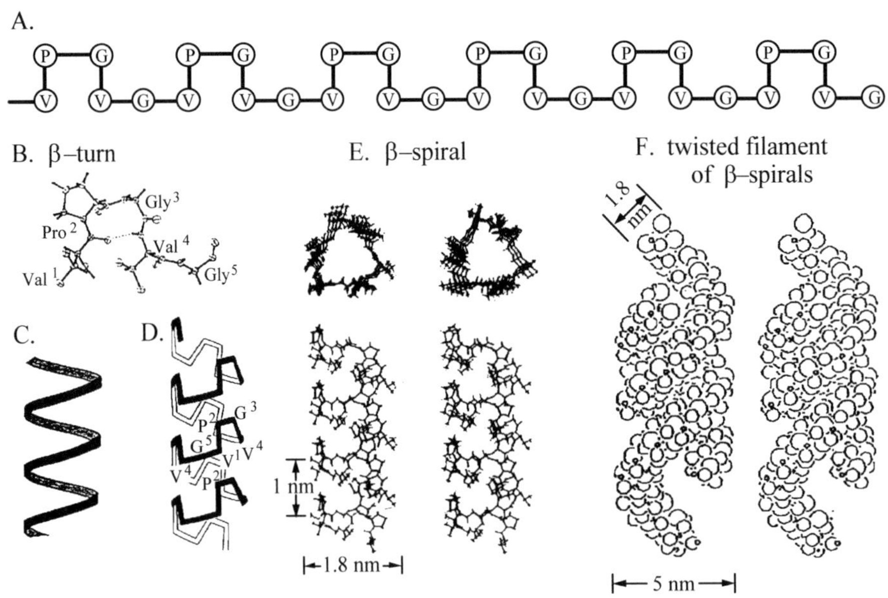

Fig. 5. Molecular structure of the elastic-contractile model protein, $(GVGVP)_n$. The structure is seen with repeating β-turns separated by dynamic suspended segments that wrap-up into associating β-spirals and exhibit simultaneous "near ideal" elasticity and phase transitional behavior from water to associate by hydrophobic interactions.
B. β-turn from the crystal structure of cyclo$(GVGVP)_3$ (Cook et al. 1980) which is the cyclic correlate of the linear β-spiral of D and E (Urry et al., 1981; Venkatachalam, et al. 1981; Venkatachalam and Urry, 1981). F. Adapted from Urry et al., 1982d.

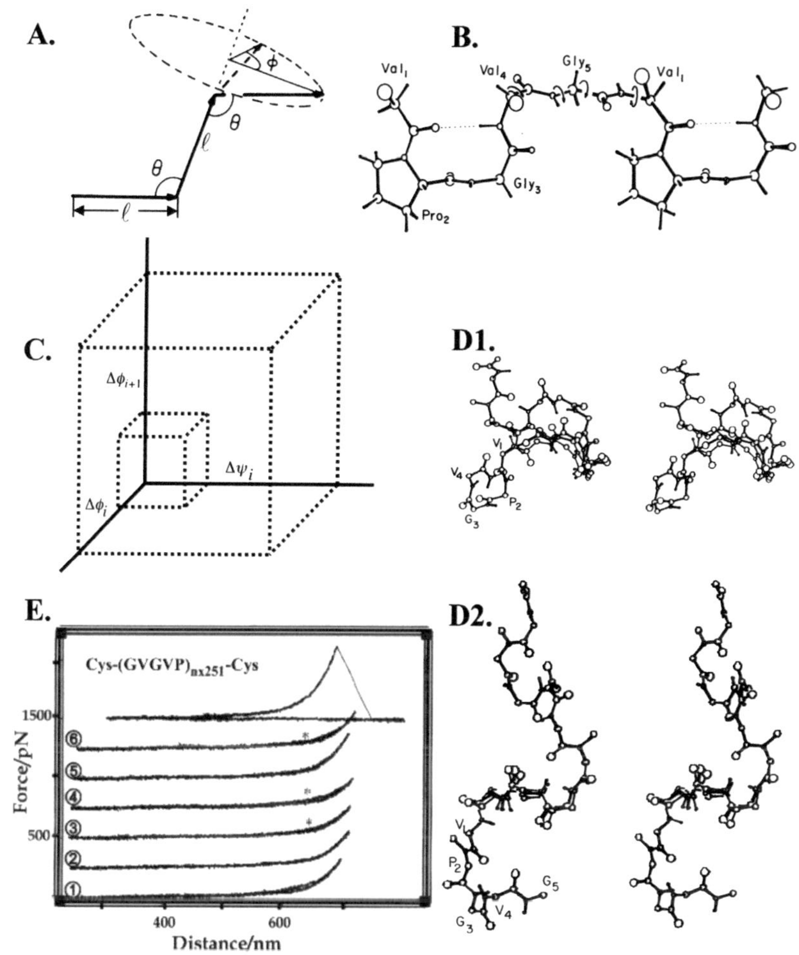

Fig. 6. Elements for understanding the nature of elasticity of the basic elastic-contractile model protein, $(GVGVP)_n$.

A. The structural elements for propagation of a rudimentary chain (Eyring, 1932). Bond length (ℓ), backbone bond angle (θ), torsion, or dihedral, angle (ϕ) due to rotation about bonds (Adapted from Urry, 1982).

B. Unrolled perspective of the β-spiral of $(GVGVP)_n$ showing suspended segments where peptide moieties are free to undergo large amplitude oscillations (peptide librations) involving the paired (ψ_4 & ϕ_5) and ψ_5 & ϕ_1 torsion angles. (From Urry, 1983)

C. Representation of entropy as a volume in configuration space, with axes plotting amplitude of torsion angle oscillations. As the volume increases due to larger torsion angle oscillations, a greater entropy can be calculated. (From Urry et al., 2010).

D1. One turn of the β-spiral with 2.7 β-turns per turn of spiral and an h of 3.5 Å showing the large torsion angle oscillations between the first and second β-turns. D2. On extension to an h of 8.0 Å, note damped oscillations. (From Urry & Venkatachalam, 1983).

E. Single-chain force-extension/relaxation curves, development of force during pulling in the z-direction of an AFM device with scans labelled from the bottom as Adapted from Urry et al., 2002.

The details of the β-turn are seen in **Figure 5B**, as obtained from the crystal structure of cyclo(GVGVP)$_3$ (Cook et al. 1980), which is the cyclic conformational correlate of the linear β-spiral conformation, as shown experimentally and computationally (Urry et al. 1981; Venkatachalam, et al. 1981; Venkatachalam and Urry, 1981). The linear β-spiral conformation is represented in increasing detail in **Figures 5C, D,** and **E** (Urry, 1990; 1991). Based on optical diffraction of negatively stained electron micrographs from incipient aggregates of dilute solutions (Volpin et al., 1976), three β-spirals of (GVGVP)$_n$ are thought to form twisted filaments as represented in **Figure 5F** (Urry et al., 1982d).

3.3 The unique properties of the basic model protein, (GVGVP)$_n$: "Near ideal" elasticity and phase transitional behavior

It is extraordinary that the basic model protein system, (GVGVP)$_n$ in water, simultaneously exhibits "near ideal" elasticity and thermally-elicited phase transitional behavior. To emphasize this unique and useful combination of properties, protein-based polymers based on (GVGVP)$_n$ have been given the descriptive name of elastic-contractile model proteins (ECMPs).

3.3.1 The "near ideal" elasticity of the basic model protein, (GVGVP)$_n$

An understanding of the "near ideal" elasticity of (GVGVP)$_n$ may be gained by discussing the component parts of Fig. 6, above. In particular, the curves of Fig. 6E utilized the basic atomic force microscope (AFM) (Hugel, 2003; Urry et al. 2002). Instead of imaging structures on a surface by rastering in the x- and y-dimensions, the cantilever tip moves in the z-direction with a long chain molecule spanning from the cantilever tip to the substrate surface to give a stress-strain curve that measures single-chain elasticity.

Ideal elasticity occurs when the plot of the force versus relaxation curve exactly overlays the force versus extension curve. Within the sensitivity (the noise level) of the measured stress-strain curves of Fig. 6E, the extension and relaxation traces of curves ② and ⑤ overlap. For curves ② and ⑤, therefore, the energy expended on extension is entirely recovered on relaxation, that is, these curves provide examples of ideal elasticity exhibited by extension and relaxation of a single-chain of Cys-(GVGVP)$_{502}$.-Cys. (Note: The Cys (cysteine, C) residues are present to achieve chemical (sulfhydryl) attachment to the cantilever tip of the AFM and for sulfhydryl attachment to the substrate surface.)

On the other hand curves ①, ③, ④, and ⑥ of Fig. 6E exhibit a higher noise level and in the original data separation is detectable as extension becomes greater than 600 nm. In these cases the extension curve is slightly higher than the relaxation curve, i.e., the cost in energy for extension is greater than the energy recovered on relaxation. Extension curves that occur at higher force levels than the relaxation curves are said to exhibit a hysteresis, which is an energy loss.

As seen in Fig. 4 of Urry et al. 2002 for Cys-(GVGIP)$_{260}$-Cys, the energy expended for extension is several times that recovered on relaxation. This is due in part to the greater hydrophobicity of (GVGIP) than of (GVGVP). The increase in the change in Gibbs free energy for hydrophobic association, ΔG_{HA}, results in a greater propensity for association with non-load bearing chain segments. A higher force on extension is required to disrupt these associations. The greater expenditure of energy to extend and disrupt these hydrophobic associations is not recovered on relaxation.

When the single-chain force-extension studies on Cys-(GVGIP)$_{260}$-Cys occur at very high dilution, however, essentially near ideal elasticity can be obtained. A slight hysteresis of each of the curves, ①, ③, ④, and ⑥ of Fig. 6E, may be due to the chain folding back on itself, as time was allowed at low extension to increase the likelihood of backfolding (Hugel, 2003). As seen in Fig. 5E, the translation along the spiral axis for each complete turn is 1nm, and one complete turn requires three pentamers. Also, note in Fig. 6D1 that it is one turn of spiral, i.e., three pentamers, that is used in the calculation of the damping of torsion angle oscillations on extension by 130% from a value of 1 nm to 2.3 nm. Using the insight of Fig. 6C and the decrease in amplitude of torsion angle oscillation on extension, the change in entropy, ΔS, can be calculated by the equation, $\Delta S = R \ln[\Pi_i \Delta\phi_i^e \Delta\psi_i^e / \Pi_i \Delta\phi_i^r \Delta\psi_i^r]$.

The total elastomeric force, f_T plotted in Figure 6E, is the sum of an entropic component of force, f_S, and an internal energy component of force, f_E, i.e., $f_T = f_S + f_E$. The entropic component of elastic force is calculated as $f_S = -T(\partial S/\partial L)_{VT}$, where ∂S is calculated from the above expression for ΔS and ∂L derives from the 130% extension as used in Fig. 6D2. The sources of f_E derive from the reversible deformation of the angle, θ, and of the bond length, ℓ, both of Fig. 6A. (See section **6.2.10**, below.)

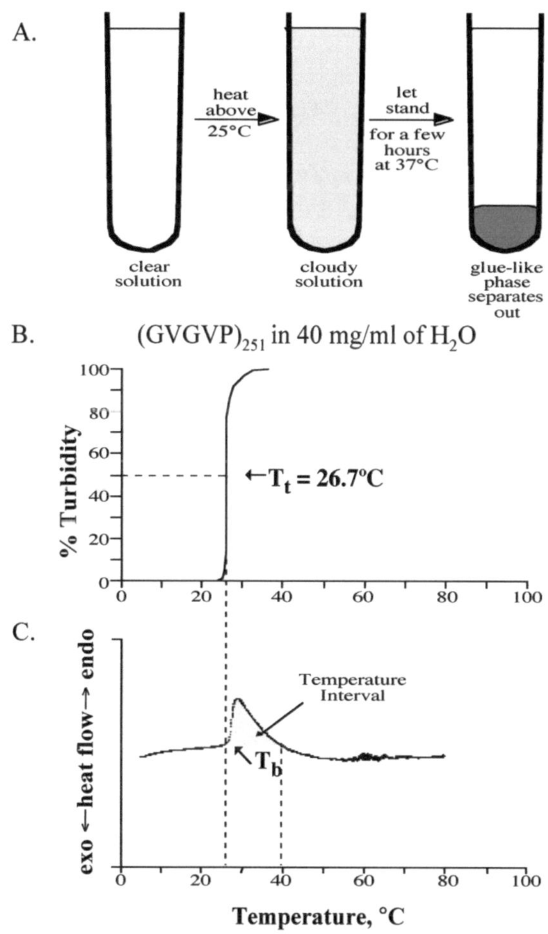

Fig. 7. Characterization of the $(GVGVP)_n$-in-water inverse temperature transition, using temperature dependence of turbidity and differential scanning calorimetry. From Urry, 1997.

3.3.2 The phase transitional behavior of the basic model protein, $(GVGVP)_n$

The basic elastic-contractile model protein, $(GVGVP)_n$, is miscible with water in all proportions below the onset temperature of the inverse temperature transition (ITT), i.e., the solutions are clear below 25°C. As depicted in Fig. 7A for concentrations of less than 400 mg/ml, on raising the temperature above 25°C, the clear aqueous solution becomes cloudy, and on standing phase separation occurs to form a viscoelastic state of 63% water and 37% model protein by weight, which constitutes a one molar concentration of pentamers, (GVGVP), of approximately 400 mg/ml (Urry et al., 1985).

As shown in Fig. 7B, when the phase separation process is followed spectroscopically, the onset of turbidity of a 40 mg/ml solution of $(GVGVP)_{251}$ begins at about 25°C. Turbidity continues on to a maximal value, 100% turbidity. The value of $\mathbf{T_t}$ is taken at 50% turbidity to give in this case a value of **26.7°C**, which value is bold-faced in this case to indicate that it is for the homopolypentapeptide, $(GVGVP)_n$, i.e., for a polypentapeptide for which the mole fraction, f_V, is one. In characterizations below, whenever values are for a general polypentapeptide, (GXGVP) as in poly[f_V(GVGVP), f_X(GXGVP)], the quantities are not bold-faced until the data for several values of f_X have been determined and then extrapolate to f_X = 1. At f_X = 1, i.e., for $(GXGVP)_n$, the values, the thermodynamic quantities for the homopolypentapeptide, are to be bold-faced. All quantities for f_X = 1 are then compared for the development of the $\mathbf{T_t}$- and $\mathbf{\Delta G_{HA}}$-Hydrophobicity Scales. The values of $\mathbf{\Delta G_{HA}}$ are similarly obtained from differential scanning calorimetry (DSC) data.

The DSC curve for $(GVGVP)_{251}$ in 40 mg/ml of water is given in Fig. 7C. The onset of the endothermic transition, indicated as $\mathbf{T_b}$, is essentially the same as that of $\mathbf{T_t}$ of Fig. 7B. While these values tend to be used interchangeably, a distinction is retained in the data of **Table 1**, below. The temperature interval over which the phase transition occurs is approximated as in Fig. 7C. If the scan could be done more slowly the temperature interval would be much narrower, as seen in Urry et al., 1985, where there was no limit on time for completion of the phase separation. The question of scan rate becomes a question of the stability and sensitivity of the DSC equipment, which presents a challenge for the low heats of the inverse temperature transition.

4. Energy conversions of designed elastic-contractile model proteins are those of living organisms!

4.1 Phenomenology of model protein-based free energy transduction.

Phenomenological demonstration of a family of 15 pair-wise energy conversions becomes possible by means of designed ECMP capable of a thermally driven inverse temperature transition to increased model protein order by hydrophobic association. The family of pair-wise energy conversions possible by designed elastic-contractile model proteins are identified by the six intensive variables of the free energy - mechanical force, pressure, chemical potential, temperature, electro-chemical potential, and electromagnetic radiation. As seen in Fig. 1, the basic sequence $(GVGVP)_n$ on γ-irradiation cross-linking of its phase separated state can be formed as elastic sheets that perform *thermo-mechanical transduction*, i.e., "pumping iron," contracting on raising the temperature from below to above that of the phase separation and relaxing on lowering the temperature from above to below that of phase separation.

On sparse replacement of one V residue, every 30 or 50 residues, by a glutamic acid or a lysine residue and cross-linking, the resulting elastic sheet performs *chemo-mechanical transduction*.

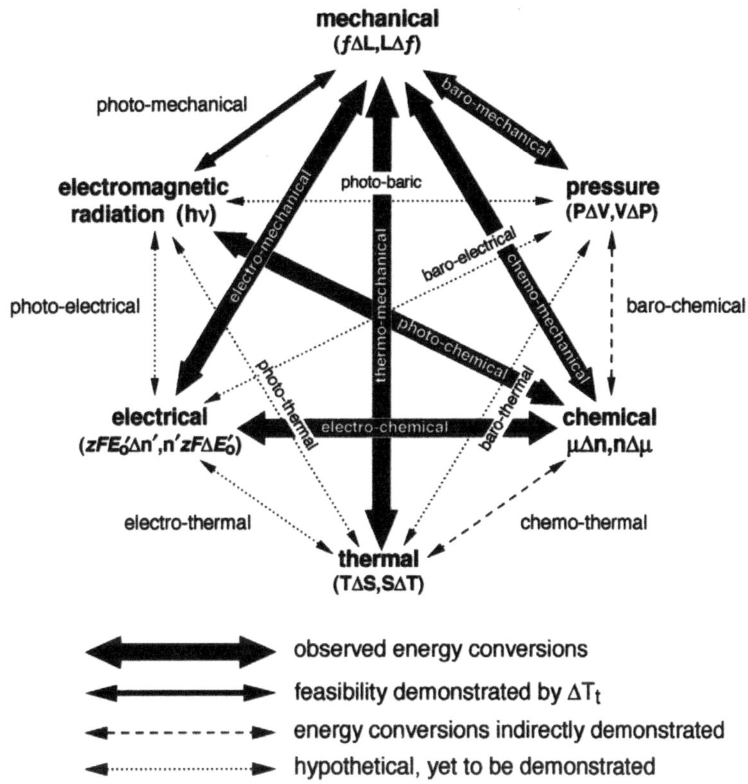

Fig. 8. The phenomenological ΔT_t-mechanism of free energy transduction using designed elastic-contractile model proteins (ECMPs). Application of pressure increases T_t and drives relaxation. Introduction of a chemical energy that neutralizes charge lowers T_t and drives contraction. Raising the temperature from below to above the transition drives contraction. Reduction of a redox function lowers T_t to drive contraction. The absorption of near ultraviolet light by a designed ECMP with attached azobenzene and cinnamide drive a *trans* to *cis* isomerization and raises the value of T_t. From Urry, 1997.

On attaching a redox group to the amino function of a lysine side chain, reduction drives contraction and oxidation effects relaxation in the performance of *electro-mechanical transduction*. Moreover, on replacing two V residues, per ECMP repeating unit, by both an acid/base function and a redox function allows that reduction of the redox function shifts the pK of the acid/base function causing the uptake of proton, thus achieving *electro-chemical transduction*. This becomes the "pumping of protons," e.g., the release of proton on oxidation, in phenomenological analogy to the electron transport chain of the inner mitochondrial membrane where oxidation of redox groups pumps protons into the inner membrane space. The examples noted here demonstrate phenomenology of three (mechanical force, chemical potential, and electrochemical potential) of the six intensive variables of the free energy for which ECMP can be designed to achieve free energy transduction.

Whenever a ΔT_t occurs and the operating temperature lies between the two values, on going from the higher value of T_t to the lower value of T_t, hydrophobic association occurs with the result of driving contraction.

4.2 Five phenomenological axioms for ΔT_t-based free energy transduction by designed model proteins

AXIOM 1: The manner in which a guest amino acid residue, or chemical modification thereof, alters the temperature, T_t, of a hydrophobic folding and/or assembly transition provides a measure of its hydrophobicity. A decrease in T_t represents an increase in hydrophobicity and an increase in T_t represents a decrease in hydrophobicity.

AXIOM 2: Raising the temperature from below to above T_t results in hydrophobic folding and/or assembly and can be used to perform useful mechanical work by contraction.

This represents the phenomenological aqueous elastic-contractile model protein heat engine of biology.

Example: Thermo-mechanical transduction

AXIOM 3: At constant temperature, lowering the value of T_t from above to below an operating temperature, i.e., increasing the hydrophobicity by changing a functional group from its more polar state to its more hydrophobic state results in contraction by hydrophobic folding and/or assembly and can be used to perform useful mechanical work, as in the lifting a weight.

Examples:	Chemo↔mechanical transduction	Electro↔mechanical transduction
	Baro↔mechanical transduction	Photo↔mechanical transduction

AXIOM 4: Any two distinct functional groups each with more and less hydrophobic states and each responsive to different variables can be coupled one to the other by being part of the same hydrophobic folding and assembly domain.

Examples:	Electro↔chemical transduction	Electro↔thermal transduction
	Baro↔electrical transduction	Photo↔voltaic transduction
	Thermo↔chemical transduction	Photo↔thermal transduction
	Baro↔thermal transduction	Baro↔chemical transduction
	Photo↔baric transduction	Photo↔chemical transduction
	Chemo↔chemical transduction	*Electro↔electrical transduction*
	Electromagnetic radiation-1 ↔ Electromagnetic radiation-2 transduction	

The italicized energy conversions represent three additional pair-wise free energy transductions for a total now of eighteen possible pair-wise free energy transductions using the above described ΔT_t-mechanism.

AXIOM 5: The energy conversions of AXIOMS 2, 3 and 4 can be demonstrated to be more efficient when carried out under the influence of more hydrophobic domains. This poising or biasing is observed in titrations by increased positive cooperativity. See Figs. 5.34 and 5.36 (Urry, 2006a), Figs. 14B and 15, and **Table** 3 below and associated discussions.

4.3 The ΔT_t-Mechanism for free energy transduction using designed elastic-contractile model proteins (ECMP) based on $(GVGVP)_n$

The temperatures, $\mathbf{T_{b(t)}}$ of **Table 1** column 2, represent either $\mathbf{T_t}$, the temperature for the onset of the inverse temperature transition (ITT) as defined by the onset of turbidity as shown in Fig. 7B, or $\mathbf{T_b}$, the temperature for the onset of the ITT as determined using differential scanning calorimetry (DSC) by the onset of the endothermic transition as

Residue X	$T_{b(t)}$°C	$\Delta G°_{HA}(\chi)$/kcal/mol pentamer
Trp	-105	– 7.00
Phe	-45	– 6.15
Tyr	-75	– 5.85
His°	-10(**T_t**)	– 4.80 (from graph)
Leu	5	– 4.05
Ile	10	– 3.65
Val	26	– 2.50
Met	15	– 1.50
His^{+}	30 (**T_t**)	– 1.90 (from graph)
Cys	30 (**T_t**)	– 1.90 (from graph)
Glu(COOH)	20 (2)	– 1.30 (– 1.50)
Pro	40	– 1.10
Ala	50	– 0.75
Thr	60	– 0.60
Asp(COOH)	40	– 0.40
$Lys(NH_2)$	40 (38)	– 0.05 (– 0.60)
Asn	50	– 0.05
Gly	55	0.00
Ser	60	+ 0.55
Arg	60 (**T_t**)	+ 0.80 (from graph)
Gln	70	+ 0.75
$Tyr(\phi\text{-}O^{-})$	140	+ 1.95
$Asp(COO^{-})$	170 (**T_t**)	≈ + 3.4 (from graph)
$Lys(NH_3^{+})$	(104)	(+ 2.94)
$Glu(COO^{-})$	(218)	(+ 3.72)
$Ser(PO_4^{=})$	860 (**T_t**)	≈ + 8.0 (from graph)

Data within parentheses utilized microbial preparations of poly(30 mers), e.g., (GVGVP GVGVP GXGVP GVGVP GVGVP GVGVP)$_n$, with n ≈ 40. The notation (from graph) indicates that the value of T_t from Table 4 was used with the sigmoid curve of Fig. 13 to estimate **$\Delta G°_{HA}$**(X). Adapted from Urry, 2004).

Table 1: Hydrophobicity Scale in terms of $\Delta G°_{HA}$, the change Gibbs free energy of hydrophobic association, for amino acid residue (X) of chemically synthesized poly[f_V(GVGVP), f_X(GXGVP)], 40 mg/ml, mw ≈ 100 kDa in 0.15 N NaCl, 0.01 M phosphate, using the net heat of the inverse temperature transition, $\Delta H_t \approx \Delta G°_{HA}$, per (GXGVP), determined at f_X = 0.2 and extrapolated to f_X = 1.

defined in Fig. 7C. In general column 2 of **Table 1** lists values for **T_b**; where **T_t** is used, it is indicated by **(T_t)** being placed to the right of the number. Poly[f_V(GVGVP),f_X(GXGVP)] gives the general ECMP composition where f_V and f_X are the mole fractions in the polymer of the defined pentamer repeats and $f_V + f_X = 1$. Experimentally, data are obtained for different low values of f_X and the data points define a straight line that is extrapolated to f_X = 1. When bold-faced, as for **T_t** or **T_b**, the result is for the homopolypentapeptide, (GXGVP)$_n$, which value may, for the more hydrophobic residues, be below 0°C or the value may be greater than 100°C for the more polar, charged residues.

For the biosynthetically prepared composition (GVGVP GVGVP GEGVP GVGVP GVGVP GVGVP)$_{36}$(GVGVP), for the glutamate state, -COO$^-$, in 0.15 NaCl and pH 2.5, T_t and T_b are 58°C, and for the glutamic acid state, -COOH, in 0.15 NaCl at pH 6.6, T_t and T_b are 22°C, i.e., $\Delta T_t = \Delta T_b = 36$°C. Using **Table 1**, the difference between the two states is 218 - 20 = 198 for (GEGVP)$_n$, which for the above composition six pentamer repeats per (GEGVP) gives a similar value, 198/6 = 33°C.

When using the γ-irradiation cross-linked matrix, X^{20}-(GVGVP GVGVP GEGVP GVGVP GVGVP GVGVP)$_{36}$(GVGVP), the elastic strip is swollen at pH 3 and 20°C and contracts on raising the temperature to 40°C; this is *thermo-mechanical transduction*. While holding the temperature constant at 40°C, however, at neutral pH the cross-linked matrix is swollen, but on lowering the pH to 3, (i.e., raising the chemical potential (inputting the chemical energy) of proton, $\Delta\mu_H$, the designed elastic-contractile model protein contracts and can perform mechanical work. This is *chemo-mechanical transduction*. The bold arrows of Fig. 8 between the mechanical and thermal energies, labeled thermo-mechanical, and between chemical and mechanical energies, labeled chemo-mechanical, represent the above-described free energy transductions.

The above represent explicit examples of the **ΔT_t**-mechanism for free energy transduction by means of designed elastic-contractile model proteins (ECMPs), as further discussed in sections **4.1** and **4.2**.

5. The genetic code provides easy access to new energy sources and improved efficiency!

All characterized life has the same Genetic Code, and it plays a central role in the thermodynamics of evolution of protein-based machines as seen through the prism of the energy converting mechanism of ECMP. Due to the Genetic Code, an ECMP, capable of performing thermo-mechanical transduction, can readily be designed to access a new energy source. For example, substitution (mutation) of a single base in 150 bases encoding for (GVGVP)$_{10}$ converts a heat engine (a machine), capable of thermo-mechanical transduction, into a chemically driven engine capable of chemo-mechanical transduction. A different single-base mutation can increase the efficiency of the designed ECMP-based machine.

The triplet codon for placing V in the sequence of a protein is fourfold redundant. Any one of four triplet codons GUU, GUC, GUA, and GUG encode for the valine (V, Val) residue. By a single-base mutation in the second position of either of two triplet codons, GUA → GAA or GUG → GAG, V is replaced by E, (glutamic acid, Glu) to introduce the carboxyl function, i.e., $-CH_2\text{-}CH_2\text{-}COO^- + H^+ \leftrightarrows -CH_2\text{-}CH_2\text{-}COOH$. Adding the proton to a carboxylate drives contraction with the consequence of *chemo-mechanical transduction*. A single-base mutation in the second position of either of another two triplet codons for V, GUU → GAU or GUC → GAC, become two ways to replace V by the D (aspartic acid, asp) residue with a slightly different carboxyl function, i.e., $-\ CH_2\text{-}COO^- + H^+ \leftrightarrows -CH_2\text{-}COOH$ similarly to drive for *chemo-mechanical transduction*.

Significantly, a single mutation in the base of the first position of the triplet codon of V, e.g., GUU → UUU, gives the much more hydrophobic F (phenylalanine, Phe) residue with the consequence of an increased efficiency of energy conversion. Also three of the V triplet codons, GUU, GUC, and GUA, by single mutations in the first position to AUU, AUC, and AUA, give rise to the slightly more hydrophobic isoleucine (I, Ile) with one CH_2 more than V to modestly increase efficiency of energy conversion. Then a single-base mutation of the isoleucine triplet codon of AUA to AAA gives the amino function of lysine, $-NH_3^+ \leftrightarrows -NH_2 + H^+$, where removal of a proton from the $-NH_3^+$ group to give $-NH_2$ provides another means with which to achieve *chemo-mechanical transduction*. But importantly, it provides a positive charge, which, with increase in hydrophobicity, increases the binding capacity of redox functions such as NAD (nicotinamide adenine dinucleotide) and FAD (flavin adenine dinucleotide) with their negative diphosphate linkage to give the *electro-mechanical* and *electro-chemical transductions* of the electron transport chain of the inner mitochondrial membrane.

These simple single-base mutations produce new and improved protein-based machines either to access new energies or to increase the efficiency of function of an existing protein-based machine. Significantly, the new or improved protein-based machines are biosynthesized *without any increase in the energy required to access a new energy source or to produce a more efficient machine.* When a single mutation with no increase in cost of energy to produce the new protein results in a new machine capable of accessing a new energy source and/or a more efficient machine, it becomes apparent why the arrow-of-time for the biological world would be one that, in the words of Toffler (1984), "... proceeds from simple to complex, from 'lower' to 'higher' forms of Life, from undifferentiated to differentiated structures."

The preceding represent elemental means whereby living organisms can naturally evolve from lower to higher forms of life using the mechanism of energy conversion identified on characterization of designed ECMP. *In fact, it would seem that the Genetic Code itself was arranged in order for the living organism to evolve from primitive to more advanced and more complex forms using the energy converting mechanism demonstrated by the elastic-contractile model proteins (ECMP) - as long as there are adequate energy sources and building materials available.* (For a more complete discussion, see Chapter 6: On the Evolution of Protein-based Machines: Toward Complexity of Structure and Diversity of Function of Urry, 2006a).

6. Designed ECMP provide the thermodynamics of protein hydration and of elasticity!

Thermodynamics of protein hydration ($\mathbf{\Delta G_{HA}}$ and ΔG_{ap}) and of elasticity (the internal energy, f_E, and entropy, f_S, components of force) can be established by designed ECMP. The thermodynamics of protein hydration are obtained using differential scanning calorimetry (DSC) of the phase separation process. Whereas acid/base and redox titrations, as a function of V $\rightarrow$ F increases in hydrophobicity, provide means of measuring ΔG_{ap}. DSC data provide the heat of the phase transition, $\mathbf{\Delta H_t}$, which on derivation, using the relationship at the phase transition of $\mathbf{\Delta H_t} \approx T_t\mathbf{\Delta S_t}$, yields the critical quantity, the change in Gibbs free energy for hydrophobic association, $\mathbf{\Delta G_{HA}}$, as the difference, $\mathbf{\Delta H_t}$**(GVGVP)** - $\mathbf{\Delta H_t}$**(GXGVP)**, where X is the substitution being characterized (Urry, 2004; 2006a). The result is the $\mathbf{\Delta G_{HA}}$-based Hydrophobicity Scale obtained for all amino acid residues, including their different functional states, where relevant, and for many biological prosthetic groups in their different functional states. As the hydrophobic R-groups are completely hydrated before the transition and essentially fully dehydrated after phase separation, the values of $\mathbf{\Delta G_{HA}}$, so obtained, may be considered to be maximal values. And this needs to be taken into consideration when interpreting a particular structural change attending protein function, e.g., in the extent of hydrophobic association experienced.

Thus, the thermodynamics of protein hydration have been obtained as the free energies of interaction, $\mathbf{\Delta G_{HA}}$ and ΔG_{ap}, and have been established by physico-chemical characterizations of designed ECMP.

6.1 Thermodynamics of the protein-in-water heat engine of biology

The protein-in-water heat engine of biology functions on the same general physical principle as the steam-powered heat engine of the birth of thermodynamics, which principle is the increase in entropy of water, when heat is applied at the temperature of a phase transition. Dramatic expansion of the bulk-water-to-vapor phase transition near the 100°C gives the increase in entropy of water for the steam-powered heat engine, whereas conversion of three-dimensionally interconnected pentagonal rings of water to bulk water gives the increase in entropy of water for the protein-in-water heat engine of biology. There is analogy to water dipole moments o three-dimensionally interconnected hexagonal rings of ice reorienting to become bulk water during the melting transition of relevance to Eyring's Significant Structure Theory of Water (Hobbs et al., 1966).

Significantly, while the phase transition of the melting of ice is quite fixed near 0 ℃, the transition temperature (T_t) of the inverse temperature transition (ITT) of protein-in-water can be shifted over much of the available aqueous range of water.

Lowering the temperature of the ITT utilizes non-random three-dimensional protein structures to which functional groups are bound. Conversion of functional groups by chemical or electrochemical energy input from a more-polar (e.g., charged) state to a more hydrophobic state, increases pentagonally-arranged water molecules, otherwise destructured by orientation toward charge. This lowers T_t from above to below the operating temperature and drives contraction by hydrophobic association (See Fig. 12 below). By this change in state of an attached biological functional group, the protein-in-water heat engine of biology functions as a protein-in-water chemical (or electrochemical) engine.

The physical process is competition for hydration. Nascent charges destroy pentagonal rings of hydrophobic hydration, as they recruit hydrophobic hydration for their own hydration, which raises T_t, disrupts hydrophobic association, and reverses contraction. Apolar and polar groups lower their free energies by reaching out for hydration unperturbed by the other. This expresses as an apolar-polar repulsive free energy of hydration, ΔG_{ap}.

6.1.1 Hydration of the hydrophobic CH_2 group is exothermic (Butler, 1937)

Butler examined the water solubility of the series of linear alcohols - methanol (CH_3-OH), ethanol (CH_3-CH_2-OH), n-propanol (CH_3-CH_2-CH_2-OH), n-butanol (CH_3-CH_2-CH_2-CH_2-OH), and n-pentanol (CH_3-CH_2-CH_2-CH_2-CH_2-OH) - and found the exothermic heat of dissolution to increase for each added CH_2, at the rate of $\Delta H/CH_2 = -1.5$ kcal/mol. Remarkably, dissolution of the CH_2 in water is a favorable heat releasing reaction. Hydrophobic hydration forms with the release of heat! Why then does the solubility of these linear alcohols decrease as the number of CH_2 moieties increase until n-octanol with seven CH_2 groups is insoluble? The answer is seen in the $[-T\Delta S]$ term of the Gibbs free energy of dissolution, Eqn. (4).

$$\Delta G(\text{dissolution}) = \Delta H - T\Delta S. \tag{4}$$

Namely, the $[-T\Delta S]$ term increases positively (unfavorably) for the Butler series as $[-T\Delta S]/CH_2 = +1.7$ kcal/mol. A positive ΔG(dissolution) means insolubility; too many CH_2 moieties exposed to water means insolubility. Before too many hydrophobic groups are exposed to water, however, they may compete for hydration with more polar groups. Thus, the exothermicity on forming the pentagonal rings of hydrophobic hydration (See Fig. 2) is

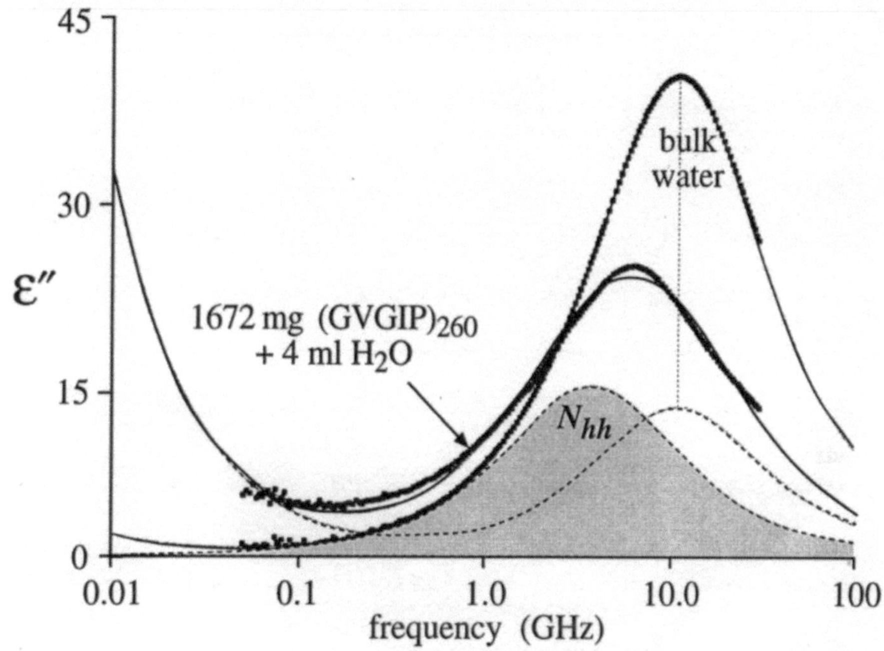

Fig. 9. Imaginary part of the dielectric permittivity of bulk water and of 1672 mg $(GVGIP)_{260}$ dissolved in 4 ml of water at 7°C. The curve for $(GVGIP)_{260}$ in water is resolved into two components, one for bulk water and one for water interacting with the model protein, $(GVGIP)_{260}$, which number of interacting water molecules is designated as N_{hh}. From Urry et al., 1997a.

the basis for a competition with polar groups for hydration. This competition for hydration is documented below in Figs. 10C and 12.

6.1.2 Hydrophobic hydration as characterized by microwave dielectric relaxation

As seen in Fig. 9, the imaginary part of the microwave dielectric relaxation spectrum of bulk water demonstrates an intense absorption just above 10 GHz. The spectrum for a solution of 1672 mg $(GVGIP)_{260}$ in 4 ml of water at 7°C, also seen in Fig. 9, has a broader absorption at lower frequency. On resolving the curve for bulk water just above 10 GHz, a second absorption occurs at a lower frequency, which represents water interacting with the ECMP, $(GVGIP)_{260}$.

The magnitude of N_{hh} is plotted in Fig. 10A as a function of dilution, which N_{hh} is found to increase until it plateaus near 5 mg/ml for both $(GVGIP)_{260}$ and $(GVGVP)_{251}$. The temperature dependence of N_{hh} in Fig. 10B, shows the magnitude of N_{hh} for $(GVGIP)_{260}$ to drop rather abruptly essentially to zero as the temperature passes through the interval of the inverse temperature transition (ITT) for $(GVGIP)_{260}$. Similarly for $(GVGVP)_{251}$, as the temperature passes through the temperature interval of the ITT for $(GVGVP)_{251}$, approximated in Fig. 7C, the magnitude of N_{hh} for $(GVGVP)_{251}$ drops to near zero. On the basis of the data of Fig. 10B, N_{hh} represents the numbers of water molecules of hydrophobic hydration (Urry et al. 1997a). The residual N_{hh} for (GVGVP) in Fig.10B is analogous to the residual pentagonal rings of water for hydrophobically associated crambin of Fig. 2.

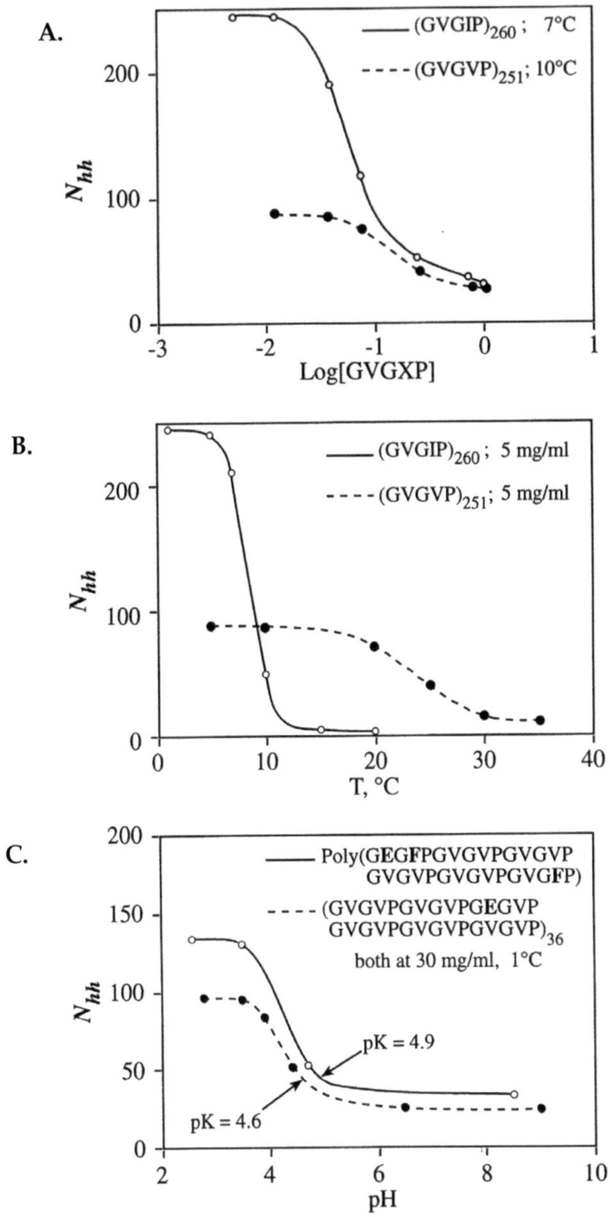

Fig. 10. Microwave dielectric relaxation studies on a unique water interacting with the basic and a designed ECMP. **A.** Demonstrates that on dilution this unique water increases. **B.** Shows that this unique water disappears as hydrophobic association develops, i.e., this identifies ***Nhh*** as hydrophobic hydration. **C.** Shows ***Nhh***, the amount of hydrophobic hydration, to decrease as charged carboxylates form. Indeed, a competition for hydration has been directly observed between hydrophobic and carboxylates, ($-COO^-$). From Urry et al., 1997a.

Having found N_{hh} to represent the number of waters of hydrophobic hydration, the issue of the competition for hydration between charged groups and hydrophobic groups is addressed in Fig. 10C at 30 mg/ml and 1°C by means of the designed ECMP, (GVGVP GVGVP GEGVP GVGVP GVGVP GVGVP)$_{36}$ and the more hydrophobic, poly(GEGFP GVGVP GVGVP GVGVP GVGVP GVGFP), with n of approximately 40. Both designed ECMPs contain the glutamic acid (Glu, E) residue, such that at low pH the functional group of the E-residue will be -COOH. As reported in Fig. 10C, on raising the pH, N_{hh}, decreases as the charged carboxylate groups, $-COO^-$, form with increased pK as 2V → 2F per 30 mer (Urry et al., 1997a).

Thus, the data of Fig. 10C directly demonstrate the competition for hydration between hydrophobic groups and charged groups. As charged $-COO^-$ form, they cooperate to disrupt the cyclically-arranged dipoles of pentagonal rings of water as the dipoles realign toward charge, not unlike charged plates reorient water dipoles in Eyring's Significant Structure Theory of Water (Hobbs et al., 1966; Jhon & Eyring, 1976). As hydrophobic and charged groups compete for limited water, they move away from each other in order to achieve hydration unperturbed by the other resulting in an apolar-polar repulsive free energy of hydration, ΔG_{ap} (Urry, 1992; 1997). In Fig. 14B ΔG_{ap} is seen as hydrophobic-induced pK shifts!

6.1.3 Thermodynamics of the inverse temperature transition

At a given temperature within the interval of the inverse temperature transition, the system of ECMP in water is at equilibrium; the chemical potential of the hydrophobically dissociated molecules in solution, μ_{HD}, is the same as the chemical potential of the phase separated, hydrophobically associated molecules, μ_{HA}. Since $\mu_{HD} = \mu_{HA}$, $\Delta G = 0$, and for the inverse temperature transition, ITT,

$$\Delta H_t \approx T_t \Delta S_t. \tag{5}$$

where the subscript, t, stands for the transition characterized in Fig. 7. Neglecting the very small heat capacity of the ECMP and of water over the temperature interval, the heat of the ITT for (GVGVP) for the transition of Fig. 7C will be given as $\mathbf{\Delta H_t}$(GVGVP), where the bold-faced $\mathbf{\Delta H_t}$ means per mole of pentamer, (GVGVP), within (GVGVP)$_{251}$.

As seen in Fig. 7C, the ITT is an endothermic transition, that is, ΔH_t is positive. Therefore, by Eqn. (5) ΔS_t is positive. The entropy increases as the phase separation occurs. Yet in this protein-in-water, two-component system, the protein becomes more ordered, i.e., ΔS_t(protein) is negative. The only other component of the ITT, water, must be the component that gives rise to the positive ΔS_t. And specifically, the hydrophobic hydration, N_{hh}, of Figs. 9 and 10B, is the water that becomes less-ordered bulk water, i.e., $\Delta S_t(N_{hh} \rightarrow$ bulk water) represents the positive entropy change that drives the ITT. During the transition the model protein becomes restricted in its motion on hydrophobic association and even becomes structured with formation of filaments, fibrils and fibers (Urry, 1992), that is, ΔS_t(model protein) is negative. So we write,

$$\Delta S_t(N_{hh} \rightarrow \text{bulk water}) > \Delta S_t(\text{total system}) + \Delta S_t(\text{model protein}) > 0 \tag{6}$$

and

$$\Delta S_t(N_{hh} \rightarrow \text{bulk water}) > \Delta S_t(\text{total system}) >> 0 \tag{7}$$

By Eqns. (5) and (6), it is apparent for the two-component ITT, that the component, ΔH_t(model protein) ≈ $T_t\Delta S_t$(model protein). Also, ΔH_t(model protein) would be an exothermic component within a larger endothermic component. Accordingly, it is interesting to report that temperature modulated differential scanning calorimetry (TMDSC) separates out such an exothermic component that is one-quarter to one-third the magnitude of the endothermic component (Rodríguez-Cabello et al., 2004). As hydrophobic association occurs, a van der Waals' interaction energy is expected to result in an exothermic component to the ITT, in this case arising largely from London's dispersion forces (See Eyring, et al. 1958).

When developing a Hydrophobicity Scale based on $\mathbf{\Delta G_{HA}}$, the change in Gibbs free energy for hydrophobic association, no attempt will be made to separate out the exothermic component, as the van der Waals' energy due principally to dispersion forces would be part of all hydrophobic associations. It is interesting to note in the plot of $\mathbf{T_b}$ versus $\mathbf{\Delta H_t}$ of Fig. 13 below that all of the amino acid residues containing aliphatic side chains and even the glycine residue where the $-CH_2-$ is in the backbone fall on a straight line, whereas the aromatic residues give a much steeper slope and a greater heat (See Fig. 13, section **6.2.4**).

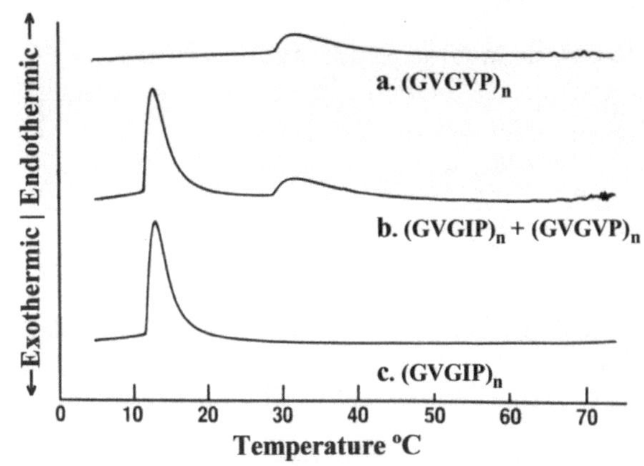

Fig. 11. Differential scanning calorimetry curves of $(GVGVP)_n$ and $(GVGIP)_n$ in water. Demonstrated are heats of $\mathbf{\Delta H_t}$(GVGVP) = 1.2 kcal/mol-(GVGVP) and $\mathbf{\Delta H_t}$(GVGIP) = 3.0 kcal/mol-(GVGIP). Scanning the aqueous mixture shows the ECMP to self associate, i.e., to demix, which suggests an hierarchical hydrophobic folding of protein. Adapted from Urry et al., 1997a.

6.2 The changes in Gibbs free energy underlying the protein-in-water heat engine of biology

6.2.1 The hierarchy of hydrophobic association demonstrated by demixing of ECMP on raising the temperature

The side chains of V, $-CH(CH_3)_2$, and I, $-CH_2-CH(CH_3)_2$ differ by a single $-CH_2-$. Similarly, the pentamers, (GVGVP) and (GVGIP) differ only by a single $-CH_2-$. In particular, the molecular weights of the pentamer units are 409 Daltons for (GVGVP) and 423 Daltons for (GVGIP). Just a 3.4% increase in molecular weight lowers the onset temperature by 15°C and increases the heat of the transition by 240%. By considering the aliphatic groups only, 2V(86D), 2G(28D), and P(42D) gives 156 Daltons for (GVGVP) and V(43D), 2G(28D), P(42D), and I(57D) gives 170 Daltons for (GVGIP), which is a small 11% increase in mass of hydrophobic moieties to be responsible for a 2.4 fold increase in heat of the transition. It was also seen in Fig. 10B that the added $-CH_2-$ of (GVGIP) resulted in a slightly larger 2.8 fold increase based on the estimated number of waters of hydrophobic hydration, N_{hh}.

Another important feature occurs in the DSC data of Fig. 11. When the temperature is scanned for an equimolar mixture of $(GVGVP)_n$ and $(GVGIP)_n$ in water, starting some 5°C or more below the lowest T_t, there is a complete demixing of the polypentamers that differ by but one $-CH_2-$ moiety per pentamer. This represents a remarkable separation capacity and surely has an impact on details of hydrophobic associations as proteins fold.

It is interesting that the inverse temperature transition, which is driven by the increase in entropy as pentagonal rings of hydrophobic hydration become higher entropy bulk water, can achieve the decrease in entropy of complete separation of a mixture of polymers as seen in the middle curve of Fig. 11. This effect has also been used in the purification of microbially-prepared ECMP (See Fig. 5 of McPherson et al., 1996). In particular, during the purification of $(GVGVP)_{251}$ from an *E-coli* lysate, an endotoxin impurity was observed with a phase separation of its own that overlapped with the temperature interval for $(GVGVP)_{251}$. The endotoxin impurity could be largely separated from $(GVGVP)_n$ by centrifugation at 23°C, facilitating later complete removal of the small amount of remaining endotoxin.

6.2.2 Experimental dependence of both T_t and ΔH_t on the amount of hydrophobic hydration, *Nhh*

In Fig. 10A and B the increase in hydrophobic hydration, N_{hh}, is a factor of 2.8 on going from $(GVGVP)_n$ to $(GVGIP)_n$ resulting from the increase of but a single CH_2 per pentamer. From yet another experiment, the endothermic heat of the transition in water for $(GVGIP)_n$, i.e., ΔH_t(GVGIP) = 2.61 kcal/mol-(GVGIP), is larger by a factor of 2.4 than that for $(GVGVP)_n$, i.e., ΔH_t(GVGVP) = 1.07 kcal/mol-(GVGVP) (Luan et al. 1990; Luan & Urry, 1991). *The magnitude of the increase in hydrophobic hydration due to the added CH_2 appears to be larger than the magnitude of the increase in the heat of the transition, which effect results in the lowered value of T_t as $\Delta H_t \approx T_t \Delta S_t$*. A similar correlation, in this case an inverse correlation, is observed between the increase in N_{hh} and the decrease in T_t for **Model protein ii** and for **Model protein I** of Fig. 12A and B.

The formation of the charged carboxylate ($-COO^-$) of the glutamic acid residue (Glu, E) decreases N_{hh}, the amount of hydrophobic hydration, as seen in Fig. 10C. In addition to the addition of a proton to neutralize the charged carboxylate, the capacity of the carboxylate charge to destructure hydrophobic hydration is moderated by ion-pairing, as with a Ca^{2+} ion. The increase in hydrophobicity, measured by the increase in number of molecules of hydrophobic hydration, N_{hh}, for a transition driven by chemical energy is demonstrated in Fig. 12 by titration of Ca^{2+} into the ECMP-water two component system. As Ca^{2+} ion-pairs with glutamic acid residues, the value of T_t decreases and the waters of hydrophobic hydration increase in an inverse relationship (See Fig. 12).

Noting the considerations of section **6.1.3**, the experimental endothermic heat given as per mole pentamer, ΔH_t(GVGVP), is comprised of two components, the *major heat absorbed* as $N_{hh} \rightarrow$ bulk water for GVGVP and the *smaller heat released* as van der Waals contacts develop as hydrophobic association proceeds between repeats of (GVGVP) (See Rodríguez-Cabello et al., 2004). These two components are given as in Eqn. (8),

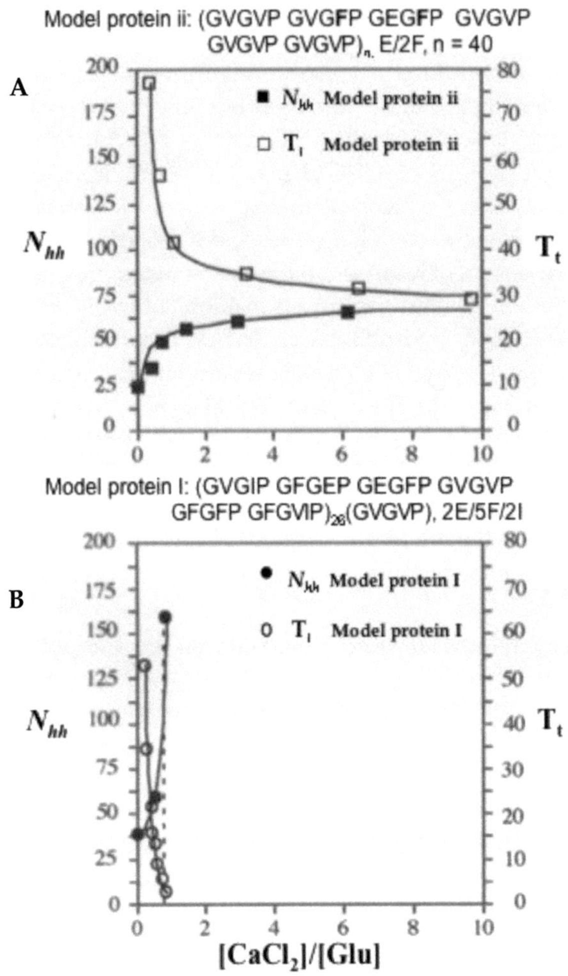

Fig. 12. Inverse correlation of T_t, the temperature of the inverse temperature transition, ITT, and $\boldsymbol{N_{hh}}$, the number of waters of hydrophobic hydration. Independent of the steepness of the $[Ca^{2+}]$ binding to carboxylate, the values for T_t mirror the values for $\boldsymbol{N_{hh}}$. ***Increases in hydrophobic hydration, N_{hh}, lower T_t, the temperature of the ITT!*** This and the correlation of ΔH_t and $\boldsymbol{N_{hh}}$ seen on comparison of data in Figs. 10 and 11 are consistent with the definition of $\boldsymbol{\Delta G_{HA}}$ of Eqn. (12) below. From Urry, 2006a.

$$\mathbf{\Delta H_t} = \mathbf{\Delta H_t}'(\text{heat absorbed as } N_{hh} \rightarrow \text{bulk water per GVGVP})$$

$$+ \mathbf{\Delta H_t}''(\text{heat released due to contacts between GVGVP}) \quad (8)$$

By Eqn. (5), $\Delta H_t \approx T_t \Delta S_t$, the $\mathbf{T_t}$(GVGVP)$\mathbf{\Delta S_t}$(GVGVP) term for the phase transition can be written,

$$\mathbf{T_t \Delta S_t} = \mathbf{T_t \Delta S_t}'(N_{hh} \rightarrow \text{bulk water per GVGVP}) +$$

$$\mathbf{T_t \Delta S_t}''(\text{contacts between GVGVP decrease entropy}) \quad (9)$$

Again there are the equivalent two components, the *larger positive entropy change* per mole pentamer, $\mathbf{\Delta S_t}'$, as hydrophobic hydration, $\mathbf{N}_{hh}$, per mole pentamer becomes less-ordered bulk water and *the smaller negative entropy change*, $\mathbf{\Delta S_t}''$, as (GVGVP) become constrained by hydrophobic association.
Keeping in mind the two equivalent per pentamer, (GVGVP), components within Eqn. (8) and Eqn. (9), Eqn. (5) can simply be written for (GVGVP) as,

$$\mathbf{\Delta H_t}(\mathrm{GVGVP}) \approx \mathbf{T_t}(\mathrm{GVGVP})\mathbf{\Delta S_t}(\mathrm{GVGVP}) \tag{10}$$

6.2.3 Derivation of the change in Gibbs free energy for hydrophobic association, ΔG_{HA}, due to a substitution of V by X

Required is an expression for the change in Gibbs free energy resulting from a change of a single R-group in the (GVGVP) pentamer. The differences in Fig. 12 between the inverse temperature transitions exhibited by GVGVP)$_n$ and (GVGIP)$_n$ represent the desired quantity, $\mathbf{\Delta G_{HA}}$, i.e., the change in the Gibbs free energy for hydrophobic association, in this example, due to the addition of a single CH_2-group. The unique position X between two G residues, GXG, allows each amino acid residue, including chemical adducts to carboxyl (e.g., Glu and Asp) and amino functions, to occur with retention of the fundamental elastic and phase transitional properties. Accordingly, Eqn (9) is restated for (GXGVP), i.e.,

$$\mathbf{\Delta H_t}(\mathrm{GXGVP}) \approx \mathbf{T_t}(\mathrm{GXGVP})\mathbf{\Delta S_t}(\mathrm{GXGVP}). \tag{11}$$

Subtraction of Eqn. (11) from Eqn. (10) gives,

$$\mathbf{\Delta H_t}(\mathrm{GVGVP}) - \mathbf{\Delta H_t}(\mathrm{GXGVP}) \approx \mathbf{T_t}(\mathrm{GVGVP})\mathbf{\Delta S_t}(\mathrm{GVGVP}) - \mathbf{T_t}(\mathrm{GXGVP})\mathbf{\Delta S_t}(\mathrm{GXGVP}). \tag{12}$$

The term on the right-hand side of Eqn. (12) expresses the sought after change in Gibbs free energy for the (GVGVP)-in-water phase transition of hydrophobic association, $\mathbf{\Delta G_{HA}}$, on substitution of V by X, that is,

$$\mathbf{\Delta G_{HA}}(\mathrm{GVGVP}\rightarrow\mathrm{GXGVP}) = \mathbf{T_t}(\mathrm{GVGVP})\mathbf{\Delta S_t}(\mathrm{GVGVP}) - \mathbf{T_t}(\mathrm{GXGVP})\mathbf{\Delta S_t}(\mathrm{GXGVP}). \tag{13}$$

Eqn. (13) represents the change in Gibbs free energy due to the increase in entropy as pentagonal rings of water become bulk water minus the smaller decrease in entropy as the model protein (GVGVP)$_n$ becomes more-ordered on hydrophobic association, as described in Eqns. (8) and (9), minus the same for (GXGVP).
As $\mathbf{\Delta S_t}$ is directly calculated from the endothermic heat, $\mathbf{\Delta H_t}$ of the experimental curve, the sought-after $\mathbf{\Delta G_{HA}}$ can be equivalently stated as,

$$\mathbf{\Delta G_{HA}}(\mathrm{GVGVP} \rightarrow \mathrm{GXGVP}) = \mathbf{\Delta H_t}(\mathrm{GVGVP}) - \mathbf{\Delta H_t}(\mathrm{GXGVP}). \tag{14}$$

It should be appreciated that such a relationship holds only for the phase transition, hence the required subscript, t.
At whatever temperature the inverse temperature transition, ITT, occurs for a particular ECMP composition within the accessible aqueous temperature range, the ECMP-in-water heat engine is based upon a phase transition in which, on heating, water undergoes a transition from lesser to greater entropy. In particular, on heating, structured hydrophobic hydration, i.e., water arranged in pentagonal rings, becomes less-ordered bulk water.

When the temperature of the transition is driven by another energy input, such as chemical energy, the particular energy input causes the responsive functional group, and hence the ECMP, to become more hydrophobic. The consequence is more hydrophobic hydration, which lowers the transition temperature, T_t, as seen in Fig. 12.

6.2.4 Plot of the temperature transition vs the change in Gibbs free energy for hydrophobic association, ΔG_{HA}

It is convenient for a hydrophobicity scale of amino acid residues to choose the most neutral residue, neither hydrophobic nor polar, as the zero reference state. This is achieved simply by replacing ΔH_t(GVGVP) of Eqn. (14) by ΔH_t(GGGVP), i.e.,

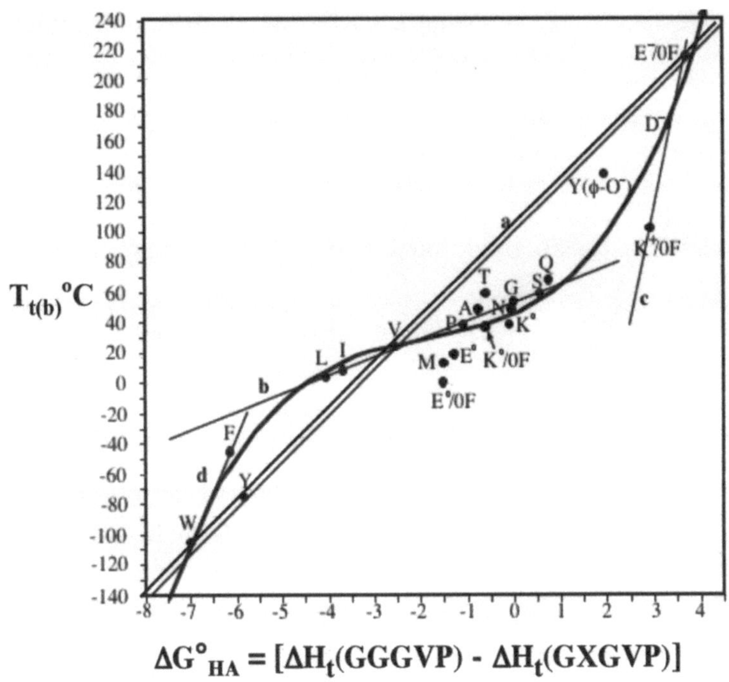

Fig. 13. The plot of the change in Gibbs free energy for hydrophobic association, ΔG°_{HA}, versus the onset temperature of the inverse temperature transition, T_t when using the midpoint of the temperature versus turbidity curve and T_b when using DSC to measure the onset of the heat of the phase transition. See text for further discussion. Adapted from (Urry, 2004).

$$\Delta G^{\circ}_{HA} \equiv \Delta H_t(\text{GGGVP}) - \Delta H_t(\text{GXGVP}), \quad (15)$$

where ΔG_{HA} has been replaced by ΔG°_{HA} to indicate that the hydrophobicity scale has chosen G as the zero reference. Eqn. (15) is used for plotting the values in Fig. 13 of $T_{t(b)}$ versus ΔG°_{HA}, where the data point for the G (glycine, Gly) is zero.

The doubled diagonal line, **a**, gives the T_t-based Hydrophobicity Scale approximation of the relationship between $T_{t(b)}$ and ΔG°_{HA}, when in reality it is a sigmoid relationship. The sigmoid plot in Fig. 13 allows that a given value of T_t can be read from the sigmoid curve for a better approximation of ΔG°_{HA}. This was done for the ΔG°_{HA} values of **Table 2**.

All of the amino acid residues with aliphatic hydrophobic groups - G, A, P, V, I, and L - fall on the straight line, **b**, i.e., they all exhibit the same linear temperature dependence of the inverse of the entropy, $(\Delta S_t)^{-1}$. All of the aliphatic hydrophobic side chains of amino acid residues, as well as the CH_2 in the backbone of G, have the same thermal stability of hydrophobic hydration, but the thermal stability of aliphatic hydrophobic hydration appears greater than that of the aromatic residues.

The temperature dependence of the amino acid residues with charged side chains, line **c**, gives the steepest slope. And E^-, D^-, and K^+ follow the relative capacity, in decreasing order, of charged side-chains to disrupt hydrophobic hydration.

The temperature dependence of the amino acid residues with aromatic side chains, line **d**, is much steeper than that of the aliphatic residues, which, as the slope is $(\Delta S_t)^{-1}$, suggests less thermal stability for the hydrophobic hydration of aromatic groups.

As seen in **Table 1** and Fig. 13, the biggest change in $\Delta G°_{HA}$ occurs on ionization of the carboxyl side chain of glutamic acid, i.e., $-CH_2-CH_2-COOH \rightarrow -CH_2-CH_2-COO^- + H^+$. From **Table 1**, the change in free energy of hydrophobic hydration, $\Delta G°_{HA}$ = [(GE°GVP) $\rightarrow$ (GE$^-$GVP)] = is 5.22 kcal/mol-(GEGVP). Also, as seen in Fig.10C, this ionization disrupts 75% of the hydrophobic hydration, N_{hh}. Furthermore, in Fig. 12 partial neutralization of E^- by ion-pairing with calcium ion markedly increases N_{hh} in a manner that mirrors the decrease in T_t. This demonstrates how putting chemical energy into a protein-in-water heat engine drives *chemo-mechanical transduction*. It does so by increasing the amount of hydrophobic hydration, and thereby lowering T_t, until the phase transition falls sufficiently below the operating temperature to have completed the hydrophobic association of contraction. Just as with the dissolution of the alcohol series of Butler (1937), as more CH_2 moieties are added, more exothermic hydrophobic hydration builds up. This continues until the unfavorable, positive $[-T\Delta S]$ term of $\Delta G(\text{dissolution}) = \Delta H - T\Delta S$ becomes greater than the favorable, negative ΔH term, and solubility is lost. The thermodynamics is such that hydrophobic hydration builds until the tipping point (the phase transition to hydrophobic association) has been reached. Once the temperature of the system is some 10°C above the onset temperature for the transition, contraction is nearly complete.

The cumulative effect of neutral residues might also be noted. For example, when the hydrophobicity of a domain or structure of a protein is being considered, the $\Delta G°_{HA}$ of glutamine, Q, of 0.75 in **Table 1** and Fig. 13 can have a disruptive effect on hydrophobic association. The cumulative effect of some half dozen glutamine residues give a $\Delta G°_{HA}$ of 4.5 kcal/6 mol-Q, which contributes more than a single glutamate where from **Table 1**, $\Delta G°_{HA}$ reads 3.7 kcal/mol-E^-.

6.2.5 (T_t and $\Delta G°_{HA}$)-Hydrophobicity scales for biological functional groups attached to amino acid side-chains

A number of biological functional groups have been attached to amino, carboxyl, and -OH functions of designed ECMP. The T_t values are listed in **Table 2** and the $\Delta G°_{HA}$ values have been approximated using the sigmoid curve of Fig. 13. Of immediate interest are the redox functional groups and most particularly N-methyl nicotinamide, which has been used to determine the effect of increased hydrophobicity on reduction potential and positive cooperativity. As will be shown in section **6.2.8** the effects of increased hydrophobicity on reduction potential and cooperativity of redox functions parallel the effects of increased hydrophobicity on pK and positive cooperativity of the carboxyl and amino functions.

Residue X	$\Delta G°_{HA}$ from **Figure13** (kcal/mol)[g]	T_t linearly extrapolated to $f_X = 1$
Lys (dihydro NMeN)[b,d]	− 7.0	−130°C
Glu(NADH)[c]	− 5.5	−30°C
Lys (6-OH tetrahydro NMeN) [b,d]	− 3.5	15°C
Glu($FADH_2$)	− 2.5	25°C
Glu(AMP)	+ 1.0	70°C
Ser(-O-SO_3H)	+ 1.5	80°C
Thr(-O-SO_3H)	+ 2.0	100°C
Glu(NAD)[c]	+ 2.5	120°C
Lys(NMeN, oxidized) [b,d]	+ 2.5	120°C
Glu(FAD)	+ 2.5	120°C
Tyr(-O- SO_3H)[e]	+ 3.0	140°C
Tyr(-O-NO_2^-)[f]	+ 3.5	220°C
Ser($PO_4^{=}$)	+ 8.0	860°C

[a]The usual conditions are for 40mg/ml polymer, 0.15N NaCl, 0.01M phosphate at pH 7.4.
[b]NMeN, N-methyl nicotinamide, attached to a lysyl side chain, i.e., N-methyl-nicotinate attached by amide linkage to the e-NH_2 of Lys. The most hydrophobic reduced state is N-methyl-1,6-dihydronicotinamide (dihydro NMeN), and the second reduced state is N-methyl-6-OH 1,4,5,6-tetrahydronicotinamide or (6-OH tetrahydro NMeN).
[c]For the oxidized and reduced nicotinamide adenine dinucleotides, the conditions were 2.5 mg/ml polymer, 0.2M sodium bicarbonate buffer at pH 9.2.
[d]For the oxidized and reduced N-methyl nicotinamide, the conditions were 5.0 mg/ml polymer, 0.1M potassium bicarbonate buffer at pH 9.5, 0.1M potassium chloride.
[e]The pK_a of polymer bound -O-SO_3H is 8.2.
[f]The pK_a of Tyr(-O-NO_2) is 7.2.
[g]Gross estimates (e.g., ± 0.5 kcal/mol) of $\Delta G°_{HA}$ using the T_t-values in the right column in combination with the $T_{b(t)}$ versus $\Delta G°_{HA}$ values from **Figure 13**. Adapted from Table 5.2 of Urry, 2006a.

Table 2. Hydrophobicity Scales (preliminary T_t and $\Delta G°_{HA}$ values) for chemical modifications and prosthetic groups of proteins[a]. T_t is the temperature for the onset of the inverse temperature transition, and $\Delta G°_{HA}$ is the change in Gibbs free energy for hydrophobic association for poly[f_V(GVGVP),f_X(GXGVP)].

In particular, increases in hydrophobicity shift the pK and reduction potential values in a supra-linear manner toward the less polar, more hydrophobic state with the same changes in free energy that contribute to the increases in positive cooperativity.

The largest change in $\Delta G°_{HA}$ of the redox functions in **Table 2** occurs on reduction of oxidized N-methyl nicotinamide ($NMeN^+$) attached by amide linkage to the lysine side chain, i.e., $\Delta G°_{HA}[NMeN^+ \rightarrow \text{dihydro NMeN}] = -9.5$ kcal/mol-NMeN. This is greater than the change in the Gibbs free energy of hydrophobic association on reduction of oxidized nicotinamide adenine dinucleotide (NAD^+) to NADH, i.e., $\Delta G°_{HA}[NAD^+ \rightarrow NADH] = -8$ kcal/mole-NAD. Similarly for the reduction of flavin adenine dinucleotide (FAD) to $FADH_2$, $\Delta G°_{HA}[FAD \rightarrow FADH_2] = -5$ kcal/mole-FAD. These numbers are of the right order and magnitude for their biological functions. In the electron transport chain of the inner mitochondrial membrane, the oxidation of $FADH_2$ provides the energy to pump at a maximum 66% of the number of protons as does the oxidation of NADH, and 100[(5 kcal/mole-FAD)/(8 kcal/mole-NAD)] = 63%.

Also, enzymatic phosphorylation of the serine residue of the serine kinase site, RGYSLG, within designed ECMP extrapolates to a **$\Delta G°_{HA}$** of +8kcal-mol-phosphate. This finding provides evidence that the presence of phosphate disrupts hydrophobic association (See section 8.1.11 of Urry, 2006a). Also, the $K_{eq} \approx 1$ for the reaction, ATP + ECMP-S = ADP + ECMP-S-P, i.e., **$\Delta G°_{HA}$**[ECMP-S → ECMP-S-P] ≈ **- $\Delta G°_{HA}$**[ATP → ADP + P] ≈ 8 kcal/mol-P.

A.

Model protein I: (GVGIP GFGEP GEGFP GVGVP GFGFP GFGIP)$_{26}$(GVGVP) 2E/5F/2I

Model protein i: (GVGVP GVGVP GEGVP GVGVP GVGVP GVGVP)$_{36}$(GVGVP) E/0F

Model protein ii: (GVGVP GVGFP GEGFP GVGVP GVGVP GVGVP)$_{n}$, n ≈ 40 E/2F

Model protein iii:(GVGVP GVGVP GEGVP GVGVP GVGFP GFGFP)$_{39}$(GVGVP) E/3F

Model protein iv: (GVGVP GVGFP GEGFP GVGVP GVGFP GVGFP)$_{15}$(GVGVP) E/4F

Model protein v: (GVGVP GVGFP GEGFP GVGVP GVGFP GFGFP)$_{42}$(GVGVP) E/5F

Fig. 14A. Family of ECMP used below in the acid/base titration studies. **Model proteins i** through **v** are the same as in **Table 3**.

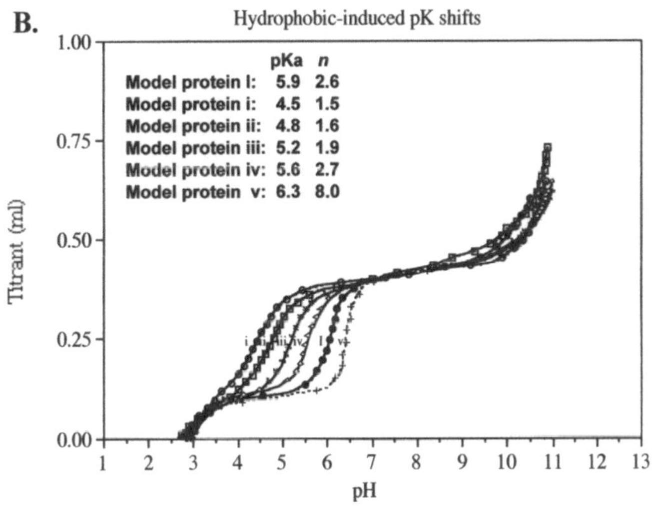

Fig. 14B. ECMP acid/base titrations exhibit supra-linear hydrophobic-induced pK & positive cooperativity shifts as V-residues are replaced by F-residues (Urry, 2006a).

6.2.6 The apolar-polar repulsive free energy of hydration, ΔG_{ap}

Three families of ECMP were designed (See **Table 3** below), each containing a different acid/base or redox function (Φ), containing a set of V → F increases in hydrophobicity, i.e., Φ/nF, where n = 0, 2, 3, 4, and 5, and V → functional groups (E, K, and NMeN). Acid/base and redox titrimetry data demonstrate systematic hydrophobic-induced pK, reduction potential, and positive cooperativity shifts. In general, five different experimental approaches used on these and other ECMPs exhibit apolar-polar group competition for hydration and quantify an apolar-polar repulsive free energy of hydration, ΔG_{ap}.

Generalized equations, containing the required terms for both apolar-polar (ap) and charge-charge (cc) repulsion, were derived for analysis of acid/base and redox titration data in order to resolve charge-charge and hydrophobic-induced pK, reduction potential and positive cooperativity shifts (Urry, 1997; 2006a). Analyses used a specialized adaptation of a Hill plot to resolve the free energies of the charge-charge (cc) and apolar-polar (ap) cooperativity interactions, $[(\partial\Delta G/\partial\alpha)_T]_{cc}$ and $[(\partial\Delta G/\partial\alpha)_T]_{ap}$, and to result in determination of the appropriate values of ΔG. The free energies due to cooperativity interactions were found to be equivalent to those free energies of corresponding pK shifts, $\Delta G_{ap} = 2.3RT\Delta pK$, of the acid/base titrations and to those of the reduction potential shifts of the redox functions, i.e., $\Delta G_{ap} = z.\mathcal{F}\Delta \mathbf{E}$.

6.2.7 Analysis of the acid-base titration curves of Fig. 14

The family of model proteins, **Model proteins i** through **v**, of Fig. 14A are poly(30 mers) in which in all five model proteins the third basic repeat, (GVGVP), has been changed to (GEGVP), and in which the first and fourth repeats remain as (GVGVP). The other three pentamers contain one or two V-residues having been replaced by one or two more-hydrophobic F-residues.

The members of this family are identified as E/0F, E/2F, E/3F, E/4F, and E/5F. Their remarkable set of acid-base titration curves are reported in Fig. 14B, where the pK and positive cooperativity shifts increase in a supra-linear manner. In particular, substitution of the first two V-residues by F-residues, as in going from **Model protein i** to **ii**, i.e., E/0F to E/2F, results in a pK shift of 0.3 pH units and the Hill coefficient changes from 1.5 to 1.6. By adding a single substitution on going from E/4F to E/5F at the higher hydrophobicity, the pK shift is 0.7 pH units and the Hill coefficient increases from 2.7 to 8 to give a change of 5.3. Thus, the effect of the initial substitution of the first F would be a pK shift of the order of 0.15, the addition of the last F-residue on going from E/4F to E/5F gives a 4.7 (= 0.7/0.15) times larger pK shift and an increase of the Hill coefficient by a factor of 106 (= 5.3/0.05).

From the **ΔG^o_{HA}**-based Hydrophobicity Scale of **Table 1**, the change in (G(HA on replacing a single V-residue by a single F-residue, found by a linear process of plotting from $f_F = 0$ to $f_F = 1$, is -3.65 kcal/mol-pentamer, yet the effect of increasing numbers of F substitutions on pK and Hill coefficient of the E/nF series is supra-linear. By what mechanism does this supra-linearity occur? And how may an understanding of the underlying physical process be found? The physical process will be found to be fundamental in understanding the function of protein-based machines of biology.

The challenge is to express accurately these hydrophobic-induced changes in pK and positive cooperativity in terms of changes in Gibbs free energy and to obtain a physical basis for the observed effects. This problem is approached below, by first deriving a generalized acid/base titration equation, and then by developing a graphical means whereby the changes in Gibbs free energies may be obtained and compared.

6.2.8 Generalized acid/base titration equations that contain apolar-polar, ap, and charge-charge, cc, repulsion terms

The familiar Henderson/Hasselbalch equation for the titration of a dilute weak acid or base has the form, $pH = pK + \log[\alpha/(1-\alpha)]$, which is modified to introduce cooperativity by introduction of the Hill coefficient, $\boldsymbol{n}$, to give the expression, $pH = pK + (1/\boldsymbol{n})\log[\alpha/(1-\alpha)]$ (Hill, 1913). The need to obtain a titration equation wherein the Hill coefficient, $\boldsymbol{n}$, could be replaced by an explicit term for the change in Gibbs free energy reflected by a change in cooperativity has been addressed over the years by Overbeek, (1948), Katchalsky & Gillis

(1949), and Harris & Rice (1954). The result for charge-charge repulsion observed in polyelectrolyte solutions would be of the form,

$$pH = pK_{cc} + \log[\alpha/(1-\alpha)] + [(\partial\Delta G/\partial\alpha)_T]_{cc}/2.3RT, \tag{16}$$

where pK_{cc} is the shifted pK due to charge-charge repulsion, and $(1/\boldsymbol{n})\log[\alpha/(1-\alpha)]$ is replaced by two terms, $\log[\alpha/(1-\alpha)] + [(\partial\Delta G/\partial\alpha)_T]_{cc}/2.3RT$.
It becomes apparent on considering the data of Fig. 14 that designed ECMP also exhibit, as discussed above, an apolar-polar repulsive free energy of hydration, ΔG_{ap}. The equation for acid/base titrations of designed ECMP requires inclusion of terms for both charge-charge and apolar-polar repulsion. The derivation may be found elsewhere (Urry, 1997; 2006a) to give the following,

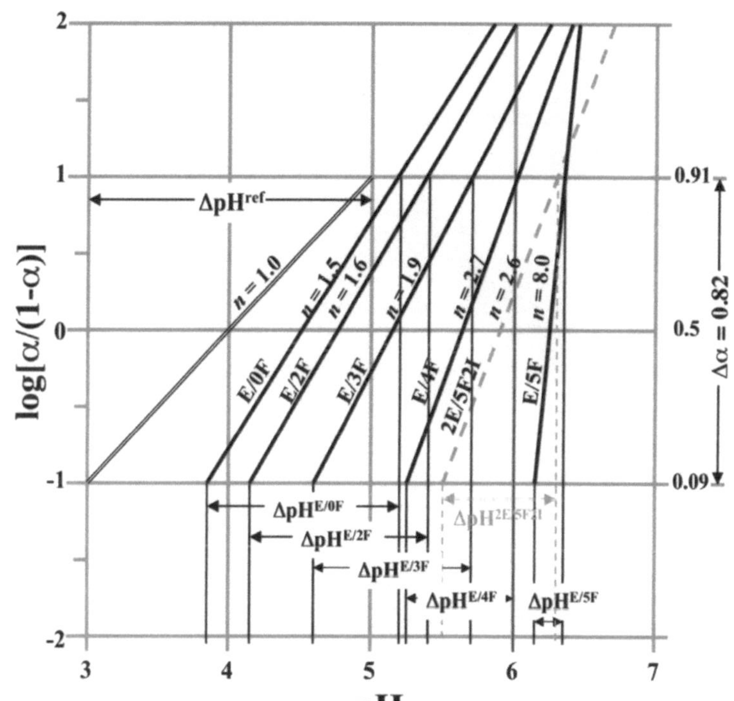

$$(\partial\Delta G^{nF}/\partial\alpha) = -2.3RT(\Delta pH^{nF} - \Delta pH^{ref})/0.82$$

Fig. 15. Hill plots of the acid/base titration data Fig. 14B has been analyzed using Eqn. (18) to calculate cooperativity values as the stand alone term $(\partial\Delta G/\partial\alpha)_T/2.3RT$ to replace the factor, $1/\boldsymbol{n}$, of the Hill coefficient, $\boldsymbol{n}$. The thermodynamic expression, $(\partial\Delta G/\partial\alpha)_T$ is equivalent to ΔG_{ap}. It is calculated by the expression, $(\partial\Delta G^{nF}/\partial\alpha)_T = 2.3RT(\Delta pH^{ref} - \Delta pH^{nF})/0.82$, where the values of ΔpH^{ref} and ΔpH^{nF} are shown in the figure. $(\partial\Delta G^{nF}/\partial\alpha)$ is positive, i.e., repulsive, for positive cooperativity and negative for negative cooperativity. For the negative cooperativity of charge-charge repulsion the acid/base titration curve is broader than and the positive cooperativity curves due to apolar-polar repulsion are steeper than given by the Henderson-Hasselbalch Eqn., pH = pK + $\log[\alpha/(1-\alpha)]$. From (Urry et al., 2010).

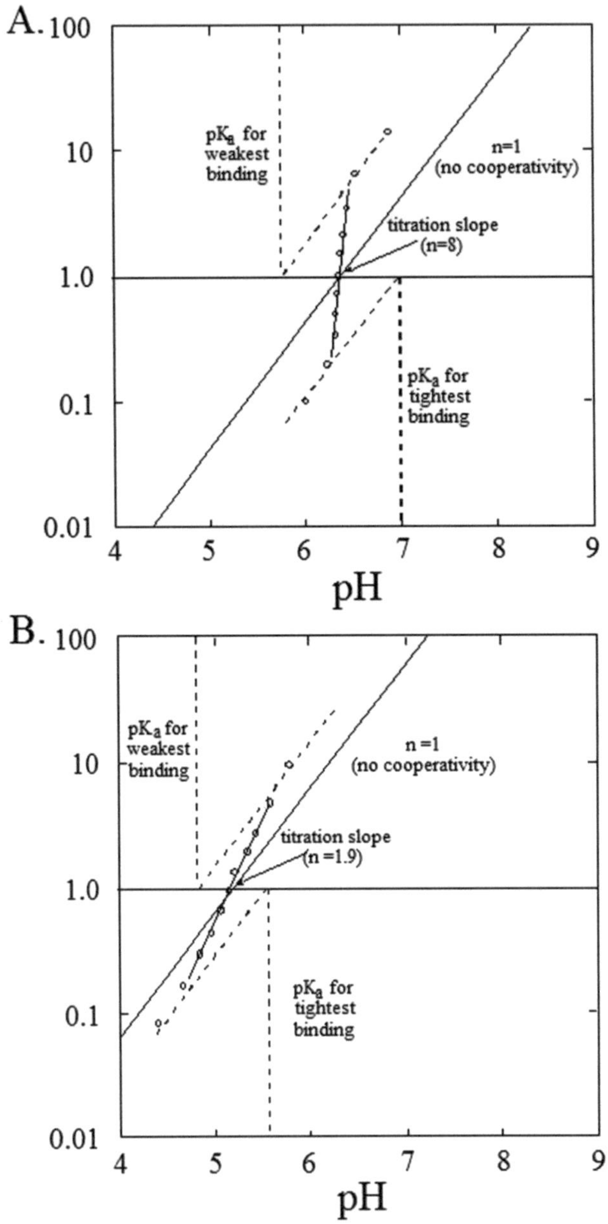

Fig. 16. Hill plots of the acid/base titration data of Fig. 14B used to estimate pK of the first and the last carboxyl to ionize, which demonstrate a residual pK shift for the last carboxyl to ionize. See text for discussion. Figure 5.31 of (Urry, 2006a).

$$pH = pK_o + \Delta pK_{cc} + \Delta pK_{ap} + \log[\alpha/(1\text{-}\alpha)] + \{[(\partial\Delta G/\partial\alpha)_T]_{cc} + [(\partial\Delta G/\partial\alpha)_T]_{ap}\}/2.3RT. \quad (17)$$

For **Model proteins i** through **v**, where the ionizable functions are separated by 90 backbone bonds plus 6 side chain bonds, charge-charge repulsion is negligible and the data for **Model proteins i** through **v** in Fig. 14B may be analyzed using the following expression,

$$pH = pK_o + \Delta pK_{ap} + \log[\alpha/(1-\alpha)] + \{[(\partial\Delta G/\partial\alpha)_T]_{ap}\}/2.3RT, \quad (18)$$

where pK_o for the unperturbed glutamic acid is taken as 4.0.

For the Hill plot, pH is plotted on the x-axis and $\log[\alpha/(1-\alpha)]$ is plotted on the y-axis as seen for Fig. 15. The $\log[\alpha/(1-\alpha)] = 0$ intercept gives pK_{exptl} and $\Delta pK = pK_{exptl} - pK_o$. The last term, $\{[(\partial\Delta G/\partial\alpha)_T]_{ap}\}/2.3RT$, measures the variation of the slope, n, from 1, which is the diagonal as plotted in Fig. 15. If $n < 1$, the slope is less steep than for $n = 1$ as occurs for negative cooperativity. If $n > 1$, the slope is steeper than for $n = 1$ as occurs for positive cooperativity.

The means, whereby $[(\partial\Delta G/\partial\alpha)_T]$ is calculated from the data as plotted in Fig. 15, is described in the footnote to **Table 3**. The results for the glutamic acid data in Fig. 14B, as calculated using Fig. 15, are given in the second and third columns of **Table 3**.

A qualitative comparison of the pK and positive cooperativity data suggest equivalent supra-linearity of hydrophobic-induced responses when plotted against linear replacement of V by F. The quantitative results of columns 2 and 3 of **Table 3** strongly suggest that the same physical process is responsible for both hydrophobic-induced pK and positive cooperativity shifts. In our view the physical process in common is the apolar-polar repulsive free energy of hydration, ΔG_{ap}. Positive cooperativity as carboxyls ionize, where ionization of subsequent carboxyls is greatly facilitated by the earlier ionizations, becomes possible due to competition for hydration between hydrophobic groups and carboxylates. The first ionization is delayed because the first carboxylate does not obtain sufficient hydration. As seen with another feature of a Hill plot, For E/5F in Fig. 16A the first carboxylate forms with of pK 7, and the last carboxylate of the titration forms with a pK of 5.7.

Accordingly, the last carboxylate of the titration of E/5F forms with a pK of 5.7 and indicates an ongoing apolar-polar repulsion, $\Delta G_{ap} = 2.3RT\Delta pK = 2.3RT(5.7 - 4.0) = 2.3$ kcal/mol-E, calculated for 298K. Even the last carboxylate of E/3F with a pK of 4.8, seen in Fig. 16B, retains a substantial apolar-polar repulsion, i.e., $\Delta G_{ap} = 2.3RT\Delta pK = 2.3RT(4.8 - 4.0) = 1.1$ kcal/mol-E. These residual apolar-polar repulsion terms reflect an extent of elastic deformation of the elastic-contractile model protein from the more nearly random chain distribution relevant to $(GVGVP)_n$ below T_t. The residual pK (and positive cooperativity) shifts warrant further discussion not possible with the present time and space limitations, but they reflect the richness of the data available from designed ECMP.

Before going on to the equation for redox titrations of designed ECMP, it may be noted that the data for **Model protein I** in Fig. 15, with both relevant charge-charge and apolar-polar repulsion terms, can be analyzed using the complete Eqn. (17). The close sequence proximity of the two E-residues in **Model protein I**, as seen in the 2E/5F/2I composition of Fig. 14A, ensures a charge-charge repulsion term that can be seen in Fig. 15. In particular, the curve for 2E/5F/2I exhibits a greater pK shift and yet a smaller slope than that of E/4F.

The first of two pieces of information that enable the resolution of all the terms is that ΔpK_{ap} and ΔpK_{cc} are of the same sign, whereas $[(\partial\Delta G/\partial\alpha)_T]_{ap}$ and $[(\partial\Delta G/\partial\alpha)_T]_{cc}$ are of opposite signs. This is apparent by comparison of the curves for E/4F and 2E/5F/2I. The second piece of information comes from the demonstration that $\Delta G_{ap} = 2.3RT\Delta pK$ equals $[(\partial\Delta G/\partial\alpha)_T]_{ap}$, as seen in **Table 3**, and the expectation that $\Delta G_{cc} = 2.3RT\Delta pK$ equals $[(\partial\Delta G/\partial\alpha)_T]_{cc}$. With this perspective, all of the terms in Eqn. (17) can be determined, i.e., $\Delta G_{ap}(2E/5F/2I) = [(\partial\Delta G/\partial\alpha)_T]_{ap}(2E/5F/2I) = 2.32$ kcal/mol-E and $\Delta G_{cc}(2E/5F/2I) = [(\partial\Delta G/\partial\alpha)_T]_{cc}(2E/5F/2I) = 0.23$ kcal/mol-E (See Urry et al., 2009; 2010). This demonstrates for the ECMP-in-water system how much more significant apolar-polar repulsion is than charge-charge repulsion, as previously shown for f_X versus pK plots for poly[f_V(IPGVG),f_E(IPGEG)] (Urry et al., 1993) and for poly[f_V(IPGVG),f_K(IPGKG)] (Urry et al., 1994). Again, while warranted, present time and space limitations do not permit further discussion here, but Figure 5.30 and associated text in (Urry, 2006a) provide further analyses.

Model protein i : [GVGVP GVGVP GΦGVP GVGVP GVGVP GVGVP]$_n$(GVGVP); Φ/0F
Model protein ii : [GVGVP GVGFP GΦGFP GVGVP GVGVP GVGVP]$_n$(GVGVP); Φ/2F
Model protein iii: [GVGVP GVGVP GΦGVP GVGVP GVGFP GFGFP]$_n$(GVGVP); Φ/3F
Model protein iv: [GVGVP GVGFP GΦGFP GVGVP GVGFP GVGFP]$_n$(GVGVP); Φ/4F
Model protein v: [GVGVP GVGFP GΦGFP GVGVP GVGFP GFGFP]$_n$(GVGVP); Φ/5F

Φ(function)	E(glutamic acid, -COOH)		K(lysine amino, $-NH_2$)		K{NMeN}-(redox)	
Energy term kcal/mol	ΔG_{ap}§ (pKa shift)	$(\partial\Delta G/\partial\alpha)$ slope; (*n*)	ΔG_{ap} (pKa shift)	$(\partial\Delta G/\partial\alpha)$ slope; (*n*)†	$z\mathcal{F}\Delta E$ redox shift	$(\partial\Delta G/\partial\alpha')$ slope; (*n*)*
Φ/0F	0.7	1.1 (1.5)	0	0 (0.9)	0	0 (1.0)
Φ/2F	1.0	1.3 (1.6)	0.3	0.3 (1.1)	0.8	0.5 (1.2)
Φ/3F	1.6	1.7 (1.9)	0.6	0.7 (1.2)	1.7	1.1 (1.5)
Φ/4F	2.2	2.2 (2.7)	1.4	1.2 (2.1)	2.5	2.7 (4.8)
Φ/5F	3.1	3.0 (8.0)	2.0	2.1 (2.7)	-----	-----

The numbers in parentheses in the table are the Hill coefficients, *n*, and Hill plots of $\log[\alpha/(1-\alpha)]$ versus pH for acid base titrations and $\log[\alpha'/(1-\alpha')]$ versus **ΔE** for the redox titrations are used, respectively, to calculate $(\partial\Delta G/\partial\alpha)$ and $(\partial\Delta G/\partial\alpha')$. $\partial\alpha$ and $\partial\alpha'$ are determined by graphical means using the y-axis intercept values of $\log[\alpha/(1-\alpha)] = 1$ (α=0.91) and -1(α=0.9), giving the graphically derived values of the divisors of 0.82 (= 0.91 - 0.09) for both $\Delta\alpha$ and $\Delta\alpha'$. The corresponding x-axis reference values are a ΔpH of 2 with $\Delta G = -2.3RT\Delta pH$ for the acid-base titrations and a **ΔE** of 59 mV with $\Delta G = -z\mathcal{F}\Delta\mathbf{E}$ for the redox titrations. The expressions for acid base titrations become $(\partial\Delta G/\partial\alpha) = -2.3RT(\Delta pH^{exptl} - \Delta pH^{ref})/0.82$ with ΔpH^{ref}= 2 and ΔpH^{exptl} being the run corresponding to the $\Delta\alpha$ = 0.82 rise for a readily defined interval of the Hill plot (See section **4.4.5**). Analogously for redox titrations, $(\partial\Delta G/\partial\alpha') = -z\mathcal{F}(\Delta \mathbf{E}^{exptl} - \Delta\mathbf{E}^{ref})/0.82$ with $\Delta\mathbf{E}^{ref}$ = 59 mV and $\Delta\mathbf{E}^{exptl}$ being the run corresponding to the rise of $\Delta\alpha'$ = 0.82 for the specified interval. As ΔpH^{exptl} approaches 2 and as $\Delta\mathbf{E}^{exptl}$ approaches 59, the slope approaches one and the terms $(\partial\Delta G/\partial\alpha)$ and $(\partial\Delta G/\partial\alpha')$ approach zero, which is ***n*** = 1. The sign on the free energies is written as the repulsive free energy, ΔG_{ap}, in which an increase in hydrophobicity increases the free energy on going from the less polar to the more polar state, i.e., to the charged or the oxidized state. This results from an increase in ΔG due to the increase in apolar-polar repulsion.

§ The pKa for the unperturbed glutamic acid is taken as 4.0.

† The Hill coefficient, *n*, for K/0F was experimentally found to be 0.9, possibly due to the effect of Cl^--NH_3^+ ion pair formation during the titration. Since the interest is the change due to addition of more hydrophobic Phe (F) residues, K/0F is taken as the reference state (Woods, TC.; Hayes, L.; Xu, J.; McPherson, DT. & Urry DW., in preparation.).

* The Hill coefficient, *n*, for the reference state composition, K{NMeN}/0F, i.e., N-methyl nicotinamide (NMeN) attached by amide linkage to lysine (K) of **Model protein** i, is taken as 1.0 after including the term, 1/z of $2.3RT/z\mathcal{F}$, for the number of electrons (2) transferred in the overall reaction (See Eqn. 19). (Hayes, LC.; Woods, TC.; Xu, J.; Gowda, DC.; McPherson, DT. & Urry, DW. In preparation.) Adapted from (Urry & Luan, 2009).

Table 3. Comparison of increases in Hydrophobicity on pK and Reduction Potential Shifts and on Positive Cooperativity for different Functional Groups in the Basis Set.

6.2.9 Generalized redox titration equation that includes apolar-polar, ap, repulsion terms

For redox titrations, the equivalent expression to the familiar Henderson/Hasselbalch equation for acid/base titrations, is the Nernst equation, $\mathbf{E} = \mathbf{E_0} + (2.3RT/z\mathscr{F})\log[\alpha'/(1-\alpha')]$, where **E** is the electrical potential in volts for dilute solutions of small non-interacting redox functions, $\mathbf{E_0}$ is the electrical potential of the reference state (the unperturbed redox function), z is the number of electrons transferred during the redox reaction, $\mathscr{F}$ is the Faraday constant of 23,060 cal/volt, and α' represents the mole fraction of the redox species being formed during the titration.

In analogy to acid/base titrations, cooperativity effects can be introduced with the Hill coefficient to give $\mathbf{E} = \mathbf{E_0} + \Delta\mathbf{E}_{ap} + (1/\boldsymbol{n})(2.3RT/z\mathscr{F})\log[\alpha'/(1-\alpha')]$ for redox titrations of designed ECMP. Your author is unaware of hydrophobic-induced shifts in reduction potentials and cooperativity effects being treated in analysis of redox titrations. **Table 3**, however, clearly shows hydrophobic-induced reduction potential and positive cooperativity effects to be larger for designed ECMP with the NMeN redox function than with the E functionality. For example, compare data for E/4F (columns 2 and 3) with those of NMeN/4F (columns 6 and 7).

Accordingly, for our purposes Eqn. (19) will be stated using equivalent terms to those of Eqn. (18), as follows,

$$\mathbf{E} = \mathbf{E_0} + \Delta\mathbf{E}_{ap} + (2.3RT/z\mathscr{F})\log[\alpha'/(1-\alpha')] + (1/z\mathscr{F})[(\partial\Delta G/\partial\alpha')_T]_{ap}, \quad (19)$$

Equation (19) neglects charge-charge interactions between redox functions, because the approximately 100 bonds between functional groups would argue charge-charge interactions to be negligible for this designed ECMP.

The values in columns 6 and 7 of **Table 3** were calculated using the titration data for the oxidation of NMeNH to $NMeN^+$ (Hayes, LC.; Woods, TC.; Xu, J.; Gowda, DC. McPherson, DT. & Urry, DW. Effect of the hydrophobicity of elastic-contractile model proteins on redox potential. In preparation.) and are plotted in analogy to the plots for titration of E/nF in Fig. 15, that is, **E** is plotted versus $\log[\alpha'/(1-\alpha')]$. As representative calculations, $[(\partial\Delta G^{NMeN/4F}/\partial\alpha')_T]_{ap} = (\Delta\mathbf{E}^{ref} - \Delta\mathbf{E}^{4F})_{ap}$ and $\Delta G_{ap}(NMeN/4F) = z\mathscr{F}\Delta\mathbf{E}_{ap}(NMeN/4F)$. For the Φ/4F compositions of **Table 3**, $z\mathscr{F}\Delta\mathbf{E}_{ap}(NMeN/4F) \approx [(\partial\Delta G^{NMeN/4F}/\partial\alpha')_T]_{ap} \approx 2.5$ kcal/mol-NMeN; $\Delta G_{ap}(E/4F) \approx 2.3RT\Delta pK \approx [(\partial\Delta G^{E/4F}/\partial\alpha)_T]_{ap} \approx 2.2$ kcal/mol-E, and $\Delta G_{ap}(K/4F) \approx 2.3RT\Delta pK(K/4F) \approx [(\partial\Delta G^{K/4F}/\partial\alpha)_T]_{ap} \approx 1.4$ kcal/mol-K (Woods, TC.; Hayes, LC.; Xu, J.; McPherson, DT. & Urry. DW. Lys-containing elastic-contractile model proteins: Biosynthesis and supra-linear increases in ΔpKa and in positive cooperativity with linear increases in hydrophobicity. In preparation).

Not only is the apolar-polar repulsive Gibbs free energy of hydration apparent for the N-methyl nicotinamide function in the ECMP composition, NMeN/4F, but it is 14% larger than for the glutamate function in the ECMP composition, E/4F, and 80% larger than the lysine function in the ECMP composition, K/4F. This might have been expected due to the large increase in hydrophobicity on reduction, i.e., $\Delta G°_{HA}(NMeN^+ \rightarrow$ dihydro NMeN) = 9.5 kcal/mol-NMeN in **Table 2**.

It seems quite apparent why biology so routinely utilizes redox functions in its energy conversion processes. In particular, the nicotinamide adenine dinucleotide (NAD), nicotinamide adenine mononucleotide (NMN) and flavin adenine dinucleotide (FAD) redox functions become reduced in the process of oxidation of foods, and especially NADH and $FADH_2$ enter into the electron transport *chain and reduce ubiquinol. Ubiquinol then becomes cyclically oxidized and reduced in the process of pumping protons across the inner mitochondrial membrane to the inner membrane space. These protons then flow back across the inner mitochondrial membrane through ATP synthase to produce 32 of the 36 ATPs formed on the oxidation of glucose, for example.*

Section **7.1** below, presents the means whereby Complex III of the inner mitochondrial membrane utilizes cyclic reduction and oxidation of its protein-bound redox functions – heme $\mathbf{b_H}$, heme $\mathbf{b_L}$ and an FeS center - in combination with the protein mechanisms discussed here - $\mathbf{\Delta G_{HA}}$, ΔG_{ap}, and single-chain elastic extension and contraction - to pump protons across the inner mitochondrial membrane. And section **7.2** below presents means whereby the F_1-motor of ATP synthase utilizes ΔG_{ap} to produce ATP from ADP and Pi(inorganic phosphate).

6.2.10 Elastic deformation on the hydrophobic association (extension) and on ΔG_{ap} (repulsion)

The elasticity of $(GVGVP)_n$ has been characterized by many physical methods with the necessity of delineating it from the random chain network theory of elasticity. The nature of ECMP elasticity is most conclusively demonstrated by means of single-chain force-extension experiments, see Fig. 6E, using the methodology of atomic force microscopy (Hugel, 2003; Urry et al, 2002). Instead of rastering in the x-y plane of the substrate surface, a single chain suspended between the substrate plane and the cantilever tip is pulled in the z-direction to demonstrate an overlay of the extension and relaxation curves. Thus, within the sensitivity of the force measurement, $(GVGVP)_n$ exhibits ideal elastic behavior. Because of a common observation of some very minor deviations of the relaxation curve from the extension curve, and because ideality is a goal approached but never quite obtained, we choose to refer to the "near ideal" elasticity of a single chain of $(GVGVP)_n$. Obviously, the random chain network theory of elasticity is not applicable to the single-chain force-extension results of Fig. 6E on $(GVGVP)_n$.

The polymer construct is central to a robust elasticity. The first mathematical expression for the propagation of a polymer (Eyring, 1932), depicted in Fig. 6A, demonstrates the two key features, the torsional, or dihedral, angle, ϕ, that provides the entropic component of elastic force, f_S, and the backbone angle, θ, and to a lesser extent the bond length, ℓ, that provide the internal energy component of elastic force, f_E.

The molecular mechanics ECEPP program due to Scheraga and coworkers (Momany et al., 1975) and molecular dynamics CHARMM program of Karplus and coworkers (Brooks, et al., 1983) both give the same value of five EU/pentamer-(GVGVP), (Urry, et al., 1982d; Urry & Venkatachalam, 1983; Chang & Urry, 1989), which satisfactorily calculate the magnitude of the entropic component of elastic force, f_S (Urry, 2006b). And the values, so obtained, for $(GVGVP)_n$ have been instructive in estimating the elastic force on extension of single protein chains on going from one functional state to another in the crystal structures of protein motors (Urry, 2006b; Urry et al, 2010). Here, the values from molecular dynamics calculations are listed in **Table 4** for the damping of torsional oscillations that occur on 130% extension of a single chain of β-spiral as described in the relaxed state in Fig. 5 and in relaxed and extended states in Figs. 6D1 and D2.

As seen in **Table 4**, the torsion angles for the suspended segment - ψ_{4i}, ϕ_{5i}, ψ_{5i}, and $\phi_{1(i+1)}$ -, indicated by rotational arrows in Fig. 6B, are the angles that exhibit large torsional oscillations in the relaxed state that become dramatically damped on extension. Fig. 6C

	Angle	Relaxed	Extended	Angle	Relaxed	Extended	Angle	Relaxed	Extended
β-turns	ψ_{16}	10.87	14.17	ψ_{26}	27.33	07.64	ψ_{36}	20.19	14.84
	ϕ_{17}	09.86	15.18	ϕ_{27}	11.71	08.33	ϕ_{37}	08.99	10.47
	ψ_{17}	47.59	46.68	ψ_{27}	11.70	13.51	ψ_{37}	21.53	32.96
	ϕ_{18}	61.70	47.41	ϕ_{28}	08.61	10.36	ϕ_{38}	11.15	10.66
	ψ_{18}	09.37	16.05	ψ_{28}	09.33	08.16	ψ_{38}	11.09	27.29
	ϕ_{19}	14.25	08.67	ϕ_{29}	09.70	07.31	ϕ_{39}	12.70	10.24
Suspended	ψ_{19}	44.09	10.99	ψ_{29}	47.32	10.48	ψ_{39}	52.00	12.50
	ϕ_{20}	41.94	09.29	ϕ_{30}	48.57	11.39	ϕ_{40}	55.88	08.37
Segments	ψ_{20}	14.50	11.15	ψ_{30}	42.56	10.62	ψ_{40}	40.67	11.08
	ϕ_{21}	27.13	24.17	ϕ_{31}	11.43	11.38	ϕ_{41}	36.44	19.06
β-turns	ψ_{21}	09.39	22.73	ψ_{31}	12.17	09.21	ψ_{41}	12.97	14.80
	ϕ_{22}	09.94	08.00	ϕ_{32}	09.90	08.93	ϕ_{42}	11.59	07.33
	ψ_{22}	11.58	16.13	ψ_{32}	15.30	10.80	ψ_{42}	11.34	13.17
	ϕ_{23}	16.37	09.33	ϕ_{33}	09.60	07.62	ϕ_{43}	09.23	09.76
	ψ_{23}	14.33	14.25	ψ_{33}	09.88	09.43	ψ_{43}	10.60	12.53
	ϕ_{24}	11.39	29.20	ϕ_{34}	11.86	09.71	ϕ_{44}	11.06	12.82
Suspended	ψ_{24}	19.53	37.87	ψ_{34}	63.80	08.36	ψ_{44}	41.89	35.22
	ϕ_{25}	25.02	23.06	ϕ_{35}	91.70	10.20	ϕ_{45}	48.98	31.31
Segments	ψ_{25}	49.32	32.10	ψ_{35}	15.03	11.51	ψ_{45}	42.05	56.89
	ϕ_{26}	31.43	27.24	ϕ_{36}	21.49	18.66	ϕ_{46}	21.55	30.33

Data from Chang & Urry, 1989, and tabulation from Urry & Parker, 2002.

Table 4. Comparison of the root mean square (RMS) fluctuations of torsion angles (ϕ and ψ) of VP(GVGVP)$_{10}$GVG for the non-extended and 130% extended states. Large decreases on extension in the amplitude of the ϕ and ψ torsional oscillations of the suspended segments are readily apparent, particularly when compared to the ϕ and ψ of the β-turns.

schematically demonstrates with the product of three torsion angle oscillations - $\Delta\phi_i \times \Delta\psi_I \times \Delta\phi_{i+1}$ – the large decrease of volume in configuration space that can occur on damping of the amplitude of three torsional oscillations. This decrease in volume represents a decrease in entropy, but instead of just three angles the computation will utilize the 31 internal torsion angles listed in **Table 4**. The equation, as written for extended, **e**, and relaxed, **r**, states, $\Delta S = R \ln[\Pi_i \Delta\phi_i^e \Delta\psi_i^e / \Pi_i \Delta\phi_i^r \Delta\psi_i^r]$, calculates the decrease in entropy due to a 130% extension to be -1.1 cal/mol-deg-residue. From looking at the decreases in amplitudes of the angles in **Table 4**, the suspended segments overwhelmingly give rise to the decrease in entropy. Using the relationship, $f_S = -T(\partial S/\partial L)_{VT}$, at 10 K for an extension, ΔL, of 3.5 to 8.0 nm per pentamer gives an f_S of 24 pN (Urry et al., 2002; Urry, 2006b). Very similar ΔS values were obtained for ΔS(GVGVP) in water (Wasserman & Salemme, 1990) using yet a third computational approach, the Kollman molecular dynamics program, for such calculations. Single-chain extension for elastomeric force development has been implicated in the function of several proteins - the Rieske Iron Protein component of Complex III of the electron transport chain discussed in section **7.2** below, e.g., the myosin II motor of muscle contraction (Urry, 2006a, 2006b), and the kinesin bipedal motor (Urry, 2005).

7. ECMP-derived thermodynamics of protein hydration and of elasticity describe function of biology's protein-based machines!

The above described thermodynamics of protein hydration ($\mathbf{\Delta G_{HA}}$ and ΔG_{ap}) and of elasticity (f_E and f_S), which with displacement give ΔG(elastic deformation), provide insight into key details of the function of biology's protein-based machines. Considered in this section are: Complex III (cytochrome bc_1/Rieske Iron Protein) of the inner mitochondrial membrane, the

F_1-motor of ATP synthase also of the inner mitochondrial membrane, and the full-length KcsA potassium ion channel of *Streptomyces lividans.* Function in each protein-based machine involves trans-membrane structures of decreasing complexity. At this stage, the thermodynamic quantities describe individual events at each site of action, but integration of events into a complete image of trans-membrane transport is in the future. Example of how this may be achieved occurs in section **8. Eyring's Absolute Rate Theory applied to biological trans-membrane transport**, whereby a single image, the free energy profile for ion passage through the monovalent cation selective Gramicidin A channel, contains all of the information required to calculate current for a chosen ion activity and trans-membrane potential.

7.1 Complex III of the electron transport chain of the inner mitochondrial membrane

There are five complexes within the inner mitochondrial membrane that achieve oxidative phosphorylation. Four complexes couple electron flow to proton translocation from the matrix side of the membrane to the inter-membrane space, i.e., cytosolic side. The fifth complex, ATP synthase, utilizes the proton concentration developed in the inner-membrane space by the first four complexes to synthesize ATP from ADP and Pi as the protons re-cross the inner membrane to the matrix side. Treated below are Complex III, of the four complexes, and the fifth complex, ATP synthase, emphasizing its F_1-motor.

7.1.1 Complex III (cytochrome bc_1/Rieske Iron Protein) of the inner mitochondrial membrane

Also referred to as the cytochrome $\mathbf{bc_1}$ complex, Complex III of yeast is a homodimeric integral membrane protein comprised of 22 subunits. The structure of Fig. 17A contains the 22 protein subunits plus a single molecule of cytochrome **c**, where it is positioned to be reduced before diffusing to the fourth complex, cytochrome **c** oxidase, to release its electron, which passes through copper ions to the two hemes, **a** and $\mathbf{a_3}$.

A stereo view of the redox functions are given in Fig. 17B, positioned exactly as in the complete complex in part A. There are two reaction sites to note, the Q_o site and the Q_i site. At the Q_o site on the cytosolic side of the membrane, ubiquinol meets and transfers one electron to the FeS center and a second electron to heme $\mathbf{b_L}$; this leaves the ubiquinol with two positive charges which it releases to the cytosolic side of the membrane as protons to become ubiquinone and the FeS center returns to the heme $\mathbf{c_1}$. The ubiquinone leaves and a second ubiqiunol enters to repeat the process. The net result at the Q_o site is a tightly coupled receipt of four electrons and the release of four protons to the cytosolic space, by means of the three thermodynamic processes presented in section **6** and as depicted in Fig. 17C.

At the Q_i site on the matrix side of the membrane, heme $\mathbf{b_H}$, having received an electron from heme $\mathbf{b_L}$, passes it on to ubiquinone. This step occurs twice to give a double negative charge to ubiquinone, and to a second ubiquinone, which pick up four protons from the matrix side of the membrane to become become two ubiquinols that can each diffuse through the lipid bilayer to enter the Q_o site.

7.1.2 The cyclic domain movement of the Rieske Iron Protein FeS center between cytochrome b and c_1

Using each of the three thermodynamic processes developed in section **6**, represented as ΔG_{HA}, ΔG_{ap}, and single chain stretching/contracting, the tightly coupled cyclic reaction at the Q_o site is depicted in three steps in Fig. 17C.

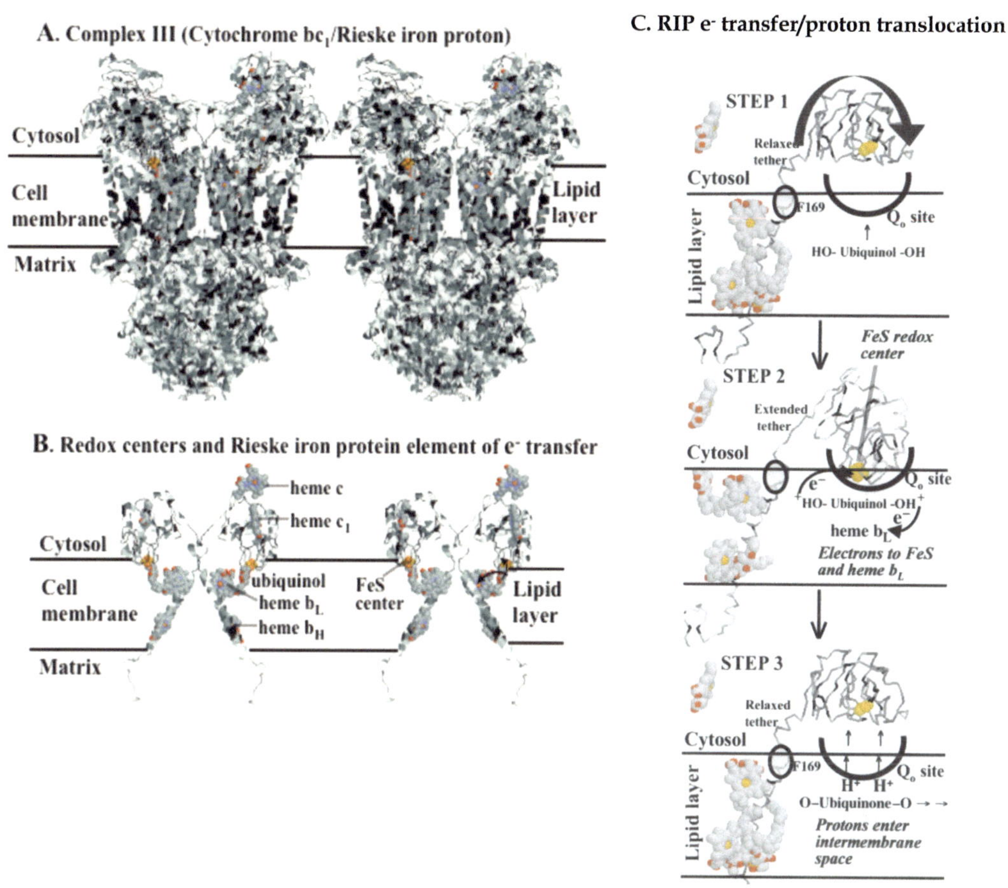

Fig. 17. **A**. Stereo pair of complete homo-dimeric structure of Complex III from yeast, plus one molecule of its substrate, cytochrome c (at front upper right). Given in ribbon representation with gray code for residues - charged white, neutral light gray, aliphatic hydrophobics gray, and aromatics black. The location of cell membrane and its lipid bilayer are indicated as well as the cytosolic (inner membrane space) and matrix sides of the membrane. Complex III functions by translocating protons from the matrix side to the cytosolic side of the membrane, as it oxidizes ubiquinol of the lipid layer at the Q_o site on the cytosolic side of the membrane and reduces ubiquinone at the Q_i site on the matrix side.
B. Stereo pair of homodimeric Rieske Iron Protein (**RIP**) with labeled redox components of Complex III (positioned exactly as in part **A**) and electron transfers at right. Note: While the globular component of RIP and its FeS center are in one monomer, its anchor in the membrane occurs with the transmembrane helices of the other monomer. **A** and **B** (Urry et al., 2010) utilize the crystal structure data of Lange & Hunte (2002). PDB accession code, 1KYO.
C. Rieske Iron Protein (RIP) and redox components of the monomer of Complex III showing 3 Steps in the cycle of electron transfer/proton translocation achieved by domain movement into and out of the Q_o site with stretching of its anchored tether as the globular tip of RIP with its FeS redox center is drawn into the Q_o site by hydrophobic association. The three steps are discussed in detail in the text, utilizing the crystal structure data of Zhang et al., 1998. Protein Data Bank, accession code 1BCC & 3BBC. From Urry et al., 2010.

STEP 1: The hydrophobic ubiquinol enters the Q_o site and, on making the site more hydrophobic, induces the globular component of RIP with its hydrophobic side and tip to roll into the Q_o site increasing its hydrophobic association as it goes. *This lowers ΔG_{HA}, the the change in Gibbs free energy for hydrophobic association (the first thermodynamic process).*
STEP 2: In the thermodynamic process of lowering ΔG_{HA} as the RIP globular component rolls into the Q_o site, the relaxed single chain tether with its end segment anchored in the cell membrane becomes stretched, *thereby utilizing the second thermodynamic process of storing energy in the stretching of the single chain.* This positions the FeS center in the Q_o site to receive one electron from ubiquinol, which transfers a second electron to heme $\mathbf{b_L}$ and on to heme $\mathbf{b_H}$ at the Q_i site. The result is a ubiquinol with two positive charges.
STEP 3: The two positive charges of ubiquinol exert an apolar-polar repulsion, *ΔG_{ap}, bringing into play the third thermodynanamic process,* that weakens the hydrophobic association, allows the stretched tether to contract and lift the globular component with FeS center to heme $\mathbf{c_1}$ for transfer of its electron. This vacating of the Q_o site opens access for release of two protons to the inner-membrane space. The resulting ubiquinone diffuses into the lipid layer, and is replaced by another ubiquinol, which returns the Q_o-site to the state for STEP 1.
*Thus, the three thermodynamic processes developed in section **6** (**ΔG_{HA}**, ΔG_{ap}, and single chain stretching/contracting) are seen to be the basis for the tight coupling of electron transport to proton translocation performed by Complex III (cytochrome bc_1/Rieske Iron Protein) of the inner mitochondrial membrane in the essential function of energy conversion in the mitochondria, the essential energy factory of the cell.*

7.2 The F_1-motor of ATP synthase of the inner mitochondrial membrane

7.2.1 Partial structure of ATP synthase as obtained by electron diffraction (Stock et al. 1999)

Fig. 18 gives a stereo view of the structure of ATP synthase of yeast mitochondria as determined by electron diffraction with residues represented as spheres. The spheres are gray-coded with charge residues white, neutral residues light gray, aliphatic hydrophobic residues gray and aromatic residues black. Missing, however, are the a-subunit, which is a partial sleeve on the F_0-rotor (the c_{10}-subunit) that completes the proton channel from cytosol to matrix, and the stator (b_2-subunit) that locks from the sleeve to the $(\alpha\beta)_3$-catalytic housing. This ensures that the F_0-rotor-couplings-$F_1(\gamma)$-rotor assembly may rotate without rotation of the $(\alpha\beta)_3$-catalytic housing.
At the top-side of Fig. 18 is the cytosolic side of the inner mitochondrial membrane, i.e., the inter-membrane space into which Complex III of Fig. 17 pumps protons. One such proton from the inter-membrane space then passes through a proton channel between the a-subunit sleeve and the 10-fold symmetric F_0-rotor to bind at the D61 carboxylate, e.g., the left most of the two white dots seen on the left-hand side of F_0-rotor just below the midline.
Drawing from the mechanism due to Fillingame and coworkers (Fillingame, 1999; Fillingame et. al., 2002), as a proton from the inner membrane space flows through the channel to bind at the dot on the left, an already protonated D61, e.g., the white dot at the right, releases a proton to the matrix side of the membrane, resulting in 36° clockwise rotation (as seen from the top) of the F_0-rotor. The translocation of 10 protons completes a 360° rotation of the F_0-rotor, which by the noted couplings of Fig. 18 drives a complete 360° rotation of the F_1-(γ)-rotor within the $(\alpha\beta)_3$-catalytic housing.

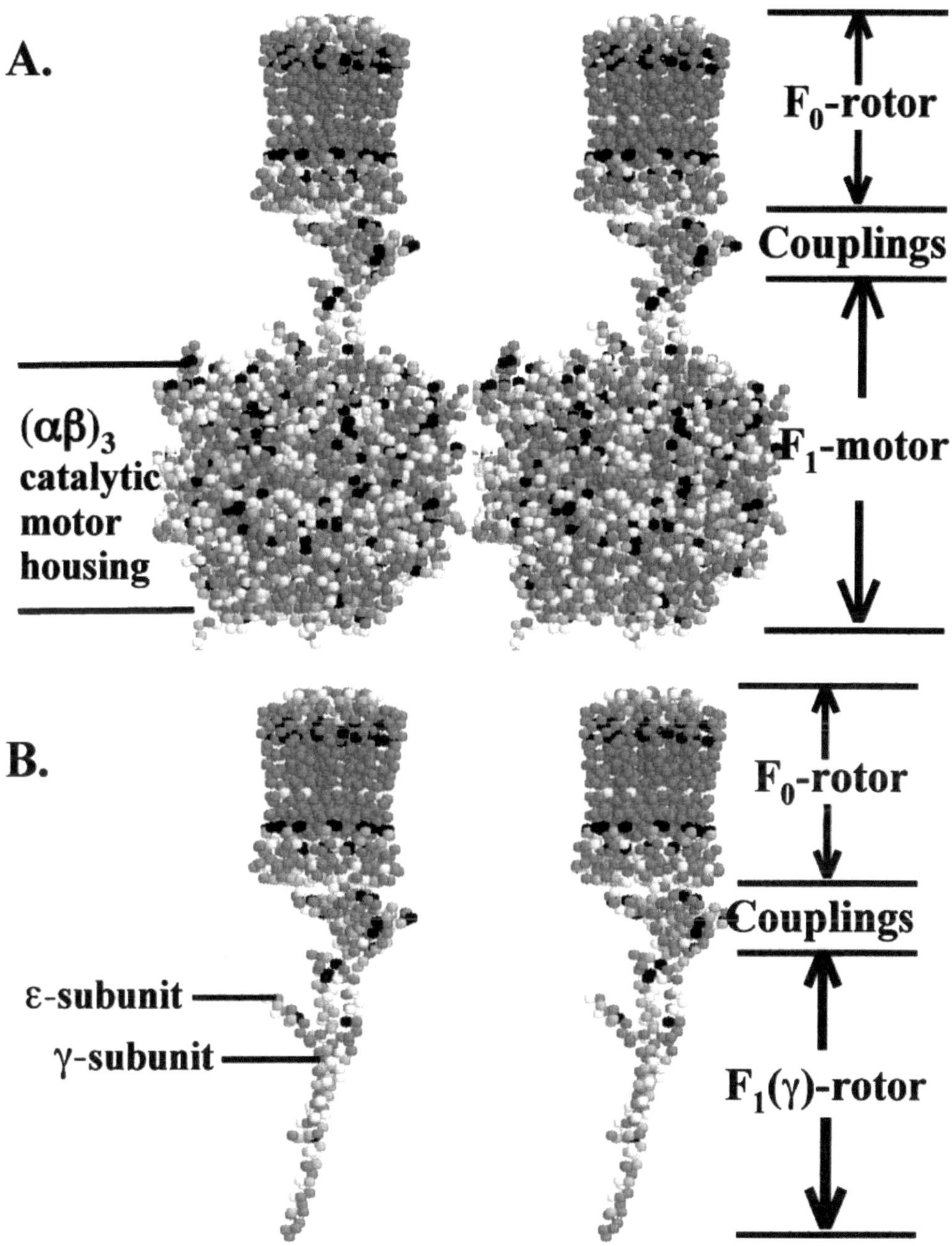

Fig. 18. Partial structure of ATP synthase obtained by electron diffraction with residues given as spheres.
A. Missing are the a-subunit and the stator (b_2-subunit).
B. The complete F_0-rotor-couplings-$F_1(\gamma)$-rotor assembly.
Protein Data Bank, accession code 1QO1. Adapted from Figure 8.23 of Urry, 2006a.

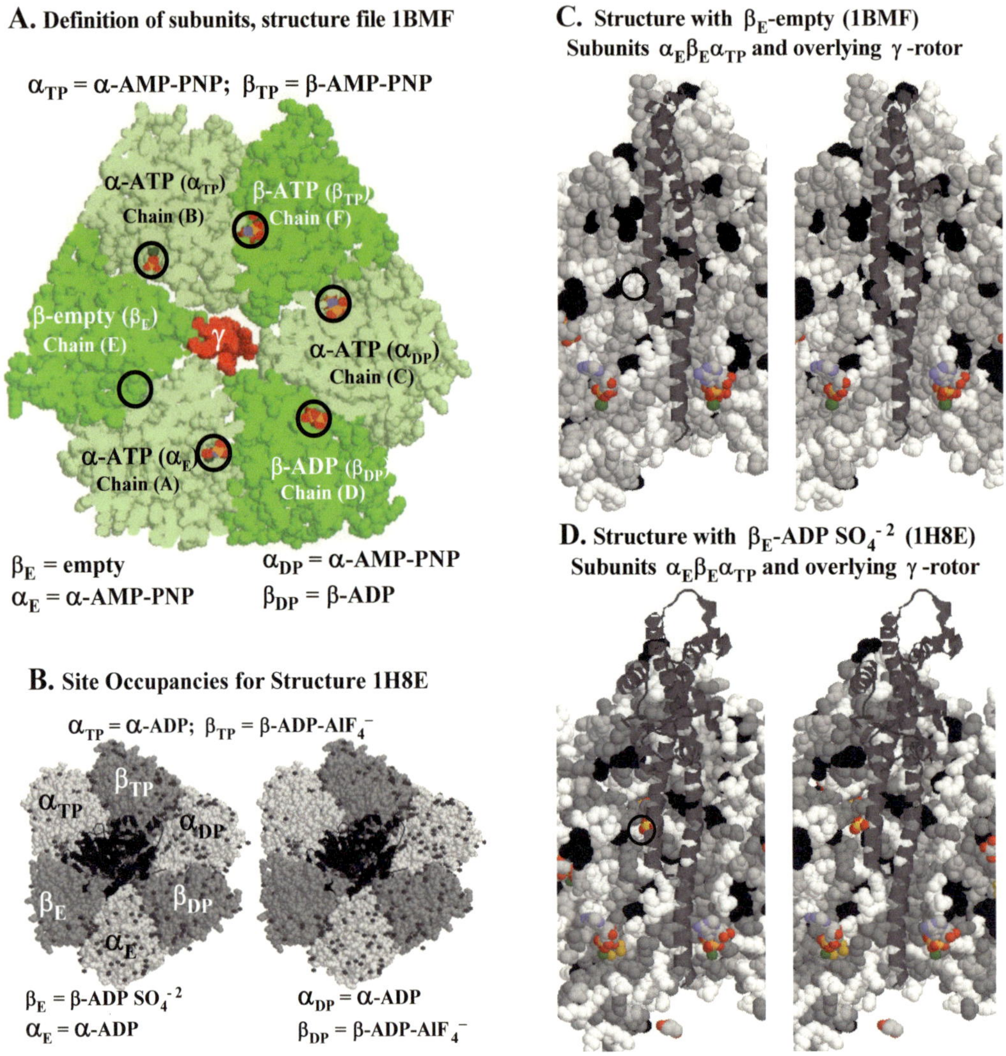

Fig. 19. **A** and **B**. Nomenclature for the α- and β-subunits of the $(\alpha\beta)_3$-catalytic housing of the F_1-motor of ATP synthase as defined from the crystal structures with Protein Data Bank accession codes, 1BMF (Abrahams et al., 1994) and 1H8E (Menz et al., 2001). The F_1-motor, when separated from ATP synthase, functions as an ATPase, named the F_1-ATPase. **C**. Looking at the inside of the catalytic housing of structure 1BMF from behind the γ-rotor toward the three subunits, $\alpha_E\beta_E\alpha_{TP}$, into a cleft, marked by small black circle, where the SO_4^{-2} analogue of HPO_4^{-2} would exist in structure 1H8E. **D**. Looking at the same internal view as in **C** to see the location of the highly charge sulfate, SO_4^{-2}, peeking out of the cleft. Adapted from Urry, 2006c.

7.2.2 Demonstration of apolar-polar repulsion, ΔG_{ap}, between phosphate analogue and the short helix of the γ-rotor

The views of Fig. 19 were developed from the structures listed in the Protein Data Bank as 1BMF (Abrahams et al., 1999) and 1H8E (Menz et al., 2001) for the purpose of demonstrating the ΔG_{ap}, the apolar-polar repulsive free energy of hydration. The strong repulsion exists, between the very polar, doubly charged sulfate group (SO_4^{-2}), an analogue for biologically potent phosphate (HPO_4^{-2}), and the very hydrophobic side of the γ-rotor. The three sides were identified as those sides facing the β-catalytic subunits as defined in Fig. 19A. By inspection the least polar, most hydrophobic β-subunit would be the empty site, β_E, the site without negatively charged nucleotides. The site containing the less charged nucleotide, ADP, is labeled β_{DP}, and the most charged site containing ATP is labeled β_{TP}. Calculations of the sides utilized the $\Delta°G_{HA}$ values of **Table 1**, and gave -20 kcal/mole for the β_E face of the rotor, +9 kcal/mol for the β_{DP} face of the rotor, and 0 kcal/mol for the β_{TP} face of the rotor. (See Urry, 2006a; 2006c). *Thus, a most hydrophobic face of the rotor was found, as necessary for ΔG_{ap}, to be an operative mechanism.*

This is the first prediction for ΔG_{ap} to be a factor in the mechanism of action of the F_1-motor of ATP synthase, namely, "***Prediction 1:*** *The rotor must be hydrophobically asymmetric.*" In what follows each "*Prediction*" will be mentioned, followed by a brief explanation.

"***Prediction 2:*** *In the static state the most hydrophobic side of the rotor faces the least polar side of the motor housing.*" As seen in Fig. 18A, the most hydrophobic side of the rotor faced the empty site of 1BMF. This should also hold for structure 1H8E, which would require that β_E = β-ADP SO_4^{-2} would be less polar than β_{DP} = β-ADP-AlF_4^{-2} and β_{TP} = β-ADP-AlF_4^{-2} as defined in Fig. 18B. On the basis of the Pauling Electronegativity Scale (Pauling, 1932; 1960) ADP-AlF_4^{-2} calculates to be more electronegative than ADP SO_4^{-2}.

"**Prediction 3:** *Role of ATP in the non-catalytic α-ATP subunits is one of triangulation of repulsive forces to lessen visco-elastic drag between rotor and housing as required for efficiency.*" The long-standing question of why is it important to have ATP in the non-catalytic α-subunits has now been answered. The ΔG_{ap} from the very polar, but not most polar occupant of the non-catalytic subunits, acts to disrupt hydrophobic associations that would otherwise cause a significant visco-elastic drag.

"***Prediction 4:*** *Negative cooperativity for ATP binding.*" The three faces of the γ-rotor are of very different hydrophobicity. The face with a $\Delta G°_{HA}$ of +9 kcal/mol would be the most attractive for the ATP to enter the apposed β-catalytic site. The next ATP would add to the β-catalytic site apposed to the face with a $\Delta G°_{HA}$ of ~0 kcal/mol. And the slowest rate of ATP addition would be to the β-catalytic site with a $\Delta G°_{HA}$ of -20 kcal/mol. Thus, there would be the appearance of a negative cooperativity of ATP binding to the β-catalytic sites. For discussion of the biochemical kinetics see Boyer, 1993: 1997.

"***Prediction 5:*** *Positive cooperativity of increased ATP occupancy of catalytic sites on rate of hydrolysis (rotation).*" Just as the ATP occupancy of the α-non-catalytic sites would decrease visco-elastic drag by disrupting hydrophobic association and increase rate of γ-rotor rotation, so too would ATP occupancy of the β-catalytic sites. Thus, the positive cooperativity of ATP binding to β-catalytic sites would be reflected in dependence of rate of hydrolysis on ATP occupancy.

"***Prediction 6:*** *Increase in distance between rotor and housing due to ΔG_{ap} repulsion, acting through water, between the most hydrophobic side of the rotor and the ADP-SO_4 analogue of the most polar state.* As seen in Fig. 20, the short α-helix of the γ-rotor seems to wrap partially around the

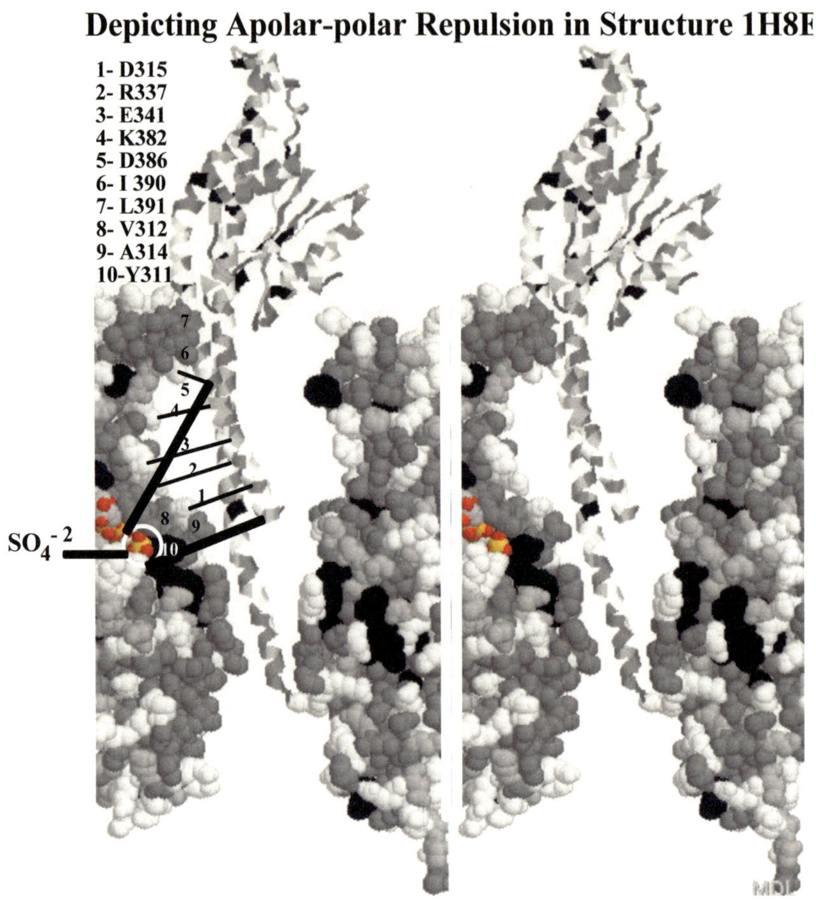

Fig. 20. Two subunits of 1H8E, as defined in Fig. 19B with β_E(β-ADP SO_4^{-2}) on the left and α_{DP}(α-ADP) on the right, to show the apolar-polar repulsive free energy of hyration between the very polar β_E(β-ADP SO_4^{-2}) on the left and the most hydrophobic face of the γ-rotor. The identified residues 8, 9, and 10 form one side of a cleft, identified by the black circle in Fig. 19 C and D. See text for further discussion. Protein Data Bank accession code 1H8E (Menz et al., 2001). Adapted from Urry, 2006c.

long α-helix of the γ-rotor, which is constrained at entry and at the bottom of the chamber. In addition to noting this, Menz et al. (2001) pointed out that the γ-rotor is displaced by an average of 3Å from the β-ADP SO_4^{-2} catalytic site. Both of these effects are remarkably well explained by ΔG_{ap}, the apolar-polar repulsive free energy of hydration, between the very polar β-ADP SO_4^{-2} catalytic site and the most hydrophobic face of the γ-rotor, as developed in section **6.2.6**.

"***Prediction 7:*** *A repulsive force acting through "waters of Thales" to store energy in elastic deformation provides the opportunity for high efficiencies of energy conversion by F_1 ATPase."* The water between hydrophobic surfaces and polar, especially charged sites has been called the "waters of Thales" in honor of Thales of Miletus, father of the Ionian Enlightenment, who in the sixth century BC asked, "What is the world made of?" and answered that water is the basis of all matter. Water, as put forward here, provides the means whereby the unique protein-in-

water heat engine of biology with hydrophobicity dependence of T_t performs the work of the cell. ΔG_{ap} between ATP binding sites and the γ-rotor, providing energy for elastic deformation, would decrease visco-elastic drag and increase efficiency of energy conversion.
"***Prediction 8:*** ***ΔG_{ap}*** *drives counter-clockwise rotation of the F_1 ATPase rotor*" follows from the relative positioning of the sulfate in the recess of a deep cleft and the orientation of γ-rotor, as seen in Fig. 19D. The rotation of the γ-rotor would be in a counter-clockwise direction due to the ΔG_{ap} even seen to cause an elastic deformation and displacement of the very polar β-ADP SO_4^{-2} catalytic site from the most hydrophobic face of the γ-rotor. This is seen more clearly in Fig. 20, where residues 8, 9, and 10 form a vein jutting out into the aqueous chamber and direct the apolar-polar repulsion, ΔG_{ap}, at the short α-helical limb of the γ-rotor. Looking down from the top, this would result in the counter-clockwise rotation of the γ-rotor. The remarkable work of Noji and coworkers (Noji, et al., 1997; Noji, 1998) demonstrated ***Prediction 8*** by directly observing the rotation by a video microscope. The rotation could be shown to occur in 120° steps as expected for a three-fold rotary motor. And the F_1-ATPase has been reported to rotate a long actin filament with an efficiency of near 100%. (Kinosita et al., 2000). See Epilogue, pages 551 - 555, of Urry, 2006a for more extensive discussion of the 8 predictions.
In your author's view, there could not be a more definitive demonstration of the prominent role of $\mathbf{\Delta G°_{HA}}$ and ΔG_{ap} in the description of the thermodynamics of protein function, and this has occurred with arguably the most important protein motor of biology that is responsible for nearly 90% of the ATP formed due to the oxidation of the representative food, glucose.

7.3 The full-length KcsA potassium-ion selective channel from Streptomyces lividans

The full-length KcsA potassium-ion selective channel from *Streptomyces lividans* has been reported recently by Uysal et al., 2009. As with the other protein structures obtained from the Protein Data Bank, KcsA has been analyzed using the program FrontDoor to Protein Explorer developed by Eric Martz (Martz, 2002).

7.3.1 Approach to the structure/function problem of the KcsA channel

The structural perspectives developed for KcsA will be in either space-filling or ribbon representation. Also a gray coding of amino acid residues will be used, where charged residues are white, neutral residues are light gray, aliphatic residues of intermediate hydrophobicity are gray, and aromatic residues, the most hydrophobic residues based on the $\mathbf{\Delta G°_{HA}}$-Hydrophobicity Scale of **Table 1**, are black.
As our interest is in visually delineating between polar, e.g., charged, groupings and hydrophobic residue or domains, the gray coding scale allows for ready visualization of the distribution and locations of these disparate groups, whereby the darkness of the domain indicates the greater hydrophobicity. In particular, the argument has been developed that there exists a competition between hydrophobic and charged residues for hydration, which competition expresses as an apolar-polar repulsive free energy of hydration, a ΔG_{ap}.

7.3.2 Relevance of the thermodynamics of protein hydration to function of the KcsA channel

In particular, hydrophobic association develops as too much pentagonally structured water exists and the [-TΔS] term of the Gibbs expression for dissolution in water, i.e.,

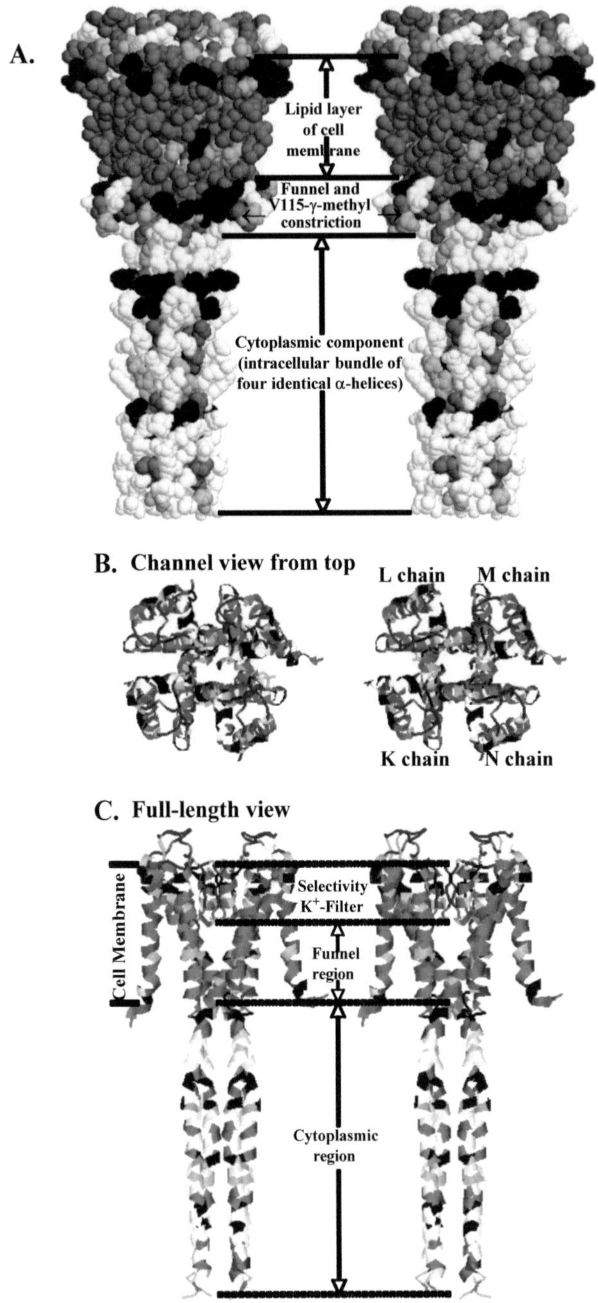

Fig. 21. KcsA potassium-ion channel of four identical chains with complete cytoplasmic component in cross-eye stereo view. **A**. In space-filling representation and gray scale where darker indicates greater hydrophobicity and white means charged residues. **B.** Channel axis view as seen from top showing four chains (KLMN) in ribbon representation. **C.** Full-length view, showing selectivity filter, funnel region, constriction due to close contact of γ-methyls of V115. PDB accession code 3EFF (Uysal et al., 2009). B. and C. from Urry et al., 2010.

Eqn. (4), ΔG(dissolution) = ΔH - $T\Delta S$, becomes too positive and solubility is lost. As a hydrophobic association fluctuates toward disassociation, too much hydrophobic hydration forms and solubility would be lost except when a charge forms during the fluctuation toward hydrophobic disassociation, wherein nascent charge recruits the pentagonal waters of hydrophobic hydration for its own hydration. The [-TΔS]-term decreases in magnitude and the fluctuation toward dissociation stands.

This perspective becomes immediately relevant to the KcsA channel due to its dependence of conduction on pH. As shown by Thompson et al., 2008, titrating the carboxyl function of glutamic acid from -COOH to -COO^- turns off conductance. With this experimental finding, one immediately looks for the formation of charged carboxylates that might disrupt hydrophobic associations that would hold the channel in the open state. So the intention is to progress from the overall representation in Fig. 21A to the details of interest.

The full structure of KcsA is given in Fig. 21 in cross-eye stereo pair and in space-filling representation with the gray coding mentioned above. Immediately apparent is the dark band with the designation of lipid layer of the cell membrane. The dark band serves to identify a ~30Å lipid-bilayer of the cell membrane that anchors the structure in the cell membrane.

By means of the ribbon representation without side chains in Fig. 21B, it is possible from the top of the channel to look from the extracellular side of the cell membrane into a continuous channel and also begin to gain a sense from this perspective of the folding of the four chains, K, L, M, and N.

The full-length view of the channel in Fig. 21C, allows a ready view of how the four chains form the superstructure of the channel. There is a specialized selectivity filter, the sequence of which provides the signature for a potassium-ion selective channel. And there is a funnel region ending in a constriction that closes the channel, and there is the lengthy protrusion into the cytoplasm of the cell. (See Uysal et al., 2009; Urry et al., 2010).

7.3.3 Carboxylate disruption of a key hydrophobic association

By means of a significant break from four-fold symmetry of the four identical chains of 3EFF (Uysal et al., 2009), the carboxyls of $E118_K$ and $E118_M$ are folded into and completely block the channel immediately below the V115 constriction. Without the apolar-polar repulsion of the $E118_K$ and $E118_M$ carboxylates, hydrophobic association occurs involving $W113_L$, interacting with $H25_L$ and $W26_L$ and nascent interaction of $W113_N$ with $H25_N$ and $W26_N$, as seen in Fig. 22B. When the $E118_L$ and $E118_N$ carboxylates are outside the channel structure, they are in a position exert a ΔG_{ap}, an apolar-polar repulsion, on $W113_K$, $H25_K$ and $W26_K$ and on $W113_M$, $H25_M$ and $W26_M$, respectively, that disrupts those hydrophobic associations as seen in Fig. 22A.

The $W113_L$ hydrophobic association with W26L and H25L, W113-(W26/H25), exerts a pull through the backbone sequence of $W113_L$-$F114_L$-$V115_L$ to draw the γ-methyl of V115 outward toward an opening of the constriction and of the selectivity filter (See Fig. 23F).

7.3.4 The effect of the $W113_L$-($W26_L$/$H25_L$) hydrophobic association on opening the Val115 γ-methyl constriction

The cross-section (slab cuts) at V115 γ-methyls are given in Fig. 23. From these the V115 γC-center to γC-center distances are obtained to give for the crystal structure 3EFF in Å, i.e., (N→L) = 6.322, (N→K) = 3.810, c(L→M) = 4.199, d(M→N) = 4.164, and e(L→N) = 4.864. By graphical analysis these distances allow estimates of channel closed and open states as shown in Fig. 23A and B, respectively.

Simply put the N and K γ-methyls of the V115 residues are in close contact, and completing them in square-planar arrangement gives the closed state in Fig. 23A, whereas the outward displacement $V115_N$ γ-methyl is made square-planar to give the open state represented in Fig. 23B.

7.3.5 Consider the hydrophobic association/dissociation (HA/HD) system, W113-(W26/H25), of the KcsA channel and its consequences

Consider, poised near 37°C, the W113-(W26/H25) hydrophobic association(HA) ↔ hydrophobic dissociation(HD) system with proximal carboxylates, **E118** and **E120**, of the KcsA channel in Fig. 24A. Decreasing charge increases the amount of hydrophobic hydration, $\mathbf{N}_{hh}$, which lowers T_t below 37°C and drives HA (as seen in the lower part of Fig. 24A where the carboxylate of **E118** is missing and in general in Fig.12). On the other hand, increasing charge (by an increase in ΔG_{ap} due to the presence of the carboxylate of **E118**) decreases $\mathbf{N}_{hh}$, which raises T_t and drives HD (as seen in the upper part of 24A and in general in Figs. 10, and 12). That the formation of carboxylates disrupt hydrophobic associations and that the removal of carboxylates allow hydrophobic association becomes apparent on studying Figs. 10C, 12, 14B, and 15, and **Table 3**, columns 2 and 3.

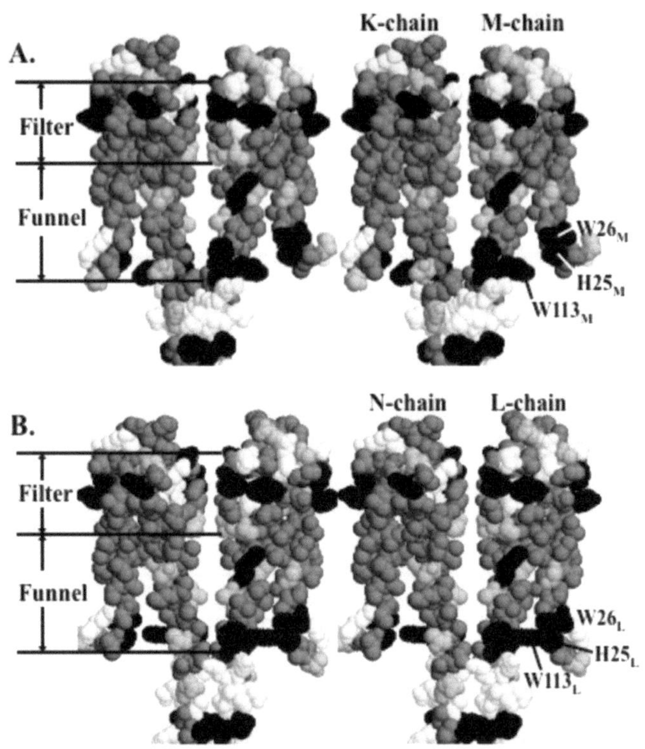

Fig. 22. Busts for the diagonally paired chains, K-M in A and L-N in B. The hydrophobic association of $W113_L$-($W26_L$/$H25_L$) pulls on the backbone sequence of $W113_L$-$F114_L$-$V115_L$ to draw the γ-methyl of $V115_L$ outward. That displacement is made square-planar to result in an open state in Fig. 23B. See text for further discussion. Protein Data Bank accession code 3EFF. From Urry et al., 2010.

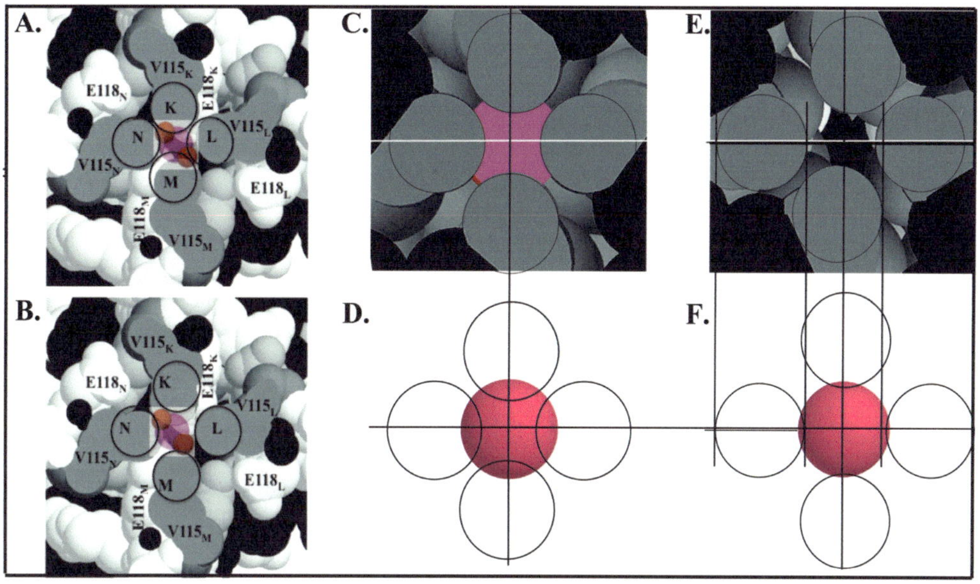

Fig. 23A and B. Cross-section at the V115 γ-methyl constriction. A. Representation of the closed state on making the N and K γ-methyl proximity square-planar. B. The open state by making the $V115_L$ γ-methyl displacement square-planar. C. A cross-section in the closed state of structure 1BL8 (Doyle et al., 1998) with representation in D. E. A cross-section at the G77 carbonyl oxygens of a closed structure, 3EFF (Uysal et al., 2009). But the displacement of the carbonyl oxygen on the left, on being made square-planar, represents an opening displacement when made in all chains, and either the lower left or the upper right oxygen pairs made square-planar give the closed state of D. See discussion.

The particular function of the W113-(W26/H25) hydrophobic association is to extend the backbone linkage $W113_L$-$F114_L$-$V115_L$ outward and thereby to move the γ-methyl of V115 outward. This outward displacement, when made square-planar for all four identical chains, opens the constriction at the base of the funnel of Fig. 21 (See Fig. 23B). A correlated outward displacement of the G77 carbonyl oxygen, also when made square-planar for all four identical chains, opens the selectivity filter as seen in Fig. 23E and F.

7.3.6 Developing the free energy profile for the open KcsA K^+-conducting channel

Fig. 24B becomes but a simple statement of the fraction of channels that will be in the open state for a given pH. Therefore, beyond a weighting of the probability of the open state, the problem of meaningful description of the pH dependence of potassium-ion conductance through the KcsA channel reduces to the state of the Gramicidin A channel discussed in section **8**.

The steps for developing a free energy profile for the KcsA channel would be similar to those listed in section **8.4.2** for the development of the free energy profile for ion passage through the Gramicidin A channel. The first challenge would be to obtain a suspension of stable lipid-bilayer membrane in which the KcsA channels could be incorporated and characterized by NMR and dielectric relaxation methods. For a detailed description of the process once the lipid-bilayer suspension of KcsA channels is obtained see section **8. Eyring's Absolute Rate Theory applied to biological trans-membrane transport** below.

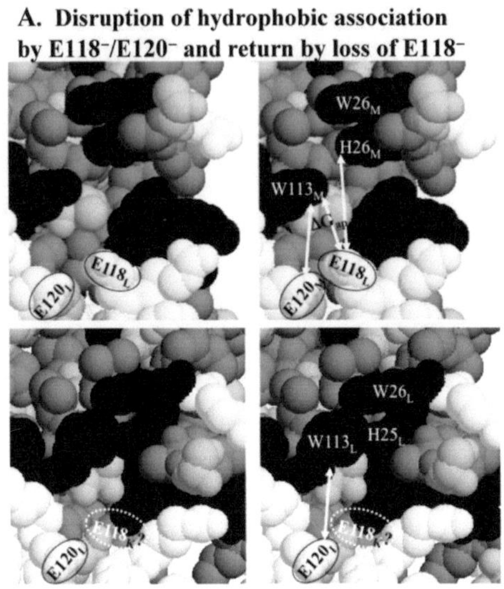

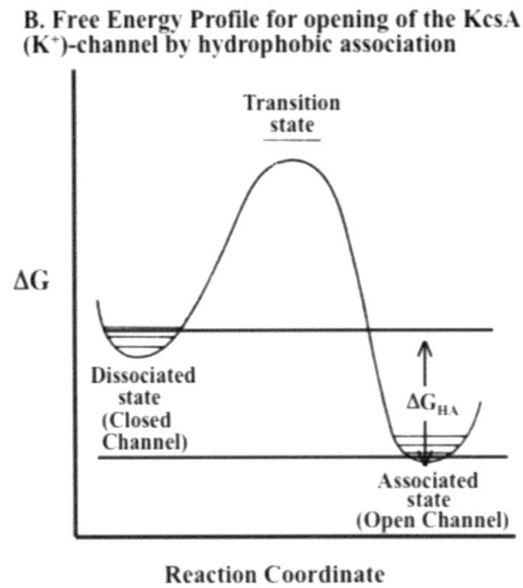

Fig. 24. A. Stereo view of hydrophobically dissociated (upper pair) and associated (lower pair) states proposed to be responsible, respectively, for the closed (Figs. 23A, C and D) and open (Figs. 23B and F) states of the KcsA (K^+)-channel. By ΔG_{ap}, the apolar-polar repulsive free energy of hydration described in section **6.2**, the charge of the carboxylates, $E118^-$ and $E120^-$, disrupt hydrophobic association, and the loss of ΔG_{ap} due to absence of $E118^-$ in the lower pair is seen to restore hydrophobic association.
B. Free energy profile representing the free energy change of hydrophobic association on protonation of the E118 carboxyl that restores the W113L,-(W26L/H25L) hydrophobic association and pulls on the backbone sequence of W113L-F114L-V115L to draw the γ-methyl of V115L outward and open the constriction. Thus, B provides a simple probability statement for the fraction of open channels. From Urry et al., 2010.

7.3.7 Looking forward to the remarkable potential of the Eyring Absolute Rate Theory to describe diverse biological trans-membrane transport processes

As will be seen in section **8**, once the image of the free energy profile is obtained, the channel conductance can be calculated as a function ion activity and trans-membrane potential. This has been demonstrated with the Gramicidin A Trans-membrane Channel as reviewed below. The entire process can be understood through the single visual image of the free energy profile of Fig. 27A. The challenge now before us is to reduce the essential and increasingly complex trans-membrane transport processes required for living organisms to visual images of free energy profiles. This potential to describe the essential energy converting trans-membrane systems of biology could result in one of the most profound legacies of the Eyring Absolute Rate Theory.

The sequence toward stepwise increasing complexity occurs in the preceding examples of the thermodynamics of biology's protein-based machines. The approach could be to begin with the KcsA channel, to proceed to Complex III of the electron transport chain of the inner mitochondrial membrane where the challenge would be in a coupled pair of visual images whereby proton flow from the matrix side would be coupled at the Q_o-site and the Q_i-site to the electron transport in the plane of the membrane, and then to proceed to ATP synthase with its pair of rotary motors, the F_0-motor and the F_1-motor, the first to use the return of proton to effect rotation of the rotor of the F_1-motor for making ATP.

8. Eyring's absolute rate theory applied to biological trans-membrane transport

8.1 The fundamental expression of Eyring's absolute rate theory

Application of the thermodynamics of Eyring's Absolute Rate Theory to essential trans-membrane transport processes of biology can result in a single figure of Gibbs free energy versus distance across the membrane that provides all of the information required to calculate transport as a function of concentration (or activity where relevant) of transported species and trans-membrane potential.

By means of Eyring's expression for the specific rate constant, k', (Eyring, 1935; Glasstone, et al. 1941; Eyring, 1969),

$$k' = \kappa kT/h \exp(-\Delta G^\dagger/RT) = \kappa kT/h \exp(-\Delta H^\dagger/RT + \Delta S^\dagger/R), \quad (20)$$

$$\Delta G^\dagger = \Delta H^\dagger - T\Delta S^\dagger, \quad (21)$$

where κ is the transmission coefficient; k is the Boltzman constant (1.38×10^{-16} erg/degree); T is the temperature in degrees Kelvin, K, which at physiological temperature of 37C would be 310K; h is Planck's constant (6.62×10^{-27} erg-sec); $\Delta G^\dagger$ is the Gibbs free energy of activation for the chemical reaction or physical process being considered, and R is the gas constant (1.987 cal/deg-mol). Importantly, Eqn. (20) expresses the activation energy in terms of the Gibbs free energy, $\Delta G^\dagger$, defined in Eqn. (21) where $\Delta H^\dagger$ and $\Delta S^\dagger$ are, respectively, the heat and entropy of activation that occur on achieving the activated state.

This is most fitting for the biophysical world where the essential processes of life occur at constant temperature and pressure, that is, where the experimentally obtained free energies are ΔG, and also where κ is commonly one. For special cases where κ differs from one, the rate process has been understood and κ calculated.

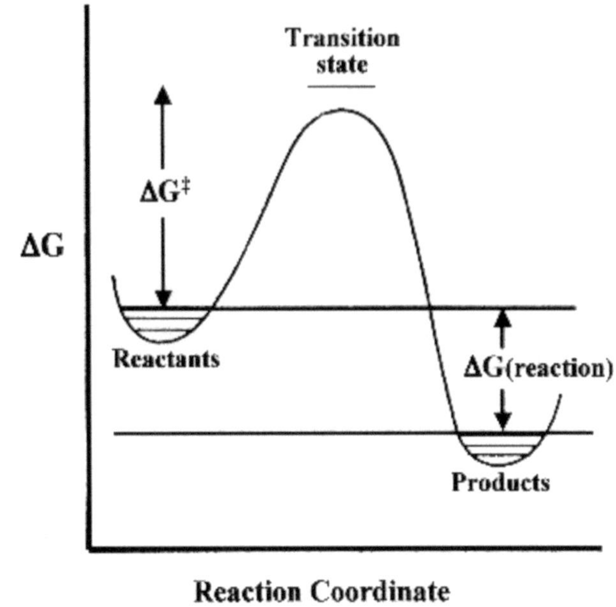

Fig. 25. The classic visual statement of Eyring's Absolute Rate Theory depicting reactants colliding to raise the free energy to that required for formation of the activated complex, now commonly called the transition state to include rate processes that may not involve chemical reactions, e.g., channel ion transport.

Fig. 25, the visual statement made by Eqn. (20), presents the archetypical image of Eyring's Absolute Rate Theory. As originally derived, it depicts the Gibbs free energy barrier that must be surmounted as reactants pass along the reaction coordinate to form products and may be called the *free energy profile* for the reaction. The reaction coordinate represents a splendid simplified plot of the distance between atoms or molecules in the gas phase flying at each other on collision course, passing over the high-energy barrier (the activated or transition state) and flying apart as product or products.

For ion transport through a trans-membrane channel the reaction coordinate is simply the length of the channel, such that plotting of the free energy profile for transit through a natural channel becomes a plot of the Gibbs free energy as a function of the actual distance along the channel. The channel state may be incorporated into a lipid-bilayer system at adequate concentrations wherein binding site locations can be determined by NMR data on individually C13-labeled carbonyl carbons of the channel. Also, by NMR methods the binding constants for single and multiple occupancy, the off-rate constants for single and multiple occupancy can be determined, and by dielectric relaxation rates for intra-channel ion translocations can even be experimentally determined.

Plotting these quantities as a function of free energy versus distance along the channel provides a free energy profile. Within a single image, the free energy profile for ion flow through the channel, the complete process of trans-membrane transport is described, and the calculated results may be compared with the single-channel currents determined by measurement of the single-channel currents in the so-called black lipid-bilayer membrane (BLM) obtained at appropriate applied potentials, and as a function of relevant ion concentrations, corrected to molal ion activity (Urry et al. 1980a; Urry et al. 1980b; Urry, 1985). Successful calculation of the experimental single-channel currents using experimentally derived distances and rate constants from a macroscopic lipid-bilayer incorporated system can be used to substantiate the detailed mechanism and, of course, to substantiate the relevance of Eyring Absolute Rate Theory.

The reported example of a natural channel for ion transport, opens the door to the development of free energy profiles for trans-membrane transport of other ions and molecules required to sustain the cell. Thus, Eyring Rate Theory provides the opportunity to obtain within a single image the visualization of many trans-membrane transport processes of biology with the potential of medically relevant interventional insights that such would provide.

8.2 The trans-membrane channel as a unique example of a free energy profile

As the objective of our first example given below, Fig. 27A contains the image of a biophysical path often-enjoyed by Henry Eyring, one of diffusion through membranes. In the present case the reaction coordinate for ion transit through a trans-membrane channel becomes the length of the channel through which the ion passes. And the plot of ion position in the channel as a function of Gibbs free energy images a *free energy profile* of ion passage over a terrain of valleys and mountain passes. The valleys locate and quantify low-energy binding sites, and the mountain passes locate and quantify high-energy barriers of the transition states. The free energy profile contains in a single figure the complete physical process of ion passage through the channel. By passing through a biological channel, ions achieve a controllable transit through an otherwise ion-impenetrable lipid membrane.

8.3 Turning channels on and off becomes central to integrating life processes.

In order to correlate and integrate the functions required for life, there must be controlled channel opening and closing. Another plot of the same form as that of Fig. 25, given above in Fig. 24, depicts the process whereby protonation of a carboxylate effects hydrophobic association to open a channel and the reverse process closes the KcsA potassium channel of living organisms. In this example, hydrophobic association pulls the channel gate open, and hydrophobic dissociation allows the channel gate to return to its closed state (See section 7.3). The equilibrium constant for this process allows calculation of the probability of the open state.

8.4 Construction of the free energy profile for the malonyl Gramicidin A trans-membrane channel and calculation of single-channel currents

8.4.1 A channel-containing lipid-bilayer state for determining structure, binding sites and rate constants

An aqueous suspension of the channel-containing lipid-bilayer state is prepared (Spisni et al. 1983; Pasquali-Ronchetti et al, 1983). This provides sufficient channel concentrations to use physical methods, such as nuclear magnetic resonance, to determine structure, channel-binding sites and rate constants. This data combines by means of Eyring rate theory to produce the free energy profile from which single-channel currents may be calculated, in a manner entirely independent of the BLM single-channel current measurement.

8.4.2 Steps in the development of a free energy profile with which to calculate single-channel currents

The steps are: **1**) Determination of structure and location of ion binding sites, as shown in Fig. 26A, B, and C (Urry et al., 1982a; Urry et al., 1982b; Urry et al., 1982c; Urry et al., 1982; Urry et al., 1983) for malonyl Gramicidin A (mal-GA). The data demonstrate the channel structure in the prepared lipid-bilayer state to be the single-stranded, head-to-head dimerized, left-handed $\beta^{6.3}$-helix. **2**) Derive the appropriate steady state equations for all occupancy states and the equation for the single-channel ion current (Urry, 1985; Urry et al., 1980a; 1980b) as given in Fig. 26D, E, and F. **3**) Construct the free energy profile for passage of ions through the trans-membrane structure by experimental determination of binding site locations and rate constants (Urry, 1985; Henze et al., 1982; Urry, et al. 1989; Urry, 1987), as in Fig. 27A, B, and C. **4**) Add to ΔG of Eqn. (20) the trans-membrane potential dependence appropriate to each rate constant with an assumed linear trans-membrane potential drop. And **5**) calculate the single-channel current as a function of ion activity and applied trans-membrane potential (Urry, 1985; Urry, 1984).

8.4.3 Comments on the structure(s) of Gramicidin A

As originally proposed in 1971 (Urry, 1971; Urry et al. 1971) and established now for three decades (Bamberg et al. 1977; Bamberg et al. 1978; Szabo & Urry 1979; Anderson & Koeppe II, 1992; Busath, 1993), the structure of the ion-conducting Gramicidin A channel in the BLM system for determining single-channel currents is a head-to-head hydrogen-bonded, single-stranded β^6-helix, a member of a new family of helices initially named π_{LD}-helices (Urry, 1971; Urry et al. 1971). The name was changed to β-helices as the hydrogen-bonded pattern between turns within the monomer is that of the parallel β-pleated sheet and between the turns at the head-to-head junction is that of the anti-parallel β-pleated sheet Urry, 1972. And the work, listed in part in the papers of reference (Urry, et al., 1982a, 1982b, 1982c, 1982e, 1983) showed the structure in lipid-bilayer suspensions (Pasquali-Ronchetti, et al. 1983; Spisni et al., 1983) to be the head-to-head hydrogen bonded, single-stranded and *left*-handed $\beta^{6.3}$-helix of Fig. 26A, B, and C, as originally proposed.

Nicholson and Cross (1989) using solid state ^{15}N NMR spectroscopy of Gramicidin A incorporated into a dimyristoyl phosphatidyl choline (DMPC) bilayer preparation, reported the same β-hydrogen-bonding pattern (β-helix) proposed in 1971, but concluded the *right*-handed rather than *left*-handed helix. We then incorporated malonyl Gramicidin A into dodecyl phosphocholine micelles, and utilized high resolution 2D NMR data to determine the handedness in this phospholipid system also to be *right*-handed (Jing & Urry, unpublished results), but with altered ion-binding interactions. Since Gramicidin A is now well-known for its many helical states (Anderson et al. 1999; Cross et al. 1999; Burkhart & Duax, 1999), the question continues to be, Is a determined Gramicidin A structure obtained at high channel concentrations relevant to the trans-membrane structure that gives rise to the single-channel current events of the (BLM) technique?

Without a methodology to observe directly the structure of a single trans-membrane channel in the BLM where the single-channel current is measured, application of Eyring's Absolute Rate Theory (ART) provides the opportunity to test a given structure by independent calculation of single-channel currents as reviewed here. Success would support the structure from which the rate constants and binding sites were derived and would be a remarkable validation of use of the Eyring rate equation for trans-membrane channel transport (with Gibbs free energies of activation for the barriers).

8.4.4 Development of the free energy profile for the malonyl Gramicidin A channel state

As depicted in the several components of Fig. 26, the malonyl-Gramicidin A trans-membrane channel - a covalent dimer with a two-fold symmetry axis perpendicular to the channel axis - contains two ion binding sites (a symmetrically positioned tight binding site for the first ion in the channel and a pair of weak binding site on entry of the second ion into the channel). Double occupancy results in repulsion between ions that weakens second ion binding. The result is two entry/exit barriers and a single central barrier for ion exchange between the two sites.

Therefore, description of ion transport through the channel requires five experimentally determinable rate constants. The following values were obtained using the lipid-bilayer suspension of malonyl GA channels: $k^t_{off} = 3x10^5$/sec, $k^t_{on} = 3x10^7$/M-sec, $k^w_{off} = 2x10^7$/sec, $k^w_{on} = 2x10^7$/M-sec and $k_{cb} = 4x10^6$/sec. The two off-rate constants were obtained by means of sodium-23 nuclear magnetic resonance (NMR) longitudinal and transverse relaxation studies. And sodium ion-induced carbonyl carbon chemical shifts (CS), and sodium ion line width (LW), excess longitudinal relaxation (ELR) rate and excess line width (ELW) data (Urry et al., 1980a, 1980b; Urry et al. 1989; Urry, 1987) were used to obtain values for the tight and weak binding constants, $K^t_b = k^t_{on}/k^t_{off}$ and $K^w_b = k^w_{on}/ k^w_{off}$, from which the on rate constants were determined. The magnitude of k_{cb} was obtained by means of thallium ion dielectric relaxation data for the ion jump inside the channel (Henze et al., 1982). Due to the two-fold symmetry and in the absence of an applied field the rate constants defined in Fig. 26D become, $k^t_{off} = k_{-1} = k_2$; $k^t_{on} = k_1 = k_{-2}$; $k^w_{off} = k_3 = k_{-4}$; $k^w_{on} = k_{-3} = k_4$; $k_{cb} = k_5 = k_{-5}$. An assumed linear applied potential drop is applied to these experimentally derived rate constants with the distances defined in Fig. 27A to be a_1 = 10.5 Å; b_1 = 13 Å; d = 15 Ă, as utilized in the current equation of Fig. 26F.

The experimental Gibbs free energies of activation were calculated, using Eqn (20) in the form of $\Delta G† = 2.3RT(12.79\text{-}\log k')$, from the experimental data to be as follows: $\Delta G†(^t{}_{off})^{exp}$ = 10.1; $\Delta G†(^t{}_{on})^{exp}$ = 7.4; $\Delta G†(^w{}_{off})^{exp}$ = 7.6; $\Delta G†(^w{}_{on})^{exp}$ = 7.6 ; $\Delta G†(cb)^{exp}$ = 8.6. These values and binding site locations are used to plot the free energy profile of Fig. 27A.

In addition, a set of five "best fit" rate constants for the experimental single-channel currents were determined by allowing all five rate constants to vary in order to optimize the fit to the experimental BLM single-channel current data, as a function of ion activity and trans-membrane potential. The "best fit" single-channel current values were found to be: $\Delta G†(^t{}_{off})^{fit}$ = 10.5; $\Delta G†(^t{}_{on})^{fit}$ = 7.3; $\Delta G†(^w{}_{off})^{fit}$ = 8.2; $\Delta G†(^w{}_{on})^{fit}$ = 8.0; $\Delta G†(cb)^{fit}$ = 7.9. The "best fit" $\Delta G†$ values for the experimental BLM single-channel current data compare favorably with the free energies of the experimentally determined rate constants using the lipid-bilayer suspension of channels without optimization.

8.5 Absolute rate theory calculates experimental single-channel currents using experimentally-derived rate constants

8.5.1 Experimental BLM single-channel currents of the malonyl Gramicidin A trans-membrane channel

We start with the malonyl Gramicidin A trans-membrane channel, the covalent dimer, formed by chemically connecting, by peptide linkage, the amino ends of two molecules of the 15 amino acid residue peptide, Gramicidin A, using the malonyl group, $\text{-OC-CH}_2\text{-CO-}$, and we measure the resulting single-channel currents in the black lipid-bilayer membrane

A. Left-handed, head-to-head dimer, $\beta^{6.3}$-helix

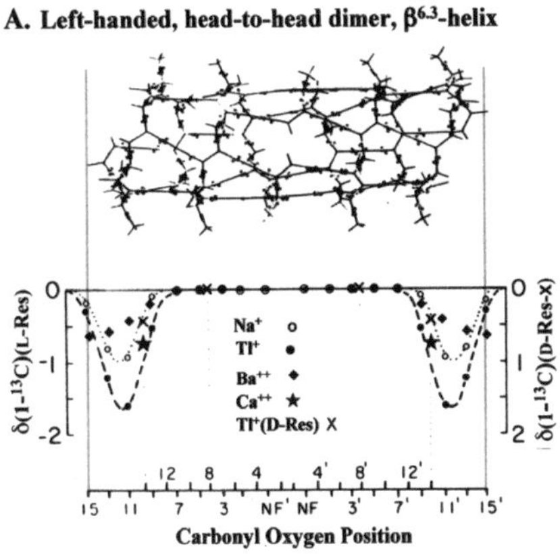

B. Right-handed, head-to-head dimer, $\beta^{6.3}$-helix

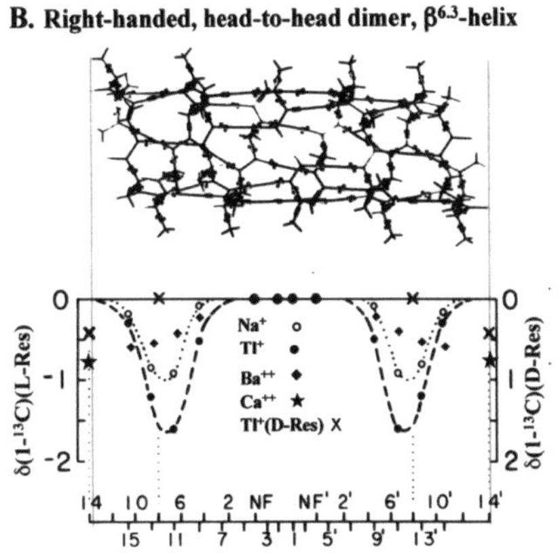

C.

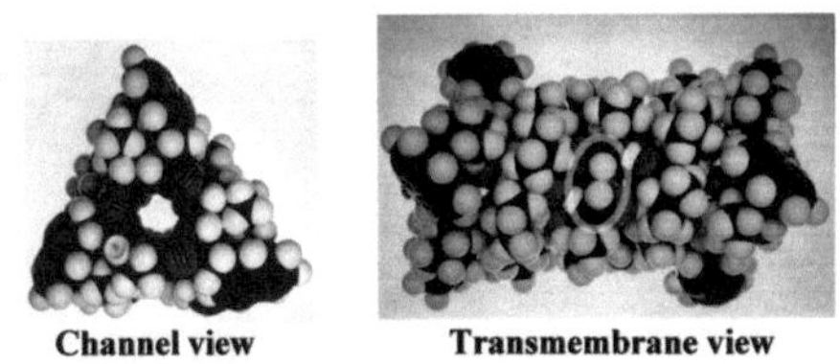

D. Occupancy states

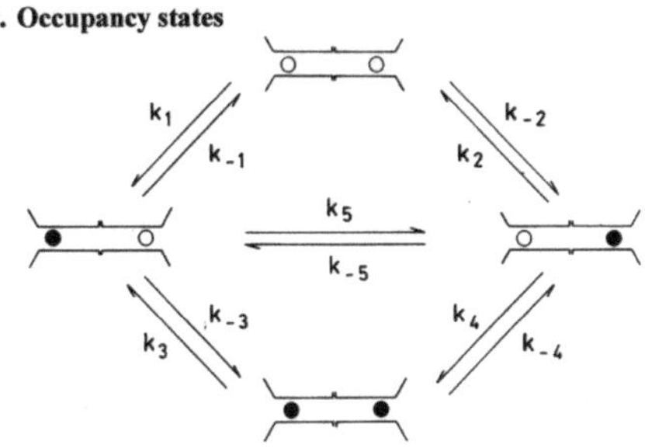

E. Steady state equations

$$\dot{\chi}(oo) = -(C'_x k_1 + C''_x k_{-2})\chi(oo) + k_{-1}\chi(xo) + k_2\chi(ox) = 0$$

$$\dot{\chi}(xo) = C'_x k_1 \chi(oo) - (k_{-1} + k_5 + C''_x k_{-3})\chi(xo) + k_{-5}\chi(ox) + k_3\chi(xx) = 0$$

$$\dot{\chi}(ox) = C''_x k_{-2} \chi(oo) + k_5 \chi(xo) - (k_2 + k_{-5} + C'_x k_4)\chi(ox) + k_{-4}\chi(xx) = 0$$

$$\dot{\chi}(xx) = C''_x k_{-3}\chi(xo) + C'_x k_4 \chi(ox) - (k_3 + k_{-4})\chi(xx) = 0$$

F. Single channel current

$$i_x = \frac{ze}{2d}\{(d - a_1)\,C_x(k_1 - k_{-2})\,\chi(oo) + [2a_1 k_5 - (d - a_1)\,(k_{-1} + C_x k_3)]\,\chi(xo) + [(d - a_1)\,(k_2 + C_x k_4) - 2a_1 k_{-5}]\,\chi(ox) + (d - a_1)\,(k_3 - k_{-4})\,\chi(xx)\}$$

Fig. 26. A. Upper image: Wire model of the single-stranded *left-handed*, head-to-head hydrogen bonded dimeric $\beta^{6.3}$-helix, indicating the β-sheet hydrogen-bonded pattern made possible by the L-D repeating sequence, namely HCO-L-Val^{1}-Gly^{2}-L-Ala^{3}-D-Leu^{4}-L-Ala^{5}-D-Val^{6}-L-Val^{7}-D-Val^{8}-L-Trp^{9}-D-Leu^{10}-L-Trp^{11}-D-Leu^{12}-L-Trp^{13}-D-Leu^{14}-L-Trp^{15}-$NHCH_2CH_2OH$, with 6.3 residues per turn of helix, as proposed in 1971 (Urry et al., 1971; Urry, 1971, 1972). From Urry et al., 1982e.
Lower part: Plot of experimental carbon-13 carbonyl carbon chemical shifts as a function of carbonyl oxygen position in the structure for the single-stranded *left-handed*, head-to-head hydrogen bonded dimeric $\beta^{6.3}$-helical structure above. Data obtained by chemical synthesis of Gramicidin A with one carbonyl carbon-13 enriched per chain synthesized for each of the 8 L-residues and the amino terminal malonyl carbonyls and for each of two D-residues, D-Val^{8} and D-Leu^{14}. Adapted from Urry et al., 1982a; 1982b; 1982c; 1982e; Urry, 1985.
B. Upper image: Wire model of the single-stranded *right-handed*, head-to-head hydrogen bonded dimeric $\beta^{6.3}$-helix, indicating the β-sheet hydrogen-bonded pattern made possible by the alternating L-D repeating sequence with 6.3 residues per turn of helix.
Lower part: Plot of experimental carbon-13 carbonyl carbon chemical shifts as a function of carbonyl oxygen position in the sequence for the *right-handed*, head-to-head hydrogen bonded dimeric $\beta^{6.3}$-helical structure immediately above. From Urry et al., 1982e; Urry, 1985.
In the plot for the *left-handed* structure in A, all of the Na^+ and Tl^+ data points and the single Ca^{++} data point consistently define the location of two binding sites, including the data points for the D-residues, D-Val^{8} and D-Leu^{14}. On the other hand, for the plot of the *right-handed* structure in B, D-Val^{8} shows no ion interaction in the middle of the binding site as defined by the L-residue carbonyl carbon chemical shifts. Furthermore, for the *right*-handed structure, the D-Leu^{14} residue carbonyl carbon chemical shifts are substantial where there is no binding site again as defined by the 8 L-residue carbonyl carbon chemical shifts. Accordingly, we believe that the structure in our bilayer preparations of the malonyl Gramicidin A is the single-stranded, *left-handed*, head-to-head hydrogen bonded dimeric $\beta^{6.3}$-helix.
C. The single-stranded *left*-handed, head-to-head hydrogen bonded dimeric $\beta^{6.3}$-helical structure of the Gramicidin A transmembrane channel: in channel view on left and in transmembrane view on the right, using CPK (Corey-Pauling-Koltun) space-filling molecular models. The oval encompasses the two formyl hydrogens at the head-to-head junction.
D. Occupancy states: The information in A and C above, define the four occupancy states, which with the ten defined elemental rate constants, provide the information required to state the steady state equations in E. From Urry, 1985.
E. Steady state equations: As the single-channel currents are determined under steady state conditions, the four steady state equations are stated here for the four occupancy states and ten elemental rate constants of D above. From Urry et al., 1980b.
F. Single-channel current: With steady state equations and defined elemental rate constants, the single-channel current may be expressed, but, if written correctly, it yields no current until a transmembrane potential is applied. With binding site locations as in Fig. 27C, with the distances defined in Fig. 27A, with the positioning of the channel in relation to the membrane as in Fig. 27B, and with the indicated fraction of the transmembrane potential applied to each experimentally determined rate constant, the single-channel current is stated as given here in F. From Urry et al., 1980b.

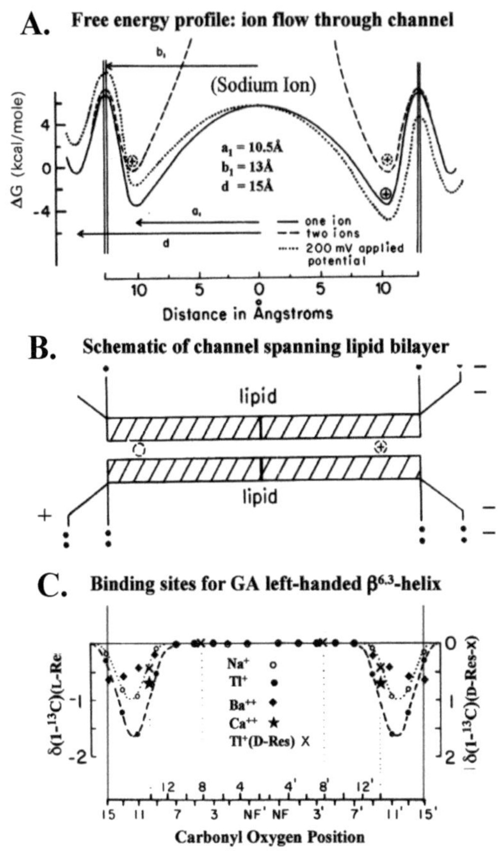

Fig. 27. A. Free Energy Profile: With five experimentally determined rate constants obtained from suspensions of transmembrane channels in bilayer preparations and the defined distances, the decrease in Gibbs free energies of binding sites and increase in Gibbs free energies of barriers are plotted here as well as the bias added by the linear applied potential drop of a 200 mV transmembrane potential, given as the dotted line for single ion occupancy, and also a similar bias added to the dashed curve for double occupancy. All the data is contained in the single image of the free energy profile with use of the expression in Fig. 26F to calculate the single-channel currents of Fig. 28A for comparison with the experimental single-channel currents given in Fig. 28B. B. Schematic representation of the Gramicidin A transmembrane channel spanning the 30Å lipid bilayer. The binding sites are noted within the channel and the charge distribution across the membrane has the left hand side indicated as positive and the right hand side as negative. Thus, the applied transmembrane potential is such that for the positive sodium ion the free energy is lower on the right hand side and raised on the left hand side of the membrane, as seen by the free energy profile in part A above. A. and B. from Urry, 1982b.
C. The lower component serves to align the binding sites in the channel and uses the outermost channel carbonyls to define the outermost reaches of the schematic representation in part B and the entrance barrier of the Gramicidin A transmembrane channel spanning the lipid bilayer in part A in order to clarify the relationship of the free energy profile of part A to the lipid bilayer being spanned by the Gramicidin A channel. Adapted from Urry et al., 1982a; 1982b; 1982c; 1982e; Urry, 1985.

(BLM) obtained at four applied potentials, 50 mV, 100 mV, 150 mV, and 200 mV, and as a function of ion concentrations, 0.1, 1, 3, and 5.5 M, corrected to molal ion activity.
It is called a black lipid membrane (BLM), because as the lipid membrane thins to a single 30Å wide lipid bilayer, the membrane is seen to go black, due to the interference of light reflected from each surface of the thin flat membrane. The 30Å wide lipid-bilayer can be spanned by a single channel-forming covalent dimer, the malonyl Gramicidin A trans-membrane channel, which allows measurement of single-channel currents from a histogram of single channel conductances.

8.5.2 Using Eyring Absolute Rate Theory to calculate the experimental BLM single-channel currents

The experimentally-derived rate constants (without any adjustment or fine-tuning of experimental values obtained on channel incorporated lipid bilayer suspensions) can be used within the formalism of Eyring Absolute Rate Theory to calculate successfully the sodium ion single-channel currents for four different trans-membrane potentials and over a wide range of ion activities, as seen in Fig. 28(a). Comparison with the directly measured sodium ion single-channel currents in Fig. 28(b) is particularly satisfying, especially for the 150 mV curve. Furthermore, also calculated were the conductance ratios for the series of alkali metal ions - Li^+, Na^+, K^+, Rb^+, and Cs^+ - and found to be within experimental error (Urry et al., 1980a; Urry et al. 1989; Urry, 1987).

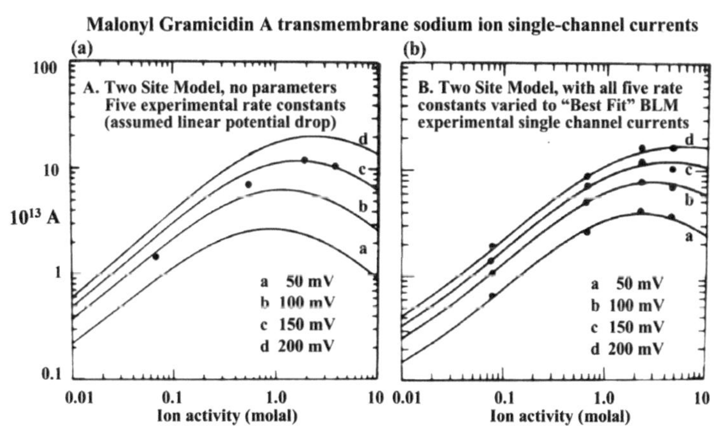

Fig. 28. Part (a): The calculated single-channel currents using the single-channel current equation of Fig. 26F with the data of the free energy profile of Fig. 27A, including a dotted curve adding the 200 mV potential bias and similarly for the dashed curve for double occupancy, that is, the five experimental rate constants and the assumed linear potential drop. The four data points for 150 mV applied potential were transferred laterally from the aligned part (b) to (a) using PowerPoint. Adapted from Urry et al., 1980a.
Part (b): Single-channel currents of Fig. 7A of Urry et al., 1980a. Data points from single channel current measurements at different ion activities and applied transmembrane potentials obtained using the black lipid membrane (BLM) approach, and solid lines are calculated "best fit" curves using a single set of five rate constants. From Urry, 1980a.

The data in Fig. 26A demonstrate the head-to-head hydrogen-bonded, single-stranded, *left-handed* $\beta^{6.3}$-helical structure of incorporated malonyl Gramicidin A transmembrane channel to occur in what has been verified as a lipid bilayer suspension (Spisni et al., 1983; Pasquali-Ronchetti et al., 1983). These findings also demonstrate the remarkable capacity of Eyring Absolute Rate Theory to construct in a single image of the free energy profile all the information with which to calculate trans-membrane transport for a number of important variables central to biological function.

The data of Fig. 26A - including the divalent ion binding data, for which the channel is impermeable - argue that the head-to-head hydrogen-bonded, single-stranded, *left-handed* $\beta^{6.3}$-helix gives rise to the experimental single-channel currents obtained in BLM experiments. *The most salient point to make here, however, is the remarkable capacity of the Eyring rate equation, Eqn. (20), to treat biologically important ion trans-membrane transport processes.*

8.6 Eyring Legacy: Absolute Rate Theory for thermodynamic description of diverse biological trans-membrane transport processes

Absolute Rate Theory is applicable to virtually all that happens in living organisms. Here we utilize Eyring's rate equation to construct a free energy profile for transit of sodium ion through a natural trans-membrane channel. Without resorting to BLM measurements of sodium ion single-channel currents through the channel, independent physical determinations of channel structure and of five separate rate constants were obtained and to each rate constant was applied an appropriate fraction of a trans-membrane potential; all of which were treated and assembled within the framework of the Eyring rate equation to construct a single free energy profile (See Fig. 27A). And the single image of the free energy profile (Fig. 27A) contains all of the information required to calculate the single-channel currents as a function of sodium ion activity and trans-membrane potential. In my mind, this stands as a demonstration of the significance and power of Eyring rate theory, embodied within Eqn. (20). To the best of my knowledge, such a capacity has been demonstrated by no other equation.

Following Henry Eyring, one might look toward, for example, quantitative description of the processes of Complex III, so central to the energy conversions that sustain living organisms. One might consider capturing the process of proton flow across the inner mitochondrial membrane in a single image by construction of the free energy profile, and similarly for electron transfer. Then there would be introduction of the coupling processes, which for the Q_o site are the naturally visual mechanical processes of hydrophobic association/dissociation and single-chain extension/relaxation. The capacity to think in such visual and comprehensive ways to convey information and understanding comes to us through the contributions of Henry Eyring.

9. Summarizing comments

9.1 The fundamental physical principle underlying the protein-in-water heat engine of biology

The steam-powered heat engine that set in motion The 19th Century Industrial Revolution heats water at the 100°C *water-to-vapor phase transition to perform mechanical work by expansion*. The fundamental physical process is the increase in entropy of water on heating at a phase transition where more-ordered (lower entropy) molecules of water become highly disordered (high entropy) molecules of water vapor.

On the other hand, the heat engine of biology heats a two-component system, protein-in-water, *to perform mechanical work by protein contraction due to hydrophobic association* (See Fig. 1). The fundamental physical process is again an increase in entropy of water, $\Delta S(H_2O)$. Heating at the phase transition causes pentagonal rings of hydrophobic hydration, N_{hh}, to become less-ordered (higher entropy) bulk water as a *much smaller decrease in entropy* occurs on the protein hydrophobic association of contraction. Due to this latter effect, the protein-in-water phase transition has been called an inverse temperature transition (ITT), wherein the protein component becomes more-ordered on raising the temperature.
The enabling feature of the protein-in-water heat engine of biology is the ease with which the state of a "biological functional group" changes the temperature of its protein-in-water phase transition, i.e., of the ITT.
A biological functional group can exist in two or more states. Invariably those states may be characterized as more-hydrophobic (m-h) and less-hydrophobic (l-h). On becoming more hydrophobic the protein of which it is part immediately develops more hydrophobic hydration. *On formation of hydrophobic hydration,* $|T\Delta S| > |\Delta H|$, *and at the phase transition,* $\Delta H_t \approx T_t\Delta S_t$. *Because of these thermodynamic relationships, the result of introducing a chemical or electrochemical energy that makes a functional group more-hydrophobic is that the phase transition shifts to a lower temperature. In particular, at the phase transition,* $T_t \approx \Delta H_t/\Delta S_t$ *and on formation of more hydrophobic hydration,* $T_t(m\text{-}h) \approx (\Delta H_t + \delta H)/(\Delta S_t + \delta S)$, *but since* $|\Delta H_t + \delta H| < |T(\Delta S_t + \delta S)|$, $T_t(m\text{-}h) < T_t(l\text{-}h)$.
Thus, instead of having to raise the temperature from below to above that of the phase transition to drive contraction by hydrophobic association, making a functional group more-hydrophobic lowers the temperature from above to below physiological temperature to drive contraction by hydrophobic association.
Finally, by way of example, because of the dependence of pK and reduction potential on hydrophobicity, the free energy transduction of electrochemical energy into chemical energy, and vice versa, occur, and because hydrophobicity induced increases in positive cooperativity occur for both acid/base and redox titrations, the result is increased efficiency, i.e., the optimization of biological free energy transduction by the proteins of biology. And so it goes for all of the pair-wise energy conversions of Fig. 8 and of section **4.2**. These (on inclusion of transient "near ideal" elastic deformation, even of single chains) constitute the dominant energy conversions that sustain Life.
Using herein-derived thermodynamic elements of the change in Gibbs free energy of hydrophobic association, ΔG_{HA}, of the apolar-polar repulsive free energy of hydration, ΔG_{ap}, and of "near ideal" elastic deformation of protein chains, this article discusses the function of key proteins involved in the trans-membrane transport energy converting processes of biology, and they point to the Eyring Theory of Rate Processes Legacy to demonstrate how the complete trans-membrane transport process may be captured in a single image of the free energy profile for transport from one side to the other of the cell membrane.

10. Acknowledgements

This work was supported in part by Bioelastics, Inc. Your author would also like to thank Edward M. Eyring of the University of Utah for obtaining references for Henry Eyring's papers, Jack Sabin of the University of Florida, Quantum Theory Project for providing a copy of (Eyring, 1969).

11. References

Abrahams, JP.; Leslie, AGW.; Lutter, R. & Walker, JE. (1994). Structure at 2.8 Å of F_1–ATPase from bovine heart mitochondria. *Nature* (London), 370, 621-628. Protein Data Bank, Structure File 1BMF.

Anderson, OS.; Apell, H-J.; Bamberg, E.; Busath, DD.; Koeppe, RE.; Sigworth, FJ.; Szabo, G.; Urry DW. & Woolley, A. (1999). Gramicidin channel controversy - the structure in a lipid environment. *Nature Structural Biology*, 6, 609.

Anderson, OS. & Koeppe II, RE. (1992). Molecular determinants of channel function," *Physiol. Rev.* 72, S89-S158.

Anfinsen, C.B. (1973). Principles that govern the folding of protein chains. *Science*, 181, 223-230.

Bamberg, E.; Apell, H.J. & Alpes H. (1977). Structure of the gramicidin A channel: Discrimination between the $\pi_{L,D}$ and the β helix by electrical measurements with lipid bilayer membranes. *Proc. Natl. Acad. Sci. USA*, 74, 2402-2406.

Bamberg, E.; Apell, H-J.; Alpes H.; Gross, E.; Morrell, JL.; Harbough, JF.; Janko K. & Läuger, P. (1978). Ion channels formed by chemical analogs of gramicidin A. *Fed. Proc.*, 2633-2638.

Boyer, PD. (1993). The binding change mechanism for ATP synthase - some probabilities and possibilities. *Biochim. Biophys. Acta*, 1140, 215-250.

Boyer, PD. (1997). The ATP synthase - a splendid molecular machine. *Annu. Rev. Biochem.*, 66, 717-749.

Brooks, BR.; Bruccoleri, RE.; Olafson, BO.; States, DJ.; Swaminathan, S. & Karplus, M. (1983). CHARMM: A program for macromolecular energy, minimization, and dynamics calculations. *J. Comp. Chem.* 4, 187-217.

Busath, DD. (1993). "The use of physical methods in determining gramicidin channel structure and function," *Ann. Rev. Physiol.* 55, 473-501.

Burkhart BM. & Duax, WL. (1999). Gramicidin channel controversy - reply. *Nature Structural Biology*, 6, 611-612.

Butler, JAV. (1937). The energy and entropy of hydration of organic compounds. *Trans. Faraday Soc.*, 33, 229-238.

Carnot, S. (1992). *Reflections on the Motive Power of Fire*, Peter Smith Publisher, ISBN-13: 978-0844618098, Gloucester, MA

Chang DK. & Urry, DW. (1989). Polypentapeptide of Elastin: Damping of internal chain dynamics on extension. *J. Computational Chem.* 10, 850-855.

Cook, WJ.; Einspahr, HM.; Trapane, TL.; Urry, DW. & Bugg, CE. (1980). Crystal Structure and Conformation of the Cyclic Trimer of a Repeat Pentapeptide of Elastin, Cyclo-(L-Valyl-L-prolylglycyl-L-valylglycyl)$_3$. *J. Am. Chem. Soc.* 102, 5502-5505.

Cross, TA.; Arseniev, A.; Cornell, BA.; Davis, JH.; Killian, JA.; Koeppe II, RE.; Nicholson, LK.; Separovic, F. & Wallace, BA. (1999). Gramicidin channel controversy - revisited. *Nature Structural Biology*, 6, 610-611.

Doyle, DA.; Morais-Cabral, J., Pfuetzner, RA,; Kuo, A.; Gulbis, JM.; Cohen, SL.; Chait, BT. & MacKinnon, R. (1998). The structure of the potassium channel: Molecular basis of K^+ conduction and selectivity. *Science*, 280, 69-77.

Dzyaloshinskii, IE.; Lifshitz, EM. & Pitaevskii, LP. (1961). General theory of van der Waals' forces. *Sov. Phys. Uspekhi*, 4, 153–176.

Eddington, A. S. (1958). The Running-Down of the Universe. Chapter IV, In: *The Nature of the Physical World*, Ann Arbor Paperbacks, The University of Michigan Press, p. 69. First published by Cambridge University Press. Also, Kessinger Publishing LLC, ISBN-10: 1417907185, ISBN-13: 978-1417907182, Whitefish, MT

Eyring, H. (1935). Activated Complex in Chemical Reactions. *J. Chem. Physics*, *3*, 107-115.

Eyring, H. (1969). Models in research. *Int. J. Quant. Chem.*, IIIs, 5-15.

Eyring, H.; Walter, J. & Kimball, GE. (1944, 1958 printing). *Quantum Chemistry*, Chapter XVIII, 18a. van der Waals' forces, pp 351-355.

Fillingame, RH. (1999). Molecular rotary motors. *Science*, 286, 1687-1688.

Fillingame, RH.; Angevine CM. & Dmitriev, OY. (2002). Coupling proton movements to c-ring rotation in F1F0 ATP synthase: Aqueous access channels and helix rotations at the a-c interface. *Biochim. Biophys. Acta*, 1555, 29-36.

Glasstone, S., Laidler, KJ. & Eyring, H. (1941). *The Theory of Rate Processes, The Kinetics of Chemical Reactions, Viscosity, Diffusion and Electrochemical Phenomena*, McGraw-Hill Book Company, Inc. New York and London.

Harris, FE. & Rice, SA. (1954). A chain model of polyelectrolytes. *J. Phys. Chem.* 581 725-732.

Henze, R.; Neher, E.; Trapane TL. & Urry, DW. (1982). Dielectric relaxation studies of ionic processes in lysolecithin-packaged Gramicidin channels. *J. Membr. Biol.*, 64, 233-239.

Hill, AV. (1910). The possible effect of the aggregation of hemoglobin on its dissociation curves, *Proc. Physiol. Soc, J. Physiol.* 40 iv-vii; Hill, AV. (1913). *J. Biochemistry* 7 471-480.

Hobbs, ME.; Jhon, MS. & Eyring, H. (1966). The Dielectric Constant of Liquid Water and Various Forms of Ice According to Significant Structure Theory. *Proc. Natl. Acad. Sci., USA*, 56, 31-38.

Hugel, T. (2003). Towards synthetic molecular motors interfaced by AFM. Dissertation, LMU München, Faculty of Physics.

Jhon MS. & Eyring, H. (1976). Liquid Theory and the Structure of Water. *Ann. Rev. Phys. Chem.*, *27*, 45-57.

Katchalsky, A. & Gillis, J. (1949). Theory of the potentiometric titration of polymeric acids. *Recl. Trav. Chim. Pays-Bas.* 678 879.

Kinosita Jr., K.; Yasuda, R.; Noji, H. & Adachi, K. (2000). A rotary motor that can work at near 100% efficiency. *Phil. Trans. R. Soc. Lond. B*, 355, 473-489.

Lange, C. & Hunte, C. (2002). Crystal structure of the yeast cytochrome bc_1 complex with its bound substrate cytochrome c, *Proc. Nat. Acad. Sci.* USA 99 2800-2805. Protein Data Bank, accession code, 1KYO.

Luan, C-H.; Dean Harris, R.; Prasad, KU. & Urry, DW. (1990). DSC studies of the inverse temperature transition of the polypentapeptide of elastin and its analogues. *Biopolymers*, 29, 1699-1706.

Luan, C-H. & Urry, DW. (1910). Solvent deuteration enhancement of hydrophobicity: DSC study of the inverse temperature transition of elastin-based polypeptides. *J. Phys. Chem.* 95, 7896-7900.

Martz E. (2002). FrontDoor to Protein Explorer; http://proteinexplorer.org; 1.982Beta.

McPherson, DT. Xu, J. & Urry, DW. (1996). Product purification by reversible phase transition following *E. coli* Expression of genes encoding up to 251 repeats of the elastomeric pentapeptide GVGVP. *Protein Expression and Purification* 7, 51–57.

Menz, RI.; Walker, JE. & Leslie, AGW. (2001). Structure of bovine mitochondrial F_1-ATPase with nucleotide bound to all three catalytic sites: Implications for mechanism of rotary catalysis. *Cell* 106, 331-341. Protein Data Bank, Structure File 1H8E.

Momany, FA.; McGuire, FF.; Burgess, AW. & Scheraga, HA. (1975). Energy parameters in polypeptides. VII. Geometric parameters, partial atomic charges, nonbonded interactions, hydrogen bond interactions, and intrinsic torsional potentials for the naturally occurring amino acids. *J. Phys. Chem.* 79, 2361-2381.

Nicholson, LK. & Cross, TA. (1989). Gramicidin cation channel: an experimental determination of the right-handed helix. *Biochemistry, 28,* 9379-9385.

Noji, H.; Yasuda, R.; Yoshida, M. & Kinosita, K. (1997). Direct observation of the rotation of F_1–ATPase. *Nature (London),* 386, 299-302.

Noji, H. (1998). AMERSHAM PHARMACIA BIOTECH & SCIENCE PRIZE: The Rotary Enzyme of the Cell: The Rotation of F_1-ATPase. *Science,* 282, 1844–1845.

Overbeek, JThG. (1948). The dissociation and titration constants of polybasic acids. *Bull. Soc. Chim. Belg.* 57 252-261.

Pasquali-Ronchetti, I.; Spisni, A.; Casali, E.; Masotti, L. & Urry, DW. (1983). Gramicidin A induces lysolecithin to form bilayers. *Biosciences Reports,* 3, 127-133.

Pauling, L. (1932). The energy of single bonds and the relative electonegativities of atoms. *J. Amer. Chem. Soc.,* 54, 3570-3582.

Pauling, L. (1960). *The Nature of the Chemical Bond and the Structure of Molecules and Crystals: An Introduction to Modern Structural Chemistry.* Third Edition, Cornell University Press, Ithaca, New York, 1960, page 93.

Prigogine, I. & Stengers, I. (1984a). *Order Out of Chaos: Man's New Dialogue with Nature.* Bantam Books, ISBN: 0553340824, New York, p. 103.

Prigogine, I. & Stengers, I. (1984b). *Order Out of Chaos: Man's New Dialogue with Nature.* Bantam Books, ISBN: 0553340824, New York, pp 12-13.

Rodríguez-Cabello, J.C.; Reguera, J.; Alonso, M.; Parker, T.M.; McPherson, D.T.; & Urry, D.W. (2004). Endothermic and exothermic components of an inverse temperature transition for hydrophobic association by TMDSC. *Chem. Phys. Letters* 388, 127-131.

Sanger, F. (1952). The arrangement of amino acids in proteins. *Adv. Protein Chem.,* 7, 1-66.

Sanger, F. & Thompson, EOP. (1953a). The amino-acid sequence in the glycyl chain of insulin. 1. The investigation of lower peptides from partial hydrolysates. *Biochem. J.,* 53, 353-366.

Sanger, F. & Thompson, EOP. (1953b). The amino-acid sequence in the glycyl chain of insulin. 2. The investigation of peptides from enzymic hydrolysates. *Biochem. J.,* 53,366-374.

Schrödinger, E. (1944a). *WHAT IS LIFE?: The Physical Aspect of the Living Cell with Mind and Matter and Autobiographical Sketches,* Cambridge University Press, 1944, (1967, edition, p. 68).

Schrödinger, E. (1944b). *WHAT IS LIFE?: The Physical Aspect of the Living Cell with Mind and Matter and Autobiographical Sketches,* Cambridge University Press, (1967, edition, p. 76).

Smith, CW. (1977). William Thomson and the creation of thermodynamics: 1840–1855. *Arch. Hist. Exact Sci.,* 16, 231-288. DOI: 10.1007/BF00328156.

Spisni, A.; Pasquali-Ronchetti, I.; Casali, E.; Lindner, L.; Cavatorta, P.; Masotti, L. & Urry, DW. (1983). Supramolecular organization of lysophosphatidylcholine-packaged Gramicidin A. *Biochim. Biophys. Acta*, 732, 58-68.

Stock D,; Leslie AGW, & Walker JE, (1999). Molecular architecture of the rotary motor in ATP synthase. *Science*, 286, 1700-1705. Protein Data Bank, Structure File 1QO1

Stackelberg, Mv. & Müller, HR. (1951). Zur Struktur der Gashydrate. *Naturwissenschaften*, 38, 456- .

Stackelberg, Mv. & Müller, HR. (1954). Feste Gashydrate II: Struktur und Raumchemie. *Zeitschrift für Elektrochemie*, 54, 25-39. (became *Berichte der Bunsengesellschaft für physicalische Chemie*).

Szabo, G. & Urry, DW. (1979). N-Acetyl Gramicidin: Single-Channel Properties and Implications for Channel Structure. *Science*, 203, 55-57.

Teeter, MM. (1984). Hydrophobic protein at atomic resolution: Pentagonal rings of water molecules in crystals of crambin. *Proc Natl Acad Sci USA*, 81, 6014–6018.

Thompson, AN.; Posson, DJ.; Parsa, PV. & Nimigean, C.M. (2008). Molecular mechanism of pH sensing in KcsA potassium channels. *Proc. Natl. Acad. Sci.* USA, 105 6900-6905.

Toffler, A. (1984). Forward, page xx. In: Prigogine, I. & Stengers, I. (1984). *Order Out of Chaos: Man's New Dialogue with Nature.* Bantam Books, ISBN: 0553340824, New York.

Urry, DW. (1971). The Gramicidin A trans-membrane channel: A proposed $\pi_{(LD)}$ Helix," *Proc. Natl. Acad. Sci.* USA, 68, 672-676.

Urry, DW. (1972). A Molecular Theory of Ion Conducting Channels: A Field-Dependent Transition Between Conducting and Nonconducting Conformations. *Proc. Natl. Acad. Sci. USA*, 69, 1610-1614.

Urry, DW. (1982a). Henry Eyring (1901-1981): A 20th Century Physical Chemist and His Models. *Mathematical Modelling* 3, 503-522.

Urry, DW. (1982b) On the Molecular Structure and Ion Transport Mechanism of the Gramicidin Transmembrane Channel, In: *Membrane and Transport*, Martonosi, A., (Ed.), 2, 285-294, Plenum Publishing Corporation, New York, New York.

Urry, DW. (1983). What is elastin; What is not. *Ultrastruct. Pathol.* 4, 227-251.

Urry, DW. (1984). Ionic mechanisms of the Gramicidin trans-membrane channel: cation nuclear magnetic resonance, dielectric relaxation, carbon-13 Nuclear Magnetic Resonance, and Rate Theory calculation of single-channel currents, In: *NATO Advanced Study Institute Series C, Spectroscopy of Biological Molecules*, Sandorfy, C. & Theophanides T., (Eds.), 511-538, D. Reidel Publishing Company, ISBN: 9789027718495, Dordrecht, Holland.

Urry, DW. (1985). Chemical Basis of Ion Transport Specificity in Biological Membranes. In *Topics in Current Chemistry*, (F.L. Boschke, Ed.) 128, 175-218, Springer-Verlag, Heidelberg, Germany

Urry, DW. (1987). NMR relaxation studies of alkali metal ion interactions with the Gramicidin A trans-membrane channel. *Bull. Magn. Reson.*, 9, 109-131.

Urry, DW. (1990). Elastomeric Polypeptide Biomaterials: Structure and Free Energy Transduction. *Mat. Res. Soc. Symp. Proc.* 174, 243-250.

Urry, DW. (1991). Protein Folding Controlled by Chemically Shifting the Temperatures of Inverse Temperature Transitions. In: *Proteins: Structure, Dynamics and Design*, Renugopalakrishnan, V.; Carey, PR.; Smith, ICP.; Huang, SG. & Storer, AC., (Eds.), Escom Sci. Publ. B.V., Leiden, The Netherlands, pp. 352-360.

Urry, DW. (1992). Free Energy Transduction in Polypeptides and Proteins Based on Inverse Temperature Transitions. *Prog. Biophys. Molec. Biol.*, 57, 23-57.

Urry, DW. (1995). Elastic Biomolecular Machines: Synthetic chains of amino acids, patterned after those in connective tissue, can transform heat and chemical energy into motion. *Scientific American*, January 44–49.

Urry, DW. (1997). Physical chemistry of biological free energy transduction as demonstrated by elastic protein-based polymers. (invited FEATURE ARTICLE) *J. Phys. Chem.B*, 101, 11007-11028.

Urry DW. (2004). The change in Gibbs free energy for hydrophobic association: derivation and evaluation by means of inverse temperature transitions. *Chem. Phys. Lett.*, 399, 177–183.

Urry DW. (2005). Hydrophobic and elastic mechanisms in Complex III/Rieske Iron Protein (RIP), walking protein motors and protein-based materials. In: *Peptide Science 2004*, Shimohigashi Y. (Ed.), The Japanese Peptide Society (2005), Proceedings of Asian Pacific International Peptide Symposium, APIPS-JPS 2004, pp. 115-118. (ISSN-1344-7661)

Urry DW. (2006a). *What Sustains Life? Consilient mechanisms for protein-based machines and materials.* Springer (Birkhauser), LLC, New York, ISBN: 08176 4346 X.

Urry DW. (2006b). Deciphering engineering principles for the design of protein-based nanomachines. In: *Protein-Based Nanotechnology*, Renugopalakrishnan, V.; Lewis, R. & Dhar PK. (Eds.), Springer, Chapter 9, 141-188.

Urry DW. (2006c). Function of the F_1-motor (F_1-ATPase) of ATP synthase by apolar-polar repulsion through internal interfacial water. *Cell Biol. Intl.*, 30, 44–55.

Urry, DW.; Goodall, MC.; Glickson, JD. & Mayers, DF. (1971). The Gramicidin A trans-membrane channel: Characteristics of head-to-head dimerized $\pi_{(LD)}$ helices. *Proc. Natl. Acad. Sci. USA, 68*, 1907-1911.

Urry, DW.; Long, MM. & Sugano, H. (1978). Cyclic Analog of Elastin Polyhexapeptide Exhibits an Inverse Temperature Transition Leading to Crystallization. *J. Biol. Chem.* 253, 6301-6302.

Urry, DW.; Venkatachalam, CM.; Spisni, A.; Bradley, RJ.; Trapane, TL. & Prasad, KU. (1980a). The malonyl Gramicidin Channel: NMR-derived rate constants and comparison of calculated and experimental single-channel currents. *J. Membr. Biol.*, 55, 29-51.

Urry, DW.; Venkatachalam, CM.; Spisni, A.; Lauger, P. & Khaled, MA. (1980b). Rate Theory Calculation of Gramicidin Single-channel Currents using NMR-derived Rate Constants, *Proc. Natl. Acad. Sci. USA*, 77, 2028-2032.

Urry, DW.; Trapane, TL.; Sugano, H. & Prasad, KU. (1981). Sequential polypeptides of elastin: Cyclic conformational correlates of the linear polypentapeptide. *J. Am. Chem. Soc.* 103, 2080-2089.

Urry, DW.; Prasad, KU. & Trapane, TL. (1982a). Location of monovalent cation binding sites in the Gramicidin channel. *Proc. Natl. Acad. Sci. USA, 79*, 390-394.

Urry, DW.; Trapane, TL. & Prasad, KU. (1982b). Molecular structure and ionic mechanisms of an ion-selective trans-membrane channel: Monovalent versus divalent cation selectivity. *Int. J. Quant. Chem.: Quant. Biol. Symp.*, 9, 31-40.

Urry, DW.; Trapane, TL.; Walker JT. & Prasad, KU. (1982c) On the relative lipid membrane permeability of Na^+ and Ca^{2+}: A physical basis for the messenger role of Ca^{2+}. *J. Biol. Chem., 257*, 6659-6661;

Urry, DW.; Venkatachalam, CM.; Long M.M. & Prasad, K.U. (1982d). Dynamic β-Spirals and A Librational Entropy Mechanism of Elasticity," In: *Conformation in Biology*, Srinivasan R. & Sarma, RH. (Eds.) G.N. Ramachandran Festschrift Volume, Adenine Press, USA, 11-27.

Urry, DW.; Walker JT. & Trapane, TL. (1982e). Ion interactions in (1-^{13}C)D•Val^8 and D•Leu^{14} analogs of Gramicidin A, the helix sense of the channel and location of ion binding sites. *J. Membr. Biol.*, 69, 225-231.

Urry, DW.; Trapane, TL. & Prasad, KU. (1983). Is the Gramicidin A trans-membrane channel single-stranded or double-stranded helix? A simple unequivocal determination. *Science*, 221, 1064-1067.

Urry, DW. & Venkatachalam, CM. (1983). A librational entropy mechanism for elastomers with repeating peptide sequences in helical array. *Int. J. Quant. Chem.: Quant. Biol. Symp.* 10, 81-93.

Urry, DW.; Trapane, TL.; Prasad, KU. & McMichens, RB. (1989). Ion interactions at membranous polypeptide sites using NMR: Determining rate and binding constants and site locations, In: *Methods in Enzymology*, Fleischer, S. (Ed.), 171, 286-342, ISBN-10: 0121820742, Academic Press, Inc., New York, New York.

Urry, DW.; Peng, S. & Parker, TM. (1993). Delineation of electrostatic-and hydrophobic-induced pka shifts in polypentapeptides: The glutamic acid residue. *J. Am. Chem. Soc.*, 115, 7509-7510.

Urry, DW.; Peng, S.; Gowda, DC.; Parker, TM. & Harris, RD. (1994). Comparison of electrostatic- and hydrophobic-induced pka shifts in polypentapeptides: The lysine residue. *Chem. Phys. Ltrs*, 225, 97-103.

Urry, DW.; Luan, C-H.; Harris, CM. & Parker, T. (1997a). Protein-based materials with a profound range of properties and applications: The elastin ΔT_t hydrophobic paradigm, In: *Protein-Based Materials*, McGrath K. & Kaplan, D., (Eds.), Birkhauser, Boston, 133–177.

Urry, DW.; Peng, S.; Xu J. & McPherson, DT. (1997b). Characterization of waters of hydrophobic hydration by microwave dielectric relaxation. *J. Amer. Chem. Soc.*, 119, 1161–1162.

Urry, DW.; Hugel, T.; Seitz, M.; Gaub, H.; Sheiba, L.; Dea, J.; Xu J. & Parker, T. (2002). Elastin: A representative ideal protein Elastomer. *Phil. Trans. R. Soc. Lond.* B, 357, 169-184.

Urry, DW. & Parker, TM. (2002). Mechanics of elastin: Molecular mechanism of biological elasticity and its relevance to contraction. *J. Muscle Res. Cell Motility*, 23, 541-547. In: Special Issue: *Mechanics of Elastic Biomolecules*, Granzier, H.; Kellermayer M. Jr. &. Linke, W., (Eds.)

Urry, DW.; Urry, KD.; Szaflarski, W.; Nowicki M. & Zabel, M. (2009). Function and frustration of multi-drug ABC exporter protein and design of model proteins for drug delivery using protein hydration thermodynamics. *Current Pharmaceutical Design*, 15, 2833-2867.

Urry, DW.; Urry, KD.; Szaflarski, W. & Nowicki M. (2010). Elastic-contractile model proteins: Physical chemistry, protein function and drug design and delivery. doi:10.1016/j.addr.2010.07.001; *Adv. Drug Del. Rev.*, xxx, xxx–xxx.

Uysal, S.; Vásquez, V.; Tereshko, V.; Esaki, K.; Fellouse, FA.; Sidhu, SS.; Koide, S.; Peroz, E. & Kossiakoff, A. (2009). Crystal structure of full-length KcsA in its closed conformation. *Proc. Natl. Acad. Sci.* USA, 106, 6644-6649. Protein Data Bank accession code 3EFF.

Venkatachalam, CM.; Khaled, MA.; Sugano, H. & Urry, DW. (1981). Nuclear magnetic resonance and conformational energy calculations of repeat peptides of elastin: Conformational characterization of cyclopentadecapeptide, cyclo-(L-Val_1-L-Pro_2-Gly_3-L-Val_4-Gly_5)$_3$. *J. Am. Chem. Soc.* 103, 2372-2379.

Venkatachalam CM. & Urry, DW. (1981). Development of a Linear Helical Conformation From Its Cyclic Correlate. β-Spiral Model of the Elastin Poly(pentapeptide), $(VPGVG)_n$. *Macromolecules* 14, 1225-1229.

Volpin, D.; Urry, DW.; Pasquali-Ronchetti, I. & Gotte, L. (1976). Studies by Electron Microscopy on the Structure of Coacervates of Synthetic Polypeptides of Tropoelastin. *Micron* 7, 193-198.

Wasserman, ZR. & Salemme, FR. (1990). A molecular dynamics investigation of the elastomeric restoring force in elastin. *Biopolymers* 29: 1613–1631.

Zhang, Z.; Huang, L.; Shulmeister, VM.; Chi, YI.; Kim, KK.; Hung, LW.; Crofts, AR.; Berry, EA. & Kim, SH. (1998). Electron transfer by domain movement in cytochrome bc_1, *Nature* 392 677-684. Protein Data Bank, accession code 1BCC and 3BBC.

Zhou, Y.; Morais-Cabral, JH., Kaufman, A. & MacKinnon, R. (2001). Chemistry of ion coordination and hydration revealed by a K^+ channel-Fab complex at 2.0 Å resolution. *Nature*, 414, 43-48.

2

Thermodynamics of Natural and Synthetic Inhibitor Binding to Human Hsp90

Vilma Petrikaitė and Daumantas Matulis
Laboratory of Biothermodynamics and Drug Design, Institute of Biotechnology Lithuania

1. Introduction

Heat shock protein 90 (Hsp90) is one of conserved heat shock proteins that protect, prevent aggregation, stabilize, activate, or otherwise regulate client proteins, is a component of the cellular chaperone machinery [Taipale et al., 2010, Taldone et al., 2009, Wandinger et al., 2008]. There are a number of recent developments in the understanding of the interesting and complex mechanism of Hsp90 action [Neckers et al., 2009a, Neckers et al., 2009b, Mayer et al., 2009, Walerych et al., 2009]. Hsp90 is over-expressed in cancer cells and Hsp90 inhibitors have shown selectivity for cancer cells. Therefore, small-molecule inhibitors are being developed as anticancer therapeutics [van Montfort and Workman, 2009, Sharp et al., 2007, Sgobba and Rastelli, 2009, Fukuyo et al., 2009].

Two groups of Hsp90 inhibitors have been designed based on naturally occuring inhibitors geldanamycin and radicicol. Geldanamycin has been modified to 17-AAG, while various resorcinol-bearing compounds were designed based on radicicol. Here we describe the thermodynamics of their binding to Hsp90 by isothermal titration calorimetry (ITC) and thermal shift assay (TSA). These assays yield not only the potency, i.e. the Gibbs free energy of binding, but also the enthalpy of binding, the entropy of binding, and the heat capacity of binding. This detailed thermodynamic description and the comparison between homologous compound structures, coupled with structural information of the Hsp90-inhibitor complex, provides insight into the structure-activity relationships (SAR) of the compounds. The SAR helps in the process of rational drug design [Freire, 2009].

2. The structure of Hsp90 and the comparison of human and yeast isoforms

There are several Hsp90 homologs in human, yeast, and bacteria. Human Hsp90 exists in two highly homologous isoforms - α and β. Alpha isoform is prevalent. There are no major known functional differences between the isoforms. Hsp90 homolog in yeast is named Hsc82 and also shares significant homology with human isoforms.

Figure 1 shows the structure of Hsp90 and Hsc82. The protein in solution exists in equilibrium between dimer and monomer. Furthermore, the protein is quite flexible and exists in equilibrium between at least three major conformations [Graf et al., 2009].

The full length protein consists of three major domains – the N-terminal domain (1-216 a.a.), the M-domain (262-524), and the C-terminal domain (525-709). There is also a charged linker (216-262) that did not crystallize and its structure is unknown [Ali et al., 2006]. Inhibitors such as radicicol and geldanamycin bind to the ATP-binding pocket of Hsp90 and are thus competitive non-covalent inhibitors.

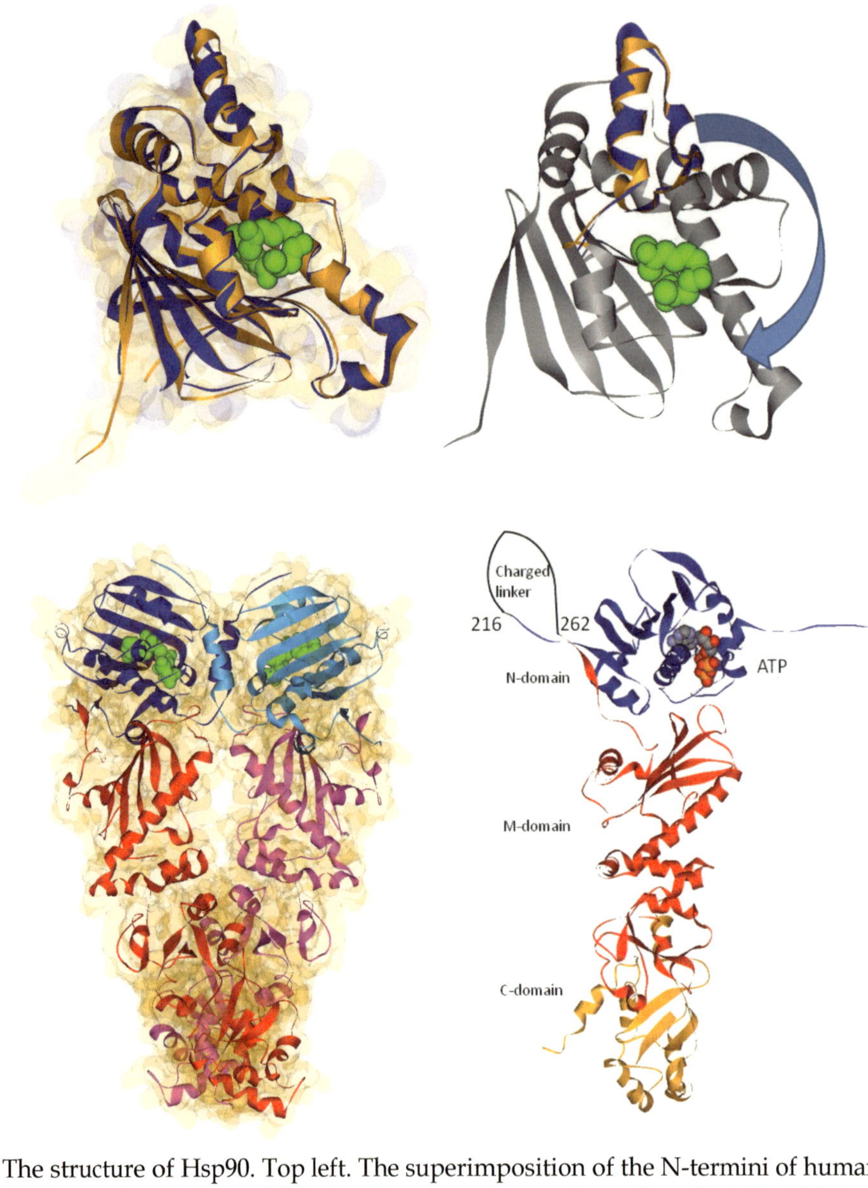

Fig. 1. The structure of Hsp90. Top left. The superimposition of the N-termini of human Hsp90αN (orange) and Hsc82N (blue) with transparent surfaces show that the fold is essentially identical. Radicicol is shown bound to yeast Hsc82N (PDB ID: 1bgq) as spacefilled green model. Top right. Rotation of the 'lid' part of the protein (94-139 a.a., 108-139 a.a. in human Hsp90). Bottom left. The structure of Hsc82 full length dimer (2cg9). One monomer is shown in dark blue-red and another in light blue-pink. Blue shows the position of the N-terminal domains. Bottom right. The structure of Hsc82 full length monomer showing the positions of three domains and the unstructured charged linker

Radicicol interactions with human and yeast Hsp90 isoforms are shown in Figure 2. Crystallographic experimental structure exists only for the yeast isoform. Therefore, radicicol binding to human isoforms was modelled computationally into the active site.

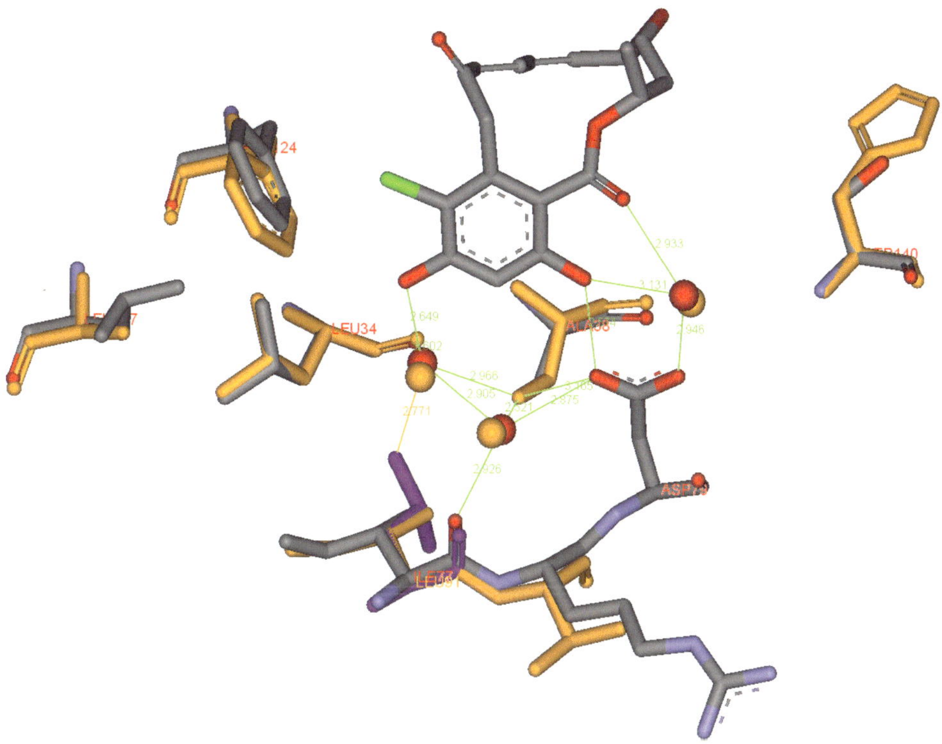

Fig. 2. Interactions of radicicol (grey substance in the middle at the top of the picture) with selected Hsp90 amino acids. Hsp90α amino acids are shown in orange (3eko) superimposed on yeast Hsc82 amino acids (colored by atom, 1bgq), and Hsp90β amino acids are shown in pink. Three important water molecules participating in hydrogen bond formation between the protein and radicicol are shown as spheres

There are very few amino acid differences between yeast and human isoforms that may have an impact on the binding thermodynamics. They are shown in the figure and listed below:

Hsp90α	Ser52	Ile91	Val92	Ala141	His154
Hsp90β	Ala47	Leu86	Val87	Ala136	His149
Hsc82	Ala38	Ile77	Arg78	Leu127	Ser140

All other amino acids in the vicinity of radicicol are identical in all three isoforms. Therefore, the difference in the binding thermodynamics should be due to the above mentioned amino acids. Furthermore, charged amino acids Arg78 and His154 point to the solvent on the protein surface different from radicicol binding site. Therefore, the difference is most likely due to the remaining three amino acids.

3. Thermodynamics of binding by isothermal titration calorimetry and thermal shift assay

Protein - ligand binding equilibrium is described by the Gibbs free energy of binding ($\Delta_b G$). More negative $\Delta_b G$ indicates a stronger binding reaction. The Gibbs free energy is sufficient to describe the equilibrium. However, several thermodynamic parameters that contribute to the $\Delta_b G$ can be correlated with structural features of the protein - ligand complex easier than the $\Delta_b G$ itself. The most important parameters are the enthalpy ($\Delta_b H$) and entropy ($\Delta_b S$) of binding:

$$\Delta_b G = \Delta_b H - T\Delta_b S \tag{1}$$

Both the enthalpy and entropy are the fist temperature derivatives (T-derivatives) of the Gibbs free energy:

$$\left(\frac{\partial \Delta_b G}{\partial T}\right)_P = -\Delta_b S \tag{2}$$

$$\left(\frac{\partial \ln K_b}{\partial \frac{1}{T}}\right)_P = -\frac{\Delta_b H}{R} \tag{3}$$

The second T-derivative of the $\Delta_b G$ (the $\Delta_b H$ T-derivative) is the heat capacity of binding ($\Delta_b C_P$). Subscript P indicates constant pressure.

$$\Delta_b C_p = \left(\frac{\partial \Delta_b H}{\partial T}\right)_p \tag{4}$$

There are other thermodynamic parameters that are pressure derivatives (P-derivatives) of $\Delta_b G$. The first P-derivative is the volume of binding ($\Delta_b V$). The second P-derivatives are the compressibility and expansion of binding. These parameters may be measured by varying the pressure of the protein - ligand system. However, they are rarely used and are beyond the scope of this chapter. Here we will concentrate on the most used, however, selected thermodynamic parameters, namely, $\Delta_b G$, $\Delta_b H$, $\Delta_b S$, and $\Delta_b C_P$.
The Gibbs free energy of ligand binding may be measured by a large variety of methods, well reviewed for carbonic anhydrase inhibitor binding in [Krishnamurthy et al., 2008]. Here we will concentrate on the application of (ITC) and the (TSA). Both methods have been described previously in detail, especially ITC [Freyer and Lewis, 2008, Landbury, 2004,

Velazquez-Campoy et al., 2004]. However, TSA is rather unconventional and underused despite its great advantages and usefulness [Zubriene et al., 2009, Cimmperman et al., 2008, Matulis et al., 2005].

ITC directly measures the heat evolved or absorbed during the binding reaction. At constant pressure, the heat is equal to the enthalpy ($\Delta_b H$) of binding. This method is the most robust and accurate way of measuring the $\Delta_b H$. However, until the isothermal titration calorimeters became commercially available in early 90s, the $\Delta_b H$ was usually estimated from the $\Delta_b G$ T-dependence using the van't Hoff relationship (3). If all contributing reactions are clearly dissected, such approach should yield the same results as titration calorimetry. However, in practice, there are many unexplained inconsistencies and only ITC provides reliable $\Delta_b H$.

However, the ITC has a number of disadvantages. Most importantly, the binding constant should be in a rather narrow range to satisfy the requirement that coefficient c is between about 5 and 500. The c is:

$$c = nM_t K_b \tag{5}$$

Where n is the binding stoichiometry, M_t is the protein molar concentration, and K_b is the binding constant defined for the reaction of $M + L \rightleftarrows ML$ as:

$$K_b = \frac{[ML]}{[L][M]} \tag{6}$$

$$\Delta G = -RT \ln K_b \tag{7}$$

In practice, ITC is useful for K_bs in the range of 10^5 to 10^9 M^{-1}. Such K_b can be usually measured by varying protein concentration. If ITC experiment is planned well, it can provide $\Delta_b G$, $\Delta_b H$, and $\Delta_b S$ in an hour. Doing the same experiment at several temperatures will yield an indirect measurement of the heat capacity $\Delta_b C_P$.

Another disadvantage of ITC is that it requires rather large amount of protein (usually more than 0.1 mg) and ligand. The protein must be well purified and soluble at micromolar concentrations.

These disadvantages can be quite easily approached using the TSA. This method is based on the observation that specifically binding ligands stabilize (sometimes destabilize) the protein. Protein solution is being heated at a constant rate in the absence or presence of a ligand and the unfolding pattern is measured by various methods such as absorbance, circular dichroism, or, most often, by fluorescence. Various fluorescent components could be followed, such as intrinsic tryptophan fluorescence or an extrinsic solvatochromic probe. Most convenient is 1,8-anilino naphthalene sulfonate (ANS). Figure 3 shows ANS fluorescence dependence on temperature upon Hsp90 unfolding. The rise in fluorescence near 50 °C is due to Hsp90 unfolding and the exposure of hydrophobic patches of the protein interior. ANS binds to such patches and its fluorescence increases.

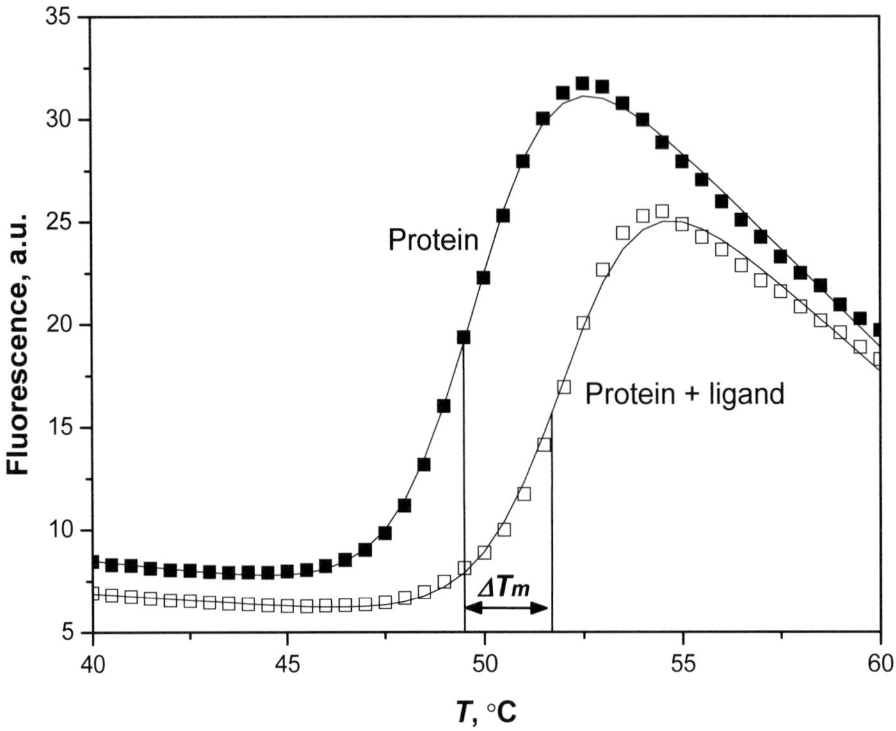

Fig. 3. Thermal shift assay protein melting curves. The midpoint of denaturation (T_m) is shifted to higher temperature upon ligand addition

Note that ANS primarily binds to cationic groups on the protein surface (first to arginine residues) [Matulis and Lovrien, 1998]. However, most such bound ANS does not fluoresce and thus we can observe the unfolding pattern of the protein. Addition of ligand stabilized the protein and shifted the curve and the midpoint of the unfolding transition (T_m) by about 5 degrees.

Protein unfolding fluorescence curves are described by the equation:

$$y = y_N + \frac{y_U - y_N}{1 + e^{\frac{\Delta_u G}{RT}}} = y_U + \frac{y_N - y_U}{1 + e^{-\frac{\Delta_u G}{RT}}} \tag{8}$$

Protein melting temperatures can be determined by fitting the protein melting curves (Figure 3) according to:

$$y(T) = y_{F,T_m} + m_F(T - T_m) + \frac{(y_{U,T_m} - y_{F,T_m}) + (m_U - m_F)(T - T_m)}{1 + e^{\left(\Delta_U H_{T_m} + \Delta_U C_p (T - T_m) - T\left(\Delta_U S_{T_m} + \Delta_U C_p \ln(T/T_m)\right)\right)/RT}} \tag{9}$$

where $y(T)$ is the calculated fluorescence as a function of temperature; y_{F,T_m} is the fluorescence of the probe bound to folded native protein before the transition at T_m; y_{U,T_m} is the fluorescence of the probe bound to the unfolded protein after the unfolding transition at T_m; m_F is the slope of the fluorescence dependence on temperature when the probe is bound to the native protein; m_U is the slope of the fluorescence dependence on temperature when the probe is bound to the unfolded protein; $\Delta_U H_{T_m}$ is the enthalpy of protein unfolding at T_m; $\Delta_U S_{T_m}$ is the entropy of protein unfolding at T_m; $\Delta_U C_p$ is the heat capacity of protein unfolding and is assumed to be temperature-independent over the temperature range studied; R is the universal gas constant; and T is the absolute temperature (Kelvin).

Ligand dosing curves (as in Figures 9 and 10) are described by the equation (10):

$$L_t = (K_{U,T_m} - 1)\left(\frac{P_t}{2K_{U,T_m}} + \frac{1}{K_{b,T_m}}\right) = \left(e^{-\left(\Delta_U H_{T_r} + \Delta_U C_p (T_m - T_r) - T_m\left(\Delta_U S_{T_r} + \Delta_U C_p \ln(T_m/T_r)\right)\right)/RT_m} - 1\right)$$

$$\times\left[\frac{P_t}{2}\frac{1}{e^{-\left(\Delta_U H_{T_r} + \Delta_U C_p (T_m - T_r) - T_m\left(\Delta_U S_{T_r} + \Delta_U C_p \ln(T_m/T_r)\right)\right)/RT_m}} + \frac{1}{e^{-\left(\Delta_b H_{T_0} + \Delta_b C_p (T_m - T_0) - T_m\left(\Delta_b S_{T_0} + \Delta_b C_p \ln(T_m/T_0)\right)\right)/RT_m}}\right]$$

L_t is the total concentration of added ligand, K_{U,T_m} is the protein unfolding equilibrium constant at T_m; P_t is the total protein concentration; K_{b,T_m} is the ligand binding constant at T_m; $\Delta_U H_{T_r}$ is the enthalpy of protein unfolding at T_r; T_r is the protein melting temperature when no ligand is added; $\Delta_U S_{T_r}$ is the entropy of protein unfolding at T_r; $\Delta_U C_p$ is the heat capacity of protein unfolding and is assumed to be temperature-independent over the temperature range studied; $\Delta_b H_{T_0}$ is the enthalpy of ligand binding at T_0; T_0 is the temperature at which the binding process is studied (usually 37 °C); is the entropy of ligand binding at T_0; and $\Delta_b C_p$ is the heat capacity of ligand binding and is assumed to be temperature-independent over the temperature range studied.

The binding constant at the physiological temperature T_0 is determined using:

$$K_{b,T_0} = e^{-\left(\Delta_b H_{T_0} - T_0 \Delta_b S_{T_0}\right)/RT_0} \tag{11}$$

TSA can be performed in the RT-PCR machine and requires only several micrograms of protein. Furthermore, there is no upper limit of the K_b to be determined. The only limit is the temperature of water boiling. Therefore, such extremely tight reactions as radicicol binding to Hsp90 can be studied by TSA. There is also no lower limit for the K_b. Therefore, millimolar and picomolar ligands can be easily measured. However, TSA does not determine $\Delta_b H$, $\Delta_b S$, and $\Delta_b C_P$. Therefore, both TSA and ITC should be used to determine the thermodynamics of Hsp90 – ligand binding.

4. Thermodynamics of Hsp90 ligand binding

Figure 4 shows the structures of Hsp90 ligands used in this study.

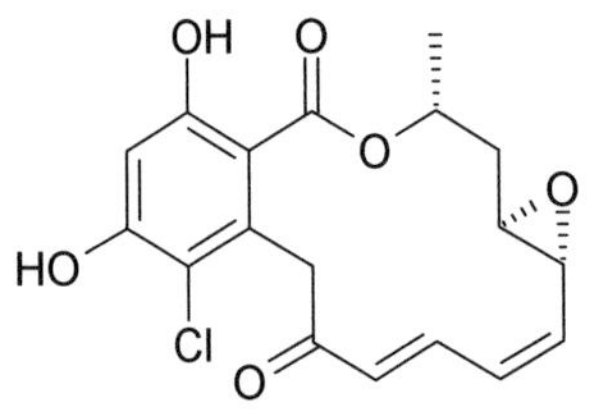

Radicicol

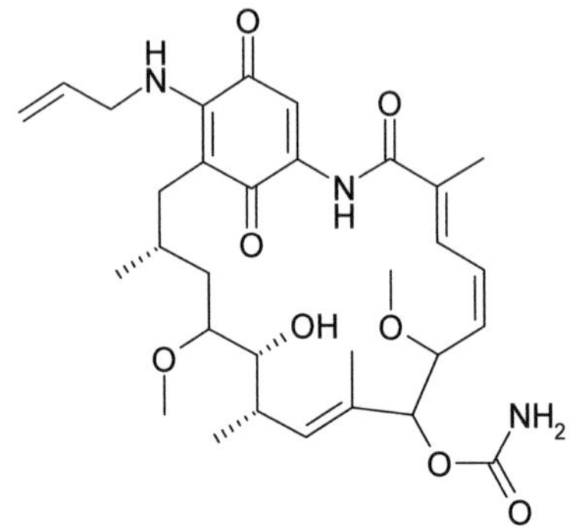

17-AAG

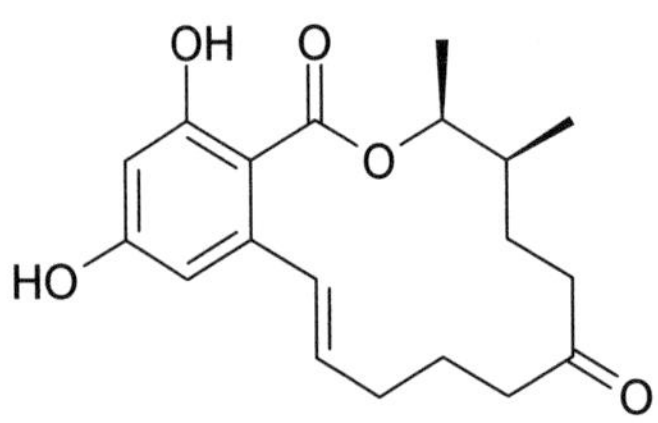

ent-35

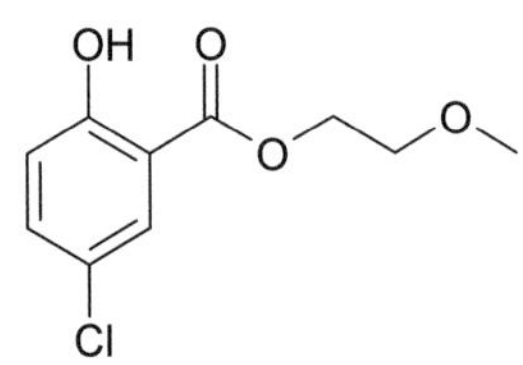

AZ1

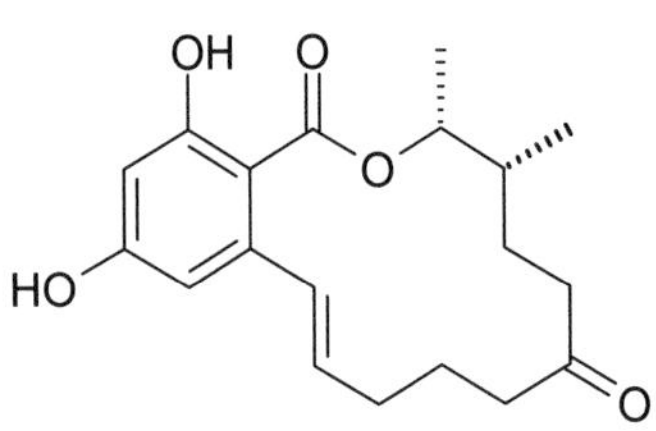

35

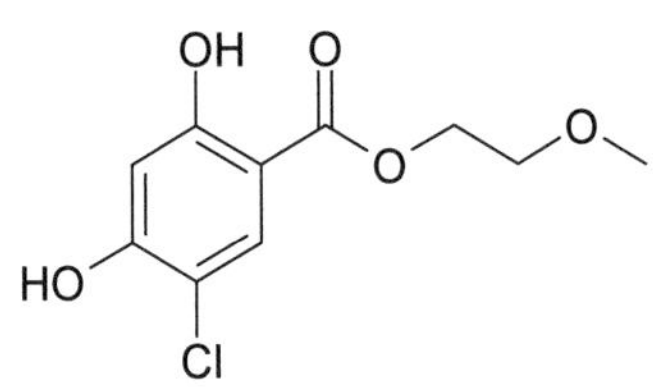

AZ3

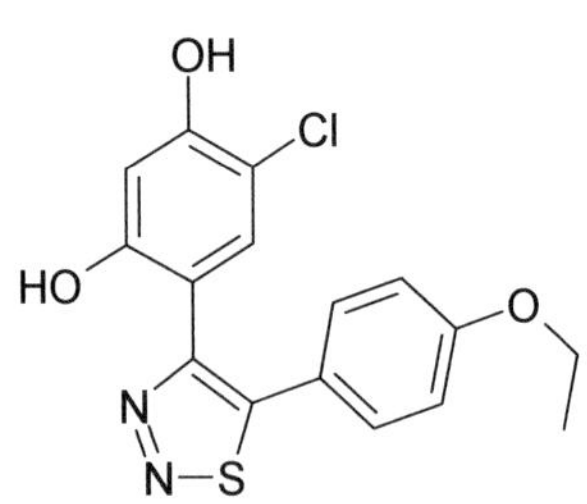

ICPD47

Fig. 4. Chemical structures of compounds discussed in this manuscript

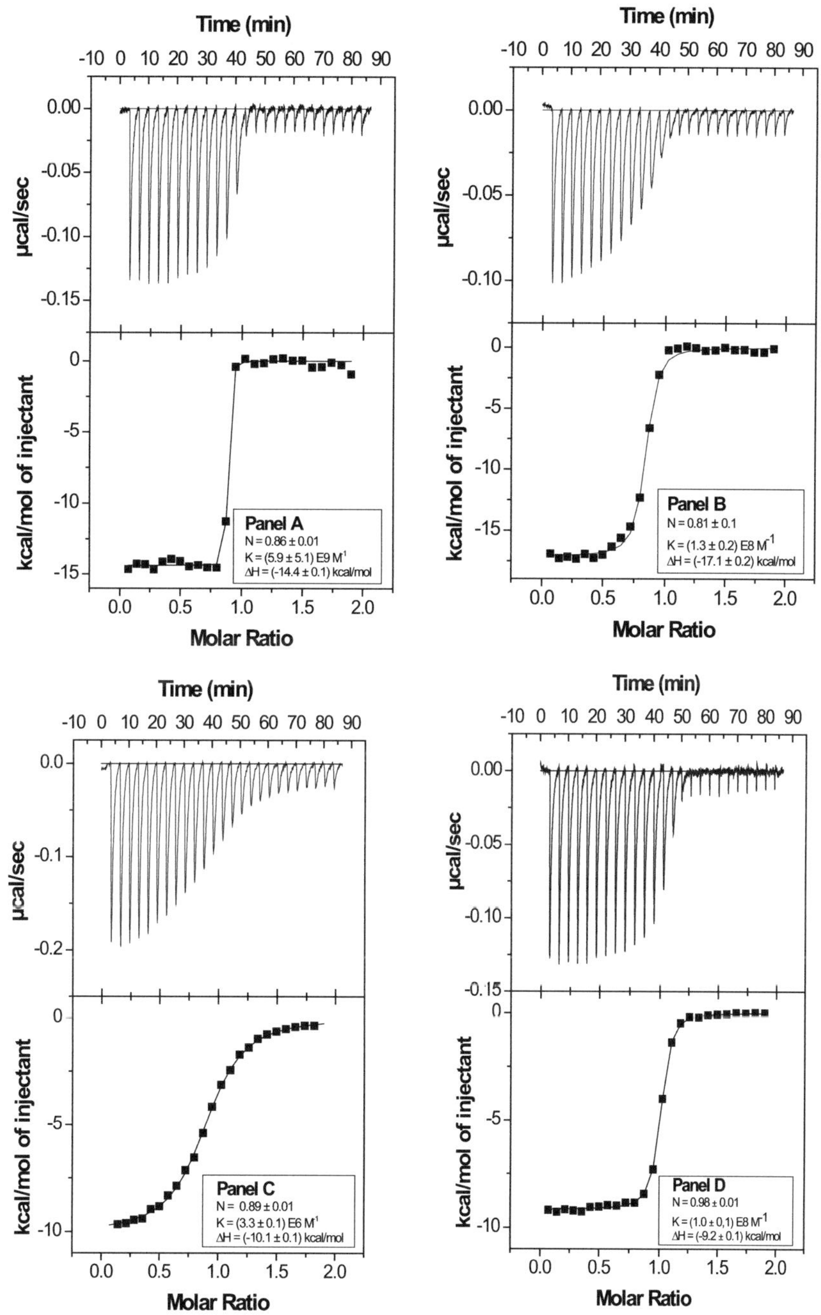

Fig. 5. A. Isothermal titration calorimetry data for radicicol binding to Hsp90αN. Upper graph - raw ITC data, lower graph - integrated ITC data with the curve fit to the standard single binding site model. The cell contained 4 µM protein, while the syringe contained 40 µM radicicol in the same buffer - 50 mM sodium phosphate, pH 7.5, 0.5% DMSO, 100 mM NaCl, at 25 °C. B. Radicicol binding ITC curve at the same conditions as in panel A except pH 8.5. C. 17-AAG binding to Hsp90αN. D. ICPD47 compound binding to Hsp90αN

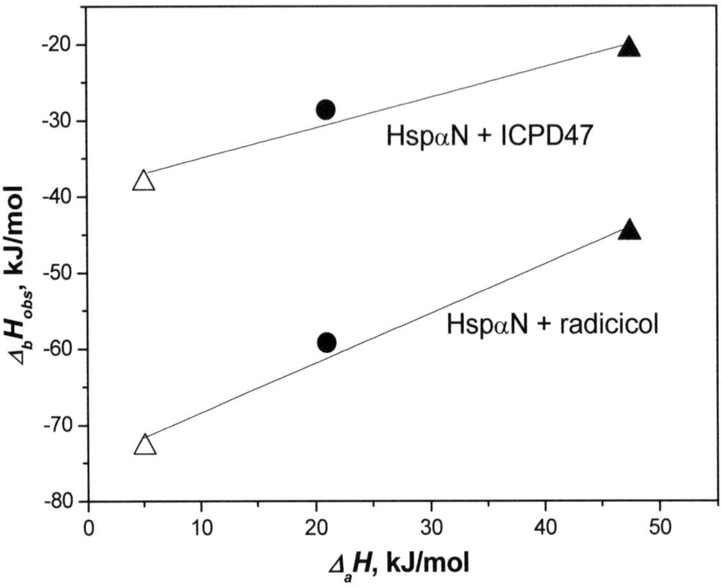

Fig. 6. The observed enthalpies as a function of the buffer deprotonation enthalpy at 25 °C temperature in various buffers: △ - phosphate, ● - Hepes, ▲ - Tris. The data points are the experimentally-observed enthalpies, and the trendlines are linear fits. Their slopes are equal to the binding-linked protonation events. Intersection with y axis is buffer-independent binding enthalpy

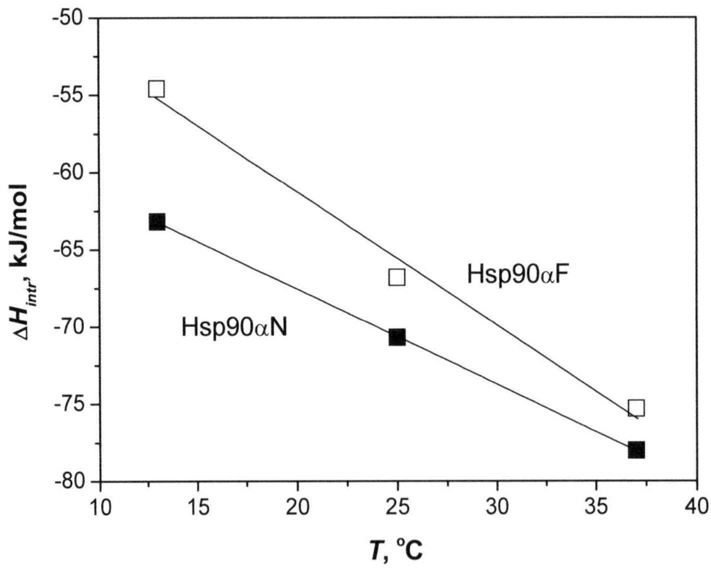

Fig. 7. Intrinsic binding enthalpies obtained after accounting for the linked protonation event as a function of temperature for radicicol binding to Hsp90αN (■) and Hsp90αF (□). The slopes are linear fits to the experimental data and are equal to the intrinsic heat capacities of radicicol binding

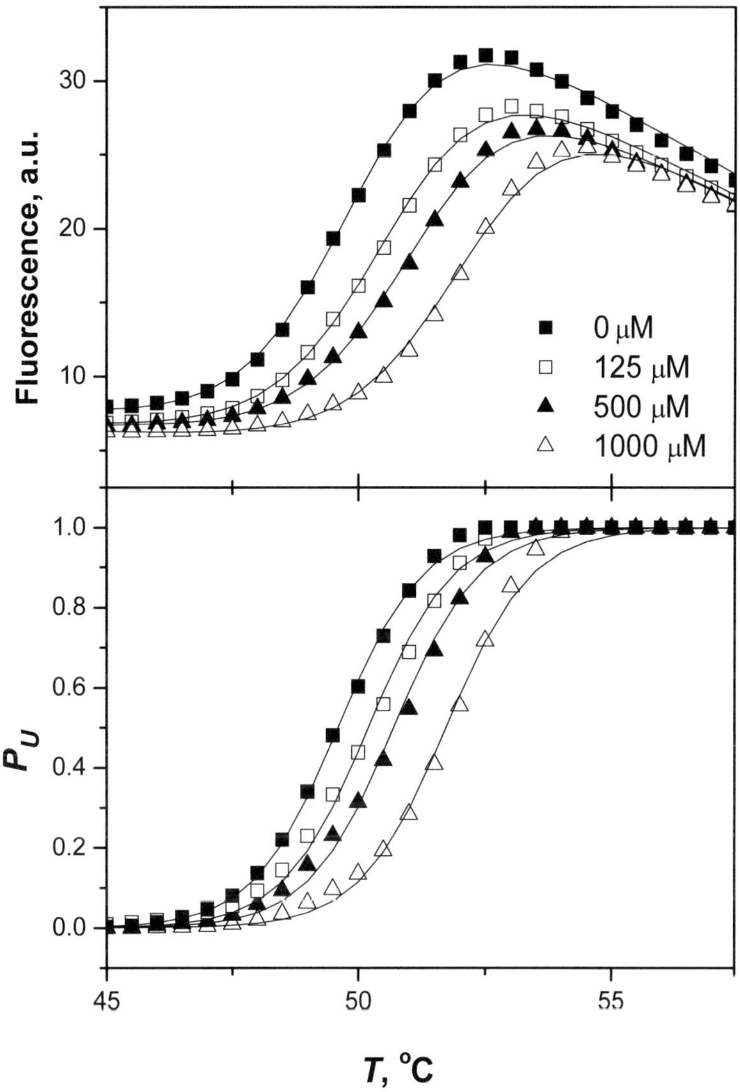

Fig. 8. The binding of AZ3 to Hsp90αN at pH 7.0 as determined by the thermal shift assay. Upper panel shows experimental fluorescence curves. The lower panel shows the same curves recalculated as probabilities to observe the protein in the denatured state

Figure 5 shows several typical ITC binding curves of the Hsp90 - ligand system. The curve in panel A is too steep, meaning that radicicol binding is too tight to be accurately determined by ITC. TSA data will be needed to determine the K_b and $\Delta_b G$ of interaction. However the $\Delta_b H$ is determined to high precision.

The $\Delta_b H$ determined in buffers with different enthalpy of protonation ($\Delta_a H$) yielded different $\Delta_b H$ (Figure 6). Therefore, the binding reaction is linked with protein or ligand protonation or deprotonation upon binding. In other words, ligand binding shifts the pK_a of ionisable groups as previous explained [Baker and Murphy, 1996].

Due the linked protonation, it is important to dissect protonation thermodynamics from binding thermodynamics in order to determine the intrinsic thermodynamics of binding. The enthalpy of, for example, Tris buffer protonation is so large (about -44 kJ/mol) that it would hide any binding enthalpies. Therefore, a series of experiments in various buffers are necessary (Figure 6).

When protonation-linkage effects are accounted for [Zubriene et al., 2010] and the intrinsic enthalpy of binding is determined, such experiments should be repeated at all temperatures of interest. Figure 7 shows intrinsic enthalpies determined at 13, 25, and 37 °C for radicicol-Hsp90α system. The full length protein bound radicicol with slightly less exothermic enthalpy. The difference was equal to ~ 4 kJ/mol (Table 1). This difference is within the standard error of the measurements. The error was greater for the full length protein because the available protein amount and concentration was lower.

ITC was useful to provide the enthalpy of binding. However, as seen in radicicol binding ITC curves in Figure 5, the binding is too tight and would require the displacement assay as described previously [Velazquez-Campoy and Freire, 2006]. Our experience shows that the TSA is much easier and yields more precise results than the displacement ITC assay [Zubriene et al., 2009].

Therefore, the binding of all ligands listed in Figure 4, was measured by the TSA. Figure 8 shows typical raw protein melting curves observed at various added ligand concentration. AZ3 bound with relatively low, millimolar affinity. Therefore, relatively large concentration of ligand had to be added in order to observe the T_m shift.

It should be noted here that the shift continues way beyond saturation of protein with ligand. This is due to the dominant entropy of mixing. If the stabilization occurred due to some kind of bond formation, then we would not observe continued stabilization past the saturation point. This is observed when an inhibitor binds covalently and irreversibly to the protein.

Figures 9 and 10 show the dosing curves for the ligands listed in Figure 4. Radicicol, the most potent binder, shifts the temperature by nearly 15 °C. 17-AAG is the average binder and shifts the temperature by about 10 °C. Some ligands, such as AZ1, barely shift the temperature even at 1 mM concentration.

Protein	K_d, nM	$\Delta_b H_{intr}$, kJ×mol^{-1}	$\Delta_b G_{intr}$, kJ×mol^{-1}	$T\Delta_b S_{intr}$, kJ×mol^{-1}	$\Delta_b S_{intr}$, J×mol^{-1}×K^{-1}	$\Delta_b C_p$, J×mol^{-1}×K^{-1}
Hsp90αN	0.04	-70.7	-59.4	-11.4	-38	-620
Hsp90αF	0.04	-66.8	-59.4	-7.5	-25	-860
Hsp90βN	0.15	-60.6	-56.1	-4.6	-15	-760
Hsc82F	0.25	-46.7	-54.8	8.1	27.1	-620
Uncertainties	±1.6-fold	±4	±2.6	±4.7	±16	±140

Table 1. Intrinsic thermodynamic parameters of radicicol binding to human ant yeast isoforms of Hsp90

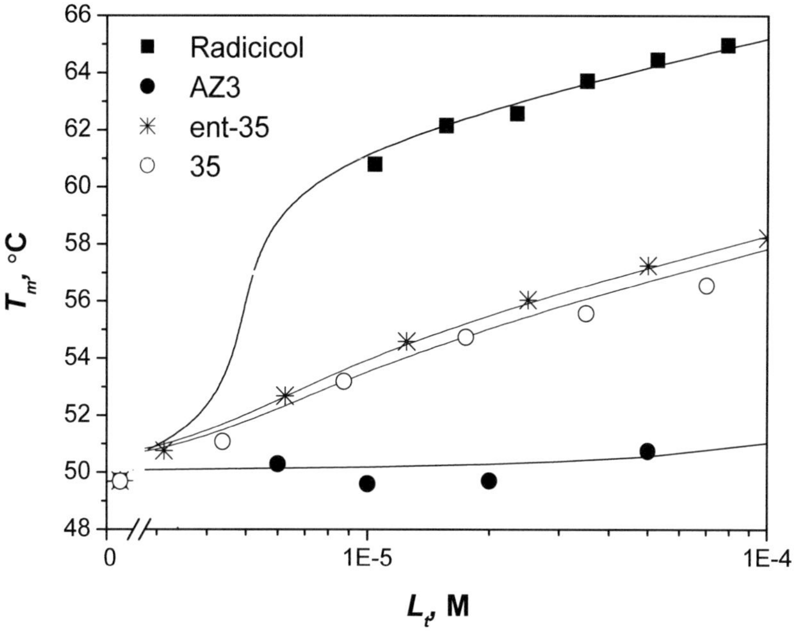

Fig. 9. The dependence of the melting temperature of Hsp90αN on the concentration of various inhibitors (ligand dosing curves) [Ugele et al., 2009]. The observed K_ds (µM) by TSA were: Radicicol - 0.00083, AZ3 - 5000, *ent*-35 - 0.27, and 35 - 0.4

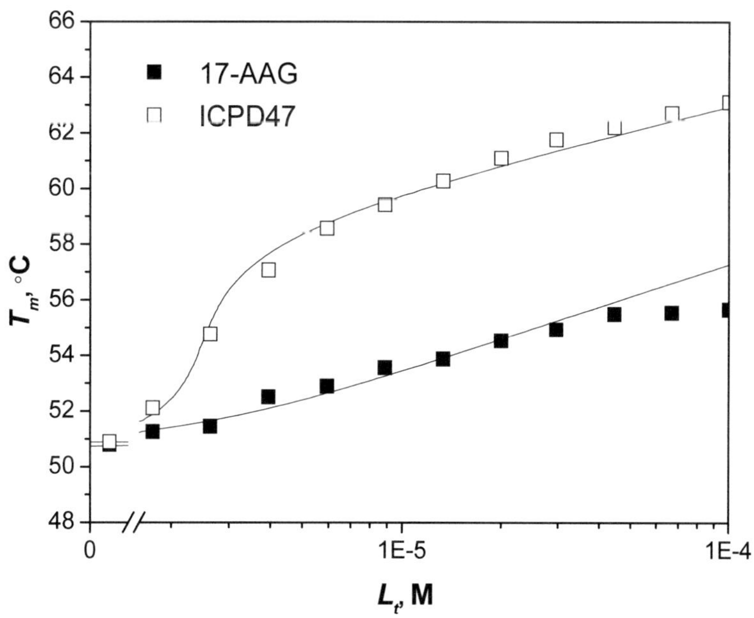

Fig. 10. The dependence of the melting temperature of Hsp90α N on the concentration of inhibitors 17-AAG and ICPD47. The observed K_ds (µM) by TSA were: 17-AAG - 0.3 and ICPD47 - 0.002

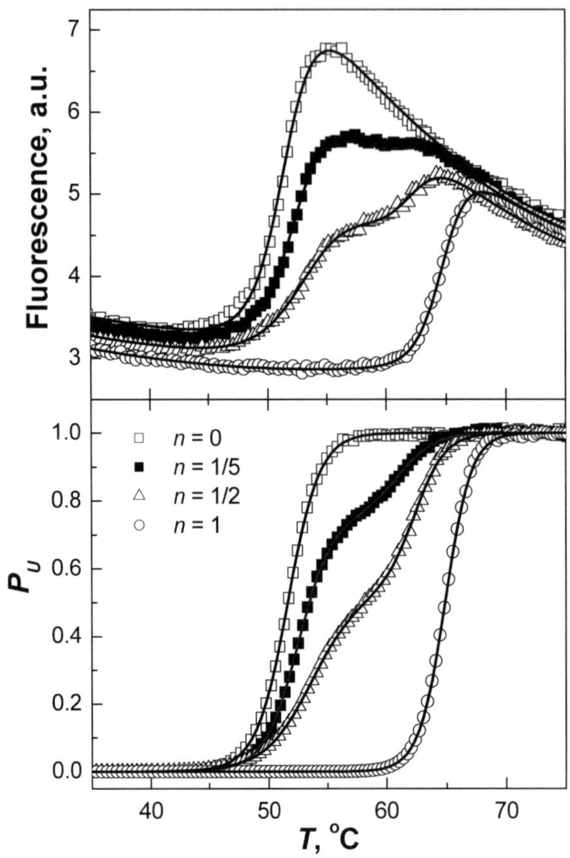

Fig. 11. Denaturation profiles of Hsp90αN in the presence of various radicicol concentrations [Zubriene et al., 2009]

Figure 11 shows an interesting phenomenon observed in TSA when the concentration of ligand is lower than protein. When the concentration of ligand is insufficient to saturate all protein binding sites, the denaturation transition splits into two transitions - the first transition is due to free protein and the second is due to the liganded protein. Relative magnitudes of both transitions are proportional to the concentrations of free and liganded protein concentrations. This phenomenon has been also observed by DSC with weakly-binding ligands at relatively high protein concentrations [Sharke and Ross, 1988].

TSA enabled determination of sub-nanomolar binding potency of naturally occuring radicicol and strongly or weakly-binding synthetic compounds where ITC does not work. However, no dissection of proton linkage was done for ligands where ITC was not feasible. Therefore, only observed K_ds are obtained for such ligands without other thermodynamic information. After dissecting the proton linkage by ITC, it was shown that radicicol binds about 4 times more strongly to recombinant human Hsp90 alpha than to Hsp90 beta isoform. This reduction in affinity is caused primarily by less favorable enthalpic rather than entropic contributions. About 90% of the binding energy comes from the favorable enthalpic contribution and small opposing entropic contribution at physiological temperature (Table 1).

Detailed proton linkage and temperature analysis had to be performed to dissect buffer and protein linked reactions from ligand binding intrinsic reaction. However, even after this detailed analysis, it is not possible to determine whether conformational change in the protein could contribute significantly to these intrinsic thermodynamic parameters. It is quite likely that some contribution comes from the rotation of the lid as shown in Figure 1. The intrinsic enthalpy of radicicol binding to Hsp90 is one of the largest enthalpies observed for any protein - small ligand binding. Note, that most of the Gibbs free energy of radicicol binding comes from the favourable enthalpic contribution. The entropy contribution is relatively small.

5. Conclusions

Radicicol and other resorcinol-bearing compound binding to Hsp90 is interesting in many respects regarding drug design. First, the binding reaction can be very tight (i.e., it has a very favorable Gibbs free energy). Radicicol stabilizes Hsp90 by 15-20°C. Second, the binding reaction has a very favorable enthalpy of binding, one of the largest for any protein - small ligand system. Third, there are few direct contacts between Hsp90 and radicicol that could account for such a large binding energy. Fourth, water molecules play an essential role in the recognition and binding. And fifth, the negative heat capacity of binding usually reflects a dominant hydrophobic origin of binding. However, hydrogen bonds are apparently essential for radicicol binding to Hsp90.

6. Acknowledgments

The project was supported in part by EEA and Norway Grants 2004-LT0019-IP-1EEE and the Lithuanian Government.

7. References

Ali, M. M., Roe, S. M., Vaughan, C. K., Meyer, P., Panaretou, B., Piper, P. W., Prodromou, C., and Pearl, L. H., 2006. Crystal structure of an Hsp90-nucleotide-p23/sba1 closed chaperone complex. *Nature*, 440(7087):1013–7.

Baker, B. M. and Murphy, K. P., 1996. Evaluation of linked protonation effects in protein binding reactions using isothermal titration calorimetry. *Biophys. J*, 71(4):2049–55.

Cimmperman, P., Baranauskiene, L., Jachimoviciute, S., Jachno, J., Torresan, J., Michailoviene, V., Matuliene, J., Sereikaite, J., Bumelis, V., and Matulis, D., 2008. A quantitative model of thermal stabilization and destabilization of proteins by ligands. *Biophys J*, 95(7):3222-3231.

Freire, E., 2009. A thermodynamic approach to the affinity optimization of drug candidates. *Chem Biol Drug Des*, 74(5):468–72.

Freyer, M. W. and Lewis, E. A., 2008. Isothermal titration calorimetry: experimental design, data analysis, and probing macromolecule/ligand binding and kinetic interactions. *Methods Cell Biol*, 84:79–113.

Fukuyo, Y., Hunt, C. R., and Horikoshi, N., 2009. Geldanamycin and its anti-cancer activities. *Cancer Lett*, 290(1):24-35.

Graf, C., Stankiewicz, M., Kramer, G., and Mayer, M. P., 2009. Spatially and kinetically resolved changes in the conformational dynamics of the hsp90 chaperone machine. *Embo J*, 28(5):602–13.

Krishnamurthy, V. M., Kaufman, G. K., Urbach, A. R., Gitlin, I., Gudiksen, K. L., Weibel, D. B., and Whitesides, G. M., 2008. Carbonic anhydrase as a model for biophysical and physical-organic studies of proteins and protein-ligand binding. *Chem Rev*, 108(3):946–1051.

Ladbury, J. E., 2004. Application of isothermal titration calorimetry in the biological sciences: things are heating up! *Biotechniques*, 37(6):885–7.

Matulis, D., Kranz, J. K., Salemme, F. R., and Todd, M. J., 2005. Thermodynamic stability of carbonic anhydrase: measurements of binding affinity and stoichiometry using thermofluor. *Biochemistry*, 44(13):5258–66.

Matulis, D. and Lovrien, R., 1998. 1-anilino-8-naphthalene sulfonate anion-protein binding depends primarily on ion pair formation. *Biophys J*, 74(1):422–9.

Mayer, M. P., Prodromou, C., and Frydman, J., 2009. The Hsp90 mosaic: a picture emerges. *Nat Struct Mol Biol*, 16(1):2–6.

Neckers, L., Mollapour, M., and Tsutsumi, S., 2009a. The complex dance of the molecular chaperone Hsp90. *Trends Biochem Sci*, 34(5):223-6.

Neckers, L., Tsutsumi, S., and Mollapour, M., 2009b. Visualizing the twists and turns of a molecular chaperone. *Nat Struct Mol Biol*, 16(3):235–6.

Sgobba, M. and Rastelli, G., 2009. Structure-based and *in silico* design of Hsp90 inhibitors. *ChemMedChem*, 4(9):1399–409.

Sharp, S. Y., Boxall, K., Rowlands, M., Prodromou, C., Roe, S. M., Maloney, A., Powers, M., Clarke, P. A., Box, G., Sanderson, S., *et al.*, 2007. In vitro biological characterization of a novel, synthetic diaryl pyrazole resorcinol class of heat shock protein 90 inhibitors. *Cancer Res*, 67(5):2206–16.

Shrake, A. and Ross, P. D., 1988. Biphasic denaturation of human albumin due to ligand redistribution during unfolding. *J Biol Chem*, 263(30):15392–9.

Taipale, M., Jarosz, D.F., Lindquist, S., 2010. HSP90 at the hub of protein homeostasis: emerging mechanistic insights. *Nat Rev Mol Cell Biol*, 11:515-28.

Taldone, T., Sun, W., and Chiosis, G., 2009. Discovery and development of heat shock protein 90 inhibitors. *Bioorg Med Chem*, 17(6):2225–35.

Ugele, M., Sasse, F., Knapp, S., Fedorov, O., Zubriene, A., Matulis, D., and Maier, M. E., 2009. Propionate analogues of zearalenone bind to Hsp90. *Chembiochem*, 10(13):2203-12.

van Montfort, R. L. and Workman, P., 2009. Structure-based design of molecular cancer therapeutics. *Trends Biotechnol*, 27(5):315–28.

Velazquez-Campoy, A. and Freire, E., 2006. Isothermal titration calorimetry to determine association constants for high-affinity ligands. *Nat Protoc*, 1(1):186–91.

Velazquez-Campoy, A., Ohtaka, H., Nezami, A., Muzammil, S., and Freire, E., 2004. Isothermal titration calorimetry. *Curr Protoc Cell Biol*, Ch. 17:Unit 17. 8.

Walerych, D., Olszewski, M. B., Gutkowska, M., Helwak, A., Zylicz, M., and Zylicz, A., 2009. Hsp70 molecular chaperones are required to support p53 tumor suppressor activity under stress conditions. *Oncogene*, 28(48):4284–94.

Wandinger, S. K., Richter, K., and Buchner, J., 2008. The Hsp90 chaperone machinery. *J Biol Chem*, 283(27):18473–7.

Zubrienė, A., Gutkowska, M., Matulienė, J., Chaleckis, R., Michailovienė, V., Voroncova, A., Venclovas, Č., Zylicz, A., Zylicz, M., and Matulis, D. 2010. Thermodynamics of radicicol binding to human Hsp90 alpha and beta isoforms. *Biophys Chem*. *In press*.

Zubriene, A., Matuliene, J., Baranauskiene, L., Jachno, J., Torresan, J., Michailoviene, V., Cimmperman, P., and Matulis, D., 2009. Measurement of nanomolar dissociation constants by titration calorimetry and thermal shift assay - radicicol binding to Hsp90 and ethoxzolamide binding to CAII. *Int J Mol Sci*, 10(6):2662–2680.

3

Enthalpy, Entropy, and Volume Changes of Electron Transfer Reactions in Photosynthetic Proteins

Harvey J.M. Hou
University of Massachusetts Dartmouth
USA

1. Introduction

Photosynthesis involves a series of electron transfer steps to convert the sunlight energy into the chemical energy in green plants, algae, and cynobacteria. The three-dimensional structures of photosynthetic reaction centers at the high resolution uncover the binding sites and precise orientation of cofactors and their interaction with proteins and provided an excellent model to investigate the mechanism of electron transfer reaction. To understand fully an electron transfer reaction, it is necessary to understand not just its kinetics, but also thermodynamics. However, in contrast to the kinetics of electron transfer mechanisms, thermodynamic information is far less accessible. Thermodynamics reveals the energy levels of reactants and products, as well as the driving forces in the reaction. The driving force of the chemical reaction is the Gibbs free energy, which is composed of enthalpic and entropic components.

Pulsed photoacoustics provides direct measurements of enthalpy and volume changes of electron transfer. Using pulsed photoacoustics, the volume change and enthalpy of electron transfer reaction were measured in the photosynthetic reaction center complex of *Rb. sphaeroides*. A large entropy was calculated based on these measurements. Further photoacoustic measurements indicated that the entropy change of electron transfer in photosystem I from *Synechocystis* sp. PCC 6803 is similar to that in bacterial center. The deconvolution fit of the photoacoustic waves distinguished thermodynamic parameters of a large negative enthalpy change and large volume change for $P_{700}^* \rightarrow A_1$ step and a positive enthalpy change and a small volume change for $A_1^- \rightarrow F_{A/B}$ step. To explore the specific role of protein matrix, the *menA* and *menB* genes were inactivated by molecular genetics and showed the altered thermodynamics of electron transfer. Inactivating the *menG* gene causes 2-phytyl-1,4-naphthoquinone (Q) to be presented as a quinone acceptor. The fit by convolution of *menG* photoacoustic waves resolved a large volume contraction for the P700* $\rightarrow$ Q step and a positive volume change for the $Q^- \rightarrow F_{A/B}$ step. The photoacoustic data of the bacterial reaction center, *menA/B* mutant, *menG* mutant, and wild type phtosystem I show significant positive entropy. In contrast, electron transfer in photosystem II is accompanied by a small negative entropy change. *In vivo* photoacoustic measurements confirmed the difference in entropy between photosysmte I and photosystem II. We conclude that apparent entropy may play a vital role in photosynthetic electron transfers.

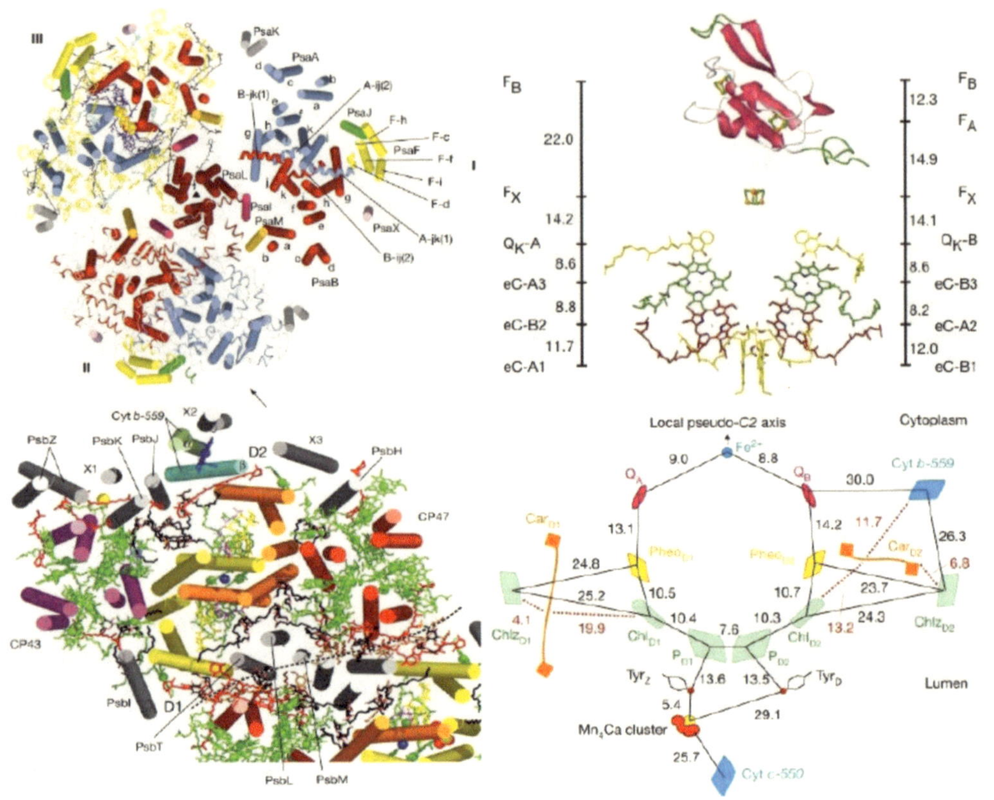

Fig. 1. Structures and arrangement of cofactors in PS I (upper) (Jordan et al., 2001) and PS II (lower) (reproduced with permission from Nature publishing Group (Loll et al., 2005)).

In this chapter we will present an overview of the protein structure and the electron transfer pathways of photosynthetic reaction centers and highlight the recent advances of thermodynamic aspects of these photosynthetic electron transfer reactions on the microsecond and nanosecond time scales using photoacoustics. The principle and experimental procedures of the pulsed photoacoustic methodology will be presented. The main focus of the chapter will be placed on the determinations, interpretations, and future endeavors of the thermodynamic parameters including the enthalpy, entropy, and volume changes of electron transfer reactions in the photosynthetic reaction centers of purple bacteria, cyanobacteria, and green plants *in vitro* and *in vivo*. The enthalpy, entropy, and volume changes of electron transfer in proteins provide novel insights into the role of the membrane environment and cofactors *in vitro* and *in vivo* in terms of a broadly thermodynamic view. These unique features offer the potential for a deeper understanding of the mechanisms of electron transfer in chemical and biological systems. They also present a framework to modify and improve the existing Marcus electron transfer theory as well as the formulation of a comprehensive electron transfer theory.

2. Protein structures, kinetics and thermodynamics in photosynthesis

Photosynthesis is the only process for large-scale solar energy storage in nature, in which the light photon energy is harvested by specific photosynthetic ligtht harvesting protein complexes in thylakoid membranes and transfered to the reaction center. The key step of energy conversion occurs in the photosynthetic reaction centers. The three-dimensional structures of both Type I (Jordan et al., 2001; Amunts et al., 2007) and Type II (Ferreira et al., 2004; Loll et al., 2005) photosynthetic reaction centers have been determined at the atomic to molecular level. Figure 1 shows the structures and arrangement of cofactors of cyanobacterial photosystems I (PS I) at 2.5 Å resolution (Jordan et al., 2001) and of cyanobacterial PS II at 3.0 Å resolution (Loll et al., 2005). These structures reveal the binding sites and precise orientation of cofactors and their interaction with proteins and provide a solid basis to interpret results of photoacoustic studies at an atomic level.

On the basis of the high-resolution structural information, the kinetics and thermodynamics of electron transfer reactions in biological systems are required for understanding precise mechanisms. During the past decade the kinetics of electron transfer steps in reaction centers of anoxygenic and oxygenic photosynthesis has been thoroughly investigated over the timescale of femtosecond to second (Brettel, 1997; Dekker & Van Grondelle, 2000; Brettel and Leibl, 2001; Gobets & van Grondelle, 2001) as shown in Figure 2. These works revealed the electron pathway and identified almost the entire electron transfer intermediates in both photosystems. However, the thermodynamics of electron transfer steps in photosynthesis, such as volume change, enthalpy and entropy, is far less well understood. There are at least two reasons to measure these thermodynamic parameters accurately. First, knowledge of the thermodynamic parameters of electron transfer reactions allows one to gauge the efficiency of energy conversion, Efficiency refers to the amount of the solar energy stored in the photosynthetic organisms. The second motivation for investigating thermodynamics is to separate the free energy into its enthalpy and entropy components, which provides more details and deeper understanding of the driving force of electron transfer mechanisms.

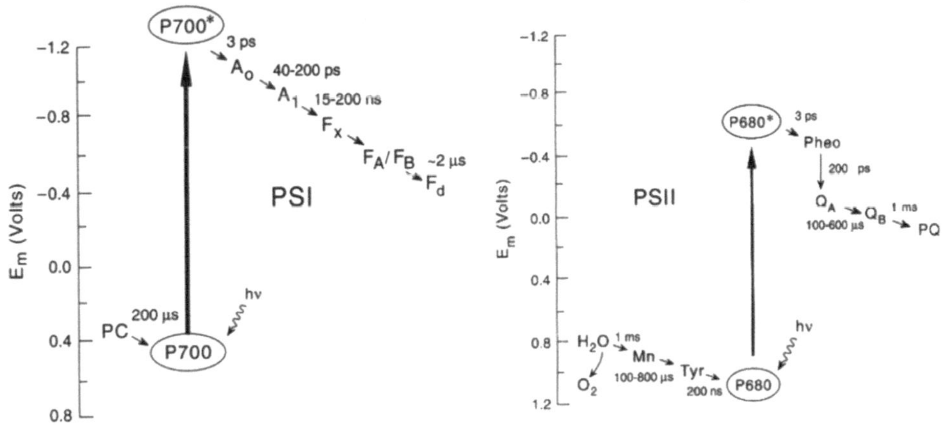

Fig. 4. Pathways and rates of electron transfer reactions in PS I (left panel) and PS II (right panel).

3. Principle of pulsed photoacoustics

Pulsed time-resolved photoacoustics provides an unique way to probe the molecular mechanism of electron transfer and proton transfer events in chemical reactions (Braslavsky & Heibel, 1992; Chen & Dibold, 1996; Borsarelli & Braslavsky, 1999; Feitelson & Mauzerall,

2002; Andres et al., 2006; Crovetto et al., 2006; Davies et al., 2008) and photosynthetic systems (Malkin, 2000; Herbert et al., 2001; Delosme, 2003; Mauzerall, 2006). For a given photoreaction, the accessible parameters include the molecular volume change secondary to conformational change or electrostriction, and enthalpy changes. With the measurement of enthalpy change by pulsed photoacoustics, the reaction entropy can be calculated when free energy is known (Feitelson & Mauzerall, 1996; Edens et al., 2000; Hou & Mauzerall, 2006).

3.1 Molecular volume change by electrostriction and structural conformational changes

Figure 3 shows the experimental setup of the pulsed photoacoustic apparatus, which enables one to determine the volume change and enthalpy of photoreactions on the nanosecond to microsecond time scales. A Nd:YAG laser and OPO are used to produce light of 680 nm and 700 nm with 1-mm light path. Consider the different response time of film, the PA detector containing a 128 mm piezoelectric film and 1-cm thickness of cell are used for the microsecond time scale photoacoustic experiments. A 28 μm film and 1-mm path cell are used for the nanosecond measurements. A 5-cm dielectric mirror was used according to the design of Arnaut *et al.* (Arnaut et al., 1992).

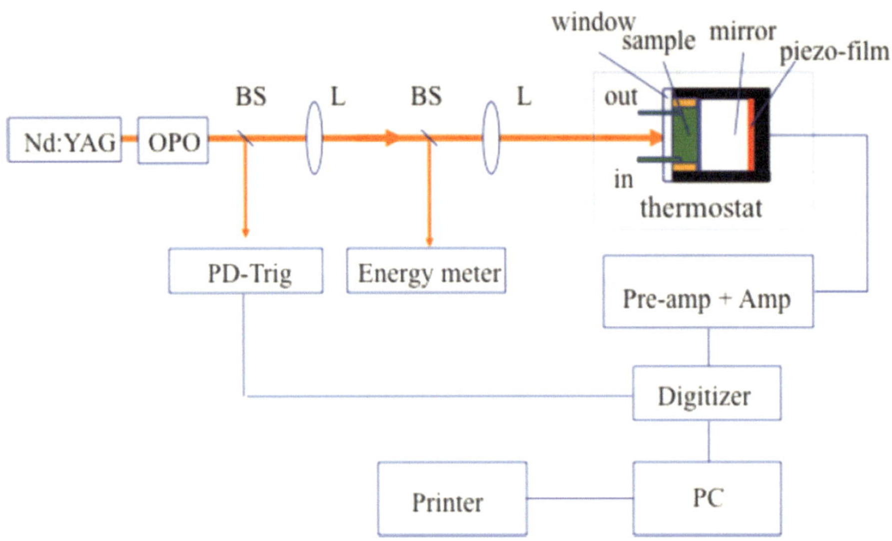

Fig. 3. Block diagram of pulsed photoacoustic system on the nanosecond-microsecond time scale. BS = beam splitter, L = lens, Amp = amplifier, OPO = optical parametric oscillator.

The advantage of using the mirror in the photoacoustic system is two-fold: (1) to increase light absorption and (2) to generate the time delay to eliminate the electric artifact in photoacoustic measurements. The volume change or heat of a photochemical reaction generates a pressure in a photoacoustic cell. The pressure is sensed by a piezo film or a microphone and recorded as a form of sound wave, designed photoaoustic wave or photoacoustic signal. To calibrate the photoacoustic signal, a reference must be used. A photoacoustic reference is a sample that degrades all the absorbed energy to heat within the resolving time interval of the apparatus. Two kinds of photoacoustic references are used to ensure the accuracy of the experimental data. The first is an external black ink reference, which is the supernatant fraction of high-quality commercial black carbon ink suspension after centrifugation. The other is an internal reference—the light-saturated photosynthetic material. The photoacoustic signal generated by a reference is described by Equation 1:

$$PA_{ref} = \frac{F \cdot \alpha'}{\kappa} E_a \cdot I(t) \tag{1}$$

where E_a is photon energy absorbed by the sample, F is piezo film sensitivity, α' is an abbreviation for thermal expansivity/heat capacity ' density, k is compressibility, and $I(t)$ is the impulse response of the system.

The photoacoustic signal is produced by the active photosynthetic reaction centers, which form a charge-separated radical pair upon light excitation, which undergo light-induced structural conformational change. The signal typically consists of two components (Equation 2).

$$PA_{RC} = \frac{F}{\kappa}\left[\alpha' \cdot Q_{RC} + \Delta V_{RC}\right] \cdot I(t) \tag{2}$$

where Q_{RC} is the heat output, which includes the enthalpy change of the reaction and other rapidly released heat, and DV_{RC} is the molecular volume change of the reaction.

When the temperature of the sample in aqueous solution is above the maximum density temperature (~4 °C), the heat output (Q_{RC}) of a photochemical reaction induces the volume expansion of solution. When the temperature is below ~4 °C, the heat induces the volume contraction. At ~4 °C, the heat component (Q_{RC}) is zero as the expansibility of water is zero. The photoacoustic signal at ~4 °C is simply the molecular volume change (ΔV_{RC}). To determine the precise values of the volume change resulting from charge separation in the photosynthetic materials, two different approaches are used: (1) volume yield measurements, and (2) saturation measurements.

In the first approach, the flash yield ($\Delta V/E$) is the volume change per unit photon energy absorbed by the system. The flash yield of the photoacoustic (DV/E) signal is dependent on the excitation photon flux, following Equation (3):

$$\frac{\Delta V_{RC}}{E} = N \Lambda V_Y \frac{(1 - e^{\Phi \sigma E})}{N \sigma E} \tag{3}$$

Assuming ΔV does not change over a small temperature range, ΔV_{RC} is obtained in the limiting low pulse energy region (the linear region of Equation 3) by normalizing to the reference photoacoustic signal, converting PA_{ref} to volume *via* a' at 25 °C and correcting for the change in the compressibility of water between T_m and 25 °C:

$$\Delta V_{RC} = \frac{PA_{RC}^{Tm}}{PA_{ref}^{25}} \times \frac{\kappa^{Tm}}{\kappa^{25}} \times \Delta V_{ref} \tag{4}$$

where $DV_{ref} = \alpha'E$, is the thermal volume change of the reference at its temperature. Since the system is linear, one can calculate ΔV_{ref}, the thermal volume change at 25 °C, for each absorbed photon at the excitation wavelength. At low energy, one obtains the volume per center multiplied by the quantum yield, $\Phi \cdot \Delta V_y$. However, if the energy is too low, the photoacoustic signal-to-noise ratio is poor. In the case of photosynthetic reaction centers the value of volume change can be obtained by fitting a curve to Equation 3 and extrapolating to zero excitation energy.

In the second approach, every photosynthetic reaction center is excited to obtain the maximum photoacoustic signal. Fitting of the photoacoustic data to Equation 5 allows one to obtain the light saturated volume change, ΔV_s:

$$\Delta V = N \Delta V_S (1 - e^{-\Phi \sigma E}) \tag{5}$$

where ΔV_S is the volume change produced per photosynthetic reaction center; s is the optical cross section per photosynthetic reaction center; Φ is the quantum yield of the photochemical reaction, and E is the excitation photon flux.
In this method, one must calculate the number of photosynthetic reaction centers in the illuminated volume of the cell (~0.34 mL), N, to obtain the real volume change ΔV_s. The effective cross-section ($\Phi \cdot \sigma$) can also be obtained from the fit of the curves. This approach measures the absolute number of photosynthetic centers, calculated without assumptions of the quantum yield assuming all centers are successfully "hit" with enough energy. Although these two methods use the same set of data, the calculation of ΔV is differently weighted and completely different.

3.2 Quantum yield of photoreaction

The photoacoustic measurement includes the enthalpy or volume changes times the quantum yield, which can be determined using the light saturation curve of photoacoustics. The light saturation function at 4 °C (where there is no thermal signal) contains the photochemical quantum yield. For a simple system with one cross section, the photoacoustic signal is described by the cumulative one-hit Poisson distribution (Equation 6).

$$PA = N \cdot PA_0 (1 - e^{-\Phi \cdot \sigma \cdot E}) \tag{6}$$

where N is the numbers of centers in the sample, PA_0 is the photoacoustic signal produced per successful hit of the reaction centers; s and Fs are the optical cross section and effective optical cross section; F is the quantum yield, E is the photon energy absorbed by the sample, and A is the absorbancy of solution.
The effective cross section can be obtained by the curve fit of the photoacoustic saturation curve. As the optical cross section, s with units of area per reaction center, can be calculated from the chlorophyll content and ratio of chlorophyll to the primary electron donor, the quantum yield of chemistry can be determined.

3.3 Enthalpy and entropy changes of photoreaction

The enthalpy change of electron transfer reactions in photosynthetic reaction centers can be calculated by the Equation 7:

$$\Delta H = (E_{h\nu} - E_{trap}) - \frac{\left(\frac{d(PA \cdot \kappa)_{RC}}{d\alpha}\right)}{\left(\frac{d(PA \cdot \kappa)_{ref}}{d\alpha}\right)} \times E_{h\nu} \tag{7}$$

As discussed above, pulsed photoacoustics directly measures the enthalpy and volume changes of the reactions on the nanosecond and microsecond time scales. With the given free energy, the entropy change is calculated from $\Delta G = \Delta H - T\Delta S$.

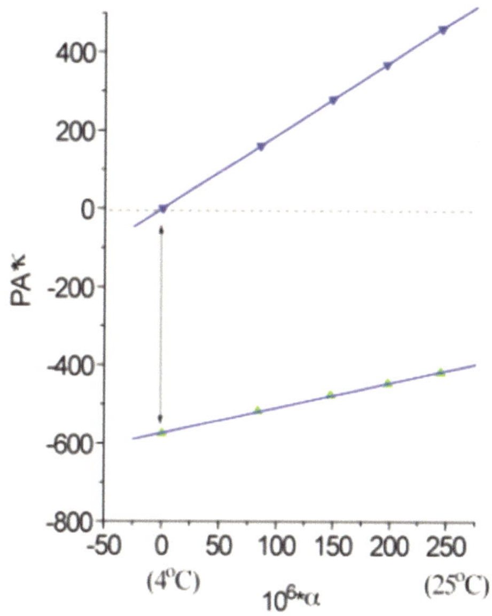

Fig. 4. Typical plot of the product of the photoacoustic signal and water compressibility (k) versus water expansivity (a) for ink reference (blue symbols) and PS I complex (green symbols).

To determine the reaction enthalpy and volume changes, the photoacoustic waves of ink references and photoactive samples such as PS I complexes are recorded at five temperatures, 4, 10, 15, 20, and 25 °C. A typical data analysis is plotted in the product of the photoacoustic intensity (peak to peak value) and water compressibility versus the water expansivity (Figure 4). The deconvolution analysis on photoacoustic waves allows one to time-resolve the thermodynamic parameters of each individual step of photoreaction (Rudzki et al., 1985; Feitelson & Mauzerall, 1996, 2002). The photoacoustic pressure wave is caused by the rate of the heat release. The two or more steps of a photochemical reaction will change the shape (wider or narrower) of photoacoustic wave than the photoacoustic reference. The linear fit of the experimental points gives intercept and slope of ink reference and PS I samples, respectively. The molecular volume change of PS I on the 1 μs time scale can be easily obtained by inserting the intercepts values of the linear fit of ink reference and PS I sample to the Equation 4. The enthalpy change of the photochemical reaction in PS I on the 1 μs time scale may be found *via* the Equation 7 using the slopes of the fit.

4. Enthalpy, entropy, and volume changes of photosynthetic reactions

The pulsed photoacoustic technique gives a direct measurement of the enthalpy change of photosynthetic electron transfer reactions (Feitelson & Mauzerall, 1996; Edens et al., 2000). A microphone may detect the photoacoustic waves *via* the thermal expansion in the gas phase. The 50-fold larger thermal expansion of a gas over liquid makes the microphone more sensitive. A gas-coupled microphone in a closed chamber is used as a detector on the ms time scale. This method allows measurements of the photosynthetic thermal efficiency, or energy storage, and of the optical cross-section of the light harvesting systems (Canaani et al., 1984; Mauzerall, 1990; Braslavsky & Heibel, 1992). However, the time resolution is low, typically about 30 μs, and inappropriate for obtain thermodynamic parameters on the microsecond and nanosecond time scales. The use of piezoelectric films acoustically coupled to a liquid sample and a pulsed laser light source increased the time resolution of the photoacoustic technique to the microsecond scale (Nitsch et al., 1988; Mauzerall et al., 1995). Photoacoustic thermodynamic studies have been carried out on isolated photosynthetic reaction centers from bacteria *Rb. sphaeroides* (Arata & Parson, 1981), on PS I from cyanobacteria (Delosme et al., 1994), and on PS II from spinach and *Chlamydomonas reinhardtii* (Delosme et al., 1994).

Measuring the energetics of photobiological reactions *in vivo* is of more interest because the local environment of photosystems may exert a prevailing consequence on their kinetics and thermodynamics. Pulsed photoacoustics *via* a microphone detector revealed the oxygen evolution and oxygen uptake *in vivo* on the millisecond time scale (Mauzerall, 1990). Using a piezo film detector, the *in vivo* enthalpy and volume changes of photosystem I and photosystem II of *Synechocysits sp. PCC 6803* were obtained on the microsecond time scale (Hou et al., 2001; Hou et al., 2001). The contribution of the two photosystems was distinguished by excitation at two different wavelengths, 625 nm for predominant excitation of PS I and 680 nm for PS II, respectively. The difficulty in the photoacoustic measurements of intact cells is the heterogeneous or "cell" artifact at the temperature of maximum density of near 4 °C. To correct the "cell" artifact, five measurements are needed instead of usual three (see reference (Hou et al., 2001) for details). The enthalpy and volume changes of intact cells of *Synechocystis sp. PCC 6803* were in good agreement with those of isolated complexes within the experimental errors.

Theoretical calculations of electron transfer have often assumed the reaction entropy to be zero. For example, the standard formulation of Marcus theory assumes that the vibrations coupled to electron transfer have the same frequency in the reactant and product states (Marcus & Sutin, 1985). Marcus theory introduces reorganization energy to interpret the reaction rate of electron transfer reactions. The reorganization energy can have two contributions: a vibration term and a solvent term. Treatments of the temperature dependence of the rate of electron transfer often assume that the free energy is independent of temperature (Gunner & Dutton, 1989). However, these assumptions are called into question by recent work. For instance, the entropy change often neglected in an artificial photosynthetic system was actually determined to be large (Rizzi et al., 2008). A fit of Marcus reorganization energy cannot interpret the observed volume change of electron transfer reaction.

Using pulsed photoacoustics, the volume change and enthalpy of electron transfer reaction were measured in aqueous solution (Feitelson & Mauzerall, 1996) and in the photosynthetic reaction center complex of *Rb. sphaeroides* (Edens et al., 2000). A large entropy was calculated based on these measurements. Further photoacoustic measurements revealed that the entropy change of electron transfer in PS I trimer from *Synechocystis* sp. PCC 6803 on the microsecond time scale was the same as that in bacterial centers (Hou et al., 2001). The volume contraction of reaction centers of PS I, which results directly from the light-induced charge separation forming $P_{700}^{+}F_A/F_B^{-}$ from the excited state P_{700}^{*}, was determined to be −26 $Å^3$. The enthalpy of the above electron transfer reaction was found to be −0.39 eV. Taking the free energy of the above reaction as the difference of their redox potentials *in situ* allows one to calculate an apparent entropy change ($T\Delta S$) of +0.35 eV. In contrast, electron transfer in PS II core complexes from *Synechocystis* sp. PCC 6803 is accompanied by a small negative entropy change (Hou et al., 2001). At pH 6.0, the volume contraction of PS II was determined to be −9 Å3, and the enthalpy change −0.9 eV for the formation of the state $P_{680}^+Q_A^-$ from P_{680}^*. The ΔV of PS II, smaller than that of PS I and bacterial centers, is assigned to electrostriction and analyzed using the Drude-Nernst equation. To explain the small ΔV for the formation of $P_{680}^+Q_A^-$ or $Y_Z^•Q_A^-$ we proposed that fast proton transfer into a polar region is involved in this reaction. These observations were confirmed using intact living cells of the same organism (Boichenko et al., 2001). These thermodynamic parameters are summarized in Figure 5. The enthalpies for the formation of states $P_{700}^+F_{AB}^-$ from P_{700}^* and $Y_Z^•P_{680}Q_A^-$ from P_{680}^* *in vivo* were estimated to be about −0.3 eV and −1 eV, respectively. Comparison of these values with the free energies of the reactions indicates a significant contribution of the apparent entropy changes $T\Delta S$, +0.4 and −0.24 eV for the formation of ion-radical pairs in PS I and PS II, respectively.

In order to obtain detailed information on intermediates in the PS I reactions, we measured the volume change and enthalpy change on the nanosecond time scale. The time constant of charge transfer from A_1^- to $F_{A/B}$ is reported to be 20 to 200 ns (Brettel & Leibl, 2001). However the modeling analysis of the electron transfer reactions in PS I by electron

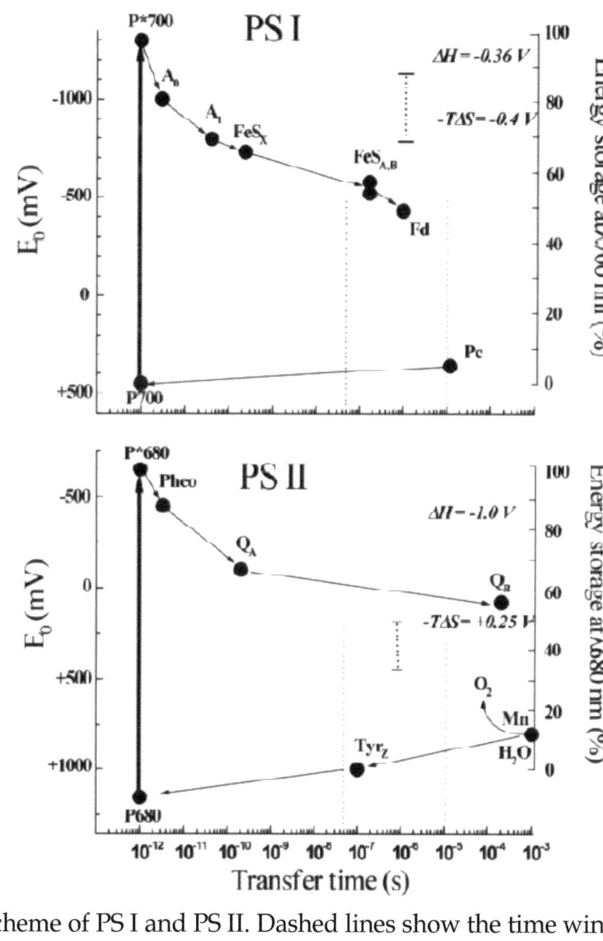

Fig. 5. Energetic scheme of PS I and PS II. Dashed lines show the time window of the PA measurements (reproduced with permission from the American Chemical Society (Boichenko et al., 2001)).

tunneling theory indicate complex equilibria between the various species (Santabarbara et al., 2005). The enthalpy and volume changes associated with this reaction and with the charge separation of the $P_{700}^* \rightarrow A_1$ reaction seem unknown. The photoacoustic waves of PS I may involve the contributions of the initial and subsequent electron transfer reactions (Figure 4). The deconvolution of the time derivative of the volume or heat release function with the apparatus response function provided by the reference signal enables us to resolve the fast and slow photoacoustic components (Feitelson & Mauzerall, 1996; Zhang & Mauzerall, 1996; Strassburger et al., 1997). It is possible to time-resolve the thermodynamic parameters of individual steps in PS I by the deconvolution analysis.

As shown in Figure 6, deconvolution analysis of photoacoustic signals on microsecond time scales resolves enthalpy and volume changes of two steps: (1) a prompt component (<10 ns)

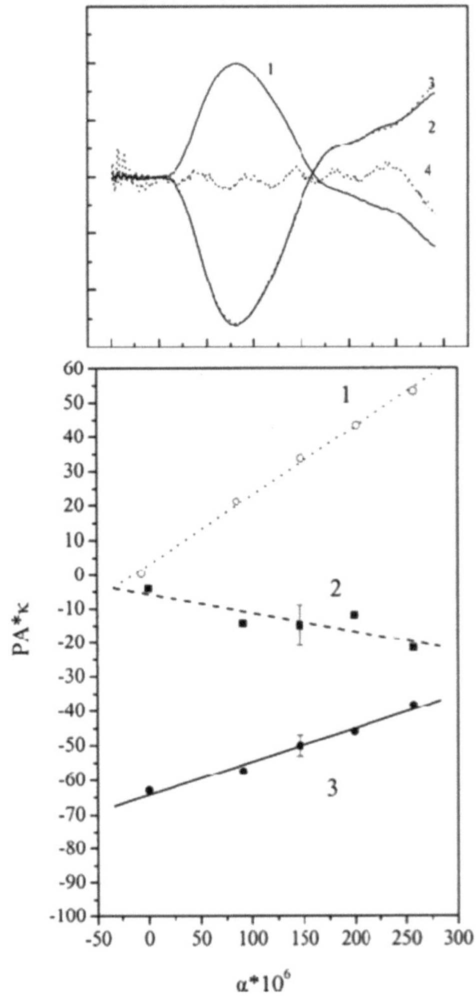

Fig. 6. Deconvolution analysis of a typical photoacoustic pressure wave to time-resolve the enthalpy and volume changes of two individual electron transfer steps in PS I of *Synechocystis* sp. PCC 6803 (reproduced with permission from the American Chemical Society (Hou & Mauzerall, 2006)).

with large negative enthalpy change (−0.8 ± 0.1 eV) and large volume change (−23 ± 2 A^3), which are assigned to the $P_{700}^* \rightarrow A_1^-F_X$ step, and (2) a component with ~200 ns lifetime, which has a positive enthalpy change (+0.4 ± 0.2 eV) and a small volume change (−3 ± 2 A^3) which are attributed to the $A_1^-F_X \rightarrow F_{A/B}^-$ step. This parameters were confirmed by a similar analytsis of photoacoustic waves on the nanosecond time scale. For the fast reaction using the redox potentials of A_1F_X (−0.67 V) and P_{700} (+0.45 V) and the energy of P_{700}^* (1.77 eV), the free energy change for the $P_{700}^* \rightarrow A_1^-F_X$ step is −0.63 eV. Thus, the entropy change ($T\Delta S$, T = 25 °C) is −0.2 ± 0.3 eV. For the slow reaction, $A_1^-F_X \rightarrow F_{A/B}^-$, taking the free energy of −0.14 eV (Santabarbara et al., 2005), the entropy change ($T\Delta S$) is positive, (+0.54 ± 0.3 eV). The redox thermodynamics of many FeS proteins (ferrodoxins) have been determined. Interestingly, most of the $Fe_4S_4(Cys)_4$ proteins have positive enthalpies of reduction of +0.3 to +0.4 eV, while the others have very small negative enthalpies. The authors explain the enthalpies in terms of electrostatic interactions with the protein dipoles. All the reduction entropies are negative as anticipated from the increase in charge and seem to vary in magnitude opposite to the enthalpies. Thus the positive enthalpy of the $A_1^-F_X$ to $F_{A/B}$ reaction may be assigned in part to the FeS cluster reduction. The positive entropy may have contributions from the freeing of oriented polar groups on quinone anion oxidation. The previous step has negative entropy as expected.

To understand more details of electron transfer thermodynamics, it is of interesting to determine the thermodynamic parameters of electron transfer in the photosynthetic electron transfer mutants on both the nanosecond and microsecond time scales. It has been generally believed that phylloquinone is difficult to dislodge from the A_1 binding site in PS I, in contrast to the ubiquinone (Q_A) in bacterial centers from *Rb. Sphaeroides* that can be easily replaced by a wide variety of different quinones (Gunner & Dutton, 1989; Xu & Gunner, 2001). A biological method to remove phylloquinone was recently devised. Targeted inactivation of the *menA* and *menB* genes that code for phytyl transferase and napthoate synthase in the phylloquinone biosynthetic pathway (Johnson et al., 2000) precludes its availability for incorporation in the A_1 site. Yet, in spite of the demonstrated absence of phylloquinone, the *menA* and *menB* null mutants grow photosynthetically. EPR measurements show that plastoquinone-9 (A_P) is recruited into the A_1 site (Johnson et al., 2000)and functions as an efficient 1-electron electron carrier (Zybailov et al., 2000). Time-resolved optical studies indicate the forward electron transfer from A_1^- to F_X is slowed 1000-fold, to 15 and 300 μs, compared to 20 and 200 ns in wild-type PS I (Semenov et al., 2001). Given the altered kinetics of electron transfer, it will be of interest to investigate the effect of these mutations on the thermodynamics of electron transfer in PS I. These thermodynamic parameters reveal that the driving force in the photosynthetic reactions may be both enthalpic and entropic.

Figure 7 summarizes the volume changes, free energies, enthalpy and entropy changes on *menA/B* PS I in comparison with those on the wild-type PS I. Opened arrows are the early step forming $P_{700}^+A_1^-$ from P_{700}^* for the wild-type PS I or $P_{700}^+A_P^-$ from P_{700}^* for the mutants, and solid arrows are the number of the following reaction: $P_{700}^+A_P^- \rightarrow P_{700}^+F_{A/B}^-$. As shown in Panel A, the volume contraction of early step of photoreaction in the mutants (−17 Å^3) is smaller than that in the wild type (−21 Å^3). Similarly, the enthalpy change (−0.7 eV) of the early step in the mutants is smaller than that (−0.8 eV) in wild type PS I (Figure 7A and 7B). Assuming a redox potential of −0.6 V for plastoquinone-9 in the A_1 site (Semenov et al., 2001), the free energy (−0.7 eV) of this early reaction in the mutants is larger than the value

(−0.6 eV) in the wild type as indicated in Figure 7C. Taking the difference of free energy and enthalpy change in the mutants, the apparent entropy change of the early step in mutants is zero. In contrast, the apparent entropy change in the wild-type PS I is calculated to be +0.2 eV. Since the apparent entropy change for the overall reaction of the generation of P_{700}^{+} $F_{A/B}^{-}$ from P_{700}^{*} is +0.35 eV (Hou et al., 2001), it implies that the latter reaction in the mutants, *i.e.*, the P_{700}^{+} A_P^{-} $F_{A/B}$ → P_{700}^{+} A_P $F_{A/B}^{-}$ reaction, is almost completely entropy-driven ($T\Delta S$ = +0.4 eV and ΔG= −0.1 eV) (Figure 7D). Therefore, based on our experimental results we propose that the foreign quinone (A_P) in PS I does affect the thermodynamics of charge separation in the early steps in PS I with a smaller volume and enthalpy changes, a large free energy and zero entropy change.

The observed thermodynamic data of the *menA* and *menB* null mutants show little difference. This is expected because the recruited quinone (A_P) is the same in both mutants. Volume change of mutant PS I following charge separation on both time scales is −16 ± 2 Å³. The quantum yield of charge separation in PS I of the mutants is slightly lower (85 ± 10%) than that of the wild-type PS I (96 ± 10%). The observed reaction is assigned to the formation of $P_{700}^{+}A_P^{-}$ $F_{A/B}$ from $P_{700}^{*}A_P$ $F_{A/B}$. The enthalpy change (ΔH) of about −0.69 ± 0.1 eV in mutant PS I was obtained for this reaction. In contrast, a large enthalpy change of ~ −1.0 eV for the formation of $P_{700}^{+}A_1^{-}$ from P_{700}^{*} in the wild-type PS I was observed. These results strongly suggest that not only the kinetics but also the thermodynamics of electron transfer reactions in PS I is significantly affected by the recruitment of the foreign plastoquinone-9 into the A_1 site (Hou et al., 2009).

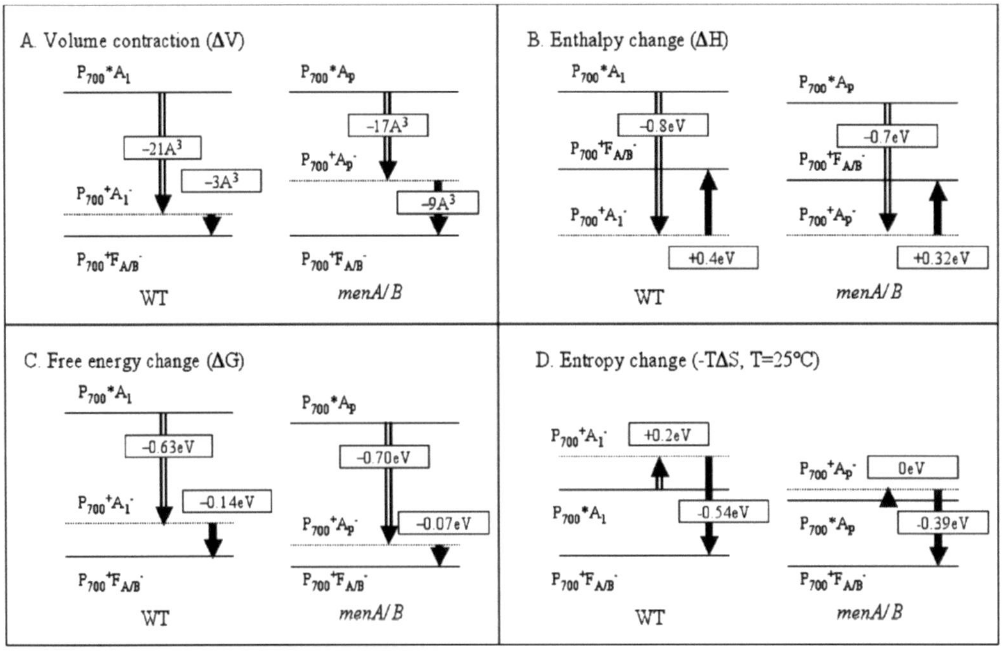

Fig. 7. Thermodynamic parameters of charge separation in *menA/B* PS I and wild-type PS I from *Synechocystis* 6803 (reproduced with permission from the American Chemical Society (Hou et al., 2009)).

5. Limitations and possible solutions

The thermodynamic parameters of electron transfer in photosynthetic reaction center complexes and in whole cells of *Synechocystis* have been determined using pulsed photoacoustics (Boichenko et al., 2001; Hou et al., 2001; Hou et al., 2001). However, the parameters in *Synechocystis* PS II were inconsistent with those obtained using *Chlamydomonas* and spinach chloroplast (De Vitry et al., 1991; Delosme et al., 1994)). The *in vitro* measurements, in particular, showed a significant difference from the *in vivo* values. These discrepancies in purified complexes from different organisms such as *Synechocystis, Chlamydomonas,* and spinach chloroplasts must be clarified.

We previously determined the volume changes and enthalpy changes of electron transfer in *Synechocystis* PS I complexes on the microsecond time scale (Hou et al., 2001) and then time-resolved two individual steps in the complexes on the nanosecond time scale (Hou & Mauzerall, 2006). A significant entropic component was found. In future work, we intend to determine the thermodynamic parameters in PS I *in vitro* and *in vivo* using photosynthetic electron transfer mutants to uncover the roles of the protein matrix in electron transfer steps.

In contrast, for the volume changes in PS I and bacterial reaction centers with an reasonable number within the error margins, the volume change of PS II charge separation is in the range from −2 to −16 Å^3 (Hou et al., 2001). For example, Boichenko *et al.* (Boichenko et al., 2001) reported a −2 Å^3 in the whole cells of cyanobacterium *Synechocystis*, and Delosme *et al.* (Delosme, 2003) observed −9 Å^3 from *Chlamydomonas* to −16 Å^3 from spinach. This obvious inconsistency needs to be investigated and clarified. The pH, species of organisms, redox potentials of electron donor/acceptors, and cell artifacts in the photoacoustic measurements may cause the discrepancy.

The discrepancy might arise from the fact that the measurements were performed at different pH values. The volume change of our previous photoacoustic measurements showed a strong pH dependence of volume change in PS II core complexes. A volume change of −3 Å^3 was observed at pH 9.0 and −9 Å^3 at pH 6.0 (Hou et al., 2001). Future work may determine the volume change of charge separation in PS II complexes using our previous photoacoustic procedures over a range of pH. The volume change will be measured by the yield and saturation method, respectively. As Delosme et al. pointed out, the discrepancy of the volume change may be due to the use of different organisms: cyanobacteria *versus* green plants (Delosme, 2003). We will test this possibility by using the photosystem II preparations from cyanobacteria, green algae, and higher plants. These PS II preparations can be purified from cyanobacterium *Synechocystis* 6803, from green algae *Chlamydomonas reinhardii*, and from spinach chloroplasts following our established method (Hou et al., 2000). The volume change of charge separation in PS II should be determined and evaluated against the data from different groups.

The inconsistency in volume changes and enthalpies of electron transfer could be due to the difference in intactness: *in vitro versus in vivo*. Most of the *in vitro* preparations gave around −10 Å^3 compared to −2 Å^3 for the *in vivo* system. A number of factors *in vivo* are different from the *in vitro* preparations. The overlap of absorption of PS I and PS II in the whole cell preparations is significant. We estimated 30% PS II contributions and 70% PS I at 625 nm. In contrast, the contribution of PS I is 20% PS I and of PS II is 80% at 690 nm, respectively (Boichenko et al., 2001). Use of the PS I-depleted mutant may be required to address this problem. Based on preliminary unpublished data Delosme *et al.* argued that the thermal efficiency of purified PS II core from *Thermosynechococcus elongatus* strongly depends on the

experimental conditions (Delosme, 2003), such as the electron donor or acceptor used, and could be significantly higher than our previous data. This discrepancy needs to be clarified.
A number of photosynthetic electron transfer mutants in addition to chemically modified the reaction centers altered the kinetics of photosynthetic reactions. For example, the photosynthetic electron transfer mutants, including *menA*, *menB*, and *menG* null PS I mutants and A_0 to F_X, F_A, and F_B mutants, are available. Pulsed photoacoustics may probe these mutants to explore the effects of mutation on thermodynamic parameters of electron transfer in PS I. The focus should be placed on the effects of the mutations on thermodynamics of photoreactions in photosynthesis. However, the entropic contribution may suffer from the limited published data reporting free energy values. To overcome these obstacles, the electrochemical measurements to determine the redox potentials *in situ* and computational study to calculate the redox potentials of the cofactors will be beneficial.
It has been a long-term goal to reveal the thermodynamics of PS II oxygen evolution. However, the time window of current photoacoustics is on the microsecond time scale, which is too fast to determine the millisecond step of PS II oxygen evolution cycle. Recently, using photothermal beam deflection techniques the enthalpies of PS II water oxidation were reported (Krivanek et al., 2008). Using a novel photopressure cell that enables one to obtain the volume and enthalpy changes on the microsecond to second time scales (Liu et al., 2008), the thermodynamic parameter of reactions involved in PS II oxygen evolution might be determined.

6. Conclusion

To fully understand a chemical reaction, it is necessary to understand not just its kinetics, but also its thermodynamic parameters. However, in contrast to the kinetics of electron transfer mechanisms, thermodynamic information is far less accessible. The driving force of the chemical reaction is the Gibbs free energy, which is composed of enthalpic and entropic components. The theory of electron transfer in chemical and biological system was developed by Marcus (Marcus & Sutin, 1985) and played a key role in advancing the understanding electron transfer mechanisms, including photosynthetic systems (Turro et al., 1996; Renger et al., 1998; Mayer et al., 2006; LeBard et al., 2008). However, one weakness of the theory is the omission of entropic contribution in the electron transfer steps.
The recent experimental results pointed out the crucial role of entropy change in chemical and biological systems (Feitelson & Mauzerall, 1996; Edens et al., 2000; Xu & Gunner, 2000; Crovetto et al., 2006; Hou & Mauzerall, 2006). The follow-up and extensive thermodynamic measurements especially by photoacoustics is worth while. Pulsed photoacoustics provides novel insights on the entropic contribution to electron transfer in proteins and is able to probe the role of the membrane environment and cofactors *in vitro* and *in vivo* in terms of a broadly thermodynamic view. These unique features of photoacoutstics studies provide the potential for a deeper understanding of the mechanisms of electron transfer and proton-coupled electron transfer in chemical and biological systems. They also provide a framework to modify and improve the existing Marcus electron transfer theory as well as the formulation of a comprehensive electron transfer theory. The results of photoacoustics experiments also provide direct information about the function of membrane proteins central to photosynthesis. In particular, the photosynthetic membrane proteins are excellent models for elucidating electron transfer and proton transfer mechanism in other membrane proteins.

7. References

Amunts A, Drory O, Nelson N (2007) The structure of a plant photosystem I supercomplex at 3.4 A resolution. Nature 447: 58-63

Andres GO, Martinez-Junza V, Crovetto L, Braslavsky SE (2006) Photoinduced Electron Transfer from Tetrasulfonated Porphyrin to Benzoquinone Revisited. The Structural Volume-Normalized Entropy Change Correlates with Marcus Reorganization Energy. J Phys Chem A 110: 10185-10190

Arata H, Parson WW (1981) Enthalpy and volume changes accompanying electron transfer from P-870 to quinones in *Rhodopseudomonas sphaeroides* reaction centers. Biochim Biophys Acta, 636: 70-81

Arnaut LG, Caldwell RA, Elbert JE, Melton LA (1992) Recent advances in photoacoustic calorimetry: theoretical basis and improvements in experimental design. Rev Sci Instrum 63: 5381-5389

Boichenko VA, Hou J-M, Mauzerall D (2001) Thermodynamics of Electron Transfer in Oxygenic Photosynthetic Reaction Centers: Volume Change, Enthalpy, and Entropy of Electron-Transfer Reactions in the Intact Cells of the Cyanobacterium *Synechocystis* PCC 6803. Biochemistry 40: 7126-7132

Borsarelli CD, Braslavsky SE (1999) Enthalpy, Volume, and Entropy Changes Associated with the Electron Transfer Reaction between the 3MLCT State of $Ru(Bpy)_3^{2+}$ and Methyl Viologen Cation in Aqueous Solutions. J Phys Chem A 103: 1719-1727

Braslavsky SE, Heibel GE (1992) Time-resolved photothermal and photoacoustic methods applied to photoinduced processes in solution. Chem Rev 92: 1381-1410

Brettel K (1997) Electron transfer and arrangement of the redox cofactors in photosystem I. Biochim Biophys Acta, 1318: 322-373

Brettel K, Leibl W (2001) Electron transfer in photosystem I. Biochim Biophys Acta 1507: 100-114

Canaani O, Barber J, Malkin S (1984) Evidence that phosphorylation and dephosphorylation regulate the distribution of excitation energy between the two photosystems of photosynthesis in vivo: Photoacoustic and fluorimetric study of an intact leaf. Proc Natl Acad Sci U S A 81: 1614-1618

Chen, H.X.; Dibold, G. (1996) Production of the photoacoustic effect and transient gratings by molecular volume changes. J Chem Phys, *104*, 6730-6741.

Crovetto L, Martinez-Junza V, Braslavsky SE (2006) Entropy changes drive the electron transfer reaction of triplet flavin mononucleotide from aromatic amino acids in cation-organized aqueous media. A laser-induced optoacoustic study. Photochem Photobiol 82: 281-290

Davies KW, Maivald D, Grabowski JJ (2008) A photoacoustic calorimetric characterization of the reaction enthalpy and volume for the preparation of a reactive intermediate from $CpMn(CO)_3$. J Photochem Photobiol A 197: 335-341

De Vitry C, Diner BA, Popot JL (1991) Photosystem II particles from *Chlamydomonas reinhardtii*. Purification, molecular weight, small subunit composition, and protein phosphorylation. J Biol Chem 266: 16614-16621

Dekker JP, Van Grondelle R (2000) Primary charge separation in Photosystem II. Photosynth Res 63: 195-208

Delosme R (2003) On some aspects of photosynthesis revealed by photoacoustic studies: a critical evaluation. Photosynth Res 76: 289-301

Delosme R, Beal D, Joliot P (1994) Photoacoustic detection of flash-induced charge separation in photosynthetic systems. Spectral dependence of the quantum yield. Biochim Biophys Acta, 1185: 56-64

Edens GJ, Gunner MR, Xu Q, Mauzerall D (2000) The Enthalpy and Entropy of Reaction for Formation of $P^+Q_A^-$ from Excited Reaction Centers of *Rhodobacter sphaeroides*. J Am Chem Soc 122: 1479-1485

Feitelson J, Mauzerall D (1996) Photoacoustic Evaluation of Volume and Entropy Changes in Energy and Electron Transfer. Triplet State Porphyrin with Oxygen and Naphthoquinone-2-Sulfonate. J Phys Chem 100: 7698-7703

Feitelson J, Mauzerall D (2002) Enthalpy and Electrostriction in the Electron-Transfer Reaction between Triplet Zinc Uroporphyrin and Ferricyanide. J Phys Chem B 106: 9674-9678

Ferreira KN, Iverson TM, Maghlaoui K, Barber J, Iwata S (2004) Architecture of the photosynthetic oxygen-evolving center. Science 303: 1831-1838

Gobets B, van Grondelle R (2001) Energy transfer and trapping in photosystem I. Biochim Biophys Acta 1507: 80-99

Gunner MR, Dutton PL (1989) Temperature and $\Delta G°$ dependence of the electron transfer from $BPh^{\cdot -}$ to Q_A in reaction center protein from *Rhodobacter sphaeroides* with different quinones as QA. J Am Chem Soc 111: 3400-3412

Herbert SK, Han T, Vogelmann TC (2001) New applications of photoacoustics to the study of photosynthesis. Photosynth Res 66: 13-31

Hou HJM, Mauzerall D (2006) The A^-F_X to $F_{A/B}$ Step in *Synechocystis* 6803 Photosystem I Is Entropy Driven. J Am Chem Soc 128: 1580-1586

Hou HJM, Shen G, Boichenko VA, Golbeck JH, Mauzerall D (2009) Thermodynamics of Charge Separation of Photosystem I in the *menA* and *menB* Null Mutants of *Synechocystis* sp. PCC 6803 Determined by Pulsed Photoacoustics. Biochemistry 48: 1829-1837

Hou JM, Boichenko VA, Diner BA, Mauzerall D (2001) Thermodynamics of electron transfer in oxygenic photosynthetic reaction centers: Volume change, enthalpy, and entropy of electron transfer reactions in manganese-depleted photosystem II core complexes. Biochemistry 40: 7117-7125

Hou JM, Dejonghe D, Shan JX, Li LB, Kuang TY (2000) Orientation of pigments in the isolated photosystem II sub-core reaction center CP47/D1/D2/Cyt b-559 complexes: A linear dichroism study. J Integr Plant Biol 42: 1211-1214

Hou JM, Boichenko VA, Wang YC, Chitnis PR, Mauzerall D (2001) Thermodynamics of electron transfer in oxygenic photosynthetic reaction centers: a pulsed photoacoustic study of electron transfer in photosystem I reveals a similarity to bacterial reaction centers in both volume change and entropy. Biochemistry 40: 7109-7116

Johnson TW, Shen G, Zybailov B, Kolling D, Reategui R, Beauparlant S, Vassiliev IR, Bryant DA, Jones AD, Golbeck JH, Chitnis PR (2000) Recruitment of a foreign quinone into the A_1 site of photosystem I. I. Genetic and physiological characterization of phylloquinone biosynthetic pathway mutants in *Synechocystis* sp. pcc 6803. J Biol Chem 275: 8523-8530

Jordan P, Fromme P, Witt HT, Klukas O, Saenger W, Krauss N (2001) Three-dimensional structure of cyanobacterial photosystem I at 2.5 A resolution. Nature 411: 909-917

Krivanek R, Dau H, Haumann M (2008) Enthalpy changes during photosynthetic water oxidation tracked by time-resolved calorimetry using a photothermal beam deflection technique. Biophys J 94: 1890-1903

LeBard DN, Kapko V, Matyushov DV (2008) Energetics and kinetics of primary charge separation in bacterial photosynthesis. J. Phys. Chem. B 112: 10322-10342

Liu Y, Edens GJ, Grzymski J, Mauzerall D (2008) Volume and Enthalpy Changes of Proton Transfers in the Bacteriorhodopsin Photocycle Studied by Millisecond Time-Resolved Photopressure Measurements. Biochemistry 47: 7752-7761

Loll B, Kern J, Saenger W, Zouni A, Biesiadka J (2005) Towards complete cofactor arrangement in the 3.0 A resolution structure of photosystem II. Nature 438: 1040-1044

Malkin S (2000) The photoacoustic effect in leaves and its applications. Probing Photosynthesis: 484-524

Marcus RA, Sutin N (1985) Electron transfers in chemistry and biology. Biochim Biophys Acta, 811: 265-322

Mauzerall D (2006) Thermodynamics in photosystem I. In J. Golbeck (ed): Photosystem I: The Light-Driven Plastocyanin: Ferredoxin Oxidoreductase Vol 24: Springer, Dordrecht, pp 571-581

Mauzerall D, Feitelson J, Prince R (1995) Wide Band, Time-Resolved Photoacoustic Study of Electron Transfer Reactions: Difference between Measured Enthalpies and Redox Free Energies. J Phys Chem 99: 1090-1093

Mauzerall DC (1990) Determination of Oxygen Emission and Uptake in Leaves by Pulsed, Time Resolved Photoacoustics. Plant Physiol 94: 278-283

Mayer JM, Rhile IJ, Larsen FB, Mader EA, Markle TF, Dipasquale AG (2006) Models for Proton-coupled Electron Transfer in Photosystem II. Photosynth Res 87: 3-20

Nitsch C, Braslavsky SE, Schatz GH (1988) Laser-induced optoacoustic calorimetry of primary processes in isolated photosystem I and photosystem II particles. Biochim Biophys Acta, 934: 201-212

Renger G, Christen G, Karge M, Eckert HJ, Irrgang KD (1998) Application of the Marcus theory for analysis of the temperature dependence of the reactions leading to photosynthetic water oxidation: results and implications. J Biol Inorg Chem 3: 360-366

Rizzi AC, van Gastel M, Liddell PA, Palacios RE, Moore GF, Kodis G, Moore AL, Moore TA, Gust D, Braslavsky SE (2008) Entropic Changes Control the Charge Separation Process in Triads Mimicking Photosynthetic Charge Separation. J Phys Chem A 112: 4215-4223

Rudzki JE, Goodman JL, Peters KS (1985) Simultaneous determination of photoreaction dynamics and energetics using pulsed, time-resolved photoacoustic calorimetry. J Am Chem Soc 107: 7849-7854

Santabarbara S, Heathcote P, Evans MC (2005) Modelling of the electron transfer reactions in Photosystem I by electron tunnelling theory: the phylloquinones bound to the PsaA and the PsaB reaction center subunits of PS I are almost isoenergetic to the iron-sulfur cluster F_X. Biochim Biophys Acta 1708: 283-310

Semenov AY, Vassiliev IR, van Der Est A, Mamedov MD, Zybailov B, Shen G, Stehlik D, Diner BA, Chitnis PR, Golbeck JH (2001) Recruitment of a foreign quinone into the A_1 site of photosystem I. Altered kinetics of electron transfer in phylloquinone biosynthetic pathway mutants studied by time-resolved optical, EPR, and electrometric techniques. J Biol Chem 275: 23429-23438

Strassburger JM, Gartner W, Braslavsky SE (1997) Volume and enthalpy changes after photoexcitation of bovine rhodopsin: laser-induced optoacoustic studies. Biophys J 72: 2294-2303

Turro C, Zaleski JM, Karabatsos YM, Nocera DG (1996) Bimolecular electron transfer in the Marcus inverted region. J Am Chem Soc 118: 6060-6067

Xu Q, Gunner MR (2000) Temperature Dependence of the Free Energy, Enthalpy, and Entropy of $P^+Q_A^-$ Charge Recombination in *Rhodobacter sphaeroides* R-26 Reaction Centers. J Phys Chem B 104: 8035-8043

Xu Q, Gunner MR (2001) Trapping Conformational Intermediate States in the Reaction Center Protein from Photosynthetic Bacteria. Biochemistry 40: 3232-3241

Zhang D, Mauzerall D (1996) Volume and enthalpy changes in the early steps of bacteriorhodopsin photocycle studied by time-resolved photoacoustics. Biophys J 71: 381-388

Zybailov B, van der Est A, Zech SG, Teutloff C, Johnson TW, Shen G, Bittl R, Stehlik D, Chitnis PR, Golbeck JH (2000) Recruitment of a foreign quinone into the A_1 site of photosystem I. II. Structural and functional characterization of phylloquinone biosynthetic pathway mutants by electron paramagnetic resonance and electron-nuclear double resonance spectroscopy. J Biol Chem 275: 8531-8539

4

Thermodynamics of Supramolecular Structure Formation in Water

Tadashi Mizutani
Doshisha University, Kyoto
Japan

1. Introduction

Hydrophobic interactions play important roles in various aspects in physics, chemistry, and biology. In particular, binding of biological molecules to proteins, polynucleotides, and cell membranes is driven by hydrophobic interactions as well as other forces such as London's dispersive interactions, van der Waals interactions, electrostatic interactions, charge-transfer interactions and hydrogen bonding (Ben-Naim & Marcus, 1984; Schneider & Yatsimirsky, 2000; Southall et al., 2002; Mayer et al., 2003; Houk et al., 2003). Hydrophobic interactions are also essential in formation of self-assemblies with a nanometer size, and construction of supramolecular structures on the interface between two different phases (Chandler, 2005). In this chapter we discuss thermodynamics of binding of a non-polar molecule to a receptor molecule/polymer, with particular attention to the hydrophobic/hydrophilic environment of the binding region as a critical factor to direct the mechanism of hydrophobic interactions. In particular, we show that enthalpy-entropy compensation would work only weakly or even non-compensation was observed when the hydrophobic interactions are extracted from various binding forces.

Origin of the hydrophobic interactions was attributed to the formation of hydrogen bonds of water molecules on the surface of non-polar molecules. To understand the thermodynamics of binding of a molecule to a biopolymer, however, other dynamics such as conformational changes in polymer and direct intermolecular interactions between the molecule and the polymer should be taken into consideration in addition to the dynamics of hydrogen bonding formation/destruction between water molecules.

The association constants of ligands to proteins for 160 protein-ligand pairs were distributed with an average of $10^{9.3}$ M^{-1} (Kunz et al., 1999). Thermodynamic studies of binding of simple molecules to synthetic receptors such as cyclodextrins and cyclophanes have been carried out to shed light on the molecular picture of hydrophobic interactions. The association constants of various ligands to cyclodextrins were smaller than those of ligands to proteins and the average was $10^{2.4}$ M^{-1} (Rekharsky & Inoue, 1998). Much tighter binding to proteins implies that proteins have been highly developed as a receptor to have both high affinity to the ligand and high selectivity. Understanding of the mechanisms of binding is needed and much work has been performed to clarify the molecular mechanism of host-guest interactions.

According to the classical theory of hydrophobic interactions (Frank & Evans, 1945), the driving force of the attractive interaction should be entropic gain due to increased freedoms of water molecules. It is puzzling, however, that hydrophobic effects are enthalpy-driven in some cases (Ross & Subramanian, 1981; Gelb et al., 1981; Smithrud et al., 1991; Spencer et al., 1995; Arena et al., 2000) and entropy-driven in others (Hooley et al., 2007; Iwamoto et al., 2007; Whitesides & Krishnamurthy, 2005, Sgarlata et. al., 2010). Jencks proposed that hydrophobic interactions can be classified into two categories, (1) classical hydrophobic interactions with an entropy term as a driving force and (2) non-classical hydrophobic interactions with an enthalpy term as a driving force (Jencks, 1969). According to his proposal, the classical hydrophobic interactions operate for a non-polar host-guest complex, while the non-classical hydrophobic interactions operate for a relatively polar host-guest complex.

We have made systematic studies on thermodynamics of binding of an alkyl group to synthetic receptors consisting of a hydrophobic porphyrin framework and poly(ethylene oxide) auxiliary groups of varying degrees of polymerization (Mizutani et al., 2003; Iwamoto et al., 2007; Matsumoto et al., 2009). We revealed that the driving force of binding of an alkyl group changed from enthalpy to entropy as the fraction of poly(ethylene oxide) moieties increased. We proposed that water accessibility to the host-guest interacting region affects the driving force of hydrophobic interactions: the binding is entropy-driven if the binding pocket is sequestered from water, while it is enthalpy-driven if the binding pocket is open to water. Thermodynamic parameters of binding of alkyl groups to proteins also showed very diverse compensation temperatures, indicating that hydrophobic interactions are characterized by the non-compensation effects of enthalpy and entropy. The findings are important to interpret the thermodynamic data of binding of biological interest, and also to rational design of drugs based on protein structures (Talhout et al., 2003; Malham et al., 2005).

2. Hydrophobic interactions

2.1 Hydrophobic interactions

Water is a solvent used in many chemical and biological reactions, and the unique feature of water is dynamic behavior of formation of hydrogen bonding between water molecules. Formation and destruction of hydrogen bonds occur very rapidly and frequently in liquid water, so that the dynamics of hydrogen bonding affects the reactions and equilibria in water to a significant way. One of such effects in water is called hydrophobic effects or hydrophobic interactions, and they are driving forces of association of non-polar molecules in water. Hydrophobic interactions operate in diverse fields such as binding of organic molecules to proteins, a folding process of proteins, double helix formation of polynucleotides, action of detergents on the interfaces, and formation of cell membranes.

In particular, hydrophobic interactions play an important role in affinity and selectivity of binding of an organic molecule to protein, and understanding of the molecular mechanism of hydrophobic interactions is quite important for the rational design of drugs, for instance. However, the hydrophobic interactions are statistical effects of the solvent, and the protein molecule has many freedoms of motion. These two factors make the molecular picture of binding mechanism of biomolecules through hydrophobic effects very complex.

2.2 Iceberg formation on non-polar molecules in water

Hydrophobic interactions originate from the solvation of non-polar molecules. Frank and Evans pointed out that, in a solution of noble gas or a non-polar molecule in water, the water molecules near the surface of non-polar molecule have greater „crystallinity" (Frank & Evans, 1945). These non-polar molecules build a microscopic iceberg around them. Formation of iceberg can explain the fact that low solubility of non-polar molecules in water originates from the unfavorable entropy. Nemethy and Scheraga calculated the statistics of water molecules which has four hydrogen bonds, three hydrogen bonds, two hydrogen bonds, and one hydrogen bond (Nemethy & Scheraga, 1962a, 1962b). The water molecule with four hydrogen bonds is most stable in terms of enthalpy (the potential energy), but is most unstable in terms of entropy since formation of multiple hydrogen bonds restricts the translational and rotational motion of the water molecule. In the opposite sense, the water molecule with one hydrogen bond is most unstable in terms of enthalpy, but is most stable in terms of entropy. The thermodynamic equilibrium will be established where the two competing terms, enthalpy and entropy, compensate to reach the minimum of the free energy.

2.3 Enthalpy and entropy changes upon transfer of alkane in water to alkane in the liquid state

Thermodynamic data of transfer of a series of alkanes from water to alkanes in the liquid state give us valuable insight into the mechanism of hydrophobic interactions. According to the iceberg structure formation theory near the surface of a non-polar molecule, thermodynamic parameters of the hydrophobic solvation or hydrophobic interactions should be proportional to the surface area in contact with water. Proportionality of the free energy and the surface area of non-polar moiety of the molecules, i.e., the solvent accessible surface area, has been confirmed for a number of thermodynamic processes such as solubilization of alkanes in water, partition of vaious molecules between water and organic solvent, protein unfolding, and protein-ligand binding (Hermann, 1972; Reynolds et al., 1974; Sharp et al., 1991a; 1991b; Richards, 1977; Böhm, 1994; Cohen & Connors, 1970; Harris et al., 1973).

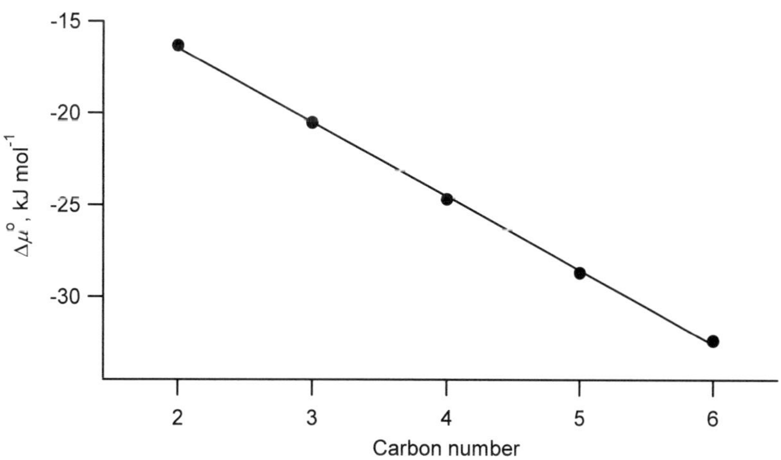

Fig. 1. Plot of the chemical potential changes in transfer of alkanes from water to pure liquid alkane, $\Delta\mu^{o}$, against the carbon number of alkanes (ethane, propane, butane, pentane, and hexane). $\Delta\mu^{o} = \mu_{HC}^{o} - \mu_{W}^{o}$. The standard states are defined as follows: chemical potential of alkane in water: $\mu_{W} = \mu_{W}^{o} + RT \ln X_{W} + RT \ln f_{W}$, where X_{W} is mole fraction of alkanes, and f_{W} is the activity coefficient. μ_{HC} denotes chemical potential of alkane: $\mu_{HC} = \mu_{HC}^{o} + RT \ln X_{HC} + RT \ln f_{HC}$, where X_{HC} is mole fraction of alkanes, and f_{W} is the activity coefficient. Least squares line fit gave $\Delta\mu^{o} = (-4.03 \pm 0.03)n - (8.4 \pm 0.3)$, where n is the carbon number. Data were taken from Tanford (1973)

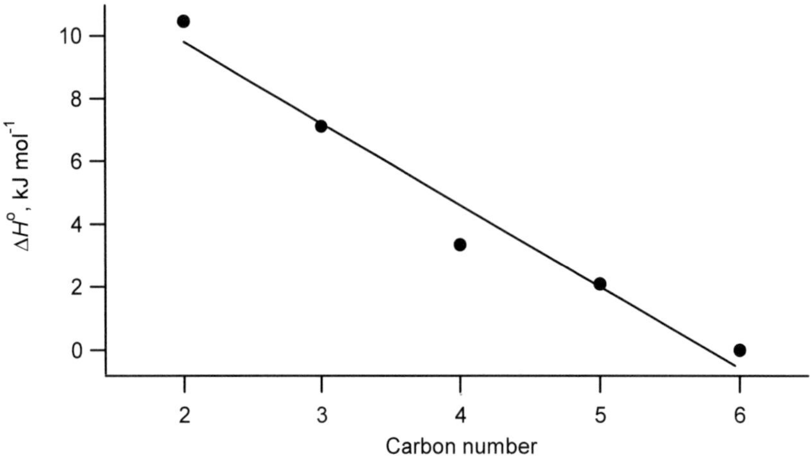

Fig. 2. Plot of the enthalpy change in transfer of alkanes (ethane, propane, butane, pentane, and hexane) from an aqueous solution to pure liquid alkane. Least squares line fit gave $\Delta H^{\circ} = (-2.6 \pm 0.3)n - (15 \pm 1)$, where n is the carbon number. Data were taken from Tanford (1973)

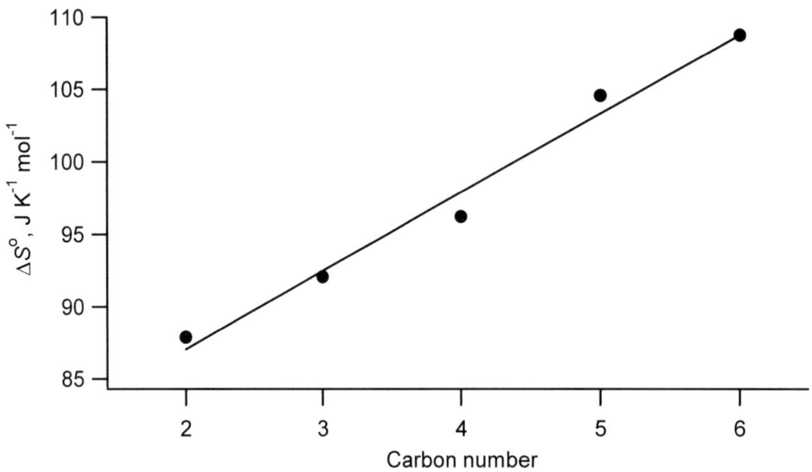

Fig. 3. Plot of the entropy change in transfer of alkanes (ethane, propane, butane, pentane, and hexane) from an aqueous solution to pure liquid alkane. Least squares line fit gave $\Delta S^{\circ} = (5.4 \pm 0.4)n - (76 \pm 2)$, where n is the carbon number. Data were taken from Tanford (1973)

First, we discuss the process of transfer of linear alkane ($CH_3(CH_2)_{n-2}CH_3$) in water to alkane in its liquid state. Figures 1-3 illustrate plots of free energy, enthalpy, and entropy changes in transfer of alkanes from water to pure alkane in the liquid phase against the carbon numbers of alkanes, n. All these thermodynamic parameters showed linear correlation with the carbon number. From the slopes of the lines we estimated the free energy change, the enthalpy change, and the entropy change per one CH_2 group as –4.0 kJ mol^{-1}, –2.6 kJ mol^{-1}, and +5.4 J K^{-1} mol^{-1}, respectively. The negative sign of the free energy change indicates that the transfer of alkane from water to alkane is a favorable process, and that hydrophobic effects, or hydrophobic forces, are attractive interactions. It is interesting to note that both the enthalpy term and the entropy term contributed favorably to the free energy changes. The favorable contribution from the entropy term is explained by cleavage of hydrogen bonds of water upon transfer of alkanes, that is, dissolution of the iceberg structure. The large negative enthalpy change implies that attractive interactions between alkane molecules in the liquid state such as attractive van der Waals forces (London's dispersive forces) make a significant contribution to the overall thermodynamics of the process.

2.4 Enthalpy and entropy changes upon transfer of alkane in water to gaseous alkane

In several studies, hydrophobic interactions have been discussed based on the thermodynamic parameters in the transfer processes of alkane or noble gas in the gaseous state to those in water (Abraham, 1982; Blokzijl & Engberts, 1993). On the basis of the solubitity of linear alkanes with carbon numbers 2-8, contribution of one methylene group to ΔG°, ΔH° and $T\Delta S^{\circ}$ of the transfer process are estimated to be -0.75 kJ mol^{-1}, +2.80 kJ mol^{-1}, and +3.56 kJ mol^{-1}, respectively. The enthalpy term and the entropy term drive the equilibrium to the opposite directions and the free energy changes are dominated by the positive entropy term. When the thermodynamic parameters of the transfer of alkane in water to liquid alkane are compared with those to gaseous alkane, the enthalpic change is negative in the former case. It implies that the attractive van der Waals interactions between alkane molecules in the liquid state play an important role.

2.5 Molecular picture of hydrophobic interactions

These thermodynamic parameters lead to the picture of classical hydrophobic interactions: the association of non-polar molecules in water is driven by the extinction of hydrogen bonded water molecules on the non-polar surface, which should be entropically favorable process where the freedom of water molecules increases. The positive slope of the plot of the entropy changes vs. carbon number for the transfer of alkane from water to organic solvent (Figure 3) reflects the increased freedom of water molecules after non-polar molecules are removed from water. If only the destruction of the iceberg structures is considered, the process should be enthalpically unfavorable since some of the hydrogen bonds in the iceberg are broken. Figure 2 shows that the slope of the plot of the enthalpy changes vs. carbon number is, however, negative. Therefore, the attractive interactions between alkanes in the liquid state overcome the enthalpic cost to cleave the hydrogen bonds in the iceberg.

The interactions involved in the protein-substrate complex formation are composed of several forces such as hydrogen bonding, hydrophobic interactions, polar interactions, ionic interactions, and so on. Thus, diverse interactions should contribute to the thermodynamic parameters of binding of ligands to proteins.

We focus on the binding of alkyl groups to a synthetic receptors or proteins to clarify the molecular picture of hydrophobic interactions in detail.

3. Enthalpy – entropy compensation

Binding of guest to host is determined by the free energy changes of the equilibrium, and the free energy can be separated into two terms: the enthalpy term and the entropy term.

Physical meaning of the two terms is quite different, the enthalpy term originates from the potential energy of the system while the entropy term from the motional freedom of the system. Two terms, however, often counteract each other to ensure stability of the thermodynamic equilibrium. In a number of processes, the potential energy term tends to restrict the motional freedom of the system to lead to smaller entropy. The entropy term tends to disturb the system to a disordered state, and this would result in larger enthalpy. The balance of these two terms finds a thermodynamic equilibrium point, where the free energy becomes minimum.

Before discussing the mechanism of hydrophobic interactions, we summarize here enthalpy-entropy compensation found in a number of reactions and equilibria. Particularly, binding of various molecules to biopolymers or other receptor molecules is characterized by an enthalpy and entropy compensation (Lumry & Rajender, 1970; Gilli et al., 1994). For a series of host-guest paris with one of these structures being systematically altered, the enthalpy changes and the entropy changes in binding are often linearly correlated. The plot of $\delta\Delta S^o$ vs. $\delta\Delta H^\circ$ gives a line with a positive slope $1/T_c$.

$$\delta\Delta S^\circ = (1/T_c)\delta\Delta H^\circ + \beta \tag{1}$$

T_c is called a compensation temperature. The positive slope indicates that the enthalpy-entropy compensation operates, while the negative slope indicates that the enthalpy term and the entropy term cooperate to result in large free energy changes. When the equilibrium is established at T_c, the free energy change can be given by equation (2).

$$\delta\Delta G^\circ = \delta\Delta H^\circ - T_c\delta\Delta S^\circ \tag{2}$$

Equation (2) is reduced to $\delta\Delta G^\circ = \beta$ by replacing $\delta\Delta S^\circ$ with equation (1). Therefore the entropy changes and the enthalpy changes due to the structural changes balance to lead to constant free energy changes at T_c. When T_c is close to room temperature, the strong compensation works, and the binding affinity remains the same even if the structures of ligands or drugs are altered. To find out a specific ligand to a target protein should be very difficult if this is the case.

Alternatively, in many literatures, $T\Delta S^\circ$ is plotted against ΔH° to show enthalpy-entropy compensation. In this plot, the slope of the line is T/T_c, where T is the temperature at which the binding data are collected.

In a number of instances of binding in water, T_c is close to room temperature. Lumry and Rajender suggested that various processes in water such as solvation of ions and nonelectrolytes, hydrolysis, and protein reactions follow the enthalpy-entropy compensation relationship with the compensation temperature in a relatively narrow range, from 250 to 315 K (Lumry & Rajender, 1970). Enthalpy-entropy compensation observed for organic reactions was reviewed by Leffler & Grunwald (Leffler & Grunwald, 1963). Dunitz suggested that enthalpy-entropy compensation is a general property of weak intermolecular interactions, and the enthalpy term and the entropy term should nearly balance out for a hydrogen bond at 300 K (Dunitz, 1995).

The two characteristic parameters, compensation temperature, T_c, and the entropy changes at zero enthalpy changes (β) as shown in equation (1) have been determined experimentally for binding of various ligands to receptors. Inoue and coworkers proposed that the compensation temperature reflects the flexibility of the host-guest system: the high compensation temperature means rigid host-guest system where binding causes minimal structural changes (Rekharsky & Inoue, 1998). The ΔS° term reflects the solvation energy where large ΔS° means large changes in solvation during host-guest complexation.

It should be noted, however, that enthalpy-entropy compensation was not observed for the transfer of alkane from water to pure alkane as shown in Figures 1-3. Increasing the carbon number caused the negative free energy, negative enthalpy and positive entropy changes. Similar non-compensation was observed for the binding of an alkyl group to synthetic receptors (see below).

4. Perturbation approach to access hydrophobic interactions: Synthetic receptors

We can determine the free energy changes, the enthalpy changes and the entropy changes from the experiments either based on the equilibrium constants obtained with various physical or spectroscopic techniques or directly determine the heat flow using calorimetry. However these thermodynamic data contain several interactions and several motional freedom changes associated with different elementary processes. For instance, binding of substrates to proteins is driven by several interactions such as hydrophobic interactions, hydrogen bonding, electrostatic interactions (salt bridges between ions and dipolar interactions), van der Waals interactions, and coordinative interactions. These attractive or repulsive forces also cause freeze of some of the motional freedoms in host and guest. Translational and rotational motion of guest and internal bond rotation of host and guest would be frozen to some extent to result in the negative entropy changes. It is generally very difficult to extract hydrophobic interactions in a pure form from these data.

Large entropic cost to freeze the translational and rotational freedom of guest upon binding should be paied by a number of weak interactions (Knox, 1971; Tabushi et al., 1978; Mizutani et al., 1994). In this case, which pairwise interaction paid the entropic cost is not clear. To extract the interaction in a pure form, we could use a relatively strong interaction, which is strong enough to freeze the translational and rotatinal freedom of guest, and then additional weak interactions can be evaluated without interference from the enthalpy-entropy compensation, since the translational and rotational entropic costs have been already paid by the strong interactions. We designed synthetic receptors to evaluate hydrophobic interactions based on the strategy. The structures of the receptors are shown in Figure 4. The receptors have the zinc ion and the zinc acts as a Lewis acid to bind Lewis bases with an exothermic reaction (Kirksey et al., 1969; Mikros et al., 1988; Mizutani et al., 1999). The Lewis acid-Lewis base interactions between the zinc and pyridine are strong enough to pay the entropic cost of freezing of translational and rotational freedoms, and a guest with Lewis base is bound to the receptor with a fixed orientation. This could allow us to evaluate the hydrophobic interactions as an additional force, and this would eliminate interference from the compensation effects of translational and rotational entropy of guest. If we determine the thermodynamic parameters of binding of various Lewis bases with varying hydrophobicity, we can determine the contribution of hydrophobic interactions to the binding enthalpy and entropy.

The receptors have hydrophobic alkyl groups and the hydrophobic porphyrin core. These moieties are very non-polar so that the molecule is not soluble in water by itself. We attached poly(oxyethylene) groups at the terminals of the alkyl groups to solubilize the receptors in water. Then the receptors have a unique structure where the non-polar environment is preserved in water. In such non-polar environment, Lewis acidic zinc can bind Lewis base and the Lewis acid-Lewis base interaction occurs without interference from water solvation. These receptors bind 4-alkylpyridines in the binding pocket as schematically shown in Figure 5. The binding constants, K/M^{-1}, as defined in equation (3) were determined by the electronic spectral changes in the Soret band upon addition of the guest in 0.1 M pottasium phosphate buffer at pH 7 at 298 K.

$$K = \frac{[\text{receptor} \bullet \text{guest}]}{[\text{receptor}][\text{guest}]} \tag{3}$$

The free energy changes were calculated according to the equation:

$$\Delta G^{\circ} = -RT \ln K. \tag{4}$$

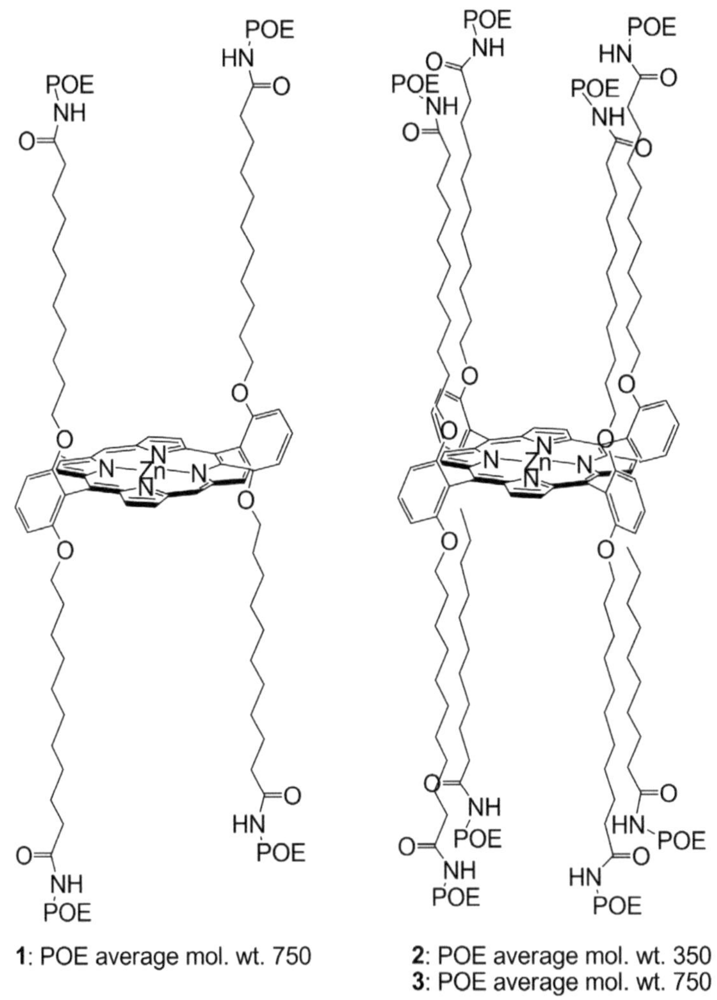

Fig. 4. Structures of receptors **1-3**

The enthalpy changes and the entropy changes were determined by either isothermal calorimetiric titrations or van't Hoff analysis of the binding free energy. For receptors **2** and **3**, both calorimetric experiments and the van't Hoff analysis based on the spectral changes were performed to give consistent results. Aggregation behavior of receptor **1**, however, prevented reliable measurements of isothermal calorimetry, so that only the thermodynamic parameters determined by the van't Hoff analysis were used. Receptor **1** and receptor **3** have poly(oxyethylene) groups with molecular weight of 750 at each terminal of the alkyl groups.

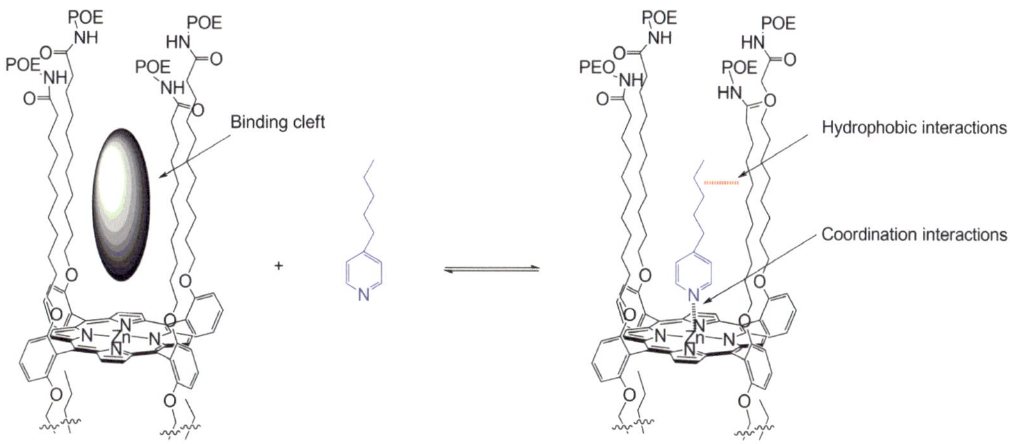

Fig. 5. Binding of alkylpyridines to receptors **1-3**

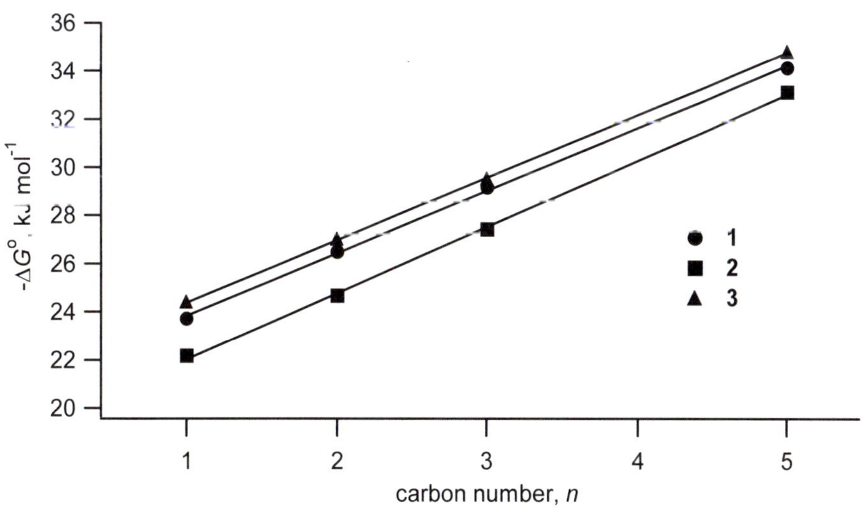

Fig. 6. Plot of the binding free energy of 4-alkylpyridines to **1**, **2**, and **3** against the number of CH_2 groups (n) in the guest. For **1**, $-\Delta G^o = (2.60 \pm 0.06)n + (21.2 \pm 0.2)$. For **2**, $-\Delta G^o = (2.75 \pm 0.06)n + (19.3 \pm 0.2)$. For **3**, $-\Delta G^o = (2.60 \pm 0.02)n + (21.78 \pm 0.6)$

Receptor **2** has poly(oxyethylene) groups with molecular weight of 350. Receptor **1** has only four alkyl chains to form the hydrophobic binding pocket while receptors **2** and **3** have eight alkyl chains to form the hydrophobic binding pocket. These differences produced the different environment of the binding pocket and influences the mechanism of hydrophobic interactions significantly.

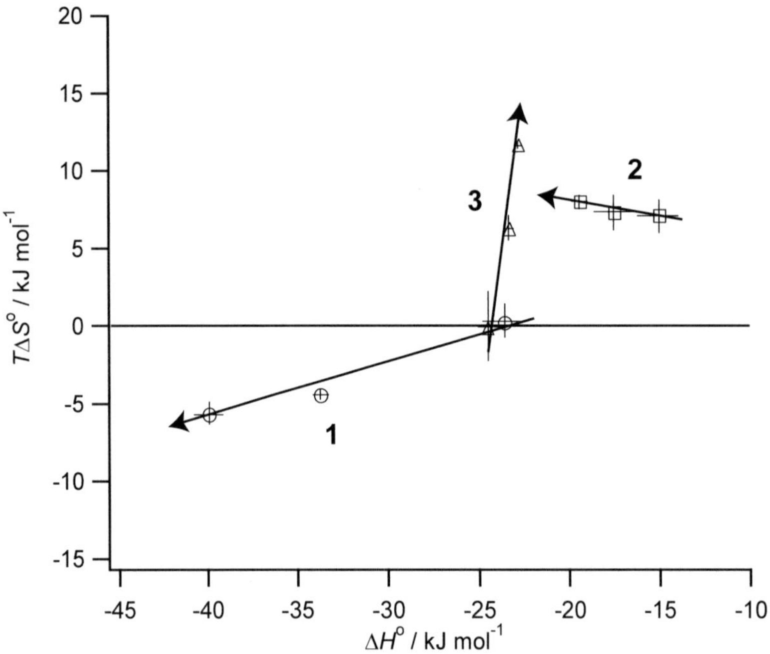

Fig. 7. Plot of $T\Delta S^o$ (T = 298 K) against ΔH^o for binding of 4-alkylpyridines to **1**, **2**, and **3**. Arrows indicates the direction of changes as the length of the alkyl group, and thus the binding free energy, $-\Delta G^o$, increases. Bars indicate standard deviations

In Figure 6 are plotted the free energy changes of binding of 4-alkylpyridines to receptors **1**–**3** against the alkyl chain length of the guest. The binding free energy decreases *ca.* 2.6–2.8 kJ/mol per one CH_2 group, to confirm that hydrophobic interactions operate effectively. All these receptors bind alkylpyridines in a similar fashion in terms of free energy. In Figure 7 are plotted the enthalpy changes and the entropy changes for these host-guest pairs. The arrows indicate the direction of increasing alkyl chain length of the guest. The plot indicates that the binding mechanisms of alkyl groups recognition, i.e., hydrophobic interactions, are quite different among three receptors **1-3**. For receptor **1**, the alkyl group binding is driven by the enthalpic term: the longer alkyl groups are favored because of the exothermicity of the equilibrium. For receptor **3**, the alkyl group binding is driven by the entropic term: the longer alkyl group binding is favored because of the randomness of the system increases. For receptor **2**, both of the two terms, enthalpy and entropy, are favorable for longer alkyl groups.

The $T\Delta S^o - \Delta H^o$ plot in Figure 7 is discussed from the standpoint of enthalpy-entropy compensation. For receptors **1** and **3**, only weak enthalpy-entropy compensation was observed. The slopes are different from unity, indicating that the compensation temperatures are much different from room temperature. For receptor **2**, the negative slope revealed that both enthalpy and entropy drive the hydrophobic interactions. For receptor **1** and **2**, the free energy changes were almost exclusively determined by the enthalpic term. For receptor **3**, the free energy changes were determined by the entropy term. The non-compensation was similar to the enthalpy-entropy changes in the transfer of alkane in water to liquid alkane, as discussed in Section **2.3**.

In independent experiments, we evaluated the water accessibility to the binding pocket. The ratios of binding constants of pyridine to imidazole are listed in Table 1 for receptors **1-3**. Receptor **3** favors pyridine while receptor **1** does not disfavor imidazole as receptor **3**. Imidazole is hydrogen bonded to water while bound to the zinc since imidazole has two nitrogens. Therefore, a hydrophobic binding pocket where water cannot penetrate into the pocket should favor pyridine and disfavor imidazole. The water accessibility to the binding pocket is high for receptor **1**, medium for receptor **2**, and low for receptor **3**. The interpretation is also reasonable considering the structures of receptors: receptor **1** has only four alkyl chains so that water penetration into the binding cleft could be frequent, while receptor **3** has eight alkyl chains to protect the binding cleft from water penetration. For receptor **1**, the binding cleft could be too small to accomodate the long alkyl group of 4-alkylpyridines. Before the guest was bound, the alkyl groups of receptor **1** may be folded to avoid contact with water, but binding of 4-alkylpyridine caused conformational changes in the alkyl groups of receptor **1** to expose the non-polar surface of the alkyl groups to water to a larger extent than the uncomplexed receptor. This could induce the iceberg structure near the complex, by which enthalpy becomes negative.

	K(4-methylpyridine)/*K*(*N*-methylimidazole)
1	20
2	96
3	124

Table 1. Ratios of binding constants of pyridines to imidazoles as a probe for the polar/nonpolar environment of the binding pocket

The hydrophobic recognition of the guest alkyl group by receptors **1–3** seems similar in terms of free energy, but the enthalpic and entropic changes revealed that the binding mechanisms are quite different. For receptor **3** having well protected binding pocket, hydrophobic interactions are driven by an entropic force. This is classical hydrophobic interactions where water gains motional freedom upon released from the non-polar surface. In contrast, for receptor **1** having less protected binding pocket, hydrophobic interactions are driven by an enthalpic force. The origin of the enthalpic force can be either van der Waals interactions or the enhanced hydrogen bonding among water, i.e., iceberg formation, where more water can be exposed to non-polar surface upon binding of alkyl group. Receptor **2** with intermediate water accessibility binds the alkyl group by both enthalpy and entropy driving forces.

5. Perturbation approach to access hydrophobic interactions: Protein-ligand binding

5.1 Entropy-driven binding of an alkyl group to protein

We then discuss the binding of the alkyl group of *p*-alkylbenzamidinium chloride to trypsin, (Talhout et al., 2003). Molecules with an amidinium group ($-C(NH_2)=NH_2^+$) are known to be an inhibitor of trypsin. The amidinium group of the ligand is bound to Asp189, Gly219, Ser190 and internal water through hydrogen bonding and the alkyl group is bound to a hydrophobic pocket made by Trp215 and Leu199. The free energy changes, the enthalpy changes and the entropy changes are plotted against the carbon number in the alkyl group of the ligands in Figures 8-10, respectively. Figure 8 shows that the increase in the carbon number leads to the increased binding affinity, $-\Delta G^{\circ}$. As shown in Figures 9 and 10, enthalpy-entropy compensation was observed: the entropy term becomes favorable with increasing carbon number, while the enthalpy term becomes unfavorable. In this study, the binding of alkyl group to trypsin is driven by the entropic term.

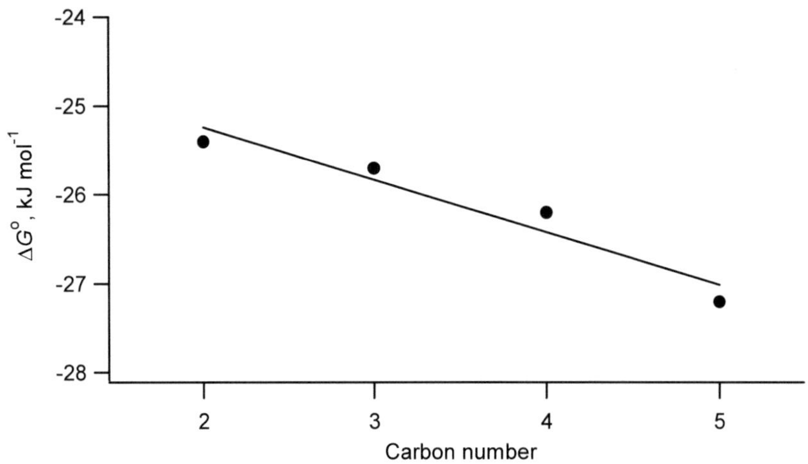

Fig. 8. Plot of the free energy change in binding of alkylbenzamidinium chlorides to trypsin at 25 °C at pH 8. Least squares line fit gave $-\Delta G^{\circ} = (-0.6 \pm 0.1)n - (24.1 \pm 0.4)$, where n is the carbon number of the alkyl group. Data were taken from Talhout et al. 2003

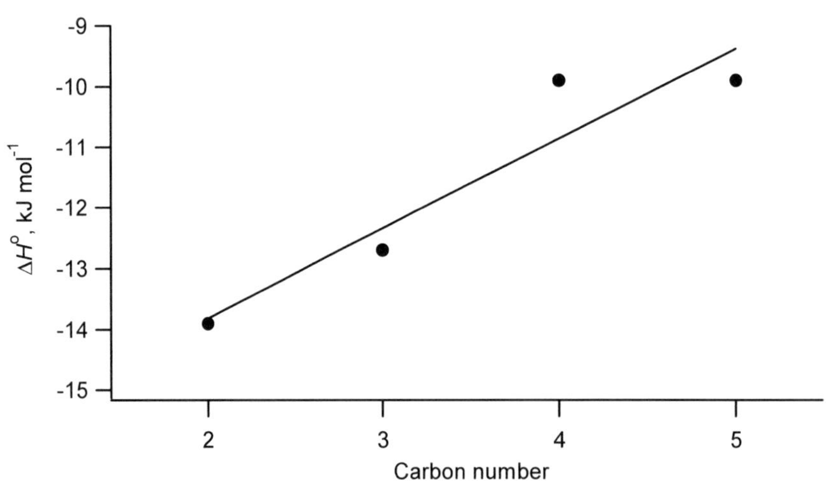

Fig. 9. Plot of the enthalpy change in binding of alkylbenzamidinium chlorides to trypsin at 25 °C at pH 8. Least squares line fit gave $\Delta H^{\circ} = (1.5 \pm 0.4)n - (16.8 \pm 1.3)$, where n is the carbon number of the alkyl group. Data were taken from Talhout et al. 2003

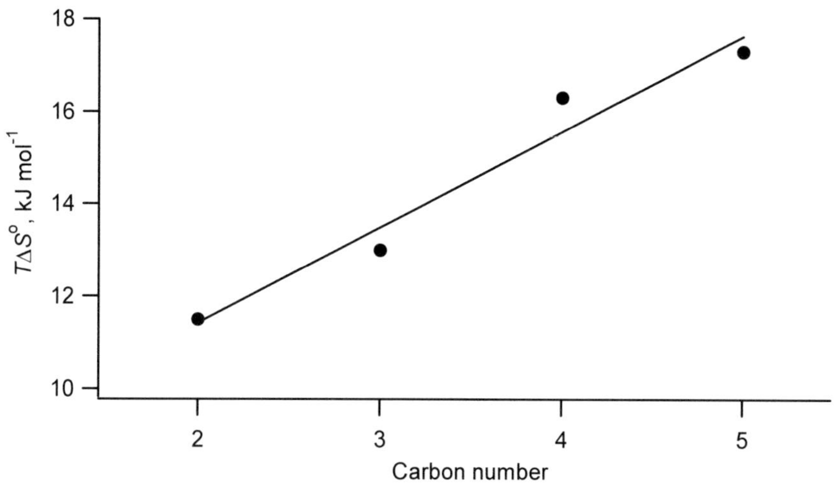

Fig. 10. Plot of $T\Delta S^o$ in binding of alkylbenzamidinium chlorides to trypsin at 25 °C at pH 8. Least squares line fit gave $T\Delta S^o = (2.1 \pm 0.3)n + (7.3 \pm 1.1)$, where n is the carbon number of the alkyl group. Data were taken from Talhout et al. 2003

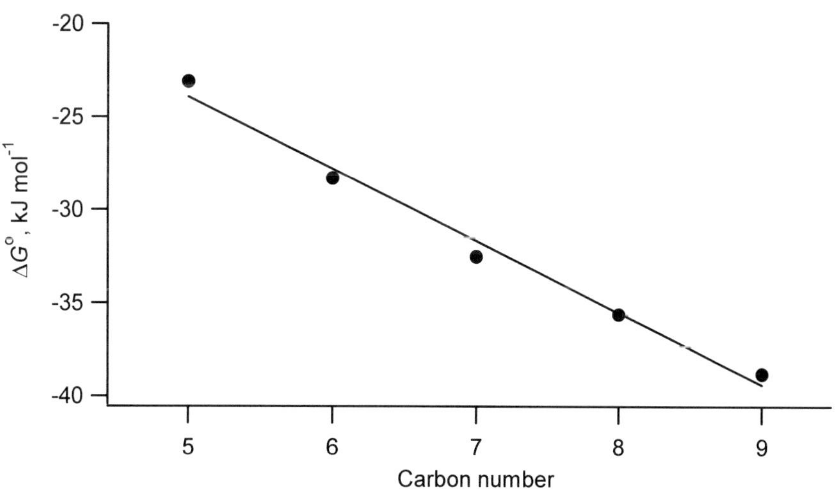

Fig. 11. Plot of the free energy change in binding of alcohols to MUP-1 at 300 K at pH 7.4. Least squares line fit gave $\Delta G^o = (-3.87 \pm 0.26)n - (4.6 \pm 1.9)$, where n is the carbon number of the alkyl group. Data were taken from Malham et al. 2005

5.2 Enthalpy-driven binding of an alkyl group to protein

Binding of alcohols to the major urinary protein (MUP) was studied by Malham et al. (Malham et al., 2005). The primary hydroxy group of alcohol was hydrogen bonded to Tyr, and the alkyl group is bound to the hydrophobic pocket formed by Leu and Phe. The free energy changes, the enthalpy changes and the entropy changes are plotted against carbon number of alcohols in Figures 11-13, respectively. Figure 11 shows that the binding becomes tighter as the alcohol is longer: the alkyl groups were bound to the protein through hydrophobic interactions. The driving force of binding was enthalpic, and the entropic term counteracted the enthalpy term, i.e., enthalpy-entropy compensation was again observed (see Figures 12 and 13). Malham et al. suggested that dispersive interactions between alcohol and the hydrophobic side chains of MUP are the origin of the negative enthalpy changes.

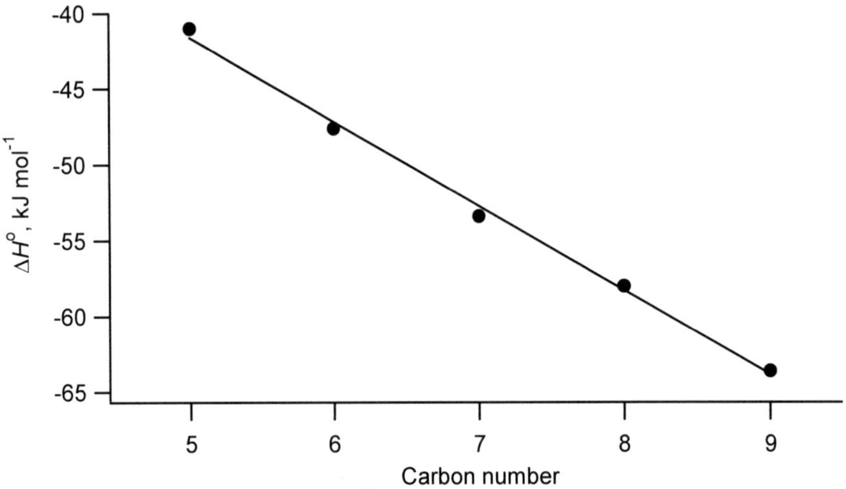

Fig. 12. Plot of the enthalpy change in binding of alcohols to MUP-1 at 300 K at pH 7.4. Least squares line fit gave $\Delta H^{\circ} = -(5.6 \pm 0.2)n - (13.8 \pm 1.4)$, where n is the carbon number of the alkyl group. Data were taken from Malham et al., 2005

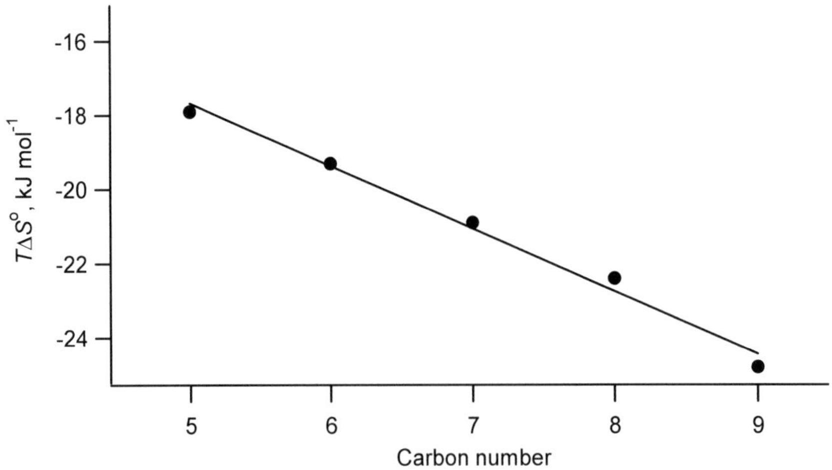

Fig. 13. Plot of $T\Delta S^{\circ}$ in binding of alcohols to MUP-1 at 300 K at pH 7.4. Least squares line fit gave $T\Delta S^{\circ} = -(1.7 \pm 0.1)n - (9.2 \pm 0.8)$, where n is the carbon number of the alkyl group. Data were taken from Malham et al. 2005

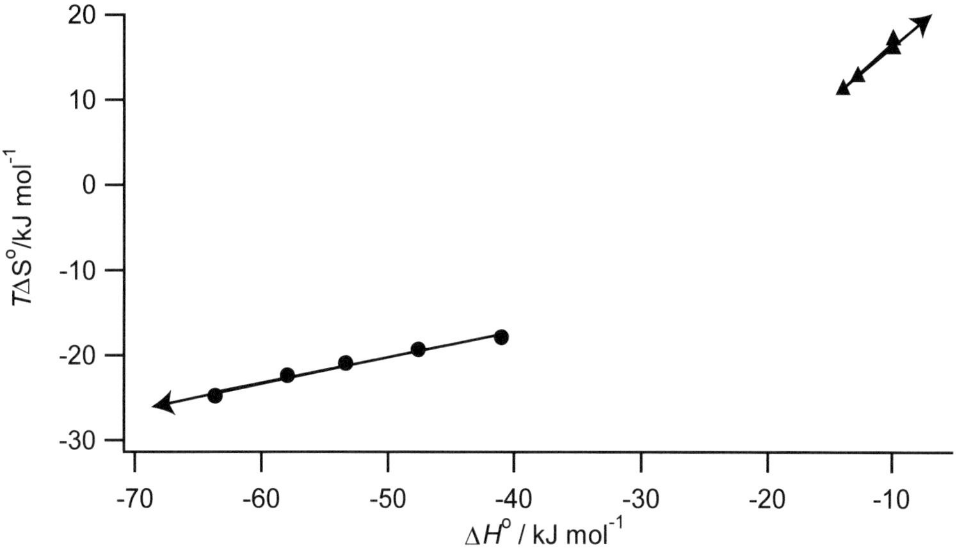

Fig. 14. Plot of $T\Delta S^{o}$ (T = 298 K) against ΔH^{o} for binding of alkylbenzamidinium chloride (filled triangle) or alcohols (filled circle) to proteins. Arrows indicates the direction of changes as the length of the alkyl group, and thus the binding free energy $-\Delta G^{o}$, increases.

6. Comparisons of thermodynamic parameters of binding of alkyl group to synthetic receptors and proteins

In Figure 14, values of $T\Delta S^{o}$ are plotted against ΔH^{o} for both binding of alkylbenzamidinium chloride and alcohol to proteins. The compensation temperature T_c was 224 K for the binding of alkylbenzamidinium chlorides to trypsin and 986 K for the binding of alcohol to MUP. In the former case, the temperature at which the binding was studied is much higher than T_c so that the entropy term dominates the overall thermodynamics. In the latter case, the temperature at which the binding was studied is much lower than T_c, to result in the enthalpy driven binding. These two studies on the thermodynamics of binding of alkyl groups to protein demonstrate that the binding of the alkyl groups can be driven either by enthalpic or entropic driving forces, and T_c's are much different from room temperature.

The compensation temperature of the binding of 4-alkylpyridines to receptor **1** was 800 K, and that of the binding to receptor **3** was 45 K based on the plot shown in Figure 7. The compensation temperature of the binding to receptor **2** was negative. These studies including both proteins and synthetic receptors demonstrated that T_c of binding of alkyl groups could be much different from room temperature, or even negative, and the origin of the diverse values of T_c should be the non-compensating properties of hydrophobic interactions.

7. Conclusion

The enthalpic and entropic contributions to the hydrophobic interactions were discussed. The transfer of alkane in the liquid state to water was driven by both enthalpic and entropic terms. The enthalpy-entropy compensation was not observed in the process. Binding of an alkyl group to synthetic receptors with varying degrees of water accessibility to the binding pocket showed that the hydrophobic interactions between the alkyl group of the guest and the alkyl chains of the receptor were driven by the entropic term if the binding pocket was sequestered from water and by the enthalpic term if the binding pocket was exposed to water. For the binding of the alkyl group to the synthetic receptors, strong enthalpy-entropy compensation was not observed. We suggest that hydrophobic interactions are driven by both enthalpic and entropic driving forces, and the relative contribution to the free energy was determined by the water accessibility to the binding cleft. Thermodynamic data for the binding of alkyl groups to proteins also showed much diverse compensation temperatures, 224 K for one case and 986 K for the other case. Non-compensating enthalpy and entropy may be a general property of hydrophobic interactions.

8. References

Abraham, M. H. (1982). Free energies, enthalpies, and entropies of solution of gaseous nonpolar nonelectrolytes in water and nonaqueous solvents. The hydrophobic effect. *J. Am. Chem. Soc.* 104, 2085, ISSN: 0002-7863.

Arena, G.; Contino, A.; Fujimoto, T.; Sciotto, D.; & Aoyama, Y. (2000). ^{1}H NMR and Calorimetric Studies of the Inclusion of Trimethylammonium Cations into Water Soluble Calixresorcinarenes. *Supramol. Chem.* 11, 4, 279-288, ISSN: 1061-0278.

Ben-Naim, A. & Marcus, Y. (1984). Solvation thermodynamics of nonionic solutes. *J. Chem. Phys.* 81, 4, 2016-2027, ISSN: 0021-9606.

Blokzijl, W. & Engberts, J. B. F. N. (1993). Understanding organic reactions in water: from hydrophobic encounters to surfactant aggregates. *Angew. Chem. Int. Ed.* 32, 1545-1579, ISSN: 1433-7851.

Böhm, H.-J. (1994). The development of a simple empirical scoring function to estimate the binding constant for a protein-ligand complex of known three-dimensional structure. *J. Comput.-Aided Mol. Des.* 8, 243-256, ISSN: 0920-654X.

Chandler, D. (2005). Interfaces and the driving force of hydrophobic assembly. *Nature* 437, 7059, 640-647, ISSN: 0028-0836.

Cohen, J. L. & Connors, K. A. (1970). Stability and structure of some organic molecular complexes in aqueous solution. *J. Pharm. Sci.* 59, 9, 1271-1276, ISSN: 0022-3549.

Dunitz, J. D. (1995). Win some, lose some: enthalpy-entropy compensation in weak intermolecular interactions. *Chemistry & Biology* 2, 11, 709-712, ISSN: 1074-5521.

Frank, H. S. & Evans, M. W. (1945). Free Volume and Entropy in Condensed Systems III. Entropy in Binary Liquid Mixtures; Partial Molal Entropy in Dilute Solutions; Structure and Thermodynamics in Aqueous Electrolytes. *J. Chem. Phys.* 13, 11, 507-532, ISSN: 0021-9606.

Gelb, R. I.; Schwartz, L. M.; Cardelino, B.; Fuhrman, H. S.; Johnson, R. F. & Laufer, D. A. (1981). Binding Mechanism in Cyclohexaamylose Complexes. *J. Am. Chem. Soc.* 103, 1750-1757, ISSN: 0002-7863.

Gilli, P.; Ferretti, V.; Gilli, G. & Borea, P. A. (1994). Enthalpy-Entropy Compensation in Drug-Receptor Binding. *J. Phys. Chem.* 98, 15, 1515-1518, ISSN: 0022-3654.

Harris, S. M. J.; Higuchi, T. & Rytting, J. H. (1973). Thermodynamic group contributions from ion pair extraction equilibriums for use in the prediction of partition coefficients. Correlation of surface area with group contributions. *J. Phys. Chem.* 77, 22, 2694-2703, ISSN: 0022-3654.

Hermann, R. B. (1972). Theory of Hydrophobic Bonding. II. The Correlation of Hydrocarbon Solubility in Water with Solvent Cavity Surface Area. *J. Phys. Chem.* 76, 19, 2754-2759, ISSN: 0022-3654.

Houk, K. N.; Leach, A. G.; Kim, S. P. & Zhang, X. (2003). Binding affinities of host-guest, protein-ligand, and protein-transition-state complexes. *Angew. Chem. Int. Ed.* 42, 40, 4872-4897, ISSN: 1433-7851.

Iwamoto, H.; Mizutani, T. & Kano, K. (2007). Thermodynamics of Hydrophobic Interactions: Entropic Recognition of Hydrophobic Moiety by Poly(Ethylene Oxide)-Zinc Porphyrin Conjugates. *Chem. Asian J.* 2, 10, 1267-1275, ISSN: 1861-4728.

Jencks, W. P. (1969). *Catalysis in Chemistry and Enzymology.* McGraw-Hill Book Company, ISBN: 0486654605, New York, USA.

Kirksey, C. H.; Hambright, P. & Storm, C. B. (1969). Stability Constants and Proton Magnetic Resonance Studies of Zinc α, β, γ, δ-Tetraphenylporphin and Substituted Pyridines. *Inorg. Chem.* 8, 10, 2141-2144, ISSN: 0020-1669.

Knox, J. H. (1971). *Molecular Thermodynamics, An Introduction to Statistical Mechanics for Chemists.* John Wiley & Sons, Ltd., ISBN: 0471996211, New York, USA.

Kuntz, I. D.; Chen, K.; Sharp, K. A. & Kollman, P. A. (1999). The maximal affinity of ligands. *Proc. Natl. Acad. Sci. U S A* 96, 18, 9997-10002, ISSN: 0027-8424.

Leffler, J. E. & Grunwald, E. (1963). *Rates and Equilibria of Organic Reactions.* John Wiley and Sons, Inc., ISBN: 9780486660684, New York, USA.

Lumry, R. & Rajender, S. (1970). Enthalpy-Entropy Compensation Phenomena in Water Solutions of Proteins and Small Molecules: A Ubiquitous Property of Water. *Biopolymers* 9, 1125-1227, ISSN: 0006-3525.

Malham, R.; Johnstone, S.; Bingham, R. J.;.Barratt, E.; Phillips, S. E. V.; Laughton, C. A. & Homans, S. W. (2005). Strong Solute-Solute Dispersive Interactions in a Protein-Ligand Complex. *J. Am. Chem. Soc.* 127, 48, 17061-17067, ISSN: 0002-7863.

Matsumoto, S.; Iwamoto, H. & Mizutani, T. (2010). Water Accessibility to Binding Cleft as a Major Switching Factor from Entropy-Driven to Enthalpy-Driven Binding of an Alkyl Group by Synthetic Receptors. *Chem. Asian. J.* 5, 5, 1163-1170, ISSN: 1861-4728.

Meyer, E. A.; Castellano, R. K. & Diederich, F. (2003). Interactions with Aromatic Rings in Chemical and Biological Recognition. *Angew. Chem. Int. Ed.* 42, 1210-1250, ISSN: 1433-7851.

Mikros, E.; Gaudemer, A. & Pasternack, R. (1988). Interactions of water-soluble zinc porphyrin with amino acid. *Inorg. Chim. Acta* 153, 199-200, ISSN: 0020-1693.

Mizutani, T.; Ema, T.; Tomita, T.; Kuroda, Y. & Ogoshi, H. (1994). Design and Synthesis of a Trifunctional Chiral Porphyrin with C_2 Symmetry as a Chiral Recognition Host for Amino Acid Esters. *J. Am. Chem. Soc.* 116, 10, 4240-4250, ISSN: 0002-7863.

Mizutani, T.; Wada, K. & Kitagawa, S. (1999). Porphyrin Receptors for Amines, Amino Acids and Oligopeptides in Water. *J. Am. Chem. Soc.* 121, 49, 11425-11431, ISSN: 0002-7863.

Mizutani, T.; Kozake, K.; Wada, K. & Kitagawa, S. (2003). An Efficient Recognition Motif for an Alkyl Moiety in Water. *Chem. Commun.* 23, 2918-2919, ISSN: 1364-548X.

Nemethy, G. & Scheraga, H. A. (1962a). Structure of water and hydrophobic bonding in proteins. I. A model for the thermodynamic properties of liquid water. *J. Chem. Phys.* 36, 3382-400, ISSN: 0021-9606.

Nemethy, G. & Scheraga, H. A. (1962b). Structure of water and hydrophobic bonding in proteins. II. Model for the thermodynamic properties of aqueous solutions of hydrocarbons. *J. Chem. Phys.* 36, 3401-17, ISSN: 0021-9606.

Reynolds, J. A.; Gilbert, D. B.; & Tanford, C. (1974). Empirical Correlation Between Hydrophobic Free Energy and Aqueous Cavity Surface Area. *Proc. Nat. Acad. Sci. USA* 71, 8, 2925-2927, ISSN: 0027-8424

Richards, F. M. (1977). Areas, volumes, packing, and protein structure. *Annu. Rev. Biophys. Bioeng.* 6, 151-176, ISSN: 0084-6589.

Ross, P. D. & Subramanian, S. (1981). Thermodynamics of Protein Association Reactions: Forces Contributing to Stability. *Biochemistry* 20, 3096-3102, ISSN: 0006-2960.

Sgarlata, C.; Mugridge, J. S.; Pluth, M. D.; Tiedemann, B. E. F.; Zito, V.; Arena, G. & Raymond, K. N. (2010). External and Internal Guest Binding of a Highly Charged Supramolecular Host in Water: Deconvoluting the Very Different Thermodynamics. *J. Am. Chem. Soc.* 132, 3, 1005-1009, ISSN: 0002-7863.

Sharp, K. A.; Nicholls, A.; Friedman, R. & Honig, B. (1991). Reconciling the Magnitude of the Microscopic and Macroscopic Hydrophobic Effects. *Science* 252, 106-109, ISSN: 0036-8075.

Sharp, K. A.; Nicholls, A.; Fine, R. F. & Honig, B. (1991). Extracting Hydrophobic Free Energies from Experimental Data: Relationship to Protein Folding and Theoretical Models. *Biochemistry* 30, 9686-9697, ISSN: 0006-2960.

Schneider, H.-J. & Yatsimirsky, A. (2000). *Principles and Methods in Supramolecular Chemistry*. John Wiley & Sons, Ltd., ISBN: 978-0-471-97253-2, Chichester, UK.

Smithrud, D. B.; Wyman, T. B. & Diederich, F. (1991). Enthalpically driven cyclophane-arene inclusion complexation: solvent-dependent calorimetric studies. *J. Am. Chem. Soc.* 113, 14, 5420-6, ISSN: 0002-7863.

Southall, N. T.; Dill, K. A. & Haymet, A. D. J. (2002). A View of the Hydrophobic Effect. *J. Phys. Chem. B* 106, 521-533, ISSN: 1520-6106.

Spencer, J. N.; DeGarmo, J.; Paul, I. M.; He, Q.; Ke, X.; Wu, Z.; Yoder, C. H.; Chen, S. & Mihalick, J. E. (1995). Inclusion Complexes of Alcohols with α-Cyclodextrin. *J. Solution Chem.* 24, 601, ISSN: 0095-9782.

Tabushi, I.; Kiyosuke, Y.; Sugimoto, T. & Yamamura, K. (1978). Approach to the Aspects of Driving Force of Inclusion by α-cyclodextrin. *J. Am. Chem. Soc.* 100, 3, 916-919, ISSN: 0002-7863.

Talhout, R.; Villa, A.; Mark, A. E. & Engberts, J. B. F. N. (2003). Understanding Binding Affinity: A Combined Isothermal Titration Calorimetry/Molecular Dynamics Study of the Binding of a Series of Hydrophobically Modified Benzamidinium Chloride Inhibitors to Trypsin. *J. Am. Chem. Soc.* 125, 35, 10570-10579, ISSN: 0002-7863.

Tanford, C. (1973). *The Hydrophobic Effect: formation of micelles and biological membranes.* Wiley, ISBN: 0471048933, New York, USA.

Rekharsky, M. V. & Inoue, Y. (1998). Complexation Thermodynamics of Cyclodextrins. *Chem. Rev.* 98, 1875-1917, ISSN: 0009-2665.

Whitesides, G. M. & Krishnamurthy, V. M. (2005). Designing ligands to bind proteins. *Quarterly Rev. Biophys.* 38, 4, 385-395, ISSN: 0033-5835.

5

Role and Applications of Electrostatic Effects on Nucleic Acid Conformational Transitions and Binding Processes[1]

Jeff D. Ballin and Gerald M. Wilson
Department of Biochemistry and Molecular Biology, University of Maryland School of Medicine
USA

1. Introduction

Processes involving nucleic acids are in general highly salt concentration dependent. Equilibrium binding constants (K_{obs}, defined in terms of the concentrations of the reactants and products) describing the association of proteins or multivalent ligands with nucleic acids (NAs) exhibit large negative power dependences on salt concentration. For the reaction of a ligand L with charge Z_L forming a complex with a DNA strand having Z_D phosphates,

$$L^{Z_L+} + DNA^{Z_D-} \xleftrightarrow{K_{obs}} complex\ , \qquad (1)$$

the binding constant is proportional to an approximately constant power of the univalent salt concentration [MX] present in solution,

$$K_{obs} \propto [\mathrm{MX}]^{SK_{obs}}\ . \qquad (2)$$

Experimentally, the exponent SK_{obs} typically ranges from $\approx 0.7Z_L$ to $\approx 0.9Z_L$ for single-stranded DNA (ssDNA) and double-stranded DNA (dsDNA), respectively, when $Z_D >> Z_L$ (Lohman & Mascotti, 1992; Mascotti & Lohman, 1990; Record et al., 1976). Helix formation between two complementary single-stranded NAs is also a binding process. Though the binding constant for this process is difficult to measure directly, the positive dependence of the thermal denaturation temperature on ln[salt] indicates that the strand binding constant must have a very strong positive power dependence on salt concentration (Record et al., 1998). Numerous detailed experimental and theoretical studies have characterized the binding of various positively charged ligands to polymeric NAs (Record et al., 1998). Those interactions are biologically significant in their own right and are of interest as models for protein-nucleic acid interactions which play critical roles in gene expression and replication.

[1] The preparation of this chapter was supported by NIH-NCI grant R01 CA102428 (awarded to G.M.W.) and based in part on Ballin, J.D. (2003) "Binding studies of model oligopeptides to nucleic acids: DNA length dependence, coulombic end effect, and amino acid composition effects," PhD dissertation, University of Wisconsin-Madison.

Much of the current research into protein-nucleic interactions focuses on specific complexes associated with these various processes. With the increasing ease and cost effectiveness of oligonucleotide and oligopeptide synthesis, many research laboratories use oligomers (*e.g.*, "cognate sequences," "binding motifs," and "recognition modules") to investigate binding interactions of larger *in vivo* systems (*e.g.*, plasmids, chromosomal DNA, etc.). Do these oligomers accurately model the behavior of the polymeric biological ensemble?

To answer this question, we must first ask how the binding behavior of a short oligonucleotide compares to that of longer NA lengths found *in vivo*. If differences exist between short *vs* long NAs, length dependences should be seen in the properties and processes involving short *vs* long NAs. Theory predicts and experiments show length-dependent effects influence a variety of nucleic acid properties including denaturation stability of double helices and binding of cationic oligopeptides (see Section 3). These length dependences can be ascribed to "end" effects which exert a greater effect on the observable (*i.e.*, average) properties of the nucleic acid as its chain length is reduced. In contrast, if a polymer is long enough, the terminal regions represent an insignificant fraction of the lattice and any observed interaction or process will be dominated by the interior region of the polymer (unless the phenomenon studied is specific to the ends). Therefore, if the behavior of the ends differs from that of the central region, these differences should become evident at shorter lengths.

Two kinds of end effects have been investigated for nucleic acid helices: a base stacking end effect and a coulombic end effect. A base stacking end effect in native DNA, either as a two-stranded duplex or a single-stranded hairpin, stems from a reduced stabilization of a base pair at the terminus because it has only one base stacking interaction while all interior bases have two. As a result, the stability per base pair at a given salt concentration, measured in terms of a change in standard free energy ($\Delta G°$) or the thermal denaturation temperature (T_m), decreases with decreasing chain length for a series of homologous sequences. In melting experiments, this base stacking end effect is exhibited for long NAs as a "fraying" of the terminal regions which occurs prior to denaturation of the interior double helix. These end stacking effects are contrasted with coulombic end effects caused by a difference in salt ion-NA interactions which manifest as a change in the extent of per phosphate counterion accumulation and coion exclusion[2] near the ends *vs* the NA lattice interior. While there is no dispute over the existence and consequences of a base stacking end effect on stability, the possibility of observable coulombic differences near the NA termini relative to the NA interior has been a more contentious issue. There are two sides to this debate: those who conclude that coulombic interactions with the terminal *vs* interior DNA regions (and thus between the averaged properties of "short" versus "long" polyions) have dramatic consequences on experimentally observable phenomena and those who hold that these coulombic differences have no significant impact on processes within the salt concentration range used in most experiments. Both camps can point to experimental and theoretical results that support their arguments. Is there a coulombic "end effect"? Does a charged species interact differently with the terminal regions *vs* interior of a polyion?

In order to understand the significance and differences in the behavior of polyelectrolytes attributable to coulombic end effects, one must understand how systems behave in the absence of end effects. While a survey of all relevant studies is beyond the scope of this

---

[2] "Counterion" and "coion" of a salt are defined relative to the (negative) nucleic acid charge. Therefore, "counterions" represent the positively charged salt cations and "coions" are the salt anions.

review, important experimental and theoretical methods that define electrostatic processes with be introduced, summarizing seminal historical studies and referencing current literature which provides technical details on the use these approaches. Where appropriate, the reader will be referred to review articles dedicated to subtopics touched upon here if more information is desired. We begin with a brief description of polyelectrolyte theories and experiments characterizing extremely long, effectively infinite, polyions. Oligoelectrolyte phenomena are discussed next, including the theories developed to explain similarities or differences observed relative to polyions. We will conclude with an overview of challenges still ahead and how they might be approached.

2. Polyelectrolyte studies

NAs are nearly ideal models of oligo- and polyelectrolytes for experimental and theoretical studies. NAs are biochemically relevant, relatively easy to manipulate experimentally, stable under a wide variety of conditions (especially DNA) and available as a monodisperse species (*i.e.*, with a defined chain length or number of monomers). This section will summarize prevailing polyelectrolyte theories to provide context for experimentally observed phenomena of nucleic acid polymers, including their fundamental properties (*e.g.*, small ion NMR relaxation rates, electrophoretic mobility) and processes involving NAs (*e.g.*, thermal denaturation, small ligand binding).

Preferential interaction coefficients. The theories described below characterize the small electrolyte ion distribution around a polyion. These ion distributions fundamentally impact NA properties and interactions but have generally eluded direct measurement of spatial distribution in solution[3]. A linkage is necessary to connect theory to experiment: preferential interaction coefficients (Γ) correlate ion concentration gradients to thermodynamic properties and therefore provide a convenient conduit from calculation to experimental observable and vice versa (Anderson & Record, 1993). Within this framework, the thermodynamic interactions of a given component or ion (*e.g.*, counterion) with some other component or species (*e.g.*, DNA) can be described simply.

Considering equilibrium dialysis as an example, the preferential interaction coefficient, Γ, characterizing salt-polyelectrolyte component interactions is defined as

$$\Gamma \equiv \lim_{C_3 \to 0} \left(\frac{\partial C_3}{\partial C_u} \right)_{T,\mu_1,\mu_3} , \tag{3}$$

where C_3 is the salt ion concentration, C_u is the molar concentration of singly charged monomer "units" (*e.g.*, NA phosphate charges), T is temperature, and μ_1 and μ_3 are the chemical potentials of the solvent and salt respectively (Anderson & Record, 1995). Preferential interaction coefficients (or if a process is considered, the stoichiometrically-weighted differences in Γ for the products and reactants) are easily related to the effects of salts and uncharged solutes on properties such as equilibrium dialysis coefficients or osmotic coefficients for processes including nucleic acid thermal denaturation studies and NA-ligand binding of biopolymers (Anderson & Record, 1983; Anderson & Record, 1993). These relationships will be discussed as their use becomes relevant.

---

[3] Recent developments in small-angle x-ray scattering (SAXS) and buffer equilibration-atomic emission spectroscopy (BE-AES) have yielded insights into the arrangements and distributions of ions around nucleic acids. See Section 3.2.2.

How is Γ determined? Preferential interaction coefficients for NA-containing systems can be calculated directly from grand canonical[4] Monte Carlo (GCMC) simulations performed at constant salt activity, a_3 (Mills et al., 1986). After evaluating C_3 for a series of GCMC simulations with varying C_u, Γ can be obtained from the slope of $C_3 - C_3^o$ *vs* C_u, where C_3^o is the salt concentration in the absence of the polyion (Olmsted et al., 1989; Olmsted et al., 1991) (See Section 2.2.3). Alternatively, single-ion preferential interaction coefficients expressed per polyion charge (with subscript *u* for "unit") may be obtained by integrating the ion distributions over the cell volume (V) containing the polyion (Ni et al., 1999):

$$\Gamma_{s,u} = \frac{1}{|Z_p|}\int (C_s - C_{s,bulk})dV\,, \tag{4}$$

where C_s is the local coion concentration and Z_p is the total polyion charge[5]. If a coion with valence z_s is assumed to obey a Boltzmann distribution dictated by a mean-field electrostatic potential, φ, the local coion concentration C_s can be obtained by

$$C_s = C_{3,bulk} e^{-\frac{\varphi z_s}{kT}}\,. \tag{5}$$

Using cylindrical symmetry coordinates together with eq 5, eq 4 becomes

$$\Gamma_{s,u} = 2\pi b \int C_{3,bulk}\left(e^{-\frac{\varphi z_s}{kT}} - 1\right) rdr\,, \tag{6}$$

where b is the average charge separation on the polyion. The potential φ is commonly modeled by the Poisson-Boltzmann equation, eq 11 (Misra & Draper, 1999; Ni et al., 1999; Sharp, 1995; Sharp et al., 1995; Shkel et al., 2000; Stigter & Dill, 1996) (see Section 2.2.1).

Note that the Γ in eq 3 describes the interaction of the electroneutral salt (*e.g.*, NaCl) with the polyelectrolyte (*e.g.*, NA), while eq 4 and eq 6 refer to specifically $\Gamma_{s,u}$, the coion (*e.g.*, Cl^-) preferential interaction coefficient expressed per polyion charge (indicated by the "*u*" subscript). By definition, when the polyelectrolyte and salt share a common ion, the coion $\Gamma_{s,u}$ and the salt component Γ are identical. Experimentally, $\Gamma_{s,u}$ is negative because coions are excluded from the vicinity of the NA polyanion. In an equilibrium dialysis experiment[6],

---

[4] Grand canonical Monte Carlo calculations are performed at constant temperature, mean ionic activity, and volume.

[5] The subscripts "*s*" and "*o*" for Γ specify the ion described: "*s*" indicates the coion, and "*o*" the counterion. These can be mnemonically related to "same" for coion and "opposite" for the counterion. Sometimes Γ_+ and Γ_- are used instead of Γ_o and Γ_s, respectively, when considering NA polyelectrolytes. If the polymer were a polycation, the coion obviously would be positive and the counterion negative. In this case, the subscript labels "+" and "-" should be transposed.

[6] An equilibrium dialysis chamber is separated into two compartments (α and β) by a semipermeable membrane through which small solutes (*e.g.*, salt ions) can pass but larger species (e.g., a nucleic acid) cannot. The system is allowed to equilibrate at constant temperature and pressure after the nucleic acid is added to compartment α, and the concentration of electroneutral salt is measured in each compartment. The Donnan coefficient, Γ^{exp}, is the quantitative measure of the difference in electroneutral salt concentrations per concentration of the larger species: $\Gamma^{exp} = \frac{m_{salt,\alpha} - m_{salt,\beta}}{m_{NA}}$. For

the coion would be excluded from the compartment containing the NA. Electroneutrality of the solution requires that nucleic acid charge must be neutralized by the accumulation of counterions and the exclusion of coions. Therefore, the $\Gamma_{o,u}$ describing the interaction of excess univalent salt with NA is easily determined since the valence-weighted sum[7] of single ion preferential interaction coefficients are related by

$$\Gamma_{o,u} - \Gamma_{s,u} = 1 \quad (7)$$

when expressed per polyion charge (denoted by the subscript "u"). Just as $\Gamma_{s,u} < 0$ reflects coion exclusion from the surface of the polyion, $\Gamma_{o,u}$ is a positive quantity, indicating that counterions are accumulated near the polyion. Because $\Gamma_{s,u}$ is readily accessible via the electroneutrality condition (*e.g.*, eq 7), subsequent discussion will generally describe phenomena in terms of coion interactions ($\Gamma_{s,u}$) by convention, since $\Gamma_{s,u}$ is equivalent to the Donnan coefficient[6]. Draper and coworkers provide a well-presented description of equilibrium dialysis and survey direct experimental approaches to measure Γ in RNA processes (Leipply et al., 2009).

Range of values of $\Gamma_{s,u}$. If there were no polyion-ion interactions in a monovalent salt solution, electroneutrality and random mixing (*i.e.*, entropy) dictate that $\Gamma_{s,u} = -0.5$ and $\Gamma_{o,u} = 0.5$. At the other extreme, strong counterion-polyion interactions which drive association of one univalent cation with every polyion charge (*e.g.*, NA phosphate) would exhibit $\Gamma_{s,u} = 0$ and $\Gamma_{o,u} = 1$. In practice, helical DNA at 0.0095 M sodium bromide has an experimental Donnan coefficient of -0.122 ± 0.004 (Strauss et al., 1967), reflecting the high nonideality of NAs in which a significant fraction of the counterions behave as if associated with the polyanion ($\Gamma_{o,u} \approx 0.9$). The nonideality decreases as salt concentration increases: $\Gamma^{exp} = -0.272 \pm 0.008$ at 0.23 M NaBr, and $\Gamma^{exp} = -0.538 \pm 0.0037$ at 0.98 M NaBr (Strauss et al., 1967).

2.1 Polyelectrolyte theory

2.1.1 Polyelectrolytes are surrounded by steep local ion concentration gradients

All polyelectrolyte theories for infinitely long highly charged polyelectrolytes predict the existence of steep salt ion concentration gradients (counterion accumulation and coion exclusion) in the radial direction near the surface of the polyion (*Fig.* 1). These concentration gradients are maintained by the electrostatic potential field resulting from a high axial density of structural charges (Anderson & Record, 1995). At 10.6 mM salt, native dsDNA with an axial charge density of two phosphates per 3.4 Å has a local[8] counterion concentration of about 3 M, while the coion concentration is approximately 20 μM (Ni et al., 1999). At 0.21 M salt, the local counterion concentration approaches 5 M, while the coion concentration is only 10 mM (Ni et al., 1999). These concentration gradients decay radially to the bulk salt concentration over a distance of ~ 50 Å. Gradients become less pronounced as the bulk solution salt concentration is increased. The concentration of salt ions near the

---

sufficiently dilute nucleic acids, Γ^{exp} converges to Γ (eq 3). Under all conditions considered in this chapter, the preferential interaction coefficient Γ is numerically equivalent to the Donnan coefficient Γ^{exp} (Record et al., 1998).

[7] For a polyanion in the presence of a 2:1 salt (*i.e.*, with a divalent cation and univalent anion), electroneutrality requires $2\Gamma_{2+,u} - \Gamma_{-,u} = 1$. If the polyanion solution contains a mixture of univalent and divalent cations with a shared univalent anion, charge balance requires $\Gamma_{+,u} + 2\Gamma_{2+,u} - \Gamma_{-,u} = 1$.

[8] "Local" concentration is the concentration of a given species at the surface of the polyion.

surface of the DNA increases slowly with increasing bulk salt concentration (Mills et al., 1986; Ni et al., 1999). As the bulk salt concentration increases, the local/bulk ion concentration ratio decreases and the thermodynamic nonideality contributed by salt-DNA interactions therefore diminish.

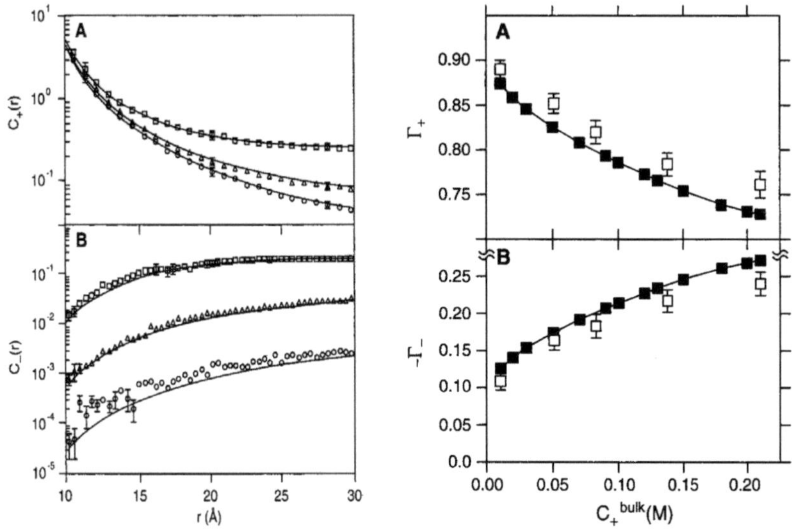

Fig. 1. (*left panel*) Dependence of monovalent cation ($C_+(r)$) and anion ($C_-(r)$) molar concentrations as a function of radial distance, r, from axis of a cylindrical model of dsDNA as derived from Monte Carlo (open symbols) and Poisson-Boltzmann calculations (smooth curves) for bulk salt concentrations of 10.6 mM (○), 51.0 mM (Δ), and 210 mM (□). (*right panel*) Canonical Monte Carlo (□) or Poisson-Boltzmann (■) calculated preferential interaction coefficients for a univalent cation (Γ_+) or univalent anion (Γ_-) with dsDNA as a function of the bulk cation concentration (C_+^{bulk}). (Figures adapted from Ni et al., 1999)

2.1.2 Important parameters in polyelectrolyte theory

When NAs are modeled as "primitive" cylindrical polyanions possessing a uniform charge distribution, polyelectrolyte behavior can be predicted using only a small number of structural parameters. The two most important parameters are the distance of closest approach between ion center and polyelectrolyte axis, *a*, and the average spacing of adjacent charges projected on the polyion axis, *b*. Often, primitive polyelectrolytes are described by the reduced axial charge density, ξ:

$$\xi = \frac{e^2}{\varepsilon b k T}. \tag{8}$$

The electron charge *e* and Boltzmann factor *k* are constants, and the product of temperature *T* and dielectric constant ε is approximately constant in water. As a result, the reduced axial charge density is almost exclusively determined by the projected axial charge spacing, *b*, with ξ = 7.14 Å/*b* in water at 25 °C. In the limit of infinite polyion dilution, counterion condensation theory (*c.f.*, Section 2.2.2) relates ξ to the preferential interaction coefficient (Manning, 1969):

$$\Gamma^{LL} = -\frac{1}{4\xi}\xi, \qquad \xi > 1, \tag{9}$$

where "LL" stands for "limiting law." Γ^{LL} is determined only by *b*: for dsDNA in water at 25 °C, $\Gamma^{LL} = -0.06$. The effective polyion radius, *a*, figures prominently in Poisson-Boltzmann and Monte Carlo calculations and is sometimes adjusted to account for hydration or penetration of small ions into helical grooves (see Sections 2.2.1, 2.2.2 and 3.1.2).

Correlation of the Debye parameter κ *to solution behavior.* The Debye parameter, κ, describes the charge screening characteristics of the solution containing the polyelectrolyte. It is dependent on the concentrations of all charged species including the polyion[9] through the ionic strength parameter, *I*:

$$\kappa^2 = \frac{8\pi N_A e^2 I}{1000\varepsilon kT}, \tag{10}$$

where N_A is Avogadro's number and *I* is ionic strength ($I = 0.5\Sigma\, z_i^2 C_i$ in which z_i is the charge of the *i*th ionic species and C_i is its molar concentration). At 25 °C in water, $\kappa = (1/3)\, I^{0.5}$Å^{-1}. For an infinitely dilute polyion in solution with a univalent salt possessing a common counterion, ionic strength is the same as the bulk salt concentration. Determining κ provides the necessary information to perform PB calculations on solutions of z_+:z_- symmetrical salts (eq 12). If employing the Debye-Hückel approximation (eq 15) or counterion condensation theory, κ parameterizes calculations for solutions containing salt ions of any valence.

2.2 Implementations of polyelectrolyte theory

Poisson-Boltzmann (PB) theory, counterion condensation (CC) theory, and Metropolis Monte Carlo (MC) simulations are the most common theoretical methodologies used to describe polyelectrolyte behavior. Each has advantages and limiting assumptions and all are still heavily used to describe and predict NA phenomena. After a brief overview of the concepts behind these approaches, their predictions will be compared to experimental studies of polyelectrolytes.

2.2.1 Poisson-Boltzmann equation

The PB formalism is the union of the Poisson equation with the Boltzmann distribution. For any arrangement of macroscopic or atomic charges, the Poisson equation relates the local electric potential, $\varphi(\vec{r})$, to the local charge density[10], $\rho(\vec{r})$, as a function of distance $\vec{r}$ from the polyion:

$$\nabla^2\varphi(\vec{r}) = \frac{4\pi\rho(\vec{r})}{\varepsilon}, \tag{11}$$

[9] The polyion concentration is defined as the concentration of its charged units. For example, if the polyion is a NA, the concentration is determined in terms of NA phosphate concentration.

[10] $\rho(\vec{r})$ is the net charge density of the surrounding ion atmosphere, determined by their concentrations and valences.

where ε is a spatially invariant dielectric constant[11]. The Boltzmann distribution relates local ion concentrations at a given radial distance from the polyion axis to the bulk solution ion concentration far from the polyion, based on the reduced electrostatic potential experienced at that given position. For a solution containing a polyion at infinite dilution in the presence of a symmetrical electrolyte (z_+:z_-) described by the Boltzmann distribution, the Poisson equation requires

$$\frac{\partial^2 y}{\partial r^2} + \frac{1}{r}\frac{\partial y}{\partial r} = \kappa^2 \sinh y \tag{12}$$

with the "Gauss' law" boundary conditions

$$\left.\frac{\partial y}{\partial r}\right|_{r=a} = -\frac{2z\xi}{a}, \qquad \left. r\frac{\partial y}{\partial r}\right|_{r=\infty} = 0 \tag{13}$$

when using cylindrical coordinates and the reduced potential $y \equiv ze\phi/kT$ in terms of the electron charge, e (Anderson & Record, 1990). Exact analytical solutions for the cylindrical PB equation (suitable for a stiff polyion like DNA) are known only for salt-free DNA solutions (*i.e.*, no added salt) in which all counterions have the same valence. However, by assuming that the potential of the mean force is approximated by the mean potential, the Poisson-Boltzmann equation can be applied to systems with a wide range of salt concentrations. This so called "PB approximation" has been justified on theoretical grounds for a cylindrically modeled DNA polyion at univalent salt concentrations below 0.1 M (Fixman, 1979). However, the PB approximation is likely to be more broadly applicable since numerous nonlinear Poisson-Boltzmann (NLPB) calculations using this approximation have made quantitative thermodynamic predictions that agree within $\approx$10% of experimental values, well inside of error, for RNA and DNA processes in up to 1 M salt (Bond et al., 1994; Korolev et al., 1998; Shkel et al., 2000; Shkel & Record, 2004).

Thermodynamic consequences can be predicted once the position-dependent electrostatic potential in hand. For example, ion gradients, measured as ratios of local (surface) to bulk concentrations, are determined by the exponential of the reduced potential. For a negatively charged polyion in solution with a 1:1 salt,

$$\frac{{}^{loc}C_+}{{}^{bulk}C_+} = \frac{{}^{bulk}C_-}{{}^{loc}C_-} = e^{y} \tag{14}$$

where ${}^{bulk}C_+ \cong {}^{bulk}C_-$ in excess salt. The reciprocal nature of the local versus bulk concentration ratios is a consequence of the sign inversion in y reflecting the ion charge z for cations *vs* anions. Eq 14 indicates that counterions are accumulated and coions are excluded from the polyelectrolyte surface (Ni et al., 1999). An increase in bulk salt concentration results in a much larger fractional increase in ${}^{bulk}C_+$ and ${}^{bulk}C_-$ than the increase in ${}^{loc}C_+$ and ${}^{loc}C_-$. Other thermodynamic parameters which are accessible via the reduced potential

[11] Distance-dependent dielectrics, $\varepsilon(\vec{r})$, have been used to model solvent effects (*e.g.*, dielectric saturation). Additional details and approaches are reviewed by (Grochowski & Trylska, 2008; Wong & Pollack, 2010; Wu & Morikis, 2006).

include preferential interaction coefficients (Γ) (Bond et al., 1994; Shkel et al., 2006; Shkel & Record, 2004), osmotic (ϕ) and activity coefficients (γ) (Anderson & Record, 1983; Record et al., 1978), the salt dependence of DNA denaturation (Bond et al., 1994; Korolev et al., 2002; Shkel & Record, 2004), and the thermodynamics of ligand binding phenomena (Rouzina & Bloomfield, 1997; Sharp et al., 1995; Stigter & Dill, 1996; Zhang et al., 1996). As mentioned above, experimental results have been very faithfully predicted in NLPB studies using both primitive and 3-D models of NAs. Numerous reviews detail the history, application, and comparisons of PB theory to experiment (Anderson & Record, 1995; Draper, 2008; Fogolari et al., 2002; Jayaram & Beveridge, 1996; Koehl, 2006; Wu & Morikis, 2006).

Development of the PB equation. Initial efforts to describe polyelectrolytes using PB theory relied on linearization of the potential distribution around a symmetric cylinder, an approximation valid only for very low charge density polyelectrolytes or at high salt concentration (see Section 2.2.2). The linearized PB solution is accurate when the reduced potential is small ($|y| < 1$), but does not describe dsDNA at experimental salt concentrations (*e.g.*, $y \approx 3$ at 0.1 M salt). In the decades since development of the Debye-Hückel limiting law, many asymptotic series expansion solutions have been developed to account for the high surface potential of NAs[12]. More recently, asymptotic solutions of the cylindrical PB equation for both high salt ($\kappa a \geq 1$, or more than 0.1 M univalent salt for B-DNA) (Shkel et al., 2000) and low salt ($\kappa a < 1$, or less than 0.1 M 1:1 salt for B-DNA) (Shkel et al., 2002; Trizac & Téllez, 2007) have been derived, providing explicit analytical expressions for the electrostatic potential and for the resulting preferential interaction coefficients, covering the entire experimental salt concentration range accessible to the PB formalism. These analytical expressions have calculated Γ for single, double, and even triple-stranded NAs with error small enough for comparison with experiments[13]. For example, the Shkel (2000) asymptotic predictions of Γ were within 5% of numerical NLPB calculations for dsDNA at 0.1 M univalent salt, with error decreasing as salt concentration increased. Most recently, a single analytical expression has been developed which predicts the electrostatic free energy of a polyelectrolyte within 2% of NLPB calculations for either a primitive cylinder or for 3-D models of B-DNA obtained from the Protein Data Bank (Shkel, 2010).

Difference between "primitive" cylindrical and structurally detailed 3-D models in NLPB calculations. Early polyelectrolyte studies by necessity considered only "primitive" representations of NAs (*e.g.*, lines of charge, cylinders, planes, etc.). With the advent of more detailed biomolecular structural information, more powerful computational resources and the availability of commercial and open-source PB calculation packages (such as DelPhi, APBS, and MEAD; see (Koehl, 2006) for additional details), Poisson-Boltzmann analyses using detailed 3-D models have become widespread in the literature. Not surprisingly, NLPB calculations employing NA models with atomic resolution have come into vogue as well (Chen & Honig, 1997; Grochowski & Trylska, 2008; Koehl, 2006; Misra et al., 1994; Sharp et al., 1995). With these developments, the minimum detail necessary to accurately describe experimentally relevant systems has come under debate. Proponents of highly detailed models have implied that such detail is necessary to predict phenomena accurately (Chen & Honig, 1997; Jayaram & Beveridge, 1996). Advocates of the "stripped down" (*e.g.*,

[12] Shkel *et al.* (2000) reviews several asymptotic approaches to the Poisson-Boltzmann equation.

[13] "Error" is defined in this case as the difference $\Gamma_{analyticalPB} - \Gamma_{NLPB}$ calculated for a cylindrical or spherical model.

cylindrical) approach (Anderson & Record, 1995) claim that simplified polyelectrolyte models capture the essence of thermodynamic phenomena under typical experimental conditions and often predict results with quantitative agreement to experiment. They point out that the lack of structural details available for some systems[14] may result in inconsistent modeling due to differences in the level of detail incorporated for the components in a given study.

To assess how inclusion of additional structural complexity might impact experimentally accessible predictions, calculations of Γ were compared for cylindrical and all-atom polyions in the presence of 0.1 M univalent salt (Sharp, 1995). Relative to a primitive cylinder model, a detailed 3-D model of DNA exhibited ~5% less counterion accumulation around the DNA due to the helical charge arrangement and groove structure. The reduced counterion accumulation resulted in a small but significant difference in the Donnan coefficient (Γ_{3D} - $\Gamma_{cyl} \cong$ -0.028 at 1 mM salt) which decreased $\lesssim$ 2-fold as salt increased in the 1 – 100 mM salt concentration range considered[15]. The salt dependence of Γ was otherwise comparable using either the cylindrical or structurally detailed DNA models. In a subsequent paper, NLPB calculations modeling the binding of DAPI with DNA as a cylinder-sphere interaction were compared with those employing an all-atom model for the complex (Sharp et al., 1995). The predicted salt dependence of binding, SK_{obs} (eq 20), from the two models were within 3% of each other (-1.9 *vs* -1.95 respectively[16]). More recently, comparison of free energy calculations of the cylindrical free energy with that calculated for the all-atom structural model of linear B-DNA found that that the cylindrical model is completely sufficient for 1:1 (monovalent) salt concentrations above 0.01 M (Shkel, 2010).

On the other hand, study of extreme phenomena such as highly complex nonperiodic geometries, large dynamic macromolecular structural changes, or very high salt concentrations, may require structural detail. For example, NLPB thermodynamic predictions of DNA operator binding by the λ repressor N-terminal domain (λbd) yielded significantly different results when comparing a cylinder-sphere model *vs* using the λbd-DNA complex crystal structure coordinates: SK_{obs} = -1.9 *vs* -4.7, respectively, while the experimentally measured SK_{obs} = -4.2. However, λbd is an ampholytic protein with 24 basic residues and 22 acidic residues, with most of the cationic amino acids near the DNA binding interface and the anionic residues on the other face (Sharp et al., 1995). A 10 Å sphere with a net +2 charge is a poor model for a protein with 46 charges and ignores, as the authors noted, the "polyelectrolyte" character of the protein. In another study, the non-uniform topology of DNA (*e.g.*, the presence of grooves) becomes important at very high salt or close proximity to the DNA surface (Montoro & Abascal, 1998). In perspective, the level of fine detail required in a model is dictated by the range of the phenomena that are to be studied. Whereas atomic detail in the model may be important in a computational study investigating the role of specifically bound Mg^{2+} in ribozyme catalysis or stability, for

---

[14] Biological systems lacking 3-D coordinates include ssDNA and the highly flexible regions of many biomolecules.

[15] This difference was determined via interpolation from Sharp's (1995a) Figure 3b. Finite-difference 3-D PB calculations on dsDNA predict $\Gamma_{3D} \cong$ -0.11 *versus* the primitive cylinder-based Γ_{cyl} = -0.083 at 1 mM *versus* $\Gamma_{3D} \cong$ -0.19 and Γ_{cyl} = -0.21 at 100 mM salt. By comparison, idealized (infinite dilution) Γ^{LL} = -0.069 (eq 9).

[16] The experimental values were SK_{obs} =-2 for DAPI-poly[d(AT)]$_2$ and -2.3 for DAPI-poly[d(GC)]$_2$ (Wilson et al., 1990).

denaturation studies and nonspecific binding events, the primitive models predict experimental results well within error. Primitive models offer great advantages when applied to appropriate biomolecules and processes: (i) results provide accurate characterization of phenomena, often using only a few, readily understood parameters; (ii) solutions are formulated analytically and are thus easy to use; (iii) and the predicted consequences offer a generalized framework which can be applied to other systems.

Limitations of Poisson-Boltzmann approaches. Standard implementations of PB neglect ion-ion correlations between salt ions, interactions between the ions and solvent, and consider ion size (*i.e.*, volume) only as a distance-limit of closest approach to the polyion surface. The solvent is usually described with a uniform dielectric continuum or with a parameterized expression for a varying, distant-dependent dielectric[17], in either case idealizing its molecular character. At lower salt ($\lesssim$ 0.01 M 1:1 salt), the calculated potentials near the polyion surface increase strongly, numerical error becomes significant, and the calculations become more time consuming. Numerous approaches addressing these issues with varying degrees of success have been developed and are the subject of several reviews (Anderson & Record, 1990; Anderson & Record, 1995; Grochowski & Trylska, 2008; Jayaram & Beveridge, 1996; Koehl, 2006; Record et al., 1998; Tan & Chen, 2009; Wu & Morikis, 2006). Finally, standard PB approaches rely on a static model of the polyion and thus do not consider polyelectrolyte dynamics (*e.g.*, if the polyion is a writhing single-stranded nucleic acid). The increasingly active efforts to describe polyion dynamics and their interactions with surrounding ion distributions are reviewed elsewhere (Prabhu, 2005; Slater et al., 2009; Viovy, 2000).

2.2.2 Counterion condensation theory

Counterion condensation theory (Manning, 1969; Manning, 1978) was originally derived within the context of a double limit: (i) the polymer is infinite in length and therefore end effects are ignored; and (ii) the concentrations of all ionic species are made infinitely small by dilution with water but that salt is in excess over the polyion charges. CC theory simplifies the continuous ion concentration gradients near the polyion by proposing the existence of a coion-free surface region where counterions are "condensed" or "territorially bound" in close proximity to the polyion backbone but still allowed to translate along it. Counterions with a valence z_c are predicted to condense if the reduced axial charge density ξ (eq 8) of the polyion exceeds the critical value $1/z_c$ (Manning, 1969). After condensation, the actual charge density of the polyelectrolyte is reduced to an effective charge density of $\xi_{eff} = 1$. Therefore, double stranded B-DNA, with $\xi = 4.2$, is predicted to exhibit significant counterion condensation.

Counterions which are not condensed are in the "diffuse ion atmosphere" where they experience a lower electrostatic potential governed by the Debye-Hückel approximation: for a given ion configuration, the electrostatic field of ion ensemble is simply the sum of the isolated electrostatic potentials associated with each ion. However, the ion atmosphere around each charged species, *i*, causes the potential to decay more sharply with distance, creating a "screened coulomb potential" analogous to that in linearized spherical PB theory (a.k.a., Debye-Hückel theory):

---

[17] Typically, a distance-dependent dielectric is invoked to take into account dielectric saturation of the solvent in close proximity to charges.

$$\varphi_i = \frac{e^{-\kappa r}}{r}. \tag{15}$$

In later CC approaches (Manning, 1977; Manning, 1978), this screened potential is assumed to describe the interactions of small ions with the polyion and its condensed counterions. At distances greater than the Debye length outside the volume enclosing the polyelectrolyte and its condensed ions, the polyion looks like a line of charges whose effective charge is reduced by ξ. At distances less than the Debye length, electrostatic interactions are not fully screened. In other words, ξ determines the onset and extent of counterion condensation, while the Debye length κ^{-1} (eq 10) defines the distance over which the ion distribution affects solution properties surrounding the polyelectrolyte.

Extended CC theory. From this theoretical framework, limiting law values for colligative properties such as preferential interaction coefficients (Γ), osmotic coefficients (ϕ), and activity coefficients (γ) are easily expressed in terms of ξ (Manning, 1969). Researchers found that limiting-law predictions of experimental values still appeared numerically accurate even at salt concentrations beyond the constraints in which CC theory was derived. In particular, the effective polyelectrolyte charge fraction, $1/(z_c\xi)$, was found to be relatively constant versus salt concentration in many experimental studies (see references in (Manning, 1977; Manning, 1978)). To explain this observation, Manning (1977, 1978) extended CC theory by assuming an analytical free energy expression, w, defined by the "charging" energy of the DNA phosphates and the ideal mixing free energy of the various species. The fraction of ions, θ, condensed around the polyion and the volume, V, which contains them are obtained by minimization of the free energy expression, w, with respect to θ. The volume depends on the polyion characteristics ξ and b, the valences of the salt ions, and only weakly on the salt concentration itself. The cylindrical condensation volume for B-DNA is predicted to be V = 720 ml/mol-phosphate, with an outer condensation radius of approximately 17 Å for a polyion radius of 10 Å (Manning, 1977). The fraction of condensed ions, $\theta = 1 - 1/(z_c\xi)$, contained within this volume is effectively considered to be a part of the polyion and is no longer electrophoretically or osmotically active. Therefore, the condensation of θ = 0.76 univalent counterions per dsDNA phosphate reduces the magnitude of the "effective" B-DNA charge to 0.24 per phosphate. The concentration of counterions in this condensed layer exceeds 1 M (on a per mol phosphate scale, 0.76 mol counterions in a 0.72 liter volume). CC theory was recently the subject of an extensive review (Spasic & Mohanty, 2008).

Counterion condensation theory has been used to describe a wide range of salt concentration-dependent nucleic acid phenomena because its simplified picture of polyelectrolyte systems has successfully predicted at least the qualitative details of polyion thermodynamic behavior (Manning, 1978). CC theory has been applied to Donnan equilibrium (Manning, 1969), osmotic coefficients (Manning, 1969), electrophoretic mobility (Sections 2.3.2 and 3.2.3), DNA denaturation (Sections 2.3.3, 3.2.4 and 3.2.5), and ligand binding (Sections 2.3.4 and 3.2.6). Manning and coworkers (Manning, 1977; Manning, 1978) have claimed that CC theory can be successfully applied with bulk monovalent salt concentrations approaching 1 M. However, several labs dispute the accuracy of this claim (Allison, 1994; Fenley et al., 2010; Record et al., 1998; Sharp et al., 1995; Stigter, 1995). With the increased availability of powerful computational resources, more sophisticated numerical methods such as NLPB (discussed above), and molecular simulation techniques such as Monte Carlo and molecular dynamics have become increasingly more prevalent.

2.2.3 Monte Carlo simulations

Monte Carlo (MC) methods determine an equilibrium distribution of component species by making a series of stochastic rearrangements to find the energetic minimum of the system. Both the "depth" and the "broadness" of the free energy minimum are important in determining the equilibrium condition. Monte Carlo simulations have significant flexibility in system modeling and require a minimum of assumptions:

i. simulations can be as large (*e.g.*, number of species in simulation) or complicated (*e.g.*, 3-D models, discrete solvent, etc.) as desired, limited only by the amount of time and computer resources available to the researcher;
ii. MC accounts for the ion-ion correlations ignored by Poisson-Boltzmann theory;
iii. electrostatic forces are calculated directly based on the coulombic potential of discrete charges, unlike counterion condensation theory which relies on significantly simplifying assumptions (Debye-Hückel approximation to describe mobile ions; counterion condensation hypothesis).

However, the error associated with calculated ion distributions scales with the number of species (*e.g.*, ions, water molecules, etc.) due to increased stochastic noise. Early MC simulations were typically limited to less than 0.1 M salt (Olmsted et al., 1989; Olmsted et al., 1991; Olmsted et al., 1995), although more recent simulations have modeled higher salt concentrations (*e.g.*, 0.2 M (Ni et al., 1999), and 4.5 M (Montoro & Abascal, 1998)). In general, the ratio of ions/polyion or waters/polyion is limited by the cell size of the calculation.

Monte Carlo simulations make a series of random changes to a given distribution of species. The decision to keep or reject the new configuration is based on the change in system energy

$$P = e^{-\frac{E_{new} - E_{old}}{kT}} , \qquad (16)$$

where E_{new} and E_{old} are the system energies of the new and old configurations respectively. If P is greater than 1, the new arrangement is kept and a new set of random moves is made. If the energy ratio P is less than 1, it is compared to a random number between 0 and 1. If P is greater than this number, the configuration is kept, otherwise it is rejected. In either case, another random move is performed. This randomized acceptance of higher energy arrangements allows the system to move out of local energy minima. Eventually, assuming "enough" iterations are executed, the ensemble average of the energy *vs* the number of steps stabilizes, reflecting a convergence to equilibrium.

Grand canonical Monte Carlo (GCMC) simulations require that the temperature, electrolyte activity, and volume of the system be specified before the simulation begins. After the number of small ions in the cell comes to equilibrium, the electrolyte concentration can be easily determined as a statistical average over fluctuations. Preferential interaction coefficients (Γ) can be extracted from these energy minimized ion distributions (Jayaram & Beveridge, 1991; Olmsted et al., 1989; Olmsted et al., 1991; Olmsted et al., 1995).

The earliest applications of GCMC simulations were used to determine activity coefficients for simple electrolytes. Since then, MC calculations have been used to determine mobile ion distributions (Montoro & Abascal, 1998; Ni et al., 1999), ion-size effects on ion-DNA interactions (Wang et al., 2007), thermodynamic coefficients (Jayaram & Beveridge, 1991; Korolev et al., 1998; Mills et al., 1986; Olmsted et al., 1991; Paulsen et al., 1987) and ligand binding (Ni et al., 1999; Olmsted et al., 1995). Because MC simulations are considered one of the more "realistic" computational methodologies available, they have been used for comparisons with experimental results and as a reference for alternative theoretical frameworks (*e.g.*, PB or CC).

2.2.4 Comparisons between PB, CC, and MC theories with experiment

All three theories can be applied to the same structural model of polyion and solution, using either a cylinder or full 3-D model for DNA, with either a uniform dielectric or variable local dielectric continuum for the solvent (*cf.*, Table 1). Many labs have made comparisons between the three theoretical frameworks and experimental data.

In general, CC theory agrees with PB calculations at very low salt concentrations (submillimolar) or where phenomena of interest are occurring far from the polyion. PB and CC determinations of the Donnan coefficient agree with one another for infinitely dilute polyions in the presence of excess salt ($\kappa a \lesssim 10^{-3}$, or ≈100 nM univalent salt) (Anderson & Record, 1980). Analytical expressions of polyion colligative properties (*e.g.*, preferential interaction coefficients, osmotic coefficients, and activity coefficients) derived from PB theory were found to be in agreement with counterion condensation theory at low but finite salt concentrations (Klein et al., 1981). Aspects of CC phenomena predicted by Manning are seen in PB calculations for the "no added salt" limit and appear to hold true even in the presence of added salt (Le Bret & Zimm, 1984). CC and PB theory agree at radial distances greater than a Debye length from the polyion (*e.g.*, cation concentrations are indistinguishable at 60 Å at 0.01 M 1:1 salt with a = 5 Å) (Stigter, 1995). CC theory is typically effective in predicting behavior of properties which depend on long range ion-polyion interactions, such as the salt distribution observed in a Donnan equilibrium experiment. CC theory also works reasonably well for polyion processes such as nonspecific ligand binding and especially its salt dependence, both of which rely more on the overall electroneutrality of the ion distribution than the specifics of its spatial organization. Myriad examples of CC analysis applied to binding interactions are described in reviews (Anderson & Record, 1982; Manning, 1978; Spasic & Mohanty, 2008).

On the other hand, CC predictions disagree with those of PB and MC at higher salt concentrations in numerous studies which compared CC to other computational and/or experimental results (Anderson & Record, 1983; Anderson & Record, 1995; Fenley et al., 2010; Korolev et al., 1998; Stigter, 1995). For example, the number of counterions (β) within an annular shell a distance Δ from the DNA cylinder surface were compared from PB and CC calculations for $[Na^+]/[P]$ concentration ratios comparable to those typically used in NMR studies ($1.2 \leq [Na^+]/[P] \leq 9.0$) (Mills et al., 1986). Mills *et al.* found that CC theory reasonably approximates the magnitude of β within the condensation layer but fails to reproduce the increase of β with increasing salt. On the other hand, even though the lack of ion-ion correlation effects in PB calculations caused a systematically lower β prediction relative to MC, the difference ($\beta^{MC(\Delta)}$ - $\beta^{PB(\Delta)}$) was relatively independent of salt concentration. As such, PB mirrored MC-predicted distance-dependent changes in ion distributions. Murthy and coworkers (Murthy et al., 1985) noted that PB underestimates the local Na^+ concentration around an infinite polyion by about 15% relative to MC and hypernetted chain (HNC) calculations done between 1 - 10 mM NaCl. However, PB, HNC, and MC predictions become indistinguishable at long distances from the DNA (~120 Å at 10 mM and ~360 Å at 1 mM NaCl) (Murthy et al., 1985).

Consideration	CC theory	PB calculation	MC simulation
Ions	considered either condensed or free; no explicit consideration of identity	no ion-ion interactions; ion radius only considered as part of polyelectrolyte cylinder radius, *a*	ions have volume; ion-ion correlations; discrete ion electrostatics
Solvent	constant dielectric continuum	constant dielectric; distance-dependent dielectric field	constant or varying dielectric; explicit water molecules
Polyion model	cylinder; line of charges	cylinder; 3-D model	cylinder; 3-D model
Electrostatic forces	linearized Debye-Hückel; minimization of free energy	continuum electrostatics (explicit charges for polyion and ligands, charge density for mobile ions) (Shkel, Trizac)	coulombic (discrete charges)
Salt concentration range	exact at 0 M univalent salt; < 1 mM salt verified; some argument for < 1 M	≤ 0.1 M theoretical limit (Fixman); ≤1 M univalent salt experimentally validated (Bond, Korolev)	< 0.1 - 0.2 M common; higher concentrations available (e.g., 4.5 M univalent (Montoro))
Approach to thermodynamic output	predict from simple picture of ion-polyion interactions; system energy is known	find potential field φ; calculate ion distribution; get thermodynamics via Γ	yields ion distribution directly; derive thermodynamics from Γ
Ease of use	very easy	moderate	moderate
Advantages	explicit analytical expression via basic parameters (ξ, κ)	well-defined distribution law; well-characterized model	Uses minimal assumptions; allows arbitrary complexity; provides estimation of error
Caveats	coarse-grained description, potentially large discrepancy for near-field phenomena	large errors when electric fields are high (*e.g.*, at high [salt]); ignores ion correlations (an issue for multivalent ions)	computationally demanding; stochastic noise increases with increasing [salt] and number of solvent molecules
Linkage to experiment	colligative properties (Γ, γ, ϕ); NA thermal denaturation; ligand binding; electrophoretic mobility	colligative properties (Γ, γ, ϕ); NA thermal denaturation; ligand binding	colligative properties (Γ, γ, ϕ); NA thermal denaturation; ligand binding

Table 1. Feature comparison of polyelectrolyte theoretical descriptions

PB and MC estimations of measurable parameters commonly agree with experiment, especially when solution conditions are amenable to PB assumptions. Paulsen *et al.* (1987) found that Γ values predicted by PB and MC calculations agree with experimental Donnan coefficients measured between 1 - 31 mM DNA phosphate and 0.2 - 111 mM NaCl.

Similarly, PB and MC approaches yielded equivalent results for Γ and mean ionic activity coefficients for a primitive DNA model in the presence of 2 - 36 mM 1:1 salt (Mills et al., 1986). Using primitive cylinders to describe polyions and lattice statistics to model flexible ligands (such as oligopeptides), NLPB calculations made accurate predictions of measured equilibrium binding constants and their salt dependences for a range of binding processes including: (i) Mg^{2+} to polynucleotides, (ii) $Co(NH_3)_6^{3+}$ to calf-thymus dsDNA, (iii) polyamines to T7 DNA, (iv) oligolysines to polyU and polyA, and (v) tripeptides to heparin, a highly charged polysaccharide (Stigter & Dill, 1996). MC and PB predictions were in excellent agreement for the change in electrostatic free energy for duplex DNA helix-coil transitions across a range of NaCl and $MgCl_2$ concentrations (Korolev et al., 1998).
However, omission of ion-ion correlations significantly decreases the predicted accumulation of multivalent cations near the polyion surface. Comparisons of PB vs MC found ~20 - 30% discrepancies in predicted local ion concentrations and Γ for systems containing divalent cations (Ni et al., 1999; Pack et al., 1999). Differences between PB and MC determinations of Γ_{2+} for polyelectrolyte solutions containing single or mixed 1:1 and 2:1 salts were attributable to a dramatically increased local cation accumulation within 3 Å of the dsDNA cylinder surface (*Fig.* 2). The net difference $\Gamma_{2+}^{MC} - \Gamma_{2+}^{PB}$ is relatively invariant across the ~10 - 200 mM univalent salt concentration range studied (Ni et al., 1999), consistent with the roughly invariant $\beta^{MC}(\Delta) - \beta^{PB}(\Delta)$ discussed previously (Mills et al., 1986). Fortunately, the relatively constant differences between predicted ion distributions cancel out for thermodynamic processes that are described by stoichiometric-weighted changes in Γ. This may explain the remarkably accurate NLPB thermodynamic predictions of diverse processes such as RNA folding (Draper, 2008), nucleic acid thermal denaturation (Shkel & Record, 2004), and ligand binding to NAs (Anderson & Record, 1995; Misra & Draper, 1999; Record et al., 1998; Stigter & Dill, 1996).

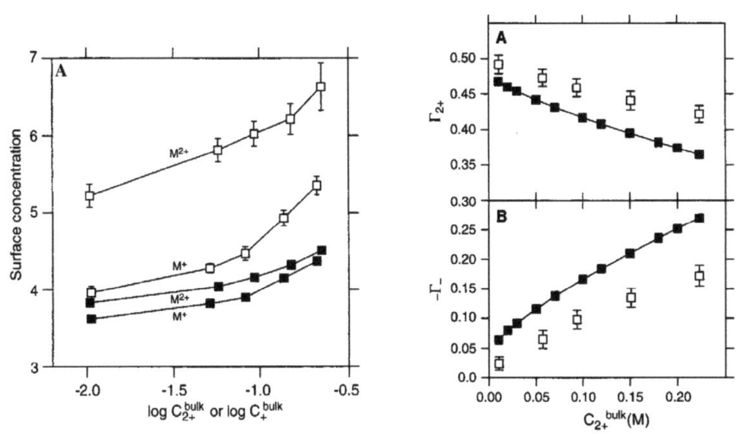

Fig. 2. PB (□) and MC (■) predictions of ion surface concentration (*left*) or preferential interaction coefficients Γ_i (*i* = -, 2+) (*right*) for 2:1 salt–dsDNA interactions. (Ni et al., 1999)

Nevertheless, emerging interest in phenomena very close to polyion surfaces has spurred development of modified NLPB approaches to account for neglected interactions that become important in high electric fields and large ion concentrations (Grochowski & Trylska, 2008; Tan & Chen, 2009; Wong & Pollack, 2010). Alternatives include PB expressions incorporating excluded volume effects (Bai et al., 2007; Gruziel et al., 2008) and/or ion correlations (Chu et al., 2008; Ramanathan & Woodbury, 1982), and a hybrid model which explicitly considers ions near the polyelectrolyte surface but uses a continuum description further away (Tan & Chen, 2009).

In summary, CC theory works well at very low salt concentrations ($\lesssim$ 1 mM) and in cases where the details of the ion concentration gradient do not affect behavior but breaks down if these conditions are violated[18]. On the other hand, comparisons of PB calculations to experiments and MC simulations confirm that NLPB approaches effectively and often quantitatively describe systems with up to 1 M univalent salt concentrations (Montoro & Abascal, 1998; Ni et al., 1999). Although differences are evident in ion distributions very close to the polyion, overall determinations of Γ and any closely related thermodynamic phenomena have been shown to be very reliable. At very high salt concentrations (*e.g.*, $\gtrsim$ 2.5 M) or at distances very close to the polyion surface (*i.e.*, $\lesssim$ 3Å for dsDNA), differences between PB and MC calculations become quantitatively significant (Montoro & Abascal, 1998; Ni et al., 1999) at least partially as a result of the greater importance of ion-ion correlations which are present in MC but not in PB.

2.3 Polynucleotide experimental studies

Experimental studies defining polyelectrolyte behavior have predominantly focused on single stranded and double stranded nucleic acid. This section will highlight seminal studies which characterized archetypal polyelectrolyte phenomena: the linear cation NMR relaxation rates in the presence of polynucleotides, the invariance of free solution electrophoretic mobilities with respect to length of high molecular weight NAs, the linear log-salt dependence of thermal denaturation of duplex DNA, and the large salt dependences seen for ligand-nucleic acid interactions.

2.3.1 Small ion NMR of polymeric nucleic acid solutions

^{23}Na NMR has been used to characterize ion-polynucleotide interactions with sonication-derived polydisperse length distributions approximately 700 bp long and various monodisperse DNA samples ranging from 125 - 165 bp in size (Anderson & Record, 1990). ^{23}Na longitudinal and transverse relaxation rates (R_{obs}) are independent of NA sequence and much larger in "salt-free" DNA-containing samples than in otherwise comparable solutions lacking DNA (Braunlin, 1995). The extent of Na^+ association is temperature independent within the 6 - 33 °C range tested, as would be expected if accumulation was due to exclusively electrostatic interactions (Bleam et al., 1983). R_{obs} has been described in terms of a two-state model in which sodium ions are considered either "bound" to the DNA or "free." Bound Na^+ ions rapidly exchange with free Na^+ on the millisecond NMR-timescale and are defined as those ions close enough to DNA that the local electric field gradients perturb $^{23}Na^+$ relaxation rates (R_B). The "free" state includes all other species which possess an average relaxation rate (R_{free}) indistinguishable from R_{obs} for an otherwise comparable DNA-free solution.

---

[18] For example, Korolev (1998) compared PB and CC predictions with experimental data on the influence of Mg^{2+} on dsDNA thermal denaturation. In this study, CC predicted that Mg^{2+} has a larger nonelectrostatic interaction with ssDNA than dsDNA, the opposite of what was found in PB calculations and in seeming contrast to Raman spectroscopy of DNA which showed only weak differences in the spectra of denatured DNA in the presence *vs* absence of Mg^{2+}.

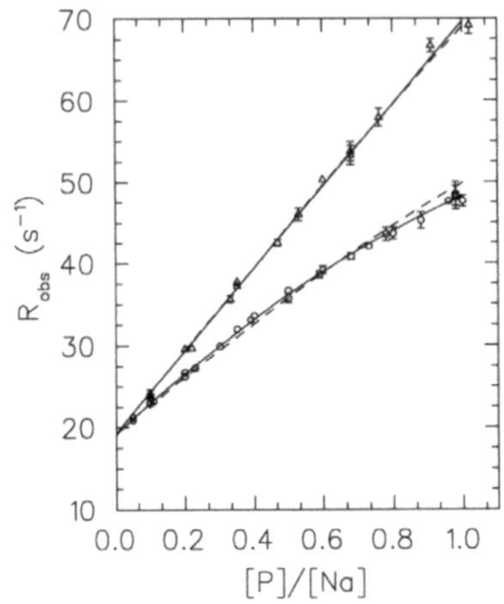

Fig. 3. Two-state analysis of ^{23}Na longitudinal relaxation rates for 160 bp dsDNA (Δ) and 20 bp dsDNA (O) titrated with Na^+. Solid lines are empirical linear and quadratic fits, respectively, and the dashed lines represent the global fit of 20 bp and 160 bp data to $R_{obs} = R_{free} + \alpha C^+_{|Z|}(a)$ [P]/[Na^+], with $\alpha = 28$ and $C^+_{|Z|}(a)$ obtained from GCMC simulations. (Stein et al., 1995)

In general, polymeric DNA has a linear relaxation rate with respect to the concentration ratio of DNA phosphate *vs* total Na^+ within $1 \lesssim [Na^+]/[P] \lesssim 20$ (Stein et al., 1995). This ratio spans DNA phosphate concentrations between 2.5 - 15.1 mM and [Na^+] between 0.003 - 1.3 M. In *Fig.* 3, the y-intercept (*i.e.*, $[Na^+] \rightarrow \infty$ as $[P]/[Na^+] \rightarrow 0$) in a R_{obs} *vs* [P]/[Na^+] plot corresponds to R_{free} (Bleam et al., 1980). At 1:1 salt conditions (no added salt), $R_{obs} = 69 \pm 1$ s^{-1} is ≈3.6-fold greater than R_{free} (19 ± 1 s^{-1}). By contrast, 20 bp dsDNA has a reduced and significantly nonlinear R_{obs} over the same [P]/[Na^+] range (see Section 3.2.1 and *Fig.* 3). R_{obs} *vs* [P]/[Na^+] data for short and long dsDNA was described with a single fitting parameter, α, weighted by $C^+_{|Z|}(a)$, the axially averaged local Na^+ concentration predicted by GCMC for an appropriately charged cylindrical dsDNA model (*Fig.* 3, *dashed lines*). Interpretations of NMR-measured cation-polyion association data and its comparison to polyelectrolyte theories (*e.g.*, CC, PB, and MC calculations) are reviewed elsewhere (Anderson & Record, 1990; Braunlin, 1995).

2.3.2 Electrophoretic mobility of polymeric nucleic acids

The electrophoretic mobility (μ) of a given nucleic acid depends on the ratio of its total effective charge (q_{eff}) and the frictional coefficient (f) it experiences while moving through the electrophoretic medium,

$$\mu = \frac{q_{eff}}{f}. \tag{17}$$

The independence of electrophoretic mobility on DNA length for long (> 400 bp) DNA strands is well documented in the literature (Olivera et al., 1964; Stellwagen et al., 1997). For example, at 0.01 M NaCl, μ was invariant with DNA length for molecular weights between 1 - 130 MDa (1.5 - 200 kbp) (Olivera et al., 1964). In a series of experiments spanning

0.0001 - 3 M NaCl, poly(styrene sulfonate) strands possessing ~150 - 4500 monomer units exhibited constant μ at fixed [NaCl] with respect to polymer length in capillary electrophoresis or electrophoretic light scattering experiments (Hoagland et al., 1999). Therefore, the independence of μ on DNA length requires that changes in the per monomer effective charge compensate any variation in the per monomer frictional coefficient for long DNAs. *Fig.* 4A demonstrates the length independence of μ for long NAs. Experimental and theoretical studies of polyion electrophoresis phenomena have been reviewed extensively (Allison et al., 2007; Hoagland et al., 1999; Slater et al., 2009; Slater et al., 2002; Viovy, 2000).

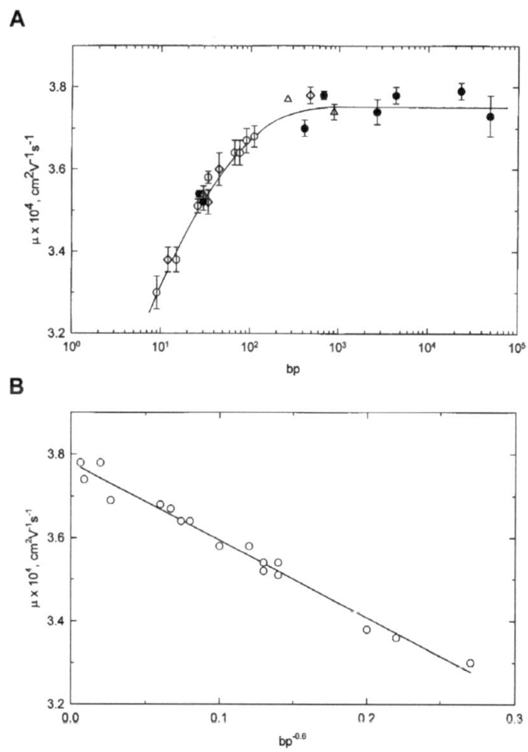

Fig. 4. *A*. dsDNA electrophoretic mobility (μ) measured by capillary electrophoresis of monodisperse DNA (●) and plasmid restriction digest fragments (O, *Msp*I of pBR322; ◇, *HpH*I of pBR322; △, *Taq*I of pUC19) as a logarithmic function of DNA size (in base pairs). The horizontal line indicates the average mobility of long DNA fragments; the curved line is a guide for the eye. *B*. Low molecular weight fragments from (*A*) re-plotted versus the -0.6-power of dsDNA length (in bp). (Stellwagen et al., 1997)

2.2.3 Polymeric melting studies

In general, DNA thermal stability increases with increasing 1:1 salt concentration up to ~1 M. For < 0.2 M salt, the T_m (the midpoint temperature of the thermal transition, where $K_{obs} = 1$) is linear in log[salt]. T_m increases 10–20 °C per decade increase in univalent salt concentration. This behavior is exhibited by both natural and synthetic DNA duplexes and appears to be primarily a coulombic effect. The salt dependence of T_m can be quantitatively interpreted by

$$\frac{\partial T_m}{\partial \ln a_\pm} = -\frac{2RT_m^2}{\Delta H_m^o}\Delta\Gamma_u, \tag{18}$$

where R is the gas constant and ΔH_m^o is the enthalpic change per nucleotide for the melting transition. Polyelectrolyte Poisson-Boltzmann calculations find that the per-nucleotide $\Delta\Gamma_u$ of nucleic acid transitions has a broad maximum around 0.01 - 0.2 M salt, explaining the relatively constant behavior of $\partial T_m/\partial \ln a_\pm$ (Bond et al., 1994; Shkel et al., 2002). NA order-disorder transitions are fully described using experimental ΔH_m^o and T_m values with $\Delta\Gamma$ determined by NLPB on primitive charged cylinders (Fig. 5).

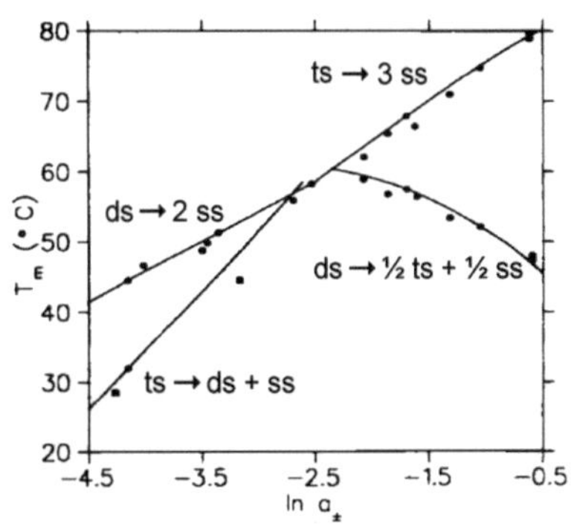

Fig. 5. The dependence of NA melting temperatures, T_m, on the logarithm of NaCl activity ($\ln a_\pm$). Symbols denote calorimetric or spectroscopic measurements, while solid lines indicate predictions (eq 18) from NLPB thermodynamic analysis. (Bond et al., 1994)

Researchers have used DNA melting studies to assess oligomer thermodynamics relative to those of polynucleotides. One common approach compared experimental $\partial T_m/\partial \ln a_\pm$ values for a given oligonucleotide relative to that predicted for a polymer with the same fractional GC content. As background for later discussion, some of the empirical $\partial T_m/\partial \ln a_\pm$ models are surveyed here.

Empirical predictive models for the salt dependence of polymeric nucleic acid denaturation. $\partial T_m/\partial \ln a_\pm$ increases linearly with increasing percentages of AT base pairs from 11 - 13 °C at 0%AT to 18 - 20 °C at 100%AT (Blake & Haydock, 1979). The reason behind the GC dependence of $\partial T_m/\partial \ln a_\pm$ is not understood, but it is well characterized. Early estimations of $\partial T_m/\partial \ln a_\pm$ relied on empirically established linearity between $\partial T_m/\partial \ln a_\pm$ and f_{GC}, the fraction of GC base pairs in the DNA duplex, f_{GC} (Blake & Haydock, 1979; Frank-Kamenetskii, 1971). These simple linear relationships are relatively successful, typically predicting $\partial T_m/\partial \ln a_\pm$ within ~10% error. However, nearest-neighbor based calculations (Blake et al., 1999; Breslauer, 1986; Delcourt & Blake, 1991) correlate better with experiment and have become the *de facto* standard in many arenas. For example, the salt dependence of thermal denaturation for a d(AT) homopolymer is underestimated by 20% using the method of Frank-Kamenetskii and by 10% using Blake and Haydock's formula but is predicted within 5% of the experimental value by the Blake (1999) nearest neighbor approach. In general, Frank-Kamenetskii, Blake and Haydock, and Blake *et al.* all see approximately a 30% difference in $\partial T_m/\partial \ln a_\pm$ for the predicted all-d(AT) and all-d(GC) extremes. These sequence-dependent methods which predict $\partial T_m/\partial \ln a_\pm$ were mostly developed using long NAs where the thermodynamic consequences of coulombic end effects are expected to be small. Comparisons of the measured oligonucleotide $\partial T_m/\partial \ln a_\pm$ with those predicted by methods described above have led some researchers to suggest that oligomers behave similarly to polymeric NAs (Williams et al., 1989). These comparisons along with other oligonucleotide duplex melting studies will be discussed in greater detail in Section 3.2.5.

2.3.4 Binding of small ligands to polymeric nucleic acids

Polyelectrolyte behavior can also be probed via polyion binding processes with small ligands (including oligocations). Below is a brief summary of some archetypal DNA/RNA-ligand binding reactions described in the literature. Reviews covering a wider array of ligand-nucleic association phenomena are available (Anderson & Record, 1995; Lohman & Mascotti, 1992; Manning, 1978; Record et al., 1978; Record et al., 1998).

Simply speaking, when a cationic ligand with charge Z_L binds to a NA in a solution containing only monovalent salt, counterions are released into the bulk solution providing an entropic driving force for ligand-NA complex formation (Record et al., 1976; Record et al., 1978). Enthalpies observed upon NA binding by oligolysines and polyamines are small and salt-independent, consistent for an entirely entropic driving force (Braunlin et al., 1982; Lohman et al., 1980; Lohman & Mascotti, 1992). The K_{obs} for binding is dependent on the mean ionic activity $a_\pm$ of monovalent salt MX in solution,

$$S_a K_{obs} = \frac{\partial \log K_{obs}}{\partial \log a_\pm} = \Delta(|Z| + 2\Gamma)\,, \tag{19}$$

where "ΔX" is the stoichiometrically-weighted difference between products and reactants (Anderson & Record, 1993; Record et al., 1998). For the process described in eq 1 between DNA (D), ligand (L), and complex (LD), $\Delta X = X_{LD} - X_D - X_L$. It is typically assumed that the binding of the ligand to DNA effectively neutralizes Z_L charges of the DNA (Olmsted et al., 1995; Record et al., 1978).

While eq 19 should be used in general, a highly approximated expression has been found to give good results for an enormous range of systems. Using the limiting law approximations of very low salt, treating the charge of the complex as $|Z_{LD}| = |Z_D| - |Z_L|$, assuming that $\Gamma_{LD} = \Gamma_D$ and that ion release only occurs from the polymer, eq 19 simplifies to

$$\left(\frac{\partial \log K_{obs}}{\partial \log[MX]}\right)^{LL}_{T,pH} = -Z_L(1 + 2\Gamma^{LL}_D) = -Z_L\psi\,, \tag{20}$$

where $\Gamma^{LL}_D = -\frac{1}{4\xi_D}$ (eq 9) and therefore $\psi = (1 - 0.5\xi_D)$. The original derivation of eq 20 (Record et al., 1976) relied on CC theory (Manning, 1969). Subsequent NLPB evaluations of ψ performed under limiting law conditions yield equivalent results (See Sections 2.2.1 and 2.2.4). dsDNA is thus predicted to have ${}^{ds}\psi = 0.88$ (Record et al., 1976). Similar values of SK_{obs}/Z_L are found regardless of whether the ligand is $Co(NH_3)_6^{3+}$ (Plum & Bloomfield, 1988), polyamines (Braunlin et al., 1982), oligolysines (Lohman et al., 1980; Record et al., 1976), or even proteins such as *lac* repressor (deHaseth et al., 1977) or RNA polymerase (deHaseth et al., 1978) for their nonspecific binding to nonoperator DNA. SK_{obs}/Z_L has shown variation in some cases, especially when complex formation involves a DNA conformational change. For example, intercalative binding processes typically exhibit reduced SK_{obs}/Z_L ratios, as highlighted by the drug sanguinarine which binds to polymeric dsDNA with SK_{obs}/Z_L ranging between 0.40 - 0.55 (Sen et al., 1996).

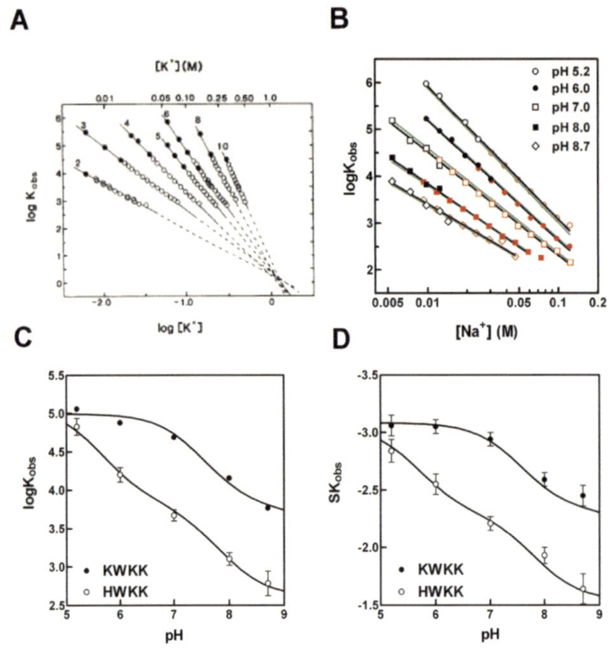

Fig. 6. Cationic peptide binding to polymeric nucleic acids is quantitatively described by limiting-law polyelectrolyte theory. (*A*) Poly(U) binding affinity of oligolysines with charge Z_L = +2 - +10 (as indicated above) exhibits a linear dependence on log[K^+] (Mascotti & Lohman, 1990). (*B*) The salt dependence of poly(U) binding affinity for oligopeptide containing two titratable groups (HWKK) decreases as pH increases from 5.2 - 8.7 (Ballin et al., 2010). (*C*) Poly(U)-XWKK binding constants (where X = histidine (H) or lysine (K)) and their salt dependence (*D*) are accurately predicted via eq 20 by accounting for peptide deprotonation with increasing pH. (Ballin et al., 2010)

The general applicability of eq 20 appears to be extensible to processes involving other polyions and/or ligands which have titratable charges. Extensive binding studies with a series of oligolysine (*Fig.* 6A) and oligoarginine peptides (Z_L = 2 - 10) with various ssDNA and ssRNA NAs found that eq 20 with $^{ss}\psi$ = 0.71 (Mascotti & Lohman, 1990) accounts for the salt dependence of these systems at low salt concentrations at low ligand charge (Z_L = 2 - 4). The full dataset is well fitted if the limiting law expression is augmented to take into account hydration changes upon binding (Mascotti & Lohman, 1990; Mascotti & Lohman, 1993). The linearity of $\log K_{obs}$ *vs* log[salt] has been reported for ligands binding to other polyelectrolytes with varying axial charge densities, ξ, including the polyanionic polysacharride heparin (Mascotti & Lohman, 1995) and polyA•$(polyU)_2$ triple stranded RNA (Record et al., 1976). Eq 20 quantitatively describes binding affinity as a function of salt concentration and pH changes (*Fig.* 6B - D) when ligand deprotonation equilibria are incorporated into the calculation of Z_L (Ballin et al., 2010; Lohman et al., 1980). Together, these studies show that polyelectrolyte theory effectively describes the electrostatic contributions to binding equilibria and that these contributions are predominantly dictated by gross physical properties such as the net charge and its spatial distribution.

3. Oligonucleotide studies

Residues within the "interior" of a polynucleotide comprise the vast majority of the NA length such that the ends contribute negligibly to the average characteristics of the polyion. Do short NAs whose ends comprise a significant fraction of their oligoion length behave any differently than their polymeric analogs? Below is a brief overview of experimental and theoretical research directly relevant to the thermodynamics of short *vs* long oligoions. This section seeks to address whether there is a coulombic end effect that differentiates the termini of any NA strand from its interior when interacting with other charged species. If such a phenomenon existed, the average per charge coulombic properties of oligoions would vary with Z_D, the total charge of the oligonucleotide until some critical threshold length was attained. The discussion of relevant theoretical work will be followed by a synopsis of experimental efforts to characterize nucleic acid properties that would be sensitive to coulombic end effects: ^{23}Na NMR relaxation rates, electrophoretic mobility, NA melting and NA-ligand interactions. Recent attempts to visualize ion distributions directly by small x-ray scattering will also be discussed.

3.1 Oligoelectrolyte theory

The evolution of oligoelectrolyte theory was similar to that of polyelectrolytes in both chronology and scope. In most cases, existing polyelectrolyte theory was adapted to account for perturbations observed in oligoelectrolyte thermodynamics. For ease of presentation, the oligoelectrolyte theory will be summarized in the following order: 1) counterion condensation theory, 2) Monte Carlo, and then 3) PB. The concepts presented here provide a context for the oligoelectrolyte experimental results presented in Section 3.2.

3.1.1 Counterion condensation and related approaches

In what may have been the first theoretical description of salt effects on an oligoion process, Elson and coworkers modeled thermal denaturation data for a series of intramolecular hairpins by calculating and then summing the discrete pairwise interactions between screened phosphate charges (Elson et al., 1970). The pairwise interactions were determined via the method of Schildkraut and Lifson which uses the linearized PB equation to estimate the screened electrostatic potential as

$$\varphi_{ij} = \frac{z_p e}{\varepsilon r_{ij}} e^{-\kappa r_{ij}} \tag{21}$$

for two phosphate charges i and j separated by a distance r_{ij} with z_p as the effective charge of the phosphates. These summed interactions were used to determine the free energy change of the helix-coil transition and to predict the experimental data using (z_p/ε) as the only fitting factor. Additional details of Elson's study and the subsequent analytical modeling of their data by Record *et al.* (1978) will be discussed further in the context of the experiments that served as their inspiration (See Section 3.2.4).

Ramanathan and Woodbury (1982) considered the BBGKY hierarchy of equations near the PB limit of a uniformly surface-charged cylinder in a continuum dielectric. The authors proved that counterion condensation occurs if ξ surpasses a critical threshold for a charged rod whose length, L, is much larger than its radius, a, and if this length is also comparable to or larger than the Debye length. The $\xi_{critical}$ is independent of salt concentration if $L \gtrsim \kappa^{-1}$. As the oligoion length decreases, the volume of the condensed layer increases, maintaining the number of ions condensed. When $L < \kappa^{-1}$, $\xi_{critical}$ varied with length and salt concentration (Ramanathan & Woodbury, 1982).
This work inspired Manning and Mohanty (1997) to adapt extended CC theory to finite polyions in the presence of monovalent salt. As discussed in Section 2.2.2, θ represents the average fraction of ions in solution which condense around a polyelectrolyte (*e.g.*, DNA) if $\xi > 1$ for a univalent salt. Manning and Mohanty reported that as the length of the polymer is traversed from the center, θ maintains the idealized (*i.e.*, polymeric) value $\theta = 1 - 1/\xi$ up to approximately κ^{-1} from the terminus. Thereafter, $\theta(s)$ continues to decrease as the end of the polymer is approached by

$$\theta(s) \approx 1 - \frac{\xi_{crit}(s)}{\xi}, \xi > \xi_{crit} \tag{22}$$

where

$$\xi_{crit}(s) = \frac{\ln(\kappa^{-1}/b)}{\ln(s/b)}, L << \kappa^{-1}, \tag{23}$$

and s is the contour distance from the terminus of the oligomer (Manning & Mohanty, 1997). In other words, the interior of an oligomer behaves just like that of a polymer for distances more than κ^{-1} from the oligoion terminus. Oligomer (or polymer) regions within κ^{-1} of the terminus will show decreasing extents of ion condensation as the end is approached. The dependence of θ on the spatial proximity to the oligoion terminus reflects a coulombic end effect. For sufficiently short oligoions, a coulombic end effect would also be evidenced as a decrease in θ with decreasing oligomer length L since the termini are always within a critical range. Indeed, the CC model predicts that oligoelectrolytes shorter than the Debye length (*i.e.*, $L < \kappa^{-1}$) will exhibit less counterion condensation than a polyion and that, unlike a polyelectrolyte, the threshold for condensation will depend on both the oligoion length and the solution salt concentration (Manning & Mohanty, 1997). In essence, κ^{-1} determines the magnitude and onset of the coulombic end effect expected by CC theory.
Fenley *et al.* (1990) evaluated CC theory predictions for a finite line charge at intermediate salt concentrations (< 0.1 M univalent salt). They found coulombic end effects at low salt which were exhibited as decreased counterion condensation relative to the polymeric limit for regions within approximately κ^{-1} of the line charge termini. To put this into context, a 20 bp dsDNA oligomer has a Debye length of 9.6 Å at 25 °C and 0.1 M univalent salt. According to the Fenley (1990) analysis, a 20 bp dsDNA will exhibit completely polymeric behavior because the DNA oligomer is approximately seven times longer than the Debye length: the predicted average fraction of condensed ions per DNA charge ($\theta = 0.740$) is close to the theoretical limit for a polyelectrolyte ($\theta = 0.764$). If instead of a finite line charge, the polyelectrolyte was modeled as a double-helical array of charges using B-DNA coordinates, the predicted fraction of condensed ions was salt-dependent, in direct conflict with the

experimentally observed salt-invariance of θ. The salt-independence of θ was restored by using an empirical distance-dependent dielectric function near the DNA surface (Fenley et al., 1990). In any case, CC theory predicts that the θ sensitivity to coulombic end effects is insignificantly small under typical experimental conditions[19]. Because θ is directly linked to experimentally measurable thermodynamic quantities, the effective equivalency of θ leads CC theory to predict that there will be no measurable coulombic end effects in NA properties or processes under typical experimentally accessible salt concentrations. These findings will be revisited in Sections 3.1.2 and 3.1.3.

3.1.2 Oligonucleotide Monte Carlo simulations

In one of the first grand canonical Monte Carlo (GCMC) studies of a finite oligoanion, B-DNA (4 - 50 bp) was represented as a right circular cylinder with 3 Å cylindrical end caps in a constant dielectric continuum at low ($a_{\pm}$ = 1.76 mM) monovalent salt (Olmsted et al., 1989). These calculations predicted a significant coulombic end effect: the surface counterion concentration increased linearly with axial position to a distance of approximately 31 Å from the ends. Beyond this distance, the interior showed a polymeric extent of counterion accumulation which was invariant with axial position. Oligomers shorter than ≈24 bp did not attain the maximal surface counterion concentration seen at the interior of polymeric DNA. These findings are in contrast to CC predictions which state that coulombic end effects should only become apparent when the Debye length is *significantly larger* than the oligomer length: even at the low salt concentration (≈2 mM) used in the Olmsted GCMC study, the Debye length κ^{-1} is only ≈2.3-fold greater than the range of the coulombic end effect they observed.

Olmsted's GCMC studies of "primitive" model DNA show that the local cation concentration in the vicinity of the polyion is much smaller at the termini than at positions in the interior (*cf., Fig.* 7A). The result is a roughly inverted parabolic ion distribution which can be approximated as a trapezoid for longer lengths (Olmsted et al., 1989; Olmsted et al., 1991; Olmsted et al., 1995). Increases in NA size beyond a critical length simply increase the length of the plateau region in the interior (*Fig.* 7A-B). If there were no coulombic end effect, the local ion concentration would be relatively invariant over the entire length of the DNA, with little difference at the ends *vs* the center of the polyion.

For B-DNA >10 bp, $\Gamma_{o,u}$ approaches the polymer limit linearly with $1/N$, a relation which holds for the entire 10 - 200 mM univalent salt concentration range considered (Olmsted et al., 1991; Shkel & Record, 2004). In other words, as the DNA strand length is increased, the average counterion accumulation per NA phosphate increases with increasing N and approaches the polymeric limit. The magnitude of the slope for $\Gamma_{o,u}$ *vs* $1/N$ increases with decreasing bulk salt concentration, indicating that larger bulk salt concentrations reduce the length dependence of ion accumulation around DNA (Olmsted et al., 1991; Shkel & Record, 2004). Olmsted *et al.* (1991) used their results to calculate $\Delta\Gamma$ values for DNA thermal denaturation which were in excellent agreement with experimental determinations.

---

[19] The predicted θ equivalency for oligomers *vs* polymers cannot be used to extrapolate comparisons of $\Gamma_{oligoion}$ *vs* $\Gamma_{polyion}$. Consider, for example, the difference between θ and $\Gamma_{o,u}$. θ represents the fraction of counterions condensed around the polyelectrolyte, whereas $\Gamma_{o,u}$ is representative of all of the counterions in solution.

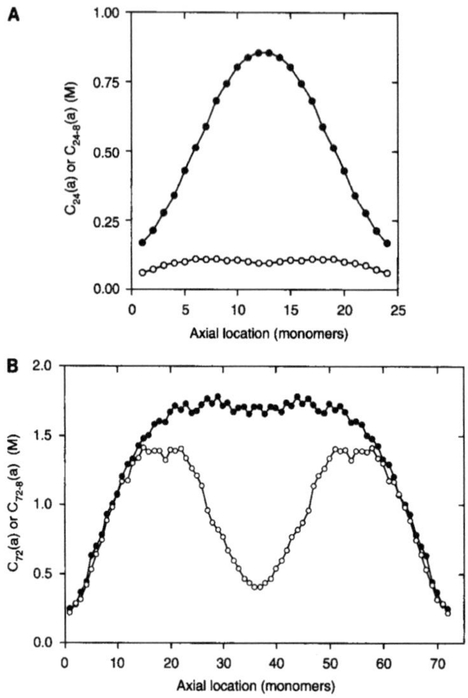

Fig. 7. GCMC predictions of surface counterion concentration $C_{|Z|}(a)$ as a function of axial location (in monomer units) along cylindrical models of uncomplexed and complexed DNA oligomers, the latter with 8 central charges eliminated to simulate binding of a ligand with Z_L = 8+. (A) 24-mer: ●, uncomplexed ($|Z_D|$ = 24); O, complexed ($|Z_D|$ = 24-8); (B) 72-mer: ●, uncomplexed ($|Z_D|$ = 72); O, complexed ($|Z_D|$ = 72-8). The axial location of the terminal charge is at position 1. (Olmsted et al., 1995)

Foreshadowing ligand-binding experiments (see Section 3.2.6), Olmsted and coworkers considered the consequences of a +8 cationic ligand binding to short-, intermediate-, and polymeric-length DNA models (Olmsted et al., 1995). Binding was represented by "turning off" 8 charges in the axial center of a given length of DNA. Binding to the center of a 24-mer severely reduced the cation surface concentration around the DNA, whereas binding to the interior of an intermediate-length model DNA effectively created two new "ends" in the interior (*Fig. 7*).

In summary, the GCMC calculations on cylindrical DNA models find that the axial distribution of counterions around the polyion is qualitatively different for short DNA (with its parabolic ion concentration profile and its nonpolymer-like interior) *vs* polymeric DNA (having a trapezoidal ion distributions with a "polymeric" interior). The coulombic end effect stems from the decreasing cation accumulation as the oligomer terminus is approached. Increasing bulk salt concentration decreases the number of terminal residues impacted by the coulombic end effect. The coulombic end effect reduces the overall average per phosphate counterion accumulation around a nucleic acid compared to the polymeric limit. Unlike CC estimations of θ, the GCMC results predict that these effects are significant at experimental salt conditions. The thermodynamics of NA processes such as S_aK_{obs} (eq 19) and $\partial T_m/\partial \ln a_{\pm}$ (eq 18) are direct repercussions of θ and Γ behavior as a function of DNA charge $|Z_D|$.

3.1.3 Oligonucleotide nonlinear Poisson-Boltzmann calculations

Characterization of the coulombic end effect by NLPB. NLPB has also been applied to short nucleic acids. Calculations on DNA lengths between 10−100 bp modeled as right circular cylinders with hemispherical caps exhibited clearly evident coulombic end effects (Allison, 1994). At all monovalent salt concentrations considered (1 - 100 mM), the reduced electric field potential, $y(a)$, at the surface of the DNA cylinder significantly decreases over a ~ 5−6 bp distance as the DNA terminus is approached. Allison (1994) found that to a first approximation, ionic strength and thus Debye length, does not strongly affect the extent of end effects on the polyion electrostatic potential. While the reduced potential $y(r)$ is not the same as the surface counterion concentration, Allison's results compare well with those of Olmsted *et al.* (1989, 1991) and Ni *et al.* (1999), probably because counterion concentration is proportional to the exponential of $y(r)$ (see eq 14). On the other hand, the large differences seen in $y(a)$, even at high salt concentrations do not correlate well with the predictions of counterion condensation theory which state that the coulombic end effect is very small under these conditions. However, θ is an integral quantity and still may attain polymeric character even if the reduced potential at or near the surface of the DNA is changing with position.

NLPB calculations on a hemispherically-capped cylindrical 80-mer DNA model evaluated the effects of salt concentration and DNA structural parameters on the axial range of the coulombic end effect (Ni et al., 1999). L_{end} is defined as the length from either oligomer terminus where the surface cation concentration reached 95% of the interior concentration. These studies found only a small salt dependence in L_{end} between 0.01 - 0.2 M univalent salt: L_{end} decreased by only 20%over a 20-fold increase in salt concentration. However, L_{end} was much more sensitive to the DNA structural parameters *a* and *b* (See Section 1.2.2). Subsequent NLPB studies discovered that above 0.1 M salt, L_{end} is determined mostly by the polyion cylinder radius, *a*, and charge separation, *b*, and only weakly by κ^{-1} (Shkel & Record, 2004).

NLPB calculations of small ligands binding to oligomers. Several NLPB analyses have considered antibiotic-DNA binding thermodynamics (Chen & Honig, 1997; Misra et al., 1994; Sharp et al., 1995). The interaction of the minor groove-binding antibiotic 4',6-diamidino-2-phenylindole (DAPI) with dsDNA exemplifies the debate on the existence of coulombic end effects. Although the original experimental binding studies of DAPI (a +2 cation) were performed using poly[d(AT)]$_2$ and poly[d(GC)]$_2$ (Wilson et al., 1990), structural data were not available for polymeric DNA. Instead, crystal structure coordinates of a DAPI–12 bp DNA complex were used to construct structurally detailed models of a 12 bp duplex oligomer for NLPB calculations (Misra et al., 1994). The results were then compared to experimental SK_{obs} values for $DAPI^{2+}$ binding to polymeric DNA. They reported that the calculated $DAPI^{2+}$-DNA SK_{obs} was the same for DAPI bound to the 12 bp Dickerson-Drew dodecamer (DDD) sequence ($(d(CGCGAATTCGCG)_2)$) versus a 57 bp DNA built around a central $d(CGCGAATTCGCG)_2$ binding site, and therefore concluded that there were no end effects. Misra *et al.* stated that the lack of difference is reasonable since DAPI binds at least 13 Å from the terminus of the Dickerson dodecamer and that their results were consistent with previous studies which reported that the coulombic end effect spans a distance of $\approx 0.5\kappa^{-1}$, or approximately 11 Å at 0.02 M salt (Olmsted et al., 1989).

However, Sharp's *et al.* (1995) NLPB studies of DAPI binding to a range of DNA lengths ($\approx$ 8−60 Å or $\approx$ 2−18 bp) predicted a discernible coulombic end effect for DNA lengths $\lesssim$ 4−5 bp at 0.1 M salt when using a structurally detailed 3-D model for the complex. For dsDNA ≥ 7 bp, the net salt-dependence showed no significant variation over the DNA lengths considered. In other words, Sharp found that DAPI-DNA complex formation requires oligomer lengths beyond the range where coulombic end effects are readily apparent for divalent ligand binding at 0.1 M salt (Sharp et al., 1995). The Misra (1994) study considered DNA lengths ≥12 bp, and therefore the predicted SK_{obs} of $DAPI^{2+}$-DDD binding should be within error of the experimentally observed SK_{obs} for binding to polymeric dsDNA.

3.1.4 Molecular dynamics simulations

In the last ~15 years, molecular dynamics (MD) simulations have become accepted to such an extent that researchers are using MD for the final stages of structure refinement (Ditzler et al., 2010). Early MD simulations of dsDNA were plagued with problems such as broken base pairing and duplex distortions (Cheatham & Kollman, 2000). However, MD can now make experimentally-consistent predictions such as the spontaneous conversion of an A-DNA duplex into the more thermodynamically stable B-DNA form, ion interactions in the grooves of B-DNA observed in NMR and X-ray studies, and sequence-dependent DNA bending. MD force fields such as AMBER, CHARMM and software packages such as Discovery Studio (Accelrys, Inc.) and GROMOS are now readily accessible[20]. Simulations commonly include full atom representations of macromolecules and many tens of thousands (sometimes millions) of solvent molecules, running simulations on nanosecond timescales or longer. Extreme examples include MD simulations of B-DNA dynamics on the microsecond timescale (Pérez et al., 2007), nucleosome core particle dynamics over a 200 ns interval (Materese et al., 2009), and tRNA entry into the ribosome during decoding (Sanbonmatsu et al., 2005). The extremely high electrostatic fields around NAs require careful consideration of ion and solvent molecules in MD simulations, with as much as 80 - 90% of the computational cost consumed by solvent-solvent interactions (MacKerell & Nilsson, 2008). Some laboratories have attempted to avoid this expense by adopting implicit methods to account for the solvent (Auffinger & Hashem, 2007), though even with these simplifications, large systems still represent a significant challenge. Conformational dynamics or binding processes (with drugs or proteins, for example) are of particular interest. MD methodology development and its applications are intensely active fields of research with reviews being published every year. Here, the basic concepts in MD will be described with salient examples as relevant. Further interest is deferred to an extensive array of available reviews (Auffinger & Hashem, 2007; Ditzler et al., 2010; Egli, 2002; Hashem & Auffinger, 2009; Koehl, 2006; MacKerell & Nilsson, 2008).

MD relies on empirical parameterizations to generate a potential field and then evaluates all atom-atom interactions between the macromolecule and the surrounding water and ions. Potential fields are generally calculated by modeling atoms as Lennard-Jones spheres possessing point charges at their centers, using harmonic spring potential energy functions describe bond lengths and bond angles, and maintaining dihedral angles within a defined torsion profile (Cheatham & Kollman, 2000; Ditzler et al., 2010). Forces acting on each atom are determined from the potential field which dictates the motions of each atom for 1 - 2 femtoseconds. From the new configuration, a new total energy (or free energy, depending on the conditions used) is calculated, and the forces and velocities are calculated again. Everything is dependent on the potential field and continual research strives to improve force field "accuracy" as judged by comparison of MD simulation results to structures

[20] An extensive list of molecular modeling and molecular dynamics software is available at http://www.netsci.org/Resources/Software/Modeling/MMMD/index.html .

predicted by *ab initio* studies or observed experimentally (Cheatham & Kollman, 2000; Hashem & Auffinger, 2009). Often, this has been done by adjustment of the parameters used in the analytical potential field equation of choice. Alternative approaches to defining atomic forces that are particularly important for NA studies include particle mesh Ewald summation (Koehl, 2006) to address the extreme electrostatic fields around the NA and hybrid quantum mechanical/molecular mechanical algorithms (Ditzler et al., 2010), often employed to describe catalytic processes in, for example, ribozymes.

Much of the discussion of ion-nucleic acid interactions in molecular dynamics simulations focuses on ion association or localization, especially in the nucleic acid grooves. Na^+ ion localization at the ApT step in the minor groove of DDD dsDNA is an archetypal example (Young et al., 1997). Since the initial Young *et al.* publication on the "ApT pocket," questions of ion-DNA site-specific binding (Egli, 2002), partial phosphate charge neutralization (Manning, 2003), and the possible repercussions of these phenomena on DNA topology (*e.g.*, bends, twist, groove widening, etc.) continue to be very active fields of research.

However, MD studies which consider ion-NA interactions and the possibility of ion-binding pockets near the NA surface seem to be focused for the most part on NA geometries and radial ion distributions. To our knowledge, very little has been reported on the axial ion distribution which would be most indicative of a coulombic end effect. That said, some studies have considered the extent of phosphate charge neutralization by the surrounding ion atmosphere. One of the earliest examples we could find analyzed the superposition of multiple DNA/counterion structures taken at regular intervals during a 1.5 ns nanosecond simulation of DDD dsDNA in the presence of 22 Na^+ ions (to maintain electroneutrality) and ≈4000 waters (Young et al., 1997). They reported that the "cloud" of counterions around the DNA under no added salt conditions is consistent with Manning's counterion condensation theory, with 76% of the ions condensed around the DNA (Young et al., 1997). A more recent MD simulation on a comparable system (DDD, 22 Na^+ ions, ≈4000 waters, 60 ns simulation) found similar results (Ponomarev et al., 2004). More recently, the net neutralization of DNA charge by core histones and surrounding counterions in the presence of ≈ 50,000 water molecules was considered after a 200 ns MD simulation (Materese et al., 2009). In this study, the authors expected that ≈24% of DNA charge should remain after accounting for the counterion and histone charges within 10 Å of the DNA surface, as anticipated by $\theta - 0.76$ from CC theory. Therefore, approximately 219 counterions were expected near the DNA surface, but only 174 were observed (Materese et al., 2009). Comparisons of radial ion distributions predicted by MD *vs* by PB calculations showed marked differences: the extreme Na^+ accumulation within 3 Å of the nucleosome surface was absent in the PB radial distribution and MD predicted three distinct ionic shells versus only one by PB calculations.

Although molecular dynamics have reproduced some experimental phenomena such as ion localization (Ditzler et al., 2010), current MD force fields generally tend to be insensitive to salt concentration. Simulations of DNA dynamics and structure in solution at nanosecond time-scales have not historically demonstrated significant differences in outcome regardless of whether they are performed in limiting salt, net-neutralizing salt, or even 1 M salt (Cheatham & Kollman, 2000). Of course, this is a serious issue, since differences in salt concentration can dramatically perturb NA structure and processes (*e.g.*, RNA folding and ribozyme function have well known salt concentration sensitivities (Draper et al., 2005)). While MD is considered relatively reliable for protein simulations, reports in 2007 (Auffinger & Hashem, 2007) detailed newly discovered issues with commonly used AMBER force field implementations. For example, some parameterizations resulted in the formation of ion clusters or aggregates when ≈ 0.25 M KCl or high concentrations of NaCl were added to NA simulations. Hidden errors which were not easily observed in typical < 10 ns simulations led to irreversible DNA backbone torsion angle transitions at long simulation times (> 100 ns) which destroyed DNA structure. Multivalent cations pose a significant challenge for MD simulations and have even been declared outside the applicability of typical force fields (Ditzler et al., 2010). Thus, as summarized in the Supplementary Content for (Krasovska et al., 2006), many researchers choose to run MD simulations with minimal salt to avoid salt-related artifacts. The ion cluster and torsion angle errors described above spurred rapid countermeasures, and MD parameterizations are continually being modified and optimized. Thus, the struggles associated with the strong electrostatic fields around highly charged NAs and multivalent ions may be resolved in time, providing another probe to assess the consequences and causes of coulombic end effects in the properties and processes of NAs.

Summary of theoretical perspectives on end effects. Counterion condensation theory states that the Debye length is the critical factor determining the size and onset of a coulombic end effect and that the significance of end effects on DNA interactions is strongly affected by the salt concentration. At or above 0.1 M univalent salt, the range of the coulombic end effect is predicted to be less than 2−3 bp. Therefore CC theory predicts no significant difference per charge between the coulombic properties of a short oligomer (approximately 5 bp, for example) and a polynucleotide at 0.1 M salt.

NLPB and MC theories predict that the axial profile of the surface counterion concentration is trapezoidal around a polymer and parabolic around a short oligomer. The length of the coulombic end effect is determined by the Debye length at very low salt ($\kappa a << 1$, or millimolar salt concentrations for B-DNA), but DNA radius, a, becomes the characteristic length determining the size of the end effect for high salt (greater than 0.1 M). Between 0.01 - 0.1 M 1:1 salt, the surface potential and counterion axial distribution show coulombic end effects which extend 5−6 bp into the dsDNA and ≈8 bases into the ssDNA interior for long enough oligomers. In both NLPB and MC calculations, preferential interaction coefficients show a linear dependence on the length of the oligomer (*e.g.*, $\Gamma = \Gamma_\infty - \gamma/N$), where both the slope (γ) and the polymeric limit (Γ_∞) are salt dependent and depend on the type (*e.g.*, structural parameters) of the DNA. These computationally derived Γ parameters are equivalent within error to those determined experimentally. Molecular dynamics simulations have found success in predicting ion binding pockets, refining coordinates in experimental structural studies, and correctly predicting some NA conformational changes and ligand interactions. However, MD still has significant difficulty with predicting properties which are sensitive to the extreme electrostatic fields surrounding NAs.

3.2 Oligonucleotide experimental studies

3.2.1 Small ion NMR of nucleic acid solutions

Unlike the linear response of the Na^+ longitudinal relaxation rate, R_{obs}, with increasing $[P]/[Na^+]$ that is observed for polymeric NAs, a short 20 bp dsDNA fragment exhibits a distinctly nonlinear dependence (Section 2.3.1 and *Fig.* 3). As the Na^+ concentration decreased relative to the DNA phosphate concentration, the difference ${}^{160bp}R_{obs}$ − ${}^{20bp}R_{obs}$ increased significantly, with the 20 bp fragment always having much less enhancement than the 160-mer at any given $[P]/[Na^+]$ ratio. The divergence of Na^+ relaxation rates between "polymeric" 160 bp DNA and "oligomeric" 20 bp DNA was particularly evident at $[P]/[Na^+] \cong 1$ where ${}^{20bp}R_{obs} = 47 \pm 1$ s^{-1} and ${}^{160bp}R_{obs} = 69 \pm 1$ s^{-1}. When compared to $R_{free} = 18.4 \pm 0.4$ s^{-1}, the 160 bp DNA had almost twice (51 *vs* 29 s^{-1}) the enhancement relative to that seen for the short oligomer (Stein et al., 1995). Since R_{obs} is indicative of the extent of cation accumulation per phosphate at the DNA surface, these results predict that the average local Na^+ concentration is lower near the surface of the 20-mer than the 160-mer at a given [P]/[Na^+] ratio. These results are consistent with the existence of a significant coulombic end effect: a reduced accumulation of counterions near nucleic acid termini would have a greater impact on the average per-phosphate counterion concentration of shorter nucleic acids since the terminal regions represent a greater proportion of the total oligoion length relative to a polyion.

3.2.2 Structural studies of oligomeric nucleic acids by small angle x-ray scattering

Structural biological applications of small angle x-ray scattering (SAXS) and its variants (*e.g.*, anomalous small-angle x-ray scattering, ASAXS; ultra-small-angle x-ray scattering, USAXS; and small-angle neutron scattering, SANS) have undergone explosive growth in the last decade. While fundamentally a low-resolution technique (typically ranging 10 - 50 Å), SAXS is unfettered by molecular weight constraints or crystallization requirements and can investigate macromolecules in diverse solution conditions with volumes as small as 15 μL and concentrations of ≈20 μM (Putnam et al., 2007). SAXS has been integrated with Monte Carlo, NLPB, molecular dynamics, atomic emission spectroscopy, and NMR applications (Bai et al., 2007; Prabhu, 2005; Rambo & Tainer, 2010). SAXS derives structural information about the population-average of all macromolecule conformations from the pattern of scattered synchrotron radiation. This intensity data is mathematically transformed into its pair distribution, *P(r)*, essentially a histogram of all inter-atomic vectors within the macromolecule. SAXS *P(r)* profiles can be directly calculated from atomic-resolution structures, but the converse is not true. The radius of gyration is directly accessible by SAXS data, thereby providing an estimation of the distribution of mass around its center and thus an indication of the relative size of the macromolecule. The hydrated macromolecular volume can also be obtained for well-folded macromolecules (Putnam et al., 2007; Rambo & Tainer, 2010). The reviews above detail these techniques and available modeling approaches (*e.g.*, rigid-body and refinement modeling) that can be additionally applied.

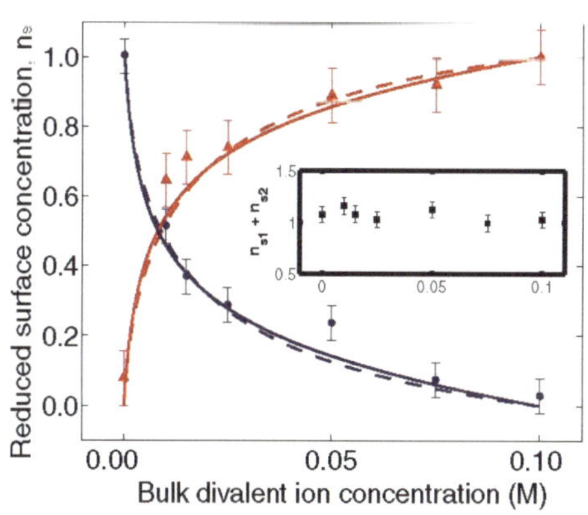

Fig. 8. The measured reduced surface concentrations of Rb^+ (●) and Sr^{2+} (▲) ions are shown as the bulk solution composition is varied. The solid lines represent a global best fit to a Boltzmann distribution of Sr^{2+} relative to Rb^+. The dashed lines represent NLPB calculations accounting for a finite ion radius with a 2 Å distance of closest approach. Inset: the sum of the reduced surface concentrations (■) equals unity within error. (Andresen et al., 2004)

Particularly salient to this review, SAXS has been used to probe NA ion distributions and their perturbation by salt concentration or ion composition changes (Andresen et al., 2004; Andresen et al., 2008; Das et al., 2003; Prabhu, 2005; Rambo & Tainer, 2010). The univalent ion atmosphere around NAs observed by SAXS is remarkably well described by NLPB calculations once finite ion-size adjustments are made (Andresen et al., 2004; Andresen et al., 2008; Das et al., 2003). SAXS studies have corroborated several PB assumptions (*Fig.* 8), including the invariance of equivalently-charged NA-ion distribution profiles and the quantitative validity of eqs 7 and 14 (Andresen et al., 2004; Bai et al., 2007; Das et al., 2003). Similar to PB-MC comparisons discussed previously (*cf.*, Section 2.2.4, *Fig.* 2A), PB calculations tend to overestimate surface cation concentrations near the DNA relative to SAXS-determined values, especially when multivalent cations are present (Andresen et al., 2004; Andresen et al., 2008; Bai et al., 2007; Das et al., 2003).

3.2.3 Oligoelectrolyte capillary electrophoresis

The capillary electrophoresis studies described in Section 2.3.3 compared free solution mobilities (μ) of a range of DNA fragment lengths (27 bp - 48.5 kbp) in tris-acetate-EDTA buffer (Stellwagen et al., 1997). The electrophoretic mobility of oligomers between 10—100 bp decreased monotonically with a -0.6 power-dependence on molecular weight (*Fig.* 4). The increase in μ for smaller DNA fragments with increasing molecular weight are in qualitative agreement with results seen for polylysines (Wilcoxon & Schurr, 1983), small polyphosphates (Grossman & Soane, 1991), and poly(styrene sulfonate) oligoions (Böhme & Scheler, 2007). However, duplexes of $\approx$ 400 bp to 45.8 kbp were independent of molecular weight, consistent with polyelectrolyte behavior (see Section 2.3.3). Does a coulombic end effect contribute to this molecular weight dependence of μ for small oligos?

Mohanty and coworkers (Mohanty & Stellwagen, 1999) derived an expression to describe μ by a series of linearizations of the field and ion transport equations and applied it to the Stellwagen (1997) experimental data. The DNA was modeled as a string of charged beads with a reduced effective charge of (1 - θ) predicted by counterion condensation theory. The condensed fraction of counterions is solely determined by axial charge density ξ (eq 8) at distances greater than the Debye length (*i.e.*, κ^{-1} < 20 Å or < 6 bp at salt concentrations >26 mM used in Stellwagen's experiments). All dsDNA fragments considered exceeded this threshold and thus coulombic end effects are predicted to be negligible. Mohanty's model qualitatively agrees with experiment in that it predicts a plateau in μ for high molecular weight DNA and decreasing with decreasing DNA length. However, the predicted μ decreases too rapidly with decreasing DNA molecular weight in comparison to experiment.

A rigorous continuum hydrodynamic-electrodynamic model describing μ found that the effective charge per base pair decreases with increasing dsDNA length (Allison et al., 2001), indicating that significantly fewer counterions were accumulated per DNA charge for short DNA oligomers than for polymeric DNA. Treating DNA as straight hemispherically capped cylinders predicted electrophoretic mobilities for DNA $\gtrsim$ 60 bp within < 9% of experimental values but overestimated μ for larger fragments (*Fig.* 9). Since the dsDNA persistence length is approximately 50 nm (~ 147 bp), the authors deduced that the increased likelihood of DNA bending for lengths $\gtrsim$ 80 bp may be contributing to the experimentally observed plateau of μ for longer NAs. Unfortunately, the impractical computational time required to sample multiple worm-like chain conformations precluded calculation of an ensemble averaged μ for a population of flexible DNA fragments. However, modeling a series of singly bent cylinders with the curvature expected for a given length of DNA predicted a μ dependence on length which qualitatively agreed with experiment (*Fig.* 9). Together, these results demonstrate that boundary element modeling (with coulombic end effects) can yield quantitatively better predictions of electrophoretic mobility for short oligomers than a CC approach (without coulombic end effects).

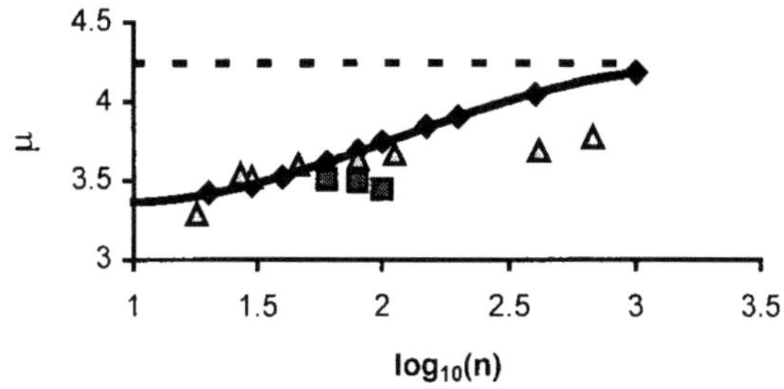

Fig. 9. Comparison of theory and experiment describing the length dependence of electrophoretic mobility, μ. Boundary element calculations modeling dsDNA as straight (◆, *solid curve*) or bent (■) primitive cylinders with hemispherically capped ends possessing phosphates set at half their full charge were compared to experimentally determined μ (Δ) for varying lengths (*n*, base pairs) of dsDNA oligonucleotides (Stellwagen *et al.*, 1997; *cf. Fig.* 4A above). The dashed line is the idealized mobility of an infinitely long straight rod. (Allison et al., 2001)

Oligomeric nucleic acid processes

Of all oligonucleotide experimental data available relevant to the question of coulombic end effects, thermal denaturation studies probably make up the most plentiful and controversial of results. In the last several years, considerable gains have been made in the characterization and predictions of oligonucleotide stability and its salt dependence. For example, a generalized set of nearest neighbor parameters are now available to describe both polymeric and short nucleic acid hairpins as well as duplex dimer complexes (SantaLucia, 1998). Recent work has generated an empirical function to quantitatively correct for the influence of divalent cations on the thermal denaturation of dsDNA oligomers (Owczarzy et al., 2008). The theoretical work of Olmsted *et al.* (1991) and Shkel & Record (2004) employ coulombic end effects to accurately describe several of the denaturation studies known at the time. However, observed and theoretical thermodynamic descriptions of NA melting led some to believe that coulombic end effects were not necessary to describe oligonucleotide melting. This brief synopsis describes some of the experimental results which seeded this controversy and concludes with a summary of oligonucleotide-ligand binding studies which more readily detected the deviation of oligonucleotide behavior from the polyelectrolyte ideal.

3.2.4 Conformational changes: denaturation of hairpin double helices (one folded strand)

In one of the first systematic melting studies of nucleic acid oligomers, Elson and coworkers investigated the thermodynamics of melting hairpin double helices formed by the intramolecular association of series of d(AT) oligomers possessing between 18 and 44 phosphate charges (Elson et al., 1970). The melting temperature T_m and its salt dependence $\partial T_m/\partial \log[\text{salt}]$ for these hairpins decreased sharply from typical polymeric values with decreasing $|Z|$, the number of oligonucleotide phosphate charges (*Fig.* 10). They used *a priori* computations based on the summation of discrete pairwise interactions of the screened DNA phosphate charges to describe their hairpin denaturation experiments (see Section 3.1.1) and found that a single value was sufficient to model all oligomer melting transitions at all salt concentrations studied. However, it was necessary to count the interactions between charges

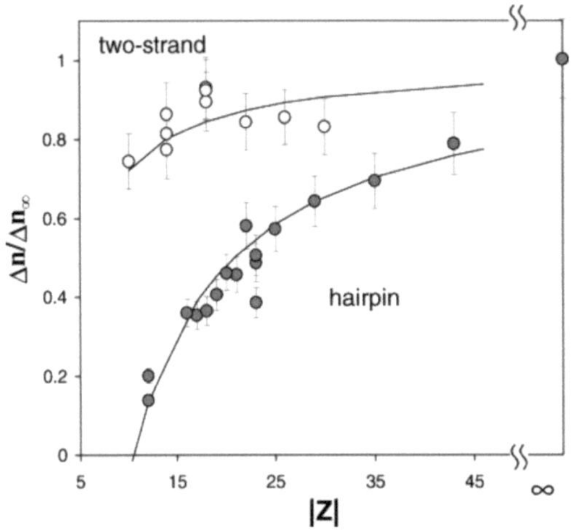

Fig. 10. The salt dependence of thermal denaturation ($ST_m = \partial T_m/\partial \ln[\text{salt}]$) as a function of NA charge $|Z|$ for (○) two-stranded and (●) hairpin NAs (including the Elson (1970) oligomers) plotted in terms of the net change in per-phosphate ion accumulation (Δn) upon thermal denaturation relative to Δn_u in the infinite polymer limit (Δn_∞). Solid lines trace $\frac{\Delta n_u}{\Delta n_{u,\infty}} = 1 - \frac{\gamma_{\text{end}}}{|Z|}$, where γ_{end} is the NLPB- determined coulombic end effect parameter quantifying the reduced fraction of ion accumulation near oligoion termini. These results relate to experiment by $\frac{ST_m}{ST_{m,\infty}} = \frac{\Delta n_u}{\Delta n_{u,\infty}} \frac{|Z|}{N_n - X} = \frac{|Z| - \gamma_{\text{end}}}{N_n - X}$, with N_n as the number of nucleotides in the native conformation and X, the number of nucleotides not contributing to the reaction enthalpy (*i.e.*, N_{loop} nucleotides in a hairpin loop). (Shkel & Record, Jr., 2004)

NA native state	γ_{end}	Valid regime	$(N_n - X)$
two-stranded duplex	2.8	$\|Z\| \geq 8$	$\|Z\|$
single-loop hairpin	10.4	$\|Z\| \geq 13$	$\|Z\| - N_h - 1$
double-loop hairpin	10.4	$\|Z\| \geq 13$	$\|Z\| - N_h + 1$

both in the helical and nonhelical segments of the hairpins to accurately model the breadth of the melting transitions. Record *et al.* (1978) developed an approximate analytical treatment of these data, using an end effect parameter in a CC-like theory to predict the observed $1/|Z|$ dependence of the $\partial T_m/\partial \log[\text{salt}]$ of hairpin melting. The methodology developed by Record and coworkers (1978) did not consider the nonhelical hairpin segments nor require any details of DNA structure other than the DNA axial charge separation *b*, yet fit the hairpin data quite well. A subsequent generalization of this approach (Olmsted et al., 1991) used GCMC simulations (*cf.*, Section 3.2.2) to predict with good accuracy the $\Delta\Gamma$ of ds → ss conversion for the Elson *et al.* (1970) hairpin studies. This same methodology, which predicts sizable coulombic end effects for short oligomers, describes hairpin denaturation, the melting transitions of dumbbell-shaped double-hairpins (Erie et al., 1987) and several of the two-stranded NA dimer denaturation studies known at the time. Shkel & Record (2004) revisited these experimental systems using NLPB calculations to develop a parameterized formula predicting $\partial T_m/\partial \log[\text{salt}]$ via consideration of the average ion accumulation near ssDNA and dsDNA termini relative to the polymeric limit (*Fig.* 10).

3.2.5 Conformational changes: two-stranded duplex ("dimer") denaturation

Some of the earlier spectroscopic, thermodynamic, and kinetic melting studies of short oligonucleotide two-stranded duplexes reported that the $\partial T_m/\partial \log[\text{salt}]$ in these systems was closer to that expected for polymeric nucleic acids with similar GC content than to analogous values seen in the hairpin melting studies described above. For example, the experimentally measured $\partial T_m/\partial \log[\text{salt}]$ for d(GCATGC) dsDNA ranged between 11 - 15 °C depending on strand concentration (Williams et al., 1989), which was consistent with a predicted $\partial T_m/\partial \log[\text{salt}]$ of 13.6 - 15.5 °C based on formulae developed to describe polymeric denaturation (Blake & Haydock, 1979; Frank-Kamenetskii, 1971). ^{1}H-NMR studies (Braunlin & Bloomfield, 1991) describing thermodynamic and kinetic salt effects on strand association for d(GGAATTCC) found that 0.13 ± 0.02 Na^+ ions are released per phosphate upon duplex dissociation, which compares well with the estimated 0.17 ions released per phosphate for polymeric DNA (Record et al., 1976). In both cases, the authors concluded that their oligonucleotide results were the same within uncertainty as those expected for the corresponding polymeric systems.

A comprehensive data review (SantaLucia, 1998) summarized both polymeric and oligomeric results across seven studies of DNA and RNA stability, including the Braunlin and Williams oligomers mentioned above. All experimental results were presented in a common format, thereby allowing direct comparison of free energies and their salt dependences. Experimental measurements of 26 NAs possessing 4 - 16 nucleotides had an average $\partial \Delta G^\circ_{37}/\partial \ln[\text{salt}] = -(0.057 \pm 0.017) * |Z_D|$ kcal/mol between 0.1 - 0.3 M NaCl. In other words, the sensitivity of NA thermal stability to salt concentration was found to be linearly proportional to the number of phosphates in the duplex, $|Z_D|$ (SantaLucia et al., 1996). Global analysis of three polymeric datasets (30 data points) yields an average $\partial \Delta G^\circ_{37}/\partial \ln[\text{salt}] = -(0.175 \pm 0.034)$ kcal/mol, independent of polymer length (SantaLucia, 1998). The larger $\partial \Delta G^\circ_{37}/\partial \ln[\text{salt}]$ for polynucleotide denaturation *vs* oligonucleotide melting in conjunction with the length dependence of $\partial \Delta G^\circ_{37}/\partial \ln[\text{salt}]$ for oligomers but not for polymers, are consistent with a coulombic end effect.

The melting thermodynamics of two-stranded oligomers appear polymeric due to a cancellation of effects: the duplex initial state denatures into two separate single strands, each with half the number of charges of the duplex. The net charge of the reactants does not change in denaturation studies ($\Delta Z = 0$) and thus[21] $\partial \Delta G^\circ/\partial \ln[\text{salt}] = -2RT\Delta\Gamma$ by eq 19. Although the length dependence of Γ for the native and denatured states of two-stranded

---

[21] Substitution of $-\Delta G^\circ/RT = \ln K_{obs}$ and $\ln[Na^+] \approx \ln a_\pm$ into eq 19 gives $\dfrac{\partial \Delta G^o}{\partial \ln[\text{Na}^+]} = -RT\Delta(|Z|+2\Gamma)$.

duplexes are significant, $\Delta\Gamma$ for the melting transition is relatively constant across a wide range of oligomer lengths. As a result, the dramatic coulombic end effects evident in NMR Na^+ relaxation experiments and hairpin thermal denaturation studies are not seen in two-strand duplex melting processes. Comparisons of oligonucleotide melting behavior of hairpins and dimer duplexes with calculations of $\Delta\Gamma$ are described in greater detail by Shkel and Record (2004). Perhaps most telling, theories which include coulombic end effects (Olmsted et al., 1991; Shkel & Record, 2004) accurately account for the melting transitions of NA hairpins and short two-stranded duplex DNA within experimental error, including d(GCATGC) (Braunlin & Bloomfield, 1991) and d(GGAATTCC) (Williams et al., 1989).

3.2.6 Oligonucleotide ligand binding

Zhang et al. (1996) performed possibly the first cationic oligopeptide-oligonucleotide binding studies to assess the difference of Z_L = 8+ peptide (KWK_6) binding to $dT(pdT)_{10}$ and poly(dT) ($dT(pdT)_{169}$). They reported that the binding affinity for KWK_6 binding to poly(dT) versus $dT(pdT)_{10}$ was about 13 times greater at 0.2 M Na^+ and 76-fold greater at 0.1 M Na^+.

The salt dependence of the equilibrium binding constant at 0.2 M Na^+ was - 6.5 ± 0.2 for poly(dT) but - 3.5 ± 0.1 for $dT(pdT)_{10}$. Poly(dT) had an extrapolated $\log K_{obs}$ at 1 M salt of - 0.6, indicating that ligand binding is unfavorable at such a high salt concentration and, more importantly, that binding is dominated by the entropic driving force of cation release characteristic of polyelectrolytes (see Section 2.3.4). $dT(pdT)_{10}$, on the other hand, has a 1 M Na^+ $\log K_{obs}$ of 0.7, still indicating that most of the poly(dT)-KWK_6 stabilization is from cation release, but with much less net thermodynamic counterion release upon complex formation. The large differences in S_aK_{obs} seen for long *vs* short nucleic acids to oligopeptides are consistent with PB (Zhang et al., 1996) and GCMC predictions (Olmsted et al., 1995). The greatly reduced salt dependence of binding is consistent with a significant change in the binding process $\Delta\Gamma$ for $dT(pdT)_{10}$ *vs* polydT (eq 19). These results conflict with the expectations of CC-based theories which maintain that little or no difference in S_aK_{obs} should be evidenced at these salt concentrations (see Section 3.1.1).

Subsequent studies of KWK_6 binding to intermediate lengths of $dT(pdT)_{|Z_D|}$, where $|Z_D|$ is the number of phosphate charges on the DNA, were later performed (Zhang et al., 1999) to define the transition from oligomeric binding of $dT(pdT)_{10}$ to polymeric binding of poly(dT). $\log K_{obs}$ and $|S_aK_{obs}|$ at 0.1 M Na^+ monotonically increased as length increased through the intermediate range $15 \leq |Z_D| \leq 69$, with values which were bracketed between the corresponding measures observed for $dT(pdT)_{10}$ and polydT. Zhang and coworkers found that S_aK_{obs} and $\Delta G°$ were inversely proportional to the number of binding sites (*Fig.* 11). Using a two-state model , they estimated that "end sites" spanning ≈10 phosphates had reduced ion accumulation relative to that expected for a ssDNA polyelectrolyte (Zhang et al., 1999). With collection of additional ssDNA-KWK_6 binding data (Ballin et al., 2004), Shkel and coworkers derived a parametric expression that quantitatively describes the full scope of nonspecific ligand-ssDNA binding studies they considered. Nonlinear least squares analysis of the experimental data indicated that coulombic end effects have a characteristic length affecting 9.0 ± 0.8 residues from each ssDNA terminus and 12 ± 1 residues from each end of dsDNA (Shkel et al., 2006). The minimum DNA length required to exhibit polymeric binding affinity and SK_{obs} for a ligand with Z_L charges is $|Z_D| = Z_L + 2N_e$ where N_e = 9.0 ± 0.8

for ssDNA and $N_e = 12 \pm 1$ for dsDNA. In other words, Mg^{2+} with $Z_L = +2$ would bind 27 bp dsDNA with polymeric affinity since polymeric binding is predicted for $|Z_D| > 2*12 + 2 = 26$.

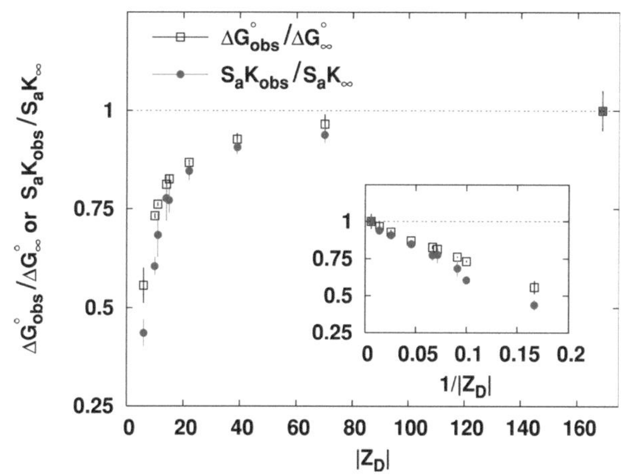

Fig. 11. ΔG°_{obs} and S_aK_{obs} for KWK_6 (Z_L = 8+) binding to ssDNA depends on DNA charge, $|Z_D|$. Values of ΔG°_{obs} at 0.1 M Na+ (□) and S_aK_{obs} (●), normalized by the corresponding quantities for binding of KWK_6 to poly(dT), are plotted against the number of DNA charges, $|Z_D|$. The inset plots the same data versus $1/|Z_D|$. (Ballin et al., 2004)

Conclusions

The experimental and theoretical results strongly support the existence of a coulombic end effect on processes involving oligonucleotides or which interact with the termini of long NAs. All experimental systems amenable to easy detection of putative coulombic end effects show such effects. Models incorporating coulombic end effects also accurately predict experimentally observed behavior for systems where end effects are expected to cancel out or be insignificantly small relative to larger effects. However, theories which neglect coulombic end effects and hence predict no difference between short *vs* long oligoelectrolytes fail to explain these phenomena and are therefore called into question. Nevertheless, theoretical systems exhibiting the greatest success in providing a unified explanation of nucleic acid properties from short to polymeric lengths account for coulombic end effects and their consequences, consideration of which is crucial for the understanding of nucleic acid phenomena in biochemistry and biotechnology.

4. Future directions

The four decades since the first reports of CC theory have witnessed tremendous advancements. Computational methods such as MC and NLPB accurately describe many polyelectrolyte thermodynamic signatures and their transitions into oligoelectrolyte behavior. Predictions are generally reliable for processes occurring at intermediate (*e.g.*, nonspecific binding events, ion distribution functions > 5 – 6 Å from the NA surface, etc.) or very far distances from the polyion (*e.g.*, colligative properties such as Donnan equilibria). These successes have encouraged researchers to tackle increasingly more difficult problems. In particular, we expect to see a growing interest in understanding processes which occur very close to the NA surface. Success in this arena will require significant improvements in the theoretical description of coulombic phenomena in extremely high field gradients, including ion-ion and ion-solvent interactions, polarizability, etc. How are these factors influenced by bent or kinked NA structures? Efficient integration of large-domain macromolecular dynamics (reptation models are of particular import for theories treating NA electrophoresis), intermediate-timescale conformational changes in RNA subdomains (essential for understanding catalytic RNA mechanisms), and a better understanding of the role of water in these processes (a universal challenge) will also become important. Of course, the excitement around SAXS and other atomic-level visualization techniques will likely continue. SAXS investigations of inter-helix DNA interactions in the presence of high-valence salts represent an active area of research. A study of the axial distribution of ions could provide enticing insights into processes involving siRNA, tRNA, and NA-drug interactions which could be compared to computational predictions.

The recent derivations of analytical solutions to the PB equation at all physiologically relevant salt concentrations may provide a viable alternative to the primitive parametric Hook's Law/Lennard-Jones potential fields used in molecular dynamics simulations. Despite their limiting assumptions, PB calculations have an enormous and growing body of evidence (SAXS, experimentally validated predictions of NA stability and its response to solution conditions, RNA folding thermodynamics, etc.) contending that PB is an effective tool for predicting electrostatic phenomena. Recent work has shown that expansion-series solutions to the PB equation yield electrostatic energy predictions that are within 2% of unapproximated calculations. The authors would like to see implementation of these simplified yet still quantitatively accurate electrostatic formulae into a MD algorithm. Such efforts could catalyze needed characterization and predictive model development describing the ***nonelectrostatic*** contributions to biomolecular processes.

5. References

Allison, S. A. (1994). End effects in electrostatic potentials of cylinders: models for DNA fragments. *J. Phys. Chem.* Vol. 98, No. 46, pp. 12091-12096, ISSN 0022-3654.

Allison, S. A.; Chen, C. & Stigter, D. (2001). The length dependence of translational diffusion, free solution electrophoretic mobility, and electrophoretic tether force of rigid rod-like model duplex DNA. *Biophys. J.* Vol. 81, No. 5, pp. 2558-2568, ISSN 0006-3495.

Allison, S. A.; Pei, H. & Xin, Y. (2007). Modeling the free solution and gel electrophoresis of biopolymers: the bead array-effective medium model. *Biopolymers* Vol. 87, No. 2-3, pp. 102-114, ISSN 1097-0282.

Anderson, C. F. & Record, M. T., Jr. (1980). The relationship between the Poisson-Boltzmann model and the condensation hypothesis: an analysis based on the low salt form of the Donnan coefficient. *Biophys. Chem.* Vol. 11, No. 3-4, pp. 353-360, ISSN 0301-4622.

Anderson, C. F. & Record, M. T., Jr. (1982). Polyelectrolyte theories and their applications to DNA. *Annu. Rev. Phys. Chem.* Vol. 33, No. 1, pp. 191-222, ISSN 0066-426X.

Anderson, C. F. & Record, M. T., Jr. (1983).The thermodynamic effects of polyelectrolyte-electrolyte interactions. In: *Structure and Dynamics: Nucleic Acids and Proteins,* E. Clementi & R. H. Sarma, (Ed.), Adenine Press, ISBN 0-940030-04-7, NY NY.

Anderson, C. F. & Record, M. T., Jr. (1990). Ion distributions around DNA and other cylindrical polyions: theoretical descriptions and physical implications. *Annu. Rev. Biophys. Biophys. Chem.* Vol. 19, No. 1, pp. 423-463, ISSN 0883-9182.

Anderson, C. F. & Record, M. T., Jr. (1993). Salt dependence of oligoion-polyion binding: a thermodynamic description based on preferential interaction coefficients. *J. Phys. Chem.* Vol. 97, No. 27, pp. 7116-7126, ISSN 0022-3654.

Anderson, C. F. & Record, M. T., Jr. (1995). Salt-nucleic acid interactions. *Annu. Rev. Phys. Chem.* Vol. 46, No. 1, pp. 657-700, ISSN 0066-426X.

Andresen, K.; Das, R.; Park, H. Y.; Smith, H.; Kwok, L. W.; Lamb, J. S.; Kirkland, E. J.; Herschlag, D.; Finkelstein, K. D. & Pollack, L. (2004). Spatial distribution of competing ions around DNA in solution. *Phys. Rev. Lett.* Vol. 93, No. 24, pg. 248103, ISSN 0031-9007.

Andresen, K.; Qiu, X.; Pabit, S. A.; Lamb, J. S.; Park, H. Y.; Kwok, L. W. & Pollack, L. (2008). Mono- and trivalent ions around DNA: a small-angle scattering study of competition and interactions. *Biophys. J.* Vol. 95, No. 1, pp. 287-295, ISSN 0006-3495.

Auffinger, P. & Hashem, Y. (2007). Nucleic acid solvation: from outside to insight. *Curr. Opin. Struct. Biol.* Vol. 17, No. 3, pp. 325-333, ISSN 0959-440X.

Bai, Y.; Greenfeld, M.; Travers, K. J.; Chu, V. B.; Lipfert, J.; Doniach, S. & Herschlag, D. (2007). Quantitative and comprehensive decomposition of the ion atmosphere around nucleic acids. *J. Am. Chem. Soc.* Vol. 129, No. 48, pp. 14981-14988, ISSN 0002-7863.

Ballin, J. D.; Prevas, J. P.; Ross, C. R.; Toth, E. A.; Wilson, G. M. & Record, M. T., Jr. (2010). Contributions of the histidine side chain and the N-terminal α-amino group to the binding thermodynamics of oligopeptides to nucleic acids as a function of pH. *Biochemistry* Vol. 49, No. 9, pp. 2018-2030, ISSN 0006-2960.

Ballin, J. D.; Shkel, I. A. & Record, M. T., Jr. (2004). Interactions of the KWK_6 cationic peptide with short nucleic acid oligomers: demonstration of large coulombic end effects on binding at 0.1-0.2 M salt. *Nucleic Acids Res.* Vol. 32, No. 11, pp. 3271-3281, ISSN 0305-1048.

Blake, R. D.; Bizzaro, J. W.; Blake, J. D.; Day, G. R.; Delcourt, S. G.; Knowles, J.; Marx, K. A. & SantaLucia, J. J. (1999). Statistical mechanical simulation of polymeric DNA melting with MELTSIM. *Bioinformatics* Vol. 15, No. 5, pp. 370-375, ISSN 1367-4803.

Blake, R. D. & Haydock, P. V. (1979). Effect of sodium ion on the high-resolution melting of lambda DNA. *Biopolymers* Vol. 18, No. 12, pp. 3089-3109, ISSN 1097-0282.

Bleam, M. L.; Anderson, C. F. & Record, M. T., Jr. (1980). Relative binding affinities of monovalent cations for double-stranded DNA. *Proc. Natl. Acad. Sci. USA* Vol. 77, No. 6, pp. 3085-3089, ISSN 0027-8424.

Bleam, M. L.; Anderson, C. F. & Record, M. T., Jr. (1983). Sodium-23 nuclear magnetic resonance studies of cation-deoxyribonucleic acid interactions. *Biochemistry* Vol. 22, No. 23, pp. 5418-5425, ISSN 0006-2960.

Böhme, U. & Scheler, U. (2007). Hydrodynamic size and electrophoretic mobility of poly(styrene sulfonate) versus molecular weight. *Macromol. Chem. Phys.* Vol. 208, No. 19-20, pp. 2254-2257, ISSN 1521-3935.

Bond, J. P.; Anderson, C. F. & Record, M. T., Jr. (1994). Conformational transitions of duplex and triplex nucleic acid helices: thermodynamic analysis of effects of salt concentration on stability using preferential interaction coefficients. *Biophys. J.* Vol. 67, No. 2, pp. 825-836, ISSN 0006-3495.

Braunlin, W. H. (1995).NMR studies of cation binding environments on nucleic acids. In: *Advances in Biophysical Chemistry,* C. A. Bush, (Ed.), Jai Press, Inc, ISBN 1-55938-978-8, Greenwich CT.

Braunlin, W. H. & Bloomfield, V. A. (1991). Proton NMR study of the base-pairing reactions of d(GGAATTCC): salt effects on the equilibria and kinetics of strand association. *Biochemistry* Vol. 30, No. 3, pp. 754-758, ISSN 0006-2960.

Braunlin, W. H.; Strick, T. J. & Record, M. T., Jr. (1982). Equilibrium dialysis studies of polyamine binding to DNA. *Biopolymers* Vol. 21, No. 7, pp. 1301-1314, ISSN 0006-3525.

Breslauer, K. J. (1986).*Thermodynamic Data for Biochemistry and Biotechnology.* H.-J. Hinz, (Ed.), pp. 402-427, Springer-Verlag, ISBN 0387163689, New York NY.

Cheatham, T. E., III & Kollman, P. A. (2000). Molecular dynamics simulation of nucleic acids. *Annu. Rev. Phys. Chem.* Vol. 51, No. 1, pp. 435-471, ISSN 0066-426X.

Chen, S. W. W. & Honig, B. (1997). Monovalent and divalent salt effects on electrostatic free energies defined by the nonlinear Poisson-Boltzmann equation: Application to DNA binding reactions. *J. Phys. Chem. B* Vol. 101, No. 44, pp. 9113-9118, ISSN 1089-5647.

Chu, V. B.; Bai, Y.; Lipfert, J.; Herschlag, D. & Doniach, S. (2008). A repulsive field: advances in the electrostatics of the ion atmosphere. *Curr. Opin. Chem. Biol.* Vol. 12, No. 6, pp. 619-625, ISSN 1367-5931.

Das, R.; Mills, T. T.; Kwok, L. W.; Maskel, G. S.; Millett, I. S.; Doniach, S.; Finkelstein, K. D.; Herschlag, D. & Pollack, L. (2003). Counterion distribution around DNA probed by solution x-ray scattering. *Phys. Rev. Lett.* Vol. 90, No. 18, pg. 188103, ISSN 0031-9007.

deHaseth, P. L.; Lohman, T. M.; Burgess, R. R. & Record, M. T., Jr. (1978). Nonspecific interactions of *Escherichia coli* RNA polymerase with native and denatured DNA: differences in the binding behavior of core and holoenzyme. *Biochemistry* Vol. 17, No. 9, pp. 1612-1622, ISSN 0006-2960.

deHaseth, P. L.; Lohman, T. M. & Record, M. T., Jr. (1977). Nonspecific interaction of lac repressor with DNA: an association reaction driven by counterion release. *Biochemistry* Vol. 16, No. 22, pp. 4783-4790, ISSN 0006-2960.

Delcourt, S. G. & Blake, R. D. (1991). Stacking energies in DNA. *J. Biol. Chem.* Vol. 266, No. 23, pp. 15160-15169, ISSN 0021-9258.

Ditzler, M. A.; Otyepka, M.; Sponer, J. & Walter, N. G. (2010). Molecular dynamics and quantum mechanics of RNA: conformational and chemical change we can believe in. *Acc. Chem. Res.* Vol. 43, No. 1, pp. 40-47, ISSN 0001-4842.

Draper, D. E.; Grilley, D. & Soto, A. M. (2005). Ions and RNA folding. *Annu. Rev. Biophys. Biomol. Struct.* Vol. 34, pp. 221-243, ISSN 1056-8700.

Draper, D. E. (2008). RNA folding: thermodynamic and molecular descriptions of the roles of ions. *Biophys. J.* Vol. 95, No. 12, pp. 5489-5495, ISSN 0006-3495.

Egli, M. (2002). DNA-cation interactions: quo vadis? *Chem. Biol.* Vol. 9, No. 3, pp. 277-286, ISSN 1074-5521.

Elson, E. L.; Scheffler, I. E. & Baldwin, R. L. (1970). Helix formation by d(TA) oligomers: III. Electrostatic effects. *J. Mol. Biol.* Vol. 54, No. 3, pp. 401-415, ISSN 0022-2836.

Erie, D.; Sinha, N.; Olson, W.; Jones, R. & Breslauer, K. (1987). A dumbbell-shaped, double-hairpin structure of DNA: a thermodynamic investigation. *Biochemistry* Vol. 26, No. 22, pp. 7150-7159, ISSN 0006-2960.

Fenley, M. O.; Harris, R. C.; Jayaram, B. & Boschitsch, A. H. (2010). Revisiting the association of cationic groove-binding drugs to DNA using a Poisson-Boltzmann approach. *Biophys. J.* Vol. 99, No. 3, pp. 879-886, ISSN 0006-3495.

Fenley, M. O.; Manning, G. S. & Olson, W. K. (1990). Approach to the limit of counterion condensation. *Biopolymers* Vol. 30, No. 13-14, pp. 1191-1203, ISSN 1097-0282.

Fixman, M. (1979). The Poisson--Boltzmann equation and its application to polyelectrolytes. *J. Chem. Phys.* Vol. 70, No. 11, pp. 4995-5005, ISSN 0021-9606.

Fogolari, F.; Brigo, A. & Molinari, H. (2002). The Poisson-Boltzmann equation for biomolecular electrostatics: a tool for structural biology. *J. Mol. Recogn.* Vol. 15, No. 6, pp. 377-392, ISSN 0952-3499.

Frank-Kamenetskii, M. D. (1971). Simplification of the empirical relationship between melting temperature of DNA, its GC content and concentration of sodium ions in solution. *Biopolymers* Vol. 10, No. 12, pp. 2623-2624, ISSN 1097-0282.

Grochowski, P. & Trylska, J. (2008). Continuum molecular electrostatics, salt effects, and counterion binding – a review of the Poisson–Boltzmann theory and its modifications. *Biopolymers* Vol. 89, No. 2, pp. 93-113, ISSN 1097-0282.

Grossman, P. D. & Soane, D. S. (1991). Experimental and theoretical studies of DNA separations by capillary electrophoresis in entangled polymer solutions. *Biopolymers* Vol. 31, No. 10, pp. 1221-1228, ISSN 1097-0282.

Gruziel, M.; Grochowski, P. & Trylska, J. (2008). The Poisson-Boltzmann model for tRNA: Assessment of the calculation set-up and ionic concentration cutoff. *J. Comput. Chem.* Vol. 29, No. 12, pp. 1970-1981, ISSN 1096-987X.

Hashem, Y. & Auffinger, P. (2009). A short guide for molecular dynamics simulations of RNA systems. *Methods* Vol. 47, No. 3, pp. 187-197, ISSN 1046-2023.

Hoagland, D. A.; Arvanitidou, E. & Welch, C. (1999). Capillary electrophoresis measurements of the free solution mobility for several model polyelectrolyte systems. *Macromolecules* Vol. 32, No. 19, pp. 6180-6190, ISSN 0024-9297.

Jayaram, B. & Beveridge, D. L. (1991). Grand canonical Monte Carlo simulations on aqueous solutions of sodium chloride and sodium DNA: excess chemical potentials and sources of nonideality in electrolyte and polyelectrolyte solutions. *J. Phys. Chem.* Vol. 95, No. 6, pp. 2506-2516, ISSN 0022-3654.

Jayaram, B. & Beveridge, D. L. (1996). Modeling DNA in aqueous solutions: theoretical and computer simulation studies on the ion atmosphere of DNA. *Annu. Rev. Biophys. Biomol. Struct.* Vol. 25, No. 1, pp. 367-394, ISSN 1056-8700.

Klein, B. K.; Anderson, C. F. & Record, M. T., Jr. (1981). Comparison of Poisson-Boltzmann and condensation model expression for the colligative properties of cylindrical polyions. *Biopolymers* Vol. 20, No. 10, pp. 2263-2280, ISSN 1097-0282.

Koehl, P. (2006). Electrostatics calculations: latest methodological advances. *Curr. Opin. Struct. Biol.* Vol. 16, No. 2, pp. 142-151, ISSN 0959-440X.

Korolev, N.; Lyubartsev, A. P. & Nordenskiold, L. (2002). Application of the Poisson Boltzmann polyelectrolyte model for analysis of equilibria between single-, double-, and triple-stranded polynucleotides in the presence of K^+, Na^+, and Mg^{2+} ions. *J. Biomol. Struct. Dyn.* Vol. 20, No. 2, pp. 275-290, ISSN 0739-1102.

Korolev, N.; Lyubartsev, A. P. & Nordenskiöld, L. (1998). Application of polyelectrolyte theories for analysis of DNA melting in the presence of Na^+ and Mg^{2+} ions. *Biophys. J.* Vol. 75, No. 6, pp. 3041-3056, ISSN 0006-3495.

Krasovska, M. V.; Sefcikova, J.; Réblová, K.; Schneider, B.; Walter, N. G. & Šponer, J. (2006). Cations and hydration in catalytic RNA: molecular dynamics of the hepatitis delta virus ribozyme. *Biophys. J.* Vol. 91, No. 2, pp. 626-638, ISSN 0006-3495.

Le Bret, M. & Zimm, B. H. (1984). Distribution of counterions around a cylindrical polyelectrolyte and Manning's condensation theory. *Biopolymers* Vol. 23, No. 2, pp. 287-312, ISSN 1097-0282.

Leipply, D.; Lambert, D. & Draper, D. E. (2009). Ion-RNA interactions: thermodynamic analysis of the effects of mono- and divalent ions on RNA conformational equilibria. *Methods Enzymol.* Vol. 469, pp. 433-463, ISSN 0076-6879.

Lohman, T. M.; deHaseth, P. L. & Record, M. T., Jr. (1980). Pentalysine-deoxyribonucleic acid interactions: a model for the general effects of ion concentrations on the interactions of proteins with nucleic acids. *Biochemistry* Vol. 19, No. 15, pp. 3522-3530, ISSN 0006-2960.

Lohman, T. M. & Mascotti, D. P. (1992). Thermodynamics of ligand-nucleic acid interactions. *Methods Enzymol.* Vol. 212, pp. 400-424, ISSN 0076-6879.

MacKerell, A. D., Jr. & Nilsson, L. (2008). Molecular dynamics simulations of nucleic acid-protein complexes. *Curr. Opin. Struct. Biol.* Vol. 18, No. 2, pp. 194-199, ISSN 0959-440X.

Manning, G. S. (1969). Limiting laws and counterion condensation in polyelectrolyte solutions. I. Colligative properties. *J. Chem. Phys.* Vol. 51, No. 3, pp. 924-933, ISSN 0021-9606.

Manning, G. S. (1977). Limiting laws and counterion condensation in polyelectrolyte solutions: IV. The approach to the limit and the extraordinary stability of the charge fraction. *Biophys. Chem.* Vol. 7, No. 2, pp. 95-102, ISSN 0301-4622.

Manning, G. S. (1978). The molecular theory of polyelectrolyte solutions with applications to the electrostatic properties of polynucleotides. *Q. Rev. Biophys.* Vol. 11, No. 02, pp. 179-246, ISSN 0033-5835.

Manning, G. S. (2003). Comments on selected aspects of nucleic acid electrostatics. *Biopolymers* Vol. 69, No. 1, pp. 137-143, ISSN 1097-0282.

Manning, G. S. & Mohanty, U. (1997). Counterion condensation on ionic oligomers. *Physica A* Vol. 247, No. 1-4, pp. 196-204, ISSN 0378-4371.

Mascotti, D. P. & Lohman, T. M. (1990). Thermodynamic extent of counterion release upon binding oligolysines to single-stranded nucleic acids. *Proc. Natl. Acad. Sci. USA* Vol. 87, No. 3142, pg. 3146, ISSN 0027-8424.

Mascotti, D. P. & Lohman, T. M. (1993). Thermodynamics of single-stranded RNA and DNA interactions with oligolysines containing tryptophan. Effects of base composition. *Biochemistry* Vol. 32, No. 40, pp. 10568-10579, ISSN 0006-2960.

Mascotti, D. P. & Lohman, T. M. (1995). Thermodynamics of charged oligopeptide-heparin interactions. *Biochemistry* Vol. 34, No. 9, pp. 2908-2915, ISSN 0006-2960.

Materese, C. K.; Savelyev, A. & Papoian, G. A. (2009). Counterion atmosphere and hydration patterns near a nucleosome core particle. *J. Am. Chem. Soc.* Vol. 131, No. 41, pp. 15005-15013, ISSN 0002-7863.

Mills, P.; Anderson, C. F. & Record, M. T., Jr. (1986). Grand canonical Monte Carlo calculations of thermodynamic coefficients for a primitive model of DNA-salt solutions. *J. Phys. Chem.* Vol. 90, No. 24, pp. 6541-6548, ISSN 0022-3654.

Misra, V. K. & Draper, D. E. (1999). The interpretation of Mg^{2+} binding isotherms for nucleic acids using Poisson-Boltzmann theory. *J. Mol. Biol.* Vol. 294, No. 5, pp. 1135-1147, ISSN 0022-2836.

Misra, V. K.; Sharp, K. A.; Friedman, R. A. & Honig, B. (1994). Salt effects on ligand-DNA binding - minor-groove binding antibiotics. *J. Mol. Biol.* Vol. 238, No. 2, pp. 245-263, ISSN 0022-2836.

Mohanty, U. & Stellwagen, N. C. (1999). Free solution mobility of oligomeric DNA. *Biopolymers* Vol. 49, No. 3, pp. 209-214, ISSN 1097-0282.

Montoro, J. C. G. & Abascal, J. L. F. (1998). Ionic distribution around simple B-DNA models II. Deviations from cylindrical symmetry. *J. Chem. Phys.* Vol. 109, No. 14, pp. 6200-6210, ISSN 0021-9606.

Murthy, C. S.; Bacquet, R. J. & Rossky, P. J. (1985). Ionic distributions near polyelectrolytes. A comparison of theoretical approaches. *J. Phys. Chem.* Vol. 89, No. 4, pp. 701-710, ISSN 0022-3654.

Ni, H. H.; Anderson, C. F. & Record, M. T., Jr. (1999). Quantifying the thermodynamic consequences of cation (M^{2+}, M^{+}) accumulation and anion (X^{-}) exclusion in mixed salt solutions of polyanionic DNA using Monte Carlo and Poisson-Boltzmann calculations of ion-polyion preferential interaction coefficients. *J. Phys. Chem. B* Vol. 103, No. 17, pp. 3489-3504, ISSN 1089-5647.

Olivera, B. M.; Baine, P. & Davidson, N. (1964). Electrophoresis of the nucleic acids. *Biopolymers* Vol. 2, No. 3, pp. 245-257, ISSN 1097-0282.

Olmsted, M. C.; Anderson, C. F. & Record, M. T., Jr. (1989). Monte Carlo description of oligoelectrolyte properties of DNA oligomers: range of the end effect and the approach of molecular and thermodynamic properties to the polyelectrolyte limits. *Proc. Natl. Acad. Sci. USA* Vol. 86, No. 20, pp. 7766-7770, ISSN 0027-8424.

Olmsted, M. C.; Anderson, C. F. & Record, M. T., Jr. (1991). Importance of oligoelectrolyte end effects for the thermodynamics of conformational transitions of nucleic acid oligomers: A grand canonical Monte Carlo analysis. *Biopolymers* Vol. 31, No. 13, pp. 1593-1604, ISSN 1097-0282.

Olmsted, M. C.; Bond, J. P.; Anderson, C. F. & Record, M. T., Jr. (1995). Grand canonical Monte Carlo molecular and thermodynamic predictions of ion effects on binding of an oligocation (L^{8+}) to the center of DNA oligomers. *Biophys. J.* Vol. 68, No. 2, pp. 634-647, ISSN 0006-3495.

Owczarzy, R.; Moreira, B. G.; You, Y.; Behlke, M. A. & Walder, J. A. (2008). Predicting stability of DNA duplexes in solutions containing magnesium and monovalent cations. *Biochemistry* Vol. 47, No. 19, pp. 5336-5353, ISSN 0006-2960.

Pack, G. R.; Wong, L. & Lamm, G. (1999). Divalent cations and the electrostatic potential around DNA: Monte Carlo and Poisson–Boltzmann calculations. *Biopolymers* Vol. 49, No. 7, pp. 575-590, ISSN 1097-0282.

Paulsen, M. D.; Richey, B.; Anderson, C. F. & Record, M. T., Jr. (1987). The salt dependence of the preferential interaction coefficient in DNA solutions as determined by grand canonical Monte Carlo simulations. *Chem. Phys. Lett.* Vol. 139, No. 5, pp. 448-452, ISSN 0009-2614.

Pérez, A.; Luque, F. J. & Orozco, M. (2007). Dynamics of B-DNA on the microsecond time scale. *J. Am. Chem. Soc.* Vol. 129, No. 47, pp. 14739-14745, ISSN 0002-7863.

Plum, G. E. & Bloomfield, V. A. (1988). Equilibrium dialysis study of binding of hexammine cobalt(III) to DNA. *Biopolymers* Vol. 27, No. 6, pp. 1045-1051, ISSN 0006-3525.

Ponomarev, S. Y.; Thayer, K. M. & Beveridge, D. L. (2004). Ion motions in molecular dynamics simulations on DNA. *Proc. Natl. Acad. Sci. USA* Vol. 101, No. 41, pp. 14771-14775.

Prabhu, V. M. (2005). Counterion structure and dynamics in polyelectrolyte solutions. *Curr. Opin. Colloid Interface Sci.* Vol. 10, No. 1-2, pp. 2-8, ISSN 1359-0294.

Putnam, C. D.; Hammel, M.; Hura, G. L. & Tainer, J. A. (2007). X-ray solution scattering (SAXS) combined with crystallography and computation: defining accurate macromolecular structures, conformations and assemblies in solution. *Q. Rev. Biophys.* Vol. 40, No. 03, pp. 191-285, ISSN 0033-5835.

Ramanathan, G. V. & Woodbury, J. (1982). Statistical mechanics of electrolytes and polyelectrolytes. II. Counterion condensation on a line charge. *J. Chem. Phys.* Vol. 77, No. 8, pp. 4133-4140, ISSN 0021-9606.

Rambo, R. P. & Tainer, J. A. (2010). Bridging the solution divide: comprehensive structural analyses of dynamic RNA, DNA, and protein assemblies by small-angle x-ray scattering. *Curr. Opin. Struct. Biol.* Vol. 20, No. 1, pp. 128-137, ISSN 0959-440X.

Record, M. T., Jr.; Anderson, C. F. & Lohman, T. M. (1978). Thermodynamic analysis of ion effects on the binding and conformational equilibria of proteins and nucleic acids: the roles of ion association or release, screening, and ion effects on water activity. *Q. Rev. Biophys.* Vol. 11, No. 02, pp. 103-178, ISSN 0033-5835.

Record, M. T., Jr.; Lohman, T. M. & Haseth, P. d. (1976). Ion effects on ligand-nucleic acid interactions. *J. Mol. Biol.* Vol. 107, No. 2, pp. 145-158, ISSN 0022-2836.

Record, M. T., Jr.; Zhang, W. & Anderson, C. F. (1998). Analysis of effects of salts and uncharged solutes on protein and nucleic acid equilibria and processes: a practical guide to recognizing and interpreting polyelectrolyte effects, Hofmeister effects, and osmotic effects of salts. *Adv. Protein Chem.* Vol. 51, pp. 281-353, ISSN 0065-3233.

Rouzina, I. & Bloomfield, V. A. (1997). Competitive electrostatic binding of charged ligands to polyelectrolytes: practical approach using the non-linear Poisson-Boltzmann equation. *Biophys. Chem.* Vol. 64, No. 1-3, pp. 139-155, ISSN 0301-4622.

Sanbonmatsu, K. Y.; Joseph, S. & Tung, C. S. (2005). Simulating movement of tRNA into the ribosome during decoding. *Proc. Natl. Acad. Sci. USA* Vol. 102, No. 44, pp. 15854-15859, ISSN 0027-8424.

SantaLucia, J. J. (1998). A unified view of polymer, dumbbell, and oligonucleotide DNA nearest-neighbor thermodynamics. *Proc. Natl. Acad. Sci. USA* Vol. 95, No. 4, pp. 1460-1465, ISSN 0027-8424.

SantaLucia, J. J.; Ilawi, H. T. & Seneviratne, P. A. (1996). Improved nearest-neighbor parameters for predicting DNA duplex stability. *Biochemistry* Vol. 35, No. 11, pp. 3555-3562, ISSN 0006-2960.

Sen, A.; Ray, A. & Maiti, M. (1996). Thermodynamics of the interactions of sanguinarine with DNA: influence of ionic strength and base composition. *Biophys. Chem.* Vol. 59, No. 1-2, pp. 155-170, ISSN 0301-4622.

Sharp, K. A. (1995). Polyelectrolyte electrostatics: salt dependence, entropic, and enthalpic contributions to free energy in the nonlinear Poisson–Boltzmann model. *Biopolymers* Vol. 36, No. 2, pp. 227-243, ISSN 1097-0282.

Sharp, K. A.; Friedman, R. A.; Misra, V.; Hecht, J. & Honig, B. (1995). Salt effects on polyelectrolyte–ligand binding: comparison of Poisson–Boltzmann, and limiting law/counterion binding models. *Biopolymers* Vol. 36, No. 2, pp. 245-262, ISSN 1097-0282.

Shkel, I. A. (2010). Coulombic free energy of polymeric nucleic acid: low- and high-salt analytical approximations for the cylindrical Poisson–Boltzmann model. *J. Phys. Chem. B* Vol. 114, No. 33, pp. 10793-10803, ISSN 1520-6106.

Shkel, I. A.; Ballin, J. D. & Record, M. T., Jr. (2006). Interactions of cationic ligands and proteins with small nucleic acids: analytic treatment of the large coulombic end effect on binding free energy as a function of salt concentration. *Biochemistry* Vol. 45, No. 27, pp. 8411-8426, ISSN 0006-2960.

Shkel, I. A. & Record, M. T., Jr. (2004). Effect of the number of nucleic acid oligomer charges on the salt dependence of stability (ΔG^{o}_{37}) and melting temperature (T_m): NLPB analysis of experimental data. *Biochemistry* Vol. 43, No. 22, pp. 7090-7101, ISSN 0006-2960.

Shkel, I. A.; Tsodikov, O. V. & Record, M. T., Jr. (2000). Complete asymptotic solution of cylindrical and spherical Poisson-Boltzmann equations at experimental salt concentrations. *J. Phys. Chem. B* Vol. 104, No. 21, pp. 5161-5170, ISSN 1089-5647.

Shkel, I. A.; Tsodikov, O. V. & Record, M. T., Jr. (2002). Asymptotic solution of the cylindrical nonlinear Poisson-Boltzmann equation at low salt concentration: analytic expressions for surface potential and preferential interaction coefficient. *Proc. Natl. Acad. Sci. USA* Vol. 99, No. 5, pp. 2597-2602, ISSN 0027-8424.

Slater, G. W.; Holm, C.; Chubynsky, M. V.; de Haan, H. W.; Dubé, A.; Grass, K.; Hickey, O. A.; Kingsburry, C.; Sean, D.; Shendruk, T. N. & Zhan, L. (2009). Modeling the separation of macromolecules: A review of current computer simulation methods. *Electrophoresis* Vol. 30, No. 5, pp. 792-818, ISSN 1522-2683.

Slater, G. W.; Guillouzic, S.; Gauthier, M. G.; Mercier, J.-F.; Kenward, M.; McCormick, L. C. & Tessier, F. (2002). Theory of DNA electrophoresis (~1999 - 2002 1/2). *Electrophoresis* Vol. 23, No. 22-23, pp. 3791-3816, ISSN 1522-2683.

Spasic, A. & Mohanty, U. (2008).*Counterion condensation in nucleic acid*. S. A. Rice, (Ed.), pp. 139-176, John Wiley & Sons, Inc., ISBN 9780470259498, Ithaca NY.

Stein, V. M.; Bond, J. P.; Capp, M. W.; Anderson, C. F. & Record, M. T., Jr. (1995). Importance of coulombic end effects on cation accumulation near oligoelectrolyte B-DNA: a demonstration using ^{23}Na NMR. *Biophys. J.* Vol. 68, No. 3, pp. 1063-1072, ISSN 0006-3495.

Stellwagen, N. C.; Gelfi, C. & Righetti, P. G. (1997). The free solution mobility of DNA. *Biopolymers* Vol. 42, No. 6, pp. 687-703, ISSN 1097-0282.

Stigter, D. (1995). Evaluation of the counterion condensation theory of polyelectrolytes. *Biophys. J.* Vol. 69, No. 2, pp. 380-388, ISSN 0006-3495.

Stigter, D. & Dill, K. A. (1996). Binding of ionic ligands to polyelectrolytes. *Biophys. J.* Vol. 71, No. 4, pp. 2064-2074, ISSN 0006-3495.

Strauss, U. P.; Helfgott, C. & Pink, H. (1967). Interactions of polyelectrolytes with simple electrolytes. II. Donnan equilibria obtained with DNA in solutions of 1-1 electrolytes. *J. Phys. Chem.* Vol. 71, No. 8, pp. 2550-2556, ISSN 0022-3654.

Tan, Z. J. & Chen, S. J. (2009). Predicting electrostatic forces in RNA folding. Vol. 469, pp. 465-487, ISSN 0076-6879.

Trizac, E. & Téllez, G. (2007). Preferential interaction coefficient for nucleic acids and other cylindrical polyions. *Macromolecules* Vol. 40, No. 4, pp. 1305-1310, ISSN 0024-9297.

Viovy, J. L. (2000). Electrophoresis of DNA and other polyelectrolytes: physical mechanisms. *Rev. Mod. Phys.* Vol. 72, No. 3, pg. 813, ISSN 0034-6861.

Wang, K.; Yu, Y. X.; Gao, G. H. & Luo, G. S. (2007). Preferential interaction between DNA and small ions in mixed-size counterion systems: Monte Carlo simulation and density functional study. *J. Chem. Phys.* Vol. 126, No. 13, pp. 135102-135108, ISSN 0021-9606.

Wilcoxon, J. P. & Schurr, J. M. (1983). Electrophoretic light scattering studies of poly(L-lysine) in the ordinary and extraordinary phase. Effects of salt, molecular weight, and polyion concentration. *J. Chem. Phys.* Vol. 78, No. 6, pp. 3354-3364, ISSN 0021-9606.

Williams, A. P.; Longfellow, C. E.; Freier, S. M.; Kierzek, R. & Turner, D. H. (1989). Laser temperature-jump, spectroscopic, and thermodynamic study of salt effects on duplex formation by dGCATGC. *Biochemistry* Vol. 28, No. 10, pp. 4283-4291, ISSN 0006-2960.

Wilson, W. D.; Tanious, F. A.; Barton, H. J.; Jones, R. L.; Fox, K.; Wydra, R. L. & Strekowski, L. (1990). DNA sequence dependent binding modes of 4',6-diamidino-2-phenylindole (DAPI). *Biochemistry* Vol. 29, No. 36, pp. 8452-8461, ISSN 0006-2960.

Wong, G. C. L. & Pollack, L. (2010). Electrostatics of strongly charged biological polymers: ion-mediated interactions and self-organization in nucleic acids and proteins. *Annu. Rev. Phys. Chem.* Vol. 61, No. 1, pp. 171-189.

Wu, J. & Morikis, D. (2006). Molecular thermodynamics for charged biomacromolecules. *Fluid Phase Equilib.* Vol. 241, No. 1-2, pp. 317-333, ISSN 0378-3812.

Young, M. A.; Jayaram, B. & Beveridge, D. L. (1997). Intrusion of counterions into the spine of hydration in the minor groove of B-DNA: fractional occupancy of electronegative pockets. *J. Am. Chem. Soc.* Vol. 119, No. 1, pp. 59-69, ISSN 0002-7863.

Zhang, W.; Bond, J. P.; Anderson, C. F.; Lohman, T. M. & Record, M. T., Jr. (1996). Large electrostatic differences in the binding thermodynamics of a cationic peptide to oligomeric and polymeric DNA. *Proc. Natl. Acad. Sci. USA* Vol. 93, No. 6, pp. 2511-2516, ISSN 0027-8424.

Zhang, W. T.; Ni, H. H.; Capp, M. W.; Anderson, C. F.; Lohman, T. M. & Record, M. T., Jr. (1999). The importance of coulombic end effects: experimental characterization of the effects of oligonucleotide flanking charges on the strength and salt dependence of oligocation (L^{8+}) binding to single-stranded DNA oligomers. *Biophys. J.* Vol. 76, No. 2, pp. 1008-1017, ISSN 0006-3495.

6

Tandem DNA Repeats: Generation and Propagation in the Microgene Polymerization Reaction and *in vivo*

Mark Itsko[1,4], Eitan Ben-Dov[2], Avinoam Rabinovitch[3] and Arieh Zaritsky[1]
Departments of [1]Life Sciences and [3]Physics, Ben-Gurion University of the Negev, PO Box 653, Be'er-Sheva 84105, Israel,
[2]Achva Academic College, MP Shikmim, 79800, Israel
[4]Present address: Laboratory of Molecular Genetics, National Institute of Environmental Health Sciences, Research Triangle Park, North Carolina 27709, USA

1. Introduction

Short and long-motif periodicities are omnipresent in genomes of both eukaryotes (Britten & Kohne 1968) and prokaryotes (Hofnung & Shapiro, 1999). Repetitive DNA can amount to more than half of higher eukaryotic genomes, specifically in *Homo sapience* (International Human Genome Consortium, 2001) and *Zea mays* (Meyers et al., 2001). Even in prokaryotes it can account for about 6% of the total genome, specifically in *Mycoplasma pneumoniae* (Ruland et al., 1990) and *Neisseria meningitides* (Parkhill et al., 2000). Two major types of repetitive DNA exist (Dogget, 2000): tandem (head-to-tail contiguous) and interspersed (non-contiguous) with a specific pattern scattered all over the genome and lengths varying up to several hundred nucleotides (nt).

Emergence and propagation of interspersed repeats such as REP (Repetitive Extra-genic Palindrome) found in bacteria and *Alu* repeats that are abundant in the human genome are putatively attributed to reproduction processes mediated by transposons (Gilson et al., 1984) or retroviruses (Ullu & Tschudi, 1984). These are multi-step complex enzymatic processes and will not be considered here. This chapter deals with tandem repeats and their generation *in vitro* by the Microgene Polymerization Reaction (MPR) (Itsko et al., 2008a; 2008b; 2009), and proposes a mechanism for their *in vivo* generation as well.

Most tandem repetitive DNA sequences in higher eukaryotes are located near the chromosomal telomers or centromers where they play important roles in maintaining genome integrity (Blackburn, 1991) and segregation (Catasti et al., 1994). In addition to this un-transcribed but evidently functional repetitive DNA, the human genome contains many apparently non-functional repetitive DNA sequences in the forms of micro-satellites (a variety of di-, tri-, tetra-, and penta-nucleotide tandem repeats) and mini-satellites (30-35 bp long, with variable sequences and conserved cores of 10-15 bp) (Dogget, 2000). The number of repeats is prone to expand during replication because its constituent strands can slide over each other between the multiple complementary regions (Wells, 1996). Thermodynamically unfavorable structures bulging out from DNA duplex that accompany the strand sliding process can be stabilized by the inner base-pairs which facilitate the expansion (Kang et al., 1995). Expansion of DNA repeats is associated with a variety of human hereditary diseases (Mirkin, 2007).

The conceptual views on repetitive DNA were changed from considering it as parasitic DNA (Orgel & Crick, 1980) up to necessary organizer of genomic information (Shapiro & von Sternberg, 2005). Furthermore, numerous periodicities encrypted in encoded proteins may reflect the evolution of modern coding sequences based on primordial oligomeric repeats (Ohno, 1987). Unused preexisted long repetitive sequences may yield coding frames expressing unique enzymes even currently (Ohno, 1984).

Different *in vitro* systems are used to study mechanisms underlying genomic repeat expansion and possible evolutionary aspects of this process. These systems include double- or single-stranded DNA with relatively short (4-8 nt) repetitive units and thermophilic DNA polymerases operating under isothermal conditions. Staggered re-annealing of constituent strands is putatively involved in expansion of repetitive DNA duplexes (Tuntiwechapikul & Salazar, 2002; Fig. 1A) whereas hairpin elongation was proposed for extension of single-stranded DNA with palindromes (Ogata & Miura, 2000; Ogata & Morino, 2000; Fig. 1B). Applying temperatures near the melting point of the hairpin-coil transition in the former or in the starting duplex in the latter profoundly facilitate such expansions. Total repetitive DNA synthesis is greatly accelerated if hairpin elongation is combined with endonucleolytic digestion of obtained long repetitive products (Liang et al., 2004; Fig. 1C).

The MPR, which was initially developed to produce artificial proteins containing repetitive motifs (Shiba et al., 1997), can be used to study not only DNA repeat propagation but their enigmatic generation as well. In this system, a medium-sized (40-50 nt) *non-repetitive homo-duplex* DNA (HD) evolves into multiple head-to-tail repetitive products during heat-cool cycles in the PCR.

In this chapter we consider this reaction, in strict terms of physical and polymer chemistry, to decipher the steps composing it and the mechanisms underlying their consecutive development. Several unorthodox views and consistent experimental results are described that obey basic thermodynamic rules, and analogous *in vivo* reactions are discussed in light of repeated motifs observed among subspecies of an entomopathogenic bacterial species.

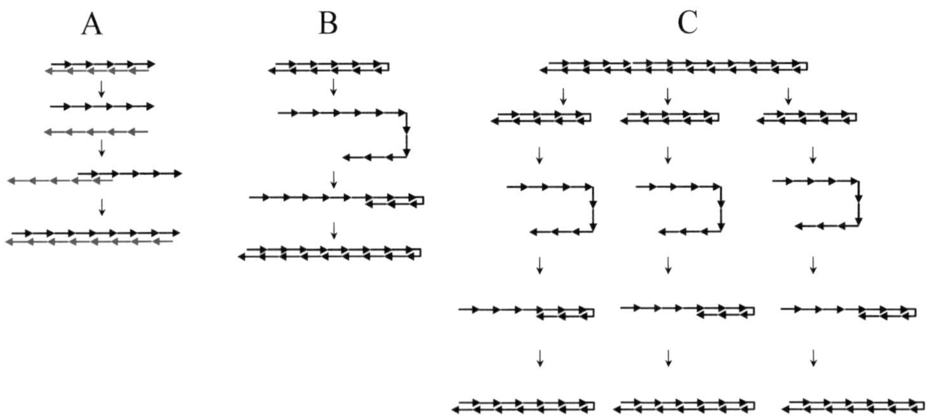

Fig. 1. *in vitro* systems for repeat expansion of a homo-duplex (**A**), single stranded (ss) DNA, (**B**) and ss-DNA with administered endonuclease (**C**). Black arrows, forward repeating units; grey arrows, reverse repeating units

<table>
<tr><th>Name</th><th>Sequence[a]</th><th>GC (%)</th><th>T_m[b] (°C)</th><th>ΔG_{OS}[c] (kcal mol^{-1})</th></tr>
<tr><td>NOMUL</td><td>5'-GG**AA**ATAGAAGAACTTAAATCTTTATTAG**AG**ATTAAC**ACA**GC-3'
| | | | | | | | | | | | | | | | | | | | | | | | | | | | | | | | | | | | | | | | | |
3'-CC**TT**TATCTTCTTGAATTTAGAAATAATC**TC**TAATTG**TGT**CG-5'</td><td>28.6</td><td>67 (68)</td><td>-43</td></tr>
<tr><td>NOMU</td><td>5'-GGTGATAGAAG**AA**CTTAAATCTTTATTAGGAATT**AAC**CTGGC-3'
| | | | | | | | | | | | | | | | | | | | | | | | | | | | | | | | | | | | | | | | | |
3'-CCACTATCTTC**TT**GAATTTAGAAATAATCCTTAA**TTG**GACCG-5'</td><td>33.3</td><td>68 (68)</td><td>-45</td></tr>
<tr><td>EVNA</td><td>5'-GGTGATAGAAG**TG**CTTAAATCTTTATTAGGAATT**GCT**CTGGC-3'
| | | | | | | | | | | | | | | | | | | | | | | | | | | | | | | | | | | | | | | | | |
3'-CCACTATCTTC**AC**GAATTTAGAAATAATCCTTAA**CGA**GACCG-5'</td><td>38.1</td><td>71 (72)</td><td>-48</td></tr>
<tr><td>EVNAH</td><td>5'-GGTGATAG**TGC**TGCTT**TG**ATCTTTATTAGGAATTGCTCTGGC-3'
| | | | | | | | | | | | | | | | | | | | | | | | | | | | | | | | | | | | | | | | | |
3'-CCACTATC**ACG**ACGAA**AC**TAGAAATAATCCTTAACGAGACCG-5'</td><td>42.9</td><td>73 (74)</td><td>-50</td></tr>
</table>

[a] Bold type letters indicate differences in composition between EVNA and NOMU. Underlined bold type letters indicate differences in composition between NOMUL and NOMU and between EVNA and EVNAH.

[b] Experimental and UNAFold-predicted (in brackets) melting temperature of the corresponding *OS* types.

[c] Association free energy (Visual OMP6-predicted) of complementary strands into *OS* at 37°C.

Reprinted from Itsko et al., 2008b with permission from Elsevier.

Table 1. Original homo-duplexes, called Original Singlets (*OS*) used in MPR research

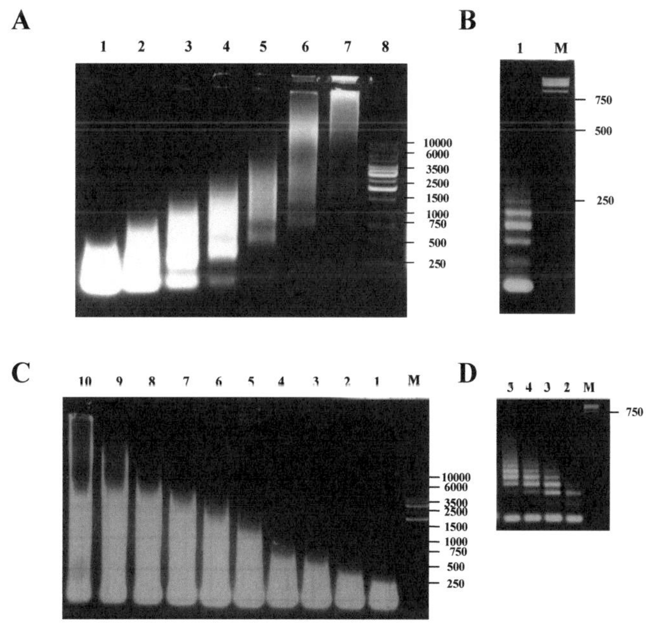

Fig. 2. Demonstration of MPR with EVNA *OS* (Table 1) and its concentration-dependence on the reactants. Change of product length with concentrations of EVNA with Vent (**A, B**) and of dNTP (**C** with Taq, **D** with Vent) on 0.8% (**A, C**) and 2.5% (**B, D**) agarose gels. (**A, B**) Lanes 1-7, 10.4, 5.2, 2.6, 1.3, 0.64, 0.32, 0.16 µM of EVNA, respectively. Lanes 8 or M, DNA ladder. (**C, D**) The following concentrations (in µM) of each dNTP were used: lane 1, 100; lane 2, 200; lane 3, 300; lane 4, 400; lane 5, 500; lane 6, 600; lane 7, 700; lane 8, 800; lane 9, 900; lane 10, 1,000. Concentration of the EVNA homo-duplex, 5.6 µM. Lanes M, DNA ladder. Reprinted from Itsko et al., 2008a with permission from Elsevier

2. Demonstration of the MPR

MPR can easily be demonstrated by applying long (up to 64 cycles) end-point-detection PCR heat/cool cycling conditions on reaction mixtures containing DNA in the form of homo-duplex (HD) of length 40-50 bp, the sequence of which does not include any repetitive motif (e.g., Table 1). The PCR product is composed of long DNA stretches of heterogeneous length that is visualized as a smear on loose agarose gels and resolvable into discrete bands of HD multiples on dense gels (Fig. 2).

3. Overall scheme of MPR

The MPR is kinetically divided into three stages: initiation, amplification and propagation (Fig. 3), each is subdivided into a number of steps that are considered later in the chapter. Here they are formulated in a simplified way with the following variables and parameters:

HD_i - homo-duplexes of DNA containing a variable number i repeats. Correspondingly, HD_1 is original non-repetitive homo-duplex called also OS (original singlet) with which the MPR starts, and HD_2 is doublet of homo-duplexes designated as D, including initial doublet (ID) that triggers propagation.

k_I is the constant rate of MPR initiation.

k_{Ampl} is the constant rate of ID amplification by OS.

k_{Pr} is the constant rate of MPR repeat propagation per PCR cycle, assumed to be independent of polymer length (measured in repeat units n).

The symbols embraced in square brackets designate concentrations of corresponding DNA species.

Initiation

The minimal repetitive unit that is prone to expand by staggered re-annealing and replication of overhangs (Fig. 3B) is D. Propagation is therefore initiated by generation of a so-called initial doublet (ID). The equation formulating the simplified process of initiation (Fig. 3 A) is:

$$2OS \xrightarrow{k_I} ID . \quad (1)$$

The mechanism of this reaction and its molecularity will be discussed in sections 5 and 6.

Amplification

The initiator (ID) can be amplified by the OS (Fig. 3 B) according to:

$$OS + ID \xrightarrow{k_{Ampl}} 2D . \quad (2)$$

Since ID and D are the same molecular species generated by different mechanisms, the reaction describes an autocatalytic process, in which the mass concentration of the initiator is rapidly brought to that of OS (see section **7** below). If $k_{Ampl} = k_{Pr}$, the amplification stage is kinetically included in the following propagation stage.

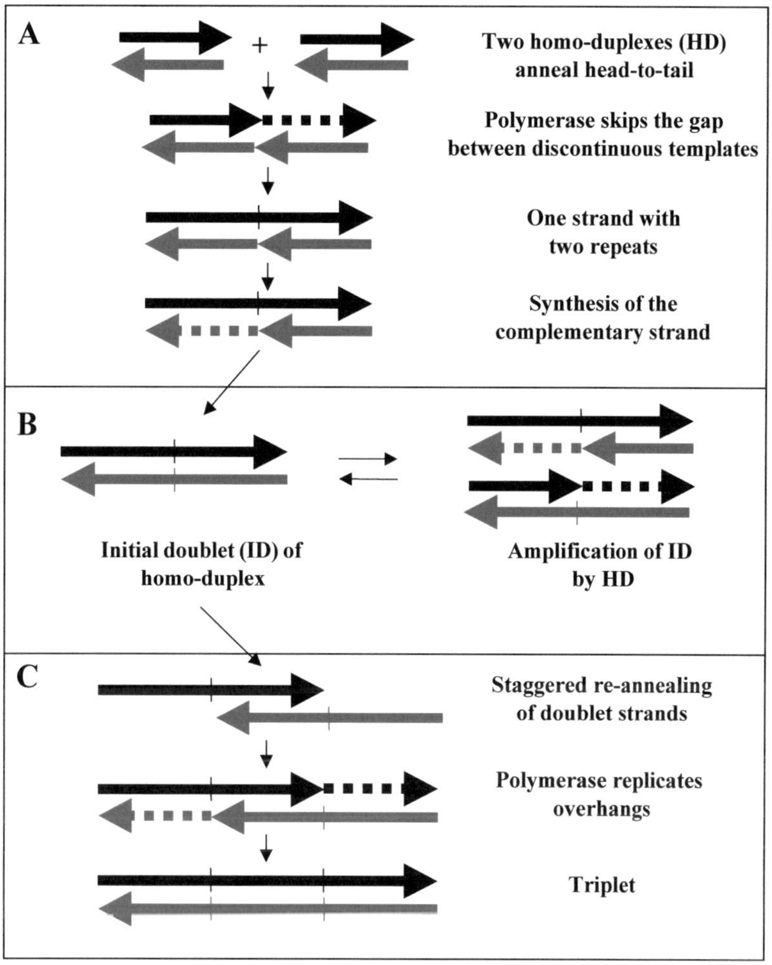

Fig. 3. A model for MPR. (**A**) Initiation. (**B**) The initiator (ID) and its amplification. (**C**) The propagation of ID in the first heat/cool cycle. Reprinted from Itsko et al., 2009 with permission from Elsevier

Propagation

After generation of doublets, the number of repeats in DNA is expanding according to

$$HD_i + HD_j \xrightarrow{k_{pr}} 2HD_n , \tag{3}$$

where $n \le i + j - 1$, since at least one repeat is always hidden in the overlap paired region (Fig. 3C).

Propagation is also an autocatalytic process resulting in an exponential growth of the number of repeats per one polymer molecule $\langle n \rangle_N$. This is justified as follows. In the extension reaction (eq. 3), two molecules of lengths n_1 and n_2 ($n_1 \le n_2$) yield two molecules with lengths in the range between n_1+1 and n_1+n_2-1. Assuming that the population of such repetitive products is uniformly distributed with a common difference of one repeat unit, the average length of the product $\langle n\rangle = ((n_1+1)+(n_1+n_2-1))/2 = n_1 + n_2/2$. Averaging over all possible reactions of this sort yields

$$\langle n\rangle_{N+1} = \langle n_1\rangle_N + \langle n_2\rangle_N/2\,. \tag{4}$$

Since $\langle n_1\rangle_N = \langle n_2\rangle_N = \langle n\rangle_N$ (belonging to the same distribution), the average length of the polymer increases by a factor of 1.5 at each cycle:

$$\langle n\rangle_{N+1} = 3\langle n\rangle_N\,/\,2 \text{ or } \langle n\rangle_N = \langle n\rangle_0 \times 1.5^N\,. \tag{5}$$

Propagation is finished when nucleotides (dNTPs) are depleted.

4. Extent of polymerization in MPR

The final polymer length in chain-growth polymerization reactions is determined by the "kinetic chain length" defined as the number of monomer units consumed in the propagation stage per active center produced in the initiation stage "(Flory, 1953; Atkins, 1994) as visualized in Fig. 4. Accordingly, the mean MPR product size increases with decreasing *OS* and increasing dNTP concentrations (Itsko et al., 2008a). The nucleotide concentration [*dNTP*] presented in the reaction mixture determines the extent of propagation, whereas that of the [*ID*] (equal to $[OS]_0$ due to amplification (eq. 2)) represents active centers. Thus, in MPR-produced multiple repetitive DNA, the final length

$$\langle n\rangle = (1/m)\times[dNTP]/[OS]_0 \tag{6},$$

where *m* is the number of nucleotides composing one *OS*. Eq. 6 is concordant with the experimental results (Fig. 2).

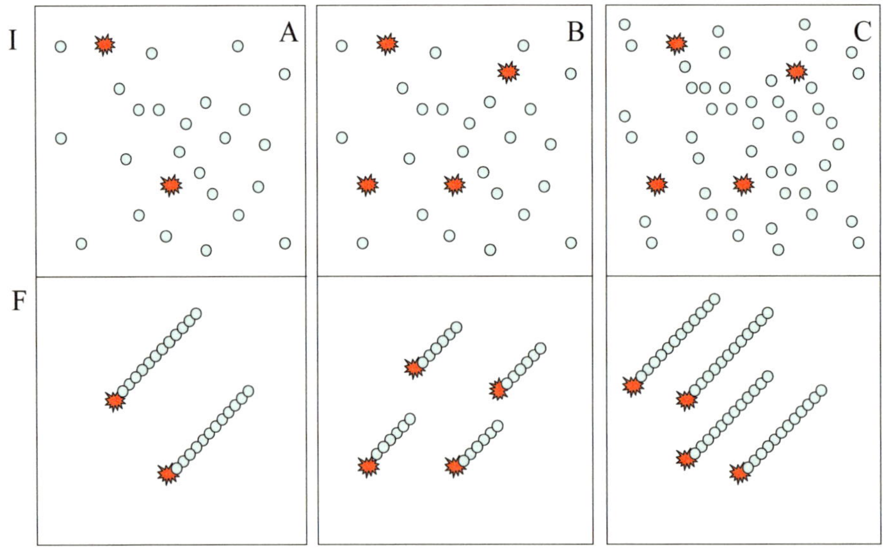

Fig. 4. Extent of polymerization in the MPR. Illustrative demonstration of its dependence on concentrations of dNTP (blue spots) and ID (red stars), at given concentrations of the reactants (**A**), at twice the concentration of ID (**B**) and at twice the concentrations of both (**C**). I, initial reactants; F, final products

5. Molecularity of the initiation stage

Progress of MPR can be followed by Real-Time PCR (Fig. 5). The initial concentration of *OS* ($[OS]_0$) determines the cycle number (the operative threshold cycle) *N*th (in cycle units *N*) at which the signal (the microgene expansion process) is first detected. A 2-fold decrease in $[OS]_0$ delayed *N*th by approximately 4 cycles (i.e., Δ*N*th = 4; Fig. 5).

Following the initiation process, total MPR products (*Tot*) are propagated exponentially formulated by

$$[Tot]_N = [ID] \times (1+E)^N \qquad (7),$$

where *E* is the amplification efficiency per cycle. Average *E* determined experimentally is about 0.7 (between 0.8 and 0.57), larger than that calculated from general considerations (eq. 5). The generation-rate of the initial doublet *ID* is assumed to be proportional to the power *m* (molecularity) of the concentration of *OS* $d[ID]/dN \propto [OS]^m$ (Fig. 3, A, B). The rate of amplification that has exponential nature (eq. 2) is much higher than that of the initiation, and hence the consumption of *OS* molecules in the initiation process is negligible compared to that during the amplification stage. The initiation process would therefore not affect $[OS]_0$ hence $[ID]$ would be $\propto N \times [OS]_0^m$. The MPR initiation is reasonably assumed (Itsko et al., 2008a) to occur in all $[OS]_0$ at the first cycle $N = 1$ (though the low sensitivity of RT-PCR detects the products at a later stage, i.e., at the *N*th). The experiments were performed with a series of 2-fold-$[OS]_0$ values (Fig. 5). The predicted ratio between $[ID]$ of two successive dilutions is thus

$$\frac{[ID_2]}{[ID_1]} = \frac{\left(2 \times [OS]_0\right)^m}{[OS]_0^m} = 2^m \quad . \qquad (8)$$

The exponential propagation of *Tot* following generation of *ID* depends on the number of cycles and amplification efficiency (eq. 7). If *E* (derived from the slope in each line of Fig. 5) remains constant, the ratio between successive $[ID]$ values in the RT-PCR experiments with two-fold different initial concentrations of *OS* may be derived at points with equal amount of $[Tot]_N$, as for example the threshold points (Fig. 5) according to

$$\frac{[ID_2]}{[ID_1]} = (1+E)^{(Nth_1 - Nth_2)}, \qquad (9)$$

and the molecularity m of the reaction can be derived by equating eq. 8 with eq. 9:

$$m = (Nth_1 - Nth_2) \times \frac{\ln(1+E)}{\ln 2} . \qquad (10)$$

The observed values of *E* however, do vary slightly (0.8–0.6) between successive lines (Fig. 5). A molecularity of 3.1 (ranging between 2.6 and 3.4) for the initiation process was estimated, using the average *E* value (0.71). It can thus be concluded that three *OS* must somehow interact to initiate doublet formation in the MPR.

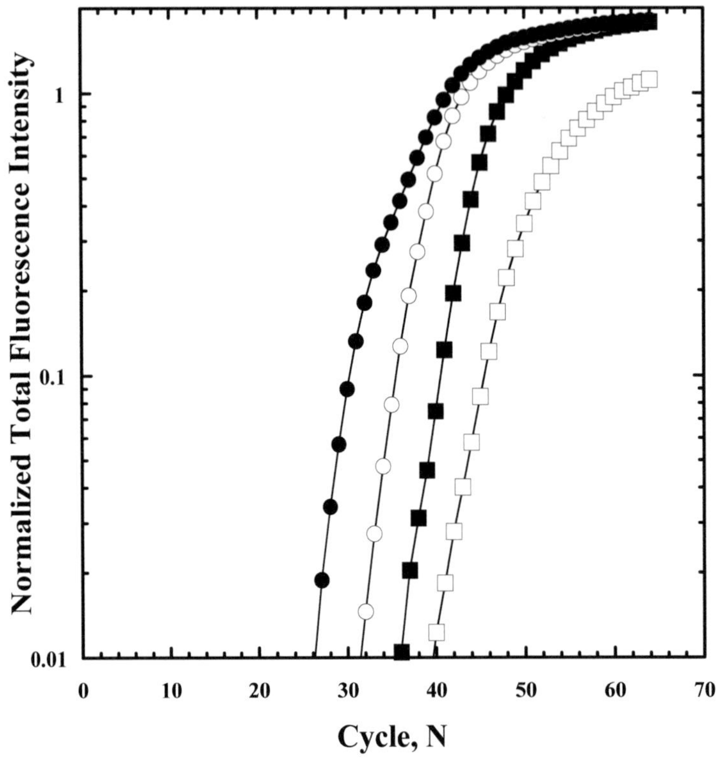

Fig. 5. Relative total MPR products generated by RT-PCR with the following initial EVNA homo-duplex concentrations (in μM): closed circles, 0.32; open circles, 0.16; closed squares, 0.08; open squares, 0.04. Reprinted from Itsko et al., 2008a with permission from Elsevier

6. Kinetics and thermodynamics of initiation

Third-order kinetics of initiation leads to a simple mechanism for the generation of *ID*: a rare and reversible association between three *OS* generates a nucleation complex (*NC*), (Fig. 6), which converts to *ID* according to:

$$3OS \xleftrightarrow{I} NC \xrightarrow{II} ID + 2OS \quad (11),$$

where *I* and *II* denote first (non-enzymatic fast equilibrium) and second (enzymatic rate-limiting) stages in the MPR initiation process.

One of these three *OS* (labeled in grey) aligns and bridges the other two, fixing them in the required proximity for the DNA polymerase to skip the inter-template gap while displacing the confronting non-template strand of the adjacent *OS*. This bridging occurs putatively through occasional Watson-Crick bonds between aligning and aligned homo-duplexes. Such putative bridging complexes are not covalently bonded, very unstable and have not been demonstrated directly. Experimental system for their revelation has still not been elaborated, but they can be predicted using the following software packages:

1. Visual OMP6 visualizes different 2nd-rank structures that can be potentially formed from constituent strands of *OS* (Fig. 7).

2. UNAfold is not so illustrative but much more comprehensive in providing schematic landscape of all hybridization patterns possible between complementary strands of *OS* and their temperature dependence (Fig. 8).

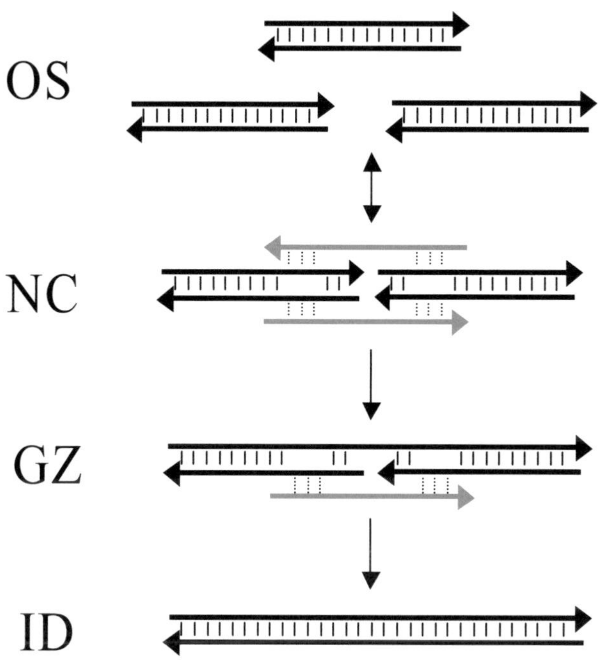

Fig. 6. A model for the MPR initiation. *OS*, original singlet *HD*s (a pair of complementary primers); *NC*, nucleation complex (an arrangement of three *OS*); grey strands correspond to the *OS* bridging the gap; *GZ*, gap zipper (conformation composed of half *NC* (*hNC*)) and a DNA polymerase-generated (by gap skipping) single strand of *ID*. Adapted from Itsko et al., 2008b with permission from Elsevier

UNAfold probability dot plots demonstrate that increasing temperature increases probabilities of generation of overall 2nd-rank structures and those that can be involved in *NC* (Fig. 8) among them. The van't Hoff equation that formulates this tendency is:

$$\frac{d\ln K_{1\to2}}{d(1/T)} = -\frac{\Delta H_{1\to2}}{R}, \tag{12}$$

where $K_{1\to2}$ and $\Delta H_{1\to2}$ are, respectively, the equilibrium constant and the standard enthalpy for the transition between more stable 1st-rank (*OS*) and less stable 2nd-rank (*NC*) structures in I part of eq. 11. $\Delta H_{1\to2} > 0$ because the stability of 2nd-rank structures is lower. Therefore, as the temperature increases ($1/T$ decreases), ln $K_{1\to2}$ and hence $K_{1\to2}$ must increase, reflecting a rise in the probability of the 2nd-rank conformations. Generation of *ID* from *NC* (Fig. 6, part II of eq. 11) requires keeping inside of *NC* 1st-rank structure of *OS* for DNA polymerase to skip the inter-template gap by extending the pre-hybridized *OS* constituent strands. On the other hand, the equilibrium constant of the formation of this 1st-rank structure from separated complementary DNA strands will decrease with temperature according to the same van't Hoff equation (eq. 12) in which $\Delta H < 0$ due to exothermic property of the hybridization. Decrease in this constant results in impeding this polymerization step and overall amplification and propagation stages. Thus temperature rise increases the rate of initiation but decreases the rates of amplification/ propagation.

NOMUL ΔG_{hNC} **(67°C) ≈ 1.80 kcal/mol**

```
                    5'-GGAAATAGAAGAACTTAAAT-------CTTTATTAGAGATTAACACAGC-3'
                            ||||                      ||||
3'-CCTTTATCTTCTTGAATTTAGAAATAATCTCTAATTGTGTCG-5' 3'-CCTTTATCTTCTTGAATTTAGAAATAATCTCTAATTGTGTCG-5'
```

NOMU ΔG_{hNC} **(68°C) ≈ 2.78 kcal/mol**

```
                      5'-GGTGATAGAAGAACTTAAAT-------CTTTATTAGGAATTAACCTGGC-3'
                                    |||:|                  |:|||
3'-CCACTATCTTCTTGAATTTAGAAATAATCCTTAATTGGACCG-5' 3'-CCACTATCTTCTTGAATTTAGAAATAATCCTTAATTGGACCG-5'
```

EVNA ΔG_{hNC} **(71°C) ≈ -1.34 kcal/mol**

```
                     5'-GGTGATAGAAGTGCTTAAAT-----CTTTATTAGGAATTGCTCTGGC-3'
                                |||  ||||                 |||  ||||
3'-CCACTATCTTCACGAATTTAGAAATAATCCTTAACGAGACCG-5' 3'-CCACTATCTTCACGAATTTAGAAATAATCCTTAACGAGACCG-5'
```

EVNAH ΔG_{hNC} **(73°C) ≈ -0.36 kcal/mol**

```
                     5'-GGTGATAGTGCTGCTTTGAT-----CTTTATTAGGAATTGCTCTGGC-3'
                                   ||||:|                  :||||
3'-CCACTATCACGACGAAACTAGAAATAATCCTTAACGAGACCG-5' 3'-CCACTATCACGACGAAACTAGAAATAATCCTTAACGAGACCG-5'
```

Fig. 7. Suggested structures of *hNC* (half *NC*) of *OS* types with approximate ΔG_{hNC} of their formation from single strands (at corresponding T_m), predicted by Visual OMP6. Conventional (|) and G:T pairings. Reprinted from Itsko et al., 2008b with permission from Elsevier

Following MPR kinetics by RT-PCR at different temperatures can test the above reasoning. Backward extrapolation of the amplification curves with different types of *OS* (next paragraph) to the first cycle retrieves the ratio $[ID]/[OS]_0$ and the value of $\ln(1+E)$ designated as amplification rate (*A*). The variability in changes of *A* (Fig. 9A) and $[ID]/[OS]_0$ (data not shown) with temperature is exceedingly high and not amenable for analysis. However these two parameters are strictly related in an exponential mode (Fig. 9B).

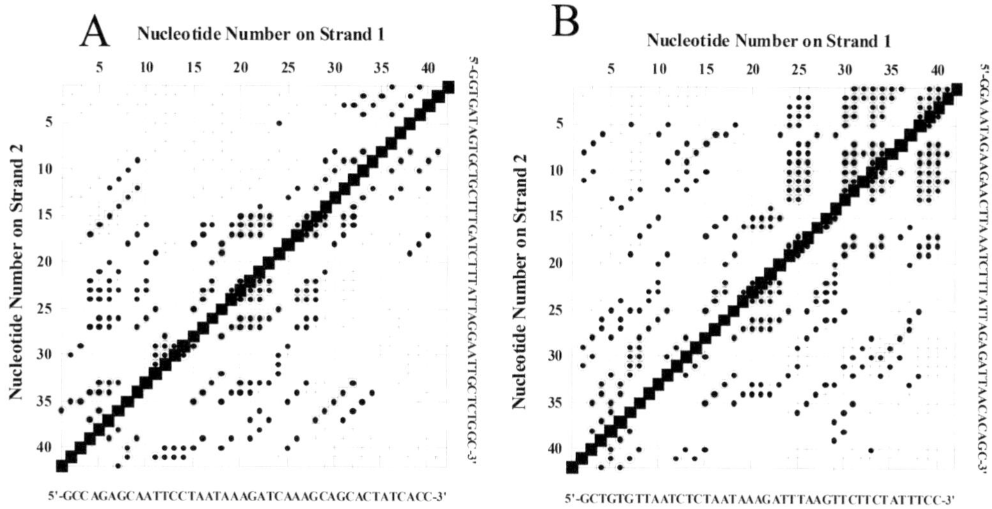

Fig. 8. Hybridization-probability dot-plot of EVNAH (**A**) and NOMUL (**B**). Each symbol designates pairing between strands' nucleotides. Filled-in squares designate pairs existing in *OS* at all tested temperatures with probability close to 1. Open circles (enclosing dots), 2nd-rank pairs at 68°C for NOMUL and 74°C for EVNAH. Dots, pairs existing at 71°C for NOMUL and 77°C for EVNAH, some of which (enclosed in open squares) form bridging structures (*hNC*) leading to *NC* (Fig. 7). Crosses, conformations existing at 74°C for NOMUL and at 80°C for EVNAH. Reprinted from Itsko et al., 2008b with permission from Elsevier

As *A* rose, the calculated ratios $[ID]/[OS]_0$ exponentially declined over seven decades. Increased *A* corresponds to a decrease in the applied temperature (Fig. 9A); hence, formation of *ID* is accelerated as temperature rises (Fig. 9B), consistent with the above reasoning. The high variability in the above parameters when they are plotted against temperature (Fig. 9A) seems to result from limited accuracy of the temperature maintenance by the heat block of the RT-PCR apparatus for the duration of each cycle (repeatedly launched by rapid cooling from 95°C to the desired temperature). The actual temperature that acts on each multiple sample may therefore differ from the registered one. The temperature-dependent amplification rate *A* (Fig. 9B) was used alternatively as a more stable indicator because it reflects the average temperature during the cycle. At each calculated *A*, the ratio $[ID]/[OS]_0$ for NOMUL was lower than that for EVNAH, consistent with lower stability of the putative bridging structure involved in NC formation of NOMUL than that for EVNAH (Fig. 7). In addition, at A=0.46 for NOMUL and 0.51 for EVNAH, the ratio lines were drastically bent (Fig. 9B), justified by lowering the fraction of *hNC* structures (Fig. 7) in overwhelmed number of emerging alternative pairings between constituent *OS* single strands at relatively high temperatures (Fig. 8, crosses) that correspond to lower amplification rates.

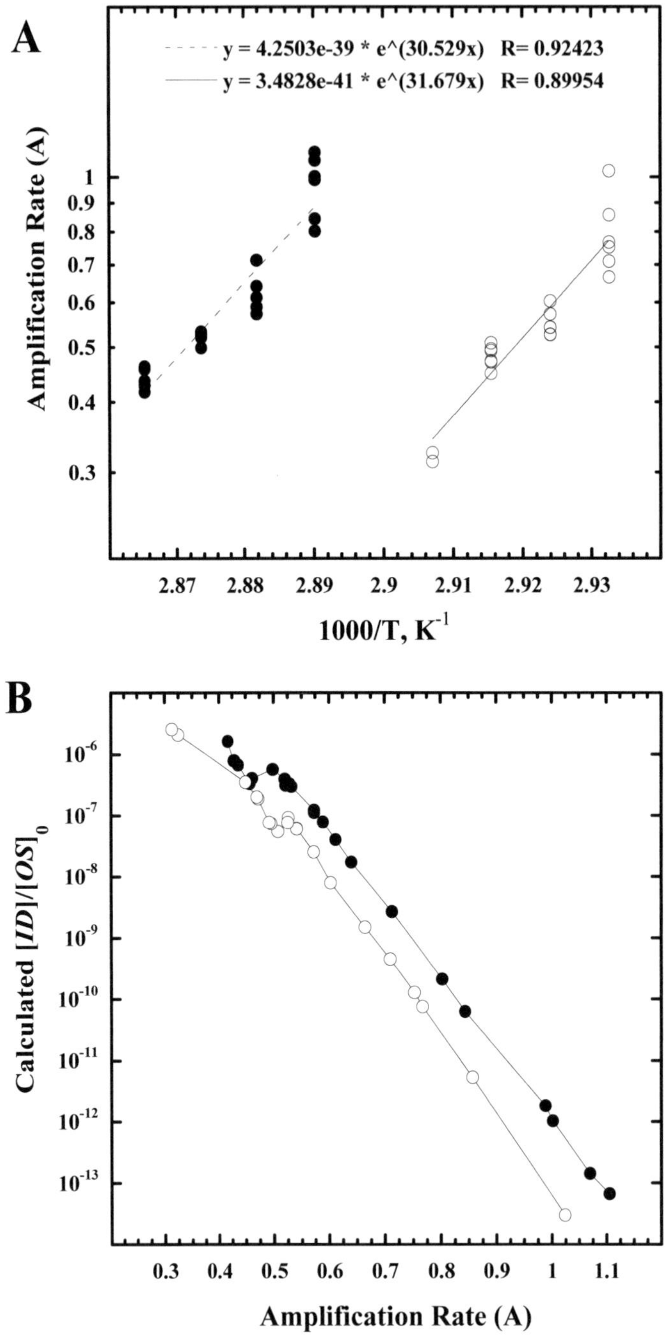

Fig. 9. **A**, Arrhenius plots of MPR amplification rate, with NOMUL (open circles) and EVNAH (filled-in circles). **B**, Dependence of calculated $[ID]/[OS]_0$ on MPR amplification rate, with NOMUL (open circles) and EVNAH (filled-in circles). Reprinted from Itsko et al., 2008b with permission from Elsevier

7. Kinetics of amplification

Kinetics of the MPR amplification stage demonstrates biphasic behavior for three of the four types of *OS* used (Fig. 10). Single-burst kinetics was seen with NOMU at all tested temperatures and with the rest of *OS*s at temperatures exceeding by 2-4°C the melting point of corresponding *OS*. The biphasic kinetics is explained by two parallel processes (with

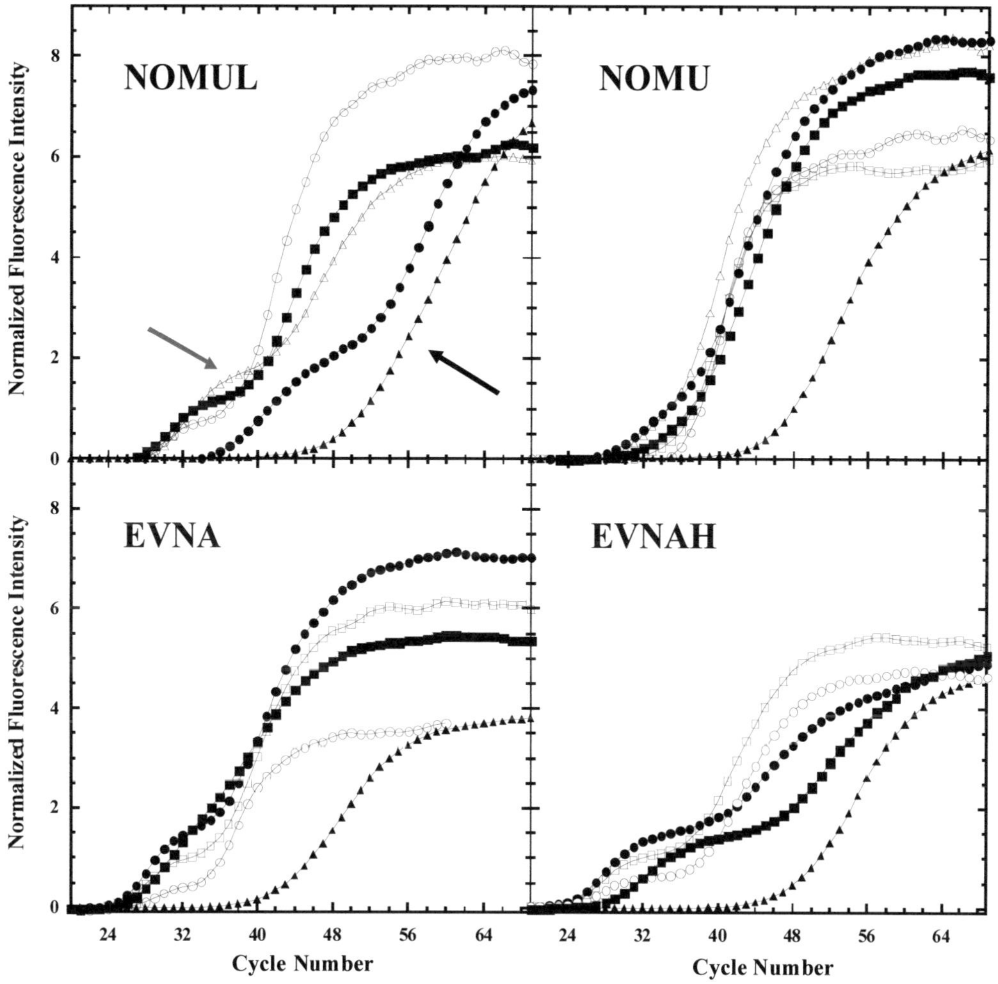

Fig. 10. MPR kinetics with four OS types: NOMUL (open circles, 68°C; filled-in squares, 69°C; open triangles, 70°C; filled-in circles, 71°C; filled-in triangles, 72°C). NOMU (open circles, 68°C; open squares, 69°C; open triangles, 70°C; filled-in circles, 71°C; filled-in squares, 72°C; filled-in triangles, 73°C). EVNA (open circles, 71°C; open squares, 72°C; filled-in circles, 73°C; open diamonds, 74°C; filled-in triangles, 75°C). EVNAH (open circles, 71°C; open squares, 72°C; filled-in circles, 73°C; filled-in squares, 74°C; filled-in triangles, 75°C). Arrowheads point at biphasic (grey) and single-burst (black) kinetics. Reprinted from Itsko et al., 2008b with permission from Elsevier

distinct kinetic parameters) involved in the MPR expansion, amplification of the *ID* by *OS* and propagation of doublets (*D*) to triplets (*T*) and higher-level multiples through staggered conformations (*SD*) (Fig. 3C) Existence of an amplification stage was suggested for the first burst in the progressively increased fluorescence intensity during MPR (Fig. 10).
The amplification (Fig. 11) can be formulated by the chemical equation

$$OS + D \overset{I}{\longleftrightarrow} 2H \overset{II}{\longrightarrow} 2D\,. \tag{13}$$

where I and II denote first (fast equilibrium) and second (rate-limiting) steps of the process. The rate of amplification (II) can be expressed as $d[D]/dt = k[H]$, where k is the rate constant of the rate-limiting step and H is the hetero-duplex composed of hybridized OS and D (Fig. 11). At temperatures higher than T_m of OS (and obviously of H) and lower than that of D, step I involves melting of OS, fraying the ends of D and hybridization of the single strands of the former with the latter.
Assuming that this step occurs near its multi-state equilibrium with $K_H = [H]/[OS][D]$, the overall process is expressed by

$$d[D]/dt = kK_H[OS][D]\,. \tag{14}$$

k rises with temperature, but K_H decreases due to the dissociation of H into D and single stranded OS. In the reaction described by eq. 13, k rises less than the drop in K_H, so that the overall value kK_H decreases with temperature. That is reflected in the highly negative value (around -62 kcal mole^{-1}) of "$\Delta G^{\ddagger}_{Ampl}$ derived from the Arrhenius plots for NOMUL and EVNAH (Fig. 9A). In other chemical processes as well, an exothermic fast-equilibrium stage leads to negative $\Delta G^{\ddagger}$ of the overall process. Solving eq. 13 (for detailed derivation see Itsko et al., 2008b) yields an expression for total fluorescence (Flu_{Tot}):

$$Flu_{Tot} = \alpha[OS]_0\left(1 + (\gamma - 1)\frac{(e^{AN} - 1)}{R + e^{AN}}\right) \tag{15},$$

where $A = kK_H\varepsilon([OS]_0 + [D]_0)$, ($[OS]_0$ and $[D]_0$ are initial concentrations of OS and D correspondingly where $[D]_0$ is actually $[ID]$), N is the number of cycles and ε is the cycle period, $R = [OS]_0/[ID]$, α is the arbitrary coefficient expressing the fluorescence brightness of OS; γ is the ratio between the values of the fluorescence brightness of D and OS. Eq. 15 was used to approximate the first burst in fluorescence intensity of MPR curves (Fig. 10) and retrieve $[ID]/[OS]_0$ ratio and amplification rate A from them.

8. Thermodynamics of transition from amplification to propagation

Denatured single strands of doublet (*ssD*) can be hybridized in fully aligned manner generating doublet homo-duplex (*D*, right in Fig. 12A) or in staggered manner generating staggered doublet (*SD*, left). Transition from amplification to propagation is mediated by the appearance of *SD*. Obviously *SD* has lower stability than *D* due to half number of hydrogen bonds ($\Delta G_{SD \to D} < 0$). *SD* cannot readily return to *D* due to the energetic barrier ($\Delta G^{\ddagger}_{SD \to D}$) between these conformations, the magnitude of which is determined by the stability (ΔG_{WSD}) of the additional 2nd-rank intermediate structure called weak staggered doublet (*WSD*) (Fig. 12B). The more stable it is, the more easily *SD* switches to *D* due to smaller $\Delta^{\ddagger}G_{SD\rightarrow D}$. *SD* switching to *D* prevents the propagation process and results in the biphasic mode of expansion (Fig. 10). Disappearance of the biphasic kinetics (Fig. 10) at temperatures higher than 71°C for NOMUL, 73°C for EVNA and 76°C for EVNAH (Table 1) is explained by increased probability of *SD* with temperature rise. UNAFold also predicts formation of *SD* at temperatures higher than: 73°C for NOMUL, 78°C for EVNA and 80°C for EVNAH (Fig. 13A). Formation of *SD* as a structure that is less stable than fully aligned hybridized *D* is stimulated by rising temperature according to the van't Hoff equation (eq. 12).

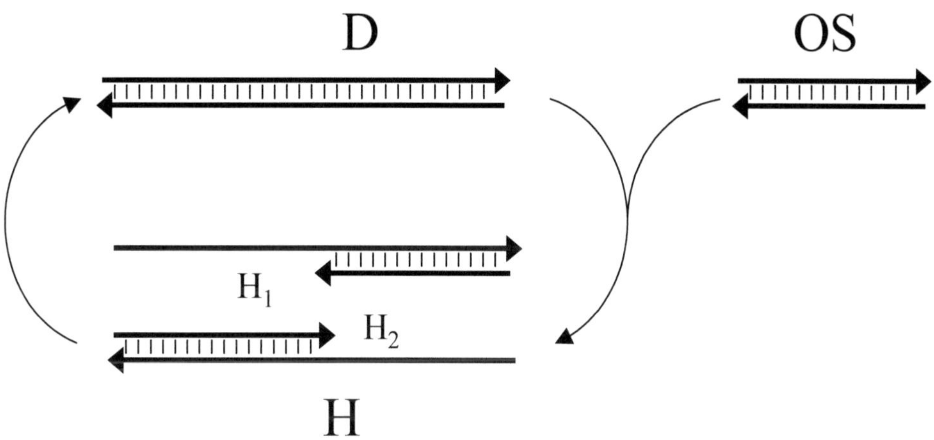

Fig. 11. The model for amplification: **OS**, Original Singlet *HD* (a pair of complementary primers), **D**, Doublet. $\mathbf{H_1}$ and $\mathbf{H_2}$, hybridized denatured *D* and *OS*. Adapted from Itsko et al., 2008b with permission from Elsevier

Melting the doublets at high temperature and their quick-cooling afterwards can demonstrate *SD* conformations. Quick cooling prevents the *D* constituent strands from finding the final most stable fully-aligned conformation when they encounter each other but entrap them in first available conformation, mostly a staggered one. Such procedure was accomplished on MPR end products generated from all four *OS* types, with concentrations of dNTP and *OS* that limit the extent of expansion (Eq. 6 and Fig. 4), and revealed that out of the four, the *OS* of NOMU was the only one to diminish (bottom band, lane 5 in Fig. 13B), most likely due to its pairings with overhangs formed by staggered structures of multiple-repeated DNA. Thus, among all used *OS*s staggering of *D* of NOMU is the most facilitated process. It is also consistent with the observation that only this *OS* displayed single burst kinetics at all tested temperatures in RT-PCR (Fig. 10).
The facilitation of staggering and of transition from amplification to propagation stage in the NOMU case is reasoned by the least stability (highest ΔG_{hNC}; Fig. 7) of its bridging structure among all used *OS*s, that would reflect also the least stability of its *WSD* (Fig. 12B) because the latter is just a staggered tandem *OS*.

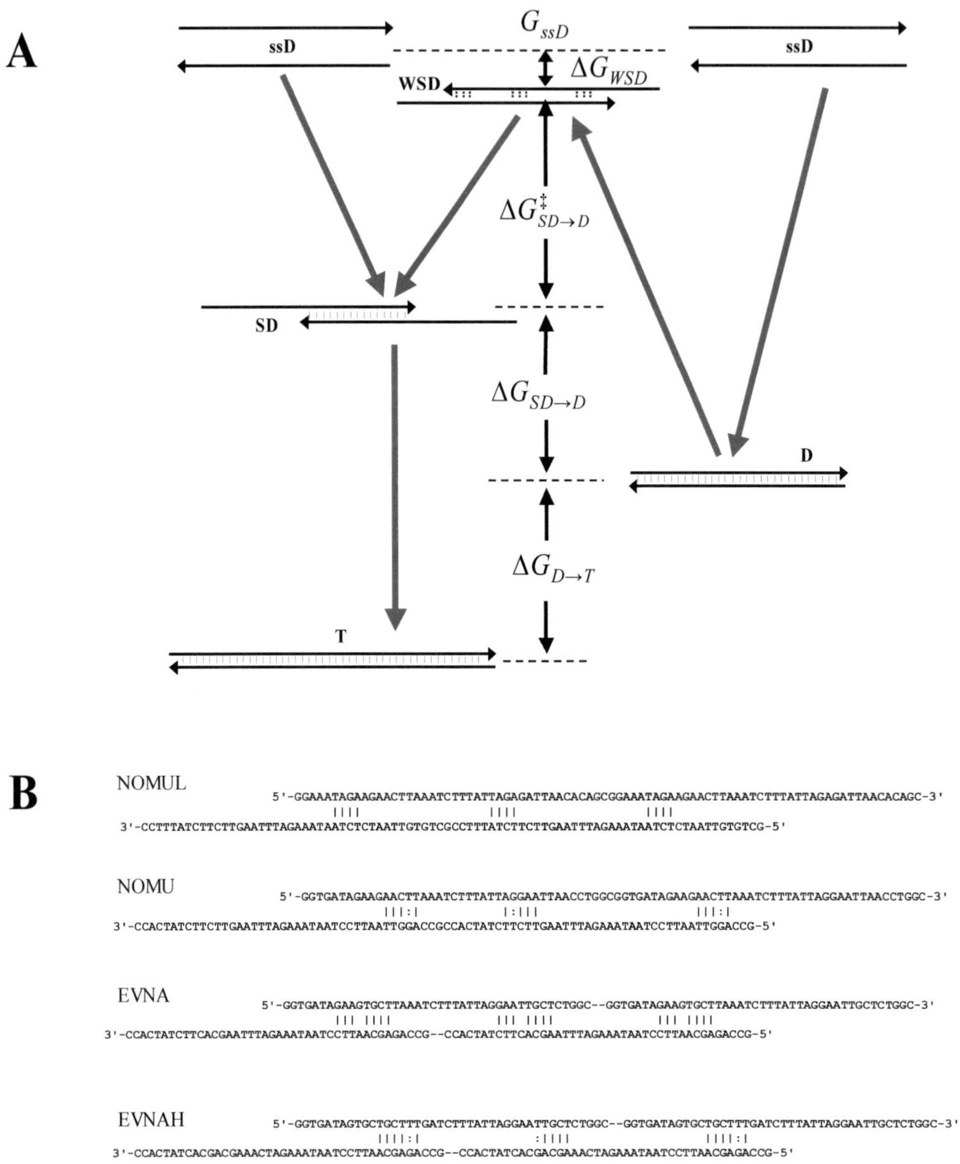

Fig. 12. Schematic free energy diagram of the various conformations that participate in the suggested transition from amplification to propagation. Dashed lines denote energy levels. G_{ssD} and $\Delta G_{D \to T}$ are, respectively, the free Gibbs energy of single stranded *D* and of the difference between *T* and *D*. **B**, *WSD* conformations (tandem staggered *OS*) of the four *D* types (related to Fig. 7). (Conventional (|) and G:T pairings). Adapted from Itsko et al., 2008b with permission from Elsevier

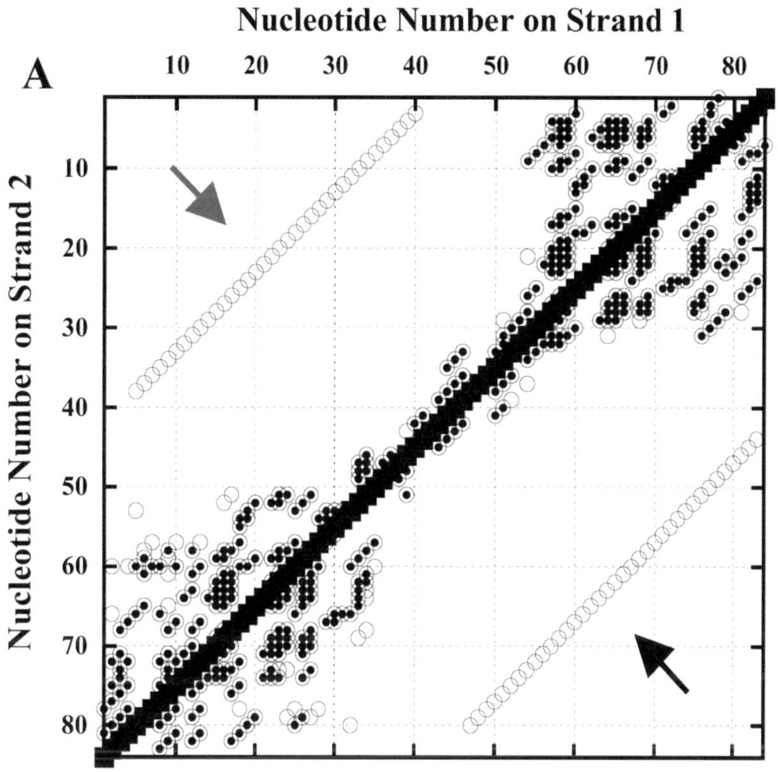

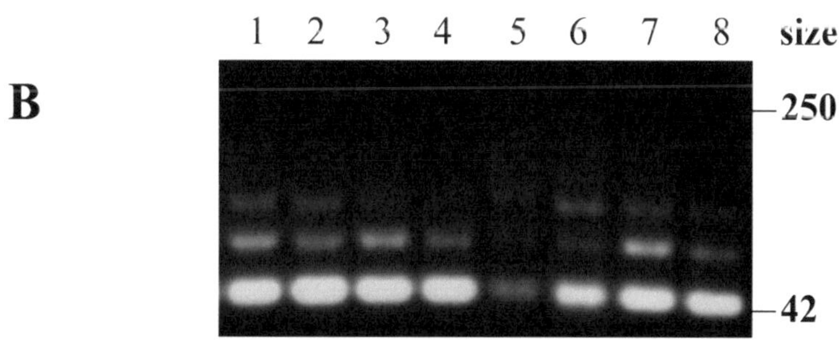

Fig. 13. Revealing SD structures *in silico* (A) and *in vitro* (B). **A**, Probability dot plot of hybridization of EVNAH's doublet. Each symbol designates pairing between strands' nucleotides. Filled-in squares designate pairs existing in *D* at all tested temperatures with probability close to 1. Open circles (enclosing dots or not, respectively) designate 2nd-rank pairs at 80°C or 81°C, with probabilities of approximately 10^{-6}. Arrows point at two lines corresponding to two *SD*, effective for propagation (black) and not (grey), **B**, MPR-generated products after 65 PCR cycles from 4 *OS* types, separated on 2.5% agarose gel. Lanes 1 and 2, with EVNAH; 3 and 4, EVNA; 5 and 6, NOMU; 7 and 8, NOMUL. Odd numbers designate samples denatured (10 min at 95°C) and then cooled rapidly. Reprinted from Itsko et al., 2008b with permission from Elsevier

9. Kinetics of propagation

The overall process of MPR propagation includes the following recurring set of three steps:

1. Aligning A_i forward and B_j reverse complementary DNA strands containing i and j repeats in staggered mode according to the equation:

$$A_i + B_j \xleftrightarrow{k_1, k_{-1}} A_iB_j, \tag{16}$$

where k_1 and k_{-1} are rate constants of alignment and of melting, respectively.

2. Association of DNA polymerase with above staggered homo/hetero-duplex:

$$A_iB_j + Enz \xleftrightarrow{k_1, k_{-1}} A_iB_jEnz, \tag{17}$$

where k_2 and k_{-2} are rate constants of association and of dissociation, respectively.

3. DNA-polymerase mediated filling-in overhangs according to:

$$A_iB_jEnz \xrightarrow{k_3} A_n + B_n + Enz, \tag{18}$$

where k_3 is the rate constant of the enzyme-driven polymerization (turnover number) for filling in $(n - i)$ repeats.

The mathematical descriptions of these steps are:

1. $$\frac{d[A_iB_j]}{dt} = k_1[A_i][B_j] - k_{-1}[A_iB_j] - k_2[A_iB_j][Enz] + k_{-2}[A_iB_jEnz], \tag{19}$$

2. $$\frac{d[A_iB_jEnz]}{dt} = k_2[A_iB_j][Enz] - k_{-2}[A_iB_jEnz] - k_3[A_iB_jEnz], \tag{20}$$

3. $$\frac{d[A_{n(i,j)}]}{dt} = k_3[A_iB_jEnz], \tag{21}$$

where square brackets designate concentrations of corresponding species and $A_{n(i,\ j)}$ is A_n generated from given i and j.

The rate of generation of intermediates such as $[A_iB_j]$ and $[A_iB_jEnz]$ is taken close to zero according to the assumption of steady state kinetics for them (Atkins, 1994). The final expression for the above chemical reaction is:

$$\frac{dA_{n(i,j)}}{dt} = \frac{k_3[Enz_{tot}][A_i][B_j]}{Enz_{tot}\,k_3/k_1 + K_D^{-1}K_M + [A_i][B_j]} \tag{22}$$

(for derivation see Itsko et al., 2009), where $K_M^{app} = K_M^{app*} + K_D^{-1}K_M$, $K_M^{app*} = Enz_{tot}\,k_3/k_1$, Enz_{tot} is the total concentration of enzyme, $K_D = k_1/k_{-1}$ is the equilibrium constant for duplex formation, $K_M = (k_{-2} + k_3)/k_2$ is the Michaelis-Menten constant.

Assuming $t = N\varepsilon$, where N is the number of cycles and ε is the cycle period, yields

$$\Delta A_{n(i,j)} = \frac{k_{Pol}[Enz_{tot}][A_i][B_j]}{Enz_{tot}\,k_3/k_1 + K_D^{-1}K_M + [A_i][B_j]}, \tag{23}$$

where $k_{Pol} = k_3\varepsilon$, and ε = 240 sec cycle^{-1} and ΔA_n [= $A_n(N+1) - A_n(N)$] is increments in $A_n(N)$ after one cycle.

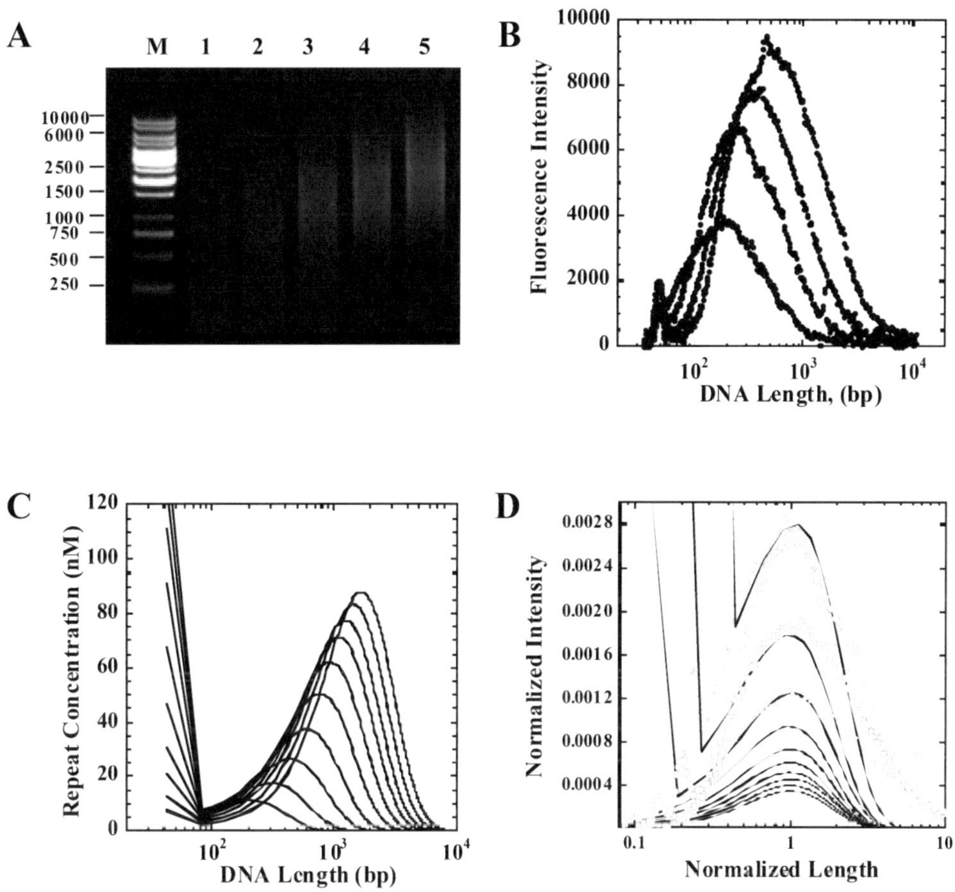

Fig. 14. Experimental versus model distributions. (**A**) Consecutive samples run on 0.8% agarose gel: M, DNA size markers; lane 1, cycle 16; lane 2, cycle 17; lane 3, cycle 18; lane 4, cycle 19; lane 5, cycle 20. (**B**) Scanned and digitized samples of cycles 17-20 (from **A**). (**C**) Matlab simulations. Final distributions of total MPR product at cycles 14-23. Model parameters: k_{Pol} = 0.09 repeat/cycle ($k_3 = 3.75\times10^{-4}$ repeat/sec), K_M^{app*} = 1 nM^2 (Eq. 22); $k_I = 10^{-24}$ nM^{-3} sec^{-1} (Eq. 1); E = 0.735 (Eq. 7). (**D**) Juxtaposition of the model (full lines for cycles 14-23) with four (cycles 17-20) experimental distributions (crosses, open circles, open squares and open diamonds, respectively), each integrally normalized. Reprinted from Itsko et al., 2009 with permission from Elsevier

Calculation of the overall change A_n during a PCR cycle requires consideration of inflow to the A_n category from all the possible combinations of A_i and B_j that together allow the generation of such product and outflow of A_n to a group of longer lengths:

$$\Delta A_n = k_{\mathrm{Pr}}\left(\sum_{i=1}^{n-1}\sum_{j=n-i+1}^{\infty}\frac{\Delta A_{n(i,j)}}{i+j-1} - \sum_{j=2}^{\infty}\frac{(j-1)\Delta A_{n(n,j)}}{n+j-1}\right) \tag{24}$$

A similar equation may be formulated for ΔB_n. Since the initial concentrations of the complementary DNA strands dealt with here are identical, $A_1(0) = B_1(0)$, the propagation kinetics of A_n and B_n are identical, hence $B_i = A_i$.
The above set of difference equations (limited to 3,000) describing kinetics of MPR was solved numerically using Matlab 7 (MathWorks, Natrick, MA (http://www.mathworks.com)). The experimental results were compared with the model (Fig. 14) and satisfactory resemblance was found between them. A discrepancy was revealed between the skewness of the experimental and modeled distributions of final MPR length. The experimental distributions are skewed to longer lengths whereas the model predicts negative skewness. This means that for every (A_i, B_j) involved in the extension step (see eqs. 3 and 18) the generated product is much shorter than the expected maximum ((i+j-1)). This is consistent with lower processivity of the Vent polymerase (Kong et al., 1993).

10. MPR as an evolutionary process

MPR can be viewed as an evolutionary process to develop biological polymers such as DNA. The repeat generation aspect included in the MPR brings this process even closer to a common scenario of Darwinian evolution, namely, spontaneous appearance of the physico-chemical trait (*ID*) that is selected (amplified) since it directly contributes to its own reproduction in the form of repeats propagation. Theoretically, two variants of *ID* can be generated in the reaction mixture from a given *OS* with comparable probabilities, cis and trans (Fig. 15). Only the cis form is prone to propagation; the trans form is "infertile" since it cannot be aligned in a staggered manner. MPR polymers indeed contain only direct repeats and never alternating inverse repeats.

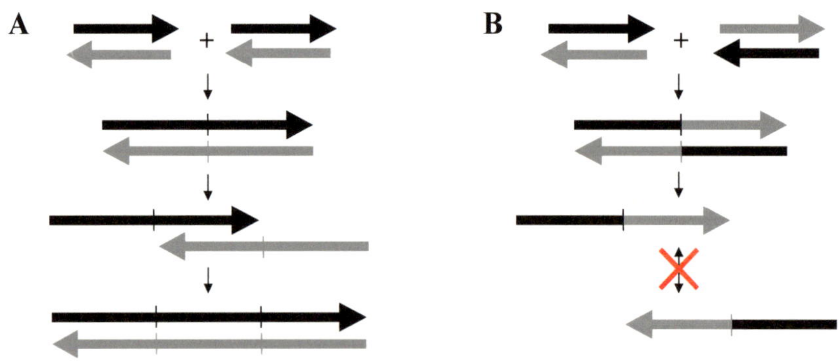

Fig. 15. Cis (**A**) and Trans (**B**) species of *ID* putatively generated in MPR initiation, and their potency in the subsequent propagation. Black and grey regions are complementary to each other. Arrowheads denote 3′

The thermodynamic stabilities of second-rank structures of different *OS*s involved in *NC* formation vary depending on sequence hence they have different propensities for propagation. In MPR conducted with a mixture of different *OS*s, the final polymers would contain predominantly the repetitive motif corresponding to *OS* with the most stable *NC* structure. This prediction is confirmed by the finding (Saito *et al.*, 2007) that the most abundant among 3 microgenic units in the combinatorial polymers generated from an equi-molar mixture is the one with the lowest free energy of formation of the *NC*-promoting 2nd-rank structure (analysis not shown), thus lending strong support to the model proposed and described here. This feature can be exploited to design special types of proteins with desired repeated motifs *in vitro* and perhaps *in vivo* as well.

11. MPR and *in vivo* expansion

Repeating motifs in natural proteins are widely documented and described (http://www.ncl-india.org/trips/), and their possible activity mechanisms discussed (Katti et al., 2000). The fact that they derive of encoding genes is sometimes not sufficiently considered. During studies with the entomopathogenic *Bacillus thuringiensis* subspecies, we encountered several such cases, where DNA repeats exist, reflecting their encoded protein motifs. Specifically, helper Orf2 proteins of some *cry* operons include different numbers of repetitive motifs. For example, Orf2$_{cry2Aa}$, encoded by *orf2* of the *cry2Aa* operon (Widner & Whiteley, 1989), contains 11 tandem repeats of a 15-amino-acids motif and seems to be involved in Cry2Aa crystallization (Ge et al., 1998). This motif is encoded by a 45-nt repeat in its gene *orf2*$_{cry2Aa}$ (Fig. 16A) that includes only 6 transitions and a single transversion among the 450 nt composing the first 10 repeats. All 5 additional changes are located at the 3' end of the last (11th) repeat. *orf2*$_{cry2Aa3}$ (Sasaki et al., 1997; Fig. 16B) includes 13 repeats with the same basic unit as *orf2*$_{cry2Aa}$. In this gene, the first 9.5 repeats are separated from the rest by a 12 nt long 'linker', i.e., a non-homologous insertion.
A second type of repeat unit among the known *orf2* genes is found in *orf2*$_{cry2Ca}$ (Wu et al., 1991) and in *orf2*$_{cry2Ac}$ (Accession # AY007687). It differs from the repeat of *orf2*$_{cry2Aa}$ by two triplets that surround a conserved T(A/G)C found in both kinds of repeats and by an extra adjacent triplet. A string of 991 nt in *orf2*$_{cry2Ca}$ encoding almost 21 repeats of a 16-mer motif in Cry2Ca contains only 35 transitions (Fig. 16C), whereas the shorter (7 repeats) string in *orf2*$_{cry2Ac}$ is much less conserved and contains many deletions and insertions (Fig. 16D).

A. *orf2*$_{cry2Aa}$ (Access. # M23723)

```
ACG TAT AAT CAA AGT CAG AAT GTA TGC CCA CAA GAT TTA GTT GAT
ACG TAT AAT CAA AGT CAG AAT GTA TGC CCA CAA GAT TTA GTT GAT
ACG TAT AAT CAA AGT CAG AAT GTA TGC CCA CAA GAT TTA GTT GAT
ACG TAT AAC CAA AGT CAG AAT GTA TGC CCA CAA GAT TTA GTT GAT
ACG TAT AAT CAA AGT CAG AAT GTA TGC CCA CAA GAT TTA GTT GAT
ACG TAT AAC CAA AGT CAG AAT GTA TGC CCA CAA GAT TTA GTT GAT
ACG TAT AAC CAA AGT CAG AAT GTA TAC ACA CAA GAT TTA ATT GAT
ACG TAT AAC CAA ACT CAG AAT GTA TGC CCA CAA GAT TTA GTT GAT
ACG TAT AAC CAA AGT CAG AAT GTA TGC CCA CAA GAT TTA GTT GAT
ACG TAT AAC CAA AGT CAG AAT GTA TGC CCA CAA GAT TTA GTT GAT
ACG TAT AAC CAA AGT CAG AAT GTA TGC CCA CAA GAT TTG AAT GTA
```

B. *orf2*$_{cry2Aa3}$ (Access. # D86064)

```
ACG TAT AAC CAA AGT CAG AAT GTA TGC GCG CAA GAT TTA GTT GAT
ACG TAT AAC CAA AGT CAG AAT GTA TGC ACA CAA GAT TTA GTT GAT
ACG TAT AAC CAA AGT CAG AAT GTA TGC ACA CAA GAT TTA GTT GAT
ACG TAT AAC CAA AGT CAG AAT GTA TGC ACA CAA GAT TTA GTT GAT
ACG TAT AAC CAA AGT CAG AAT GTA TGC ACA CAA GAT TTA GTT GAT
ACG TAT AAC CAA AGT CAG AAT GTA TGC GCG CAA GAT TTA GTT GAT
ACG TAT AAC CAA AGT CAG AAT GTA TGC ACA CAA GAT TTA GAA GAT
ACG TAT AAC CAA AGT CAG AAT GTA TGC ACA CAA GAT TTA GTT GAT
ACG TAT AAC CAA AGT CAG AAT GTA TGC GCG CAA GAT TTA GTT GAT
ACG TAT AAC CAA AGT CAG AATGGTGTCTACAAG CAA GAT TTA GTT GAT
ACG TAT AAC CAA AGT CAG AAT GTA TGC GCG CAA GAT TTA GTT GAT
ACG TAT AAC CAA AGT CAG AAT GTA TGC ACA CAA GAT TTA GTT GAT
ACG TAT AAC CAA AGT CAG AAT ACT TAT ATA CAA GAT TTA ATT GAT
AAG TAT AAT CAA AGT CAG AAT
```

C. *orf2*$_{cry2Ca}$ Access. # X57252)

```
ACG TAT AAC CAA AGC CAA AATTGGCTG    TAC GCG CAA GAT TTA GTG GAT
ACG TAT AAC CAG AGC CAA AAT GGT GCC TGC GCG CAA GAT TTA ATG GAT
ACG TAT AAC CAA AGC CAA AAT GGT GCC TAC GCG CAA GAT TTA GTG AAT
ACG TAT AAC CAA AGT CAG AAT GGT GCC TAC GCG CAA GAT TTA GTG GAT
ACG TAT AAT CAA AGT CAG AAT GGT GCC TGC GCG CAA GAT TTA ATG GAT
ACG TAT AAC CAA AGC CAA AAT GGT GCC TAC GCG CAA GAT TTA GTG AAT
ACG TAT AAC CAA AGT CAG AAT GGT GCC TAC GCG CAA GAT TTA GTG GAT
ACG TAT AAT CAA AGT CAG AAT GGT GCC TGC GCG CAA GAT TTA ATG GAT
ACG TAT AAC CAA AGC CAA AAT GGT GCC TAC GCG CAA GAT TTA GTG AAT
ACG TAT AAC CAA AGT CAG AAT GGT GCC TAC GCG CAA GAT TTA GTG GAT
ACG TAT AAT CAA AGT CAG AAT GGT GCC TAC GCG CAA GAT TTA GTG GAT
ACG TAT AAC CAA AGT CAG AAT GGT GCC TAC GCG CAA GAT TTA GTG AAT
ACG TAT AAC CAA AGT CAG AAT GGT GCC TAC GCG CAA GAT TTA GTG AAT
ACG TAT AAC CAA AGT CAG AAT GGT GCC TAC GCG CAA GAT TTA GTG GAT
ACG TAT AAT CAA AGT CAG AAT GGT GCC TAC GCG CAA GAT TTA GTG GAT
ACG TAT AAC CAA AGT CAG AAT GGT GCC TAC GCG CAA GAT TTA GTG AAT
ACG TAT AAC CAA AGT CAG AAT GGT GCC TAC GCG CAA GAT TTA GTG GAT
ACG TAT AAT CAA AGT CAG AAT GGT GCC TAC GCG CAA AAT TTA GTG GAT
ACG TAT AAT CAA AGT CAG AAT GGT GCC TGC GTG CAA GAT TTA GTG AAT
ACG TAT AAT CAA AGT CAG AAT GGT GCC TGC GTG CAA GAT TTA GTG GAT
ACG TAT AAC CAA AGT CAG AAT GGT GCC TAC G
```

D. *orf2*$_{cry2Ac}$(Access. # AY007687)

```
                                 GT GCCTTAC GCG CAAAGAT TTA GTG GAAT
ACGGAAT AACCCAAAAGT CAG AAT GGT GCCNTAC GCG CAANGAT TTA AGT GAAT
ACG TATTAANCCAA AGT CAG AAT GGT GCC TAC GCG CAA GAT TTA AGT GNAT
ACGGTATAATC CAA AG  CAG AAT GGG CCC TAC GCG CAA AAT TTA GTG GAT
ACG TAT AAT CAA AGT CAG AAT GGT GCC TGC GTG CAA GAT TTA GTG AAT
ACG TAT AAT CAA AG  CAG AAT GGT GCC TGC GTG CAA GAT TTA GTG GAT
ACG TAT AAC CAA AGT CAG AAT GGT
```

E. *cry11Bb2* (Access. # HM068615)

```
            CAATAAT ACA AGC AGT GGG TAT GAG CAA GGA TAT AAC GAT
AAT TAT AAC CAA AAT ACA AGT AGT GGG TAT GAG CAA GGA TAT AAC GAT
AAT TAT AAC CAA AAT ACA AGT AGT GGG TAT GAG CAA GGA TAT AAC GAT
AAT TAT AAC CAA AAT ACA AGT AGT GGA TAT GAG CAA GGA TAT AAC GAT
AAT TAT AAC CAA AAT ACA AGC AGT GAG TAT GAG CAA GGA TAT AAC GAC
AAT TAT AAC CAA AAT ACA AGT AGT GGA TAT GAG CAA GGA TAT AAC GAT
AAT TAT AAC CAA AAT ACA AGT AGT GGA TAC GAG CAA GGA TAT ATT GAT
AAT TAT AGG CCA
```

Fig. 16. Individually aligned repeated motifs found in different *B. thuringiensis* (Bt) subspecies. **A.** *orf2*$_{cry2Aa}$ from Bt subsp. *kurstaki* HD-1; **B.** *orf2*$_{cry2Aa3}$ from Bt subsp. *sotto*. **C.** *orf2*$_{cry2Ca}$ from Bt S_1; **D.** *orf2*$_{cry2Ac}$ from Bt; **E.** *cry11Bb2* from Bt strain K34. Fonts: red, pyrimidine transitions; blue, purine transitions; underlined, transversions. Highlights: green, the conserved triplet in the middle of the motif; yellow, region of varied length (modulus 3); grey, sequence homology between *cry11Bb2* and *orf2*; red, non-homologous insertions

Repetitive motifs are encountered in certain *cry* genes themselves. The newly discovered Cry11Bb2 (Melnikov *et al.*, 2010) contains 7 tandem repeats of a 16-amino acids motif, the role of which is still to be determined. Seven out 16 triplets constituting the repetitive motif of Cry11Bb2 (highlighted grey in Fig. 16E) are identical/homologous in composition and location to those found in previously described *orf2* genes (Fig. 16, A-D). Repeated blocks have been found in other mosquito larvicidal δ-endotoxins (de Maagd et al., 2003): Cry11Ba1 (four repeats), Cry11Bb1 (5 repeats), Cry20Aa (eight), Cry27Aa (8 repeats).

The spectrum of replacements in the repeats of $orf2_{cry2Aa}$ and $orf2_{cry2Ca}$ is reciprocal on two sides of the conserved T(A/G)C triplet: in its 5′ part there is a bias to pyrimidine transitions whereas purine transitions are exclusively observed at its 3′ part. Intriguingly, such pattern of changes may point to occurrence of double strand break (DSB) in the middle of the repeat. Repair of DSB is mediated by trans-resection of opposite strands surrounding it in a way that transiently exposes generated ssDNA stretches to increased mutagenesis with characteristic pattern of transitions encompassing the DSB (Yang, 2008). Moreover, much evidences has accumulated in eukaryotic cells that point to blunt dsDNA ends as key intermediates in the process leading to gene amplification (Messer & Arndt, 2007; Pace et al., 2009; Mondello et al., 2010).

To survive DSB, cells exploit two major processes: homologous recombination (HR) and non-homologous end-joining (NHEJ). HR retrieves lost genetic information in error-free way from undamaged homologous template using Rad50/Mre11 complex in yeasts (Wyman & Kanaar, 2006) or RecBCD in *Escherichia coli* (Dillingham & Kowalczykowski, 2008). In contrast, NHEJ break repair is a ligation-like non-template-directed process that occurs between the non-homologous termini of a DSB and is therefore considered to be more error-prone. In the course of NHEJ, a DSB is repaired by attracting the Ku70/80 hetero-dimer, which recruits the ligase IV complex (comprised of ligase IV, XRCC4 and XLF) to seal the DNA ends (Mahaney et al., 2009). NHEJ, regarded as exclusive prerogative of eukaryotic cells, has recently been found in prokaryotes by identifying bacterial homologues of Ku protein (Doherti et al., 2001; Brissett & Doherty, 2009). Moreover, a functional NHEJ repair pathway is essential for spore viability in *Bacillus subtilis* under conditions that yield DSBs (Weller et al., 2002; Wang et al., 2006; Moeller et al., 2007). It is noteworthy that many of the bacteria that contain the Ku ligase system are capable of sporulation (*B. subtilis, Streptomyces coelicolor*) or spend long periods of their life cycle in the stationary phase (*Mycobacterium tuberculosis, Mesorhizobium loti, Sinorhizobium loti*) (Weller et al., 2002). The sporulating *B. thuringiensis* may contain NHEJ repair pathway as well.

Exogenous oligonucleotides complementary to the broken ends can efficiently target DSB for repair in yeasts (Storici et al., 2006). We propose that partial complementarity between sequences flanking the broken ends may assist their sealing and contribute to repeat generation via MPR-like slipped structures (Fig. 17, panels A-H). An example involving putative 2nd rank complementarity that can participate in the process to seal the broken ends (Fig. 17I) was revealed in Cry2Bb2 (Melnikov et al., 2010).

Once the initial doublet has been generated inside the chromosome, the repeats may propagate during DNA replication since the constituent strands may slide over each other between the multiple complementary regions. Second rank complementary structures similar to WSD in MPR (Fig. 12B) may stabilize such slipped structures. The alternative mechanism for the propagation is possible recurrence of DSBs in the middle of the generated repeats. Bridging the broken ends by occasional Watson-Crick bonds due to partial complementarity (Fig. 17D) can bring about nucleotides insertions/deletions during fill-in by DNA polymerase similar to those occurring during conversion of the *NC* to the *ID* in MPR (Fig. 6 and Itsko *et al.*, 2008a). Selection for functionality of a given coding region can restore the lost frame by generating an additional triplet or dropping an existing one. This can result in the two/three triplets varied region (highlighted yellow in Fig. 16) surrounding the conserved T(A/G)C triplet in the 4 *orf2* versions and in *cry11Bb2* listed here (highlighted green in Fig. 16). The inserted linker in that region as in $orf2_{cry2Aa3}$ (highlighted red in Fig. 16B) is also consistent with this explanation.

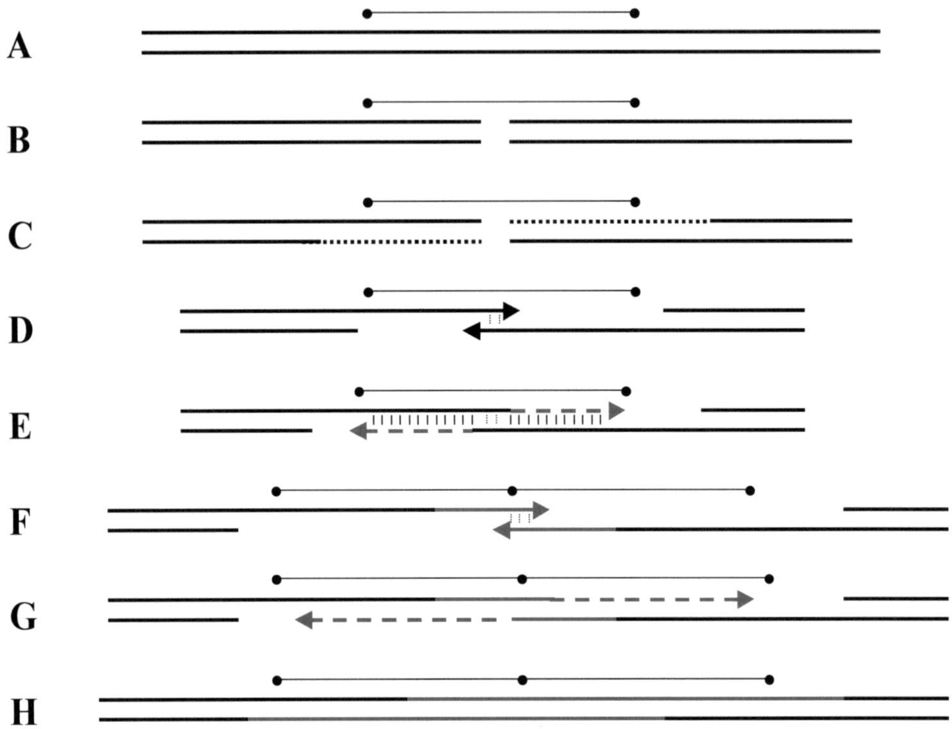

I

```
5'-AATACAAGTAGTGGATATGAGCAAGGATATAACGATAATTATAACCAA-3'
   ||||||||||||||||||||||||||||||||||||||||||||||||
3'-TTATGTTCATCACCTATACTCGTTCCTATATTGCTATTAATATTGGTT-5'


5'-AATACAAGTAGTGGATATGAGCAAGGATATAACGATAATTATAACCAA-3'
                           |||||| |        |||||| |
           3'-TTATGTTCATCACCTATACTCGTTCCTATATTGCTATTAATATTGGTT-5'


           5'-AATACAAGTAGTGGATATGAGCAAGGATATAACGATAATTATAACCAA-3'
                           |||||| |        |||||| |
3'-TTATGTTCATCACCTATACTCGTTCCTATATTGCTATTAATATTGGTT-5'
```

Fig. 17. Generation of sequence duplication due to DSB repair via NHEJ-like repair pathway. Chromosome region (**A**) suffers a DSB (**B**). **C**. Trans resection of DNA strands flanking the break by a dedicated exonuclease. **D**. Bridging the broken ends in staggered mode by occasional Watson-Crick bonds due to partial complementarity. **E**. Filling-in single stranded regions by DNA polymerase. **F**. Back slippage of the broken ends under repair that is mediated by a weak 2nd rank complementarity. **G**. Filling-in single stranded regions by DNA polymerase. **H**. Restored integrity of the chromosome with the generated duplicated region. Dumbbells designate schematic location of the sequence undergoing duplication. Newly synthesized DNA is designated in grey. **I**. Fully-aligned and two slipped structures existing between constituent strands of *cry2Bb2* repeat motif that may assist to bridge the DSB ends. Bold and underlined regions designate two sets of complementarity

It is intriguing that the AGT triplet in $orf2_{cry2Ca}$ (Fig. 16C) is always adjacent to CAG whereas its 5′- replaced AGC is adjacent to CAA, both combinations encode the same amino acids, Serine by AGT/C and Glutamine by CAG/A. This coincidence cannot be explained by selection for suppression of one transition by the other. Such replacements do affect stability of slipped structures presumably emerging during replication of this region. Without the replacements, almost triplet repetition AAT CAA AGT CAA AAT (Fig. 16C) would be explicitly observed that could putatively propagate in uncontrolled way through slipped structures, bringing about genetic instability. Strong selection for silent transitions that save triplet coding but prevent deleterious expansion by disrupting repetition pattern would thus be anticipated to stabilize the genome.

12. Concluding remarks

The MPR was dissected to sub-reactions and their thermodynamics and kinetics were analyzed. Different propensities of various HDs to expand into multiple repeating units were justified in terms of different stabilities of *NC*s engaged in MPR initiation. The proposed models, thermodynamic for initiation and kinetic for propagation, agree satisfactorily with experimental results.

The MPR with non-repetitive HD presents an optional chemical evolutionary system in which the thermodynamic advantage of very weak interactions results in biased proliferation of a certain reaction product. The learned approach to optimize MPR is necessary in protein engineering. The importance of studying this phenomenon lies far beyond applied interest; it may reflect primordial molecular evolution of primitive DNA sequences into complex genomes.

Molecular participants in the repeat expansion process and reaction conditions accompanying it *in vivo* obviously differ from those existing in MPR mixtures. The facilitating factor in the *in vitro* repeat expansion, temperatures manipulations above the T_m of expanding oligonucleotide *HD*, is impossible *in vivo*, where an ensemble of various enzymes is operating. DNA/protein complexes can select for generation of specific functionality-relevant hybridized species of DNA molecules out of competing structures. This principal difference affects only the rate of the processes and their efficiencies but not their thermodynamic feasibilities. Even strongly unfavorable energetic interactions (high ΔH values) between DNA species (putatively bringing to repeat generation and their propagation) can occur in both systems when overcome by entropic component of their multiplicity and redundancy ($T\Delta S$) (Harvey, 1997). Thus, the basic thermodynamic consideration about stabilities of 2nd rank structures underlying repeat propagation propensity can be common to the repeat expansion phenomenon in both arenas.

13. Acknowledgments

This investigation was partially supported by a grant from the United States-Israel Binational Science Foundation (BSF, number 2007-037), Jerusalem, Israel (to A.Z.), and a Levi Eshkol scholarship (to M.I.) from the Israeli Ministry of Science, Culture and Sports.

14. References

Atkins, P.W. (1994). *Physical Chemistry*. 5th ed., Oxford University Press, ISBN 0198557329, Oxford, UK, pp. 861–959.

Blackburn, E.H. (1991). Structure and function of telomeres. *Nature*, 350, 569-573.

Brissett, N.C. & Doherty, A.J. (2009). Repairing DNA double-strand breaks by the prokaryotic non-homologous end-joining pathway. *Biochem Soc Trans*, 37, 539-545.

Britten, R.J. & Kohne, D.E. (1968). Repeated sequences in DNA. *Science*, 161, 529-540.

Catasti, P.; Gupta, G.; Garcia, A.E.; Ratliff, R.; Hong, L.; Yau, P.; Moyzis, R.K. & Bradbury, E.M. (1994). Unusual structures of the tandem repetitive DNA sequences located at human centromeres. *Biochemistry*, 33, 3819-3830.

de Maagd, R.A.; Bravo, A.; Berry, C.; Crickmore, N. & Schnepf, E. (2003). Structure, diversity and evolution of protein toxin from spore-forming entopathogenic bacteria. *Annu Rev Genet*, 37, 409–433.

Dillingham, M.S. & Kowalczykowski, S.C. (2008). RecBCD enzyme and the repair of double-stranded DNA breaks. *Microbiol Mol Biol Rev*, 72, 642-671

Dogget, N.A. (2000). Overview of human repetitive DNA Sequences. *Current Protocols in Human Genetics*, APPENDIX 1B, DOI: 10.1002/0471142905.hga01bs08

Doherty, A.J.; Jackson, S.P. & Weller, G.R. (2001). Identification of bacterial homologues of the Ku DNA repair proteins. *FEBS Lett*, 500, 186-188.

Flory, P.J. (1953). *Principles of Polymer Chemistry*. Cornell University Press, ISBN 0801401348, Ithaca, New York.

Ge, B.; Bideshi, D.K.; Moar, W.J. & Federici B.A. (1998). Differential effects of helper proteins encoded by the *cry2A* and *cry11A* operons on the formation of Cry2A inclusions in *Bacillus thuringiensis*. *FEMS Microbiol Lett*, 165, 35-41.

Gilson, E.; Clément, J.M.; Brutlag, D. & Hofnung, M.A. (1984). Family of dispersed repetitive extragenic palindromic DNA sequences in *E. coli*. *EMBO J*, 3,1417-1421.

Harvey, S.C. (1997). Slipped structures in DNA triplet repeat sequences: entropic contributions to genetic instabilities. *Biochemistry*, 36, 3047-3049.

Hofnung, M. & Shapiro, J. (1999). *Res Microbiol* 150 (special issue on bacterial repeats).

International Human Genome Consortium (2001). Initial sequencing and analysis of the human genom. *Nature*, 409, 860-921.

Itsko, M.; Zaritsky, A.; Rabinovitch, A. & Ben-Dov E. (2008a). Initiation of the microgene polymerization reaction with non-repetitive homo-duplexes. *Biochem Biophys Res Commun*, 368, 606-613.

Itsko, M.; Zaritsky, A. & Rabinovitch A. (2008b). Thermodynamics of unstable DNA structures from the kinetics of the microgene PCR. *J Phys Chem B*, 112, 13149-13156.

Itsko, M.; Rabinovitch. A. & Zaritsky, A. (2009). Kinetics of repeat propagation in the microgene polymerization reaction. *Biophys J.*, 96, 1866-1874.

Kang, S.; Jaworski, A.; Ohshima, K. & Wells, R.D. (1995). Expansion and deletion of CTG repeats from human disease genes are determined by the direction of replication in *E. coli*. *Nat Genet*, 10, 213-218.

Katti, M.V.; Sami-Subbu, R.; Ranjekar, P. K. & Gupta, V.S. (2000). Amino acid repeat patterns in protein sequences: their diversity and structural-functional implications. *Protein Sci*, 9, 1203-1209.

Kong, H.; Kucera, R.B. & Jack, W.E. (1993). Characterization of a DNA polymerase from the hyperthermophile archaea *Thermococcus litoralis*. Vent DNA polymerase, steady state kinetics, thermal stability, processivity, strand displacement, and exonuclease activities, *J Biol Chem*, 268, 1965–1975.

Liang, X.; Jensen, K. & Frank-Kamenetskii, M.D. (2004). Very efficient template/primer-independent DNA synthesis by thermophilic DNA polymerase in the presence of a thermophilic restriction endonuclease. *Biochemistry*, 43, 13459-13466.

Mahaney, B.L.; Meek. K. & Lees-Miller, S.P. (2009). Repair of ionizing radiation-induced DNA double-strand breaks by non-homologous end-joining. *Biochem J*, 417, 639-50.

Melnikov, O.; Baranes, N.; Einav, M.; Ben-Dov, E.; Manasherob, R.; Itsko, M. & Zaritsky, A. (2010). Tandem repeats in a new toxin gene from a field Isolate of *Bacillus thuringiensis* and in other *cry11*-like genes. submitted

Messer, P.W. & Arndt, P.F. (2007).The majority of recent short DNA insertions in the human genome are tandem duplications. *Mol Biol Evol*, 24, 1190-1197.

Meyers, B.C.; Tingey, S.V. & Morgante, M. (2001). Abundance, distributionand transcriptional activity of repetitive elements in the maize genome. *Genome Res*, 11, 1660-1676.

Mirkin, S.M. (2007). Expandable DNA repeats and human disease *Nature*, 447, 932-940.

Moeller, R.; Stackebrandt, E.; Reitz, G.; Berger, T.; Rettberg, P.; Doherty, A.J.; Horneck, G. & Nicholson, W.L. (2007). Role of DNA repair by non-homologous end joining (NHEJ) in *Bacillus subtilis* spore resistance to extreme dryness, mono- and polychromatic UV and ionizing radiation. *J Bacteriol*, 189, 3306–3311.

Mondello, C.; Smirnova, A. & Giulotto, E. (2010). Gene amplification, radiation sensitivity and DNA double-strand breaks. *Mutation Res*, 704, 29–37.

Ogata, N. & Miura, T. (2000). Elongation of tandem repetitive DNA by the DNA polymerase of the hyperthermophilic archaeon *Thermococcus litoralis* at a hairpin-coil transitional state: a model of amplification of a primordial simple DNA sequence. *Biochemistry*, 39, 13993-14001.

Ogata, N. & Morino, H. (2000). Elongation of repetitive DNA by DNA polymerase from a hyperthermophilic bacterium *Thermus thermophilus*. *Nucl Acids Res*, 28, 3999-4004.

Ohno, S. (1984). Birth of a unique enzyme from an alternative reading frame of the preexisted, internally repetitious coding sequence. *Proc Natl Acad Sci USA*, 81, 2421-2425.

Ohno, S. (1987). Evolution from Primordial Oligomeric Repeats to Modern Coding Sequences. *J Mol Evol*, 25, 325-329.

Orgel, L.E. & Crick, F.H. (1980). Selfish DNA: the ultimate parasite. *Nature*, 284, 604-607.

Pace, J.K. II; Sen, S.K.; Batzer, M.A. & Feschotte, C. (2009). Repair-mediated duplication by capture of proximal chromosomal DNA has shaped vertebrate genome evolution. *PLoS Genet*, 5(5):e1000469.

Parkhill, J.; Achtman, M.; James, K.D.; Bentley, S.D.; Churcher, C.; Klee, SR.; Morelli, G; Basham, D.; Brown, D.; Chillingworth, T.; Davies, R.M.; Davis, P.; Devlin, K.; Feltwell, T., et al. (2000). Complete DNA sequence of a serogroup A strain of *Neisseria meningitidis* Z2491. *Nature*, 404, 502-506.

Ruland, K.; Wenzel, R. & Herrmann R. (1990). Analysis of three different repeated DNA elements present in the P1 operon of *Micoplasma pneumonia*: size, number and distribution on the genome. *Nucl Acids Res*, 19, 637-647.

Saito, H.; Minamisawa, T. & Shiba, K. (2007). Motif programming: a microgene-based method for creating synthetic proteins containing multiple functional motifs. *Nucleic Acids Res*, 35(6):e38, doi:10.1093/nar/gkm017.

Sasaki, J.; Asano, S.; Hashimoto, N.; Lay, B.W.; Hastowo, S.; Bando, H. & Iizuka T. (1997). Characterization of a cry2A gene cloned from an isolate of *Bacillus thuringiensis* serovar *sotto*. *Curr Microbiol*, 35: 1-8

Shapiro, J.A. & von Sternberg, R. (2005). Why repetitive DNA is essential to genome function. *Biol Rev*, 80, 227-250.

Shiba, K.; Takahashi, Y. & Noda T. (1997). Creation of libraries with long ORFs by polymerization of a microgene. *Proc Natl Acad Sci USA*, 94, 3805-3810.

Storici, F.; Snipe, J.R.; Chan, G.K.; Gordenin, D.A. & Resnick, M.A. (2006). Conservative repair of a chromosomal double-strand break by single-strand DNA through two steps of annealing. *Mol Cell Biol*, 26, 7645-7657.

Tuntiwechapikul, W. & Salazar, M. (2002). Mechanism of in vitro expansion of long DNA repeats: effect of temperature, repeat length, repeat sequence, and DNA polymerases. *Biochemistry*, 41, 854-860.

Ullu, E. & Tschudi, C. (1984). Alu sequences are processed 7SL RNA genes. *Nature*, 312, 171-172.

Wang, S.T.; Setlow, B.; Conlon, E.M.; Lyon, J.L.; Imamura, D.; Sato, T.; Setlow, P.; Losick, R. & Eichenberger, P. (2006). The forespore line of gene expression in *Bacillus subtilis*. *J Mol Biol*, 358, 16–37.

Weller, G.R.; Kysela, B.; Roy, R.; Tonkin, L.M.; Scanlan, E.; Della, M.; Devine, S.K.; Day, J.P.; Wilkinson, A.; di Fagagna, F.D. *et al.* (2002). Identification of a DNA non homologous end-joining complex in bacteria. *Science*, 297, 1686–1689.

Wells, R.D. (1996). Molecular basis of genetic instability of triplet repeats. *J Biol Chem*, 271, 2875-2878.

Widner, W.R. & Whiteley. H.R. (1989). Two highly related insecticidal crystal proteins of *Bacillus thuringiensis* subsp. *kurstaki* possess different host range specificities. *J Bacteriol*, 171, 965-974.

Wu, X.; Cao, X.L.; Bai, Y.Y. & Aronson, A.I. (1991). Sequence of an operon containing a novel N-endotoxin gene from *Bacillus thuringiensis*. *FEMS Microbiol Lett*, 81, 31-36.

Wyman, C. & Kanaar, R. (2006). DNA double-strand break repair: all's well that ends well. *Annu Rev Genet*, 40, 363-83.

Yang, Y.; Sterling, J.; Storici, F.; Resnick, M.A. & Gordenin, D.A. (2008). Hypermutability of damaged single-strand DNA formed at double-strand breaks and uncapped telomeres in yeast *Saccharomyces cerevisiae*. *PLoS Genet*, 4(11):e1000264.

7

The Second Law of Thermodynamics and Host-tumor Relationships: Concepts and Opportunities

Joseph Molnar[1], Zoltán G. Varga[1],
Elysia Thornton-Benko[2] and Barry S. Thornton[3]
[1]*Institute of Medical Microbiology and Immunobiology, University of Szeged,*
[2]*Royal Prince Alfred Hospital Sydney, currently Research Associate, Key Centre for Health Technologies University of Technology Sydney*
[3]*School of Physics, University of Sydney*
[1]*Hungary*
[2,3]*Australa*

1. Introduction

The human organism continuously takes up nutrients to build up macromolecules and functionally active structures. On the basis of the first law of thermodynamics, the energy of nutrients is transformed, proceeding toward minimum entropy production and the release of heat and "waste products" in the open system. Part of the heat produced during exothermic processes supports the optimal efficiency of the endothermic biological reactions in the organism, and the rest is dissipated in the environment on the basis of the Second Law of Thermodynamics. In contrast with the living state, cancers live on the energy and mass of the organism as a parasite metabolizing and destroying the tissues of the organism according to the Second Law of Thermodynamics.

Schröedinger put the question of whether life is based on the laws of physics. He suggested that: chemical processes lead to entropy production, which is a direct measure of molecular disorder (1). At the end of his book, he summarized his view: "We must therefore not be discouraged by the difficulty in interpreting life by the ordinary laws of physics" (1). Where processes under conditions far from equilibrium correspond to an interplay between chance and necessity, between deterministic laws and fluctuations (2, 3).

During tumor growth, there are many differences between the healthy tissues and the growing tumor, including metabolic, structural and thermodynamic differences, for heat production. Both structural differences and heat formation can be used to follow the entropy differences between cancerous and nomal tissues. These entropy changes can be followed by various methods, e.g. histology and thermography. In general, the calculation of entropy production due to thermal fluxes based on temperature differences and, similarly, differences in other entropy-producing processes as driving forces are promising as potential targets for tumor demarcation (2-4).

Entropy production results from the transport of heat and matter between the two phases of the system and also from the chemical reactions taking place. Thus, the rate of increase of entropy production $\frac{dS}{dt}$ is a bilinear form of the rates of the irreversible processes and some functions of state, which may be called "affinities" or "generalised forces" as explained in Appendix A. S is entropy, T Temperature, V volume, Gibbs function, A affinity, n, number of species, I, II locations.

$$\frac{d_iS}{dt}=\left(\frac{1}{T^I}-\frac{1}{T^{II}}\right)\frac{d_i^IG}{dt}-\sum_{\gamma}\left(\frac{\mu_\gamma^I}{T^I}-\frac{\mu_\gamma^{II}}{T^{II}}\right)\frac{d_0n_\gamma^I}{dt}+\frac{A^I\mathrm{v}^I}{T^I}+\frac{A^{II}\mathrm{v}^{II}}{T^{II}}\geq 0$$

The above can be replaced by Equation (3) from the Appendix and is as follows:

$$Sd\mathrm{T} = -\,dG + Vdp + \sum_i \mu_i dn_i$$

A relatively simple form of the entropy production per unit time was expressed by Prigogine (3) below:
If the entropy changes in the two phases (at the border of the tumorous and the normal organ tissues) are compared as cancers' velocity, then the ratio between the two forms of tissue may be characteristic of the borderline, where an overlap in entropy production rate may be found between the normal tissues and the tumor. This can be considered a new target for demarcation (5).
Dissipative structures are useful models for *in vivo* experimental studies of the onset of tumors (3, 4), where individual tumour cell dynamics is seen as a fluctuation between molecular products smaller than a critical size and others larger than critical size, uncontrollably and permanently able to appear and develop through replication. The cell is then confronted with a population of cells that either succeed in destroying it or not. Cytotoxic cells can confuse dead tumour cells with living cells and as a result, the destruction of cancer cells becomes increasingly difficult. Here, the outcome of instability through fluctuation is growth occurring irreversibly in time, producing an evolution of a tumour whith the tumor developing a stealthy growth (6).
This simplest example of this kind of evolution is associated with the concept of structural stability: the competition determines the threshold of stability. The the new series of reactions then enter into competition with the previous mode of functioning of the system(7). However, chemical processes lead to entropy production, where processes under conditions far from equilibrium correspond to an interplay between chance and necessity, between deterministic laws and fluctuations (3).
If the structural fluctuation imposes itself, whereby the kinetics innovators is fast enough for the latter to invade the system instead of being destroyed, then whole system will adopt a new mode of functioning and its activity will be governed by a new syntax (4, 8).
Living organisms metabolize basic nutrients, converting their chemical energy to other forms of biomolecules and heat is produced as a by-product or waste energy that is dissipated to the environment. In the complex metabolic process maintained by the concentration gradients of the compounds, the transformed chemicals become incorporated into tissues of the organism. Leo Szilárd presumed, that an organism may behave differently from a mechanical system, causing a permanent decrease of entropy and thus violation of the Second Law (9). He then realised the possibility that the proposed arrangement

threatens the validity of the Second Law. It was thought that such a simple inanimate device could achieve the same result as that attained by the intervention of an intelligent being. The biological phenomenon of a non-living, theoretical device (Maxwell's "demon") that generates the same quantity of entropy as required by thermodynamics was examined along the same line of thought but the Second Law remains valid (9).

Thermodynamic properties such as entropy are of predictive value for further reactions and for the direction of the reactions. Organisms and their compartments are thermodynamically open systems; they have been analyzed by means of structural entropy. Such morphological studies give information about the entropy changes from a directed exchange with the environment, as considered by Kayser (10). When a solid tumor is embedded in healthy tissues, the entropy current is equivalent to the amount of entropy exported or imported through a boundary defined at a thermodynamic distance; this permits the introduction of the entropy concept to the processing of information (10).

Direction of Informational entropy

The Second Law of Thermodynamics does not allow the creation of entropy barriers, but it does not exclude changing the direction of some particular components of entropy flow from a tumor to the normal tissues in the close environment. The reversal of entropy flow in coexisting normal and tumor tissues may halt the tumor development due to reversal of the signal transmission as informational entropy in the tumor-host entity (11). The different entropy production rate between the normal and cancerous cells determines the direction of informational entropy current between normal and cancerous cells.

Energy requirements and time

The changes in the universe are continuous or permanent including the entropy increase content in the world, but "constant" things are transient in time. To preserve things in a given state, constant energy is required and the energy requirements for such preservation increase with time. One simple example is the system of melting ice (which can serve as an analog for very limited conditions for life). To preserve the transient state of coexisting ice cubes in water, we have to invest energy that will keep the temperature at around 4 °C, in order to avoid complete ice formation or complete melting of the ice cubes in the water. There is a permanent energy investment or requirement to maintain life, when the living object extracts energy from its environment. As in the case of melting ice cubes or completely frozen water, the system cannot be repeated or reinitiated. Similarly, the life processes cannot be reinitiated if the transient state is once interrupted by the cessation of energy expenditure. The instability of the melting ice system gives a good example of the opportunity for us to change the stability of the system by changing the direction of entropy flow by an increase of temperature. This opportunity does not mean that hyperthermia is applicable as a cancer cure: our previous calculations have demonstrated that the thermogenesis produced by cancerous tissue plays an insignificant role in the entropy production of cancer (12). However, the water-ice transition example shows that there are opportinities for modification of the direction of entropy flow in a thermodynamically open system.

There was a historical objection against the universal validity of the Second Law of Thermodynamics. This objection was embodied in the notion of Maxwell's demon, which appears again and again according to Szilárd. (The Maxwell demon catches the fast molecules and lets the slow ones pass, to gain energy at the expense of heat in a hypothetical engine with no mechanical structures that could lose energy in the form of heat). Thus it was stated that, in a system left to itself, no perpetuum mobile of the second kind could operate, in spite of the fluctuation phenomena (9). No successful objection has ever been made so far. Thermodynamic differences between healthy and cancerous tissues suggest new strategies to contribute to improvement of the perspectives for the therapy of solid tumors and to reduce the development of their metastases (7).

The occasional question of what entropy actually implies has been answered given a very simple answer that entropy is the "shadow of energy" (13). The shadow of light is an illusion that can be conceived as a mirror-reflected picture of disordered material or chaos. The shadow of light is an illusion but the products of cancer in an organ can be considered functionally as inactive tissues that progressively reduce the specific physiological function of the organ and this process continuously develops, constantly gaining more space from the host and preventing the organ's function.

The development of the host-cancer relationship is a one-way road; it is an irreversible process in the coexisting cancer and host tissues once cancer begins. Cancer grows toward the maximum entropy according to the Second Law of Thermodynamics and lives on the negative entropy of the host in a superparasitic relationships if no external force field is applied (7). At the same time, the host develops toward the entropy minimum that is characteristic of the living state. At the border of the tumour and the normal tissues, the order-disorder transition area, the basic thermodynamic and biochemical changes occur. The developing tumor changes the minimum entropy production of the tissues toward maximum entropy production in this part of the organism functioning as a time delayed bomb. However, thermodynamic aspects of the composition and direction of entropy flow between cancerous and healthy tissues living in close proximity can be analyzed in reasonable detail. The questions arise as to what the known components of entropy production and what factors determine the direction of entropy flow from cancerous to normal tissues or *vice versa.* The main components are now listed below:

3. Components of the entropy program

1.	Cancer grows by invading into normal tissues but the opposite never occurs: normal tissues never invade cancer tissues.
2.	Tumors produce heat that is dissipated from the cancer to the normal tissues and can be regarded as a symptom.
3.	The low pH of cancer cells and the neutral pH of healthy cells produces a pH gradient between the two forms of tissues.
4.	At acidic pH, the ATP synthase activity of the tumor is further reduced in a vicious circle.
5.	The membrane potential differences between normal and cancer cells are responsible for maintaining some type of information flow.
6.	The electrogenesitic features of normal and cancerous tissues differ (cortical electrogenesis, muscle, retina, etc.).
7.	Enzymes and other biologically active compounds produced by the tumour cells which invade and survive carry information from cancerous to normal cells and mediate the nutrients and immunomodifiers.
8.	Dielectric structures are formed after transition from normal to tumorous cells (14, 15). In general, the direction of the entropy flow produced by the cancerous and healthy tissues is driven by concentration gradients, potential gradients, conduction, convection, diffusion, blood circulation, etc. (Table 1).

Thermodynamic forces drive fluxes:
1. temperature differences drive thermal fluxes. a. convection movement of biological tissues. b. conduction-related heat transfer. c. conduction-related matter transfer in the opposite direction (4. Glansdorff P. and Prigogine I. Thermodynamic Theory of Structure, Stability and Fluctuations, Wiley Interscience, New York, p.17, 1978,)
2. chemical potential differences drive diffusion currents: (disordered cytoskeleton, disordered mitochondria).
3. electric potential differences drive electric currents.
4. Gibbs energy differences drive chemical reactions, *e.g.* ATP can be synthesized from glucose depending upon the anaerobic glucose fermentation or on the respiration chain. Thus, the energy produced by one mole of glucose varies between 37 and 301 kcal/mol. At any rate, the high rate of glucose utilization of a tumor is greater than that of the normal tissues and this is exploited in tumor diagnosis by PET.
5. Viscosity differences make medium deformations.
6. Application of external forces can induce different entropy production in tumor and healthy tissue
7. Entropy differences drive entropy fluxes by determining the directions of processes. If different fluxes flow in the same direction, they result in flows or currents of information.
8. Chemotactic constraints in coexisting populations drive nutrient fluxes (Lotka Voltera type processes).

Table 1. Differences between cancerous tissues and normal tissues drive fluxes

The entropy flow carries information from cancerous tissues to the healthy tissues, which involves various fluxes carring information from the cancerous to the normal tissues, and also negative entropy from the healthy tissues toward the cancerous tissues. The entropy flow production and intensity are related to the accompanying convection to the biological tissues. The conduction relating to heat transport only can be neglected due to the homogeneity of temperature in the human organism. Conduction in the opposite direction to the transport of matter (cancer cells), may be called negative entropy. Since entropy flow is the flow of information from cancerous to normal tissues, only the contribution from items 1 and 3 listed above are now considered. The other items are considered later in this report.

Cancer can be considered as a type of energy parasitism where cancerous tissues divert or extract energy from the normal tissues (utilizing glucose, proteins and fats in various ways). In this way, the cancerous tissues transform the pathways of normal energy production and dissipate the excess energy as waste material. An aspect of practical importance in chemotherapy is the kinetic resistance: the chemotherapy acts mainly on proliferating cells whereas the resting cells survive. This reflects the intrinsic resistance of the DNA of the non-proliferating cells to drug-induced strand breaks. Pharmacokinetic resistance is manifested when drugs are administered at the "wrong time", when tumor cells are dormant, or are going through mitosis. The historically related criticism of ineffectiveness of cancer chemotherapy indicated the need to optimise treatment so as to avoid the continuous

administration of chemotherapy when tumor cells are in a resting or dormant state while the cells of the immune system will be eradicated as a consequence of the simultaneous adverse effects of tumor chemotherapy because these cells are renewing continuously.
A tumour comprises non-functional abnormal tissues, accepted by the immune system of the host as its own and not a foreign tissues but is a stealthy invader of the healthy tissues (15, 16). Tumour cells and tissues differ from normal cells in electric potential, electrogenesis, dielectric properties, electric conductivity, permittivity, impedance. For example disorder in the brain caused by a tumour means an entropy increase, and differences in entropy production rates between the two types of tissues determine the direction of entropy flow between the tumor and the healthy brain tissues around the tumor. The growing tumour competes for available energy sources. This may result in EEG changes in non-operated on individuals with a brain tumor. The focal slowing of the normal electrical rhythms close to the tumor are revealed as slow alpha and theta waves appearing in the projection area of the tumor. These changes are related to the high entropy production and the consequent import of negative entropy from the normal tissues. The growing tumor attenuates cortical electrogenesis and the electrical activity of the brain around the tumor tissues declines (17). Impaired electrogenesis means that a cancer has an electron deficiency which continuously reduces the electron pool of normal living tissues by oxidizing the living state, resulting in cancer. (In dielectric insulators or dielectric layers, the charges are fixed and are not free to move).
The extracellular matrix is less conductive than fibers or cytoskeleton components inside the cells. An applied electric field tends to restore the orientation of the dipoles and produces a field inside the dielectric which opposes the applied field. The conductivity describes the ability of the system to transport charge and the electric charge moves relatively freely. Permittivity characterises the ability of the material to trap or store charges. Factors tending toward the entropy maximum simultaneously result in a malignant state by invading host tissues, healthy tissues never invading cancerous ones.

4. What can be changed and how?

The key question is how the process of cancer development can be changed? One possibility is to overcome the kinetic resistance of a solid tumor or multiple tumors by consideration of the Gompertz growth model with chemotherapy included. The chemotherapy plus multi-drug resistance (MDR) -reversal compounds can halt the indefinite growth of cancer through the use of selective anticancer agents and selective inhibitors of ABC transporters (18). Since the Pgp17o glycoprotein responsible for the MDR of tumor cells also has physiological functions, it was reasonable to search for specific MDR-reversing compounds and highly selective antitumor drugs by using stereoisomers of certain active compounds. As a result of such experiments, tissue or organ-specific MDR inhibitors, combined with a tumor-specific anticancer drug result in strong synergy (18). The direction of entropy flow from tumorous to normal tissues in close proximity in the host was presumed to be reversed by such external forces based on the difference in entropy production of the two cell populations.
To maintain the normal signal transmission by particular protease inhibitors or by inhibitors of angiogenesis is another possibility. The blockade of various components of entropy production that mediate harmful by-products causing or inducing molecular or tissue disorders is a promising perspective. The other side of the coin is disruption of the quorum sensing mechanism (a population signal transmission) that triggers carcinogenesis (19). Nevertheless, we suppose that in the case of the host–tumor entity, the differences in

entropy production induced by various external interventions are able to change the directions of fluxes (Table2).

Effects of external force field on entropy dissipation
The rate of entropy production in tumor/normal cell is 1.1 without an external force field.
Application of a static electronic force field induces lower entropy production in thumor than in normal tissues.
A square wave electric potential increases the entropy differences by continuous switch on and off. The increase of frequency enhances entropy production. Applied square wave electric pulses produce more entropy in normal cells then in tumor cells. Tumor/normal ratio = 0.33
The effect has been exploited in the electrochemotherapy of skin cancer. (Rodriguez Cuevas Sergio et al. Archives Medical Res. 32,, 273, 2001Tisdale M J: J.Supportive Oncol 1, 159 (2003)
Delivery of electric pulses to solid tumors increases the susceptibility to anticancer agents due to transiently increased membrane permeability (Sersa M et al Cancer Res 55,3450, 1995. Belehradek M. et al Cancer 72, 3694, 1993).
Ultrasound-mediated chemotherapy with application of low output intensity 1 MHz, 2 W/ch2) increased the chemosensitivity of solid tumors to bleomycin (Tomizawa M et al Cancer Letters 173, 31, 2001).
The mechanism of action of cavitation, poration and re-sealing of cell membrans by acoustic pressure increases the intra-cytoplasmic concentrations (Luo L. et al . Diagnostic Pathology 2006, 1:43 doi:10.1186/1746-1596-1-43.)

Table 2. The possible effects of external forces on entropy dissipation from normal to tumorous tissues.

The entropy differences are maintained by different fluxes.

(I)	Temperature differences associated with thermogenesis and hyperthermia can produce an entropy maximum in tumours, but the normal tissues are more sensitive than tumour cells.
(II)	The diffusion current is driven by a chemical potential gradient, where a number of subcellular compartments are greatly reduced in cancer cells due to the dielectric layers formed.
(III)	Chemical reaction are driven by Gibbs energy, glycolysis being the main energy source[14] for cancer cells, which metabolise 10 times more glucose than normal cells. The roles of glucose metabolism and glycosylation are known. Anaerobic glycolyis provides the energy for cancer and the increased glucose catabolism is used as a basis of the identification of tumors by PET. The glucosylation of ABC transporters including MDR1, proceeds in cancer cells and in physiologically important transporters, such as the blood-brain barrier. The glucose-regulated proteins modify the drug resistance in cancer when upregulated (20). The organ-specific, *e.g.*, brain metastasis formation of breast cancer requires posttranslational glycosylation.
(IV)	The velocity gradient is driven by viscous stress. The change in electrochemical potential in cell membrane can change the dielectric structure and permittivity[14]

	beween the internal membrane surfaces. These structures are unable to provide free conduction electrons or can act as insulators.
(V)	The dissipation of entropy is due to the work performed by an external force field. An example is the expositure of cancerous tissues to an electric field, which may increase the entropy production of the normal tissues above that of the tumorous tissues. This would mean that this intervention can reverse the direction of entropy current to enable its flow from normal tissues toward cancerous tissues. Another element of entropy production is the intensity of the individual fluxes and the sum of the "collected" entropy flows. Those effects determine the influence of the main pathway of interaction between the two co-existing entities. Heat dissipation from a tumor is mediated by the thermal flux driven by temperature differences. The heat production contains several parameters, one of which is diagnostic: it gives the metabolic rate of the lesion. Chemicals can diffuse in both directions, depending on their structures. The diffusion current between tumorous and normal tissues is driven by a chemical potential gradient. Chemical reaction rates are always driven by a Gibbs energy decrease: a velocity gradient coupled with viscous stress due to internal friction differences or dissipation due to work performed by external force fields.

5. The role of pH

With a pH gradient coupling with H potential differences, tumor cells preferentially convert glucose to lactic acid *in vitro*. Low intracellular pH values were demonstrated when pH electrodes were inserted into tumor tissues: the tumor was significantly more acidic than the normal tissues (8, 9, 21, 22). The membrane potential gradient is driven by potential differences between the cancerous and normal cell membranes. When tumour cells preferentially converts glucose and other compounds to organic acids, the intracellular pH increases. We can consider several examples of therapies:

a. Ultrasound

The pH differences between tumorous and normal tissues can be exploited in the reversal of the direction of entropy flow in the therapeutic application of the ultrasound irradiation of solid tumors when the entropy production of normal tissues may exceed that of tumorous tissues.

The aim of a good cancer outcome could be to change the direction of entropy flow based on pH, electric potential differences, *etc.*, where the most important contribution to the ultrasound absorption is related to the chemical relaxation and not shearing motions of medium molecules (10) and viscous stress or the heat loss due to conduction mechanisms because chemical relaxation (23) the basic contribution to the ultrasound absorption based on the pH differences (11, 12, 24).

In general, the pH measured in normal cells microelectronically is 7.2-7.6 while that in tumor cells is 5.6-7.2 which leads to healthy tissues displaying stronger ultrasound absorption than cancerous tissues. If we considered the relative difference between them to be 10%, then the difference in entropy production rate between tumorous and normal cells is 8×10^{-7} erg/degree/sec for an ultrasound power of 1 watt/cm on two types of cells, and lower for cancer at 2×10^{-7} erg/degree/sec (24). Naturally, decreasing the acidity in cancer tissue or increasing the acidity in normal cells would change the relative magnitudes of the entropy production rate of the two tissues and cause the reversal of entropy flow (24).

Acidic pH changes the conductance and dielectric properties of cancer cells, as discussed by Albert Szent-Györgyi (25).

In the host, the relationship, between healthy tissues and malignant tissues can be described by the Second Law of Thermodynamics. This defines the direction of physical and biological processes resulting in tumor growth. In the two living systems, the entropy dissipation is encoded in their genes; the exception is the response rate of dissipation to stress conditions affecting the organisms. As previously mentioned the negative entropy uptake compensating entropy production differs during life. Entropy production is a thermodynamic quantity for a living system. It depends on the quality of life conditions such as nutrition, physical exercise and the frequency of stress exposure. Entropy always increases for any non-equilibrium system due to the more disordered sub-cellular structures of the cancerous cells, with their different subcellular structures, following the Second Law of Thermodynamics (11, 12, 15). The information inherent in a cancerous cell is different from that in a normal cell. In living healthy cells, the free energy is high (increases) and their entropy is low (decreases) due to the altered metabolism at the expense of the environment (12, 16).

b. Electronic/Dielectric

Dielectric structures are formed in cells after malignant transformatiom, as shown by Thornton, who used inversion of Raman spectra to compare normal living cells and cancer cells[14]. With regard to induced dielectric structures, it was proved that chemical or other agents -which can change the electrochemical potential locally at the outer membrane- will produce internal dielectric zones. However, the transformation then subsequently requires that the polarisation P and membrane potential V change progressively to the Frölich metastable point in order for the transition to cancer occur. For example, a normal cell may not exhibit dielectric zones but P and V may never reach the metastable state and the cell will be benign. The progression to the metastable point involves successive mutations, (often quoted as about five) which could take a considerable time[14] . However, it does give time to set to act to avoid this progression.

The entropy production of healthy cells is lower than that of cancerous cells if no external energy input is applied to the tissues. However, when appropriate external energy is applied, the rate of entropy production of normal tissues may exceed that of cancerous tissues. Cancerous tissues develop toward the maximum entropy, but maximum entropy involves the total amount of energy not being available for work in the cells (11, 17).

As stated at the beginning of this report, we aim to review the essentials of the theoretical background of thermodynamics to introduce the reader to the application of the concept, with some examples of the combination of theory with multiple parameters of entropy production and the differences between healthy and cancerous tissues. Cancer and its surrounding normal tissues live together in a competition where the cancer cells are numerically inferior to the host cells. A small group of transformed malignant cells in a particular organ change the stability constraints of mass and energy flows in a thermodynamically non-equilibrium state. Conditions which favor the cancer cells or limit the growth or survival of their normal neighbors have therefore been studied.

In a breast cancer example, when normal human breast MCF10 cells or MRC5 fibroblasts were cultured *in vitro* in the presence of MCF7 human breast cancer cells in the absence of serum, the normal cells fed the cancer cells. In some experiments, serum as growth-promoting factor was replaced by the two sets normal cells. The results were defined by the competition factor F. The larger the competition factor, the stronger the exclusion effect between the two types of cells. It was found that the competition exclusion between two kinds of normal cells is stronger than between normal and cancer cells *in vitro* in the presence of serum nutrients. That means that F is in the parasitism region and the cancer cells live in parasitic coexistence with normal cells. The case is different during starvation: in the absence of serum, F is lower than zero, *i.e.* this is a superparasitism region[7].

The competition between the two cell populations for nutrients may be dependent on activation energy, the intake of nutrients and the metabolism by cell populations. The differences in the network in signal transmission, the direction of conductance and subcellular compartmentisation may play a role in the coexistence of cancer, stromal and normal cells. In summary, there is some type of competition between the different cell populations. This can be explained partly by competition exclusion, if cancer cells have a smaller least resource requirement than normal cells[26, 27].

6. The role of nutrition

We presume that the entropy and energy contents of nutrients are also of importance in cancer development in open systems. The life processes need a continuous ingestion of low entropy and high energy-containing nutrients. The nutrients are converted to high entropy and low Gibbs energy-containing waste products that are excreted by the host. If the living state is maintained by the thermodynamically determined exothermic processes, and the non-equilibrium state by the continuous supply of the described nutrients, the question arises, whether the effects of highly processed food with a low free energy content and a high entropy state contribute to the increasing frequency of cancer, or whether the increasing incidence of cancer is a consequence solely of the increased and accumulating environmental pollution itself modifying the thermodynamic processes maintaining the normal living state of mankind.

We suppose that tumour cells are selected on the basis of their nutrient requirements. Cannibalism appears via autophagy which can occur for cancer cells to stay alive during dormancy. Cancer cells have different energy requirements than normal cells and finish or complete their development at an evolutionarily lower stage than complete differentiation. However, cell population systems operate at thermodynamic non-equilibrium, which continuously acquires free energy, heat and work from the surrounding. The life of cancer is ordered at the expense of disordering its surroundings, when cancer becomes a parasite of the living state. The mitochondrial activity decereases continuously in cancer. However, life is considered a physical phenomenon in which negative entropy is essential for maintaining the self-organization compensating the spontaneous entropy production according to the Second Law[15, 28].

This is an ideal parasitic relationship for cancer as a commensural type of co-existence whereby cancer and healthy cells can survive and grow. There is a tendency for the host to halt the parasite by the immune system from one side, and there is a tendency for the cancer to accommodate and to adapt the host and for example, avoid the immune system. Cancer colonises the host without provoking a response as functionally and immunologically inactive tissues propagate themselves and kill their host by increasing the ratio of functionally inactive tissues at the expense of the functionally intact tissues in the host. In the end stage the specific functions are reduced in the host as multi-organ system. The question is, what is the driving force of this process? Cancer cells can grow on a wide scale of nutrients from small molecules to dead cells or living cells, whereas healthy cells are able to survive and grow only on a limited range of specific nutrients. The nutrient specificity of normal cells is limited to maintain the differentiated state, while cancer cells survive and grow under minimal nutrient conditions and in addition they are able to gain energy directly from the healthy cells and indirectly with the help of the stroma of the tumor. The simplest explanation is that the existing nutrient concentration differences and those of a large scale of substrates favor the slow overgrowth of cancer in the host when a transient immune deficiency can enhance the chance for tumor initiation and promotion. As a result, cancer has a selective advantage for growth due to the dissipation of the working structures and excess entropy. Another contributing factor can be the transforming growth factor produced by tumor cells inhibiting the growth of normal cells for their promotion. At a cellular level, after the initiation, deactivation of the immune system, the developing stroma and finally the metastatic invasion result in cancer. At the level of electrons, an electron acceptor blocks the electron donation from a purine or other molecules and the inhibited electron migration disturbs the conductance, resulting in restructuralization in the cells. Macrophage-like tumor cells are attracted, leading to repulsion in the absence of attractive forces.

There are changes in electromagnetic effects such as from the formation of dipole-dipole interactions and production of dielectric layers are produced and membrane and other oscillations are disturbed. The available energy does not satisfy the energy needs and the loosely bound cells continuously divide and subdivising can result because of the discrepancies between needs and availability. The high entropy production of tumor cells cannot be fully compensated by importing the negative entropy from the environment, and consequently the tumor growth is based upon the decomposition of normal cells and protein–lipid mobilisation by specific factors produced by the tumour[7].

7. Driving forces

The role of driving forces in cancer progression is unquestionable. The direction of entropy flow during parasitism is well defined. The irreversibility of the process can be explaind by competition for least resource requirements. This process consists of several components, such as accelerated glycolysis. Damage to the mitochondria, -or more exactly in the respiratory chain in tumor cells, - results in a compensatory increase in glycolytic ATP production in the malignant cells, together with a decreased level of oxidation of NADH-linked substrates. The mitochondrial defects contribute to tumor progression in several ways, such as by modified energy production, free radical generation and programmed cell death[28, 29, 30].

Proteolysis and lipid mobilization are energy sources in the host. The result is that the large majority of available energy sources in the host are governed and used by the tumor to support its survival and growth. As mentioned previously, this process can be considered as energy parasitism when the entropy production of cancer is compensated by energy withdrawal from the normal tissues in the form of negative entropy and high energy-containing macromolecules of the host. Consequently, the minimum entropy production of the living state becomes no longer sustainable. Enhanced glycolysis is the characteristic metabolism in the tumor[12].

The proteolysis-inducing factor produced by tumor cells appeared to fulfill the function of triggered muscle proteolysis, resulting in severe muscle protein degradatios[31]. Tumours also produce a lipid-mobilizing factor, which is associated with the progressive depletion of adipose fat tissues and the depletion of skeletal muscle, and with changes in the body composition via induced hydrolysis of fats and triglycerides in adipocytes, protein degradation and protein degradation[32]. Also the tumor-produced lipid mobilizing factor causes stimulation of the protein synthesis. Different components of fluxes are essential in entropy production. The entropy flow from normal tissues to cancerous tissues carries information concerning the normal tissues, while entropy flow in the opposite direction carries information for the cancerous to the healthy tissues. As stated at the beginning of the report, the Second Law of Thermodynamics does not allow us to set up entropy barriers, but it does not exclude changing the direction of components of entropy flow from tumor to normal tissues. Entropy production in living cells contains more than 5 terms[12, 16].

In general, the relation between entropy and information quantity was defined by Schrödinger who revised the statistical meaning of entropy on the basis of the investigations of Boltzmann and Gibbs in statistical physics and expressed by entropy=k log D, where k is the Boltzmann constant and D is a quantitative measure of the atomistic disorder of the body in question. The differences in basic factors of entropy production between normal and cancerous cells determine the direction of fluxes, and those can therefore be regarded as targets of specific interventions in cancer therapy.

The rates of entropy production in healthy and cancer cells have been compared[12]. The differences were analysed via the comparison of selected factors, as follows:

1. The thermal flux driven by temperature differences, the production of energy and entropy loss due to heat dissipation.
2. The diffusion current driven by chemical potential gradients.
3. The chemical reaction rates driven mostly by decreasing Gibbs energy (affinity and glycolysis).
4. The velocity gradient coupled with viscous stress.
5. The dissipation of energy to the external field, resulting from the external work completed.

Many other factors contributing to entropy production were not included in those comparative studies, *e.g.* : the pH gradient associated with the higher acidity of cancer cells than of normal tissues due to the glucose metabolism, the membrane potential gradient driven by the different membrane potentials of normal and cancer cells, possible interactions between cells with different electric charges or resting membrane potentials; the information flux by conductance is driven by dielectric permittivity differences, where electronic energy bands are saturated, proteins are non-conductant; entropy currents driven by differences in entropy production by various processes.

The toxic effect of entropy flow plays a key role in the competition between tumour and host cells and determines the rate of tumor progression and the rate of cachexia. The cancer invasion has the first priority, while healthy tissues are not able to invade into cancerous tissues. Consequently reversal of the direction of entropy propagation can be a realistic goal as a therapeutic possibility. The intracellular pH of cancer cells is much lower than that of normal cells. The increased glycolysis resulting in metabolic acidosis was first described by Warburg 1956.

Dielectric layers have been demonstrated in proliferating cancer cells (14). The main point to be noted is that, the differences in the basic factors of entropy production between normal and cancer cells determine the direction of fluxes and consequently can be targets of specific interventions in cancer therapy.

8. Reversing the flow

What is the solution? How can we reverse the entropy flow? If we were able to increase the entropy flow from normal tissues, then the informational entropy of the normal tissues would counteract the informational entropy of the cancerous tissues.

1.	The extreme extent of glucose metabolization and heat production and the effects of an external force field on the studied 5 entropy production terms Appendix eq.3 have been discussed and evaluated from practical point of view in the case of external energy input, when the entropy production of normal cells can exceed that of cancer cells[10,11,12,15]. The direction of entropy flow would then be reversed and that leads to a blockade of the propagation of harmful information from the cancerous to the healthy tissues. Reversal of the directions of entropy flows by exposure to external forces appears to be possible.
2.	The artificial modification of acidity for cancerous cells can change the relative entropy production rates of the two kinds of cells and may lead to reversal of the direction of entropy flow. It can be supposed that modification of the basicity of a tumour leads to a standard free energy decrease and therefore entropy production would be lowered when the pH increases in a tumour. Modification of the glucose metabolism of tumors seems to be most difficult to achieve.
3.	Electronic energy bands are saturated and proteins are non-conductant in cancer. The conduction change due to the formation of dielectric layers in cancer cells is another possible factor for intervention[15]
4.	The conductivity differences between normal and cancerous tissues can be exploited as targets for intervention. The different sensitivities of cancer and normal cells to electric fields, *e.g.* square wave electric impulses, may result in changes in the polarisation of cancer cells, whereby they become more sensitive to external foces such as ultrasound or chemotherapy[11, 12].
5.	The application of low-frequency (1 MHz) and low-intensity ($<1\ W/cm^2$) ultrasound can destabilize the connections of tumor tissues with their environment. Ultrasound absorption increases entropy production in normal tissues more efficiently than in tumorous tissues as a consequence of the more acidic nature of the tumour. Recent publications[23, 24] show that intervention is able to reduce the entropy flow from the cancer to the healthy tissues[23, 24].
6.	Alternating electromagnetic fields several volt/cm applied at low frequency appeared to have caused the reversal of entropy flow[24].
7.	An alternating magnetic field (10,000 Gs) induces additional entropy production in tumor cells by changing the entropy production threshold[15, 24]
8.	Modification of angiogenesis by an inhibitor reduces the nutrient support.
9.	Increasing temperature also appears to reverse the entropy flow direction somewhat and is thought to reduce the rate of entropy production from the heat flux, meaning some temporary benefit in therapy in some cases. However, mathematical and physics calculations have shown that entropy production due to heat is only a very small fraction of the total entropy production, and consequently the efficiency of hyperthermia as therapy is very low[12, 15]

Table 3. Some Possible Methods

9. Optimism and caution

The orderliness encountered in the unfolding of life springs from a different source. It appears that there are two different 'mechanisms' by which orderly events can be produced: the 'statistical mechanisms' which produce 'order from disorder', and a new one, producing 'order from order'. The Second Law appears to give a simple explanation. That is why physicists were so proud to have fallen in with the 'order-from-disorder' principle, which is actually followed in Nature and which alone conveys an understanding of the great line of natural events, and primarily of their irreversibility. We cannot expect that the 'laws of physics' derived from it automatically suffice to explain the behavior of living matter, whose most striking features are visibly based to a large extent on the 'order-from-order' principle. We would not expect two entirely different mechanisms to bring about the same type of law as -we would not expect a single solution of cancer treatments. We must therefore not be discouraged by the difficulty in interpreting life by the ordinary laws of physics. For that is just what is to be expected from the knowledge we have gained of the structure of living matter[2, 3, 13]. We must be prepared to find a new type of physical law. Or are we to term it a non-physical one, not to say a super-physical law, as the question was ironically put by Schrödinger[1]. This assertion, cannot fail to arouse contradiction. There are phenomena whose conspicuous features are visibly based directly on the 'order-from-order' principle and have nothing to do with statistics or molecular disorder. These calculations do not imply any statistics; they are based solely on Newton's law of universal attraction. The mechanical events seem to follow distinctly the 'order-from-order' principle, and if we say 'mechanical', the term must be taken in a wide sense[2, 13].

In a paper witten by Max Planck on the topic 'The Dynamic and the Statistical Type of Law' ('Dinamische und Statistische Gesetzmässingkeit'), the distinction is precisely labeled as 'order from order' and 'order from disorder'. The object of that paper was to show how the interesting statistical type of law is constituted from the 'dynamic' laws put forward to govern the interactions of single atoms and molecules[2, 7], where the understanding and the identification of the tumor escape mechanism are easier[16].

By combining morphology with kinetics of tumor development and textural, biological properties of tumor, the equations from thermodynamics can be used to illustrate the differences between healthy and cancer cells. The data allow appropriate calculations to be performed which induce the entropy concept to the processing of information collected by the analysis of various contributing factors. Healthy tissues display an efficient and stable way to perform the biological functions so as to maintain the lowest entropy level. Deviations from these values are reflected in differences in structures and functions between neighboring tissues in the thermodynamically open system of the host-tumor entity[12, 13, 16].

The entropy current is equivalent to the amount of entropy exported through the boundary between tumorous and healthy tissues. This is an appropriate measure of thermodynamic theorems[4, 26, 28] based on the evidence that the normal healthy state develops toward minimum entropy production, while a cancer moves towards the entropy maximum[1, 14, 15]. mc^2 is negative in nuclear fission, due to a decrease in mass, but positive for a growing mass of living organism[27, 28]. The difference between the two is the energy dissipation due to mass. The test of this came with the advent of nuclear fission, where energy appears to come from nowhere, but a term provided by Einstein readily maintains the validity of the conservation because the change in mass of the system, Δm is negative since the mass of the system decreases. While the First Law applies equally to living and to dead matter, the Second Law does not.

The First Law states that energy is conserved, while the Second Law states that, within the framework of conservation, one cannot have it any way one might like. It is a matter of everyday experience that certain processes do and certain processes do not occur in the world. Heat cannot be caused to flow from a cooler to hotter body without producing some other effect. Processes occur so as to progress from non-equilibrium toward equilibrium states; the direction of processes is therefore defined for a heat engine or a biological object operating irreversibly. For a heat pump or a heat engine or the diffusion of molecules through cell membranes, we may write irreversible when the total entropy changes are greater than zero. This means that the Secon Law is a sweeping generalization to the effect that, for any process, the sum of all the entropy changes occurring as a result of the process is greater than zero, and approaches zero in the limit as the process becomes reversible.

10. Mitochondria of cancer cells

One of the first attempts to destroy cancer was the idea of the selective effect of mitochondrial toxin for tumor cells. Papers have described mitochondrial toxins selective for tumor cells. The concept of a mitochondrial basis for killing tumor cells has existed since the early part of last century, based upon metabolic studies demonstrating general differences in metabolic control between cancer and normal cells[29, 30, 32]. Indeed, the metabolic disturbance in tumor cells is known. Metabolic disturbance due to abnormally high rates of glycolysis is demonstrated on a daily basis in the clinic through the fluoro-deoxyglucose PET scanning diagnosis of recurrent or residual malignant disease[33]. It was initially posited that mitochondrial oxidative phosphorylation was defective in tumor cells, and this initial insult led to gradual and compensatory increases in glycolytic ATP production as the central event of cell transformation. It was predicted that treatment producing a mitochondrial injury of a general nature would strike a greater blow against cancer cells, due to their already compromised respiratory state, than against normal cells. There is also the possibility that high rates of glycolysis may suppress respiration, known as the Crabtree effect, without implicating defective mitochondria[23, 24]. Interventions designed to attack metabolic susceptibilities of tumor cells have included glycolytic inhibitors (2-deoxyglucose), membrane pore activators (lonidamine), and thiol-active chemicals (arsenicals). More recently, screening of a combinatorial library of compounds based on natural benzopyran inhibitors of mitochondrial complex-I yielded several novel compounds with pronounced cytostatic effects against a cancer cell line panel[34]. It was calculated that cancer cells extract a much higher amount of glucose from the organisms than do healthy cells[12, 24]. In addition glucose-regulated proteins play an important role in the induction of MDR[20]. The mitochondrial susceptibility in cancer cells is interwoven with cellular metabolic control mechanisms. The mitochondrial uptake of particular toxins (F16 and other DLCs) also reflects the distribution across the plasma membrane, according to the plasma membrane potential, $\Delta\psi_p$, which may vary between cancer and normal cell populations[30,35,36].

New developments in chemotherapy are encouraging but the immunosuppression of chemotherapy is another important question considering the discontinuous tumour growth with reduced entropy production and the increasing import of negative entropy from healthy tissues. Attention to timing of therapies including chemotherapies is particularly important. In a dormant or resting phase of a tumor, the growth of the tumor cells is not really sensitive to the cytotoxic chemotherapy, but the toxicity toward bone marrow and other immune cells is continuous[15, 16, 33].

Living cells create an electric potential force between their various phases by three distinct mechanisms. Charge separation creates a potential of 120 to 145 mV between cytoplasmic and mitochondrial phases by unbalanced proton expulsion powered by the redox energy of the respiration chain. The resting potential of cells, which varies from -85 mV in the heart to -4.5 mV in red cells does not appear to result from an unbalanced flow of ions. Movement of an ion between phases entails three types of energy. The concentration work is that required to move an ion between phases containing different concentrations of ions. The electric work is that required to move an ion between phases with differing electric potentials. The osmotic work term is small and can generally be ignored. The measured resting potential between extra- and intracellular phases is approximately -85 mV, depending on the Na^+ or K^+ current flow. In the liver, the resting potential ranges from -28 mV to -40 mV, while in red cells the resting potential is about -4.5 mV. The resting potential between extra- and intracellular phases of cells should be thought of not as a diffusion potential, but rather as a measure of the electrical work. The use of intracellular KCl electrodes to measure resting electrical potential results from the existence of a mono-ionic Gibbs-Donnan near-equilibrium system between extra- and intracellular phases with the energy of ATP hydrolysis[34, 36]. In our experiments on resting cells in the presence of glucose concentrations increasing from 0.05 to 2.0% the fluorescence of R123 accumulation or uptake in MDR cells did not vary. MDR cells growing in the presence of various glucose concentrations displayed a somewhat elevated R123 uptake relative to the control. Acid diffuses from tumor to normal tissues as one of the component of entropy flow, making the normal tissues more sensitive to tumor progression and invasion.

10.1 Membranes and electrical potentials

The electric potentials of the membranes of resting (normal) and tumor cells are lower than those of normal cells. The membrane potential of proliferating tumour cells is somewhat lower than that of resting cells. The potential differences create a non-equilibrium state. These unique situations at the borderline of the two charged tissues and the resulting charge will be dissipated as bioelectric current, facilitating tumor invasion, or in other cases the ionized atmospheric electric weather fronts induce sensitivity to migraine in sensitive patients. If a larger area of the two cell populations with a larger difference in electric potentials is equalized in charges due to conductance, when the different charges are not isolated or the conductance is enhanced between them, this electric excitation may lead to epileptic seizures in a susceptible individual. The question arises of the consequences of entropy dissipation, the synchronisation of chaos, or the first step of the formation of a new highly ordered system. This is another example of entropy dissipation due to different extents of bioelectricity generation. In the case of brain tumors, the disorder in the brain caused by the growing tumour increases the entropy and reduces the nutritive proceses of the brain tissue. Differences in entropy production between the two kinds of tissues determine the direction of entropy flow from the tumor to the healthy brain tissues, with local slowing of the normal electrical rhythm in the neighborhood of the tumour, slow alpha and theta waves appearing in the projection area of the tumor. This finding should be related to the high rate of entropy production, and a negative entropy import from the normal healthy brain tissues can contribute to the attenuated cortical-electrogenesis around the tumorous tissues[21]. This *in vivo* observation can be supported by the superparasitic nature of cancer cells in *in vitro* experiments[7]. Bioelectricity and conductance are well known and measurable in diagnosis via the heat function by ECG (in mV), in electromyography (EMG) (in mV), or in EEG (in μV) (36). At a cellular level, the differences in electric potential are much smaller and energy dissipation in the human central nervous system is 10^{-9} W per neuron. However, for comparison, the typical dissipation for a transistor has been estimated to be as small as 10^{-1} W [17, 37].

These differences can drive the bioelectric currents, and can contribute as a well-defined component of the entropy production, mediating information between healthy and tumorous tissues.

11. Combination treatments and need for maths and physics

Combination treatment of cancer may consist of electric treatment plus a chemotherapeutic agent, radiotherapy or surgery. The chemotherapy for recurrent or metastatic tumors results in poor response rates due to the MDR and/or the relative impermeability of cell membranes to chemotherapeutics. Combined treatment consisting of pulsed electric fields and chemotherapeutic agents has been used recently as electrochemotherapy[38, 39]. This treatment procedure relies on the physical effects of locally applied electric fields to destabilise cell membranes in the presence of chemotherapeutics. The electric pulses are used in attempts to deliver drugs locally to the cell interior and retain them there.

Topics in discontinuation can be demanding in mathematics and physics such as inverse problems for radiation treatment and in the dynamics of molecular motion in membrane penetration, Tumour growth is discontinuous and it is supposed that a discontinuity in entropy production takes place during tumor development. Discontinuity increases and decreases in the entropy indicate the stages where the tumor has to start importing negative entropy in order to survive and continue to grow. A sequence of progressive levels of density in tumors has been demonstrated by use of a new technique for the analysis of mammograms or computer tomography with graphic displays, reported together with new implications for therapy timing. The entropy production can be defined as heat dissipation, diffusion current glucose utilization, electric potential differences, membrane polarization differences and sensitivity to the application of external work, such as square wave electric potential exposure[40]. It is known that the thermodynamic entropy of a tumour is different from that of normal tissues due to a disordered structure of the cytoskeleton in the cancer cell the information contents of the two tissues must therefore be considered when the direction of entropy flows is planned to be changed, including mass transport or metastasis formation across the border between the normal and cancerous tissues[41, 42]. Informational entropy is a complex phenomenon in physics but can be defined logically with the aid of Information Theory. The large quantity of information includes some coincidences. Among the coincidences, there are some sequences. The sequences can transiently associated. Closely related associations can format premises. The overlap or condensation of related sequences can result in the formation of generalised analogies. The comparison of similarity analogies may lead to distinctions of differences. The disorder of tumors and the long-range correlations in ordered tissues are in competition with each other. Tumour cells break the ordering of processes in host and resulting propagation in existing structures. Their difference in entropy development increases with time. We presume that the entropy for the two different living systems increases monotoniously with time from zero to infinity, but the differences are constant in time. The process of informational entropy flow thus introduces the arrow of time into cancer dynamics. There are opportunities to enhance the effectivity of chemotherapy, *e.g.* by using electrochemotherapy[38, 39], when a high intensity of electric pulses facilitates the uptake of chemotherapeutics into the cells[40, 43].

There is experience with the application of extracorporeal high-intensity focused ultrasound in producing the thermal ablation of solid carcinomas[43] and the microwave coagulation of solid tumours[44]. The electronic sensitisation of cancer cells to ultrasound, is an example of the combination of various physical effects for the treatment of solid tumours[45]. It has been shown that the exposure of tissue to high-intensity electrical pulses of short duration can induce reversible electropermeation and irreversible structural changes in the cancer cell membrane, in this way facilitating the uptake of normally impermeable substances into the cell [40, 43].

12. Conclusion

Under natural conditions, various entropy-producing processes exist in the four different types of cell populations coexisting in cancer cells, the stromal cells, immune cell and normal tissue constituents. Entropy proceeds from the cancerous toward the normal tissues. The large differences between the entropy of these four cell populations are the driving forces of the complex cancerous information flow from the tumor to the normal tissues. The numerous information fluxes coincide; among them, there are sequences based on the tendency of ordering toward energy utilization or mobilization from their environments. As the examples of external force fields suggest, there are opportunities to change the entropy production rates of coexisting cancer and normal cell populations (Fig. 1).

Alterating metabolic pathways of tumours shows promising approaches to achieve cancer therapies and publications have recently been reviewed[46]one new opportunity is being developed in Australia[47], particularly relevant to melanomas. It involves changing membrane potentials of tumour cells before reaching the metastic state so as to enter a new metabolic pathway, which, under pH control, can enter a further pathway chosen to either avoid a metastatic state or delay it to allow other therapies to gain advantage. Another opportunity is to induce cellular immunity, or an antipromotion-like mechanism to demarcate the cancerous tissue or stroma formation, resulting in the inhibition of negative entropy inflow as nutrient import from the healthy tissues.

In a reverse mode what are the prospects for the reversal of the entropy flow from tumorous to normal cells? Could we enhance the entropy production of the normal tissues around the tumour, to provide entropy flow from the normal tissues to the cancerous tissues while carrying the information from the healthy tissue toward the tumour. If, for example, the ultrasound absorption and entropy dissipation were made greater in normal than in cancerous tissues, the direction entropy might be reversed. Application of external forces may therefore give us hope that we can halt the progression of cancer or at least halt progression to metastasis for new therapies to be effective.

13. Perspectives to reverse the direction of entropy flow

As mentioned above, let us consider an entropy flow from normal tissues to tumorous tissues carries the information on the healthy tissues toward the tumor.

The first tems of entropy production contribute to a higher rate of entropy production for cancerous cells than for healthy cells; however, the situation is different for the chemical effects term, on application of an external force field.

Ultrasound absorption and dissipation in normal tissues is greater than in cancerous tissues, and consequently the direction of entropy flow can be reversed by exposure to an external force field such as an electric field or ultrasound. Low-frequency and low-intensity ultrasound is weakly absorbed in biological tissues (when the induced temperature increase can be neglected): consequently the effect of entropy production on non-damaged cells can be followed independently from temperature effects.

Entropy production due to exposure to a square wave electric potential is higher for normal tissues than for cancerous tissues (pulse frequency and field strength). Square wave electric potential-induced entropy production is comparable to chemical reaction-induced entropy production.

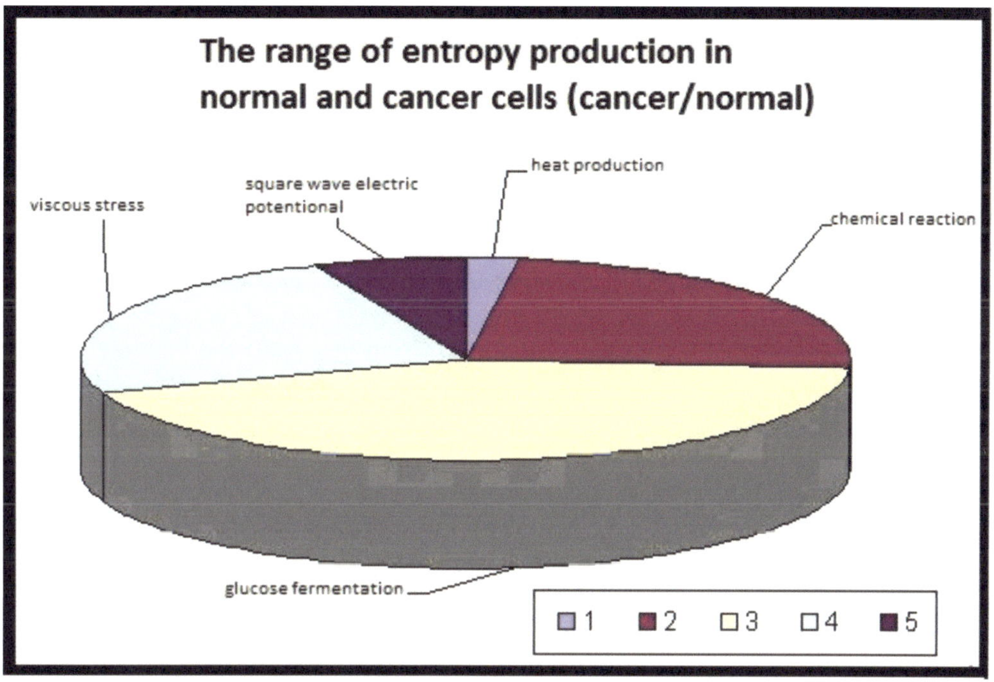

Figure 1. The ratio of entropy production between normal and cancer cells

Entropy production results from the transport of heat and matter between the two phases of the system and also from the chemical reactions taking place. Thus entropy production is a bilinear form of the rates of the irreversible processes and some functions of state, which may be called "affinities" or "generalised forces".

A relatively simple form of the entropy production per unit time was expressed by Prigogine (Prigogine I: Introduction to Thermodynamics of Irreversible Processes, Interscience Publishers, Division of John Wiley and Sons, New York, London).

If the entropy changes in the two phases (at the border of the tumorous and the normal organ tissues) are compared regarding the difference in velocity of change, the ratio between the two forms of tissue may be characteristic of the borderline, where an overlap in entropy production rate may be found between the normal tissues and the tumor. If so, this can be considered a new target for demarcation.

14. Appendix

14.1 The second law and biochemistry

Gibbs "free energy" (thermodynamic potential) G is the function of most use in biochemistry and is the part of the energy which can be used for performance of work.

$$G = \mathrm{H} - \mathrm{T}S \tag{1}$$

The "total" energy H (heat content or "enthalpy") is the sum of the "free" energy, G and a function $\mathrm{T}S$ of the entropy S (ie the bound or unavailable energy which is present because of the chaotic motion of atoms and molecules.

The Second Law of Thermodynamics states the universal tendency of energy to become unavailable. Equation (1) defines G for an isothermal system at constant pressure which may gain or lose heat energy. The Second Law requires a net increase in entropy for any process in such a system together with its evniromnent. Because entropy can be transferred from the system to the environment the actual system can then have a decrease in entropy. However, an isothermal system at constant pressure must suffer a decrease in free energy. In an equilibrium state free energy has a minimum value and the difference ΔG in free energy between the system's actual state and its equilibrium state is, in effect, a measure of the system's tendency to change. At any given temperature T the difference in entropy ΔS and, the change in ΔG and $\Delta \mathrm{H}$ are related from equation (1) as $\Delta H/\mathrm{T} = \Delta G/\mathrm{T} + \Delta S$.

The Second Law requires an overall production of entropy from the environment and system, ie

$$\Delta S(e) + \Delta S > 0 \tag{2}$$

Where $\Delta S(e)$ is the increase in entropy of the environment.

It is a straightforward matter to define E, volume V and chemical potential $\partial G/n_i$ in terms of G,T and pressure p together with the number of chemical species n_i as

$$Sd\mathrm{T} - dG + Vdp\sum_i \mu_i \, dn_i \tag{3}$$

Reference: Bray H.G. and White K., Kinetics and Thermodynamics in Biochemistry, p77 (J2A Churchill) London, 1966.

14.2 Dissipative structures and the second law

With regard to the basic equation (3) it is of importance to note that any state of non-equilibrium associated with the system can be described by the same variables as used in an equilibrium state and they are governed by the same equation of state ie local equilibrium hypothesis simply means the validity of the Gibbs Law for irreversible processes. The summation term $\sum \mu_i dn_i$ where μ_i is the chemical potential of the component i . For each chemical reaction the affinity $\mathrm{A}j = -\sum \frac{v}{ij} \mu_i$ where the v_{ij} are stoechiometric coeffecients [see Glansdoff P., Energy evolution of complex networks of reactions, p41-54 in Living Systems as Energy Converters, (Eds R. Buvet, M.J. Allen and J-P Massue), North Holland Publn, Amsterdam, 1977].

Auto-catalytic or feedback effects in a sequence of coupled chemical reactions may generate chemical instability. Catalytic effects in the kinetic equations can lead to a critical state beyond which a completely new behaviour may actually appear. The system still evolves in accordance with the Second Law of Thermodynamics but may not, in general, still comply with the minimum entropy production theorem. A new internal organisation in the system can become more orderly than those of the thermodynamic branch and are then called dissipative structures including spatio - temporal "chemical clocks".

15. Acknowledgement

The authors thank to Prof. Shlomo Ben Effraim and Prof. Liaofu Luo for discussions on questions the differences in entropy production of normal and cancerous tissues, and discussions on relationships concerning modification of the direction of entropy flow from aspects of electrobiochemistry. This work was supported by the Szeged Foundation of Cancer Research.

16. References

[1] Erwin Schrödinger: What is life? *The physical aspects of the living cell.* pp. VI.57-58, Cambridge University Press, 1948.

[2] Prigogine I. Stengers I: Order out of Chaos, *Flamingo*, UK, Harper Collins Publishers p.188, 1985.

[3] Prigogine I.: *Introduction to thermodynamics of irreversible processes.* Interscience Publishers Division of John Wiley and Sons, New York and London.

[4] Glansdorff P. Prigogine I.: *Thermodynamic Theory of Structure, Stability and Fluctuations,* Wiley Interscience, New York, p17, 1978.

[5] Garay R.P., Lefever R: A kinetic approach to the immunology of cancer: stationary states properties of effector-targer cell reactions. *J.of Theoretical Biology* 73,, 417-438, 1978.

[6] Bergmann-Leitner E.S, Abrams S.I. *Cancer Immunology and Immunotherapy* 49, 193-2007, 2000.

[7] Molnar Joseph, Luo Liaofu, and Gyémánt Nora, Mucsi Ilona, Vezendi Klára, Ocsovszki Imre, Szőkefalvi-Nagy Elisabeth, Thornton Barry S.: Cancer growth is superparasitism in the host: a predator-prey relationship. *Acta Scientiarum Naturalium Universitatis NeiMongol* 38/1, p.44-57, 2007.

[8] Allen P.M. (Darwinian evolution and predator-prey ecology) Evolution of population stability, *Proc.of Natl. Acad. Sci.* 73/3, 665-668. 1976).

[9] Szilárd Leó: On the decrease of entropy in a thermodynamic system by intervention of intelligent beings, Maxwell's Demon, page 124-133. *Zeitschift für Physic,* 32, 753-1925

[10] Kayser Klaus, Hans-Joachim Gabius: The application of thermodynamic principles to histochemical and morphometric tissue research: principles and practical outline with focus on the glycosciences. *Cell Tissue Res.* 296,, 443-455,, 1999.

[11] Molnár J., Thornton B.S:, Thornton-Benko E., Amaral L., Schelz Zs., Novák M.: Thermodynamics and electrobiologic prospects for therapies to intervene in cancer progression. *Current Cancer Therapy Rev.* 5, 158-169, 2009.

[12] Luo L, Molnar J. Ding H. Lv X, Spengler: Physicochemical attack against solid tumors based on the reversal of direction of entropy flow: an attempt to introduce thermodynamics in anticancer therapy. *Diagn. Pathol.* 2006: 1,:43.

[13] ,..„socratus" israelsad@bezeqint.net (1a

[14] Thornton B.S. : Inversion of Raman spectra of living cells indicates dielectric structure related to energy control. *Physics Letters* A 106/4, 198-202, 1984.

[15] Molnar J., Thornton B.S., Molnar A., Gaál D., Luo L., and Bergmann Leitner E.,: Thermodynamic aspects of cancer: Possible role of negative entropy in tumor growth, its relation to kinetic resistance. *Letters in Drug Design and Discovery* 2, 429-438 (2005).

[16] Bergann-Leiner Elke S., H.Huncan Elisabeth and W.Leitner Wolfgang: Identification and targeting of tumor escape mechanisms: A new hope for cancer therapy? *Current Pharmaceutical Design* 2003, 9, 2009-2023.

[17] Járdánházy Anette, Járdánházy T., Luo L. and Molnár Joseph: Brain tumor induced textural changes and impaired cortical electrogenesis. *Acta Scientiarum Naturalium Universitatis Neimongol* 38,/4, 390-402, 2007.

[18] Molnár J, Szabo D., Mándi Y., Mucsi I., Fisher J., Varga A., König S., Motohashi N.: Multidrug resistance reversal in mouse lymphoma ceélls by heterocyclic compounds, *Anticancer Res.* 18, 3033-3038, 1998.,

[19] Agur Z. Kogan Y. Levi L, Harrison H. Lamb R., Kirnasovsky OU, Clarke RB.: Disruption of quorum sensing mechanism triggers tumorigenesis: a simple discrete model corroborated by experiments in mammary cancer cells. *Biol.Direct* 2010 Apr 20:5(1), 20.

[20] Zhang L., Wang S., Wandtao, Wang Y. Wang J. Jiangli, Lisheng. Hu, and Wang Qi: Upregulation of GRP78 and GRP94 and its function in chemotherapy resistance to VP-16 in human lung cancer cell line SK-MES-1. *Cancer Investigation* 1-6, 2009, DOI:10.10.80/073579008002527239.

[21] Arnold D.L. Shoubridge E.A. Villemure J.G., Feindel W. Proton and phosphorus magnetic resonance spectroscopy of human astrocytomas in vivo. *NMR in Biomed* 4, 184- (1990),

[22] Griffiths J.R. Are cancer cells acidic? *Br.J. of Cancer* 64, 425-427, 1991.

[23] Luo-L., Molnár J, Ding H., Lv Xg, Spengler G: Ultrasonic absorption and entropy production in biological tissue. *Diagnostic Pathology* 1,:35, 2006

[24] Luo Liaofu: Entropy production in a cell and reversal of entropy flow as an anticancer therapy. *Frontiers of Physics in China* 112-131 (2008)

[25] Szent-Györgyi A.: A submolecular biology of cancer, Elsevier, North Holland, *Ciba Foundation Symposium 67, 1967, and Electronic Biolgy and Cancer,* Marcell Dekker 1976, page 80.

[26] Luo L. W.L..:A mathematical model on competition of several species for one resource. *Acta Sci.Nat. Universitatis Intramongoliae* 25/6, 645-651, 1994.

[27] Terrell L. Hill: *Free energy transduction and biochemical cycle kinetics.* Springer Verlag ISBN 3-540-96836-9.

[28] Van Ness H. C.: *Understanding themodynamics*. McGraw_Hill Book Company, New York, 1969
[29] Carew Jennifer S. and Huang P.: Mitochondrial defects in cancer. *Molecular Cancer*, Biomed Central, 2002, I:9. p.1-12.
[30] Hockenbery David M.: A mitochondrial Achilles' heel in cancer. *Cancer cell*: July 2002 page 1-2,
[31] Watchorn T.M., Waddell I., Dowidar N., Ross J. A.: Proteolysis-inducing factor regulates hepatic gene expression via the transcription factors NF-kB and Stat3, *The FASEB Journal* 15.562-564, 2001.
[32] IslamAli BS and Tisdale M.J.: Effect of a tumor-produced lipid mobilizing factor on protein synthesis and degradation,.*British J.of Cancer* 84/12, 1648-1655, 2001.
[33] Warburg O.: On the origin of cancer cells, Science 123, 309-314, 1956.
[34] Sussmann I., Erecinska M., Wilson DF.: Regulation of cellular energy metabolism: the Crabtree effect. *Biochim.Biophys.Acta* 591, 209-223, 1980.
[35] Nicolaou K; Pfefferkorn J; Schuler F; Roecker A; Cao G; Casida J.: Combinatorial synthesis of novel and potent inhibitors of NADH: ubiquinone oxidoreductase. *Chemistry & biology* 2000;7(12):979-92.
[36] Veech R.L., Kashiwaya Y., King M.T.: The resting potential of cells are measures of electrical work, *Integr.Physiol., Behav. Sci.*(0) 30, 283-307, 2010.
[37] Neuman John and Neumann Clara, *The computer and the brain* .New Haven and London, University Press 1963,
[38] Jaroszeski M.J., Gilbert R., Heller R.: Electrochemotherapy: an emerging drug delivery method for the treatment of cancer. *Advances in Drug Delivery Reviews* 26, 185-197, 1997.
[39] Rodriguez-Cuevas S., Barrose-Bravo S. Almanza-Estrada J, Cristobal-Martinez L., González-Rodriguez E.: Electrochemotherapy in primary and metastatic skin tumors: Phase II trial using intralesional Biomyon. *Arch. Med. Res.* 32,273-276, 2001
[40] Larkin J. Soden D., Collins C., Tangney M. Preston J.M., Russell L.J., McHale A.P. Dunne C. and O'Sullivan GC: Combined electric field and ultrasound therapy as a novel antitumor theatment. *European J. Cancer* 41,, 1339-48 2005.
[41] Engi Helga , Gyémánt Nóra, Motohiro Ohkoshi, Amaral Leonard, Molnar Joseph: Modelling of tumor-host coexistence in vitro in the presence of serine protease inhibitors. *In Vivo* 23/5, 711-716, 2009
[42] Ohkoshi Motohiro and Sasaki J.:Antimetastatic activity of synthetic serine protease inhibitor FOY 305 (Foypan). *In Vivo* 19, 133-136, 2005,
[43] Madersbacher S., Pedevilla M. Vingers L.et al: Effect of high intensity focused ultrasound on human prostate in vivo. *Cancer Res.* 55, 3346-3351, 1995.
[44] Umeda T., Abe H. Kurumi Y, Naka S. Shiomi H. Hanasawa K, Morikawa S., Tani T.: Magnetic resonance-guided percutaneous microwave coagulation for liver metastases in a case. *Breast cancer* 2005,, 12/4, 317-321.
[45] Rollan Haro AM, Smyth A. Hughes P., et al. Electro-sensitization of mammalian cells and tissues to ultrasound: a novel tumor treatment modality. *Cancer Letters* 2005, 222,/1, 49-55
[46] Tennant D., Durban R. and Gottlieb E., Targetting metabolic transformation for cancer therapy, *Nature Reviews Cancer*, 10, 267-277, April 2010.

[47] Thornton B.S. and Thornton-Benko E. Properties of cell membrane potentials in biophysics suggest altering tumour cell metabolic pathways to avoid entry to the metastactic state. Opportunities for melanoma therapies, Foundation for Australian Resources, *Internal Report* 2006, C/- University of Technology, Sydney Australia.

8

Thermodynamics of the Heart

Uehara, Mituo and Sakane, Kumiko Koibuchi
Universidade do Vale do Paraíba
Brazil

1. Introduction

It is well known that heart and respiratory rates increase with increasing the level of physical exercise. The heart rate (heart beats per minute) is related to the cardiac output (volume of blood pumped by the heart per unit time) and the respiratory rate is related to the rate of oxygen consumption. Therefore there is relation between cardiac output and oxygen consumption, as confirmed by many experiments (Milnor, 1990).
Adolph Fick, German physiologist, was the first to relate, mathematically, cardiac output to gas exchange in the lungs. In 1870, he derived the equation (Acierno, 2000; Shapiro, 1972)

$$Q = \frac{q}{[O_2]_{pv} - [O_2]_{pa}}, \tag{1}$$

where Q is the cardiac output, q the rate of oxygen consumption, $[O_2]_{pv}$ the O_2 concentration in the pulmonary veins, and $[O_2]_{pa}$ the O_2 concentration in pulmonary arterial blood.
Fick´s equation is a landmark of the history of cardiology because it provided the theoretical basis for the measurement of cardiac output in intact animals and people. The first measurement of cardiac output in humans was made in 1930 (Acierno, 2000), sixty years after the publication of Fick´s theoretical derivation, which marks the beginning of modern cardiology.
The cardiac output is related to the work that the heart does in pumping blood to the vascular system. A thermodynamic approach has been applied to estimate energy gain and the work performed by the heart (Blick & Stein, 1977). Recently, by viewing the heart as a thermodynamic engine, a new equation, which relates cardiac output to oxygen consumption, has been derived (Uehara et al., 2008). In this work, the new equation is applied to derive the behavior of important physiological quantities of the cardiovascular system during exercise, such as the efficiency index of the heart, and to discuss the conditions for the linearity of the relation between cardiac output and oxygen consumption. It is shown that the efficiency index of the heart decreases with increasing the exercise level. Furthermore, the referred equation is generalized to include pathological situations such as, for example, ischemic heart disease.

2. The cardiovascular system

The cardiovascular system is diagrammatically represented in Figure 1. The heart consists of a double pump, the left side of the heart or the left heart and the right side of the heart or the right heart, connected in series by two independent circulatory systems, the pulmonary circulation and the systemic circulation. The left heart pumps blood through the systemic circulation and the right heart pumps blood through the pulmonary circulation. Each heart has two chambers, the atrium and the ventricle, which periodically contract and relax. The movements of contraction and relaxation are, respectively, *systole* and *diastole*. The systole and diastole of the atria and of the ventricles are synchronized so that when the atria are contracting the ventricles are relaxed, and vice-versa. The atrium receives and stores blood during the ventricular contraction, and blood flows from the atrium to the ventricle during ventricular relaxation. The right ventricle propels blood through the blood vessels of the lungs (the pulmonary circulation) and the left ventricle propels blood to all other tissues (the systemic circulation). The blood conveys nutrients and oxygen to the cells. Inside the cells, oxygen and nutrients react chemically with release of energy, which is used by the cells, and production of water and carbon dioxide (CO_2) that are disposed by the cells to the blood, so that the concentration of oxygen in the blood decreases and the concentration of carbon dioxide increases as the blood flows through the systemic circulation. In the pulmonary circulation oxygen from the pulmonary alveoli is added to blood and carbon dioxide comes out from the blood and gets into the pulmonary alveoli to be exhaled. The freshly oxygenated blood flows from the pulmonary circulation to the left side of the heart, fills the left ventricle, and a new cardiac cycle begins.

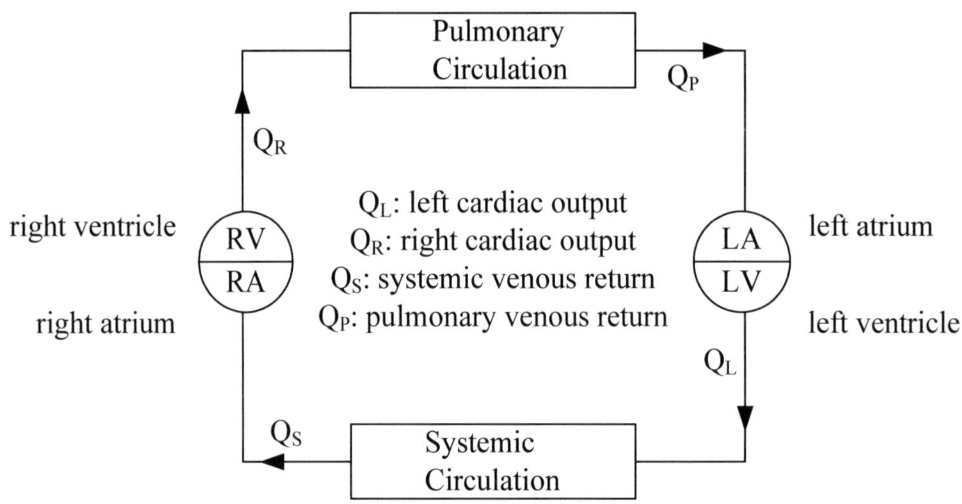

Fig. 1. Diagrammatic representation of the cardiovascular system

In Figure 1, Q_L denotes the cardiac output of the left heart, Q_R the cardiac output of the right heart, Q_P the blood flow from the pulmonary circulation, and Q_S the venous return (blood flow from the systemic circulation). A system of valves ensures that blood flows in the direction indicated in Fig. 1. We consider time-averaged values of blood flows over a cardiac period. During transient phenomena, these time-averaged blood flows are functions of time and satisfy a set of differential equations that are useful for discussing the stability of the system (Uehara and Sakane, 2003). In a steady state of the cardiovascular system the time-averaged blood flow is the same throughout the system, that is

$$Q_L = Q_R = Q_P = Q_S = Q \tag{2}$$

In this work we shall consider only steady-states of the cardiovascular system and time-averaged values of time dependent quantities.
The cardiac output of a ventricle is given by

$$Q = f\,S \tag{3}$$

where f is the cardiac rate (number of heart beats per minute) and S the *stroke volume* (volume of blood ejected by the ventricle in a cardiac cycle).
The blood arterial pressure oscillates periodically between a maximum called *systolic* pressure and a minimum called *diastolic* pressure. The mean arterial pressure in the systemic circulation is approximately given by (Berne&Levy, 2001)

$$P_{sa} = P_{sd} + \frac{P_{ss} - P_{sd}}{3} \tag{4}$$

where P_{sa} is the mean arterial pressure, P_{sd} the diastolic arterial pressure and P_{ss} the systolic arterial pressure.
The left ventricle pumps blood into the arteries of the systemic circulation, which already contain blood at pressure P_{sa}. The blood pressure falls along the systemic circulation at the end of which there are two large veins (venae cavae). These large veins convey blood to the right atrium. The blood flow to the atrium, denoted by Q_S, is called *systemic venous return*. The term "return" refers to the fact that the blood is returning to the heart and the term "venous" indicates that the blood vessels that convey blood to the right atrium are veins. The venous pressure P_{sv} at the venae cavae is very low as compared to the systemic arterial pressure P_{sa}. The systemic vascular resistance R_S is defined by (Hoppensteadt & Peskin, 1992)

$$R_S = \frac{P_{sa} - P_{sv}}{Q_S}\,. \tag{5}$$

Since the venous pressure P_{sv} is much smaller than the arterial pressure Psa, it can be neglected, so that usually R_S is calculated as $R_S = P_{sa} / Q_S$.

3. Relation between cardiac output and oxygen consumption

We shall derive a relation between cardiac output and oxygen consumption by considering the efficiency of the heart in pumping blood to the vascular system.

3.1 Energy consumption

Principles and concepts of thermodynamics can be applied to the heart in order to determine the energy expended in pumping blood. By considering all known sources of energy gain or loss by the heart, it can be shown that many terms are negligibly small (Blick & Stein, 1977). In this work we shall consider only the dominant terms.
Energy provided by food is stored in the body as internal energy. Among the various factors that can change the heart internal energy, the dominant one is the metabolic energy change. This contribution arises from the oxidation of carbohydrates (starch and sugars), fats, and proteins. For glucose ($C_6H_{12}O_6$), a form of sugar used in intravenous feeding, the oxidation equation is (Cameron et al., 1999)

$$C_6H_{12}O_6 + 6O_2 \rightarrow 6CO_2 + 6H_2O + 686\ \text{kcal} \tag{6}$$

where 686 kcal (kcal = 4.184 kJ) is the energy released by one mole of glucose.

Coronary blood flowing through the myocardium delivers oxygen and other substances to the cells of the cardiac muscle. In the interior of the cells, nutrients react chemically with oxygen and energy is released.

From Equation (6) it can be seen that the energy released per liter of oxygen consumed, in the oxidation of glucose, is 5.1 kcal/(liter of oxygen). For a typical diet the energy released is about 5.0 kcal/(liter of oxygen) (Milnor, 1990; Cameron et al., 1999).

The metabolic process we have just described is called *aerobic* metabolism, which occurs in the presence of free oxygen. Energy can also be provided by *anaerobic* metabolism, which occurs in the absence of free oxygen.

The rate of oxygen consumption by the heart (Ω_H) is a fraction of the total rate of oxygen consumption (Ω) by the body and can be written as

$$\Omega_H = c\Omega \tag{7}$$

where $c < 1$.

Assuming that the energy consumed by the heart is provided only by aerobic metabolism, we write

$$P = Ec\Omega \tag{8}$$

where P is the power consumed by the heart and E is the metabolic energy released per liter of oxygen consumed. Part of the power consumed by the heart is used to perform external work in the form of ventricular contractions to pump blood.

The assumption that the energy consumed by the heart is generated by aerobic metabolism is supported by experimental observations. Vary and collaborators have analyzed the aerobic and anaerobic processes of energy production in the cardiac muscle and concluded that the aerobic metabolism corresponds to more than 90% of the total energy produced (Vary et al., 1981). In the transition from the rest condition to exercise the ventricles must generate sufficient power to maintain blood pressure and organ perfusion, which requires tight coupling of the transfer of chemical energy to mechanical power. This controlled energy transfer can require up to fivefold increase in cardiac power generation and oxygen consumption, yet appears to occur without anaerobic metabolism (Zhou et al., 2006; Sharma et al., 2005).

3.2 Work done by the heart

Among the various forms of work that are performed by the heart, many of them give rise to terms that are negligibly small (Blick & Stein, 1977). In this work we shall consider only the work in pumping blood, which is the dominant term.

Each ventricle pumps blood to the corresponding arterial system, which already contains blood at a pressure that oscillates periodically. The maximum and minimum arterial pressures are called the systolic and diastolic pressure, respectively. The process of filling the left ventricle and the process of blood ejection to the vascular system can be described by a pressure-volume diagram as shown in Fig.2, which illustrates the cardiac cycle of the left ventricle (Burkhoff & Sagawa, 1986). The line DA describes the filling of the ventricle, in which the volume increases from V_{ES} (end systolic volume) to V_{ED} (end diastolic volume). The vertical line AB describes the isometric increase of the ventricular pressure, during which the volume remains constant, because the inflow and the outflow valves are closed.

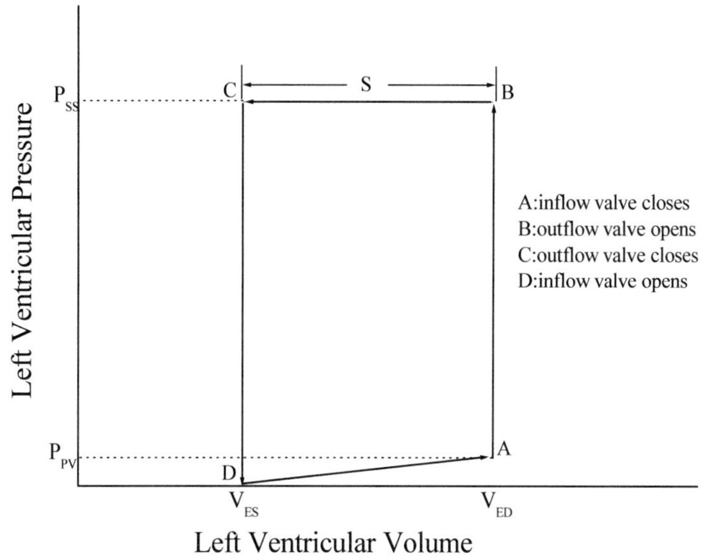

Fig. 2. Pressure-volume diagram of the left ventricle. P_{ss}: systemic systolic pressure; P_{pv}: pulmonary venous pressure; V_{ES}: end systolic volume; V_{ED}: end diastolic volume; S: stroke volume.

The horizontal line BC describes the ejection of blood to the vascular system. The volume of blood ejected $S = V_{ED} - V_{ES}$ is the *stroke volume*. V_{ES} is the volume of blood that remains in the left ventricle in the end of the ejection process. The descending vertical line CD describes the isometric decreases of the ventricular pressure.

During the filling process, described by the line DA, blood flows from the pulmonary veins into the left ventricle, which is relaxed. Thus, the filling pressure is equal to the pulmonary venous pressure P_{pv}, which is much smaller than the arterial pressure of the systemic circulation. During the ejection process, described by the line BC, the ventricular pressure must be larger than the arterial pressure of the systemic circulation, which varies between its systolic and diastolic values. The time-average of the ventricular pressure, during the ejection process, is approximately equal to the systemic systolic pressure P_{ss}. In Figure 2, for simplicity, the ventricular pressure is assumed to be constant during the ejection process.

From the pressure-volume diagram of the left ventricle it can be seen that the external work done by the left ventricle in a cardiac cycle is approximately given by (Uehara et al., 2008; Burkhoff & Sagawa, 1986)

$$W_L = S\,P_{ss} \tag{9}$$

where W_L is the work done by the left ventricle, S the stroke volume, and P_{ss} is the systemic systolic pressure of blood in the large arteries of the systemic circulation.

The pressure-volume diagram of the right ventricle is similar to that of the left ventricle, but the time average of the right ventricle pressure during the ejection of blood is approximately equal to the pulmonary systolic pressure P_{ps}, which is about seven times smaller than the systolic systemic pressure P_{ss} (Higginbotham et al., 1986; Milnor, 1990). Thus, the external work done by the right ventricle in a cardiac cycle can be approximately given by

$$W_R = S P_{ps}, \tag{10}$$

so that the total external work done by the heart in a cardiac cycle is

$$W = W_L + W_R = S(P_{ss} + P_{ps}) \tag{11}$$

and the average power generated by the heart is

$$Wf = (P_{ss} + P_{ps})Sf \tag{12}$$

where f is the heart rate.

The mechanical energy output of the heart to the systemic and pulmonary blood is subsequently converted to thermal energy throughout the body by the action of viscous dissipation of the blood.

3.3 Heart efficiency

The heart can be viewed as a thermodynamic engine that transforms part of the expended power P into external power Wf, given by Equation (12). The mechanical efficiency of the heart can be defined in several ways. It is reasonable to define it as the mechanical power generated by the heart to pump blood divided by the metabolic energy rate consumed by the heart muscle. Thus we write

$$\eta = \frac{Wf}{P} \tag{13}$$

where η is the heart efficiency in pumping blood to the vascular system.

Equations (3), (8), (12), and (13) yield

$$Q = \frac{\eta c E \Omega}{P_{ss} + P_{ps}} \tag{14}$$

Equation (14) is useful to investigate the cardiopulmonary response to exercise. For numerical calculations, we shall rewrite the equation in a more convenient form. Since the systolic pulmonary pressure P_{ps} is seven times smaller than the systolic systemic pressure P_{ss} (Higginbotham et al., 1986; Milnor, 1990), we write Equation (14) as

$$Q = \frac{rE\Omega}{1.14 P_{ss}} \tag{15}$$

where

$$r = \eta c\,. \tag{16}$$

We shall call r *efficiency index* of the heart.

The energy released in metabolic reactions, for a typical diet, is E = 5.0 kcal/(liter of O_2) (Milnor, 1990). Introducing this value into Equation (15) we obtain

$$Q = \frac{1.38 \mathrm{x} 10^5\, r\Omega}{\{P_{ss}\}} \tag{17}$$

where $\{P_{ss}\}$ is the numerical value of P_{ss} in mmHg (1 mmHg = 0.133 kPa). For example, if P_{ss} = 120 mmHg, then $\{P_{ss}\}$ = 120.

3.4 Cardiopulmonary response to exercise

Exercise provides a powerful tool that permits the study of the regulation of the cardiovascular system under controlled and reproducible conditions. Exercise more than any other stress taxes the regulatory ability of the cardiovascular system. More is learned about how a system operates when it is forced to perform than when it is idle.

We shall use Equation (17) and experimental data published in the literature for calculating the efficiency index of the heart r.

Higginbotham and collaborators (Higginbotham et al., 1986) have measured several quantities of the cardiovascular dynamics, during upright exercise in normal man. The study population consisted of 24 healthy males aged 20-50 years. Subjects varied from 63 kg to 103 kg in weight and from 173 cm to 185 cm in height; body surface area (a quantity that depends on the mass and height of the person, according to the empirical formula given below) ranged from 1.65 m^2 to 2.10 m^2. Those investigators have not published the values of cardiac output but the values of *cardiac index*, which is defined by

$$I = \frac{Q}{A} \tag{18}$$

where I is the cardiac index, Q the cardiac output, and A the body surface area. Cardiologists use the cardiac index to compare the pumping capability of a patient to that of a healthy person, irrespective of the body size.

The body surface area, which depends on the mass and the height of the person, is calculated from the empirical formula (Cameron et al., 1999)

$$A = 0.202 M^{0.425} H^{0.725} \tag{19}$$

where A is the body surface area in m^2, M the mass in kilogram, and H the height in meters. Table 1 shows experimental data from Higginbotham et al. (1986).

	P_{ss} (mmHg)	Ω (liter/min)	I (liter.min^{-1}.m^{-2})
Rest	136	0.33	3.0
Peak Exercise	220	2.55	9.7

Table 1. Experimental data from Higginbotham et al. (1986). P_{ss} is the systemic systolic pressure, Ω the rate of oxygen consumption, and I the cardiac index.

Higginbotham and collaborators have not calculated the mean body surface area but they have just reported that the body surface area varied from 1.65 m^2 to 2.10 m^2, so that we shall take A = (1.65+ 2.10) $m^2/2$ = 1.88 m^2 as the mean value for the considered population. With the data shown in Table 1 and Equations (17) - (18) we have calculated the cardiac output Q and the efficiency index $r = \eta_C$. The results are shown in Table 2.

The results show that the efficiency index r is not constant but varies with the rate of oxygen consumption Ω. In order to determine the form of the function $r(\Omega)$, let us consider the *rate-pressure product* that is defined as the heart rate times the systemic systolic pressure (fP_{ss}).

	Ω (liter/min)	Q (liter/min)	10^2r
Rest	0.33	5.6	1.7
Peak Exercise	2.55	18.2	1.1

Table 2. Rate of oxygen consumption Ω, cardiac output Q and efficiency index $r = \eta c$.

This quantity is clinically useful in case of ischemic heart disease. It has been experimentally observed that the rate-pressure product correlates linearly with the rate of oxygen consumption (Kitamura et al., 1972).
Equations (3) and (15) yield

$$fP_{ss} = \frac{rE\Omega}{1.14S} \tag{20}$$

Experiments show that the stroke volume S increases in the transition from rest to exercise and then remains practically constant (Uehara et al., 2008; Rowell, 1986). On the other hand r is a function of Ω, so that differentiating Equation (20) we obtain

$$\frac{d(fP_{ss})}{d\Omega} = \frac{rE}{1.14S} + \frac{E\Omega}{1.14S}\frac{dr}{d\Omega} \tag{21}$$

where we have neglected the variation of the stroke volume. The rate-pressure product correlates linearly with Ω if its derivative is constant. Thus, we write

$$\frac{rE}{1.14S} + \frac{E\Omega}{1.14S}\frac{dr}{d\Omega} = K \tag{22}$$

where K is a constant. Equation (22) can be written as

$$r + \Omega\frac{dr}{d\Omega} = 1.14S\frac{K}{E} = b \tag{23}$$

where b is practically constant insofar as S is practically constant. Integrating Equation (23) we obtain

$$r = \frac{a}{\Omega} + b \tag{24}$$

where a is an integration constant. The constants a and b can be determined from experimental data.
From Equations (21) and (23) we obtain

$$\frac{d(fP_{ss})}{d\Omega} = b\frac{E}{1.14S}, \tag{25}$$

which is practically constant during exercise, so that Equation (20) describes the nearly linear correlation between fP_{ss} and Ω that has been experimentally observed (Kitamura et al., 1972).
Equation (24) and the data shown in Table 2 yield

$$a = 0.0023 \frac{\text{liter}}{\text{min}} \tag{26}$$

and

$$b = 0.010 \,. \tag{27}$$

By Equation (23), r is practically constant and equal to b for values of Ω such that

$$\Omega >> \frac{a}{b} \,. \tag{28}$$

Thus, from Equations (26)-(27) it can be seen that for Ω >> 0.23 liter/min. the efficiency index of the heart is practically constant. Since the value of Ω for the subject in exercise is Ω > 1.00 liter/min., we conclude that the efficiency index is practically constant during exercise.
The relation between cardiac output and the rate of oxygen consumption is practically linear, as shown by many experiments (Beck et al., 2006). Thus, assuming the linear relation

$$Q = A + B\Omega \tag{29}$$

we obtain, from the data in Table 1 and Table 2

$$A = 3.7 \frac{\text{liter}}{\text{min}} \tag{30}$$

and

$$B = 5.7 \,. \tag{31}$$

The value of the slope B = 5.7 corresponds to the value expected for healthy individuals (Skarvan, 2000). Lower values indicate an inadequate augmentation of cardiac output in relation to oxygen demand.
For comparison, we shall estimate the efficiency index of the heart considering experimental data published by other investigators. In order to calculate the efficiency index r from Equation (17), it is necessary to measure the cardiac output Q, the rate of oxygen consumption Ω, and the systemic systolic pressure P_{ss}. Poliner and collaborators (Poliner et al., 1980) have studied the left ventricle performance at rest and during multilevel exercise in seven normal subjects (six man and one woman) with a mean age of 26 years and a mean body surface area of 1.83 m^2. Exercises were performed in a bicycle ergometer. They have measured the cardiac output Q and the systemic systolic pressure P_{ss}, but they have not measured the rate of oxygen consumption Ω, so that in order to calculate the efficiency index r = ηc from Equation (16), it is necessary to provide the lacking data from other sources. Table 3 shows data from Poliner and collaborators (Poliner et al., 1980).
Beck and collaborators (Beck et al., 2006) have experimentally studied the relation between cardiac output and oxygen consumption during upright cycle exercise in healthy humans. They have submitted 72 healthy subjects, aged 20 to 40 years, to a cycle-ergometer exercise test. They have measured the cardiac output and the rate of oxygen consumption, but they have not measured the systemic systolic pressure. The mean height of the subjects was 173 cm and the mean weight was 72.1 kg. Introducing these values into Equation (19) we obtain the value A = 1.85 m^2 for the mean body surface area. Since the mean body surface area for the population studied by Beck and collaborators differs in less than 1% from the

corresponding value for the population studied by Poliner and collaborators, we shall use the data of both publications to complement each other. Thus, for the same value of the cardiac output we shall use the systemic systolic pressure P_{ss} measured by Poliner and collaborators and the rate of oxygen consumption Ω measured by Beck and collaborators. For each value of the cardiac output given by Poliner and collaborators (Poliner et al., 1980) we have obtained the rate of oxygen consumption Ω from Figure 3 of the paper published by Beck and collaborators (Beck et al., 2006). Table 4 shows the experimental data we have obtained from Beck and collaborators.

	Rest	Stage I	Stage II	Peak exercise
Q (liter/min.)	4.8	10.4	15.1	18.0
Heart rate f (beats/min.)	89 ± 5	124 ± 4	165 ± 4	182 ± 2
P_{ss} (mmHg)	125 ± 5	161 ± 7	190 ± 8	204 ± 8

Table 3. Cardiac output Q, heart rate f, and systemic systolic pressure P_{ss} from Poliner et al. (1980).

	Rest	Stage I	Stage II	Peak exercise
Q (liter/min.)	4.8	10.4	15.1	18.0
Ω (liter/min.)	0.19	1.18	2.00	2.51

Table 4. Cardiac output Q and oxygen consumption Ω from Beck et al. (2006)

By using Equation (17) and the values of Q and P_{ss} given in Table 3, and Ω given in Table 4, we have calculated the efficiency index r. The results are shown in Table 4. It can be seen that the efficiency index decreases with the level of exercise.

	Rest	Stage I	Stage II	Peak exercise
Ω (liter/min.)	0.19	1.18	2.00	2.51
Efficiency index r	0.023	0.010	0.010	0.011

Table 5. Efficiency index r as a function of the rate of oxygen consumption Ω

Assuming that the efficiency index r is a function of the rate of oxygen consumption of the form expressed by Equation (24), we have determined by the least -squares method the values of a and b that correspond to the best fitting to the experimental data given in Table 5. We have found

$$a = 0.0027 \frac{\text{liter}}{\text{min}} \tag{32}$$

and

$$b = 0.009 \tag{33}$$

with a correlation factor equal to R = 0.9898.
Figure 3 shows r as a function of Ω, given by Equation (24), with a and b given, respectively, by Equation (32) and Equation (33).

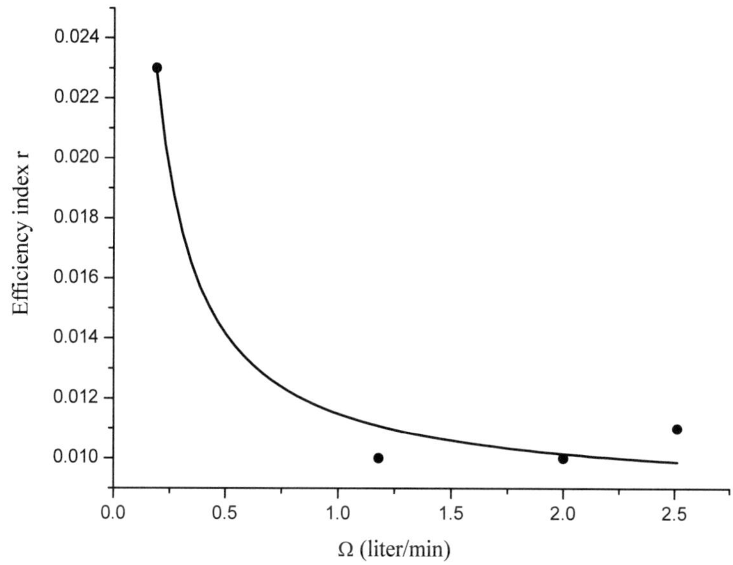

Fig. 3. Efficiency index r as a function of the rate of oxygen consumption. The experimental points correspond to the values of r given in Table 5.

The results (32) - (33) show that the values of a and b, obtained from data published by Poliner and collaborators (Poliner et al., 1980) and Beck and collaborators (Beck et al., 2006), are not very different from that ones obtained from Higginbotham 's data (Higginbotham et al, 1986). Therefore, we conclude that, according to experimental data, the efficiency index of the heart r is larger for the subject at rest than in exercise, and is practically constant for higher levels of exercise.
Measuring the fraction, denoted by c, of oxygen consumed by the heart relatively to the total oxygen consumed by the body, the efficiency of the heart can be calculated as $\eta = r/c$. For a subject consuming oxygen at the rate $\Omega = 0.25$ liter/minute, the heart consumes oxygen at the rate $\Omega_H = 0.035$ liter/minute (Rowell, 1993), so that $c = \Omega_H/\Omega = 0.14$. With these data and Equations (16), (24), (32)-(33) we obtain $\eta = 0.14$, which agrees with the result $\eta = 0.13$ reported in the literature (Blick & Stein, 1977).

3.5 Linearity of the relation between cardiac output and oxygen consumption

The relationship between cardiac output and oxygen consumption during exercise has generally been assumed to be linear. Beck and collaborators have made experimental studies to test that assumption (Beck et al., 2006). They have used a statistical approach for analyzing the experimental data. We shall use Equation (15) to discuss the question of the linearity of the relation between cardiac output and oxygen consumption.
Assuming the linear relation expressed by Equation (29) and by using the least- squares method we have determined the values of A and B that best fit the experimental data shown in Table 3. We have found

$$A = 3.7 \frac{\text{liter}}{\text{min}} \tag{34}$$

and

$$B = 5.7 \tag{35}$$

with a correlation factor equal to R = 0.9999. We note that the values of A and B are the same as those ones, given by Equations (30)-(31), obtained by using Higginbotham´s data (Higgibotham et al., 1986). Figure 4 shows cardiac output Q as a function of the rate of oxygen consumption Ω.

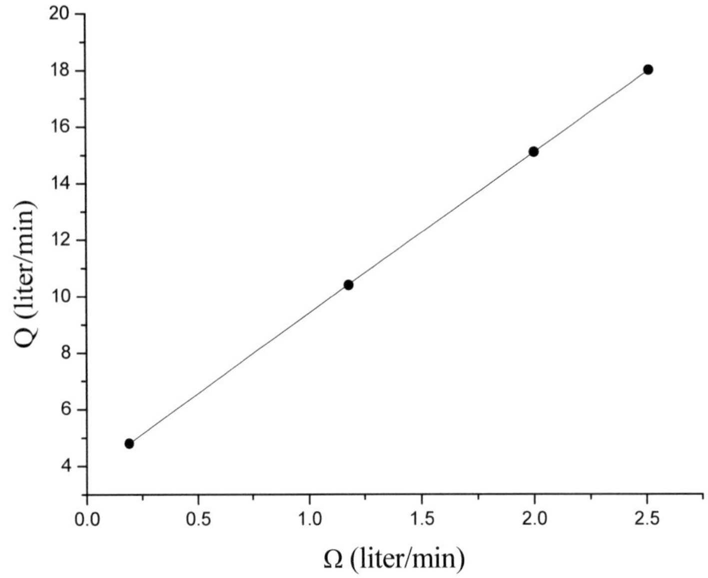

Fig. 4. Cardiac output Q as a function of the rate of oxygen consumption Ω.

Let us now consider the behavior of the systemic systolic pressure P_{ss}. Equations (3),(20), (24) and (29) yield

$$P_{ss} = \frac{E(a + b\Omega)}{1.14(A + B\Omega)}, \tag{36}$$

from which we obtain

$$\frac{dP_{ss}}{d\Omega} = \frac{E(bA - aB)}{1.14(A + B\Omega)^2} \tag{37}$$

and

$$\frac{d^2P_{ss}}{d\Omega^2} = -\frac{2BE(bA - aB)}{1.14(A + B\Omega)^3} \tag{38}$$

Equations (37) and (38) show that if $(bA - aB) > 0$ then $d P_{ss} / d\Omega > 0$, and $d^2 P_{ss} / d\Omega^2 < 0$, so that in this case the systemic systolic pressure P_{ss} would be an increasing function of the

rate of oxygen consumption Ω with a downwards concavity. On the other hand if $(bA - aB) < 0$, then P_{ss} would be a function that decreases with Ω, which contradicts experimental data. Therefore, we must have $(bA - aB) > 0$ or, equivalently, $A/B > a/b$, which is satisfied by the values of a, b, A, and B, given, respectively, by the results (26),(27), (30), and (31).
For a typical diet, the caloric equivalent per liter of consumed oxygen is E = 5.0 kcal/(liter of O_2) (Milnor, 1990; Cameron et al., 1999). Introducing this value into Equations (36)-(37) we obtain

$$\{P_{ss}\} = 1.38\text{x}10^5 \frac{(a + b\Omega)}{(A + B\Omega)} \tag{39}$$

and

$$\frac{d\{P_{ss}\}}{d\Omega} = 1.38\text{x}10^5 \frac{(bA - aB)}{(A + B\Omega)^2} \tag{40}$$

where $\{P_{ss}\}$ is the numerical value of the systemic systolic pressure P_{ss} in mmHg. Equations (39)-(40) are convenient for numerical calculations. Figure 5 shows P_{ss} as a function of the rate of oxygen consumption Ω.

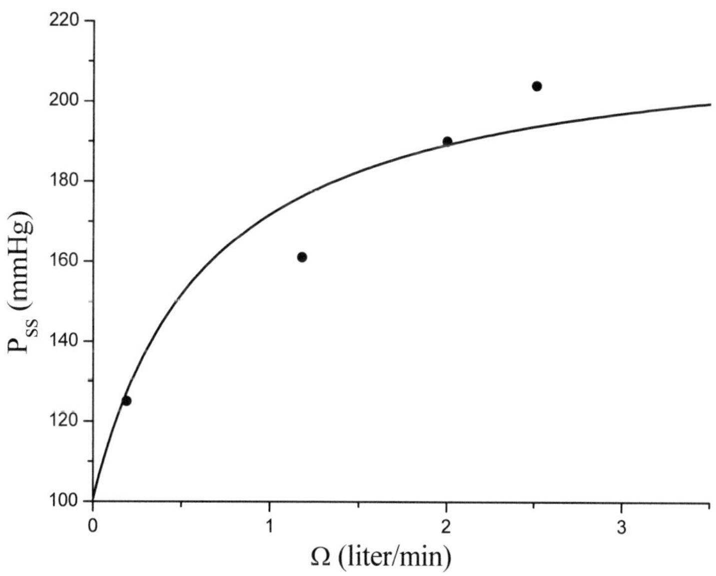

Fig. 5. Systemic systolic pressure P_{ss} as a function of the rate of oxygen consumption Ω. The experimental points correspond to the values given in Table 3.

For large values of Ω such that $B\Omega >> A$, we have also $b\Omega >> a$, because $A/B > a/b$. In this case Equation (39) can be reduced to

$$\{P_{ss}\} = 1.38\text{x}10^5 \frac{b}{B} \tag{41}$$

Introducing the values of b and B given, respectively, by (27) and (31) we get P_{ss} = 242 mmHg.

Assuming the linear relation $Q = A + B\Omega$ between cardiac output Q and the rate of oxygen consumption Ω, and considering the behavior of the efficiency index of the heart r, expressed by Equation (24), we have shown that during exercise the systemic systolic pressure P_{ss} is an increasing function of Ω, which has a derivative that decreases with the level of exercise, so that the systemic systolic pressure approaches a limiting value with increasing the exercise level.

3.6 Ischemic heart disease and essential hypertension

Blood supply to the *myocardium* (cardiac muscle) via coronary arteries is normally abundant. Inadequate blood supply to the heart (*myocardial ischemia* or *coronary ischemia*) leads to the activation of anaerobic processes for generating energy (Chandler et al., 2002; Stanley et al., 1997). Easily measured predictors of myocardial blood flow are necessary to evaluate objectively functional improvement in patients with ischemic heart disease. The rate-pressure product can be a useful parameter for this purpose, since by Equation (20) it is an index of myocardial oxygen consumption. It has been observed that in case of *angina pectoris* (chest pain caused by myocardial ischemia) the pain occurs during exercise, in a given patient, at a characteristic value of the rate-pressure product (Kitamura et al., 1972; Gobel et al., 1978).

In the presence of ischemic heart disease, Equation (8), which gives the power consumed by the heart, must be generalized to include anaerobic processes for generating energy. Thus, we write

$$P = Ec\Omega + \chi \,, \tag{42}$$

where X is the rate of energy provided by anaerobic processes. Equation (15) for the cardiac output is modified as

$$Q = \frac{\eta(cE\Omega + \chi)}{1.14P_{ss}} \tag{43}$$

and Equation (20) for the rate-pressure product is modified as

$$f\,P_{ss} = \frac{\eta(cE\Omega + \chi)}{1.14S} \tag{44}$$

Let us call L the value of the rate-pressure product at which occurs chest pain, during exercise, in a given patient with *angina pectoris*. Thus, we write

$$(f\,P_{ss})_L = L \tag{45}$$

so that for $fP_{ss} < L$, the rate-pressure product is given by Equation (20) and for $fPss \geq L$, it is given by Equation (44).

The rate-pressure product is related to the mechanical power Wf generated by the heart, as can be seen from Equation (12), which we write as

$$Wf = 1.14P_{ss}Sf \tag{46}$$

since P_{ps} is about seven times smaller than P_{ss}. Then if $Wf \geq 1.14LS$ chest pain occurs, indicating the activation of anaerobic processes for generating energy.

Let us write the efficiency index of the heart as $r_i = \eta c_i$ for a patient with ischemic heart disease. Since the fraction c of oxygen consumed by a normal heart is larger than the corresponding value c_i for the diseased heart, we have $r_i < r$. Therefore the efficiency index of the heart can be a clinically useful parameter to distinguish a diseased heart from a normal one. It would be interesting to make experiments for verifying this possibility.
Let us now consider *essential hypertension*, a disease in which the systemic resistance, defined by Equation (5) is higher than the normal one, due to constriction of *arterioles* (small arteries that can constrict and dilate). In this case, an arterial pressure higher than the normal is required to maintain the cardiac output at a normal value. This means that the systemic systolic pressure P_{ss} is higher than the normal pressure. Then, Equation (15) shows that the rate of oxygen consumption Ω must be higher than the normal rate to maintain the cardiac output at a normal value, so that the line that represents cardiac output as a function of the rate of oxygen consumption would be shifted towards the right relatively to the normal line shown in Figure 4..

4. Conclusion

By applying a thermodynamic approach we have derived a relation between cardiac output and the rate of oxygen consumption. Using the derived relation and experimental data published in the literature we have estimated the efficiency index of the heart. We have also derived an equation for the rate-pressure product as a function of the rate of oxygen consumption. The derived equation explains the practically linear correlation observed between the rate-pressure product and the rate of oxygen consumption by the heart. This work suggests a possible clinical application of the efficiency index of the heart as a parameter to distinguish a diseased heart from a normal one. This possibility must be investigated experimentally in order to verify if it is clinically useful. The theory presented in this work provides a theoretical basis for designing experiments to investigate the cardiopulmonary response to exercise.

5. References

Acierno, L. J. (2000). Adolf Fick: Mathematician, Physicist, Physiologist. *Clinical Cardiology*, Vol. 23, Part 5 (2000), 390-393, ISSN 0160-9289

Beck, K. C.; Randolph, L. N.; Bailey, K. R.; Wood, C. M.; Snyder, E. M. & Johnson,B. D. (2006). Relationship between cardiac output and oxygen consumption during upright cycle exercise in healthy humans. *Journal of Applied Physiology*, Vol. 101, No. 5, (May 2006), 1474-1480, ISSN 8750-7587

Blick, E. F.; & Stein, P. D. (1977). Work of the heart: A general thermodynamic analysis. *Journal of Biomechanics*, Vol. 10, No. 9, (Sept 1977) 589-595, ISSN 0021-9290

Burkhoff, D. & Sagawa, K. (1986). Ventricular efficiency predicted by an analytical model. *American Journal of Physiology*, Vol. 250, No. 6, (June 1986), p. R1021-R1027, ISSN: 0002-9513

Cameron, J. R..; Skofronick, J. G. & Grant, R. M. (1999). *Physics of the Body*, 2nd edition,p.21, Medical Physics Publishing, ISBN 0-944838-91-X, Madison, Wisconsin, USA,

Chandler, M. P.; Huang, H.; McElfresh, T. A. & Stanley, W. C. (2002). Increased monoxidative glycolysis despite continued fatty acid uptake during deman-induced myocardial ischemia. *American Journal of Physiology – Heart and Circulatory Physiology*, Vol. 282, No. 5, (May 2002), p. H1871-H1878, ISSN 1522-1539

Gobel, F. L.; Nordstrom, L. A.; Nelson, R. R.; Jorgen, C. R. & Wang, Y. (1978). The rate-pressure product as an index of myocardial oxygen consumption during exercise in patients with angina pectoris. *Circulation*, Vol. 57, No. 3, (March 1978), 549-556, ISSN 0009-7322

Higginbotham, M. B.; Morris, R. S.; Williams, P. A.; McHale, P.A.; Coleman, R. E. & Cobb, F. R. (1986). Regulation of stroke volume during submaximal and maximal upright exercise in normal man. *Circulation Research*, Vol. 58, No. 2, (February 1986), 281-291, ISSN 0009-7330 (print); ISSN 1524-4571 (online)

Hoppensteadt, F. C. & Peskin, C. S. (1992). *Mathematics in Medicine and the Life Sciences*, Springer-Verlag, ISBN 0-387-97639-6, New York, USA, p. 115

Kitamura, K.; Jorgensen, C. R.; Gobel, F. L.; Taylor, H. L. & Wang, Y. (1972). Hemodynamic correlates of myocardial oxygen consumption during upright exercise. *Journal of Applied Physiology*, Vol. 32, No. 4, (April 1972), 516-522, ISSN 0021-8987

Milnor, W. R. (1990). *Cardiovascular Physiology*, Oxford University Press, p. 129, p. 130, p. 174, ISBN 0-19-505884-4, New York, USA,.

Poliner, L. R.; Deimer, G. J.; Lewis, S. E.; Parkey, R. W.; Blomqvist, C. G.; Willerson, J. T. (1980). Left ventricular performance in normal subjects: a comparison in the upright and supine positions. *Circulation*, Vol. 62, No. 3, (September 1980), 528-534, ISSN 0009-7322

Rowell, L. B. (1986). *Human Circulation – Regulation During Physical Stress*, Oxford University Press, p. 215, ISBN 0-19-504075-9, New York, USA.

Rowell, L. B. (1993). *Human Cardiovascular Control*, Oxford University Press, p. 205, ISBN 0-19-507362-2, New York, USA.

Skarvan, K. (2000). Ventricular performance. In: *Cardiovascular Physiology*, Priebe, H.-J. & Skarvan, K., (Ed.), 27-72, BMJ Books, 2nd edition, ISBN 0-7279-1427-8, London, UK.

Sharma, N.; Okere, I. C.; Brunengraber, D. Z.; McElfresh, T. A.; King, K. L.; Sterk, J. P.; Huang, H.; Chandler, M. P. & Stanley, W. C. (2005). Regulation of pyruvate dehydrogenase activity and citric acid cycle intermediates during high cardiac power generation. *The Journal of Physiology*, Vol. 562, Pt 2, (Jan 2005), 593-603, ISSN 0022-3751

Shapiro, E. (1972). Adolf Fick - Forgotten Genius of Cardiology. *The American Journal of Cardiology*, Vol.. 30, No. 6, (Nov 1972), 662-665, ISSN 0002-9149

Stanley, W. C.; Lopaschuk, G. D.; Hall, J. L. & McCormack, J. G. (1997). Regulation of myocardial carbohydrate metabolism under normal and ischaemic conditions. Potential for pharmacological interventions. *Cardiovascular Research*, Vol. 33, No. 2, (Feb 1997), p. 243-257, ISSN 0008-6363

Uehara, M. & Sakane, K. K. (2003). Physics of the cardiovascular system: An intrinsic control mechanism of the human heart. *American Journal of Physics*, Vol. 71, No. 4, (April 2003) 338-344, ISSN 0002-9505

Uehara, M. ; Sakane, K. K. & Bertolotti, S. A. (2008). Thermodynamics of the heart: Relation between cardiac output and oxygen consumption. *American Journal of Physics*, Vol. 76, No. 6, (June 2008) 566-569, ISSN 0002-9505

Vary, T. C.; Reibel, D. R. & Neely, J. R. (1981). Control of energy metabolism of heart muscle. *Annual Review of Physiology*, Vol. 43, 419-430, ISSN 0066-4278

Zhou, F; Yu, X.; Cabrera, M. C. & Stanley, W. C. (2006). Role of Cellular Compartmentation in the Metabolic Response to Stress. *Annals of the New York Academy of Sciences*, Vol. 1080, 120-139, ISSN 0077-8923 (print); ISSN 1749-6632 (online)

9

The Protein Surface as a Thermodynamic Frontier: A Fractal Approach

Mariana Pereyra and Eduardo Méndez
Laboratorio de Biomateriales, Facultad de Ciencias, Universidad de la República, 11400 Montevideo Uruguay

1. Introduction

The objects in Nature cannot be simply described in terms of the Euclidean geometry. It is difficult to find natural objects that can be represented as perfect spheres, planes and stright lines. Unfortunatelly, when researchers are intent on describing Nature in detail, they usually fall in mathematical descriptions that are extremately complex, many times with no solution. The fractal geometry is a fascinating conceptual framework (Mandelbrot, 1982) as it possibilitates to characterize Nature irregularities with a single number, a really tempting idea *per se*. Moreover, this idea push us to a change in our minds when describing Nature, as we are used to the limited vision given by the Euclidean geometry. The word "irregular" itself remind us the idea that these objects do not fit into the Euclidean description of Nature. Fractals act as compression algorithms, as they contain minimal information and a reduced complexity (Abel & Trevors, 2006). Most of the objects in Nature are irregular, and the most fascinating thing is that irregular objects are the norm in the fractal geometry. Unfortunately, the application of the "fractal" label has been too extended, mainly because the mathematical law that rules this concept (a potential law) is versatile enough to allow that virtually all experimental data set fits the equation (Bryant et al., 1989).

The surface of a protein constitutes the spatial domain through which the proteins interact with the surroundings. A great number of processes depend on surface phenomena, which at the same time depend on a wide range of structures and geometrical patterns. The protein surface is determined by the packing achieved in the folding process. Hence, the study of the geometrical characteristics yields valuable information not only on the folding process itself, but also related to the proteins in their interactions with the surroundings. The folded structure of a protein determines two different but closely related characteristics: stability and functionality. The stability of a protein relates with the surrounding media, while its functionality relates with its capacity to interact with that media through the interfacial surface.

There is a consensus in considering the aqueous media surrounding the protein as the interacting media. In this regard, the protein surface is determined as the surface contacting the water molecules. However, this definition left behind other types of interactions that may take place, like protein-protein, protein-ligand, protein-DNA, or even, out of the natural context, protein-metallic surface. The immediate conclusion that one can take out is that the exact nature of the protein surface is highly dependent on the size of the interacting media, which may vary from something very tiny (water molecule) to something relatively big (metallic surface). In other words, the surface of a protein depends on the scale considered.

Proteins have occupied a privileged status in the application of the fractal concepts: from the first work in 1980 (Stapleton et al., 1980) until mid 2010, almost 800 papers appeared. However, the eventual fractal nature of the proteins is an issue still not resolved: neither is proved that the fractal dimension is a property of the protein, nor exist a unique fractal dimension value that characterizes all the proteins. Thirty years after the first study, one can arrive to the conclusion that the point has not been firmly concluded. In this chapter, we will briefly review the main point of views in which protein structures have been treated by a "fractal approach", and we present a new view in which we will show that a specific protein, cytochrom *c*, is not a fractal object, but forms part of a materials distribution network of fractal nature.

2. The protein surface

2.1 Definitions

The surface of a protein constitutes the first level of communication with its sorroundings. It is already recognized that the global and local roughness of the protein surface affects this communication in terms of diffusion, molecule recognition, and physical properties.

Among the different definitions of protein surface, the following are the most commons (Fig. 1).

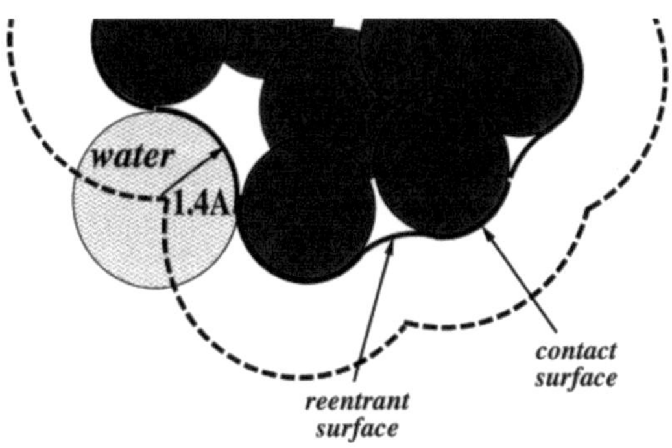

Fig. 1. Protein surfaces defined after rolling a rigid sphere along the protein. The full line represents the contour of the SES, and the dashed line represents the contour of the SAS

- **van der Waals surface (SVDW).** It is the surface resulting from considering each atom of the protein as a sphere with a van der Waals radii. The external surface results from the multiple superpositions between each spheres.
- **Solvent accessible surface (SAS).** This defintion was introduced by Lee and Richards (Lee & Richards, 1971) as the surface generated by the center of a rigid sphere of fixed radii (usually that of water, 1.4 Å) after let it roll all over the van der Waals surface.
- **Solvent excluded surface (SES).** Originally named as "molecular surface" (Richards, 1977), is the surface defined by two geometric domains: the contact surface resulting from the accessible van der Waals surface contacted by a rolling sphere of fixed radii (probe sphere), and the reentrant surface defined as the internal face of the rolling sphere when contancting simultaneously two or more atoms.

- **Hydration surface (HS).** It is defined as the second layer of water molecules sorrounding the protein (Gerstein & Lynden-Bell, 1993). This definition put the protein surface between the bulk solvent and the water molecules that strongly interact with the porotein. It is based on the fact that out of this surface, water molecules posses the same energy than those of the bulk, and are not oriented according to the dipole moment of the protein.

The protein surface definitions schematically depicted in Fig. 1, show the singular points resulting from the intersections of atoms in the SVDW. This singular points are also projected in the SAS, but resulted smoothed in the SES. The analytical and geometrical solution of these singularities are one of the main aspects in the development of algorithms for the calculation of the protein surfaces, because the singular points are not differenciables, and hence do not have analytical solution. An additional point of concern is the assessment of those atoms that do not participate of the exposed surface of the proteins, and that could be left out in the consideration in order to speed up the calculations. Hence, most of the algorithms which are continuously proposed differ in the way the singular points and the non-surface atoms are treated (Wodak & Janin, 1980, Goetze & Brickmann, 1992, Gabdoulline & Wade, 1996, Totrov & Abagyan, 1996, Flower, 1997, Can et al., 2006); their descriptions are beyond the scope of this chapter.

Among all algorithms proposed, MSDOT created by Connolly (Connolly, 1983) is considered the reference algorithm aginst which all the new ones are compared. The algorithm of Connolly calculates analytically the SAS. Of historical value is the algorithm developed by Lee and Richards (Lee & Richards, 1971), which was the first one that solved the singular points by designing a smooth surface coincident with the VDWS.

For the interest of the present work, it was necessary to have an algorithm capable of changing the size of the probe sphere. We have successfully employed the suit of programs ARVOMOL (Fernández Pacios, 1994) and SurfRace (Tsodikov et al., 2002), which were upgraded by their authors to suit the specific requierements of our work.

2.2 The protein surface and the thermodynamic stability of the folded state

The stability of the native state of a protein is governed by the sign and mangnitud of the Gibbs energy, $\Delta_{unf}G$, of the unfolding process:

$$\Delta_{unf}G = \Delta_{unf}H - T\Delta_{unf}S \tag{1}$$

where $\Delta_{unf}H$ and $\Delta_{unf}S$ represent the change in the enthalpy and entropy of the process, respectively, and T is the temperature. Eq. (1) is applicable to every process involving some kind of phase transition, and protein denaturing is one of this processes. The change in the Gibbs energy depends on the temperature. For a given temperature T different to a reference temperature T_R, the change in the Gibbs energy is:

$$\Delta_{unf}G\,(T) = \Delta_{unf}H(T_R) + \Delta_{unf}C_P\,(T-T_R) - T[\Delta_{unf}S(T_R) + \Delta_{unf}C_P \ln(T/T_R)] \tag{2}$$

where $\Delta_{unf}C_P$ is the net change in the heat capacity in the unfolding process. The thermodynamic stability of a protein can be obtained in calorimetric experiments. The interpretation of these changes in the thermodynamic properties requires the knowledge of the events taking place during the process. Of course, this does not imply that these values depends on the pathway taken during the process, as being state functions, the change in the thermodynamic properties G, H, S and C_P depends exclusively on their values in the initial and final equilibrium states. Bearing this in mind, it must be recognized that proteins are folded in an aqueous environment, and most of the interactions that are broken in the unfolding process were between the chemical groups of the aminoacids and the water molecules of the solvent. Hence, the changes in the thermodynamic properties taking place during the unfolding process are due to a balance between the interactions of the folded protein within their own aminoacids and with the solvent, and the interactions between the unfolded protein and the solvent. This can be summarized as the addition of several contributions to the total change in the Gibbs energy:

$$\Delta_{den}G = \Delta_{gen}G + \Delta_{ion}G + \Delta_{tr}G + \Delta_{other}G \qquad (3)$$

where $\Delta_{gen}G$ includes all the generic contributions associated with the formation of the secondary and tertiary structures (van der Waals, H-bonds, hydration, conformational entropy), $\Delta_{ion}G$ takes into account all the electrostatic effects in the formation of the native structure, $\Delta_{tr}G$ includes the change in the translational degree of freedom produced in the native structure in the unfolding process, and $\Delta_{other}G$ includes all the specific interactions (prosthetic groups, metals, ligands) within the protein.

The major contribution to the enthalpy change comes from the formations of intramolecular interactions, like van der Waals and H-bond, together with the desolvation of the chemical groups involved in these interactions. The enthalpy change due to these interactions is included in a generic term, $\Delta_{gen}H$, which can be expressed as a function of the atomic contributions of the changes in the salvation degree produced during the unfolding process:

$$\Delta_{gen}H = \Sigma_i\ \alpha_i(\rho)\ \Delta SAS_i \qquad (4)$$

where the sum includes all the atoms of the protein, ΔSAS_i is the change in the solvent accessible surface for each atom and $\alpha_i(\rho)$ is a coefficient that depends on each atomic type and the mean packing density in the protein.

The change in the heat capacity of the proteins in the unfolding process depends on the change of hydration of the exposed groups in the native and unfolded structures. These changes are closely related to the solvent accessibility:

$$\Delta_{unf}C_P = \Sigma_i\ \alpha_i(T)\ \Delta SAS_i \qquad (5)$$

where $\alpha_i(T)$ is a coefficient that depends on each atomic type, and it is a function of the temperature. The factor ΔSAS_i depends on the atom considered: non-polar atoms contribute negatively to $\Delta_{unf}C_P$ upon loosing the interaction with the solvent, while polar atoms contribute positively (oxygen is an exception).

Finally, the calculation of the entropy change for the unfolding process includes two main contributions: one arising from the salvation changes ($\Delta_{sol}S$) and the other due to the conformational changes ($\Delta_{conf}S$). While $\Delta_{conf}S$ is practically independent of the temperature, $\Delta_{sol}S$ dependence of the temperature is a function of the type of aminoacidic residue (polar or non-polar) in accordance to:

$$\Delta_{sol}S\ (T) = \Delta_{sol}S_{non\text{-}polar} + \Delta_{sol}S_{polar} \qquad (6)$$

$$\Delta_{sol}S\ (T) = \Delta_{sol}C_{P,non\text{-}polar} \ln(T/T^{*}_{non\text{-}polar}) + \Delta_{sol}C_{P,non\text{-}polar} \ln(T/T^{*}_{non\text{-}polar}) \qquad (7)$$

where T^{*}_i corresponds to the temperatures at which the absolute entropies of hydration of polar and non-polar residues are zero. The dependence between $\Delta_{sol}S$ and $\Delta_{sol}C_P$ (Eq. 7) establishes the dependence between the entropy changes for the unfolding process and the solvent accessible surface (viz. Eq. 5).

2.3 Calculation of the protein surface

There are several techniques to determine the surface of a protein. The method of the rolling ball yields surface area values which depend on the radius of the sphere probe. The dimension is calculated from the double logarithmic transformation:

$$D_s = 2 - (\partial \log S(r) / \partial \log r) \quad (8)$$

Another way to calculate the fractal dimension of the protein surface is based on the co-dimension rule. After rolling the probe sphere over the protein, the contour obtained after performing parallel cuts with planes are measured with a variable rule of size ε. The size of the contour is proportional to ε^{D_c}, where D_c is the fractal dimension of the contour. According to the co-dimension rule, the surface fractal dimension is (Pfeifer et al., 1985):

$$D_s = D_c + 1 \quad (9)$$

3. The fractal approach to the description of the protein structure

Proteins are heterpolymers with variable composition obtained by combination of a basic pool of 20 aminoacids. The aminoacid sequence constitutes the primary structure of the protein, and the information contained in such structure is enough to determine the three-dimensional folding of the protein. The structure of the folded protein reveals a variety of interactions that brings into contact aminoacids located far apart each other in the primary structure. Such interactions include covalent (disulfide bridges), electrostatics and weak forces. These long-range interactions lead to the formation of the compact structure of the proteins, which shape fall in the non-Euclidean geometry. Fractal geometry appears as the alternative concept that allows for the description of the protein structure.

3.1 Proteins as mass fractals

Let us consider concentric spheres of increasing radius around the centre of mass of a protein, and its mass M determined as a function of the radius r. The mass fractal dimension (D) is defined by the scaling relationship:

$$M(r) \propto r^D \quad (10)$$

If the mass fractal dimension is determined outside the centre of mass of the protein, i.e. near the surface, the values are greatly affected (Enright & Leitner, 2005). The mass fractal dimension of proteins gives values for the fractal dimension for proteins of 2.47 (Moret et al., 2005, Moret et al., 2006, Moret et al., 2009)

3.2 Proteins as line polymers

When the protein is considered from the point of view its primary structure, the fractal dimension of the protein backbone can be determined. The length of the backbone is measured by a stepwise connection of straight lines between alpha carbon atoms, being the intervals of increasing length ε. The number of steps N is then considered as a function of the variable length ε, and the fractal dimension is defined by the scaling relationship:

$$N(\varepsilon) \propto \varepsilon^{-D} \tag{11}$$

The value reported for the fractal dimension calculated according to Eq. 11 is 1.65 (Isogai & Itoh, 1984, Isvoran et al., 2001, Isvoran et al., 2008).

3.3 Proteins as roughened surfaces

A rigid sphere that rolls all over a protein is able to detect its surface irregularities. However, the degree of detail achieved in the description of the surface depends on the size r of the sphere. Let it $N(r)$ be the number of probe spheres needed to cover the whole surface. The dimension for this surface emerges from the following relationship:

$$N(r) \propto r^{-D} \tag{12}$$

The measured surface for the object is a function of the sphere probe radii, $s = s(r)$, and it is defined by the number of sphere probes multiplied by the sectional area (Torrens et al., 2001):

$$s(r) \propto N(r)\,\sigma(r) \propto r^2 r^{-D} \propto r^{2-D} \tag{13}$$

If the scale exponent is unique, the dimension D has a fractal nature. The surface dimension calculated in this way is between 2.06 - 2.17 (Craciun et al., 2009, Aqvist & Tapia, 1987, Pettit & Bowie, 1999).

3.4 But...Are protein fractal objects?

As Mandelbrot himself recognized (Mandelbrot, 1984), the fact of being him who coined the term "fractal", he cannot control its use. In fact, he does not use the term "fractal", and instead he prefers to use the term "fractal dimensionality" to refer to any anomalous dimensionality. We feel that it is important to be strict in maintaining the mathematical definition of fractal in order to discuss the eventual feasibility of a fractal description of the proteins.

Let us first consider the scale length. In order to assess the fractality of an object, the scaling behaviour should hold in several orders of magnitude. Avnir (Avnir et al., 1998) has been highly critic in this respect, questioning the wide use of the label "fractal". Most of the published works determined the fractal nature of different objects in a scale range spanning between 0.5 and 2 orders; the accomplishment of a potential equation by the data is a necessary but not a sufficient condition for applying the concept of "fractal" to that object. From a strict mathematical point of view, the arguments of Avnir are correct. However, the same author acknowledges that the practical limitation for considering an object as fractal lies on the cut off limits that exist in real objects, a fact considered by other authors in a well-known discussion on this topic held in the *Science* magazine (Mandelbrot et al., 1998). In the case of proteins, the lower cut off is ideally given by the size of an atom (1 Å), and the upper cut off by the size of the protein itself (few tens of Å), implying just two orders of magnitude in the scale. Facing this fact, Fernández-Pacios (1995) proposed the use of the term "effective dimension" to refer to the dimensionality of the proteins.

Some methodological aspects in the evaluation of the surface dimension of the proteins are now considered. The co-dimension rule gives unequivocally information of the protein surface, as strictly the surface contour is considered. Unfortunately, there is not enough

evidence to assess the quality of this method. For the case of lysozyme, the D_s obtained was 2.17. When the surface assessed through Eq. (4) is employed to calculate the fractal dimension, some discrepancies are notorious. Usually, values around 2.2 are obtained. In a single case, the calculated surface dimension of the lysozyme was 2.44 (Lewis & Rees, 1985). In this case, the radii range employed was 1 - 3.5 Å, may be so little that one can wonder if the probe sphere was able to get into the protein and evaluate its interior. In fact, when the mass fractal dimension is calculated in two different ways, one putting the increasing sphere in the centre of the protein, and the other putting the sphere towards the border of the protein, the value of the calculated surface dimension were 2.4 (Enright & Leitner, 2005, Moret et al., 2005) and 2.2 (Xiao, 1994, Enright & Leitner, 2005).
From the above discussion, arises an intuitive definition of *internal surface* and *external surface*. The interactions that take place across the internal surface are related to the intramolecular interactions between atoms and residues, while those established through the external surface would be related with the protein surroundings, like the solvent, ligands, etc. Hence, it is the external surface the one that is considered the thermodynamic surface.

4. An alternative view for the fractality of proteins

4.1 The West-Brown-Enquist theory of biological scaling laws

The biological scaling laws are known since a long time, and they were since their origin, of axiomatic nature. West, Brown and Enquist (1997) proposed a quantitative model that explain the origin of the ¼ power laws, and for the first time these scaling laws were demonstrated, based both on hydrodynamic (West et al., 1997) and geometrical considerations (West et al., 1999).
For the description of geometric forms, the authors are based on two conceptually different frameworks: the Euclidean and the fractal. In the framework of Euclidean geometry, the objects are described through an external surface S that keep inside a volume V, while in the fractal framework the objects have an exchange surface s and a total volume v, that correspond to biological active matter (Table 1). In this regard, the objects that belong to the last framework are called *biological objects*.

Variable	Euclidean framework	Fractal framework
Length	$L \propto S^{1/2} \propto V^{1/3} \propto M^{1/3}$	$l \propto s^{1/3} \propto v^{1/4} \propto M^{1/4}$
Surface	$S \propto L^2 \propto V^{2/3} \propto M^{2/3}$	$s \propto l^3 \propto v^{3/4}\ M^{3/4}$
Volume	$V \propto L^3 \propto M$	$v \propto l^4 \propto M$

Table 1. Scaling laws for length, surface and volume scaling laws for objects in the framework of Euclidean and fractal geometries. Adapted from (West et al., 1999)

For all cases, the proportional equations are resumed as $Y = Y_o\ M^b$, where Y is a scaling property, Y_o is a normalization constant, and b is the scaling exponent. The scaling exponents for biological objects are defined as:

distance: $b = 1 + \varepsilon_l$ $0 \leq \varepsilon_l \leq 1$

surface: $b = 2 + \varepsilon_s$ $0 \leq \varepsilon_s \leq 1$

volume: $b = 1 + \varepsilon_v$ $0 \leq \varepsilon_v \leq 1$ $(\varepsilon_v = \varepsilon_l + \varepsilon_s)$

where ε_l, ε_s and ε_v are the fractional scaling exponents for distance, surface and volume, respectively. These exponents correspond to the fraction above the Euclidean dimension giving the fractional exponent. According to these definitions, the following proportionalities hold between the geometrical variables:

$$l \propto s^{1/(2+\varepsilon s)} \qquad s \propto v^{(2+\varepsilon s)/(3+\varepsilon s+\varepsilon l)} \qquad l \propto v^{1/(3+\varepsilon s+\varepsilon l)}$$

The exchange surface of the biological objects is maximized in such a way that occupy as much as possible of the volume. This means that the exponent $b = (2+\varepsilon_s)/(3+\varepsilon_s+\varepsilon_l)$ must be maximum, condition achieved when $\varepsilon_s = 1$ and $\varepsilon_l = 0$. That is, the surface of a biological object may achieve a surface dimension of 3, and a volume dimension of 4. The theory explains that biological objects are shaped to optimize distribution of materials in a hierarchical structure.

One of the most interesting aspects of the West-Brown-Enquist (WBE) theory, more than 10 years after its proposal, is that is accomplished by an impressive 27 orders of magnitude for the mass of biological objects spanning from respiratory complexes to large mammals (West et al., 2002).

4.2 Proteins as biological objects

According to the studies of West, Brown and Enquist, the enzyme citochrome *c* oxidase is the smallest biological object proved to accomplish with the theory, and in this regard, it would be the smallest unity in the hierarchical network that share all biological objects. This network is of fractal nature, and this point is important in relation to proteins: *proteins are not fractal objects but biological objects, in the sense of the WBE theory, as part of a distribution network of fractal nature.*

Which are the results already obtained that support this proposal? Most of the research agrees that the structure of proteins seen from the mass fractal dimension coincides with a percolation system of dimension 2.5, and a surface dimension near 2.2. Other macromolecules, like dendrimers, show a mass dimension of 2.5 and a surface dimension of 2 (Pricl et al., 2003). It seems that large molecules show two different dimensions, depending on weather the surface of the object is considered or not. Hence, one could hypothesize that one of the internal dimension would be related to the structure of the protein, while the external dimension would relate to the protein surface, in the thermodynamic sense.

Some isolated results are in partial agreement with this view, when considering only the surface. For example, (Lewis & Rees, 1985) and (Fernández Pacios, 1995) showed that the surface of a protein displayed a high dimensional value (around 3) for a limited range of probe sphere radii (between 1.5 - 2.5 Å), and for larger radii, the show the usual value around 2.2. This high dimensional value could be related with the high dimension proposed by the WBE theory.

4.3 Test hypothesis: cytochrome *c* as a biological object

Allthough there is evidence that the surface of a protein is not fractal in the strict mathematical sense, proteins may display a fractal behaviour according to the WBE theory. This hypothesis is based on the results from (Lewis & Rees, 1985) and (Fernández Pacios, 1995), who showed that the protein surface evaluated by probe spheres of radii of ca. 2 Å display a dimensionality close to 3, the maximum value for a surface in a non-Euclidean framework.

Although these authors did not conclude about this specific value, it should be noted that the diameter of such probe sphere (4 Å) is of the order of the optimal values for interatomic distance inside the proteins. Hence, one could hypothesize that a probe sphere of such diameter would fill all the protein inside, thus "reading" a value corresponding to the maximum achievable occupancy.

The WBE theory postulates that the biological scaling laws base on the optimization of the materials distribution within the organism. This optimization is achieved when the exchange surfaces of the biological structure are maximum, and at the same time, the internal distances of the structure are minimum. The WBE theory predicts that biological objects, independently of their size, form a real distribution network of fractal nature. Is the independence from the object size that gives the fractal basis of the theory.

The accomplishment of the WBE theory by the different biological objects requires the measure of the relationship between the exchange surface and the real volume of the object. Usually, all the published works dealing with this problem employed the measure of the metabolic activity (proportional to the surface) as a function of the mass of the object, assuming a constant density (homogeneity) of the object. The smallest biological object accomplishing the WBE theory is the cytochrome *c* oxidase, being the enzymatic activity in this case the property evaluated. The fact that this enzyme is the smallest biological object measured, it was proposed as the "vital unit" of Nature.

Cytochrome *c* is the protein in charge of electron transport between the cytochrome *c* reductase and oxidase. According to the WBE theory, if what it is optimized is the materials transport in a hierarchical network, one should remember that are the electrons the ultimate material transported. Based on these considerations, cytochrome *c* would play the role of vital unit held by cytochrome *c* oxidase.

In order to probe that proteins accomplish the WBE theory, it is necessary to calculate both the surface and the volume of the protein, in order to obtain the values for ε_s and ε_l. It should be mentioned that this is the first time an object is analyzed for the accomplishment of the WBE theory by direct measure of the surface and the volume, and not employing other measures proportional to these properties.

5. Results and discussion

In order to apply the proposed methodology, it is necessary that the surface calculation programs allow for the evaluation of the surface and volumes in the widest possible sphere radii range. SurfRace and GEPOL were employed for the calculation of the SES of cytochrome *c*, and the results were compared to those of MS-DOT. The ratio of the surface obtained between GEPOL and MS-DOT was 1.01 (r^2 = 0.98), between SurfRace and MS-DOT was 1.09 (r^2 = 0.97), and between GEPOL and SurfRace was 1.06 (r^2 = 0.99), indicating that all three programs could be equally used.

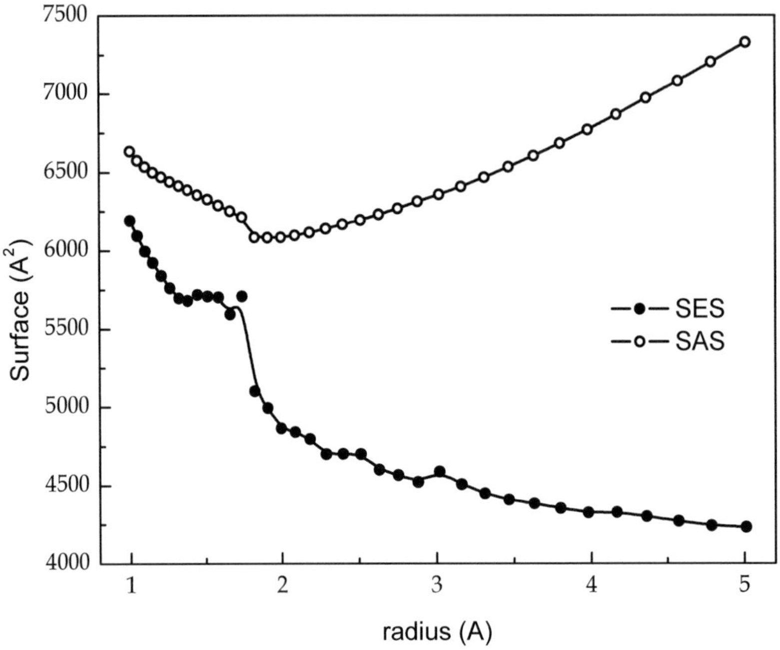

Fig. 2. Variation of the SAS and SES surfaces with the radius of the rolling sphere probe

The comparison of SAS and SES values (Fig. 2) show that for $r < 1.8$ Å, there is a monotonous decrease in the surfaces. However, at this sphere radius value, both surface show different behaviours: while SAS increases with r, SES decreases. This is because SES has less contact points with the protein as r increases, giving rise to a decrease in the surface, while SAS, due to its own definition, increases with the value of r. Below a radii of 1.8 Å, the decrease observed in SAS is due to the balance between the lesser number of contact points and the increase due to its own definition.

The divergence observed between SAS and SES for $r > 1.8$ Å suggest that below this value, the sphere probe is able to enter the protein and also evaluates the internal cavities of the protein. Let us consider a sphere of radius r which is able to enter the protein interior. If the internal distances (d) of the free spaces inside the protein are higher that the diameter ($2r$) of the sphere, it could be able to freely move within these spaces. As the sphere radius is increased, if the condition $2r < d$ is met, the contact points with the internal surface are lesser, and SES will progressively decrease. If the internal distances are homogeneous, a transition point at $2r = d$ will be verified, above which the sphere probe will not be able to enter the protein. Hence, those spheres that accomplish the condition $2r > d$ will be able to evaluate only the external surface of the protein, giving rise to a decrease in SES. In this way, the surface that can be considered as the thermodynamic surface is quantitatively defined.

The surface dimension calculated for cytochrome *c* show two clear values: one of ca. 3.0 for sphere probe radii of ca. 1.8 Å, and another in the range 2.05 – 2.20 for sphere probe radii above 2.5 Å (Fig. 3).

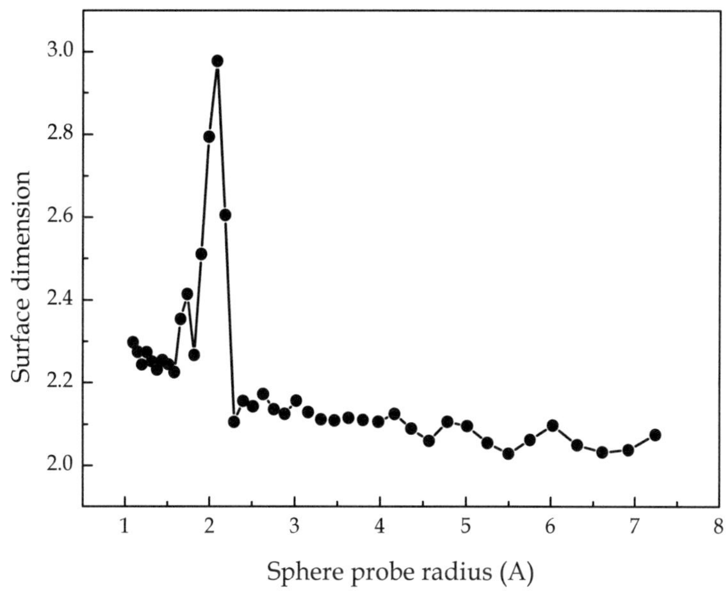

Fig. 3. Surface dimension of cytochrome *c* as a function of the sphere probe radius

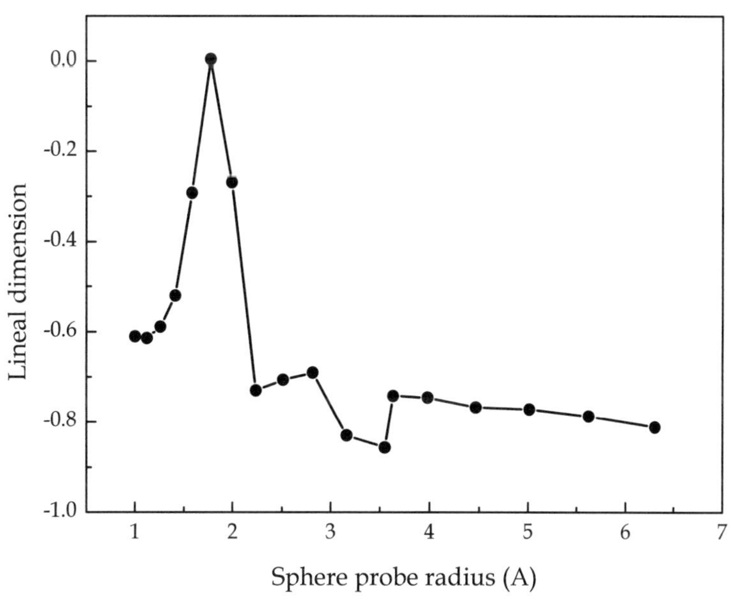

Fig. 4. Lineal dimension of cytochrome *c* as a function of the sphere probe radius

For the calculation of ε_l, the graphs of ε_l vs. r show the same behaviour to that of the surface dimension (Fig. 4). It is worth to note that the calculated values for ε_l are negative, that is, have no physical meaning. However, only for r = 1.8 Å, it is obtained the only value with physical meaning: $\varepsilon_l = 0$.

Hence, for cytochrome *c* it is demonstrated that $\varepsilon_s = 1$ and $\varepsilon_l = 0$. Notice that for establishing the accomplishment of the WBE theory for an object, it is necessary to measure the relationship between geometrical parameters, particularly surface and volume. In general, for biological objects (cells, animals), some proportional properties to these geometrical parameters are used, as the metabolic rate or enzyme activity (proportional to the surface) and the mass (proportional to the volume) (West et al., 1997). In our case, for the first time, it was possible to directly measure the surface and the volume of the object.

The results of the calculations indicate that for cytochrome *c*, a bimodal dimension was obtained. The existence of these two values is in agreement with the fact that proteins are not fractal objects in the strict sense of the definition, and accordingly, the dimension calculated is not a fractal dimension. Accordingly, it is more correct to name this dimension as *effective dimension* (Fernández Pacios, 1995). Besides, the value of the effective dimension of ca. 2.2 obtained for r > 1.8 Å has been extensively reported in the literature for proteins. Hence, it can be concluded that the thermodynamic surface of the proteins have an effective dimension of ca. 2.2.

The particular condition in which the sphere probe diameter equals the internal distances of the protein ($2r = d$), the sphere probe will completely fill all the free spaces inside the protein, it the internal distances are homogeneous. Under this condition, the maximum dimensionality is obtained, as correspond to a surface that fill all the volume of the object. The value of $2r$ = 3.6 Å is a typical value of the internal distances between the aminoacidic residues in the protein, and at these separation the interacting energies (van der Waals, Hydrogen bonds) are optimum. These optimal distances have been related to the structure of a crumpled surface, which may be a good model for protein structure (Cassia-Moura & Gomes, 2006).

6. Conclusions and future research

We have explored another view of the fractal nature of proteins, particularly focusing on the possible role of cytochrome *c* as a step in the materials distribution network that scales along 27 orders of magnitude, according to the theory of West, Brown and Enquist. For the first time, the prove of the accomplishment of this theory is based on the direct measurement of the surface and the volume of the object. Our conclusions cannot be extended to all proteins, as it was only limited to cytochrome *c*. However, our initial insights into the behaviour of other proteins showed that many proteins have a higher dimensionality at a radius value of 1.8 Å, confirming our initial suggestion that this value represent common internal distances related to long-range interactions within aminoacid residues.

Another point of interest relies on the fact that the protein surface can be considered as a thermodynamic frontier. Our approach was directed to the use of proteins in the design of biosensors. In this regard, we assumed that the maximum interaction between two surfaces is achieved upon equality of their surface dimensions. For such purpose, we have designed a modified surface based on the adsorption of 2-thiobarbituric acid onto a gold surface, for which we obtained a surface dimension of 2.17, that is, of the same order to the protein surface (Méndez et al., 2008). Upon contacting both surfaces, we have obtained a high protein coverage and a high electron transfer, indicating that indeed the surface of the protein behaves as a true thermodynamic frontier. Hence, we expect that these findings can bring new insights into biosensor design (Charcosset, 1998, Collier et al., 2001, Butala et al., 2003, Butala et al., 2004, Costanzo et al., 2004, Doke & Sadana, 2005, Doke et al., 2006, Doke & Sadana, 2006, Chodankar et al., 2007).

Finally, it is interesting to consider the reliability of the default value of 1.4 Å for the calculation of the protein surfaces. We have shown that this value may not be the best for this purpose, as the protein surface display large variations around this value. We think that, if the value of 1.8 Å is universal, this may be the best value to obtain information on the protein structure, and a value above 2.5 Å would be adequate to obtain information on the surface of the protein.

7. Acknowledgements

The authors thank CSIC (UdelaR), ANII, and PEDECIBA for financial support. Also acknowledged are Dr. Tsodikov and Dr. Fernández-Pacios for improvements in their softwares to fit the present research needs.

8. References

Abel, D. L. & Trevors, J. T. (2006). Self-Organization Vs. Self-Ordering Events in Life-Origin Models. *Physics of Life Reviews*, 3, 4, 211-228, 15710645 (ISSN)

Aqvist, J. & Tapia, O. (1987). Surface Fractality as a Guide for Studying Protein-Protein Interactions. *Journal of Molecular Graphics*, 5, 1, 30-34, 02637855 (ISSN)

Avnir, D., Biham, O., Lidar, D. & Malcai, O. (1998). Is the Geometry of Nature Fractal? *Science*, 279, 5347, 39-40, 00368075 (ISSN)

Bryant, S. H., Islam, S. A. & Weaver, D. L. (1989). The Surface Area of Monomeric Proteins: Significance of Power Law Behavior. *Proteins: Structure, Function and Genetics*, 6, 4, 418-423, 08873585 (ISSN)

Butala, H. D., Ramakrishnan, A. & Sadana, A. (2003). A Fractal Analysis of Analyte-Estrogen Receptor Binding and Dissociation Kinetics Using Biosensors: Environmental and Biomedical Effects. *BioSystems*, 70, 3, 235-253, 03032647 (ISSN)

Butala, H. D., Tan, Y. & Sadana, A. (2004). Analyte-Receptor Binding on Surface Plasmon Resonance Biosensors: A Fractal Analysis of Cre-Loxp Interactions and the Influence of Cl, O, and S on Drug-Liposome Interactions. *Analytical Biochemistry*, 332, 1, 10-22, 00032697 (ISSN)

Can, T., Chen, C. I. & Wang, Y. F. (2006). Efficient Molecular Surface Generation Using Level-Set Methods. *Journal of Molecular Graphics and Modelling*, 25, 4, 442-454, 10933263 (ISSN)

Cassia-Moura, R. & Gomes, M. A. F. (2006). A Crumpled Surface Having Transverse Attractive Interactions as a Simplified Model with Biological Significance. *Journal of Theoretical Biology*, 238, 2, 331-339, 00225193 (ISSN)

Collier, I. E., Saffarian, S., Marmer, B. L., Elson, E. L. & Goldberg, G. (2001). Substrate Recognition by Gelatinase A: The C-Terminal Domain Facilitates Surface Diffusion. *Biophysical Journal*, 81, 4, 2370-2377, 00063495 (ISSN)

Connolly, M. L. (1983). Solvent-Accessible Surfaces of Proteins and Nucleic Acids. *Science*, 221, 4612, 709-713, 00368075 (ISSN)

Costanzo, P. J., Patten, T. E. & Seery, T. A. P. (2004). Protein-Ligand Mediated Aggregation of Nanoparticles: A Study of Synthesis and Assembly Mechanism. *Chemistry of Materials*, 16, 9, 1775-1785, 08974756 (ISSN)

Craciun, D., Isvoran, A. & Avram, N. M. (2009). Does Fractal Characteristics of Hemoglobin Change from One Organism to Another? *Romanian Journal in Physics*, 54, 5-6, 569-575, 1221146X (ISSN)

Craciun, D., Isvoran, A., Reisz, R. D. & Avram, N. M. Distinct Fractal Characteristics of Monomer and Multimer Proteins. *Fractals*, 18, 2, 207-214, 0218348X (ISSN)

Charcosset, C. (1998). Fractal Analysis of Antigen-Antibody Binding Kinetics: Membrane Chromatography Applications. *Journal of Colloid and Interface Science*, 203, 2, 485-487, 00219797 (ISSN)

Chodankar, S., Aswal, V. K., Hassan, P. A. & Wagh, A. G. (2007). Structure of Protein-Surfactant Complexes as Studied by Small-Angle Neutron Scattering and Dynamic Light Scattering. *Physica B: Condensed Matter*, 398, 1, 112-117, 09214526 (ISSN)

Doke, A., Mathur, S. & Sadana, A. (2006). Fractal Binding and Dissociation Kinetics of Heart-Related Compounds on Biosensor Surfaces. *Journal of Receptors and Signal Transduction*, 26, 4, 337-357, 10799893 (ISSN)

Doke, A. M. & Sadana, A. (2006). Detection of Glucose and Related Analytes by Biosensors: A Fractal Analysis. *Journal of Receptors and Signal Transduction*, 26, 1-2, 35-59, 10799893 (ISSN)

Enright, M. B. & Leitner, D. M. (2005). Mass Fractal Dimension and the Compactness of Proteins. *Physical Review E - Statistical, Nonlinear, and Soft Matter Physics*, 71, 1, 15393755 (ISSN)

Fernández Pacios, L. (1994). Arvomol/Contour: Molecular Surface Areas and Volumes on Personal Computers. *Computers and Chemistry*, 18, 4, 377-385, 00978485 (ISSN)

Fernández Pacios, L. (1995). A Numerical Study on the Effective Dimension of Protein Surfaces. *Chemical Physics Letters*, 242, 3, 325-332, 00092614 (ISSN)

Flower, D. R. (1997). Serf: A Program for Accessible Surface Area Calculations. *Journal of Molecular Graphics and Modelling*, 15, 4, 238-244, 10933263 (ISSN)

Gabdoulline, R. R. & Wade, R. C. (1996). Analytically Defined Surfaces to Analyze Molecular Interaction Properties. *Journal of Molecular Graphics*, 14, 6, 341-353, 02637855 (ISSN)

Gerstein, M. & Lynden-Bell, R. M. (1993). What Is the Natural Boundary of a Protein in Solution? *Journal of Molecular Biology*, 230, 2, 641-650, 00222836 (ISSN)

Goetze, T. & Brickmann, J. (1992). Self Similarity of Protein Surfaces. *Biophysical Journal*, 61, 1, 109-118, 00063495 (ISSN)

Isogai, Y. & Itoh, T. (1984). Fractal Analysis of Tertiary Structure of Protein Molecule. *Journal of the Physical Society of Japan*, 53, 6, 2162-2171, 00319015 (ISSN)

Isvoran, A., Licz, A., Unipan, L. & Morariu, V. V. (2001). Determination of the Fractal Dimension of the Lysozyme Backbone of Three Different Organisms. *Chaos, Solitons and Fractals*, 12, 4, 757-760, 09600779 (ISSN)

Isvoran, A., Pitulice, L., Craescu, C. T. & Chiriac, A. (2008). Fractal Aspects of Calcium Binding Protein Structures. *Chaos, Solitons and Fractals*, 35, 5, 960-966, 09600779 (ISSN)

Lee, B. & Richards, F. M. (1971). The Interpretation of Protein Structures: Estimation of Static Accessibility. *Journal of Molecular Biology*, 55, 3, 00222836 (ISSN)

Lewis, M. & Rees, D. C. (1985). Fractal Surfaces of Proteins. *Science*, 230, 4730, 1163-1165, 00368075 (ISSN)

Mandelbrot, B. B. (1982). *The Fractal Geometry of Nature*, W.H. Freeman and Company, 0-7167-1186-9,

Mandelbrot, B. B. (1984). Fractals in Physics: Squig Clusters, Diffusions, Fractal Measures, and the Unicity of Fractal Dimensionality. *Journal of Statistical Physics*, 34, 5-6, 895-930, 00224715 (ISSN)

Mandelbrot, B. B., Pfeifer, P., Biham, O., Malcai, O., Lidar, D. A. & Avnir, D. (1998). Is Nature Fractal? [1]. *Science*, 279, 5352, 783-786, 00368075 (ISSN)

Méndez, E., Wörner, M., Lages, C. & Cerdá, M. F. (2008). Topological and Electron-Transfer Properties of the 2-Thiobarbituric Acid Adlayer on Polycrystalline Gold Electrodes. *Langmuir*, 24, 9, 5146-5154, 07437463 (ISSN)

Moret, M. A., Miranda, J. G. V., Nogueira Jr, E., Santana, M. C. & Zebende, G. F. (2005). Self-Similarity and Protein Chains. *Physical Review E - Statistical, Nonlinear, and Soft Matter Physics*, 71, 1, 15393755 (ISSN)

Moret, M. A., Santana, M. C., Nogueira Jr, E. & Zebende, G. F. (2006). Protein Chain Packing and Percolation Threshold. *Physica A: Statistical Mechanics and its Applications*, 361, 1, 250-254, 03784371 (ISSN)

Moret, M. A., Santana, M. C., Zebende, G. F. & Pascutti, P. G. (2009). Self-Similarity and Protein Compactness. *Physical Review E - Statistical, Nonlinear, and Soft Matter Physics*, 80, 4, 15393755 (ISSN)

Pettit, F. K. & Bowie, J. U. (1999). Protein Surface Roughness and Small Molecular Binding Sites. *Journal of Molecular Biology*, 285, 4, 1377-1382, 00222836 (ISSN)

Pfeifer, P., Welz, U. & Wippermann, H. (1985). Fractal Surface Dimension of Proteins: Lysozyme. *Chemical Physics Letters*, 113, 6, 535-540, 00092614 (ISSN)

Pricl, S., Fermeglia, M., Ferrone, M. & Asquini, A. (2003). Scaling Properties in the Molecular Structure of Three-Dimensional, Nanosized Phenylene-Based Dendrimers as Studied by Atomistic Molecular Dynamics Simulations. *Carbon*, 41, 12, 2269-2283, 00086223 (ISSN)

Richards, F. M. (1977). Areas, Volumes, Packing and Protein Structure. *Annual review of biophysics and bioengineering*, 6, 151-176, 00846589 (ISSN)

Stapleton, H. J., Allen, J. P., Flynn, C. P., Stinson, D. G. & Kurtz, S. R. (1980). Fractal Form of Proteins. *Physical Review Letters*, 45, 17, 1456-1459, 00319007 (ISSN)

Torrens, F., SÃ¡nchez-MarÃ-n, J. & Nebot-Gil, I. (2001). New Dimension Indices for the Characterization of the Solvent-Accessible Surface. *Journal of Computational Chemistry*, 22, 5, 477-487, 01928651 (ISSN)

Totrov, M. & Abagyan, R. (1996). The Contour-Buildup Algorithm to Calculate the Analytical Molecular Surface. *Journal of Structural Biology*, 116, 1, 138-143, 10478477 (ISSN)

Tsodikov, O. V., Thomas Record Jr, M. & Sergeev, Y. V. (2002). Novel Computer Program for Fast Exact Calculation of Accessible and Molecular Surface Areas and Average Surface Curvature. *Journal of Computational Chemistry*, 23, 6, 600-609, 01928651 (ISSN)

West, G. B., Brown, J. H. & Enquist, B. J. (1997). A General Model for the Origin of Allometric Scaling Laws in Biology. *Science*, 276, 5309, 122-126, 00368075 (ISSN)

West, G. B., Brown, J. H. & Enquist, B. J. (1999). The Fourth Dimension of Life: Fractal Geometry and Allometric Scaling of Organisms. *Science*, 284, 5420, 1677-1679, 00368075 (ISSN)

West, G. B., Woodruff, W. H. & Brown, J. H. (2002). Allometric Scaling of Metabolic Rate from Molecules and Mitochondria to Cells and Mammals. *Proceedings of the National Academy of Sciences of the United States of America*, 99, SUPPL. 1, 2473-2478, 00278424 (ISSN)

Wodak, S. J. & Janin, J. (1980). Analytical Approximation to the Accessible Surface Area of Proteins. *Proceedings of the National Academy of Sciences of the United States of America*, 77, 4 II, 1736-1740, 00278424 (ISSN)

Xiao, Y. (1994). Comment on "Fractal Study of Tertiary Structure of Proteins". *Physical Review E*, 49, 6, 5903-5905, 1063651X (ISSN)

10

Biomimetics - Thermodynamics to Study Wetting of Self-Cleaning Surfaces

Erwin Hüger, Jürgen Rost, Marion Frant, Gerhard Hildebrand, and Klaus Liefeith
Department of Biomaterials, Institute for Bioprocessing and Analytical Measurement Techniques (iba) e.V., Rosenhof, D-37308 Heilbad Heiligenstadt Germany

1. Introduction

Brilliant, recyclable and always surprising are the inventions of nature. Hence, it is no wonder that industry attempts to imitate natural environments. However, there is not a single engine that can keep up with the low-consumption of a hummingbird, and there are no paints available that are as permanently clean as a lotus leaf. The lotus effect relies on a strong water repellency. Rain drops have an almost spherical shape on the lotus leaf with minuscule contact area of low adhesion. Hence, the drops easy roll-up collecting dust and pathogens, cleaning in that way the leaf surface. One of the most important reasons for the existence of water-repellent natural surfaces is that it provides protection against pathogens supplied by free water such as bacteria or fungal spores. Therefore, water removal minimizes the chances of infection. In addition, dust particle removal from natural surfaces minimizes organism overheating or salt injury. However, surfaces of self-cleaning plants and insects are not the only example from nature for water repellency. Due to humidity, all surfaces are less or more covered by a nanoscopic thick water layer. In general, water environments are a frequent habitate on our blue planet. Hence, water repellency is long-established in nature. Feather of e.g. terrestrial birds have evolved primarily to repeal water (Rijke & Jesser, 2010).

Although nature still offers the best self-cleaning-properties by its remarkable ability for self-healing, there is a huge effort to constructing artificial biomimetic superhydrophobicity. Extreme water-repellency as well as extreme water-wetting properties are of considerable technological interest with a high-potential for energy economization. For example, extreme water-repellency is the basis for self-cleaning and anti-fouling which are the objects of a fast-growing research field with a multitude of applications in goods used every day. On the other hand, liquid-wetting and the dewetting-resistance are useful for tissue-engineering and implants where the biological cells need an aqueous environment. Accordingly, currently there is an intensified endeavor in research and development of materials with strong wetting and dewetting resistance. A huge number of studies have been carried out to produce artificial biomimetic roughness-induced superhydrophobic surfaces. A valuable recent review published by InTech (Qu & He & Zhang, 2010) is worth mentioning here.

We show that a simply constructed Gibbs-energy expression for the system composed of a liquid-droplet on material surfaces can adequately determine the water-repellency or water-absorbency in dependence of the surface structure and chemistry. The calculations do not need exceptional computational facilities and, in fact, do not even require computational knowledge. Since biomimetics has evolved into an interdisciplinary field of research, this chapter addresses scientists involved in engineering, chemistry, physics, biology, botany and related subjects, engaged in both academia or industry.

Up-to date wettability calculations are restricted to theoretically-modeled surface morphology, where the morphology is considered to be an identically repetition of one and the same asperity feature, e.g. pillars, (homogeneously dispersed) on the surface (Barbieri & Wagner & Hoffmann, 2007; Patankar, 2004; 2003; Nosonovsky & Bhushan, 2005; 2006; Ishino & Okumura, 2006). Real surfaces do not possess only one kind of asperity geometry, but a diversity of roughness features often randomly dispersed on the material surface. We apply our equations on real surfaces. For that reason we select the surface of *Colocasia esculenta* plant leaves which are evolution-optimized for self-cleaning property through natural selection. We determine experimentally the surface morphology of the *Colocasia esculenta* leaf surface with atomic force microscopy of a special setup designed for experiments in LifeScience (Huger, 2009). Using Gibbs-energy calculations we determine metastable and equilibrium states during the wetting of *Colocasia esculenta* leaf surface. However, the experiments are terestial. To make the calculations even more realistic, the gravitational force will be included in the Gibbs energy function. Liquid drops sitting or hanging from textured surfaces will be considered as well as the type of liquids. Illustrative graphical presentations as well as analytical minimization of the Gibbs-energy function as a function of the surface roughness, surface chemistry and acting forces will be provided.

This chapter focusses on the description of the Gibbs energy formula for wetting analysis with application on the evolution-optimized self-cleaning *Colocasia esculenta* leaf surface. To that purposes the chapter is organized as follows: Section (2) gives a brief description to the interaction of the liquid drop molecules to vapor (e.g. air) and material surfaces. In section (3) the Gibbs energy of a liquid drop on a material surface is constructed. Special emphasis will be given to the constituents of the Gibbs energy function that can be determined experimentally and how unavowed components can be expressed by experimentally procurable values, or avoided in order to solve the equations. Approaches will be discussed. After that, the Gibbs energy function will be built up step by step by considering flat (section (4)), rough (section (5)) and composite (section (6)) surfaces. Special attention will be given to the effect of gravitational fields. It will be shown how the surface morphology and surface chemistry enters the equations. Gibbs energy maps and Gibbs energy minimizations will illustratively explain the Wenzel (wetting) regime and the Cassie-Baxter (partial-wetting) equation. Section (6) will show how the constructed Gibbs energy function can be applied on a real system composed by the self-cleaning *Colocasia esculenta* leaf. Illustrative images will explain how the roughness parameters are extracted from the experimental data to enter the Gibbs energy function. The gravitational contribution will be included for drops resting on or hanging from the plant leaves. Our conclusion and outlook will be summarized in section (7). Limitations of the current model will be expressed.

2. Hydrophilic and hydrophobic interraction. Contact-angle.

Interaction of substances to water molecules (i.e. hydrophilicity and hydrophobicity) plays a crucial role in biological cell building and functionality, hydrophobic and hydrophilic interaction being the succesful strategy of life-fountain. Furthermore, the bond strength to water molecules plays also a role for evolution-optimized macroscopic properties of life-form such as the mentioned self-cleaning plant leafs and feather evolution.

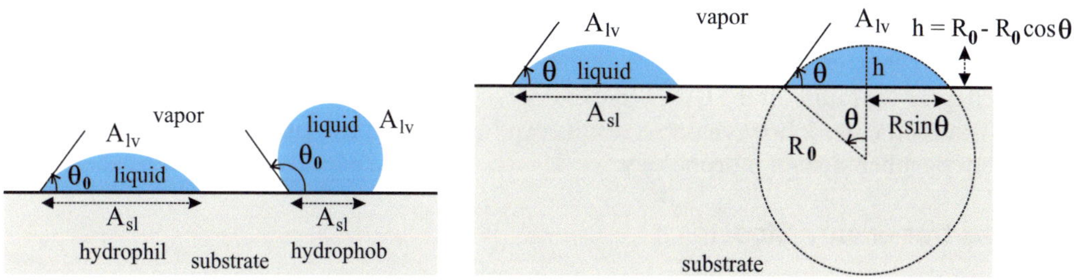

Fig. 1. Left sketch: Contact angle of a liquid drop on a substrate. Right panel: The liquid droplet is a spherical dome.

Water repellency can be conveniently expressed by the contact angle θ, defined as the angle between the tangent to the curved water surface at the point of contact with the solid surface and the plane of the surface on which the drop is resting, measured through the water (Fig. 1) (Marmur, 2010).
Hydrophilic materials "like" the contact with water due to a gain in energy at the material-water interface. In contrast, a water repellent surface (hydrophobic surface) "hates" the contact with water due to an increase in energy at the water-material interface, or a lower decrease in energy for drops in contact to hydrophobic chemistry in comparison to drops in contact with hydrophilic materials. Consequently, a water-repellent surface seeks to minimize the contact with water as far as possible, whereby a hydrophilic surface seeks to maximize its contact area with water. Hydrophilic surfaces are very common in industrial application and environment. For example, all metals and almost all oxides are intrinsically hydrophilic from nature. In contrast, hydrophobic surfaces are much more rare to be found. A hydrophobic surface is a surface with very low surface free energy. Low-energy surfaces can be obtained by saturated fluor (Hunter, 2010) and sulphur (Huger, 2008) bonds, both of them being poissoners.
The contact angle of water droplet which forms on a hydrophobic (hydrophilic) surface (see Fig. 1) is a result of energy minimization of the system built up by the surface material, liquid (e.g. water) droplet and gas environment (e.g. air) and is successfully related by the Youngs equation:

$$\cos\theta_0 = \frac{\sigma_{sv} - \sigma_{sl}}{\sigma_{lv}} \tag{1}$$

where θ_0 is the intrinsic contact angle, called also the Young contact angle, σ is the surface tension and the subscripts s, l and v denote the solid, liquid (e.g. water) and vapor (e.g. air) phase. Hydrophilic (hydrophobic) surfaces force the liquid droplet to have contact angles with more (less) than 90 degrees. Up to date there is no chemistry which allows a smooth surface to be built up at a water contact-angle larger than 120 degrees (Hunter, 2010). However, the contact angle measured on so-called superhydrophobic surfaces is larger than 150 degrees. The surface morphology and not the chemistry have to be the reason for such high contact-angles.
Super-repellent surfaces are that class of superhydrophobic surfaces on which water-droplets are not sticky. Hence, on such surfaces the droplets can easily roll-off already at a small inclination of no more than a few degrees. Such surfaces possess then self-cleaning properties, i. e. the possibility to roll-off of a water droplet under even a tiny disturbance taking the dust adhered to it cleaning in that way the surface. The cleaning procedure happens because

usually the dust adheres more strongly on the water drop than on the surface due to the surface liquid tension.
In this chapter we build up the Gibbs free energy of the system under discussion and we will show through the work how valuable results can be obtained by energy minimization on the ground of peculiar surface morphology.

3. Gibbs free energy. Model

3.1 Approach

Thermodynamics is a subject of immense practical value. One reason why thermodynamics is so valuable for physicists, chemists and engineers is that it is a theory which can be developed in its entirity, without gaps in the argument, on the basis of only moderate knowledge of mathematics (Denbigh, 1968). In this spirit, the Gibbs free energy of the system built up by a liquid droplet on a solid surface is given by the sum of the interfacial energies (Huger, 2009):

$$G = A_{lv}\sigma_{lv} + A_{sl}\sigma_{sl} + A_{sv}\sigma_{sv} \tag{2}$$

where A denotes the areas of the liquid-vapor (A_{lv}), solid-liquid (A_{sl}) and solid-vapor interfaces (A_{sv}).
Equation (2) represents a balance between each interfacial energy. Interfaces where the liquid molecules possess low bonding (i.e. hydrophobic interaction) increases the total interfacial energy. Hydrophilic interaction reduces the energy. So, the liquid drop would try to minimize its interface to air (vapor) (A_{lv}) and to maximize the solid-liquid interface (A_{sl}). The increase of the solid-liquid interface would reduce the energy, but would also change the drop shape. The drop-shape change can increase the interface to air (vapor) (A_{lv}) and, consequently, increase the energy. The system can reach such a wetting state, where the energy reduction by a growth of the solid-liquid interface is lower than the energy increase by the accompanied drop-shape change. There, the drop-shape lies in metastable or global energy minima. So, total energy minima appears for specific drop shapes and wetting behaviour. This is the aim of thermodynamics, which focuses on the description of equilibrium properties. Energy calculations are performed to analytically or numerically predictable local or global equilibrium properties (Cool & Garcia, 2010; Huger & Sob, 2010). In that spirit, we intend to find out how drop shapes and wetting behaviours influences the total interfacial energy.
Now, the interface areas can be determined. σ_{lv} can be measured, but σ_{sl} and σ_{sv} are unknown (Chaudhury & Whitesides, 1992). However, from equation 1 one can get the difference:

$$\sigma_{sv} - \sigma_{sl} = \sigma_{lv}\cos\theta_0 \tag{3}$$

So, in order to be able to calculate the Gibbs energy of the system under study it is necessary to express the Gibbs energy function to be dependent only on the difference between σ_{sl} and σ_{sv}. Fortunately, this can be done easily.
A_{sv} is the difference between the total area A_{tot} of the material surface and the area of the wetted material surface A_{sl}:

$$A_{sv} = A_{tot} - A_{sl} \tag{4}$$

From equations (2) and (4) one obtains:

$$G - A_{tot}\sigma_{sv} = A_{lv}\sigma_{lv} + A_{sl}(\sigma_{sl} - \sigma_{sv}) \tag{5}$$

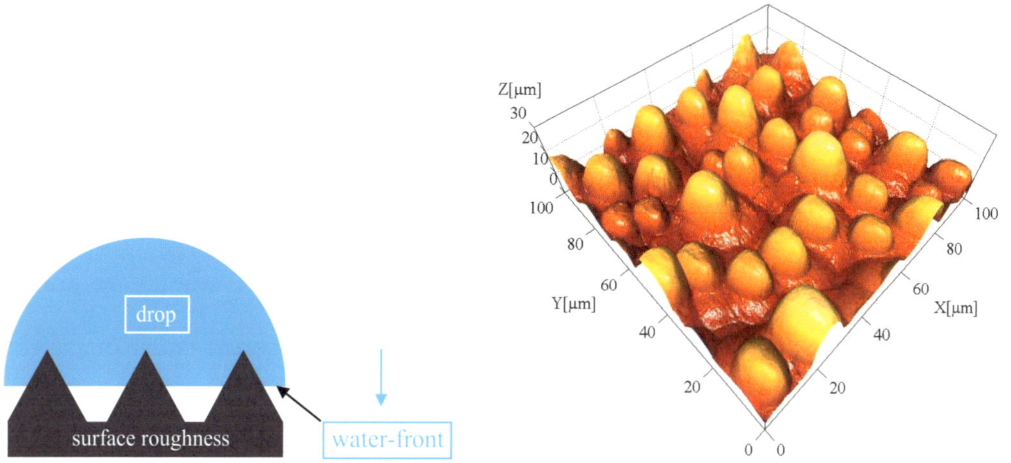

Fig. 2. Left panel: Schematic representation of a Cross-section through a liquid-droplet penetrating surface asperities. Right image: Atomic force microscopy from Colocasia esculenta leaf.

and using the equation (3) follows:

$$E = G - A_{tot}\sigma_{sv} = A_{lv}\sigma_{lv} - A_{sl}\sigma_{lv}\cos\theta_0 \qquad (6)$$

Now, the term $A_{tot}\sigma_{sv}$ is constant during the behavior of the liquid droplet on the material surface, whereas the other terms of equation (6) depends on the droplet state on the surface. Fortunately, all terms on the right part of the equation (6) can be determined. It makes no difference if the Gibbs energy G or the difference $E = G - A_{tot}\sigma_{sv}$ is used for calculations, since metastable and equilibrium drop shape are obtained by energy minimization:

$$0 = \frac{\partial E}{\partial \theta} = \frac{\partial (G - A_{tot}\sigma_{sv})}{\partial \theta} = \frac{\partial G}{\partial \theta} = \frac{\partial (A_{lv}\sigma_{lv} - A_{sl}\sigma_{lv}\cos\theta_0)}{\partial \theta} \qquad (7)$$

So, in the following we will calculate the energy difference E expressed in eq. (6). Before doing so, we will take gravity into account.

3.2 Gravity contribution

The experiments on drops done so far are terrestrial. For very small drops the gravity field can be neglected due to the larger contributions from surface tension, but, as it will be further shown, for larger drops the gravity contribution becomes relevant. The gravity contribution to the total Gibbs energy will be investigated for drops resting on surfaces or hanging from surfaces. In the following we will take into account drops whose size is much larger than the size of the surface roughness feature. The water front penetrating the surface roughness is taken as a horizontal plane, as depicted in the left sketch of Fig. 2. The right panel of Fig. 2 presents an atomic force microscopy (AFM) image of the *Colocasia esculenta* leaf surface.

The surface contains surface protrusions called papillae. The distance between the highest papillae extends up to 40 μm. So, the radius of the droplet base (see Fig. 1) should be equal at least to $R\sin\theta = 80\mu m$. In our laboratory we measured a contact angle of $\theta = 170^o$ for the leaf surface presented in Fig. 2. Hence, the radius of the liquid droplet should be some 460 μm. Consequently, the volume of the droplet considered through this work should be at least some

0.4 μl. However, in our laboratory we routinely used liquid droplets of 3 to 5 μl for contact angle measurments, the drop bases of which contacts a lot of leaf papillae.
For large droplets, the center of mass will strongly depend on the drop contact angle. A lower contact angle (θ) means that the center of mass will be lower and the systems lowers in that way its total energy. So, the θ-dependent part of the gravity is given by:

$$E_g(\theta) = mgC_z(\theta) = \rho V g C_z(\theta) \tag{8}$$

where m is the mass of the drop, V the drop volume, ρ the liquid density, g the gravity acceleration and C_z the vertical position of the drop mass-center.
In the case the drop is homogeneously filled by the liquid, the drop mass-center is given by the formula for the centroid of solids:

$$C_z = \frac{1}{V}\int_V z dV = \frac{1}{V}\int_{zmin}^{zmax} z\frac{\partial V(z)}{\partial z}dz \tag{9}$$

The drop is a spherical dome where $zmax = h$ (see Fig. 1). $zmin = 0$ can be selected. $V(z)$ is the volume of the drop from $zmin = 0$ up to the vertical value z. For $zmax = h$ the volume is given by:

$$V(zmax) = \pi R_0 h^2 - \frac{1}{3}\pi h^3 = \frac{\pi R_0^3}{3}(1-\cos\theta)^2(2+\cos\theta) \tag{10}$$

For an arbitrary value z ($0 < z \leq h$) the volume is given by:

$$V(z) = V(zmax) - V^{dome}(z) = (\pi R_0 h^2 - \frac{1}{3}\pi h^3) - [\pi R_0(h-z)^2 - \frac{1}{3}\pi(h-z)^3] \tag{11}$$

With $h = R_0(1-\cos\theta)$ (see Fig. 1) follows:

$$\begin{aligned} V(z) &= \frac{\pi R_0^3}{3}(1-\cos\theta)^2(2+\cos\theta) - \frac{\pi}{3}(h-z)^2(3R_0 - h + z) = \\ &= \frac{\pi R_0^3}{3}(1-\cos\theta)^2(2+\cos\theta) - \frac{\pi R_0^3}{3}(1-\cos\theta-\frac{z}{R_0})^2(2+\cos\theta+\frac{z}{R_0}) \end{aligned} \tag{12}$$

With $p = z/R_0$ eq. (12) can be written as:

$$\begin{aligned} V(p) &= \frac{\pi R_0^3}{3}(1-\cos\theta)^2(2+\cos\theta) - \frac{\pi R_0^3}{3}(1-\cos\theta-p)^2(2+\cos\theta+p) \\ &= \frac{\pi R_0^3}{3}(3p\sin^2\theta - 3p^2\cos\theta - p^3) \end{aligned} \tag{13}$$

and consequently:

$$dV = \frac{\partial V}{\partial p}dp = \frac{\pi R_0^3}{3}(3\sin^2\theta - 6p\cos\theta - 3p^2)dp = \pi R_0^3(\sin^2\theta - 2p\cos\theta - p^2)dp \tag{14}$$

From $z = p \cdot R_0$ follows $pmax = h/R_0 = 1 - \cos\theta$ and $pmin = 0$. So, eq. (9) becomes:

$$
\begin{aligned}
C_z &= \frac{1}{V}\int_V z dV = \frac{1}{V}\int_{pmin}^{pmax} p \cdot R_0 \frac{\partial V(p)}{\partial p} dp = \\
&= \frac{1}{\frac{\pi R_0^3}{3}(1-\cos\theta)^2(2+\cos\theta)} \cdot \int_0^{1-\cos\theta} p \cdot R_0 \cdot \pi R_0^3(\sin^2\theta - 2p\cos\theta - p^2)dp \\
&= \frac{3R_0[\frac{(1-\cos\theta)^2\sin\theta}{2} - \frac{2}{3}\cos\theta(1-\cos\theta)^3 - \frac{1}{4}(1-\cos\theta)^4]}{(1-\cos\theta)^2(2+\cos\theta)} \\
&= \frac{\frac{3R_0}{12}(5\sin^2\theta + 5\cos^2\theta - 3 + \sin^2\theta - 2\cos\theta)}{(2+\cos\theta)} = \frac{R_0}{4}\frac{(1-\cos\theta)(3+\cos\theta)}{2+\cos\theta}
\end{aligned}
\tag{15}
$$

From eq. (8) and eq. (15) follows the gravity contribution to the total Gibbs energy to be:

$$
E_g(\theta) = \frac{\rho V g R_0}{4} \cdot \frac{(1-\cos\theta)(3+\cos\theta)}{2+\cos\theta} \tag{16}
$$

With:

$$
V = \frac{\pi R_0^3}{3}(1-\cos\theta)^2(2+\cos\theta) \Leftrightarrow R_0 = \left(\frac{3V}{\pi(1-\cos\theta)^2(2+\cos\theta)}\right)^{\frac{1}{3}} \tag{17}
$$

follows finally:

$$
\begin{aligned}
E_g(\theta) = \quad & \frac{\rho V g}{4}\left(\frac{3V}{\pi(1-\cos\theta)^2(2+\cos\theta)}\right)^{\frac{1}{3}} \cdot \frac{(1-\cos\theta)(3+\cos\theta)}{2+\cos\theta} = \\
= \quad & \frac{(3^{\frac{1}{3}})\rho(V^{\frac{4}{3}})g}{4\pi^{\frac{1}{3}}} \cdot \frac{(1-\cos\theta)^{\frac{1}{3}}(3+\cos\theta)}{(2+\cos\theta)^{\frac{4}{3}}}
\end{aligned}
\tag{18}
$$

Now, ρ, V, g are constant. Hence, the gravity contribution (eq. (18)) is dependent only on the drop contact angle (θ).
Finally, the Gibbs energy including the gravity contribution is given by:

$$
\begin{aligned}
E &= G - A_{tot}\sigma_{sv} = A_{lv}\sigma_{lv} - A_{sl}\sigma_{lv}\cos\theta_0 + E_g(\theta) = \\
&= A_{lv}\sigma_{lv} - A_{sl}\sigma_{lv}\cos\theta_0 + \frac{(3^{\frac{1}{3}})\rho(V^{\frac{4}{3}})g}{4\pi^{\frac{1}{3}}} \cdot \frac{(1-\cos\theta)^{\frac{1}{3}}(3+\cos\theta)}{(2+\cos\theta)^{\frac{4}{3}}}
\end{aligned}
\tag{19}
$$

The following sections apply the Gibbs formula without (eq. (6)) and include the gravity (eq. (19)) on flat and rough surfaces.

4. Flat surfaces

4.1 Without gravity contribution

In the case of a flat surfaces, the liquid-vapor interface is considered to be a spherical cap, the curvature surface of a spherical dome (see Fig.1). The surface area of a spherical dome's ceiling is:

$$A_{lv} = 2\pi R_0 h = \pi R_0^2 (2 - 2\cos\theta) \tag{20}$$

where R_0 is the radius of curvature and θ is an arbitrary contact angle imposed to the drop and not necessarily its equilibrium contact angle (should be used in the minimization of Gibbs energy).

The area of the solid-liquid interface is equal to the area of the spherical dome base:

$$A_{sl} = \pi(R_0 \sin\theta)^2 \tag{21}$$

Introducing relation (20) and (21) into the equation (6) of the Gibbs energy follows:

$$\frac{G - A_{tot}\sigma_{sv}}{\pi R_0^2 \sigma_{lv}} = 2 - 2\cos\theta - (\cos\theta_0)\sin^2\theta \tag{22}$$

Now, a change in θ changes also R_0. In order to work with a constant value, R_0 will be expressed by the volume V of the drop which is constant during the drop shape change. From the volume of the drop (spherical dome) eq. (10) follows:

$$\frac{G - A_{tot}\sigma_{sv}}{\pi(\frac{3V}{\pi})^{\frac{2}{3}}[(1-\cos\theta)^2(2+\cos\theta)]^{-\frac{2}{3}}\sigma_{lv}} = 2 - 2\cos\theta - (\cos\theta_0)\sin^2\theta \tag{23}$$

and finally:

$$\frac{G - A_{tot}\sigma_{sv}}{\pi\sigma_{lv}(\frac{3V}{\pi})^{\frac{2}{3}}} = [(1-\cos\theta)^2(2+\cos\theta)]^{-\frac{2}{3}} \cdot (2 - 2\cos\theta - \cos\theta_0 \sin^2\theta) \tag{24}$$

In the left side of equation (24) all values up to G are constant during the behavior of a droplet on the surface. Consequently we can put:

$$G - A_{tot}\sigma_{sv} = E \tag{25}$$

and hence calculate the energy E:

$$E(\theta,\theta_0) = (\pi\sigma_{lv}(\frac{3V}{\pi})^{\frac{2}{3}})[(1-\cos\theta)^2(2+\cos\theta)]^{-\frac{2}{3}} \cdot (2 - 2\cos\theta - \cos\theta_0 \sin^2\theta) \tag{26}$$

With

$$\Phi(\theta_0) = \Phi_{smooth}(\theta_0) = \cos\theta_0 \tag{27}$$

and

$$F(\theta) = [(1 - \cos\theta)^2(2 + \cos\theta)]^{-\frac{2}{3}} \tag{28}$$

follows:

$$E(\theta,\theta_0) = (\pi\sigma_{lv}(\frac{3V}{\pi})^{\frac{2}{3}}) \cdot F \cdot [2 - 2\cos\theta - \Phi\sin^2\theta] \tag{29}$$

If we use:

$$\frac{E}{\pi\sigma_{lv}(\frac{3V}{\pi})^{\frac{2}{3}}} = \frac{G - A_{tot}\sigma_{sv}}{\pi\sigma_{lv}(\frac{3V}{\pi})^{\frac{2}{3}}} = E^\star \tag{30}$$

one can calculate the normed energy $E^\star$ which is no more dependent on the droplet volume and liquid surface tension:

$$E^\star(\theta,\theta_0) = [(1 - \cos\theta)^2(2 + \cos\theta)]^{-\frac{2}{3}}(2 - 2\cos\theta - \cos\theta_0\sin^2\theta) = F[2 - 2\cos\theta - \Phi\sin^2\theta] \tag{31}$$

For calculations we wrote a computer-code in MATLAB. The left-image of Fig. 3 depicts the energy surface in dependence of the material surface chemistry (intrinsic contact angle) and the contact angle imposed artificially by the calculation code.
The right-panels of Fig. 3 presents the path of minimal energy (upper panel) obtained from the energy map. The contact angles corresponding to the path of minimal energy is given

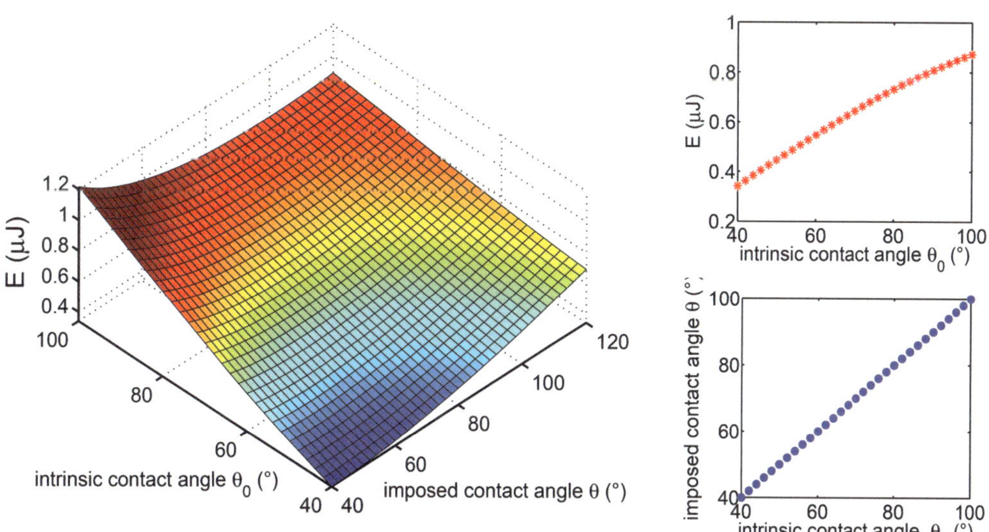

Fig. 3. Left-image: Energy map of a liquid droplet on flat surfaces with changing surface chemistry (intrinsic contact angle). Right upper panel: Path of minimal energy obtained from the left energy map. Right lower panel: The contact angles corresponding to the path of minimal energy. Drop volume $5\mu L$.

in the lower panel. It can be observed that equilibrium contact angles equals the intrinsic contact angles. This can be obtained also by the differentation of eq. (26) after θ and finding the minima by setting the diferential equal with zero:

$$0 = \frac{\partial E(\theta,\theta_0)}{\partial \theta} \Longleftrightarrow 0 = \frac{\partial E^\star(\theta,\theta_0)}{\partial \theta} = \frac{\partial [F(\theta)\cdot(2-2\cos\theta-\Phi(\theta_0)\sin^2\theta)]}{\partial \theta} =$$
$$\left(\frac{\partial F}{\partial \theta}\right)\cdot(2-2\cos\theta-\Phi\sin^2\theta)+F\cdot\left(\frac{\partial(2-2\cos\theta-\Phi\sin^2\theta)}{\partial\theta}\right) = [(1-\cos\theta)^2(2+\cos\theta)]^{-\frac{5}{3}}$$
$$\cdot(\sin\theta)\cdot[-2\sin^2\theta+2\cos\theta\sin\theta+2-3\cos\theta+\cos^3\theta-\Phi(\cos^4\theta-\sin^4\theta+2\cos\theta-3\cos^2\theta)] \quad (32)$$

There are three factors in the last expression of eq.(32). The first factor possesses a negative exponent which requires that:

$$(1-\cos\theta)^2(2+\cos\theta) \neq 0 \Longrightarrow \theta \neq 2n\pi \quad (33)$$

The second factor is $(\sin\theta)$. Considering the restriction (33), $\sin\theta$ vanishes for:

$$\theta = (2n+1)\pi \quad (34)$$

For $\theta = (2n+1)\pi$ the drop is entirely surrounded by the fluid and corresponds to an extrema, but which is a maxima instead of a minima. So, the vanishing of the last multiplier of the Gibbs-energy derivative (32) should furnish the energy minima:

$$-2\sin^2\theta+2\cos\theta\sin\theta+2-3\cos\theta+\cos^3\theta-\Phi(\cos^4\theta-\sin^4\theta+2\cos\theta-3\cos^2\theta)=0 \quad (35)$$

Equation (35) leads to:

$$\Phi = \frac{-2\sin^2\theta+2\cos\theta\sin\theta+2-3\cos\theta+\cos^3\theta}{\cos^4\theta-\sin^4\theta+2\cos\theta-3\cos^2\theta} \Rightarrow$$
$$\Phi = \frac{(-\cos\theta)(1-\cos\theta)^2}{-(1-\cos\theta)^2} \Longrightarrow \Phi = \cos\theta \quad \Longleftrightarrow \quad \cos\theta = \Phi \quad (36)$$

For smooth surfaces follows:

$$\cos\theta^{equil} = \cos\theta_0 = \Phi = \Phi_{smooth} \quad (37)$$

Equation (36) is an important expression which will find discussion also in the following sections. The next section investigates the influence of gravity on the equilibrium state.

4.2 Gravity contribution included

If we insert eq. (18) in eq. (29) one obtains the energy of the system with the gravity contribution included:

$$E(\theta,\theta_0) = (\pi\sigma_{lv}(\frac{3V}{\pi})^{\frac{2}{3}})F[2-2\cos\theta - \Phi\sin^2\theta] + \frac{(3^{\frac{1}{3}})\rho(V^{\frac{4}{3}})g}{4\pi^{\frac{1}{3}}} \cdot \frac{(1-\cos\theta)^{\frac{1}{3}}(3+\cos\theta)}{(2+\cos\theta)^{\frac{4}{3}}} \tag{38}$$

The contact-angles corresponding to the minimization of the Gibbs energy minima are obtained by:

$$0 = \frac{\partial E(\theta,\theta_0)}{\partial\theta} \Longleftrightarrow$$

$$0 = \frac{\partial\left[(\pi\sigma_{lv}(\frac{3V}{\pi})^{\frac{2}{3}}) \cdot F(\theta) \cdot (2-2\cos\theta - \Phi\sin^2\theta)\right]}{\partial\theta} + \frac{\partial\left[\frac{(3^{\frac{1}{3}})\rho(V^{\frac{4}{3}})g}{4\pi^{\frac{1}{3}}} \cdot \frac{(1-\cos\theta)^{\frac{1}{3}}(3+\cos\theta)}{(2+\cos\theta)^{\frac{4}{3}}}\right]}{\partial\theta} \tag{39}$$

According to eqs. (32,35,36) the first term in eq. (39) can be written as:

$$\frac{\partial\left[(\pi\sigma_{lv}(\frac{3V}{\pi})^{\frac{2}{3}}) \cdot F(\theta) \cdot (2-2\cos\theta - \Phi\sin^2\theta)\right]}{\partial\theta} =$$

$$= (\pi\sigma_{lv}(\frac{3V}{\pi})^{\frac{2}{3}}) \cdot [(1-\cos\theta)^2(2+\cos\theta)]^{-\frac{5}{3}} \cdot (\sin\theta) \cdot (1-\cos\theta)^2 \cdot (-\cos\theta - \Phi) \tag{40}$$

The second term of eq. (39):

$$\frac{\partial\left[\frac{(3^{\frac{1}{3}})\rho(V^{\frac{4}{3}})g}{4\pi^{\frac{1}{3}}} \cdot \frac{(1-\cos\theta)^{\frac{1}{3}}(3+\cos\theta)}{(2+\cos\theta)^{\frac{4}{3}}}\right]}{\partial\theta} -$$

$$(\frac{(3^{\frac{1}{3}})\rho(V^{\frac{4}{3}})g}{4\pi^{\frac{1}{3}}} \cdot [(1-\cos\theta)^2(2+\cos\theta)]^{-\frac{5}{3}} \cdot (\sin\theta) \cdot (1-\cos\theta)^2 \cdot \frac{(1-\cos\theta)^{\frac{2}{3}}}{(2+\cos\theta)^{-\frac{1}{3}}} \tag{41}$$

Inserting eqs. (40) and (41) into eq. (39) one obtains the contact angles which corresponds to the Gibbs energy minima by solving the equation:

$$V^2\rho^3 g^3(1-\cos\theta)^2 = 3\pi^2\sigma_{lv}^3(2+\cos\theta)(\cos\theta+\Phi)^3 \tag{42}$$

Hence, the contact angles (θ^{equil}) corresponding to the Gibbs energy minima are a function of the drop-volume (V), liquid-density (ρ), liquid-fluid surface tension (σ_{lv}) and (Φ).
Without gravitational contribution, an analytical solution can be easily achieved because eq. (37) is simple. Including the gravity field, the equation becomes of fourth order (eq. (42)). A analytical solution is possible, but more difficult. A numerical solution of eq. (42) is more convenient.
In our laboratory the volume of the water drop usualy used to measure contact angles was 5 μL. As mentioned in the previous section, the minimal volume of the droplet considered in this work should be 0.4 μL. Maximum droplets can be considered possessing a radius

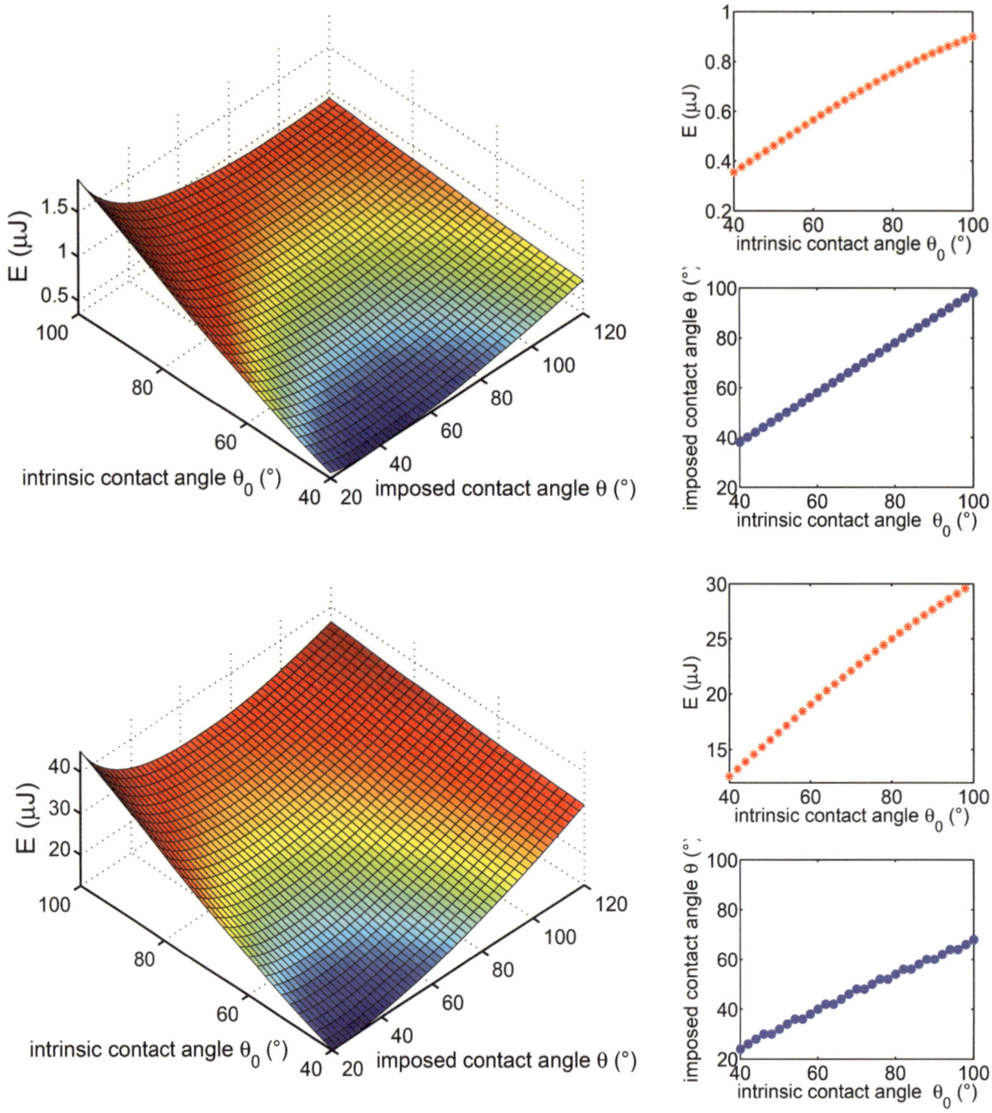

Fig. 4. Effect of gravitational field on a liquid droplet resting on a flat surface. Left images: energy maps in dependence of the surface chemistry (intrinsic contact angle) and the drop contact angle imposed by the calculation code. Gravity contribution is included for drop volumes of 5 μL (upper panel) and 523 μL (lower panel). Right panels: Path of minimal energy and corresponding contact angles of the energy surfaces depicted in the left-images.

of 5 mm, having in that way a volume of 523 μL. With $g = 9.81 m/s^2$, $\rho = 1000 kg/m^3$ and $\sigma_{lv} = 0.072 N/m$ one obtains the energy maps illustrated in the left images of Fig. 4 for water drop volumes of 5 μL (upper panel) and 523 μL (lower panel).

The corresponding paths of minimal energy and the equilibrium contact angles are presented in the right panels of Figure 4. The gravity contribution increase the energy of each state. Large

drops changes the topography of the energy map. Small drops do not change the equilibrium state whereas large drops can reduce significantly the equilibrium contact angle.
In the case a drop is hanging from (instead of resting on) a surface, the gravity contribution has to be subtracted from (instead of adding to) the Gibbs energy:

$$E(\theta,\theta_0) = (\pi\sigma_{lv}(\frac{3V}{\pi})^{\frac{2}{3}})F[2-2\cos\theta-\Phi\sin^2\theta] - \frac{(3^{\frac{1}{3}})\rho(V^{\frac{4}{3}})g}{4\pi^{\frac{1}{3}}} \cdot \frac{(1-\cos\theta)^{\frac{1}{3}}(3+\cos\theta)}{(2+\cos\theta)^{\frac{4}{3}}} \tag{43}$$

With $g = 9.81m/s^2$, $\rho = 1000kg/m^3$ and $\sigma_{lv} = 0.072N/m$ one obtains the energy maps depicted in the left images of Fig. 5 for water drops hanging from a surface for droplet volumes of 5 μL (upper panel) and 523 μL (lower panel).
The right panels of Fig. 5 presents the path of minimal energy and the corresponding equilibrium contact angles. The gravity contribution reduces the energy of each state. The topography of the energy map is strongly influenced for large drops. Small drops do not change the equilibrium state whereas large drops enhance significantly the equilibrium contact angle. Equilibrium contact angles approaching 180 degrees reflect drops which lost contact to surface. As the lowest panel of Fig. 5 shows, the equilibrium contact angle for large drops jumps to 180 degrees also for water-attractive surface chemistries (hydrophilic materials). In general, for drops hanging from the surface, the gravity contribution acts alike a non-adhesive (repulsive) field, whereas for drops resting on surfaces the gravity acts alike an adhesive field.
Figure 6 shows the path of minimal energy and Fig. 7 the corresponding equilibrium contact angles as a function of the drop volume and the surface chemistry for drops resting (left panels) and hanging (right panels) from smooth surfaces.
The next section investigates the case of rough surfaces.

5. Full-wetted rough surfaces. Wenzel regime.

This chapter analyses the case where rough surfaces beneath the drop-base are fully wetted. Textured surfaces with roughness-features smaller than the size of the drop-base will be considered.

5.1 Without gravity contribution

The area of the solid-liquid interface is given by:

$$A_{sl} = r \cdot A_{drop-base} = r \cdot \pi(R_0 \sin\theta)^2 \tag{44}$$

where r is the roughness ratio defined as the fraction between the real and projected surface. The roughness parameter r is always larger than unity. It equals 1 for flat surfaces.
The vapor-liquid interface remains the same as given in eq. (20). Introducing eq. (44) into the formula of the Gibbs-energy eq. (6) one obtains with the help of eqs. (10), (29) and (28):

$$E(\theta,\theta_0,r) = (\pi\sigma_{lv}(\frac{3V}{\pi})^{\frac{2}{3}})F[2-2\cos\theta - r(\cos\theta_0)\sin^2\theta] \tag{45}$$

The normed energy $E^{\star}$ is then given by:

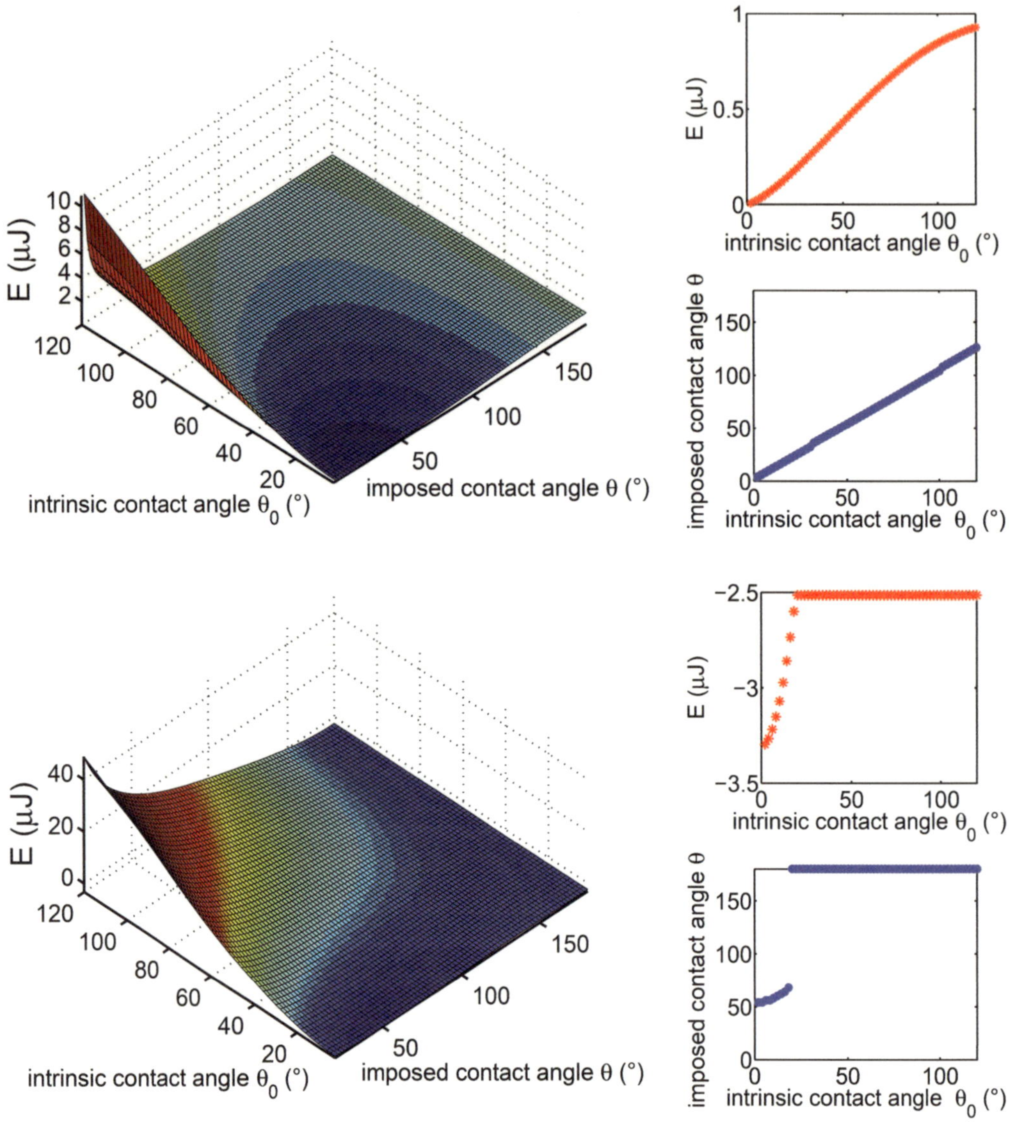

Fig. 5. Effect of gravitational field on a liquid droplet hanging from a flat surface. Left images: energy maps in dependence of the surface chemistry (intrinsic contact angle) and the drop contact angle imposed by the calculation code. Gravity contribution is included for drop volumes of 5 μL (upper panel) and 523 μL (lower panel). Right panels: Path of minimal energy and corresponding contact angles of the energy surfaces depicted in the left-images.

$$E^{\star}(\theta,\theta_0,r) = F(\theta) \cdot [2 - 2\cos\theta - r \cdot (\cos\theta_0)\sin^2\theta] \tag{46}$$

By comparing eq. (29) with eq. (45) one obtains:

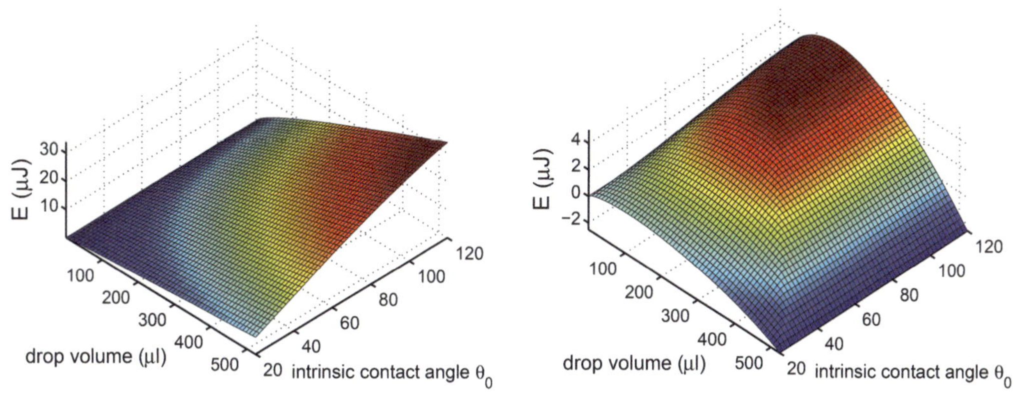

Fig. 6. Energy minima map of a liquid droplet resting on (left panel) or hanging from (right panel) a flat surface as a function of the drop volume and the surface chemistry (expressed by intrinsic contact angle), with gravity contribution included.

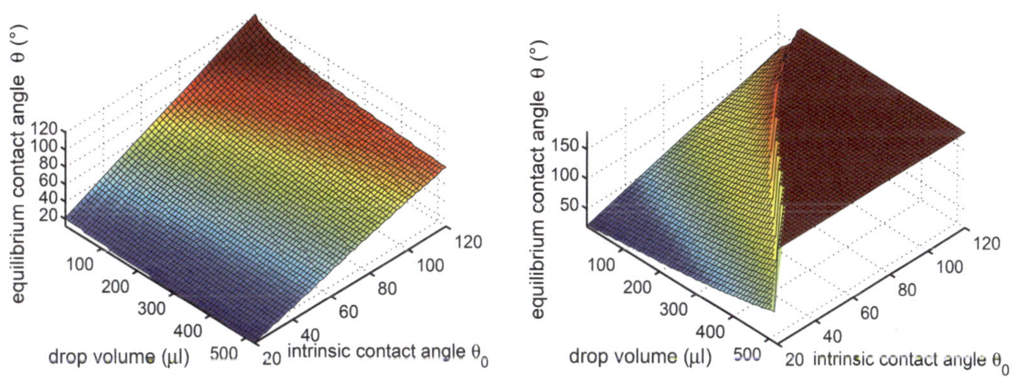

Fig. 7. Equilibrium contact angles of a liquid droplet resting on (left panel) or hanging from (right panel) a flat surface as a function of the drop volume and the surface chemistry (expressed by intrinsic contact angle), with gravity contribution included. $\theta \cong 180^o$ means that the drop lost his contact to surface. This is the case for large drops hanging from the surface.

$$\Phi(\theta_0) = \Phi_{rough}(\theta_0) = r \cdot \cos\theta_0 \tag{47}$$

From the solution (eq. (37)) of energy minimization (eq. (32)) one obtains the equillibrium contact angle of the drop on the rough surface given by:

$$\cos\theta = r \cdot \cos\theta_0 = \Phi = \Phi_{rough} \tag{48}$$

Equation (48) is the well-known Wenzel equation which provides the contact angle of a drop on a rough surface if on the drop-base the surface is fully wetted. For the case of the *Colocasia esculenta* leaf presented in Fig. 2 the roughness parameter r is 2.60. The left image of Fig. 8

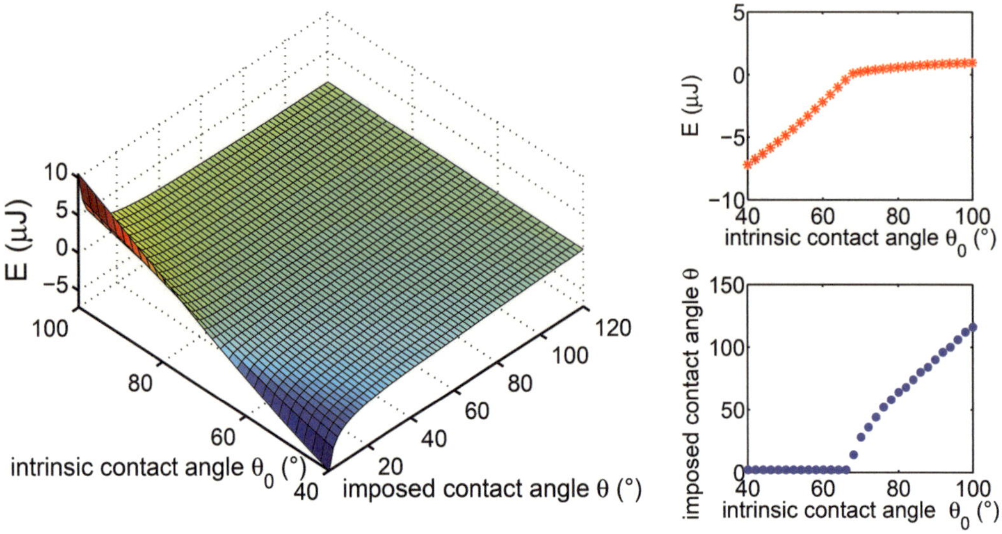

Fig. 8. Left-image: Energy map of a liquid droplet on the rough surfaces of the *Colocasia esculenta* leaf, in dependence of surface chemistry (intrinsic contact angle) and drop contact angle. Right panels: Path of minimal energy obtained from the left energy map. Corresponding equilibrium contact-angles.

illustrates the energy surface E of a drop on the *Colocasia esculenta* leaf surface as a function of the material surface chemistry (intrinsic contact angle) and the contact angle imposed artificially by the calculation code.

The paths of minimal energy and equilibrium contact-angles are presented in the right-panels. It can be observed that equilibrium contact angles are given by the Wenzel equation (48).

5.2 Gravity contribution included

If we insert eq. (18) by adding or subtracting into eq. (45) one obtains the energy of the system for drops resting on a rough surface or hanging from a rough surface with the gravity contribution included:

$$E(\theta,\theta_0) = (\pi\sigma_{lv}(\frac{3V}{\pi})^{\frac{2}{3}})F[2-2\cos\theta-\Phi\sin^2\theta] \pm \frac{(3^{\frac{1}{3}})\rho(V^{\frac{4}{3}})g}{4\pi^{\frac{1}{3}}}\frac{(1-\cos\theta)^{\frac{1}{3}}(3+\cos\theta)}{(2+\cos\theta)^{\frac{4}{3}}} \qquad (49)$$

Figure 9 presents the path of minimal energy and Fig. 10 the corresponding equilibrium contact angles as a function of the drop volume and the surface chemistry. Again, for drops hanging from the surface the gravity contribution act alike a non-adhesive (repulsive) field, whereas for drops resting on surfaces it acts alike an adhesive field.

The next section shows the energy and the equilibrium contact angles for all drop states appeared during the drop-sink into the surface roughness.

6. Partial-wetting: Cassie-Baxter

Partially wetted states have the wetting-fraction in between 0 und 1 ($0 < f < 1$). For $f = 1$ the state is defined as wetted (or Wenzel state) where on the drop-base the material surface

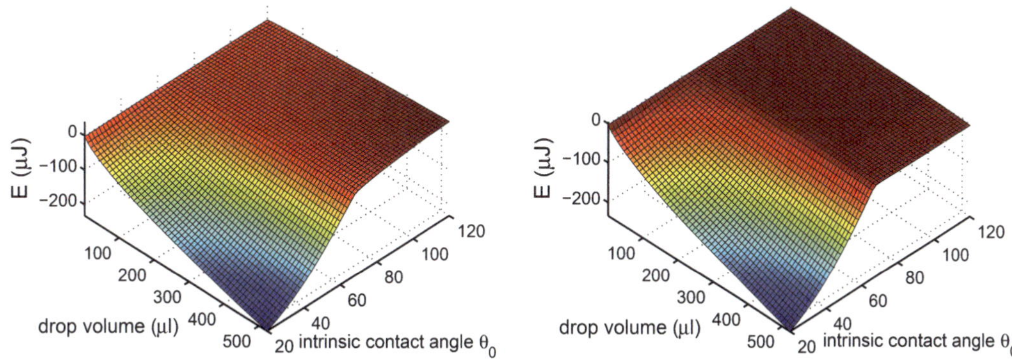

Fig. 9. Wetting regime: Energy minima map of a liquid droplet resting on (left panel) or hanging from (right panel) the surface of *Colocasia esculenta* leaf as a function of the drop volume and the surface chemistry (expressed by intrinsic contact angle), with gravity contribution included.

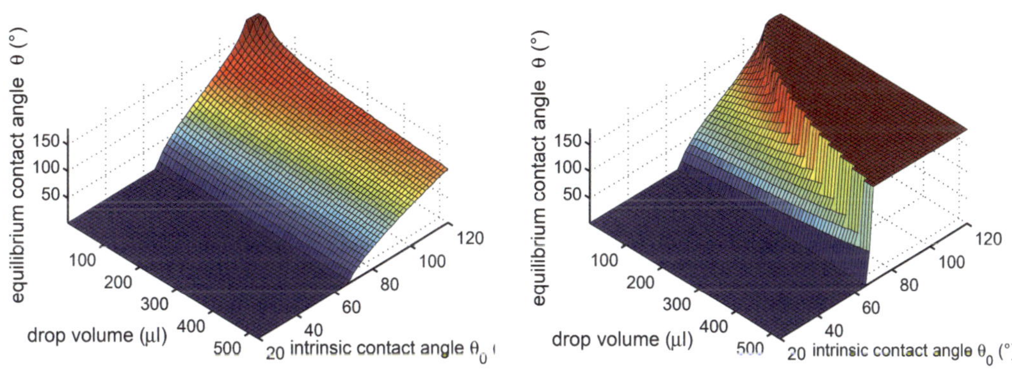

Fig. 10. Wetting regime: Equilibrium contact angles of a liquid droplet resting on (left panel) or hanging from (right panel) the surface of *Colocasia esculenta* leaf as a function of the drop volume and the surface chemistry (expressed by intrinsic contact angle), with gravity contribution included. $\theta \cong 180^o$ means lack of contact to surface, i.e. drop is falling.

is fully wetted. States for which $f = 0$ will be defined as non-bonded states. Drops hanging with $f = 0$ are falling.

The main objective of this subsection is to investigate the existence of Gibbs energy minima for states where underneath the drop-base the material surface is only partial wetted, i.e. if energy minima appears during the drop's sink into the surface roughness. This appears if there is an energy barrier between the wetting (Wenzel) state and the partial wetting (Cassie-Baxter) state. Obviously, such a energy barrier cannot be produced by a flat surface.

In the following we make the assumption that the water-front on the drop-base which sinks into the surface roughness is allways parallel to the macroscopic surface, as sketched in Fig. 2. The area of the solid-liquid interface is then given by:

$$A_{sl}(x) = r(x) \cdot f(x) \cdot A_{drop-base} = r(x) \cdot f(x) \cdot \pi(R_0 \sin\theta)^2 \qquad (50)$$

where $x = z/zmax$ is the sink depth ratio between the intermediate (z) and maximal (zmax) sink depth. The sink depth ratio x is in between 0 and 1. It is equal to 0 for the case the drop do not sink into the surface roughness and is equal to 1 if on the drop base all the surface is wetted. $f(x)$ is the fraction of the flat drop-base area which is wetted at the sink depth ratio x. Obviously also the roughness ratio is a function of the sink depth ratio $r = r(x)$.
Usually, surfaces possess one roughness-ratio (r) and one wetting-fraction (f) at a given sink-depth ratio (x). Hence, in place of expressing r and f by the sink-depth, one can directly express the roughness-ratio as a function of the wetting-fraction ($r(f)$). Figure 11 illustrates the wetted area of the papillae of *Colocasia esculenta* leaf for four sink-depths (i.e. four wetting fractions). The drop base is considered to be larger than the extent of the AFM images. The flat white regions in between the roughness corresponds to the non-wetted $(1 - f)$ part of the drop-base, i.e. the liquid-vapor interface on the drop-base. Figure 12 presents the dependence of the roughness ratio (r) on the wetting fraction (f) for the papillae of *Colocasia esculenta* leaf. With the help of this surface morphology characteristics we will compute the Gibbs energy in dependence of the surface chemistry. The results are presented in the following subsections.

6.1 Without gravity contribution

The area of the liquid-vapor interface is given by adding to eq. (20) also the non-wetted part of the drop-base area $(1 - f)\pi(R_0 \sin\theta)^2$:

$$A_{lv} = \pi R_0^2(2 - 2\cos\theta) + (1 - f(x)) \cdot \pi(R_0 \sin\theta)^2 \tag{51}$$

By introducing the eqs. (50) and (51) into the formula of the Gibbs-energy eq. (6) one obtains with the help of eqs. (10), (25) and (28):

$$G^{\star}(\theta,\theta_0,x) = F(\theta) \cdot [2 - 2\cos\theta - (r \cdot f \cdot \cos\theta_0 - 1 + f)\sin^2\theta] \tag{52}$$

where the wetted fraction f and the roughness ratio r are functions of the sink depth ratio x. From the solution (eq. (37)) of energy minimization (eq. (32)) one obtains the equilibrium contact angle of the drop on the rough surface given by:

$$\cos\theta^{equil} = r \cdot f \cdot \cos\theta_0 + 1 - f = \Phi \tag{53}$$

This equation is the well-known Cassie-Baxter equation which provides the contact angle of the drop in different partial wetting states. For $x = 0$ one have $f = 0$ from which follows that $\theta = 180^o$. For full wetting on the drop-base ($f = 1$) the Cassie-Baxter equation (53) transforms into the Wenzel equation (48) which describes the wetting regime.
The Cassie-Baxter equation (eq. (53)) does not predict if there exists a stable or metastable partial wetted state. Their existence is given by zero-points in the Gibbs energy derivatives after the sink-depth ratio:

$$0 = \frac{\partial G^{\star}(\theta,\theta_0,x)}{\partial x} \Longrightarrow 0 = \frac{\partial \Phi}{\partial x} = \frac{\partial(rf\cos\theta_0 + 1 - f)}{\partial x} = (r\frac{\partial f}{\partial x} + f\frac{\partial r}{\partial x})\cos\theta_0 - \frac{\partial f}{\partial x} \tag{54}$$

As can be observed from eq. (54), the sink-depths which corresponds to Gibbs energy extrema, depends on the surface chemistry (expressed by θ_o) and the surface roughness gradient (i.e. $\frac{\partial f}{\partial x}$ and $\frac{\partial r}{\partial x}$).

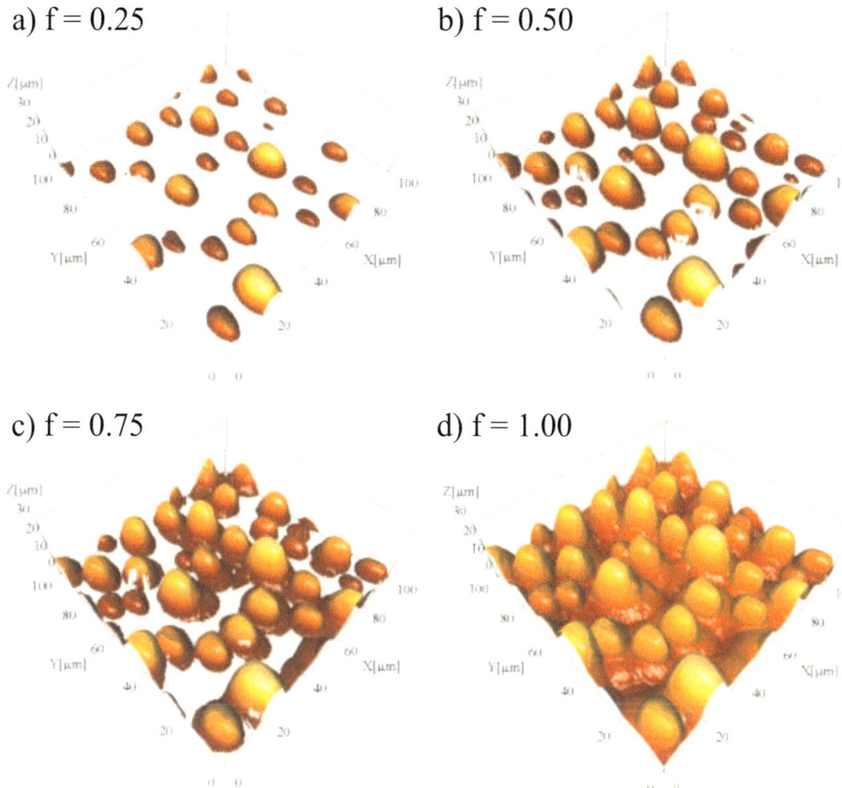

Fig. 11. The wetted area of *Colocasia esculenta* leaf surface roughness as a function of the wetting fraction f (i.e. as a function of the sink-depth). The pictures where obtained by processing with SPIP an atomic force microscopy image recorded from the *Colocasia esculenta* leaf. The wetted area of *Colocasia esculenta* for partial wetting f=0.25; 0.50; 0.75 and full-wetting f=1 are illustrated in the upper-left image (f=0.25), upper-right image (f=0.50), low-left image (f=0.75) and low-right image (f=1), respectively.

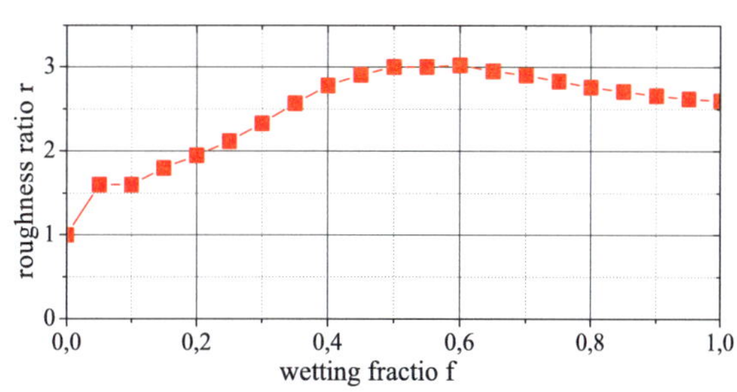

Fig. 12. Roughness ratio vs wetting fractio determined from the AFM image of *Colocasia esculenta* leaf.

In order to get the contact angles and sink-depths corresponding to Gibbs-energy minima, one can analytically solve the eqs. (54) and (53), or, alternatively, compute the Gibbs-energy for all possible contact angles (θ), all sink-depths (x) for a given surface morphology (f,r) and chemistry (θ_o): $G^\star = G^\star(\theta,\theta_0,r,f,x)$.
The second strategy is more convenient, especially, as previosly mentioned, when the gravity contribution is considered. Furtheremore, an analytical expression for f and r is not always easy to be determined, especially for random rough surfaces. Random rough surfaces are much more relevant from a practical perspective, because they are cheaper to fabricate. They can be made from a large class of materials, metallic and non-metallic. So, there is also a practical standpoint for the preference to compute the Gibbs energy for all possible contact angles, if random rough surfaces are considered.
Figure 13 presents the normed energy for a drop resting on papillaes of *Colocasia esculenta*. θ is the contact angle imposed artificially by the calculation code. For each wetting fraction f one can obtain the Gibbs energy minima and the corresponding (equilibrium) contact angle presented in Fig. 14 for intrinsic contact angles of $\theta_0 = 72^o$ (left panel), $\theta_0 = 109^o$ (middle panel) and $\theta_0 = 112^o$ (right panel).
For hydrophilic surface chemistry, Gibbs energy minima appears only in the wetted state ($f = 1$) (see Figs. 13,14). There is no energy barrier which could stop the wetting on the drop-base. So, for hydrophilic surface chemistry, always the wetting regime is obtained. This is different for hydrophobic surface chemistry. For hydrophobic surface chemistry, the *Colocasia esculenta* roughness induces a local Gibbs energy minima for partially-wetted states ($f < 1$) (Figs. 13,14). This can be observed also in Fig. 15 where the maps of the Gibbs energy minima and their corresponding maps of contact angles are presented in dependence of surface chemistry (θ_0) and wetting fraction. Firstly, for hydrophobic contact angles up to ($\theta_0 = 110^o$) the Gibbs energy minima for partial-wetting is metastable. The global minimum being the wetting-regime ($f = 1$). For intrinsic contact angles higher than 110°, the partial-wetting generates the global minimum.
The stars surrounded by circles in Fig. 16 reveal that the partially-wetted states are global minima.

6.2 Gravity contribution included

If we insert eq. (18) in eq. (52) one obtains the energy of the system with the gravity contribution included for drops resting on (+) or hanging from (-) rough composite surfaces:

$$E(\theta,\theta_0) = (\pi\sigma_{lv}(\frac{3V}{\pi})^{\frac{2}{3}})F[2-2\cos\theta-\Phi\sin^2\theta] \pm \frac{(3^{\frac{1}{3}})\rho(V^{\frac{4}{3}})g}{4\pi^{\frac{1}{3}}} \cdot \frac{(1-\cos\theta)^{\frac{1}{3}}(3+\cos\theta)}{(2+\cos\theta)^{\frac{4}{3}}} \quad (55)$$

In order to investigate the gravity contribution on the wetting behavior, we compare the case of no-gravity inclusion with the case when the gravity contribution is included. This will be done for a drop volume 523 μL resting on or hanging from the *Colocasia esculenta* leaf.
With $g = 9.81 m/s^2$, $\rho = 1000 kg/m^3$ and $\sigma_{lv} = 0.072 N/m$ one obtains the energy maps illustrated in Fig. 17 for resting drops (middle row) and for hanging drops (lower raw). Figure 18 presents the path of energy minima in dependence of the wetting-fractio (f) for drops resting or hanging in the case of no-gravity contribution and the case of gravity contribution included. The corresponding contact angles are presented in Fig. 19.
Gravity enhance (reduce) the Gibbs energy for drops resting on (hanging from) surfaces. On that way, the gravity field reduces (enlarges) the contact angle of energy minima states, for

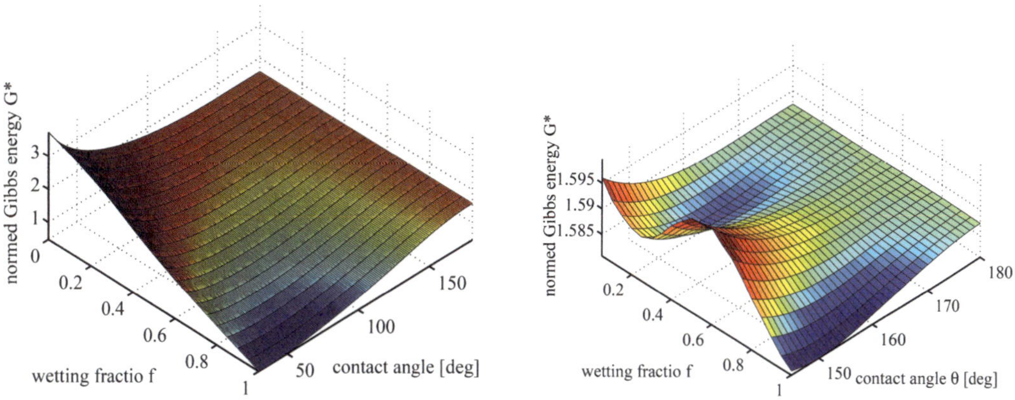

Fig. 13. *Colocasia esculenta* leaf: No gravity consideration. Normed Gibbs energy of a liquid droplet as a function of the wetting fraction f and the contact angle θ imposed artificially by the calculation code for a intrinsic contact angle of $\theta_0 = 72^o$ (left panel) and $\theta_0 = 109^o$ (right panel).

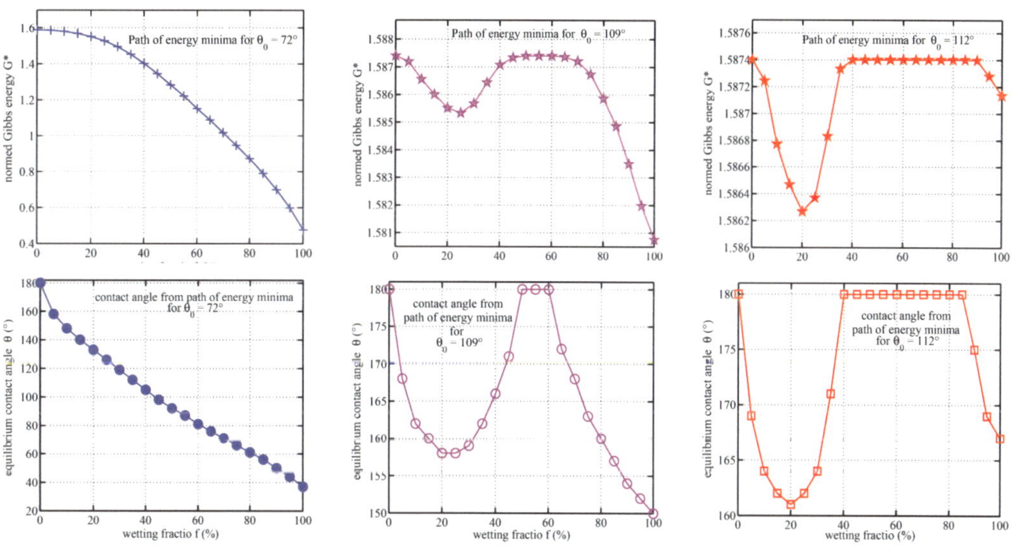

Fig. 14. *Colocasia esculenta* leaf: No gravity consideration. Upper row: Path of minimal Gibbs energy, e.g. obtained from the energy surface presented in Fig. 13 for intrinsic contact angle of $\theta_0 = 72^o$ (left panel), $\theta_0 = 109^o$ (middle panel) and $\theta_0 = 112^o$ (right panel). Lower row: The contact angles corresponding to the path of minimal energy.

drops resting on (hanging from) surfaces. Contact angles of $\theta = 180^o$ for hanging drops means drop-fall.

The dependence on the surface chemistry is presented in Fig. 20. Figure 21 presents the surface chemistry dependence of the Gibbs energy minima, their corresponding contact angles and the wetting fraction where the Gibbs energy minima appears.

The extension to liquids of different liquid surface tensions is straightforward. In that case σ_{lv} has to be changed. Of course, a different liquid yields the surface chemistry to establish a different intrinsic contact angle θ_0.

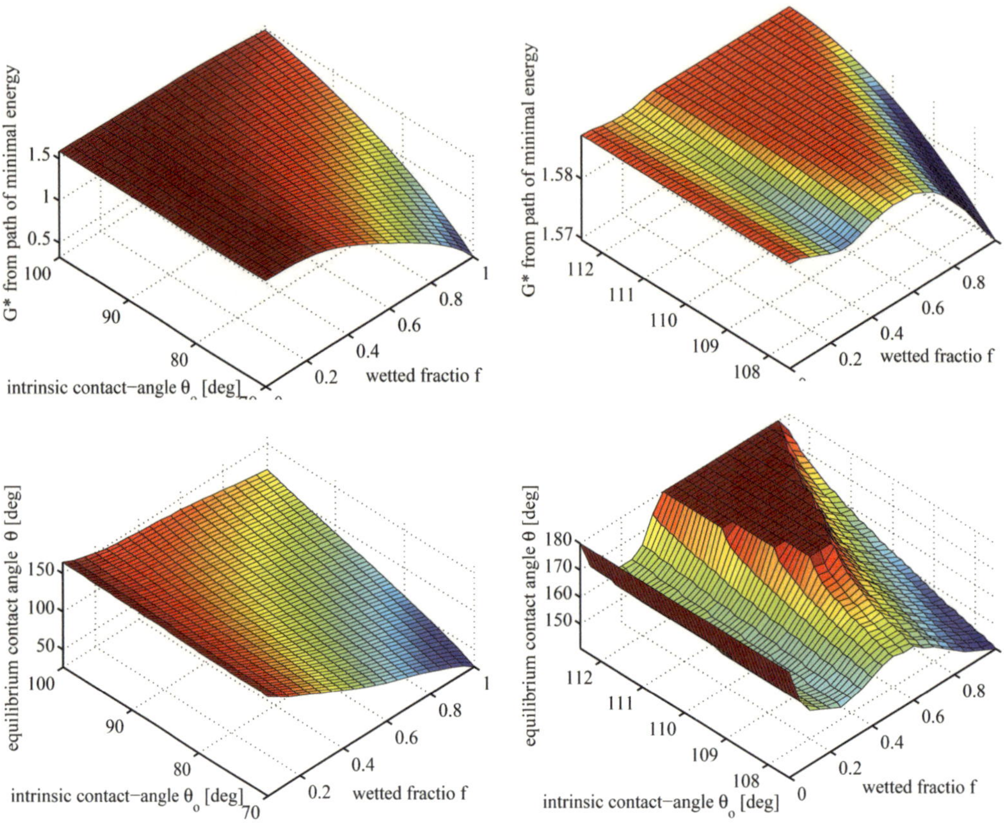

Fig. 15. *Colocasia esculenta* leaf: No gravity consideration. Upper row: Path of minimal Gibbs energy in dependence of the surface chemistry expressed by intrinsic contact angle of θ_0 (i.e. Gibbs energy minima as a function of wetting fraction and intrinsic contact angle). Lower row: The contact angles corresponding to the path of minimal energy.

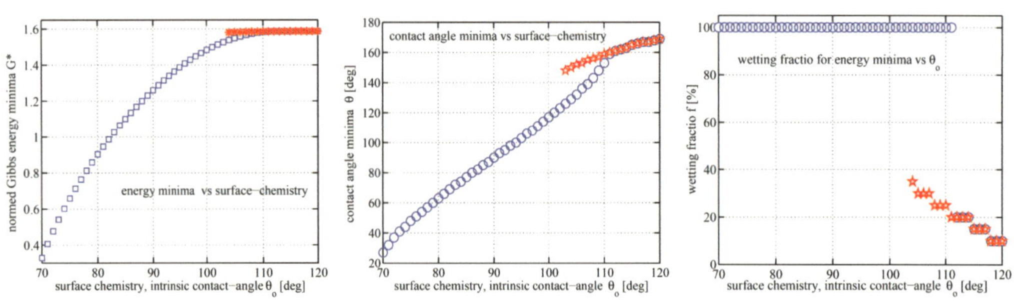

Fig. 16. *Colocasia esculenta* leaf: No gravity consideration. Left-panel: Metastable and global Gibbs energy minima in dependence of the surface chemistry expressed by the intrinsic contact angle. Stars mark partially-wetted states ($f < 1$). Circles denote the global minima. Stars surrounded by circles means that the partially-wetted states are global minima. Middle-panel: The contact angles corresponding to the Gibbs energy minima. Right-panel: The wetting fraction for the Gibbs energy minima.

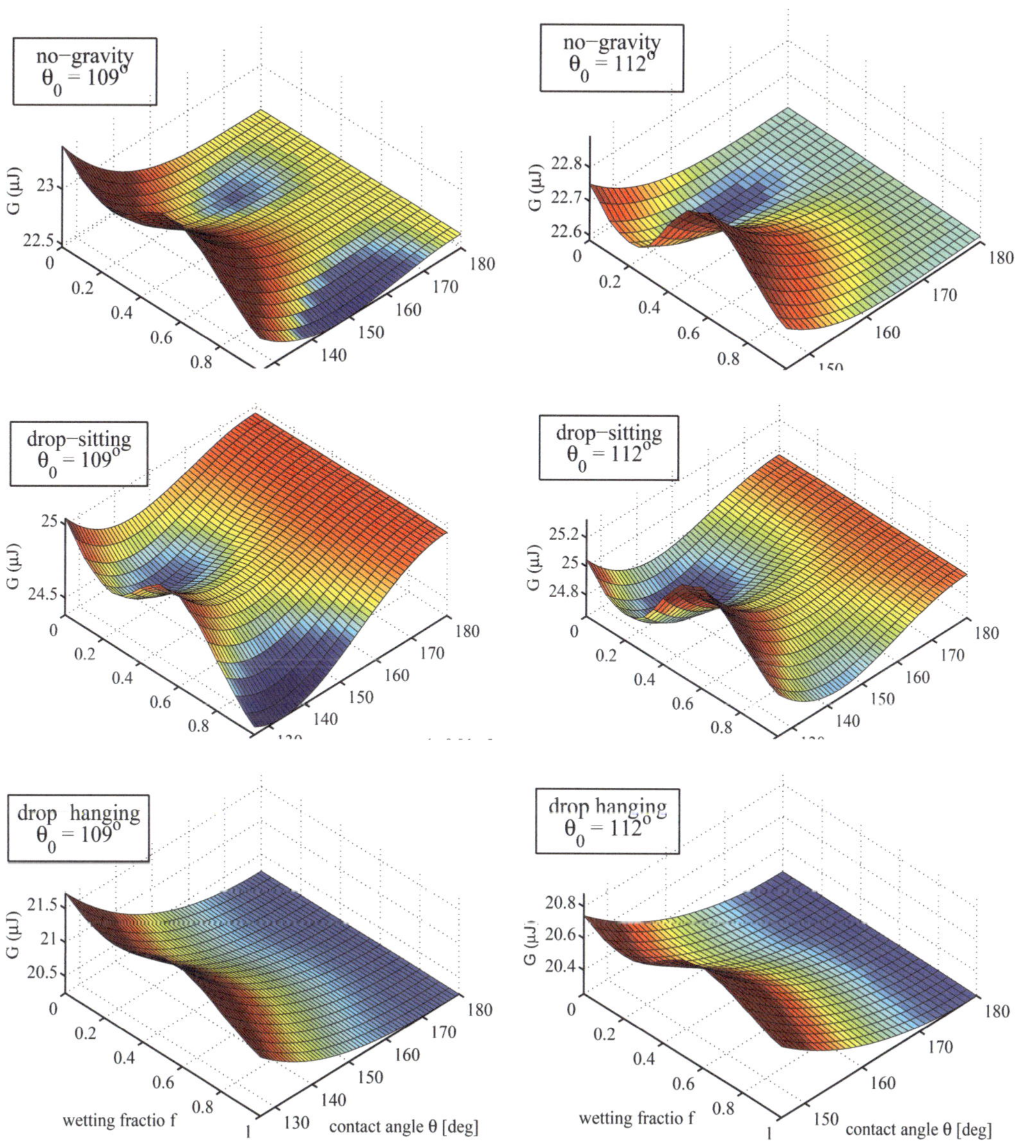

Fig. 17. *Colocasia esculenta* leaf: Effect of gravity for a drop resting (middle panels) or hanging (lower panels). Energy maps without (uppers panels) and with (middle and lower panels) gravity included as a function of the wetting fraction f and the contact angle θ imposed artificially by the calculation code. Left panels: for intrinsic contact angle of $\theta_0 = 109^o$. Right panels: $\theta_0 = 112^o$. The volume of the water droplet is 523 μL.

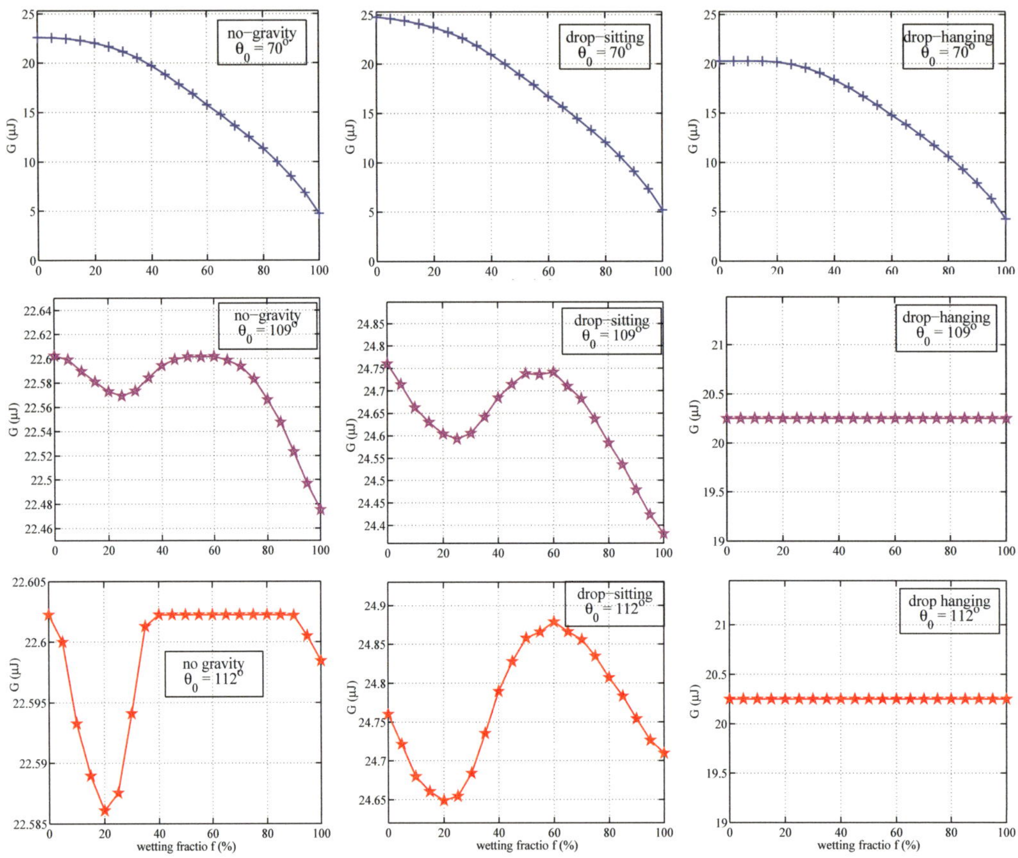

Fig. 18. *Colocasia esculenta* leaf: Effect of gravity on paths of minimal energy as a function of wetting fraction f. Paths of minimal Gibbs energy obtained from the energy surface presented in Fig. 17 with lack of gravity (left column), with gravity consideration for resting droplet (middle column) and with gravity contribution for hanging droplet (right column). Upper row: for intrinsic contact angle of $\theta_0 = 70^o$. Middle row: $\theta_0 = 109^o$. Lower row: $\theta_0 = 112^o$.

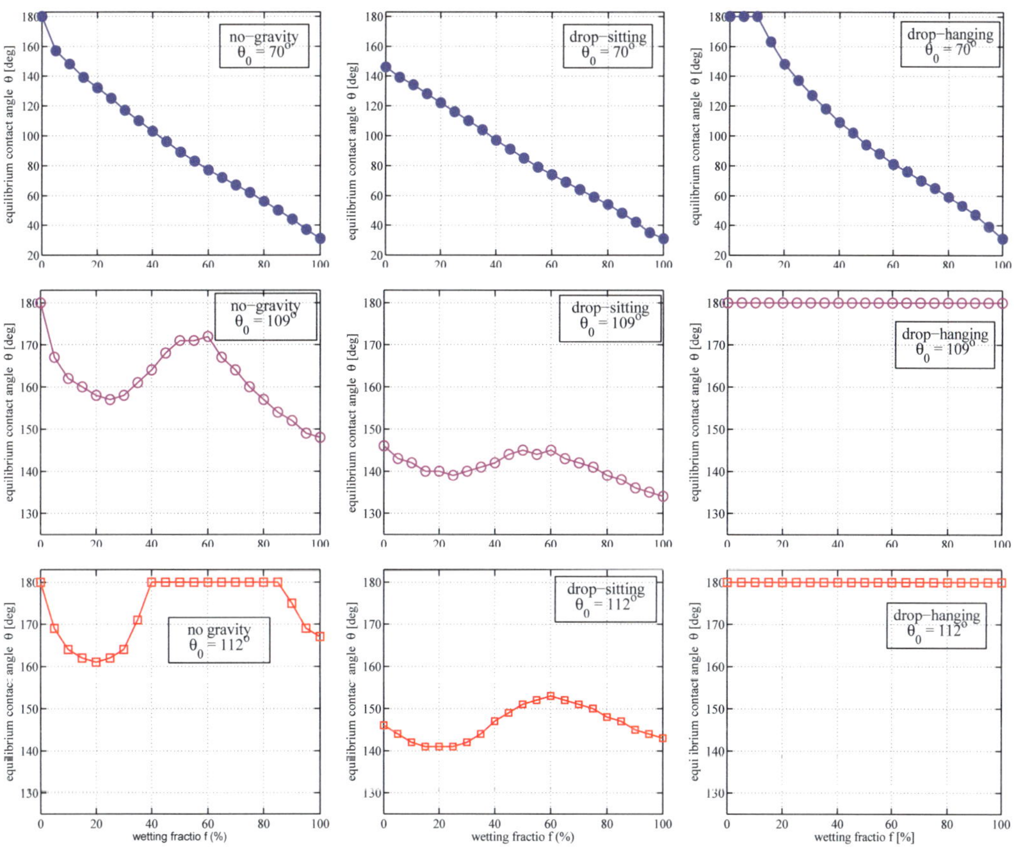

Fig. 19. *Colocasia esculenta* leaf: Effect of gravity on the contact angles corresponding to the paths of minimal energy as a function of wetting fraction f. Contact angles which corresponds to the paths of minimal Gibbs energy presented in Fig. 18. Lack of gravity (left column), with gravity consideration for resting droplet (middle column) and with gravity contribution for hanging droplet (right column). Upper row: for intrinsic contact angle of $\theta_0 = 70^o$. Middle row: $\theta_0 = 109^o$. Lower row: $\theta_0 = 112^o$.

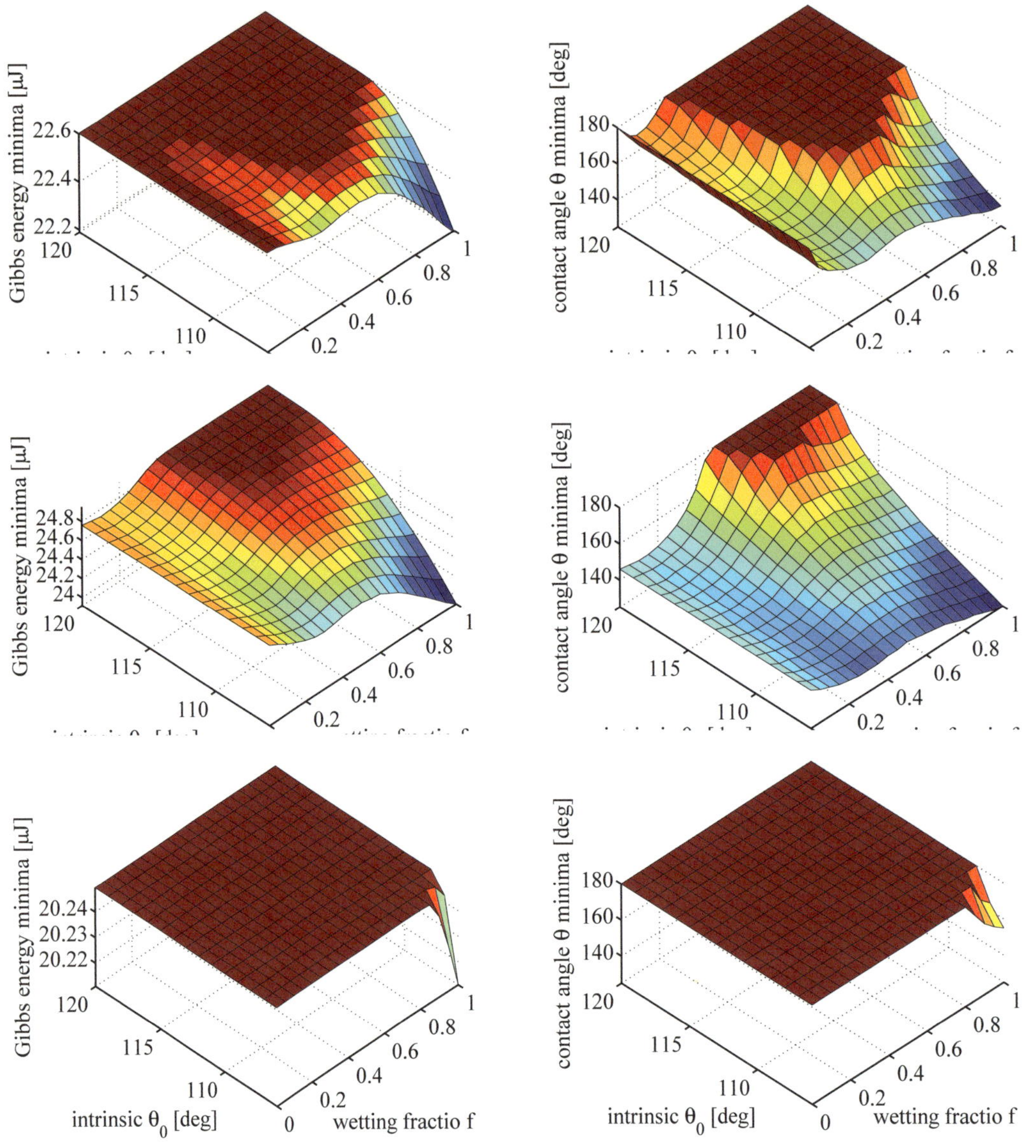

Fig. 20. *Colocasia esculenta* leaf: Effect of gravity on Gibbs energy minima (left panels) and their corresponding contact-angles (right panels) for drops resting (middle panels) or hanging as a function of the surface chemistry and wetting fraction. Upper panels: lack of gravity. Middle panels: gravity included for drop-resting. Lower panels: gravity included for drop-hanging.

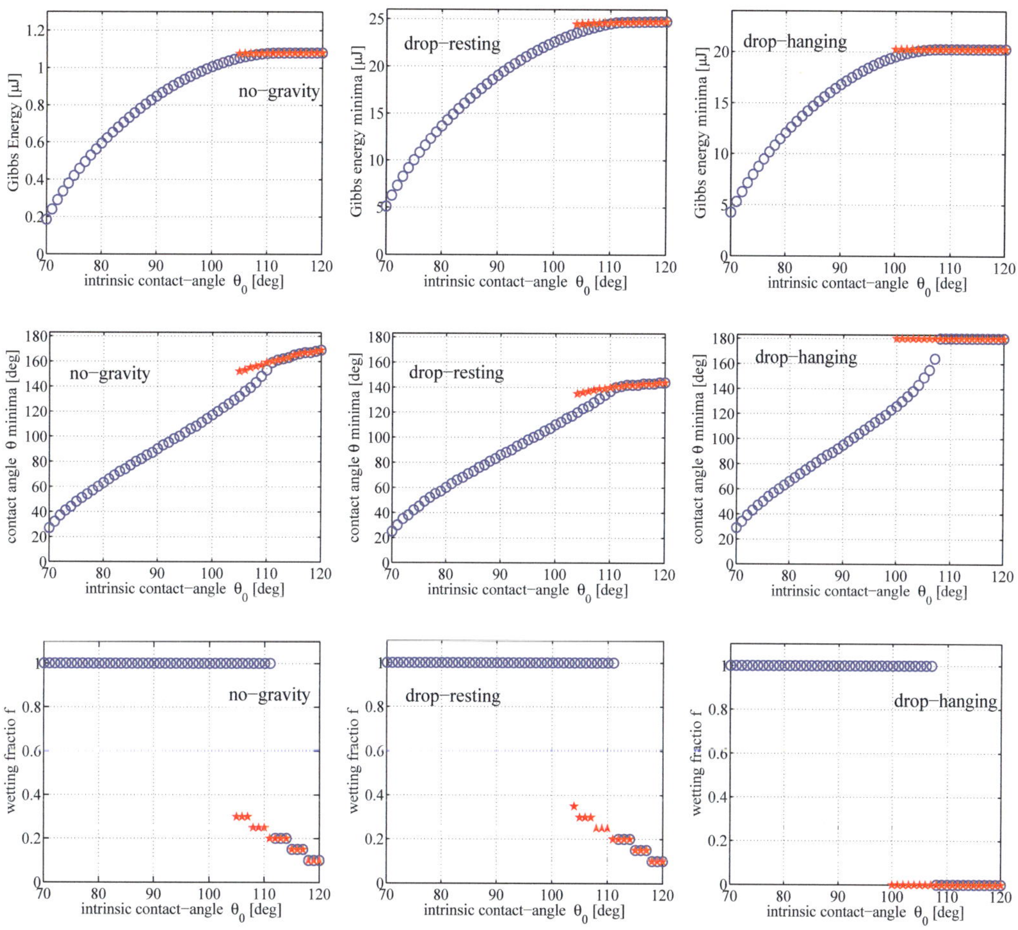

Fig. 21. *Colocasia esculenta* leaf: Effect of gravity on the values obtained for the paths of minimal energy as a function of the surface chemistry. Left-column: for no gravity. Middle-column: for drop resting. Right-column: for drop-hanging. Upper panels: Metastable and global Gibbs energy minima in dependence of the surface chemistry expressed by the intrinsic contact angle. Stars mark partially-wetted states ($f < 1$). Circles denotes the global minima. Stars surrounded by circles means that the partially-wetted states are global minima. Middle panels: The contact angles corresponding to the Gibbs energy minima. Lower panels: The wetting fraction for the Gibbs energy minima.

7. Conclusion and Outlook

We described an easy to handle Gibbs energy function to study the influence of surface chemistry and surface roughness on liquid wetting behaviour. An example how forces affect the wetting behavior was supplied by considering the gravitational field. The equations were applied on the surface morphology measured by atomic force microscopy from the evolution-optimized self-cleaning *Colocasia esculenta* plant leaf.
The input parameters for the equations are:

- surface chemistry, expressed by the Young contact angle (θ_o), i.e. the drop contact angle which the chemistry induces on a smooth surface;
- surface roughness, expressed by the roughness ratio r and the wetting fraction f. Both values are dependent on the sink-depth of the drop into the surface roughness.

The equations delivers:

1. energy of each drop status including metastable and global energy minima.
2. contact angle of metastable and global energy minima.
3. wetting fraction for non-wetted area of metastable and equilibrium drop shape status.

From that results, the following useful properties can be determined:
a.) wetting area of metastable and global energy minima:

$$A_{sl} = r \cdot f \cdot A_{base}^{drop} = \frac{r \cdot f \cdot (3V/\pi^2)^{2/3} \sin^2\theta}{(2 - 3\cos\theta + \cos^3\theta)^{2/3}} \tag{56}$$

where V is the drop volume.
b.) Energy barriers for wetting/dewetting and for transitions between energy minima. The energy barriers can be compared with drop fall experiments.
c.) Contact angle hysteresis, a key factor for drop roll-up (self-cleaning) ability, given by (Bhushan & Her, 2010):

$$\Delta\theta = \theta_{adv} - \theta_{rec} = f \cdot r \cdot \frac{\cos\theta_{r0} - \cos\theta_{a0}}{\sin\theta_{equil}} \tag{57}$$

where θ_{adv} and θ_{rec} are the advancing and receding angle for a rough surface, and θ_{a0} and θ_{r0} are the advancing and receding angle which the surface chemistry build-up on a smooth surface. θ_{equil} is the drop contact-angle on the rough surface, which corresponds to the status under study, e.g. metastable or equilibrium states.
d.) Adhesion tension $\tau^o = \gamma_{lv} \cdot \cos\theta$, a factor affecting biocompatibility (Parhi & Golas & Vogler, 2010; Vogler, 2001; 1998; 1992).
The generalization to hierarchical structures and various acting forces are straightforward.

8. References

Rijke A. M. & Jesser, W. A. (2010). The feather structure of dippers: Water repelency and resistance to water penetration. *The Wilson Journal of Orinithology*, Vol. 123, 563 – 568.

Mengnan Qu & Jinmei He & Junyan Zhang (2010). Superhydrophobicity, Learn from the Lotus Leaf, In *Biomimetics Learning From Nature*, Amitava Mukherjee (Ed.), ISBN: 978-953-307-025-4, INTECH, Available from: http://www.intechopen.com/articles/show/title/superhydrophobicity-learn-from-the-lotus-leaf

Kenneth Denbigh (1968). In: *The principles of chemical equilibrium*, Cambridge University Press, London, New-York.

Cool Th., Bartol A., Kasenga M, Modi K, & Garcia R. E. (2010). Gibbs: Phase equilibria and symbolic computation of thermodynamic properties. *Calphad: Computer Coupling of Phase Diagrams and Thermochemistry*, Vol. 34, 393 404, doi:10.1016/j.calphad.2010.07.005

Hüger E., Kana Th., & Sob M. (2010). A mechanism of inhibition of phase transitions in nano-grained close-packed Pd thin films. *Calphad: Computer Coupling of Phase Diagrams and Thermochemistry*, Vol. 34, 421 427, doi:10.1016/j.calphad.2010.07.009

Hüger E., Rothe H., Frant M., Grohmann S., Hildebrand G., & Liefeith, K. (2009). Atomic force microscopy and thermodynamics on taro, a self-cleaning plant leaf. *Applied Physics Letters*, Vol. 95, 033701-1 – 033702-3.

Barbieri L., Wagner E. & Hoffmann, P. (2007). Water wetting transition parameters of perfluorinated substrates with periodically distributed flat-top microscale obstacles. *Langmuir*, Vol. 23, 1723 – 1734.

Patankar N. A. (2004). Transition between superhydrophobic states on rough surfaces. *Langmuir*, Vol. 20, 7097 – 7102.

Patankar N. A. (2003). On the modeling of hydrophobic contact angles on rough surfaces. *Langmuir*, Vol. 19, 1249 – 1253.

Nosonovsky M. & Bhushan, B. (2005). Roughness optimization for biomimetic superhydrophobic surfaces. *Microsyst. Technol.*, Vol. 11, 535 – 549.

Nosonovsky M. & Bhushan, B. (2006). Stochastic model for metastable wetting of roughness-induced superhydrophobic surfaces. *Microsyst. Technol.*, Vol. 12, 231 – 237.

Ishino, C. & Okumura, K. (2006). Nucleation scenarios for wetting transition on textured surfaces: The effect of contact angle hysteresis. *Europhysics Letters*, Vol. 76, 464 – 470.

Bhushan B. & Her, E. K. (2010). Fabrication of Superhydrophobic Surfaces with High and low Adhesion Inspired from Rose Petal. *Langmuir*, Vol. 26, 8207 – 8217.

Chaudhury M. K. & Whitesides, G. M. (1992). Correlation between surface free energy and surface constitution. *Science*, Vol. 255, 1230 – 1232.

Hunter L. (2010). The C-F bond as a conformational tool in organic and biological chemistry. *Bellstein Journal of Organic Chemistry*, Vol. 6, 1 – 14.

Hüger E., Zeleny M, Kana T., Osuch K. & Sob, M. (2008). A peculiar bonding of sulphur at the Nb(001) surface. *Europhysics Letters*, Vol. 83, 26001-1 – 26001-6.

Marmur A. (2010). Solid-Surface Characterization by Wetting. *Annual Reviews of Material Research*, Vol. 39, 473 – 489.

Parhi P., Golas A. & Vogler Erwin A. (2010). Role of Proteins and Water in the Initial Attachment of Mammalian Cells to Biomedical Surfaces: A Review. *Journal of Adhesion Science and Technology*, Vol. 24, 853 – 888.

Vogler Erwin A. (2001). How Water Wets Biomaterial Surfaces. In: *Water in Biomaterials Surface Science*, M. Morra (Ed.), John Wiley and Sons, Ltd, New-York.

Vogler Erwin A. (1998). Structure and reactivity of water at biomaterial surfaces. *Advances in Colloid and Interfacial Science*, Vol. 74, 69 – 117.

Vogler Erwin A. (1992). Practical Use of Concentration-Dependent Contact Angles as a Measure of Solid-Liquid Adsorption. 2. Experimental Aspects. *Langmuir*, Vol. 8, 2013 – 2020.

11

Thermodynamics of Self-Assembly

L. Magnus Bergström
Royal Institute of Technology
Sweden

1. Introduction

Self-assembly of molecules to form much larger colloidal or nanosized aggregates has received a great deal of attention the last decades or so and the number of technical applications and products based on the principle of self-assembly is still rapidly increasing. Moreover, fundamental processes in life sciences such as the properties and stability of lipid membranes and their interactions with proteins, DNA etc are based on the phenomenon of self-assembly.

In this chapter we outline the fundamental principles of the thermodynamics of self-assembling molecules into self-associated colloidal aggregates and interfaces. Since the process of self-assembling molecules implies an increase of order in a system, this process must be unfavourable from an entropic point of view. Consequently, in the absence of any additional interactions or driving forces, molecules that interact favourable with solvent molecules do mix completely as one single phase of free molecules dispersed in a solvent. On the other hand, the presence of a driving force that is strong enough to overcome the unfavourable entropy may lead to a spontaneous self-assembly of molecules. In the present work, we mainly focus on the technically and biologically important case of self-assemblies formed by amphiphilic molecules, i.e. molecules composed of one hydrophilic and one hydrophobic part. The driving force for the self-association of amphiphilic molecules in an aqueous solvent is the hydrophobic effect, i.e. the principle preventing oil and water to mix in a liquid phase. As a result, under certain conditions amphiphilic molecules may spontaneously form a single disperse phase of thermodynamically stable self-associated interfaces, in addition to the options available for ordinary (non-amphiphilic) molecules, i.e. either to completely mix as free solute molecules with the solvent or to separate into two or more liquid phases.

Below it is demonstrated that, by means of combining thermodynamics of self-assembly with the bending elasticity approach of calculating the free energy of a single self-associated interface, it is possible to rationalize the geometrical size and shape of self-assembled interfaces in terms of the chemical architecture of the constituting amphiphilic molecules (size of hydrophilic and hydrophobic part, respectively, charge number, flexibility etc) as well as solution properties (salinity, temperature etc).

2. Amphiphilic molecules

Amphiphilic molecules consist of a hydrophilic head group and a hydrophobic tail and may roughly be subdived into different types: (*i*) Surfactants usually have a moderately sized tail and do form rather small globular or long wormlike or threadlike self-assembled interfaces, called micelles, in an aqueous solvent. (*ii*) Lipids have a much larger hydrophobic part as compared to the head group and form comparatively large bilayer structures. As a common notation in the present work, we refer to micelles and bilayers as self-assembled interfaces. The head groups may be either charged (ionic surfactants) or uncharged (nonionic surfactants). Many lipids are zwitter-ionic, i.e. they have a head group that consists of both positive and negative charges whereas the whole molecule lacks a net charge. The driving force needed for amphiphilic molecules to spontaneously self-assemble into micelles and bilayers is usually denoted the hydrophobic effect, i.e. the principle that hydrocarbon (oil) and water do not mix. As a result, in order to reduce the hydrocarbon-water contact in an aqueous solution the hydrophobic tails self-assemble to form liquid-like cores with the head groups located at the aggregate interface pointing towards the aqueous bulk solvent phase. Any quantitative molecular theory that accurately describes the hydrophobic effect is unfortunately not available, but it is generally believed that water molecules in the proximity of hydrophobic surfaces form entropically unfavourable ordered structures (clathrates). (Evans & Wennerström, 1994) Nevertheless, rather accurate semi-empirical expressions have been generated for the hydrophobic free energy of bringing hydrocarbon chains from an aqueous solution to the corresponding hydrocarbon liquid bulk phase using solubility data for hydrocarbons in water. (Tanford, 1980)

The hydropobic effect, together with some partly counteracting contributions to micelle formation, may be incorporated into a single free energy parameter $\Delta\mu_{mic}$ that may be interpreted as the free energy per aggregated molecule of forming a micelle out of free surfactant molecules [see Section 4 below]. The most important contributions to $\Delta\mu_{mic}$, in addition to the hydrophobic effect, are chain conformational entropy (Ben-Shaul et al., 1985; Gruen, 1985), electrostatics and excluded volume repulsive interactions among the surfactant head groups. The hydrophobic effect and chain conformational entropy are related to the surfactant tails whereas electrostatics and excluded volume repulsion are related to the surfactant head groups.

The electrostatic free energy contribution is essential for surfactant aggregates consisting of at least one ionic surfactant. It appears to be mainly a result of the unfavourable entropic effect of concentrating counter-ions in a diffuse layer outside charged aggregate interfaces. An accurate explicit expression of the electrostatic free energy have been derived, as an expansion to second order in curvature of the charged surface, from the Poisson-Boltzmann theory. (Mitchell & Ninham, 1989)

An additional repulsive free energy contribution arises as solvent molecules penetrate and mix with the hydrophilic head groups. The resulting excluded volume repulsion among the surfactant head groups increases in magnitude with head group volume. For nonionic surfactants this is the main free energy contribution related to the head groups, counteracting free energy contributions related to the tails and making the molecule amphiphilic in nature. (Bergström, 2009).

3. Entropy of self-assembly

The ideal entropy of mixing N free amphiphilic molecules with N_w solvent molecules is given by the expression

$$S_{free} = -k\left(N_w \ln\phi_w + N\ln\phi_{free}\right) \tag{1}$$

where ϕ_w and ϕ_{free} are the volume fractions of solvent and solute, respectively, and k is Boltzmann's constant. Eq. (1) is a general expression in the sense that it allows for the mixing of solute and solvent with different molecular volumes. (Guggenheim, 1952; Flory, 1974) For the special case of equal molecular volumes, volume fractions may be replaced with mole fractions in Eq. (1).

Now, consider the process of self-assembling N free amphiphilic molecules to form a single aggregate or self-assembled interface. Since the self-assembling process is associated with an ordering of molecules, there is a change in entropy as a result of this process. The entropy of mixing one single aggregate with solvent molecules is given by

$$S_{agg} = -k\left(N_w \ln\phi_w + \ln\phi_N\right) \tag{2}$$

Hence, the entropy change per aggregate formed in the process of self-assembly may be obtained by subtracting Eq. (1) from Eq. (2), i.e.

$$\Delta S_{agg} \equiv S_{agg} - S_{free} = -k\left(\ln\phi_N - N\ln\phi_{free}\right) \tag{3}$$

Usually $\ln\phi_N$ and $\ln\phi_{free}$ are about the same order of magnitude [see further below], and Eq. (3) may be approximately simplified so as to give

$$\Delta S_{agg} \approx kN\ln\phi_{free} \tag{4}$$

for the case $N >> 1$.

We note that $\Delta S_{agg} < 0$ and, as expected, the increase in order due to the process of self-association implies a reduction in overall entropy. This means that ordinary molecules that interacts more or less favourable with solvent molecules will not spontaneously self-assemble, but prefer to remain as free molecules in a close to ideal mixture. Moreover, it becomes clear from Eq. (4) that the entropic driving force of dissociating molecules increases with decreasing solute concentration ϕ_{free}. As will be further discussed below, this entropic contribution is utmost important as to determine the equilibrium size of self-assembled interfaces.

4. Multiple equilibrium

From a thermodynamic point of view, a spontaneous self-assembly process may be written as a set of reactions, one for each value of N, i.e.

$$N \text{ molecules} \rightarrow A_N \tag{5}$$

The total free energy of the processes can then be written as a sum of the "interaction" free energy of forming a micelle (= $N\Delta\mu_{mic}$) and the unfavourable free energy of self-assembling amphiphilic molecules ($\Delta G_{agg} = -T\Delta S_{agg} > 0$, where T is the absolute temperature), giving the following multiple set of equilibrium conditions

$$\Delta G_{tot} = N\Delta\mu_{mic} + \Delta G_{agg} = 0 \tag{6}$$

It follows from Eq. (6) that $\Delta\mu_{mic} < 0$ must be a necessary condition for the process to overcome the unfavourable entropy of self-association ($\Delta S_{agg} < 0$). As previously mentioned,

it is mainly the hydrophobic effect that accounts for negative values of $\Delta\mu_{mic}$, the remaining contributions discussed above having slightly counteracting positive contributions to $\Delta\mu_{mic}$. (Eriksson et al., 1985)
From Eq. (3) it now follows that

$$\Delta G_{agg} = kT\left(\ln\phi_N - N\ln\phi_{free}\right) \tag{7}$$

and by combining Eqs. (6) and (7) we may write the set of equilibrium conditions in the following way

$$\Delta G_{tot} = N\left(\Delta\mu_{mic} - kT\ln\phi_{free}\right) + kT\ln\phi_N = 0 \tag{8}$$

It will become clear throughout this chapter that Eq. (8) is the central thermodynamic relationship when the spontaneous and reversible process of self-assembling amphiphilic molecules is due to be rationalized.

5. Critical micelle concentration

In order to achieve a full equilibrium in accordance with Eq. (8), the surfactant molecules exchange more or less rapidly between the self-assemblies and as free molecules in the surrounding bulk solution. The latter may be considered as a large reservoir where environmental variables defining the solution state, e. g. temperature, pressure, chemical potentials of the free surfactant molecules etc, are kept constant during the reversible process of self-assembly. (Eriksson et al., 1985; Israelachvili, 1991) One important consequence of this thermodynamic equilibrium situation, and the multiple equilibrium approach inherent in Eq. (8), is that the entropic dissociation driving force in Eq. (4) is too large at low surfactant concentrations for self-assembled interfaces with comparatively large N to be present. As a result, self-assembled interfaces will only form above a certain concentration of surfactant, and above this concentration limit both free surfactants and self-assembled interfaces must coexist simultaneously in a dispersed surfactant solution.
The volume fraction of self-assembled interfaces (ϕ_N) and free surfactant (ϕ_{free}), respectively, are plotted against the overall surfactant volume fraction (= $\phi_{free} + \phi_N$) in Fig. 1. It is clearly seen that the aggregate concentration is virtually zero below a certain value of the total surfactant concentration, whereas above this value the aggregate concentration increases proportionally to the surfactant concentration ($\phi_{free} + \phi_N$), with ϕ_{free} approximately constant. This threshold value is called the critical micelle concentration (CMC).
The appearance of the curves in Fig. 1 may be rationalized as follows: at low surfactant concentration (i.e. below CMC), ϕ_{free} is small and $-kT\ln\phi_{free} \gg -\Delta\mu_{mic}$ (note that both quantities on either side of the inequality sign are positive quantities). As a result, the term including $\Delta\mu_{mic}$ in Eq. (8) becomes negligible and, as a consequence, $\phi_N = \phi_{free}^N \ll \phi_{free}$. As a result, the concentration of higher order aggregates (dimers, trimers etc.) rapidly becomes negligible as N increases beyond unity. As the surfactant concentration increases, the quantity of $-kT\ln\phi_{free}$ will eventually be of the same order of magnitude as $-\Delta\mu_{mic}$. However, we may neglect the term including ϕ_N, since $-NkT\ln\phi_{free} \gg -kT\ln\phi_N$ in so far $N >> 1$. As a result, Eq. (8) may be simplified so as to give

$$\Delta\mu_{mic} \approx kT \ln \phi_{free} \tag{9}$$

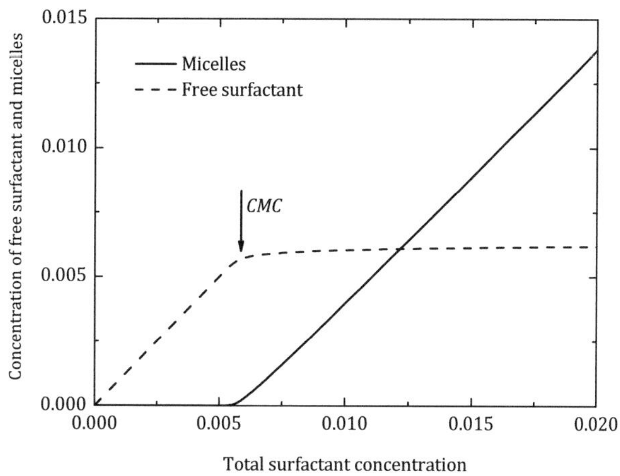

Fig. 1. The volume fraction of micelles (ϕ_N, solid line) and free surfactant (ϕ_{free}, dashed line) plotted against the overall surfactant volume fraction ($\phi_{free} + \phi_N$). The free energy was set equal to $\Delta\mu_{mic} = -5kT$ giving a $CMC \approx 0.0055$ (in volume fraction units)

Since ϕ_{free} is approximately constant above *CMC*, we obtain the following well-known and important relation between *CMC* (which is straightforward to measure) and the micelle "interaction" free energy

$$\Delta\mu_{mic} = kT \ln CMC \tag{10}$$

Eq. (10) is frequently claimed to follow from the so called "pseudo phase separation-model". However, from our present treatment it is perfectly clear that Eq. (10) follows straightforwardly from conventional solution thermodynamics as an approximation that becomes increasingly valid in the limit of large aggregation numbers ($N >> 1$). The smallest micelles observed to be formed by conventional surfactants have aggregation numbers about $N = 50$, which appears to be a sufficiently large number for Eq. (10) to be, for all practical purposes, accurate.

5.1 Dependence of CMC on molecular architecture and solution properties

CMC depends strongly on the molecular architecture of amphiphilic molecules. In particular, CMC decreases considerably with increasing size of the hydrophobic tail. For surfactants with tails consisting of aliphatic hydrocarbon alkyl chains C_nH_{2n+1}, the logarithm of CMC has been found to follow a linear relationship (sometimes referred to as the Shinoda equation), i.e.

$$\ln CMC = A - Bn \tag{11}$$

where *A* and *B* are two constants. For ionic surfactants, the slope has been experimentally found to equal $B \approx 0.7$, whereas $B \approx 1.1$ for nonionic surfactants and $B \approx 1.5$ for alkanes (corresponding to solubility rather than CMC, since alkanes are not amphiphilic). (Shinoda et al., 1963; Israelachvili, 1991; Jönsson et al., 1998) The latter value corresponds to what is expected from the empirical contribution to $\Delta\mu_{mic}$ due to the hydrophobic effect for aliphatic hydrocarbon chains ($\Delta\mu_{mic}^{hb} = -2.0 - 1.5n$). (Tanford, 1980) The reduction of B for ionic and nonionic surfactants may be rationalized as a result of the contributions due to electrostatics and excluded head group volume repulsion to $\Delta\mu_{mic}$. In particular, the electrostatic free energy increases significantly as the electrolyte concentration of free surfactant is reduced with the lowering of CMC. As a consequence, the lowering of $\Delta\mu_{mic}$ with n becomes strongly counteracted and B significantly reduced. Upon adding additional salt to an ionic surfactant mixture, B increases so as to approach the value observed for nonionic surfactants.

The reduction of electrostatic free energy with increasing electrolyte concentration also causes a strong CMC dependence of ionic surfactants on the concentration of added salt. It appears that, at moderate and high salt concentrations (c_{salt}), CMC approximately follows the linear relation (Shinoda et al., 1963)

$$\ln CMC = constant - B' \ln c_{salt} \tag{12}$$

where the constant $B' = 0.5\text{-}1.0$, depending mainly on the size of the surfactant aggregates. The approximate mathematical form of Eq. (12), as well at the value of B', may be accurately reproduced by calculations of the electrostatic free energy from the Poisson-Boltzmann theory.

The CMC of nonionic surfactants display a slight dependence on head group volume (CMC increases with increasing size of the head group) as a result of excluded volume interactions. More significant, however, appears to be the temperature dependence of CMC of nonionic surfactants consisting of an ethylene oxide based head group. The ethylene oxide group become significantly more hydrophobic upon raising the temperature and, as a result, $\Delta\mu_{mic}$ and CMC become significantly lowered upon increasing T. (Jönsson et al., 1998)

5.2 CMC in surfactant mixtures

Two or more surfactants may readily mix with each other and form mixed self-assembled interfaces. The CMC in a mixture of two surfactants (denoted Surfactant 1 and Surfactant 2, respectively) is given by the following relation

$$CMC = a_1 CMC_1 + a_2 CMC_2 = \gamma_1 x CMC_1 + \gamma_2 (1 - x) CMC_2 \tag{13}$$

where x is the mole fraction of Surfactant 1 *in the self-assembled interfaces*, and CMC_1 and CMC_2 are the CMCs of pure Surfactant 1 and 2, respectively. $a_1 \equiv x\gamma_1$ and $a_2 \equiv (1 - x)\gamma_2$ denote the activities of Surfactant 1 and 2, respectively, where γ_1 and γ_2 are the corresponding activity coefficients. In a surfactant mixture close to CMC it is crucial to distinguish between the mole fraction in the self-assembled interfaces (x) and the mole fraction based on the total surfactant concentration (y), including surfactant present as free monomers. It is straightforward to demonstrate that, in the limit of zero aggregate concentration, CMC as a function of y may be written as (Clint, 1975; Bergström & Eriksson, 2000)

$$\frac{1}{CMC} = \frac{y}{\gamma_1 CMC_1} + \frac{1-y}{\gamma_2 CMC_2} \tag{14}$$

The activity coefficients may be calculated from the dependence of $\Delta\mu_{mic}$ on aggregate composition employing the following expressions (Bergström & Eriksson, 2000)

$$\gamma_1(x) = \exp\left[\left(\mu_{ex}(x) - \mu_{ex}(x=1) + (1-x)\frac{d\mu_{ex}}{dx}(x'=x)\right)/kT\right] \tag{15}$$

$$\gamma_2(x) = \exp\left[\left(\mu_{ex}(x) - \mu_{ex}(x=0) - x\frac{d\mu_{ex}}{dx'}(x'=x)\right)/kT\right] \tag{16}$$

where the excess free energy per aggregated molecule is defined as

$$\mu_{ex} \equiv \Delta\mu_{mic} - kT\left[x\ln x + (1-x)\ln(1-x)\right] \tag{17}$$

Note that x in Eqs. (15) and (16) refers to the equilibrium mole fraction which is obtained by means of minimizing the molecular free energy with respect to the aggregate mole fraction x'. It follows from Eqs. (15) and (16) that a linear expression of $\mu_{ex}(x)$ gives $\gamma_1 = \gamma_2 = 1$ and, as a result, a linear (ideal) behaviour of *CMC* as a function of x.

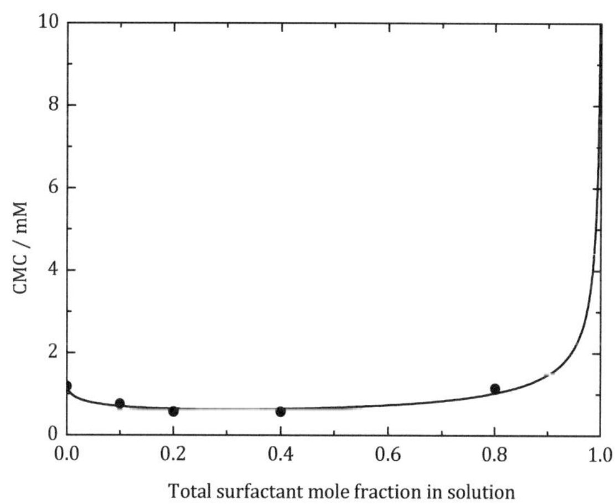

Fig. 2. The critical micelle concentration (*CMC*) plotted against the mole fraction of cationic surfactant in solution (y). The symbols represent experimentally measured values for the system sodium dodecyl sulphate (anionic surfactant)/tetracaine (cationic surfactant) in [NaCl] = 0.154 M, whereas the solid line represents calculations using the Poisson-Boltzmann theory. (Bergström & Bramer, 2008)

An ideal behaviour has usually been observed in mixtures of two rather similar surfactants, e.g. mixtures of two nonionic surfactants. On the other hand, synergistic effects, i.e. negative deviations of CMC from ideal behaviour, is frequently observed in different surfactant mixtures including ionic surfactants, for instance mixtures of an ionic and a nonionic surfactant, two oppositely charged surfactants or two surfactants with identical charge number but different tail lengths. Calculations of γ_1 and γ_2 using the Poisson Boltzmann theory usually gives accurate agreement with experiments for these cases [*cf.* Fig. 2]. (Bergström, 2001a; Bergström et al., 2003; Bergström & Bramer, 2008)

The magnitude of synergistic or antagonistic (= positive deviation from ideal behaviour) effects may be appropriately quantified with the model-independent thermodynamic synergy parameter defined as

$$\beta_{syn} \equiv 4\ln(a_1 + a_2) = 4\ln(x\gamma_1 + (1-x)\gamma_2) \tag{18}$$

$\beta_{syn} < 0$ corresponds to a negative deviation (synergism), whereas $\beta_{syn} > 0$ implies a positive deviation (antagonism) from ideal behaviour ($\beta_{syn} = 0$). β_{syn} may be calculated from experimentally obtained *CMC* values using the relations $a_1 \equiv x\gamma_1 = y\, CMC/CMC_1$ and $a_2 \equiv (1 - x)\gamma_2 = (1 - y)\, CMC/CMC_2$ for the activities. (Bergström et al., 2003)

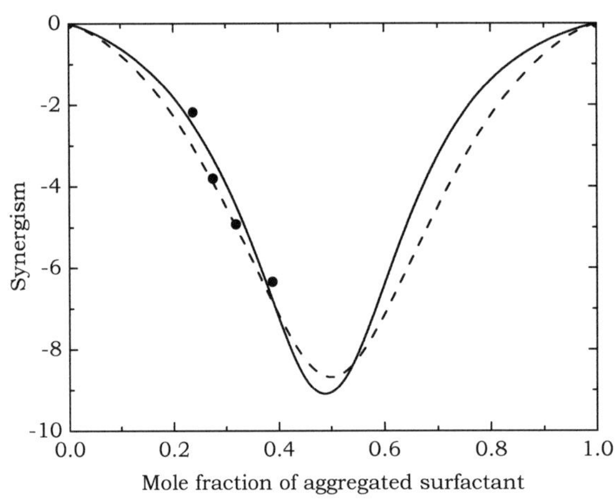

Fig. 3. The synergy parameter β_{syn} plotted against the mole fraction of self-assembled cationic surfactant (x). The symbols represent experimentally measured values for the system sodium dodecyl sulphate (anionic surfactant)/tetracaine (cationic surfactant) in [NaCl] = 0.154 M whereas the solid line is calculated from the Poisson-Boltzmann theory. The minimum is located at $x = 0.488$ where $\beta_{syn}^{min} = -9.1$. The dashed line represents the best available fit to the data obtained with the regular mixture theory giving $\beta_{syn}^{min} = \beta_{int} = -8.7$. (Bergström & Bramer, 2008)

It is common to employ the theory of regular mixtures in the evaluation of CMC data for mixed surfactant systems, and quantify synergistic effects with the model-dependent interaction parameter β_{int}. (Holland, 1992) The theory of regular mixtures is based on the difference in 1-1, 2-2 and 1-2 pair-wise molecular interactions, and the corresponding excess free energy per aggregated molecule equals

$$\frac{\mu_{ex}}{kT} = \beta_{int} x(1-x) \tag{19}$$

Inserting Eq. (19) in Eqs. (15) and (16) gives $\gamma_1 = e^{(1-x)^2 \beta_{int}}$ and $\gamma_2 = e^{x^2 \beta_{int}}$. Hence, $\beta_{syn} = 4\ln\left(xe^{(1-x)^2 \beta_{int}} + (1-x)e^{x^2 \beta_{int}}\right)$, according to the regular mixture theory, always goes through a minimum at $\beta_{syn}(x = 0.5) = \beta_{int}$ (or maximum if $\beta_{int} > 0$) [cf. Fig. 3]. As a matter of fact, the factor 4 in Eq. (18) is introduced with the purpose of achieving $\beta_{syn}^{min} = \beta_{int}$ in the particular case of the regular mixture theory. However, experimentally obtained β_{syn} vs. x usually display a minimum not located at $x = 0.5$, most evident in the cases of ionic/nonionic surfactant mixtures and ionic/ionic surfactant mixtures with equal charge numbers. (Bergström et al., 2003) An exception is mixtures of two oppositely charged surfactants with equal magnitude of charge number, for which both Poisson-Boltzmann theory and regular mixture theory predicts β_{syn} to have a minimum close to (or exactly for latter case) $x = 0.5$ [cf. Fig. 3].

According to the regular mixture interpretation of deviations from ideal behaviour, favourable 1-2 interactions (as compared to 1-1 and 2-2 interactions) would cause synergism whereas unfavourable 1-2 interactions would yield antagonism. This interpretation is in sharp contrast to the thermodynamic interpretation based on Poisson-Boltzmann mean field calculations. According to the latter, synergism arises with necessity as long as two surfactants mix readily in the self-assembled interfaces and there is a pronounced asymmetry between the two surfactants. For the particular case of monovalent and oppositely charged surfactants, the asymmetry (difference in charge number = 1 – (–1) = 2) results in a release of a pair of counter-ions for every pair of surfactants that is aggregated. As a result, the chemical potential is considerably reduced and so is CMC. On the other hand, antagonism appears as a result of the surfactants not mixing properly with each other. For instance, in the extreme case where two surfactants do not mix at all and form separate one-component self-assembled interfaces, $a_1 = a_2 = 1$ which implies that $\beta_{syn} = 4\ln 2 \approx 2.77 > 0$.

6. Size distribution of self-assembled interfaces

It appears to be convenient to introduce the free energy function (Eriksson et al., 1985)

$$\mathrm{E}_N \equiv N\left(\Delta\mu_{mic} - kT\ln\phi_{free}\right) \tag{20}$$

which allows us to rewrite the set of equilibrium conditions in Eq. (8) in the following way

$$\Delta G_{tot} = \mathrm{E}_N + kT\ln\phi_N = 0 \tag{21}$$

E_N may be interpreted as the free energy of forming a self-assembled interface out of N free surfactant molecules in the bulk solution. As previously discussed, the last term in Eq. (20), including ϕ_{free}, is virtually constant above CMC. Solving ϕ_N in Eq. (21) gives the following size distribution function of self-assembled interfaces

$$\phi_N = e^{-\mathrm{E}_N/kT} \tag{22}$$

Now it is straightforward to obtain a relation between the total volume fraction of self-assembled interfaces or micelles (ϕ_{mic}) and E_N by means of summing up ϕ_N over all N, i.e.

$$\phi_{mic} = \sum_{N=1}^{\infty} \phi_N \int_1^{\infty} e^{-E_N/kT} dN \tag{23}$$

where an integral approximation have been employed in the second equality. (Israelachvili et al., 1976; Eriksson et al., 1985) From Eq. (23) it is, in principle, possible to obtain the full size distribution for an arbitrarily shaped self-assembled interface, once the mathematical form of E_N as a function of aggregate size and shape is known.

7. Bending elasticity and curvature energy

In recent years it has become widely accepted that bending properties play a decisive role as to rationalize the behaviour of self-assembled interfaces. According to the mathematical discipline of differential geometry, the curvature at a single point on a self-assembled interface (most conveniently defined at the hydrocarbon/water interface) may be defined by considering two perpendicular curves on the interface with radii of curvature, R_1 and R_2, respectively. As a result, each point at the aggregate interface may be distinguished by two principal curvatures, $c_1 \equiv 1/R_1$ and $c_2 \equiv 1/R_2$. It is postulated that the free energy per unit area γ depends uniquely on c_1 and c_2 for a given system of surfactant and solvent at some environmental condition. Moreover, by switching variables from c_1 and c_2 to the mean curvature $H \equiv ½(c_1 + c_2)$ and Gaussian curvature $K \equiv c_1c_2$, respectively, we may introduce the function $\gamma(H, K)$. (Porte, 1992; Hyde et al., 1997)

A quantitative description of the bending properties of self-assembled interfaces is obtained as γ is expanded to second order with respect to H and K, i.e.

$$\gamma(H,K) = \gamma_0 + 2k_c(H - H_0)^2 + \bar{k}_c K \tag{24}$$

commonly referred as to the Helfrich expression. (Helfrich, 1973) The total free energy of a self-assembled interface, corresponding to the E_N-function defined in Eq. (20), is given as an integral of γ over the entire interfacial area A of the self-assembled interface, i.e.

$$E_N = \int \gamma(H,K)dA = \gamma_0 A + 2k_c \int (H - H_0)^2 dA + \bar{k}_c \int K dA \tag{25}$$

The first term on the right-hand side of Eq. (25) ($= \gamma_0 A$) represents the free energy of stretching the self-assembled interface. The second and third terms take into account effects due to the dependence of free energy on the local curvature of the self-assembled interface, usually referred as to the bending free energy.

The Helfrich expression introduces three important parameters related to different aspects of bending a surfactant monolayer interface, i.e. the bending rigidity (k_c), the spontaneous curvature (H_0) and the saddle splay constant ($\bar{k}_c$). As a common notation, we will refer to all of them below as bending elasticity constants. For thermodynamically stable objects, such as surfactant micelles and bilayers, the three bending elasticity constants may be interpreted as thermodynamic parameters, and they may be calculated from any suitable molecular model by means of minimizing the free energy per molecule of a surfactant interfacial layer at any given H and K. (Bergström, 2006a; Bergström, 2006b; Bergström, 2008a).

7.1 Bending rigidity

The bending rigidity k_c quantifies the resistance against deviations from a uniform mean curvature $H = H_0$. Positive values of k_c secures $\gamma(H, K)$ to display a minimum with respect to H, whereas negative k_c -values corresponds to γ having a maximum. As a consequence, the sign of k_c determines the stability of an amphiphilic self-assembled interface, i.e. stable interfaces may only exist for positive values of k_c. Moreover, high k_c -values are expected to favour self-assembled interfaces with small deviations from a homogenous curvature or geometry, i.e. rigid and monodisperse objects with a uniform shape [see further below]. (Porte, 1992; Bergström, 2007; Bergström, 2008a)

From a molecular point of view, k_c appears to be a measure of the amphiphilic nature of a surfactant molecule, and it displays a clear maximum when plotted against the ratio between the hydrophilic and hydrophobic (= lipophilic) parts, respectively, of an amphiphilic molecule. This maximum corresponds to the optimal hydrophilic-lipophilic balance (HLB) of a surfactant molecule [*cf.* Fig. 4]. (Bergström, 2009) For instance, attaching a point charge to the end of a C_{12} aliphatic hydrocarbon chain significantly rises k_c to positive values as the molecule become amphiphilic in nature. (Bergström, 2006a) On the other hand, if either the hydrophilic or the hydrophobic part of the molecule becomes too large or small compared to the other, k_c may become negative which means that the amphiphilic nature of the molecule is too weak for stable surfactant self-assembled interfaces to form.

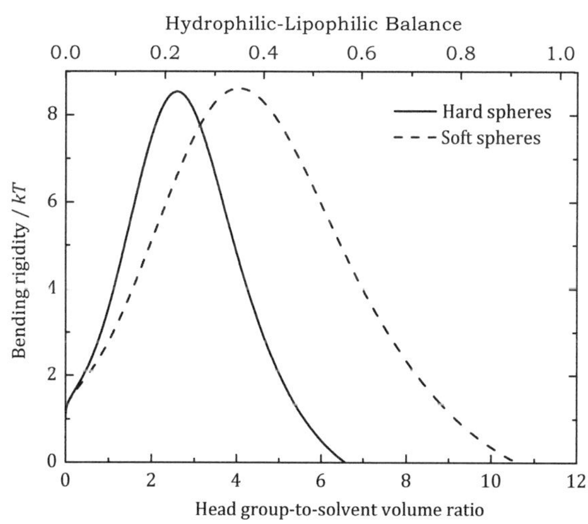

Fig. 4. Bending rigidity (k_c) plotted against the head group-to-solvent volume ratio and the hydrophilic-lipophilic balance (HLB) for a self-assembled interface formed by a single-chain nonionic surfactant with the head group treated as a hard sphere (solid line) and a soft sphere (dashed line), respectively. (Bergström, 2009)

The flexibility of surfactant tails appears to be crucial for the self-assembly process of nonionic surfactants. In accordance, it may be demonstrated that k_c must always equal zero, thus preventing self-assembly, for nonionic surfactants made up of a hydrophobic and a hydrophilic part that are both rigid. (Bergström, 2009) Ionic surfactants with a rigid hydrophobic part may have $k_c > 0$, but are predicted to behave substantially different than conventional ionic surfactants. With comparatively small k_c and high k_cH_0, they appear to be

much smaller and more polydisperse than micelles formed by an ionic surfactant with flexible tail. (Bergström, 2006a)
Moreover, in expressions derived for k_c as a function of composition in a surfactant mixture, an explicit contribution appears that is due to the mixing of surfactants in the self-assemblies *per se*. (Kozlov & Helfrich, 1992; Porte & Ligoure, 1995; Safran, 1999; Bergström, 2006b) This contribution always brings down k_c, and the magnitude of this reduction increases with increasing asymmetry between two surfactants with respect to head group charge number, volume of surfactant tail etc. The origin of this mixing dependence of k_c is as follows: the surfactant composition, as obtained by minimizing the molecular free energy at given H and K, becomes a smooth function of aggregate curvature. As a consequence, different geometrical parts of a self-assembled interface may have different surfactant compositions so as to enrich the surfactant (in an aqueous solvent) with the highest HLB in the most curved parts of the interface. As a result, the free energy penalty of distorting the aggregate curvature from its optimal value (= H_0) becomes significantly reduced as k_c decreases in magnitude upon mixing. Corresponding explicit dependences on mixing are absent for the other two bending elasticity constants, i.e. k_cH_0 and $\bar{k}_c$.

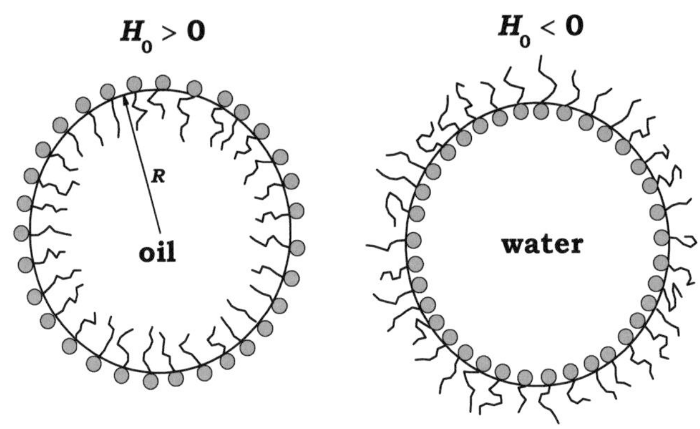

Fig. 5. Schematic illustration of a spherical oil-in-water and water-in-oil microemulsion droplet, respectively, with radius R. (Bergström, 2008a)

7.2 Spontaneous curvature

The spontaneous curvature H_0 represents the sign and magnitude of the preferential curvature of a single surfactant layer. H_0 is usually defined to be positive for a film that appears convexly curved from a position in the hydrophilic phase, like ordinary surfactant micelles in water or oil-in-water microemulsion droplets [*cf.* Fig. 5]. As a matter of fact, it appears that the product k_cH_0 is more readily to interpret from a physical and molecular point of view than H_0 itself. k_cH_0 depends on the architecture of an amphiphilic molecule in such a way that it increases with increasing hydrophilic-lipophilic balance [*cf.* Fig 6]. (Bergström, 2006a; Bergström, 2009) As a result, k_cH_0 rapidly decreases with increasing surfactant tail length whereas ionic surfactants usually have larger k_cH_0 than nonionic surfactants. Consequently, as an additional oil phase of solvent is present, surfactants with a high HLB tends to form oil-in-water microemulsion droplets whereas surfactants with low HLB values tend to form water-in-oil droplets [*cf.* Fig. 5].

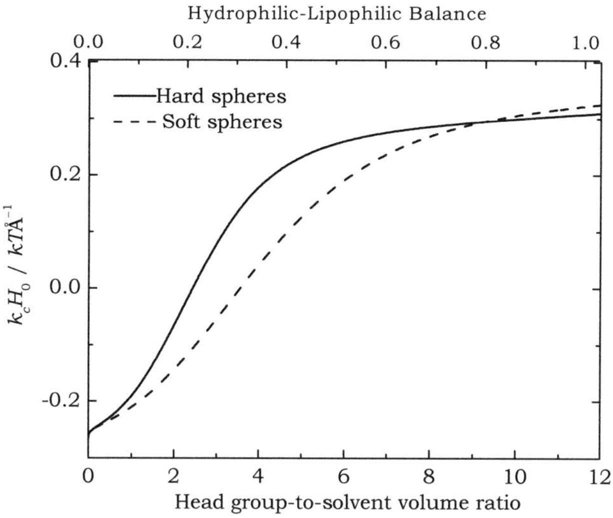

Fig. 6. Bending rigidity times the spontaneous curvature (k_cH_0) plotted against the head group-to-solvent volume ratio and the hydrophilic-lipophilic balance (HLB) for a self-assembled interface formed by a single-chain nonionic surfactant with the head group treated as a hard sphere (solid line) and a soft sphere (dashed line), respectively. (Bergström, 2009)

7.3 Saddle-splay constant

All interfacial shapes that may be generated from one another by twisting and stretching, without breaking the interface, belong to the same topology. (Hyde et al., 1997) Different topologies are characterized with the quantity genus (g), which represents the number of handles or holes present in the self-associated interface. For instance, ordinary surfactant micelles may be spherical, cylindrical or tablet-shaped (see further below) but belong to the same topology since the self-assembled interfacial layer is geometrically closed ($g = 0$).
According to the so called Gauss-Bonnet theorem,

$$\int K dA = 4\pi(1-g) \tag{26}$$

As a result, the last term in Eq. (25), for a geometrically closed interface ($g = 0$), equals $4\pi\bar{k}_c$, the quantity of which does not depend on the size of the self-assembled interface. Since $\bar{k}_c$ contributes with a size-independent term to E_N, the value of $\bar{k}_c$ indirectly determines the size of a dispersed set of geometrically closed self-assembled interfaces. Positive values of $\bar{k}_c$ means that the total free energy in the system increases as one single self-assembled interface is split up to form two or many smaller interfaces. As a result, increasing values of $\bar{k}_c$ favours large aggregation numbers. Calculations based on detailed molecular models indicate that $\bar{k}_c$ usually assumes negative values, favouring the formation of small self-assembled interfaces. (Bergström, 2006a; Bergström, 2009).
It has been demonstrated that the remaining parameter in Eqs. (24) and (25), the interfacial tension γ_0, is directly determined for a given concentration of micelles ϕ_{mic}. (Israelachvili et

al., 1976; Bergström, 2006c) As a consequence, it follows that the overall average size in a dispersion of self-assembled interfaces is determined by, in addition to the three bending elasticity constants, the total surfactant concentration ($\phi_{tot} = \phi_{mic} + \phi_{free}$). This dependence is a result of the entropy of self-assembly which, according to Eqs. (3) and (4), effectively works as a driving force tending to reduce the size of self-assembled interfaces. This effective driving force increases in magnitude with decreasing surfactant concentration and, as a consequence, surfactant micelles are occasionally find to increase in size with increasing surfactant concentration [see further below].

8. Spherical micelles and microemulsion droplets

The simplest geometry of a self-associated surfactant monolayer interface is the geometrically closed spherical shape. This is the only geometrical shape for which the curvature is equal on every single point on the self-assembled interface. The free energy as a function of the radial distance to the hydrophobic-hydrophilic interface (R), in accordance with Eq. (25), is simply obtained by inserting $c_1 = c_2 = 1/R$ in the Helfrich expression in Eq. (24) and multiplying γ with the surface area $A = 4\pi R^2$. (Safran, 1991; Bergström, 2008a).

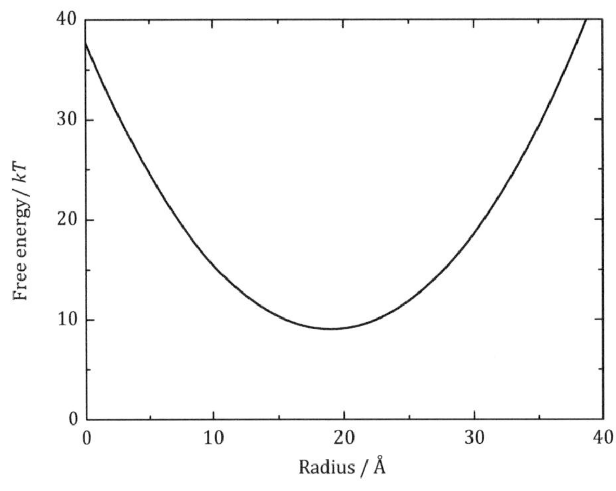

Fig. 7. The free energy E of spherical micelles as a function of radius R for the case of bending elasticity constants set to $H_0 = 0.03$ Å^{-1}, $k_c = 2kT$ and $\bar{k}_c = -kT$

As a result,

$$E(R) = 4\pi\left(2k_c + \bar{k}_c - 4k_c H_0 R + \gamma_p R^2\right) \tag{27}$$

where $\gamma_p \equiv \gamma(H = K = 0) = \gamma_0 + 2k_c H_0^2$ is the interfacial tension of a strictly planar surfactant layer. Hence, E may be considered as the sum of two contributions: (*i*) the stretching work of forming a planar interface [$= 4\pi R^2 \gamma_p$] and (*ii*) the work of bending the interface at constant interfacial area [$= 4\pi\left(2k_c + \bar{k}_c - 4k_c H_0 R\right)$].
Eq. (27) is plotted in Fig. 7 for a set of typical values of the bending elasticity constants. It follows from Eq. (27) that the free energy of a spherical micelle is proportional to the

aggregation number N in the limit $N \to \infty$, consistent with the thermodynamics of large macroscopic systems. In our treatment, however, it has become clear that non-extensive contributions to E(R) appear as a result of curvature effects and, as a consequence, a sharp minimum of E(R) is observed in Fig. 7, the location of which is mainly determined by the spontaneous curvature. The appearance of significant non-extensive contributions to the free energy is the basis of the discipline called thermodynamics of small systems. (Hill, 1963-64)

8.1 Size distribution of spherical micelles and microemulsion droplets

The size distribution of spherical micelles may be obtained by inserting Eq. (27) in Eq. (23) and switching variables in accordance with $dN = 4\pi R^2 dR/v$, where v is the molecular volume of the surfactant. As a result,

$$\phi_{mic} = \int_0^\infty \phi(R)dR = \int_0^\infty \frac{4\pi R^2}{v} e^{-4\pi\left(2k_c+\bar{k}_c-4k_cH_0R+\gamma_pR^2\right)/kT} dR \qquad (28)$$

The volume fraction density $\phi(R)$ according to Eq. (28) is plotted in Fig 8. The size distribution function appears to be close to Gaussian with a sharp peak located at the minimum of E(R).

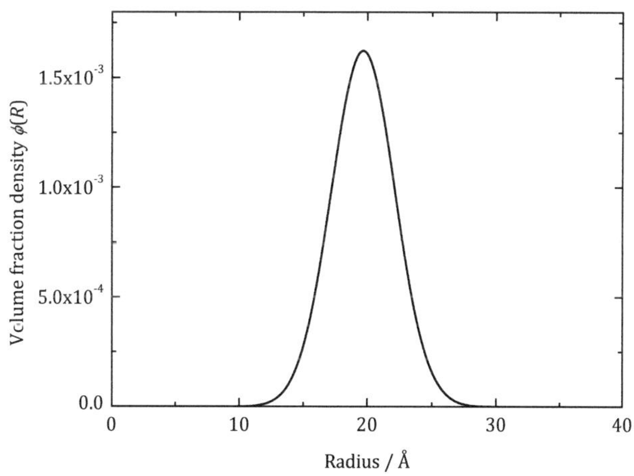

Fig. 8. Size distribution of spherical micelles for the case of bending elasticity constants set to $H_0 = 0.03$ Å^{-1}, $k_c = 2kT$ and $\bar{k}_c = -kT$

The size distribution shown in Fig. 8 is obtained by giving the spontaneous curvature a rather large value (H_0 = 0.03 Å^{-1}), implying an average micelle radius $\langle R \rangle \approx 20$ Å. This is about the largest accessible value for micelles made up of surfactants with a C_{12} aliphatic chain to retain a strictly spherical shape. Upon further decreasing H_0, the micelles tend to grow in size so as to achieve a reduced interfacial curvature. However, at some point the micelle radius starts to exceed the length of a fully stretched surfactant molecule and, in order to avoid the formation of a hole in the centre of the hydrophobic micelle core, surfactant micelles must assume some kind of non-spherical shape. Non-spherical micelles will be treated below in Section 9. Alternatively, in systems including, in addition to surfactant and water, a hydrophobic component (oil), the micelle core may be filled up with oil molecules to retain its spherical shape [*cf.* Fig 5].

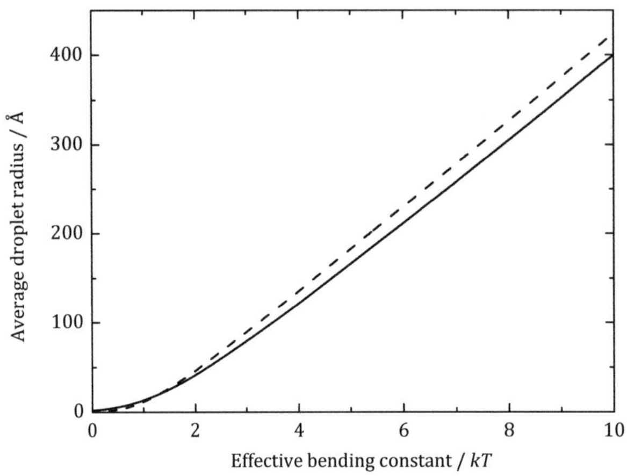

Fig. 9. The volume equivalent average droplet radius $\langle R\rangle$ for an ellipsoidal microemulsion droplet (solid line) and for a strictly spherical droplet (dashed line) plotted against the effective bending constant $k_{eff} \equiv 2k_c + \bar{k}_c$ for given values of $H_0 = 0.01$ Å^{-1} and $k_c = kT$. (Bergström, 2008a)

An explicit expression for the average size of a spherical micelle or microemulsion droplet may be derived from a given size distribution function as

$$\langle R\rangle = \frac{1}{\phi_{mic}} \int_0^\infty R\phi(R)\,dR \tag{29}$$

As a result, we obtain (Bergström, 2008a)

$$\langle R\rangle = H_0^{-1}\left(1 + \frac{\bar{k}_c + k_S}{2k_c}\right)\frac{1}{f(z)} \tag{30}$$

where $f(z)$ is a function that approaches zero as $z \rightarrow 0$ and equals unity as $z \rightarrow \infty$ and

$$z \equiv \frac{16\pi (k_c H_0)^2}{\gamma_p kT} = \frac{4\pi}{kT}\left(2k_c + \bar{k}_c + k_S\right) \tag{31}$$

The exact appearance of $f(z)$ depends on in what way shape fluctuations of the droplets are taken into account. k_S is a parameter, in units of energy, that takes into account the unfavourable entropy of self-assembling surfactant molecules in micelles (or surfactant and oil molecules in microemulsion droplets). k_S assumes negative values and equals zero as $N \rightarrow 1$ (equivalent to $k_{eff} \equiv 2k_c + \bar{k}_c \rightarrow 0$, *cf.* Fig. 9). $f(z) = 1$ whereas k_S becomes constant in the limit of large values of the effective bending constant ($k_{eff} \equiv 2k_c + \bar{k}_c$).

The average radius of swollen micelles and microemulsion droplets as a function of the effective bending constant k_{eff}, in accordance with Eq. (30), is plotted in Fig. 9. It appears that the droplet radius always vanishes in the limit $k_{eff} \to 0$. This is a general feature that emphasizes k_{eff} as a direct measure of the strength of the driving force of self-assembling amphiphilic molecules, the criterion of which is $k_{eff} > 0$. As k_{eff} turns into negative values, the hydrophobic driving force is no longer strong enough so as to counteract the entropy of self-assembly, and the surfactant is expected to be present as free molecules.
It is evident that all bending elasticity constants, as well as k_S, contribute to the final size of spherical micelles or microemulsion droplets. The droplet size is directly influenced by H_0, $\bar{k}_c$ and k_S. $\langle R\rangle$ increases as the curvature of the droplets tends to decrease with decreasing H_0, whereas the droplet size increases with increasing $\bar{k}_c$ as a result of the Gauss-Bonnet theorem discussed above. (Safran, 1991) Self-association entropy effects, taken into account by $k_S < 0$, tend to decrease the size of the micelles and droplets as it becomes increasingly more unfavourable to self-assemble surfactant and solute molecules with increasing aggregation number N. (Bergström, 2008a).
The bending rigidity influences $\langle R\rangle$ in a more indirect way. At high k_c -values, the influences of $\bar{k}_c$ and k_S are reduced since the free energy penalty to change the interfacial curvature from a uniform value equal to H_0 is large, and $\langle R\rangle \approx 1/H_0$. On the other hand, lowering k_c is expected to increase the magnitude of deviations from a spontaneous curvature. Since both $\bar{k}_c$ and k_S are usually negative, the droplet size is usually found in the regime $\langle R\rangle < 1/H_0$.
The polydispersiy, in terms of the relative standard deviation, is defined as

$$\frac{\sigma_R}{R} = \sqrt{\frac{\langle R^2\rangle}{\langle R\rangle^2} - 1} \tag{32}$$

and it may be evaluated for spherical micelles and microemulsion droplets employing the size distribution function in Eq. (28) [*cf.* Fig. 10]. It is seen that as $k_{eff} \to 0$, $\sigma_R/\langle R\rangle$ approaches a constant value that depends on how shape fluctuations are taken into account. As $k_{eff} \to \infty$, the polydispersity approaches the expression (Safran, 1991; Bergström, 2008a)

$$\frac{\sigma_R}{R} = \sqrt{\frac{kT}{8\pi\left(2k_c + \bar{k}_c + k_S\right)}} \tag{33}$$

Hence, the polydispersity appears to be a function of mainly k_{eff}. $\sigma_R/\langle R\rangle$ is primarily influenced by the bending rigidity since low values of k_c allows for large variations in droplet curvature. However, $\sigma_R/\langle R\rangle$ is also indirectly influenced by $\bar{k}_c$ through a constraint of fixed surfactant and oil concentration. The number of droplets must always increase with increasing polydispersity for a given droplet volume fraction and, as a consequence, the droplet polydispersity becomes raised as $\bar{k}_c$ is decreased. (Safran, 1991) Eqs. (30) and (33) have been derived by Safran for the case where self-association entropy effects have been neglected, i.e. with $f(z) = 1$ and $k_s = 0$ [*cf.* Fig. 10]. (Safran, 1991).
Finally, allowing for ellipsoidal shape fluctuations of micelles or droplets it may be demonstrated that deviations from a spherical shape increase with decreasing values of k_c, whereas the droplet shape is not influenced by k_cH_0 and $\bar{k}_c$ [*cf.* Fig 11]. (Bergström, 2008a).

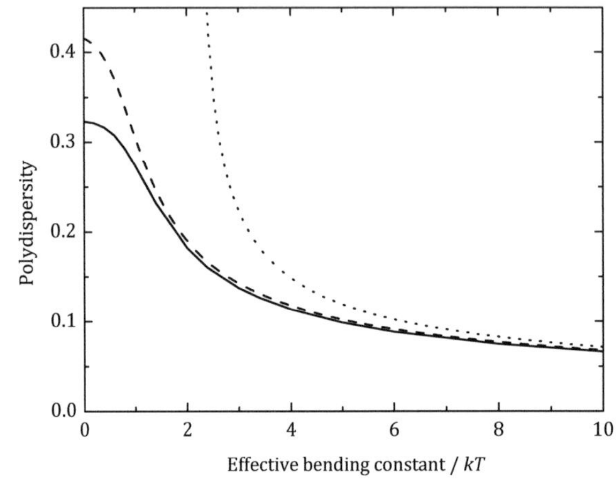

Fig. 10. The polydispersity $\sigma_R/\langle R\rangle$ of ellipsoidal microemulsion droplets (solid line) and strictly spherical droplets (dashed line) plotted against the effective monolayer bending constant $k_{eff} \equiv 2k_c + \bar{k}_c$. The dotted line represents $\sigma_R/\langle R\rangle$ as calculated from Eq. (30) with $f(z) = 1$ and $k_s = 0$. (Bergström, 2008a)

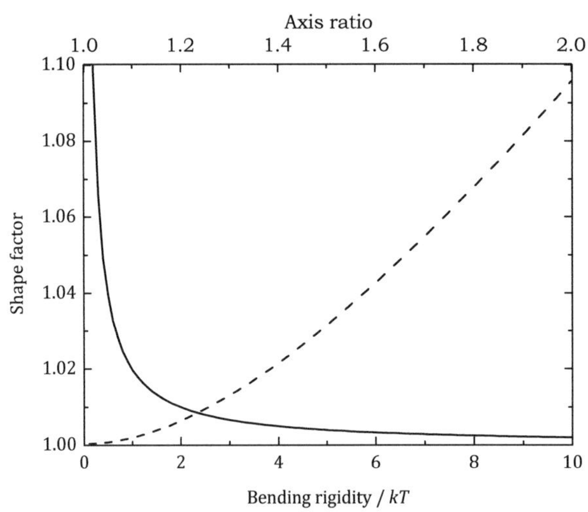

Fig. 11. The average shape factor $\langle\chi\rangle \equiv \langle A\rangle/4\pi R^2$ for ellipsoidal microemulsion droplets plotted against the bending rigidity k_c (solid line, bottom axis) for given values of $H_0 = 0.01$ Å^{-1} and $\bar{k}_c = 0$. The dashed line shows χ as a function of the axial ratio ε (top axis). (Bergström, 2008a)

9. Non-spherical micelles

In a solution with only surfactant and water present (and no oil), the strictly spherical shape of micelles corresponds to a minimum in size. As already mentioned, as micelles grow beyond a certain aggregation number they are forced to turn into some kind of non-spherical shape. As a result, they may grow in length to become shaped as long rods, worms or threads. Alternatively, they may grow both with respect to length and width to form disklike micelles or some kind of bilayer structure [see Section 10 below]. The thickness, however, must be more or less constant, approximately equalling twice the length of a surfactant molecule. The most general form of a growing surfactant micelle is the so called tablet-shaped or triaxial micelle. The tablet is characterized as having three dimensions (thickness, width and length) that may be different from one another. An example is the triaxial ellipsoid, the model of which has proven to be able to accurately fit small-angle scattering data of surfactant micelles. (Bergström & Pedersen, 1999a; Bergström & Pedersen, 1999b).

9.1 Tablet-shaped micelles

In order to investigate the influence of the various bending elasticity constants on the width and length of a tablet-shaped micelle, a theoretical model has recently been developed. (Bergström, 2007) In this model a tablet-shaped micelle is considered as composed of three geometrical parts: (*i*) a central planar bilayer surrounded by (*ii*) two straight semi-cylindrical rims and (*iii*) two semi-toroidal rims [*cf.* Fig. 12]. The planar bilayer is composed of a rectangular part with length L and width $2R$ with two semi-circular parts with radius R at each short side. The thickness of the bilayer part equals 2ξ and the radii of the cylindrical and toroidal rims equal ξ.

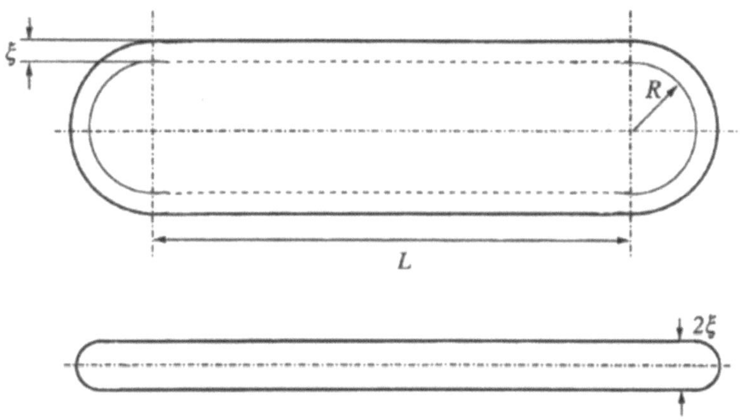

Fig. 12. Schematic illustration of a tablet-shaped micelle. (Bergström, 2007)

The free energy of a tablet-shaped micelle as a function of the dimensionless half width ($r \equiv R/\xi$) and the dimensionless length of the central bilayer part ($l \equiv L/\xi$) have been derived from Eq. (25) giving

$$\frac{E(r,l)}{kT} = \alpha + \delta\psi(r)\beta(\pi r + l) + 2r(\pi r + 2l)\lambda \qquad (34)$$

where the reduced planar interfacial tension is defined as $\lambda \equiv \xi^2 \gamma_p / kT$ and $\alpha \equiv 2\pi\left(3k_c + 2\bar{k}_c - 8\xi k_c H_0\right) / kT + 4\pi\lambda$, $\beta \equiv \pi k_c (1 - 4\xi H_0) / kT + 2\pi\lambda$ and $\delta \equiv 2\pi k_c / kT$ are three dimensionless parameters taking into account the bending free energy. The ψ-function ($0 < \psi < 1$) equals unity in the limit $r \to 0$ and zero as $r \to \infty$. (Bergström, 2007; Bergström, 2008b).

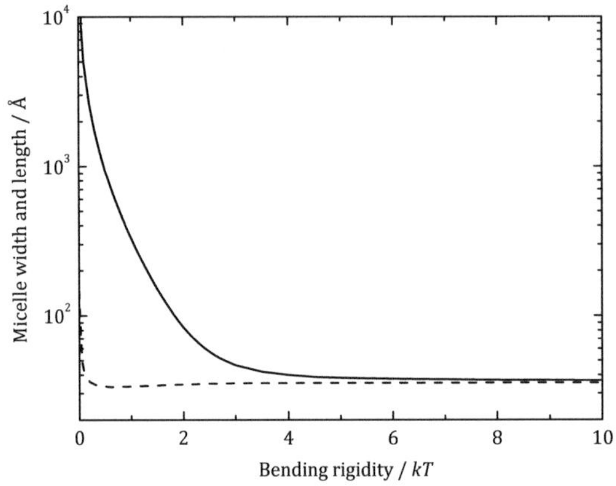

Fig. 13. The average length $\langle\Lambda\rangle$ (solid line) and width $\langle\Omega\rangle$ (dashed line) of tablet-shaped micelles plotted against the bending rigidity. The spontaneous curvature was set to $H_0 = 1/2\xi$ with $\xi = 12$ Å. β was set equal to zero corresponding to the maximum attainable size of the micelles. (Bergström, 2007)

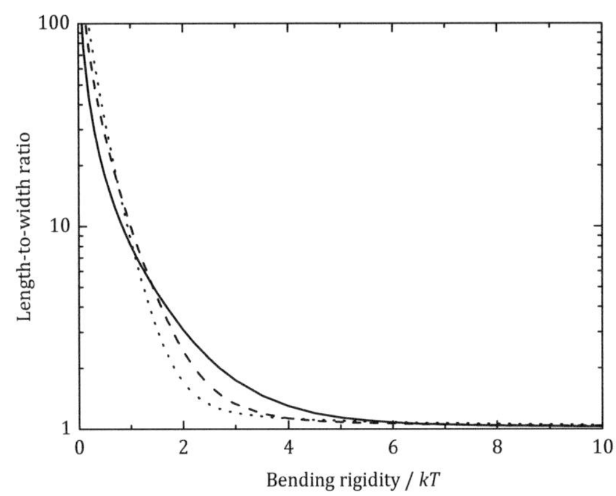

Fig. 14. The length-to-width ratio $\langle\Lambda\rangle/\langle\Omega\rangle$ of tablet-shaped micelles plotted against the bending rigidity. The spontaneous curvature was set to $H_0 = 1/\xi$ (solid line), $H_0 = 1/2\xi$ (dashed line) and $H_0 = 1/3\xi$ (dotted line) with $\xi = 12$ Å. β was set equal to zero corresponding to the maximum attainable size of the micelles. (Bergström, 2007)

The corresponding size distribution of tablet-shaped micelles has been evaluated by inserting Eq. (34) in Eq. (23). As a result, by means of integrating over l the following relation is obtained

$$\phi_{mic} = \frac{\pi\xi^6}{\upsilon^2}\int_0^\infty \frac{8r^2 + 6\pi r + \pi^2}{\beta + 4\lambda r} e^{-\delta\psi(r) - \pi\beta r - 2\pi\lambda r^2} dr \tag{36}$$

From the length and width distribution function it is straightforward to calculate the average width $\langle\Omega\rangle \equiv 2(\langle R\rangle + \xi)$ and length $\langle\Lambda\rangle \equiv \langle L\rangle + \langle\Omega\rangle$, in a dispersed phase of tablet-shaped micelles, as functions of the different bending elasticity constants. It appears that k_cH_0 mainly influence the width of the micelles ($\langle\Omega\rangle$ increases slightly with decreasing k_cH_0) whereas the length of the micelles increase significantly upon lowering the bending rigidity. As a consequence, more disklike ($l = 0$ in Eq. (34) and Fig. 12) or oblate shaped micelles are expected at high k_c -values, whereas elongated worms or threads are expected at k_c -values below about kT [*cf.* Figs. 13 and 14]. For instance, spherocylindrical micelles (two hemispheres attached to the ends of a cylinder, $r = 0$ in Eq. (34) and Fig. 12) are expected to be present at large values of k_cH_0 and low k_c. This behaviour may be rationalized as a result of the property of k_c to influence to what extent self-assembled interfaces are composed of geometrical parts with different curvatures. For instance, the large difference in curvature between the cylinder and hemispherical parts of a spherocylindrical micelle is favoured by low k_c -values. The saddle-splay constant contributes with a constant factor to the micelle size distribution and influence the size of tablet-shaped micelles (including spherocylinders and disks), but not the shape.

In the extreme case of long spherocylindrical micelles, the work of inserting an extra molecule in the cylinder part of a micelle is much lower than the work (per molecule) of forming an additional micelle with an extra pair of hemispheres. As a result, all surfactant molecules that are added will be incorporated in the existing micelles and, as a consequence, the micelles grow in length with increasing surfactant concentration in accordance with $\langle L\rangle \propto \phi_{mic}$, whereas the number of micelles is not changed. (Bergström, 2001b) Likewise, the (volume-weighted) relative standard deviation reaches a maximum value of $\sigma_L/\langle L\rangle = 1$ for spherocylindrical micelles. This behaviour is in sharp contrast to what is predicted for spherical or disklike micelles that form as the bending rigidity is large in magnitude. In the latter case, the curvature of a single micelle is changed upon inserting an extra molecule and, as a result, the work of forming an extra micelle is lower than the work of inserting the same amount of molecules in the existing micelles. Hence, adding surfactant to a system of spherical or disklike micelles increases the number of micelles, but does not change the micelle size, and the micelles turn out be nearly monodisperse. As a consequence, both growth rate and polydispersity of oblong tablet-shaped micelles are predicted to increase with decreasing k_c -values so as to finally reach the maximum values of long spherocylindrical micelles.

Moreover, Eq. (36) sets the limits for the existence of a solution of thermodynamically stable tablet-shaped micelles. (Bergström, 2007) According to Eq. (36), there is a maximum value that ϕ_{mic} may attain as the parameter β approaches zero. If the concentration of self-assembled surfactant exceeds this maximum value it is required that $\beta = r = 0$ so as to fulfil the condition of constant total surfactant concentration. There is no resistance for micelles to grow in the length direction as β and r reaches zero and, as a consequence, infinitely long cylinders ($r = 0$), rather than discrete tablet-shaped micelles, are expected to form above this critical concentration of surfactant. The surfactant concentration where the transition from discrete tablets to infinite cylinders is expected to occur decreases with increasing values of the effective bending constant $k_{eff} = 2k_c + \bar{k}_c$.

9.2 Toruslike micelles

Alternatively, toruslike micelles (cylinder that is geometrically closed in a ring shape), rather than infinite cylinders, may form at large k_{eff} -values, in the regime where tablets may not form. The free energy of toruslike micelles may be derived in a similar way as for tablets giving (Bergström, 2008b)

$$\frac{E(r)}{kT} = 2\pi r\left[\beta + \frac{\pi k_c}{kT}\left(\frac{r}{\sqrt{r^2-1}} - 1\right)\right] \tag{37}$$

where $r = R/\xi$ is the dimensionless torus radius and $\beta \equiv \pi k_c(1-4\xi H_0)/kT + 2\pi\lambda$ is defined in the same way as for tablet-shaped micelles in Eq. (34). Toruslike micelles belong to a different topology than tablets, with a single hole and a genus $g = 1$. As a result, the integral over the Gaussian curvature in Eq. (26) equals zero and, as a consequence, $\bar{k}_c$ is absent in Eq. (37). From the Gauss-Bonnet theorem, this implies that tori are favoured at the expense of tablets by large values of the saddle-splay constant.
The size distribution function for toruslike micelles equals

$$\phi_{tor} = \frac{2\pi^2\xi^3}{v}\int_1^\infty \exp\left[-2\pi\beta r - \frac{2\pi^2 k_c}{kT} r\left(\frac{r}{\sqrt{r^2-1}} - 1\right)\right] dr \tag{38}$$

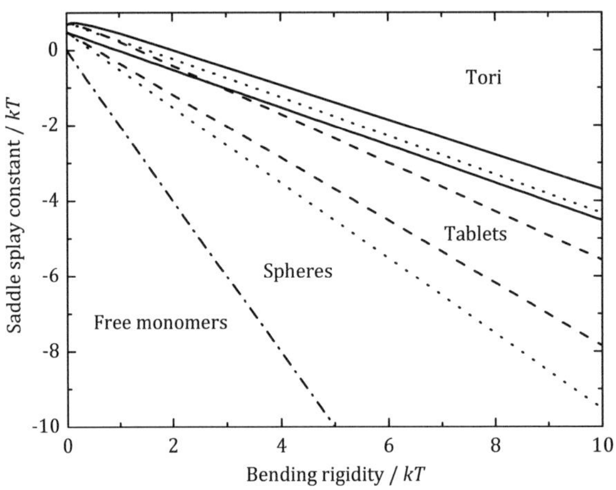

Fig. 15. Diagram showing regimes where toruslike, tablet-shaped and spherical micelles are predicted to predominate at $\xi H_0 = 1/2$ (solid lines), $\xi H_0 = 1/3$ (dashed lines) and $\xi H_0 = 1/4$ (dotted lines). Surfactants do not self-assemble in an aqueous solvent below the dash-dotted line, corresponding to $k_{eff} = 2k_c + \bar{k}_c = 0$. (Bergström, 2008b)

By setting $\phi_{mic} = \phi_{tor}$ according to Eqs. (36) and (38), we obtain the micelle predominance diagram in Fig. 15. It is seen that toruslike micelles are expected to be found at large values of k_c and $\bar{k}_c$, whereas a transition to tablets (including spherocylinders and disks) is expected as $k_{eff} = 2k_c + \bar{k}_c$ is lowered. As k_{eff} is further decreased, the tablets become smaller in size, eventually ending up as spheres. As k_{eff} falls below zero the surfactant may no longer form micelles and is expected to be present as free surfactant [*cf.* Fig. 9].
A second type of abrupt transition is predicted from Eq. (36). (Bergström, 2007) In the vicinity of $H_0 = 1/4\xi$, tablet-shaped micelles grow considerably in both width and length to

form large but discrete bilayer fragments. The width-to-length ratio is determined by the magnitude of the bending rigidity and large circular disks may be present in so far k_c is large in magnitude and $H_0 \gtrsim 1/4\xi$. As H_0 is further decreased, ϕ_{mic} rapidly approaches zero and as H_0 falls below $1/4\xi$ thermodynamically stable micelles can no longer exist. Rather than micelles, various types of bilayer structures appear to predominate as $H_0 < 1/4\xi$. A micelle-to-bilayer transition at $H_0 = 1/4\xi$ is also predicted from Eq. (38) for toruslike micelles. (Bergström, 2008b)
The behaviour of surfactant micelles as a function of the various bending elasticity constants is summarized in Table 1.

Micellar systems ($H_0 > 1/4\xi$)	Small $\bar{k}_c$	Large $\bar{k}_c$
Small k_c	Rods	Disordered Cylinders or Tori
Large k_c	Disks	Ordered Cylinders or Tori

Table 1. The structural behaviour of self-assembled micellar systems with bending rigidity and saddle-splay constant

10. Bilayers

10.1 Vesicles

As $H_0 < 1/4\xi$, amphiphilic molecules are predicted to self-assemble into various kinds of bilayer structures. In comparatively small and discrete form, bilayers tend to close themselves geometrically in order to eliminate their unfavourable edges and form vesicles (spherical bilayer shells). Vesicles may be considered as made up of an outer, positively curved ($H > 0$), and an inner, negatively curved ($H < 0$), monolayer [*cf.* Fig. 16].

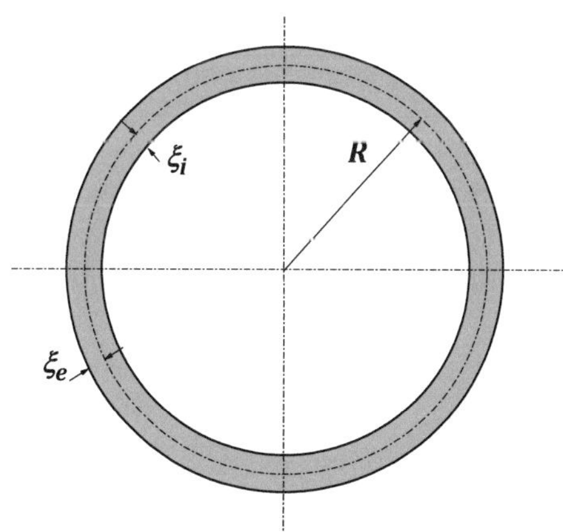

Fig. 16. Schematic representation of a unilamellar vesicle with radial distance R to the surface that divides the inner and the outer monolayers, the thicknesses of which are denoted ξ_i and ξ_e, respectively. (Bergström, 2001c)

The free energy of a vesicle may be derived in a similar way as for micelles giving (Helfrich, 1973; Bergström & Eriksson, 1996)

$$E(R) = 4\pi\left(k_{bi} + 2\gamma_p R^2\right) \quad (39)$$

where the bilayer bending constant

$$k_{bi} = 4k_c\left(1 - 2\xi H_0\right) + 2\bar{k}_c \tag{40}$$

incorporates contributions due to the bending of a planar bilayer at constant interfacial area $[= 8\pi\gamma_p R^2]$. Considering the entire bilayer as a closed interface gives a topology with genus $g = 0$. However, if we want to compare with tablet-shaped ($g = 0$) and toruslike micelles ($g = 1$), we must consider each single monolayer as an interface, and the vesicle as consisting of two closed interfaces. As a result, the genus of a unilamellar vesicle equals $g = -1$ and, consequently, the integral in Eq. (26) equals 8π, and a term equal to $8\pi\bar{k}_c$ appears in Eq. (39). The size distribution of vesicles may be written as (Bergström, 2001c)

$$\phi_{ves} = const \times \int_0^\infty R^\alpha e^{-4\pi\left(k_{bi} + 2\gamma_p R^2\right)} dR \tag{41}$$

Notably, the bending free energy ($= 4\pi k_{bi}$) is constant with respect to vesicle size, implying that high values of k_{bi} favours large vesicles. This is consistent with Eq. (41), according to which the aggregation number $\langle N\rangle \propto e^{-4\pi k_{bi}}$. Hence, rather small unilamellar vesicles are expected to form at comparatively low values of k_{bi} whereas, at higher k_{bi}, the bilayers may grow in to some more or less macroscopically large structure.

By considering Eq. (40) we may analyze the size of vesicles in terms of the three bending elasticity constants. The vesicle size is expected to slightly increase with decreasing values of k_cH_0. This may be explained as a result of the fact that the different contributions to k_cH_0 from the two oppositely curved monolayers, respectively, partly cancel one another. However, the outer layer is always more voluminous than the inner layer and, as a result, the net effect must be the vesicle size showing the same trend as a function of k_cH_0 as the outer layer. The bending rigidity has a large influence on vesicle size, in contrast to the case of spherical micelles and microemeulsion droplets. The reason for this has to do with the geometrical heterogeneous nature of vesicles. Increasing values of k_c favours a more uniformly curved geometry as the curvature of the outer and inner layers both approaches zero with increasing vesicle size. On the other hand, at low k_c -values the free energy penalty of accomplishing a large difference in curvature between the two layers is significantly reduced, allowing for comparatively small vesicles. As a matter of fact, rather small and thermodynamically stable vesicles have mainly been observed to be formed in surfactant mixtures, particularly in mixtures of two oppositely charged surfactants. (Tondre & Caillet, 2001) As previously mentioned, k_c is expected to be significantly reduced as two or more surfactants are mixed, and the magnitude of decrease is particularly pronounced in mixtures of oppositely charged surfactants. (Bergström, 2006b) Analogous to micelles, the aggregate size is expected to increase with increasing values of the saddle-splay constant, in accordance with the Gauss-Bonnet theorem.

10.2 Lamellar structures

The bending elasticity constants as the entire bilayer is regarded as an interface may be expressed in terms of the corresponding constants for a single monolayer, i.e. $H_0^{bi} = 0$, $k_c^{bi} = 2k_c$ and $\bar{k}_c^{bi} = 2\bar{k}_c - 8\xi k_c H_0$. (Porte et al., 1989; Szleifer et al., 1990; Ljunggren & Eriksson, 1992) Since H_0^{bi} is always zero for a thermodynamically equilibrated bilayer, more or less macroscopic bilayer structures, with a mean curvature of the bilayer close to zero, are expected to be present as k_{bi} assumes large values.

The most common type of lamellar structure is ordered sheets of planar lamellar bilayers (L_α-phase). Since the two principal curvatures of planar geometry both equal zero ($c_1 = c_2 = 0$), it follows that $H = K = 0$ for the L_α-phase. However, another set of morphologies of bilayers with $H = ½(c_1 + c_2) = 0$ may be obtained by setting $c_1 = -c_2$. These kinds of bilayer interfaces, distinguished as having $H = 0$ and $K = c_1 \cdot c_2 < 0$, are examples of minimal surfaces. (Hyde et al., 1997) Macroscopic minimal surface bilayers are characterized as having a topology with $g \to \infty$. As a result, minimal bilayer interfaces ($g \to \infty$, $K < 0$) are favoured at the expense of macroscopic planar layers ($g = 0$, $K = 0$) by large values of $\bar{k}_c^{bi} = 2\bar{k}_c - 8\xi k_c H_0$. (Porte et al., 1989) Notably, low (possibly negative) values of the monolayer spontaneous curvature H_0 may cause $\bar{k}_c^{bi} > 0$, although $\bar{k}_c < 0$. Minimal bilayer interfaces may be both ordered (bicontinuous cubic lamellar structure, gyroids etc) and disordered (L_3-phase or "plumbers nightmare"). (Hyde et al., 1997) It may be noted that analogous monolayer structures (L_α-phase, L_3-phase, bicontinuous cubic structure etc) may be formed in oil/water/surfactant microemulsion systems as the monolayer spontaneous curvature $H_0 \approx 0$. (Jönsson et al., 1998).

The dependence of bilayer structure on the different bending elasticity constants are summarized in Table 2.

Bilayer systems ($H_0 < 1/4\xi$)	Small $\bar{k}_c$	Large $\bar{k}_c$
Small k_c	Small Vesicles	Disordered bilayers
Large k_c	Large Vesicles	Ordered bilayers

Table 2. The structural behaviour of self-assembled bilayers as a function of bending rigidity and saddle-splay constant

11. References

Ben-Shaul, A.; Szleifer, I. & Gelbart, W. M. (1985). Amphiphile Chain Organization in Micelles of Different Geometries. In: *Physics of Amphiphiles: Micelles, Vesicles and Microemulsions*, V. Degiorgio and M. Corti (Ed.), 404-428, North-Holland, ISBN: 978-0-44-486940-1, Amsterdam

Bergström, L. M. (2006a). *Langmuir*, 22, 3678-3691, ISSN: 0743-7463

Bergström, L. M. (2006b). *Langmuir*, 22, 6796-6813, ISSN: 0743-7463

Bergström, L. M. (2006c). *J. Colloid Interface Sci.*, 293, 181-193, ISSN: 0021-9797

Bergström, L. M. (2007). *ChemPhysChem*, 8, 462-472, ISSN: 1439-4235

Bergström, L. M. (2008a). *Colloids Surf. A*, 316, 15-26, ISSN: 0927-7757

Bergström, L. M. (2008b). *J. Colloid Interface Sci.*, 327, 191-197, ISSN: 0021-9797

Bergström, L. M. (2009). *Langmuir*, 25, 1949-1960, ISSN: 0743-7463

Bergström, L. M. & Bramer, T. (2008). *J. Colloid Interface Sci.*, 322, 589-595, ISSN: 0021-9797

Bergström, M. (2001a). *Langmuir*, 17, 993 - 998, ISSN: 0743-7463

Bergström, M. (2001b). Thermodynamics of Micelles and Vesicles. In: *Handbook of Surfaces and Interfaces of Materials*, H. S. Nalwa (Ed.), 233-264, Academic Press, ISBN: 0-12-513915-2, San Diego

Bergström, M. (2001c). *J. Colloid Interface Sci.*, 240, 294-306, ISSN: 0021-9797

Bergström, M. & Eriksson, J. C. (1996). *Langmuir*, 12, 624-634, ISSN: 0743-7463
Bergström, M. & Eriksson, J. C. (2000). *Langmuir*, 16, 7173-7181, ISSN: 0743-7463
Bergström, M.; Jonsson, P.; Persson, M. & Eriksson, J. C. (2003). *Langmuir*, 19, 10719-10725, ISSN: 0743-7463
Bergström, M. & Pedersen, J. S. (1999a). *Langmuir*, 15, 2250-2253, ISSN: 0743-7463
Bergström, M. & Pedersen, J. S. (1999b). *Phys. Chem. Chem. Phys*, 1, 4437-4446, ISSN: 1463-9076
Clint, J. H. (1975). *J. Chem. Soc., Faraday Trans. 1*, 71, 1327-1333, ISSN: 0300 -9599
Eriksson, J. C.; Ljunggren, S. & Henriksson, U. (1985). *J. Chem. Soc., Faraday Trans. 2*, 81, 833-868, ISSN: 0300-9238
Evans, D. F. & Wennerström, H. (1994). *The Colloidal Domain, Chapter 3*, Wiley-VCH, ISBN: 978-0-47-124247-5, New York
Flory, P. J. (1974). *Faraday Disc.*, 57, 7-29, ISSN: 0301-7249
Gruen, D. W. R. (1985). *J. Phys. Chem.*, 89, 153-163, ISSN: 1520-6106
Guggenheim, E. A. (1952). *Mixtures*, Clarendon Press, Oxford
Helfrich, W. (1973). *Z. Naturforsch. C*, 28, 693-703, ISSN: 0341-0382
Hill, T. L. (1963-64). *Thermodynamics of Small Systems (Parts I and II)*, Dover, ISBN: 0-486-68109-2, New York
Holland, P. M. (1992). Modeling Mixed Surfactant Systems. In: *Mixed Surfactant Systems*, P. M. Holland and D. N. Rubingh (Ed.), 31-51, American Chemical Society, ISBN: 978-0-84-122468-1, Washington, DC
Hyde, S.; Andersson, S.; Larsson, K.; Blum, Z.; Landh, T.; Lidin, S. & Ninham, B. W. (1997). *The Language of Shape*, Elsevier, ISBN: 978-0-444-81538-5, Amsterdam
Israelachvili, J. N. (1991). *Intermolecular and Surface Forces*, Academic press, ISBN: 978-0-12-375182, London
Israelachvili, J. N.; Mitchell, D. J. & Ninham, B. W. (1976). *J. Chem. Soc., Faraday Trans. 2*, 72, 1525-1568, ISSN: 0300-9238
Jönsson, B.; Lindman, B.; Holmberg, K. & Kronberg, B. (1998). *Surfactant and Polymers in Aqueous Solution*, Wiley, ISBN: 978-0-47-149883-4, Chichester
Kozlov, M. M. & Helfrich, W. (1992). *Langmuir*, 8, 2792-2797, ISSN: 0743-7463
Ljunggren, S. & Eriksson, J. C. (1992). *Langmuir*, 8, 1300, ISSN: 0743-7463
Mitchell, D. J. & Ninham, B. W. (1989). *Langmuir*, 5, 1121-1123, ISSN: 0743-7463
Porte, G. (1992). *J. Phys.: Condens. Matter*, 4, 8649-8670, ISSN: 0953-8984
Porte, G.; Appell, J.; Bassereau, P. & Marignan, J. (1989). *J. Phys. France*, 50, 1335-1347, ISSN: 1155-4304
Porte, G. & Ligoure, C. (1995). *J. Chem. Phys.*, 102, 4290-4298, ISSN: 0021-9606
Safran, S. A. (1991). *Phys. Rev. A*, 43, 6, 2903-2904, ISSN: 0556-2791
Safran, S. A. (1999). *Adv. Phys.*, 48, 4, 395-448, ISSN: 0001-8732
Shinoda, K.; Nakagawa, T.; B.-I., T. & Isemura, T. (1963). *Colloidal Surfaces*, Academic Press, New York and London
Szleifer, I.; Kramer, D.; Ben-Shaul, A.; Gelbart, W. M. & Safran, S. A. (1990). *J. Chem. Phys.*, 92, 6800-6817, ISSN: 0021-9606
Tanford, C. (1980). *The hydrophobic effect*, Wiley, ISBN: 978-0-47-104893-0, New York
Tondre, C. & Caillet, C. (2001). *Adv. Colloid Interface Sci.*, 93, 115-134, ISSN: 0001-8686

12

Thermodynamics and Mesomechanics of Nanostructural Transitions in Biological Membranes as Liquid Crystals

Lev Panin
Scientific Research Institute of Biochemistry SB RAMS
Russia

1. Introduction

Biological membranes play a key role in structural-functional organization of any cells. Membranes constitute 65% of the cell dry weight, of them 40% belong to lipids and 60% to simple and conjugate proteins. Membranes separate cell into functional units (divisions); as a result, cell operates as a complicated factory in miniature. Besides, membranes integrate the operation of cell as a whole. Cell is an open system. It steadily exchanges weight, energy and information with the environment. This process also involves the biological (cell) membranes.
All biological membranes are liquid-crystalline structures. They consist of phospholipids bilayer, with simple and conjugate proteins immersed in it. These are various membrane-bound enzymes, transmembrane carriers (of glucose, Na^+ and K^+ ions, and others), and hormone receptors that serve as signal mechanisms of the cell.
Structure-forming bonds in liquid crystals are represented by covalent and hydrogen bonds, hydrophobic and electrostatic interactions. These are the low-energy binding: covalent bonds — 50-100 kcal/mol, hydrogen bonds — about 6 kcal/mol.
Between fatty acid tails of phospholipids there is a large amount of water dipoles. Such structure of the membrane provides high mobility of its structural components relative to each other. Membranes are reinforced by cholesterol molecules, which strengthen the hydrophobic interactions.
Such membrane responds to external action as a cooperative system.

2. Changes in structure and function of erythrocyte membranes in vivo

These studies were performed with thirteen participants of a Russia-Canadian transarctic ski transition "Dickson (Russia) - North Pole - Canada" within the polar day period (Observation..., 1992). In this section, we present the results of analysis of the structural-functional characteristics of erythrocyte membranes within a long (three-month) stay of humans in arctic deserts.

2.1 Methodology

Erythrocyte 'shadows' were obtained at Dickson, the North Pole, and in Ottawa. The objectives of this investigation were to study the viscosity, specific electrical conductivity, and the activation energy associated with ionic transport, to analyze the structure of the cell membrane by infrared spectroscopy, and to determine K_m and V_{max} for the Na^+, K^+ -ATPase. Observations were made over the temperature range 34-42 °C.

Viscosity and electrical conductivity were plotted against temperature, and the points of phase transition (Tc) were determined for each sample. A shift in phase transition reflects structural and functional changes in the membrane, with implications for Na^+, K^+-ATPase. The activity of this membrane-bound enzyme depends strongly upon the degree of order in the membrane (that is, upon the content of polyenoic fatty acids and cholesterol, the type of phospholipids, the presence of their lysoforms and structural transitions).

The viscosity was measured in an ultrathermostat, by traditional capillary viscosimetry. The speed of flow of a suspension of erythrocyte shadows through a capillary tube of known diameter (0.34 mm) and length (100 mm) was determined under a standard driving pressure (20 mm H_2O). Viscosity was measured at intervals of 1 °C over the 34-42 °C range, calculations being based on the equation:

$$P_x = P_o t_x / t_o \tag{1}$$

where P_o and P_x are the viscosities of the phosphate buffer and the suspension of erythrocyte shadows respectively, and t_x and t_o are the times for a fixed volume of the fluid (0.1 ml) to pass through the tube. Over the chosen temperature range, the viscosity of the buffer was 1 cP. The viscosity of the erythrocyte shadows was estimated to within 1 % relative to this standard.

The electrical conductivity of a colloidal solution depends on the disperse and the dispersive phases of the system (the number and motility of the colloidal particles, the size of their charge, and the number and charge of the ions in solution.

The charge on the particles is given by:

$$Q = z \cdot e \cdot k\,(1 + x \cdot z) \tag{2}$$

where z is the size of the particle, e is the dielectric strength, κ is the electrokinetic potential, and x is the thickness of the double electrical layer around a colloidal particle. The specific conductivity C is then given by:

$$C = Q \cdot v \cdot U_o \tag{3}$$

where v is the number of particles in 1 cm^3, as determined from the ratio:

$$v = 0.01 / \{4\pi Z^3/3\} \tag{4}$$

The term 0.01 assumes that 1% of colloidal particles are in the solution. U_o is the cataphoretic mobility, determined from the expression:

$$U_o = e \cdot k / 6\pi P \tag{5}$$

where P is the viscosity of the solution. By substitution,

$$C = 0.01 \cdot e^2 \cdot k^2 \cdot (1 + x^2)/8\pi^2 \cdot Z^2 P \tag{6}$$

From this final equation, it emerges that the electrical conductivity of a colloidal system is strongly influenced by the particle's electrokinetic potential, the dielectric strength, the particle size, the thickness of the double layer and the viscosity of the solution. The

equations presented are strictly limited to particles around 50 Å in size, but the concepts can be extended to larger particles, provided that the concentration and surface potential of the colloid, and the concentrations of the potential-forming ions in solution are all low.
Biological membranes display some of the properties of liquid crystals. The temperature dependence of conductivity of the erythrocyte shadows should thus be given by the expression:

$$C = C_o e^{-Ea/KT} \tag{7}$$

where C_o is a constant depending on the nature of the particle, K is Boltzmann's constant, Ea is the activation energy, and T is the absolute temperature. The temperature dependence should also be related to the number and motility of the carriers of current, according to the equation:

$$C = n \cdot e \cdot m_o \tag{8}$$

where n is the number of free charges, m_o is the motility of the charge, and e is the charge on an electron.
The experimental technique measured the resistance of suspensions of erythrocyte shadows when these were placed in a cuvette with electrodes of known surface, separated by a known distance. A 6-V, 2-kHz voltage was applied. Conductivity was then obtained as the inverse of specific resistance. The concentration of the shadows was 0.46 mg/ml, measured as a protein concentration, while the dispersing sodium/potassium/phosphate buffer had a pH of 7.4 and an osmolality of 310 osm.
Structural changes in the erythrocyte membrane were also evaluated by infrared spectroscopy. A suspension of the erythrocyte shadows (0.2 ml) was placed in a cuvette with a calcium fluoride base, and was vacuum dried for 1.0-1.5 h at 3-4 °C. The film of dried erythrocytes was then treated twice with phosphate buffer and was incubated at the desired temperature for 30 min before vacuum drying at 21-23 °C for a further 15 min. Spectra were determined on an(UR)-20 infrared spectrophotometer. Measurements were made of the half widths of absorption at 1660 cm^{-1} (corresponding to the C–O bonds in proteins), 1,000-1,100 cm^{-1} (the P - O - C bond in phospholipids), 1,500-1,700 cm^{-1} (C=O and NH bonds in proteins), and 2,700-3,000 cm^{-1} (the C—H bonds in proteins and phospholipids), together with the maximum intensity of absorption at 1,745 cm^{-1} (C=O bond in phospholipids) and the background intensity at 1,480 cm^{-1}.
Since the erythrocyte membrane contains many proteins, it was not possible to translate the spectral absorption into standard units of liters per mole per centimeter. Rather, the integral of the absorption peak was expressed as square centimeters per gram of erythrocyte shadow. An essentially equal total mass of erythrocytes was used for preparing test films at the three sites (Dickson, 0.50 ± 0.05, North Pole 0.49 ± 0.09, and Ottawa 0.41 ± 0.06mg).
The spectrophotometer was calibrated using a polystyrol film. The relative error of the absorption measurements was estimated at 1%.

2.2 Changes in viscosity during the ski trek

The initial viscosity of the erythrocyte "shadow" suspensions at a temperature of 34 °C was 1.85 cP. The viscosity decreased rather uniformly as temperature increased, both in Dickson and in Ottawa (fig. 1), but the curve had an anomalous region between 36 and 38 °C.
In Canada, the phase transition was displaced upwards, by 0.5 °C, and changes in viscosity of the erythrocyte shadows were more marked than in Dickson or at the North Pole (table 1). The

data thus suggest that an increase in rigidity of the erythrocyte membranes developed during the course of the trek, which reversed after completion of the mission. These changes can probably be attributed to increases in lysoforms phospholipids, and lipid hydroperoxides. Not only would such changes disturb hydrophobic interactions, making the membranes more porous, but they would presumably increase interactions between proteins and lipids, making the membrane more stable. Later, these changes may have been offset by an increase in the number of newly formed erythrocytes, giving the increased delta viscosity seen in Ottawa.

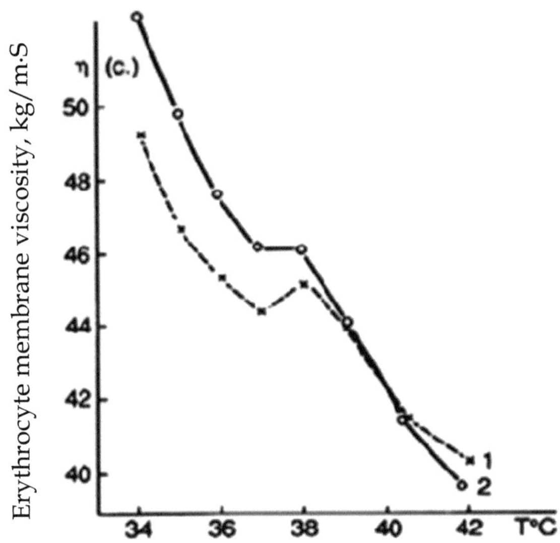

Fig. 1. Temperature dependence of viscosity of erythrocyte "shadows" on Dickson (1) and in Ottawa (2)

Parameters	Origin of the blood sample		
	Dickson	North Pole	Ottawa
Point of Phase transition, °C	36.7±1.2	36.7±1.4	37.2±1.1
Viscosity Changes, kPa	0.305±0.063	0.220±0.091*	0.439±0.136*

Table 1. Changes in viscosity of erythrocyte 'shadows' over the temperature range 34-42 °C (mean ± SD)

* - *Differences ($p < 0,05$) from Dickson*

2.3 Changes in electrical conductivity during the ski trek

At low temperatures, conductivity showed an almost exponential relationship to temperature, but at high temperatures the relationship (fig. 2). became more linear. The transition from one pattern of temperature dependence to the other was smoother than that observed in homogenous liquid crystals (table 2).

The transition points showed no significant differences between the blood samples collected on Dickson, at the North Pole and in Ottawa. However, differences in activation energy before and after the phase transition point indicate the structural change in the cell membrane at 37-38 °C. Moreover, the ratios of activation energies at low and high temperatures changed over the trek, being highest at the North Pole. The increase in activation energy at the North Pole, and subsequently in Ottawa, suggests that the small ions (H^+, K^+, Na^+ etc) had less ability to compensate for their particulate charge in a suspension of membranes with a lesser structural integrity (table 3).

Parameters	Origin of the blood sample		
	Dickson	North Pole	Canada
Point of Phase transition, °C	37.5±1.2	37.7±1.0	36.7±1.0
Changes in intrinsic electrical conductivity, $\Omega^{-1} \times m^{-1}$	0.131±0.035	0.101±0.029*	0.105±0.022*

Table 2. Changes in the phase transition point and intrinsic electric conductivity of erythrocyte shadows over the temperature range 34-42 °C (mean ± SD)
** - Differences (p < 0,05) from Dickson*

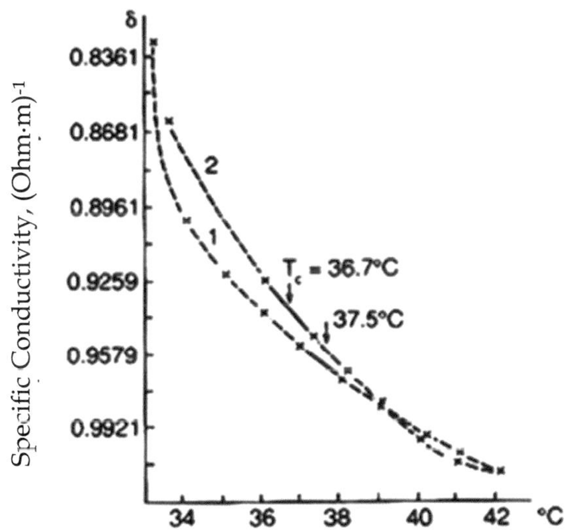

Fig. 2. Temperature dependence of electrical conductivity of erythrocyte "shadows" on Dickson (1) and in Ottawa (2)

Origin of the sample	Activation energy, cal/mg		
	Ea_1	Ea_2	Ea_1/Ea_2
Dickson	6.492 ± 2.487	5.106 ± 2.955	1.643 ± 0.475
North Pole	7.325 ± 2.707	3.148 ± 1.085	2.306 ± 1.051
Canada	11.771 ± 2.494	5.592 ± 2.045	2.108 ± 1.012

Table 3. Activation energy (Ea) of erythrocyte 'shadows' at low temperature (Ea_1) and high temperature (Ea_2) parts of the dependence curve (mean ± SD)

2.4 Changes in Infrared absorbance during the ski trek

Absorption in the regions of C=O bond (1.660 cm^{-1}) and the NH bond (1.550 cm^{-1}) decreased significantly as the temperature of the measurements was reduced, these changes being more marked in the initial blood samples from Dickson than in the final specimens taken in Ottawa. The Dickson specimens also showed a significant decrease in absorption at 1.645-1.650 cm^{-1}, and at 1.635, 1.685 and 1.696 cm^{-1} (fig. 3). The difference between the two locations again points to the development of structural changes in the proteins of the erythrocyte membrane.

As the measuring temperature was decreased, there was also a change in absorption at 1.070 cm^{-1}, reflecting changes in the P-O-C grouping of the phospholipids, the difference amounting to 9.8% at Dickson, 6.9% at the North Pole, and 7.9% in Canada (table 4). The P-O-C region showed two maxima, at 1.070 and 1.090 cm^{-1}, this being more marked in Dickson than in Ottawa. Thus, changes in phospholipids also seem to have developed over the course of the trek.

Origin of the sample	Absorption intensity, % ΔI between 40 and 37 °C		
	1.660 cm^{-1}	1.070 cm^{-1}	1.660 cm^{-1}/1.070 cm^{-1}
Dickson	8.87 ± 3.48	9.85 ± 3.67	1.82 ± 0.37
North Pole	7.52 ± 4.00	6.92 ± 2.31	2.61 ± 0.21
Canada	5.87 ± 3.12	7.91 ± 4.75	2.54 ± 0.36

Table 4. Temperature dependence of integrated intensity of the infrared spectra of erythrocyte "shadows" (mean ± SD)

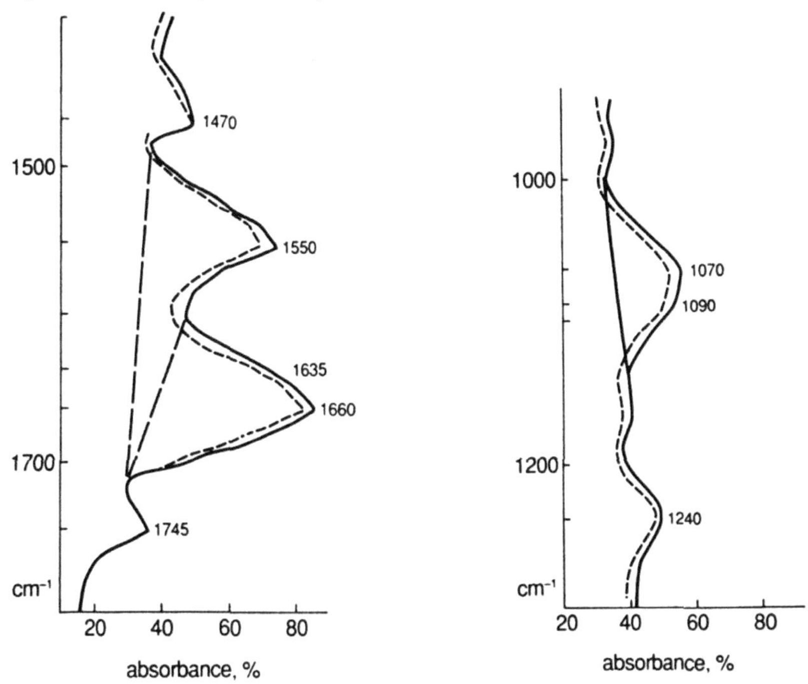

a b

Fig. 3. Infrared absorbance of erythrocyte shadows at Dickson, before trek: a. λ = 1400-1800 cm^{-1}, b. λ = 900-1300 cm^{-1}

——— 40 °C ; .– – – – 37 °C

Disarrangement of the structural proteins was further suggested by an increase in absorption at the wavelength of the C=O bond 1.660 cm^{-1}. The decrease in absorption at 1.745 cm^{-1} points to a reduced phospholipids content of the membrane at the North Pole and in Canada, relative to the Dickson baseline (table 5).
The intensity of background absorbance at 1.480 cm^{-1} depends on the water content of the sample. Given the standard conditions of vacuum drying, any alteration of this value must reflect changes in the amount of bound water. Values were 10% higher at the North Pole (fig. 4) and in Canada (fig. 5), relative to initial blood samples. This suggests that there was an increase in the hydration of the proteins and/or phospholipids.

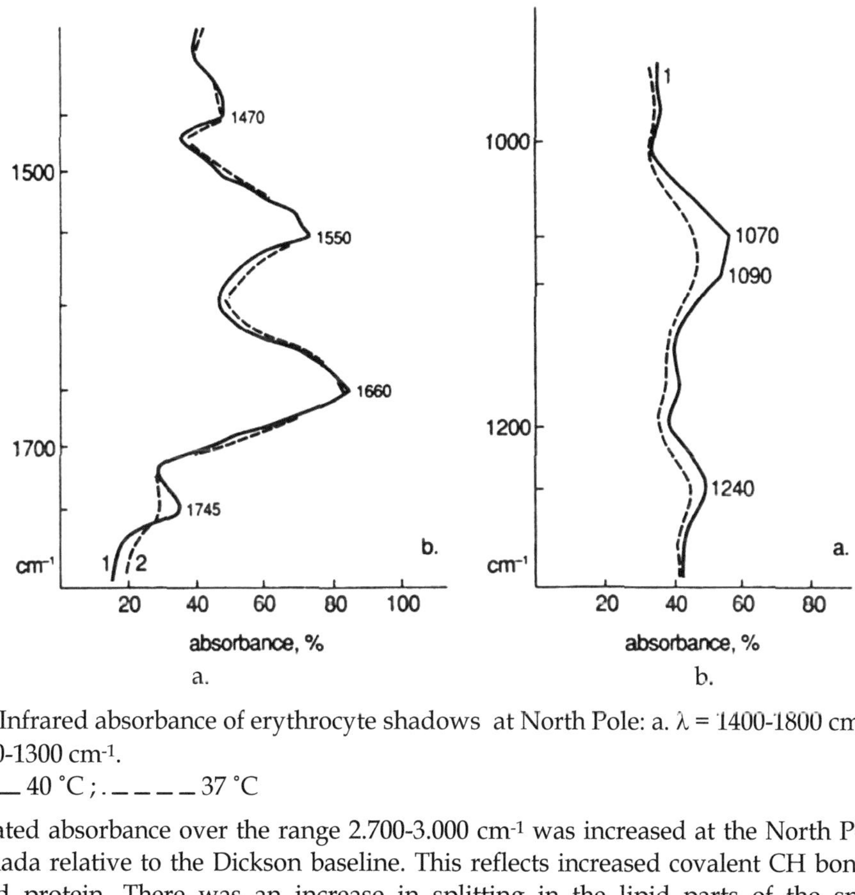

a. b.

Fig. 4. Infrared absorbance of erythrocyte shadows at North Pole: a. λ = 1400-1800 cm^{-1}, b. λ = 900-1300 cm^{-1}.
______ 40 °C ; . _ _ _ _ 37 °C

Integrated absorbance over the range 2.700-3.000 cm^{-1} was increased at the North Pole and in Canada relative to the Dickson baseline. This reflects increased covalent CH bonding of fat and protein. There was an increase in splitting in the lipid parts of the spectrum (particularly, at 1.470 cm^{-1}), suggesting deformational oscillation of the CH bonds.

Origin of the sample	Absorption intensity, %		
	1.745 cm^{-1}	1.480 cm^{-1}	1.660 cm^{-1}
Dickson	18.9 ± 5.0	33.1 ± 8.8	64.2 ± 8.6
North Pole	15.5 ± 4.5	41.8 ± 8.0	82.4 ± 10.3
Canada	13.2 ± 3.8	40.9 ± 8.8	80.7 ± 9.5

Table 5. Changes in some characteristics of the infrared spectra of the erythrocyte "shadows" during the Ski trek (mean ± SD)

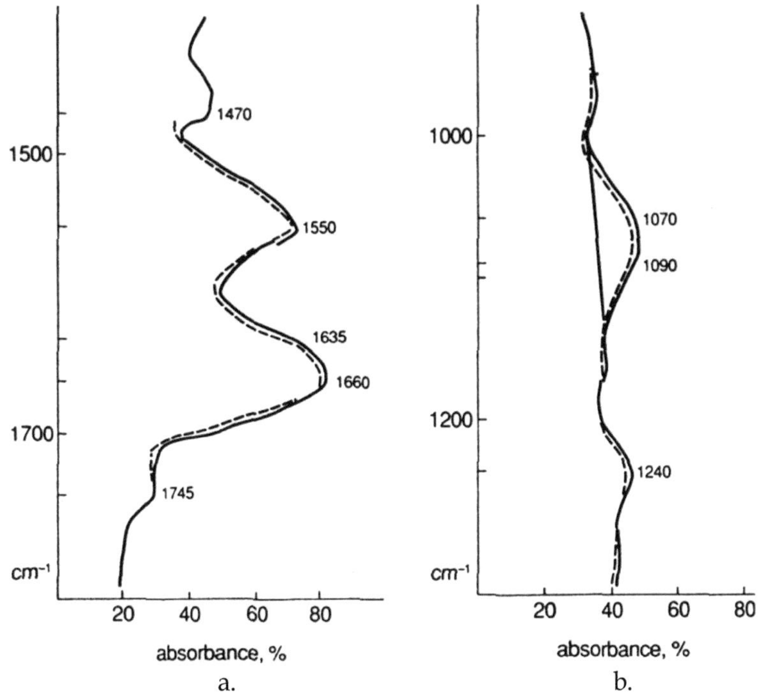

Fig. 5. Infrared absorbance of erythrocyte shadows sampled in Canada after trek: a. λ = 1400-1800 cm^{-1}, b. λ = 900-1300 cm^{-1}
_______ 40 °C ; ._ _ _ _ _37 °C

2.5 Changes in ATPase over the ski trek

The activity of ATPase and its K_m for ATP were both decreased in Ottawa relative to initial values in Dickson: from 3.50 ± 0.25 to 2.65 ± 0.32 mmol/mg/h, and from 5.64 ± 0.21 to 3.83 ± 0.72 mmol, respectively. Changes of kinetic characteristics of Na^+/K^+-ATPase of man erythrocyte "shadows' upon variation of suspension temperature are presented in table 6. In the North Pole the decrease in K_m was more then 30% in relation to Dickson. Thus, activity of Na^+/K^+-ATPase, which is responsible for maintaining the transmembrane electrochemical potential, decreased in 1.2-1.4 times. The Michaelis constant (Km), which determines the enzyme-substrate affinity, also decreased.

These changes could be related to a decreased content or an altered composition of phospholipids, together with conformational changes in the enzyme molecule itself. Probable, stress hormones (cortisol, adrenalin) also make a substantial contribution to the structural transitions. Their content in the blood serum was increased during the ski trek: cortisol before trek - 565±78 nmol/l, after trek - 927±70 nmol/l; adrenalin before trek - 0.7±0.03 mkg/l, after trek - 1.49±0.07 mkg/l.

2.6 Conclusions

Erythrocyte membranes responds to extreme conditions as cooperative system. The changes in intermolecular interactions and membrane permeability indicated by these various tests are not necessarily negative adjustments, since they enhance the rheological properties of the erythrocytes, and may also enhance the diffusion of oxygen through the cell membranes.

However, when subjects are exercising hard, there may also be changes designed to modify the oxygen-transporting ability of the red cells, microviscosity of erythrocyte membranes etc.

Origin of the sample	T, °C	$K_m/M_{Atp} \cdot 10^{-6}$
Dickson	36	4.88 ± 0.54
	37	5.64 ± 0.21
	38	4.54 ± 0.11
North Pole	36	2.89 ± 0.13
	37	3.51 ± 0.34*
	38	2.92 ± 0.12
Ottawa	36	3.31 ± 0.15
	37	3.83 ± 0.72*
	38	3.61 ± 0.15

Table 6. Changes of Michaelis constant (Km) over the course of ski trek.
** - Differences ($p < 0,05$) from Dickson.*

3. The interaction mechanism of cortisol and catecholamines with structural components of erythrocyte membranes

Physiological activity of the great majority of hormones is determined by the receptors located on plasmatic membranes of target cells. They all have high affinity for the hormone

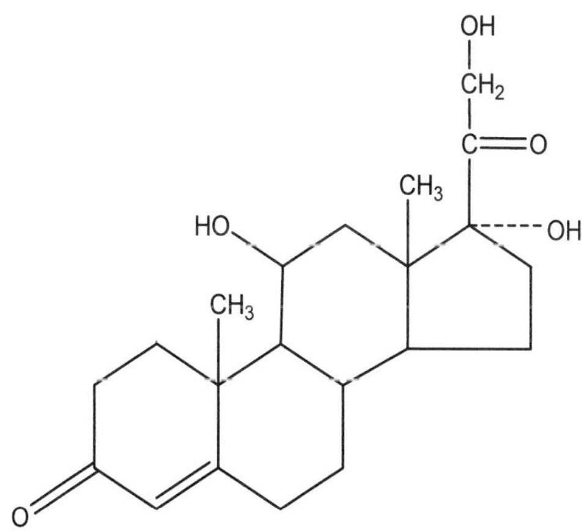

cortisol

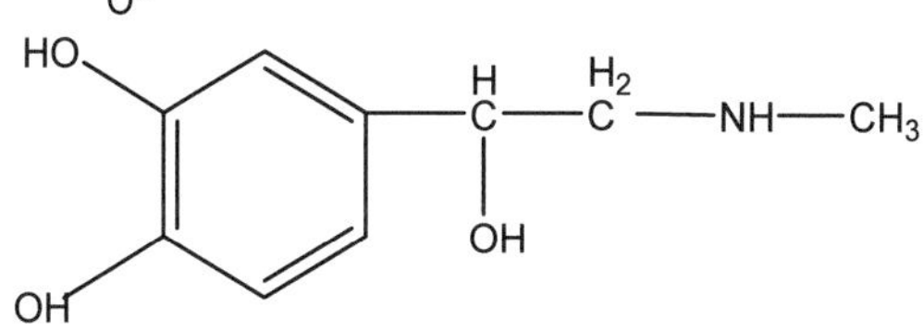

adrenaline

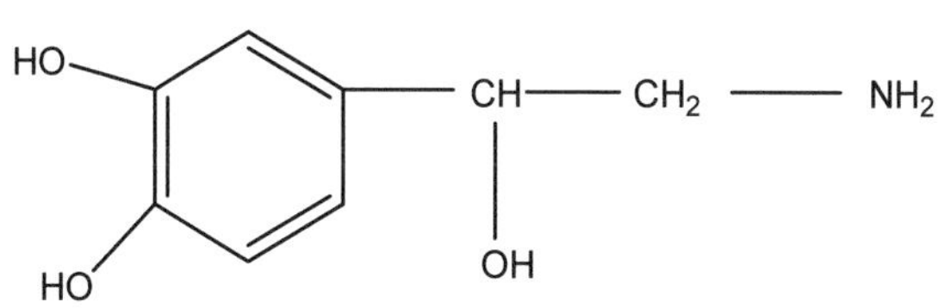

noradrenaline

ligand. Detection of a rather large amount of steroid hormones in the albumin fraction of blood serum and on plasmatic membranes of some cells (erythrocytes, leukocytes) points to the possibility of their nonspecific (non-receptor) interaction with protein and membrane structures. Some mechanisms of this interaction are discussed in available literature (Sergeev et al., 1996). Here, of particular interest are glucocorticoids and catecholamines, which blood content may increase several fold under stress. This will increase nonspecific interaction of these hormones with plasmatic membranes of blood cells.

We studied cortisol and catecholamines interaction with erythrocyte membranes, and the mechanism of structural transitions that occur in membranes under the action of these hormones.

Three stress hormones: cortisol, adrenaline and noradrenaline (Amersham) is analyzed in the work.

3.1 Methodology

3.1.1 Atomic force microscopy (AFM) of erythrocytes

Erythrocytes were obtained from fresh blood after decapitation of Wistar rats under light nembutal narcosis. Blood was diluted twofold by isotonic phosphate buffer (pH 7.35) containing 0.16 M of KH_2PO_4 and 0.18 M of Na_2HPO_4. After precipitation of cells by centrifuging at 330 g for 10 min, supernatant liquor was decanted, and the washing procedure was repeated twice more.

All the procedures were performed at 4 °C. The resulting erythrocyte suspension of 20 mcl volume was deposited onto a glass slide as a thin smear. The smear was predried for 10 min in air at 24 °C and humidity of 40%. After evaporation of excessive surface moisture, the smear was observed under a «Solver Bio» atomic force microscope (NT-MDT, Russia) at 24 °C using a semi-contact mode. An analogous procedure of obtaining red blood cells for the AFM examination was employed earlier by other authors (Wu et al., 2009). In each experiment, we first tested a control specimen without hormones, and then the experimental one. Silicon cantilevers NSG11 (NT-MDT, Russia) with a resonant frequency between 120 and 180 kHz and spring constant ~ 6 N/m were used (all of these probe parameters were offered by manufacturer). Images of the surface relief of erythrocyte membrane after absorption of hormones were obtained with the scan size 1x1 μm^2 and 1.3x1.3 μm^2.

3.1.2 IR spectroscopy of erythrocyte shadows

Erythrocyte shadows were obtained after their hemolysis in hypotonic phosphate buffer (pH 7.35) containing 2.75 mM of KH_2PO_4 and 8.5 mM of Na_2HPO_4. Shadows were precipitated by centrifuging at 5500 g, supernatant liquor was decanted. The washing procedure was repeated four more times. All operations and further storage of shadows were performed at 4 °C.

A film for taking the IR spectra of erythrocyte shadows was prepared in a cuvette with fluorite backing via slow evaporation of water under weak vacuum at a pressure of ca. 0.1 atm (ca. $0.5 \cdot 10^4$ Pa) and temperature 4±1 °C (Kunitsyn et al., 2001). Drying lasted 180 min. A suspension of erythrocyte shadows in a 0.001 M phosphate buffer with pH 7.35 and volume 60 mcl was introduced into a cuvette. This was supplemented with 30 mcl of the same buffer and 1.0 mcl of the hormone solution with concentration 10^{-6} M. Stirring and incubation lasted 10 min at 16-17 °C. The cuvette was placed horizontally on a special table of a vacuum unit.

When the film was prepared, the cuvette was transferred into an optical chamber and blown with dry air for 30 min, then the scanning unit was switched on. IR spectra were taken on a Specord-M80 spectrometer (Germany, Leipzig), sequentially experiment and control against the fluorite backing, or experiment and control to obtain a difference spectrum. Integration, determination of the spectrum band frequency, and mathematical processing were performed with special programs enclosed to the spectrometer.

3.1.3 Fluorescence analysis of erythrocyte shadows

Fluorescence measurements were performed with a Shimadzu spectrofluorophotometer RF-5301(PC)SCE. 4 ml of hypotonic phosphate buffer containing 2.75 mM of KH_2PO_4 and 8.5 mM of Na_2HPO_4 (pH 7.35), and erythrocyte shadows were poured into a quartz cuvette of size 1 x 1 x 4 cm^3. The concentration of shadow proteins was determined by the Warburg-Christian method from changes in the optical density of suspension (Dawson et al., 1986). On the average, it varied in the range of 0.100-0.250 mg/ml.
A cuvette with the shadow suspension was placed into a spectrofluorimeter thermostat for 1 hour. Getting a stationary temperature regime in the cuvette was controlled by an electronic thermometer. In all the experiments, temperature in the cuvette was 36 °C. After establishing a stationary temperature in the cuvette, intensity of the intrinsic fluorescence of tryptophan residues in protein membranes was measured. The tryptophan emission spectrum was taken in the range of 300 nm $\leq \lambda \leq$ 400 nm at the excitation wavelength 281 nm, with the maximum of emission intensity observed at 332 nm. The average value of maximum emission intensity was obtained graphically after its continuous measuring for 4 minutes. Intensity of tryptophan fluorescence fluctuated within 1%. The possible reasons include variation of temperature in the cuvette with suspension, instrumental error in determination of fluorescence intensity, and photochemical reactions occurring in the system. Spectral width of the slits was 1.5/10. The tryptophan absorption spectrum was recorded in the range of 220 nm $\leq \lambda \leq$ 300 nm at the emission wavelength λ = 332 nm. Cortisol was dissolved in a mixture of dimethyl sulfoxide (DMS) and ethanol (1 : 1, V/V). Concentration of the hormone in the initial mother liquor was 10^{-3} M. If necessary, the solution was diluted with hypotonic phosphate buffer to obtain a desired concentration.
A solution of adrenaline or noradrenaline with the concentration 10^{-6} M was prepared in hypotonic phosphate buffer. The time of hormone incubation with shadows was one hour. Absorption and emission spectra were taken, the average value of emission and absorption intensity was measured. For each hormone (cortisol, adrenaline or noradrenaline), the binding constant K_b was calculated by the method (Attallah & Lata, 1968) as well as the stoichiometric concentration of a bound hormone B_{max} and a change in free energy of the system ΔG. The interaction of cortisol and erythrocyte membrane is described by the equation

$$B + nS = S_nB, \tag{9}$$

where B is a membrane protein, S is the hormone, and n is the number of moles of hormone per a mole of proteins. The binding constant K_b was calculated by the formula

$$K_b = [S_nB]/[S]\cdot[B] \tag{10}$$

where $[S_nB]$ is the concentration of bound protein, $[B]$ is the concentration of free protein, and $[S]$ is the concentration of free hormone. It is supposed that hormone, upon binding to protein, completely quenches its fluorescence. Thus, the fluorescence intensity F will be proportional to the concentration of free protein. Let's write C for total concentration of protein in the cuvette, and x for concentration of the bound protein. Then,

$$F_{max} = \beta C; \qquad F = \beta(C - x) \tag{11}$$

where F is the intensity of tryptophan fluorescence at $\lambda = 332$ nm (the excitation wavelength $\lambda = 228$ nm), F_{max} is the intensity of tryptophan fluorescence in the absence of hormone (when the entire protein is free), β is the proportionality factor, and A_s is the stoichiometric concentration of hormone. When concentration of hormone exceeds $\boldsymbol{A_s}$, the fluorescence quenching does not increase. Dividing the first equation of set (11) by the second one gives

$$x = QC, \text{ where } Q = (F_{max} - F)/F_{max} \tag{12}$$

$[S] = A - n{\cdot}x = A - n{\cdot}Q{\cdot}C$, where A is the total concentration of hormone; $n = A_s/C$; $[B] = C - x = C(1-Q)$. Substitution of (11) and (12) into expression for binding constant (10) gives

$$K_b = Q/(1-Q)(A-nQC) \tag{13}$$

In our case, molar mass of membrane proteins is unknown, so the concentration of proteins in cuvette C is determined in mg/ml, and concentration of hormones A in mol/l. The constant n is expressed in moles of molecules of hormone per milligram of protein (M/mg), and is a ratio of the maximum concentration of bound hormone to the concentration of membrane proteins. This can be written as

$$B_{max} = A_s/C \text{ [mole/mg protein]} \tag{14}$$

Changes in Gibbs free energy ΔG of the system upon transition of hormone from aqueous medium to erythrocyte membrane are calculated by the formula

$$\Delta G = - RT \cdot \ln(K_c) \text{ [J/mole]}, \tag{15}$$

where $R = 8{,}314$ $J/K{\cdot}$mole, and T is the absolute thermodynamic temperature.
For adrenaline and noradrenaline, in formulas (11) and (12) we use the dependence of $Q_1 = (D_{max} - D)/D_{max}$ on concentration A of hormone introduced into the cuvette. Here, D_{max} and D are the intensities of tryptophan absorption in shadows at $\lambda = 228$ nm, respectively, without addition of hormone and with hormone. The emission wavelength $\lambda = 332$ nm.
The measurement errors appeared due to inaccuracy in volumetric dosing of the shadow suspension specimens and their titration against hormones. Specimens were dosed using pipette dispensers DPAOF-1000 and DPAOF-50, their relative error at (20±2)°C being 1% and 2%, respectively. 4 ml of the buffer was taken once with a DPAOF-1000 pipette, and suspension of erythrocytes and hormones was dosed twice using a DPAOF-50 pipette. The fluorescence intensity of erythrocyte shadows F is directly proportional to the concentrations of proteins C and hormones A in the specimen. Relative error E_F in measuring the F value can be estimated by the formula

$$E_F = \sqrt{(1\%)^2 + (2\%)^2 + (2\%)^2} = 3\% \tag{16}$$

Relative measurement errors for K_b and B_{max} can be obtained in the same way. They are equal to 10%.

In calculation, the values of fluorescence intensity *F* and absorption intensity *D* were corrected for dilution of suspension after the introduction of solution with hormone, for quenching of tryptophan emission by the solvent (a mixture of DMS and ethanol), for proper fluorescence of hormones, and evaporation of water from the cuvette.

3.1.4 Measurement of erythrocyte membrane microviscosity

Membrane microviscosity for translational diffusion of pyrene probe was calculated as a ratio of fluorescence intensity of the pyrene dimer to fluorescence intensity of the pyrene monomer (fig. 6). Microviscosity of erythrocyte membranes was measured also on a Shimadzu RF-5301(PC)SCE spectrofluorimeter. The experimental specimen was prepared as follows: 4 ml of hypotonic phosphate buffer containing 2.75 mM of KH_2PO_4 and 8.5 mM of Na_2HPO_4 (pH 7.35), a fluorescent pyrene probe, erythrocyte shadows and a specified amount of hormone were placed in a quartz cuvette of size 1 x 1 x 4 cm^3. Before use, all the components were stored at 4 °C. The concentration of shadow protein in the cuvette was 0.100-0.250 mg/ml; that of pyrene, $7.76 \cdot 10^{-6}$ M. Pyrene was diluted in ethanol, its initial concentration being $1.5 \cdot 10^{-3}$ M. The cuvette was placed into the spectrofluorimeter thermostat for 10 min, then the fluorescence measurements were performed at 36 °C. Before placing the specimen into the spectrofluorimeter thermostat, it was shaken vigorously for 1 min. For fluorescence measurements of shadows upon their loading with a different amount of hormones, each time a new specimen was prepared by the same procedure. Such a procedure is necessary because pyrene favors fast degradation of erythrocyte membranes.

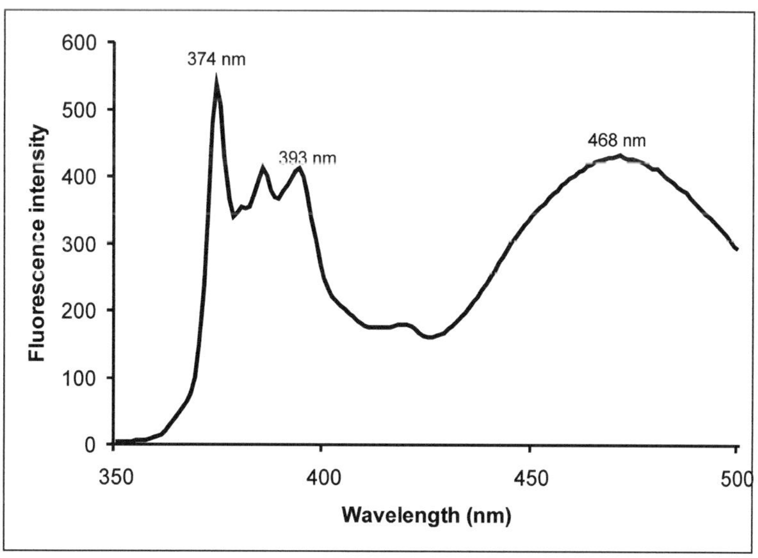

Fig. 6. Emission spectra of the pyrene in the suspension of shadows. Excitation wavelength λ = 337 nm and spectral slit width 1.5/3. At that, the maxima of emission intensity were observed at λ = 374 nm and λ = 393 nm (the vibronic emission peaks of excited pyrene monomers), and λ = 468 nm (the emission maximum of excited pyrene dimer). Concentration of shadows C = 0.128 mg protein/ml. Concentration of pyrene in the suspension is $7.7 \cdot 10^{-6}$ M, temperature of the specimens 309.1±0.1 K (36 °C), pH of the suspension 7.35

To measure the microviscosity of a lipid bilayer near proteins (the region of protein-lipid interaction), we used the excitation wavelength $\lambda = 281$ nm and spectral slit width 1.5/5. Microviscosity of a lipid bilayer far from proteins (the region of lipid-lipid interaction) was measured with the excitation wavelength $\lambda = 337$ nm and spectral slit width 1.5/3. At that, the maxima of emission intensity were observed at $\lambda = 374$ nm and $\lambda = 393$ nm (the vibronic emission peaks of excited pyrene monomers), and $\lambda = 468$ nm (the emission maximum of excited pyrene dimer). Fig. 6 shows the emission spectrum of pyrene in the shadows suspension.

The relative microviscosity of membranes was determined as a ratio $\eta(A)/\eta(0)$, where $\eta(A)$ and $\eta(0)$ are microviscosities of membranes, respectively, with and without hormone added to the shadow suspension. For the region of lipid-lipid interaction, relative microviscosity was calculated by the formula

$$\eta(A)/\eta(0) = \{F_{468}(0)/F_{468}(A)\} \cdot \{F_{393}(A)/F_{393}(0)\} \tag{17}$$

where $F_{468}(A)$ is the fluorescence intensity of pyrene at wavelength $\lambda = 468$ nm in a specimen at the hormone concentration A in suspension; $F_{468}(0)$ is the fluorescence intensity of pyrene at wavelength $\lambda = 468$ nm in a specimen with no hormone in suspension. $F_{393}(A)$ and $F_{393}(0)$ are the fluorescence intensities of pyrene at wavelength $\lambda = 393$ nm at the hormone concentration A in suspension and without hormone in suspension, respectively. The excitation wavelength is 337 nm.

For the region of protein-lipid interaction, relative microviscosity was calculated by the formula

$$\eta(A)/\eta(0) = \{(F_{468}(0) - I_{468})/(F_{468}(A) - I_{468})\} \cdot \{(F_{393}(A) - I_{393})/(F_{393}(0) - I_{393})\} \tag{18}$$

where I_{393} and I_{468} are the fluorescence intensities of tryptophan at wavelength $\lambda = 393$ nm and $\lambda = 468$ nm, respectively. The excitation wavelength is $\lambda = 281$ nm. A relative measurement error for relative microviscosity was equal to 6%.

3.2 Results

3.2.1 IR spectroscopy

The analysis of IR spectra of rat erythrocyte shadows upon cortisol addition showed a ca. 20% increase in the intensity of absorption bands of CO (1655.2 cm^{-1}) and NH bonds (1548 and 3290 cm^{-1}), the effect becoming more pronounced with increasing concentration of the hormone (table 7, figs. 7, 8). The increased intensity of the band at 1655.2 cm^{-1} points to a growing orderliness in membrane proteins caused by the structural transition tangle $\rightarrow$ α-helix. Appearance of absorption band at 1630 cm^{-1} corresponds to β-structure (tangle $\rightarrow$ β-structure transition) at the cortisol concentration 10^{-7} M (fig. 9). The interaction of hormone ligand with protein enhances intermolecular interaction between membrane proteins and phospholipids, and facilitates the formation of lipid-protein clusters.

A shift of stretching vibrations of the peptide bond (NH bond), 3308 $\rightarrow$ 3280 cm^{-1} ($\Delta v = 28$ cm^{-1}), and an increase in its intensity were found, which are caused by the formation of hydrogen bond between cortisol and NH bond of proteins. It forms most likely between the keto group in hormone A-ring ($C_3 = 0$) and NH bond of the membrane protein. Although both the keto group at C_{20} in D-ring and the OH group at C_{11} in C-ring may also participate in the formation of hydrogen bonds. Cortisol is a cholesterol derivative; however, the presence of two keto and three hydroxy groups considerably changes its interaction with the surface of erythrocyte membrane as compared to cholesterol. The latter penetrates deep into the membrane and enhances the hydrophobic interactions between chains of fatty acids in phospholipids. Cortisol acts on the surface. Of interest are the shifts of CH bond stretching vibrations: 2848 $\rightarrow$ 2852 cm^{-1} (Δv = 4 cm^{-1}) and 2930 $\rightarrow$ 2925 cm^{-1} (Δv = 5 cm^{-1}). The latter increased also its intensity under the action of hormone. A change in the intensity of this bond confirms the occurrence of structural transition order $\rightarrow$ order under the action of cortisol, but does not reveal where it takes place: in membrane proteins or in phospholipids, since CH bond resides both in proteins and phospholipids. An increase in the intensity of absorption band of C=O bond in phospholipids and its shifting from 1748 to 1740 cm^{-1} were observed. This increase in the band intensity points to an enhanced orderliness of higher carboxylic acids and a decreased entropy in phospholipids. The band shift is caused by the formation of hydrogen bond between the hormone, probably OH group at C_{21}, and CO bond in phospholipids. Such a concurrent interaction of the hormone with protein and phospholipids takes place at the interface of proteins and phospholipids, which enhances the lipid-protein interactions and leads to the formation of complex domains revealed by atomic force microscopy.

A 3 cm^{-1} shift in the frequency of P=O bond to the short-wave region and a minor increase in its intensity were observed.

This is attributed to dehydratation of membranes upon their deformation (compression) under the action of hormone. This is exactly the loss of bound water that increases the frequency of P=O bond. Displacement of water dipoles from lipid-protein domains to adjacent regions results in the appearance of meso-stripes with tensile hydrodynamic stress (see fig. 20).

It is known that spectrin accounts for 30% of membrane proteins (Leto & Marchesi, 1984). It can be expected that additional deformation of membranes would occur due to the spectrin-actin-ankyrin meshwork capable of contraction, this meshwork residing on both the internal and external surfaces of the membrane (Palek & Sahr, 1992). This mechanism of erythrocyte membrane deformation plays a more essential role upon interaction with adrenaline and noradrenaline. Thus, adrenoreceptor and cholinoreceptor were found earlier on the surface of erythrocyte membranes (Leto & Marchesi, 1984). The EPR spectroscopy study revealed that deformation (contraction) of erythrocyte membranes occurs in erythrocyte membranes (shadows) under the action of adrenaline or carbachol (the analog of choline). The contraction effect was eliminated by cytochalasin, which can stabilize spectrin. Later, the IR spectroscopy study showed that carbachol and adrenaline produce changes in the secondary structure of these proteins. In the present study we also observed a similar effect on erythrocyte membranes caused by the action of adrenaline and noradrenaline.

Upon incubation of adrenaline with concentration 10^{-9} M and shadows, we observed shifting of absorption bands of CO and NH groups of the peptide bonds. The CO band shifted from 1655.4 to 1654 cm^{-1}, whereas at $C_{adrenaline} = 10^{-7}$ M the shift was 1655.4 $\rightarrow$ 1656 cm^{-1}, accompanied by appearance of the band at 1646 cm^{-1}. In the first case, absorption band of NH bond shifted by 32 cm^{-1}, and in the second case, by 36 cm^{-1}. Splitting of the band of NH bonds was also recorded (table 7, figs. 10, 11).

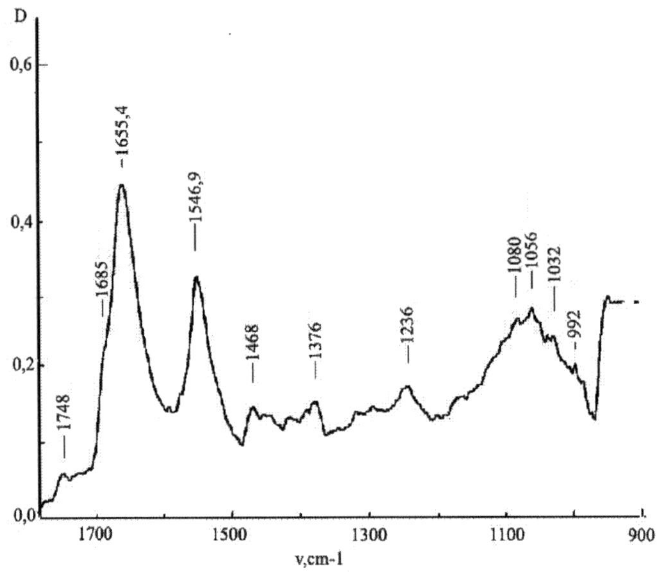

a

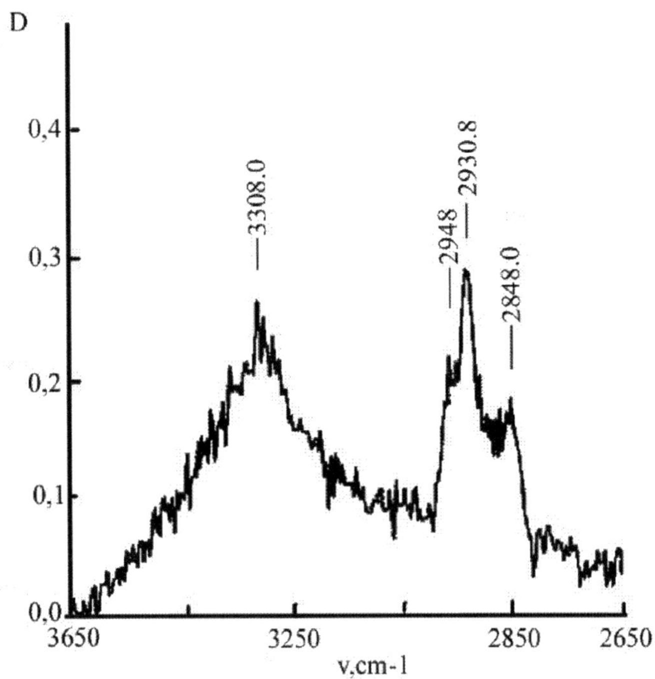

b

Fig. 7. IR spectra of rat erythrocyte membranes (control) ($C_{phosph.\ buff.}$ = 0.01 M, pH 7.35, relative humidity 0%): a) ν = 900-1800 cm^{-1}, b) ν = 2600-3700 cm^{-1}

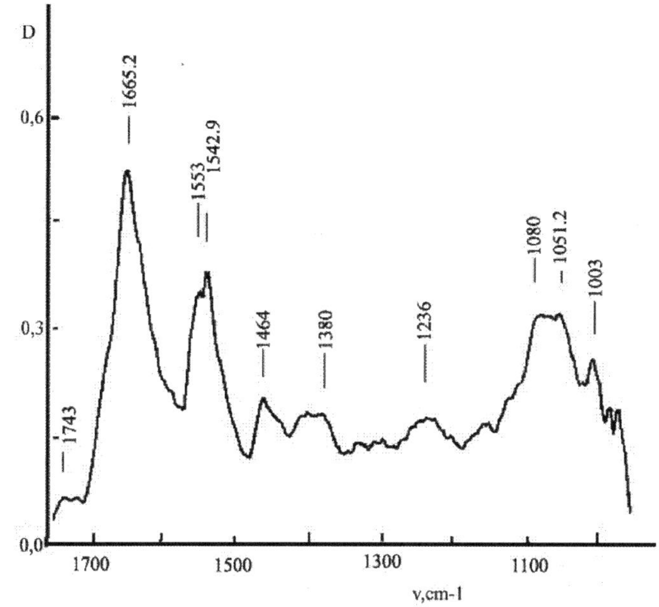

a

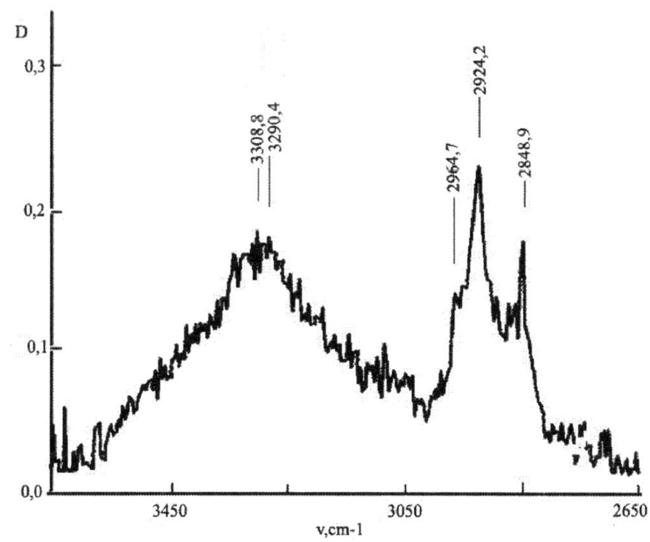

b

Fig. 8. IR spectra of rat erythrocyte membranes incubated with cortisol ($C_C = 4.44 \times 10^{-8}$ M, $C_{phosph.\ buff.} = 0.001$ M, pH 7.35, relative humidity 0%): a) ν = 900-1800 cm^{-1}, b) ν = 2600-3700 cm^{-1}

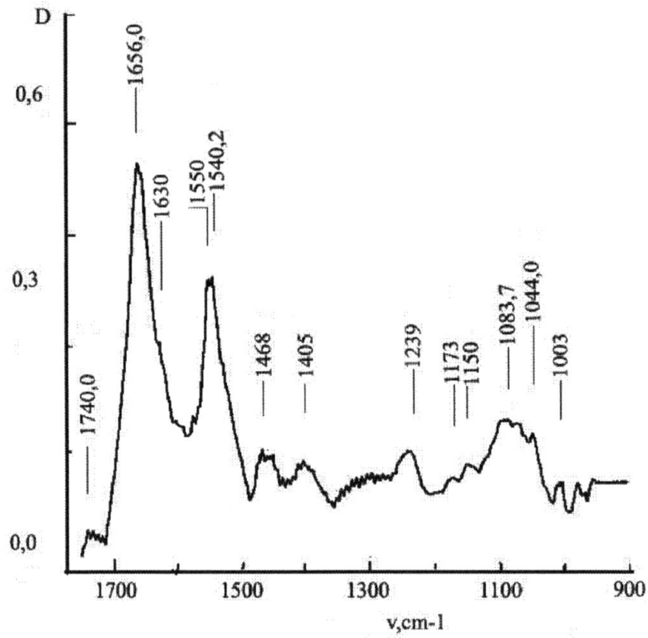

a

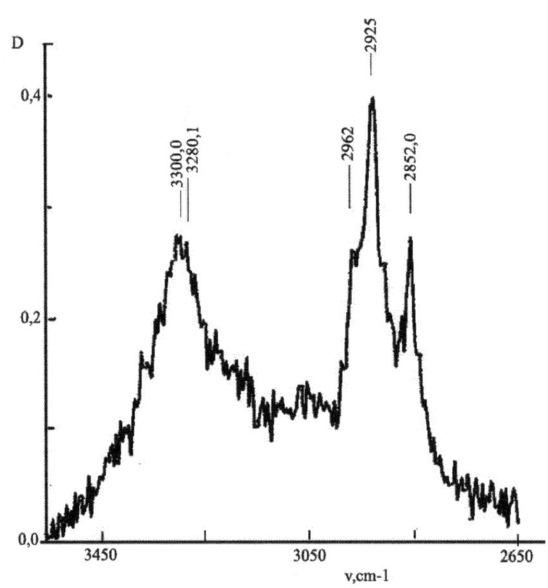

b

Fig. 9. IR spectra of rat erythrocyte membranes incubated with cortisol (C_C = 1.05 x 10^{-7} M, $C_{phosph.\ buff.}$ = 0.001 M, pH 7.35, relative humidity 0%): a) ν = 900-1800 cm^{-1}, b) ν = 2600-3700 cm^{-1}

Interestingly, the low concentration of adrenaline gave a 28 cm^{-1} shift of the band of phospholipid CO bonds, whereas with noradrenaline this value was only 18 cm^{-1}. Splitting of the band was also observed. At the same time, the absorption band of PO bond showed strong splitting (table 7, fig. 10). Its shift to the short-wave region was 22 cm^{-1}, and to the long-wave region, 16 cm^{-1}. Upon splitting, the band shift between extreme bands attained 38 cm^{-1}.

Absorption band in the region of 2928 cm^{-1} (CH stretching vibrations) showed an additional splitting. Bands shifting was also recorded in this part of the spectrum.

Incubation of noradrenaline with erythrocyte shadows was accompanied by shifting of absorption bands assigned to the peptide bond of membrane proteins, in particular, the 1655.4 → 1654 cm^{-1} shift of CO bond ($\Delta v = 1.4\ cm^{-1}$) and 3308 → 3270 cm^{-1} shift of NH bond (stretching vibrations) ($\Delta v = 38\ cm^{-1}$) (table 7, fig. 12). Shifting of the indicated absorption bands points to conformational transitions in membrane proteins: tangle → α-helix and tangle → β-structure. The bands at 1654 and 1685 cm^{-1} increased in intensity as well as the band of NH bond (amide II) at 1548 cm^{-1}. In addition, splitting in the region of NH bond (stretching vibrations) was observed. This splitting indicates the mutual interaction of intramolecular domains comprising α-helices or β-structures, which results from their ordering. However, the splitting may be related also with intermolecular interaction of proteins. Deformation or contraction of membrane under the action of noradrenaline takes place here, similar to the case of adrenaline, thus leading to pronounced changes in the structure of membrane proteins. These changes may be caused by interaction of several functional groups of noradrenaline with the receptor, in particular, amino group and OH groups. Hydrophobic interaction of the aromatic ring of noradrenaline with hydrophobic region of its receptor is also possible.

The 1748→1730 cm^{-1} band shift of phospholipid C=O bond was recorded as well as the 1236→1246 cm^{-1} shift of phospholipid P=O bond. The band of P-O-C bond also shifts by 16 cm^{-1} to the short-wave region of the spectrum. So strong shifting of these bands results from the increased ordering of phospholipids. However, this is caused also by the enhanced interaction with membrane proteins, in particular, integral proteins, protein of stripe 3, and finally with retraction proteins. Note that hydrogen bonds are formed with any of these proteins. Probably the receptor for adrenaline may have not only the protein, but also the protein-lipid nature. If this is the case, OH groups of noradrenaline will interact with C=O or P=O bonds.

Shifting of absorption band of $O_4C_4C_5O_5$ in monosaccharides was recorded (1056 → 1072 cm^{-1}, $\Delta v = 16\ cm^{-1}$). This shift indicates the formation of hydrogen bonds of noradrenaline with the functional groups of monosaccharides as well as the ordering of monosaccharides relative to each other accompanied by changes in their conformation, such as boat → chair. The chair conformation is energetically more advantageous.

Note. A_{CO} is the integral intensity of v_{CO} absorption band of peptide bond on semi-log scale.

Thus, the comparison of adrenaline and noradrenaline effect shows that action of the first hormone on phospholipids is more pronounced as compared to the second hormone. This is related with the presence of methyl radical at nitrogen due to higher electronegativity of nitrogen atom, charge of the methyl group increases, thus enhancing the effect of adrenaline on the membrane. This gives ground to suggest that amino group and methylamino group interact with phospholipids by CO and PO bonds.

The total effect of all the hormones is presented in table 7.

Compound	ν_{CO} (cm^{-1})	ν_{NH} stretch. (cm^{-1})	$\nu_{C=O}$ (cm^{-1})	$\nu_{P=O}$ (cm^{-1})	ν_{P-O-C} (cm^{-1})	N O_4C_4-C_5O_5 (cm^{-1})	ν_{CH} stretch. (cm^{-1})	A_{CO}
Shadows (control)	1655.4 1686	3308	1748	1236	1080	1056	2948 2930 2848	*1.2150E + 01*
Shadows + cortisol (A = 4.4·10^{-8} M)	1655.2	3290.4 3308.0	1743	1236	1080	1051.2	2924.2 2848.9	*1.5169E + 01*
Shadows + cortisol (A = 1.05·10^{-7} M)	1656.0 1630	3280 3300	1740	1239	1083.7	-	2962 2925 2852	*1.5640E + 01*
Shadows + cortisol (A = 2·10^{-7} M)		3280 3296					2952.5 2931.4 2920.0 2853.8	
Shadows + adrenaline (A = 10^{-9} M)	1654	3276 3292 3316	1720 1734 1744	1220 1236 1244 1258	1080 1088 1094	1066	2958 2952 2928* 2852	
Shadows + adrenaline (A = 10^{-7} M)	1656 1646	3272 3288 3304 3320	1740	1220 1236 1248 1258	1080 1096	1070 1060 1044	2956 2924 2852	
Shadows + noradrenaline (A = 1.87·10^{-7} M)	*1654*	*3271* *3300*	*1730*	*1246*	*1096*	*1072*	*2952* *2928* *2852*	

Table 7. IR spectroscopy. Frequency parameters of rat erythrocyte shadows after their interaction with hormones

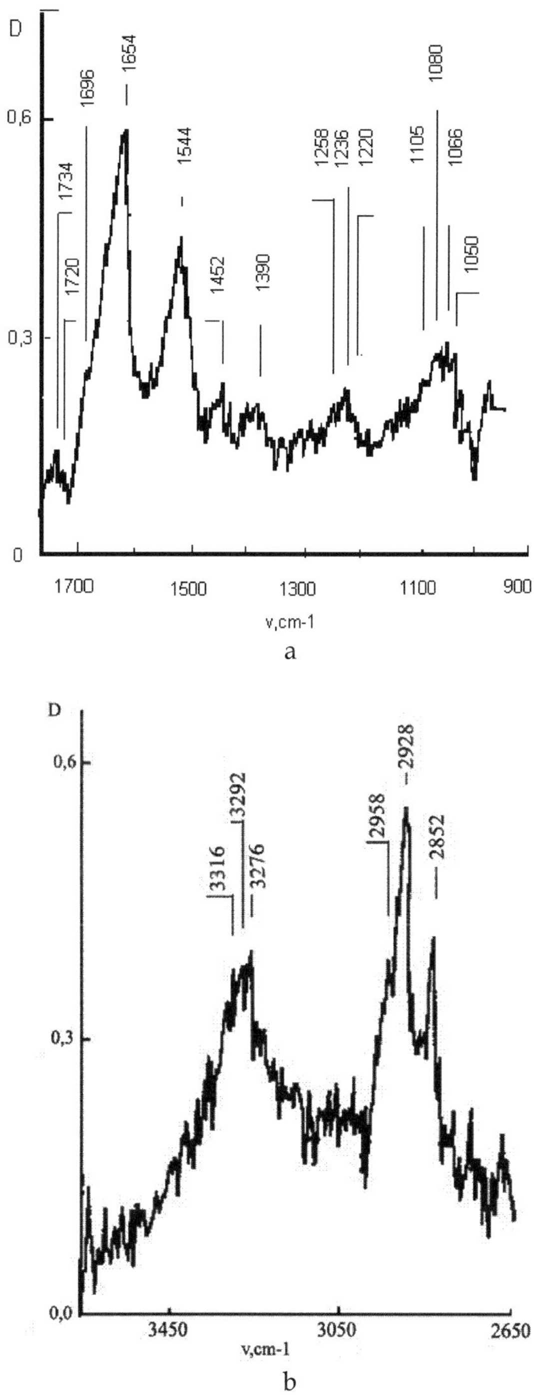

Fig. 10. IR spectra of rat erythrocyte membranes incubated with adrenaline ($C_A = 10^{-9}$ M, $C_{phosph.\ buff.} = 0.001$ M, pH 7.35, relative humidity 0%): a) $\nu = 900\text{-}1800\ cm^{-1}$, b) $\nu = 2600\text{-}3700\ cm^{-1}$

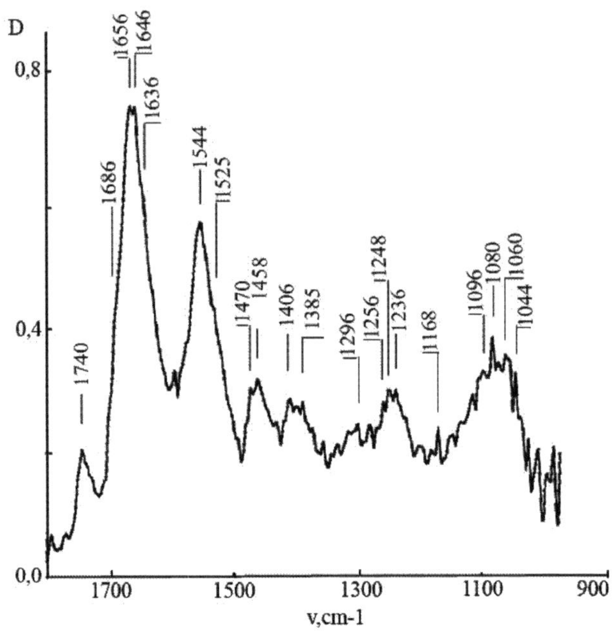

a

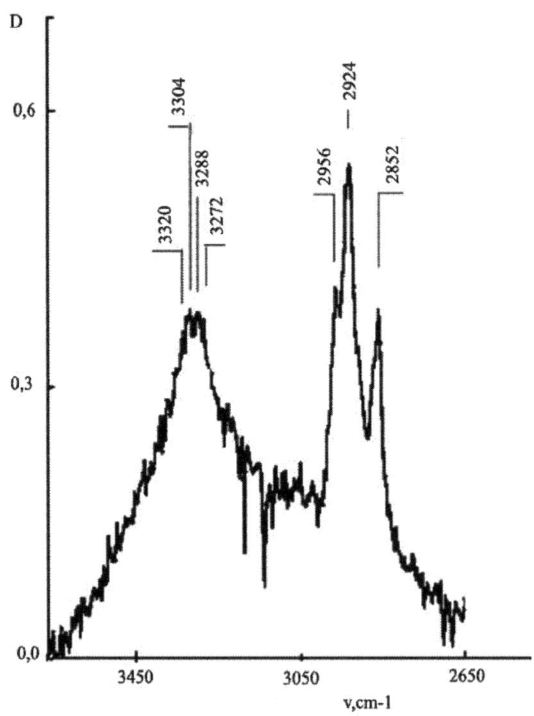

b

Fig. 11. IR spectra of rat erythrocyte membranes incubated with adrenaline ($C_A = 10^{-7}$ M, $C_{phosph.\ buff.} = 0.001$ M, pH 7.35, relative humidity 0%): a) ν = 900-1800 cm^{-1}, b) ν = 2600-3700 cm^{-1}

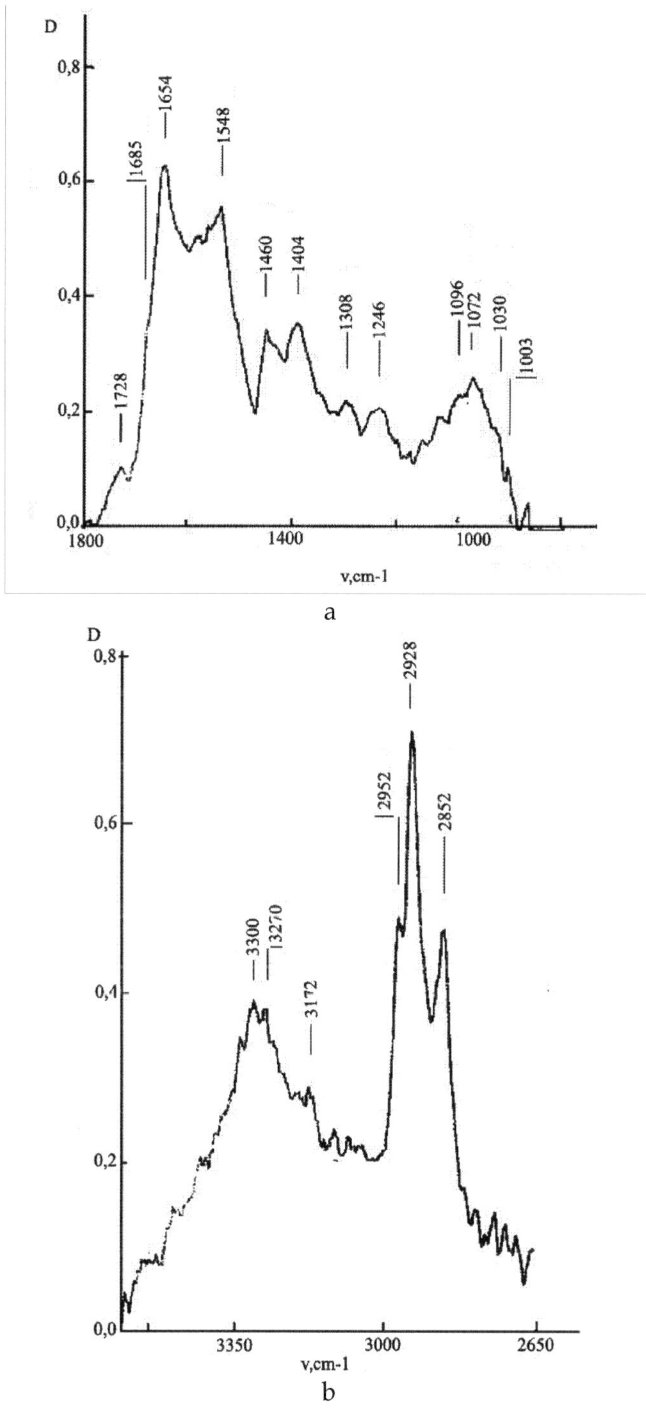

Fig. 12. IR spectra of rat erythrocyte membranes incubated with noradrenaline (C_{NA} = 1.87 x 10^{-7} M, $C_{phosph.\ buff.}$ = 0.001 M, pH 7.35, relative humidity 0%): a) ν = 900-1800 cm^{-1}, b) ν = 2600-3700 cm^{-1}

Fluorescence analysis allowed us to observe the contribution of proteins to initiation of structural transitions in erythrocyte membranes under the action of hormones. For this purpose we used both the effect of light absorption and the tryptophan fluorescence quenching when determining the protein secondary structure.

3.2.2 Fluorescence analysis

As was noted above, cortisol cannot penetrate deep into phospholipid bilayer and interacts with proteins on the surface of cell membrane. This leads to pronounced transformation of their secondary structure and is accompanied by a decrease in the absorption maximum at 228 nm. Addition of hormone to the incubation medium with the concentration $1.82 \cdot 10^{-7}$ M decreased the absorption intensity by 30 a.u., which made up 4.6% as compared to control (the hypochromic effect). The hypochromic effect is related with transition of protein molecule from static tangle to α-helix (Panin & Panin, 2007), i.e., with increased orderliness of its structure.

The absorption spectrum of macromolecules, which have an ordered structure in the excited state, reflects the interaction between excited monomeric units and strongly depends on their spatial arrangement in the molecule.

The structural transitions in proteins at cortisol interaction with erythrocyte membranes are evidenced also by the curves of tryptophan fluorescence quenching. A maximum of fluorescence decrease was observed at the hormone concentration in the incubation medium equal to $11.6 \cdot 10^{-8}$ M and the protein content of 0.256 mg/ml, ΔF was 34 a.u., which makes up 11% with respect to control (fig. 13).

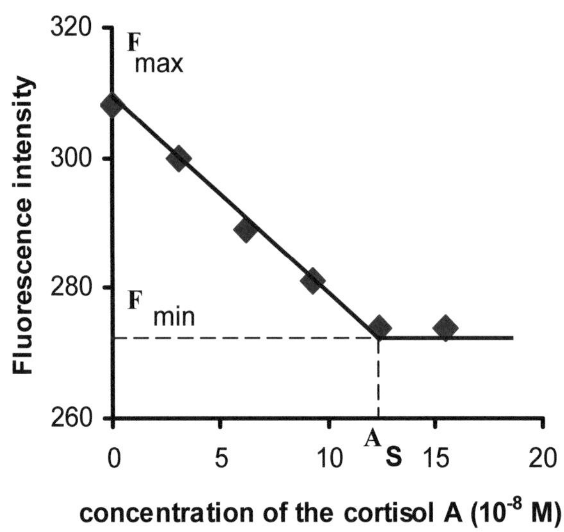

Fig. 13. The dependence of fluorescence intensity of shadows on the concentration *A* of hormone cortisol introduced into the cuvette. A_s is the stoichiometric concentration of the hormone. Concentration of membrane protein C = 0.256 mg/ml. The excitation wavelength λ = 281 nm, emission wavelength λ = 332 nm

The data on tryptophan fluorescence quenching were used to calculate the constants of hormone binding (K_b), the amount of bound hormone (B_{max}, mol/mg of protein), and a change in free energy (ΔG, kJ/mol) upon interaction of cortisol with erythrocyte membrane (table 8). It was found that K_b for cortisol is $(1.23 \pm 0.12) \cdot 10^6 M^{-1}$. A similar value for transcortin (a specific receptor) is $3 \cdot 10^7 M^{-1}$ (Panin & Panin, 2004), which is an order of magnitude higher. However, even in the norm, 6-8% of the hormone is bound to blood cells. These are mainly erythrocytes. Under stress, this value may rise considerably. Our calculations showed that the maximal saturation of erythrocyte membranes with hormone (B_{max}) in vitro is $(4.7 \pm 0.47) \cdot 10^{-10}$ mol/mg of protein, which is quite a low value (table 8). When cortisol interacts with erythrocyte membranes, free energy of the entire system (ΔG) decreases by 30.0 kJ/mol. This indicates that under the action of hormone the orderliness of structural components in erythrocyte membranes increases, whereas their entropy decreases.

Hormone	Binding constant K_b (M^{-1})	Amount of bound hormone B_{MAX} (mol/mg protein)	*Changes in Gibbs free energy* ΔG *(kJ/mol)*
Cortisol	$(1.23 \pm 0.12) \cdot 10^6$	$(4.69 \pm 0.47) \cdot 10^{-10}$	*- 36.0*
Adrenaline	$(6.3 \pm 0.63) \cdot 10^6$	$(1.60 \pm 0.16) \cdot 10^{-11}$	*- 40.2*
Noradrenaline	*$(1.70 \pm 0.17) \cdot 10^6$*	*$(8.10 \pm 0.81) \cdot 10^{-10}$*	*- 36.9*

Table 8. Parameters of cortisol binding to erythrocyte membrane based on tryptophan fluorescence quenching of membrane proteins and parameters of adrenaline and noradrenaline binding to erythrocyte membrane based on the hypochromic effect of tryptophan in membrane proteins

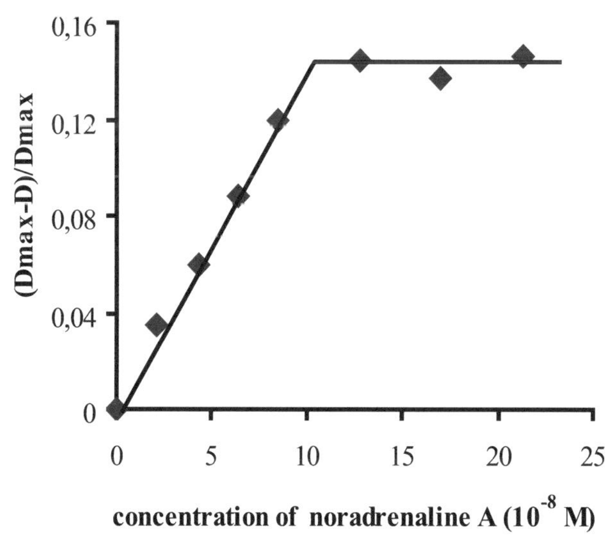

Fig. 14. The dependence of $(D_{max}-D)/D_{max}$ on the concentration A of hormone noradrenaline introduced into the cuvette. Here, D_{max}, D - are the intensities of tryptophan absorption in shadows at $\lambda = 228$ nm, respectively, without addition of hormone and with hormone. Concentration of membrane protein C = 0.124 mg/ml. The emission wavelength $\lambda = 332$ nm

Similar results were obtained in our study of two other stress hormones, adrenaline and noradrenaline. With both hormones, the hypochromic effect is even more pronounced than with cortisol (table 8, figs. 14, 15). This suggests an increased orderliness of the membrane proteins structure due to transition tangle → β-structure → α-helix. A decrease in absorption may be related with changes in the dipole moments direction of quantum transitions of monomeric protein residues accompanying their transition into another conformation.

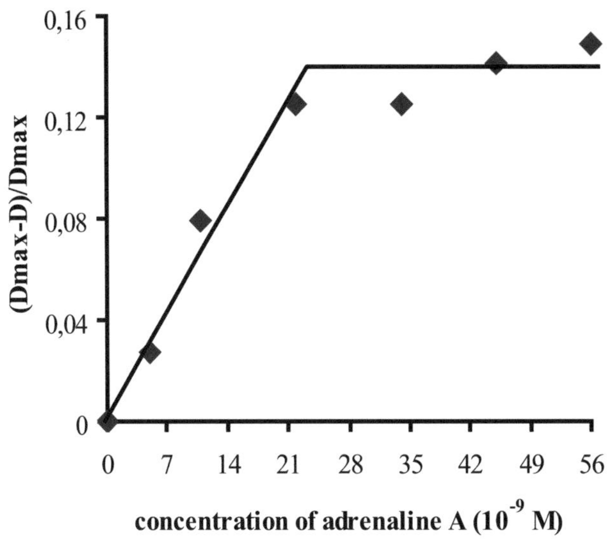

Fig. 15. The dependence of $(D_{max}-D)/D_{max}$ on the concentration A of hormone adrenaline introduced into the cuvette. Here, D_{max}, D - are the intensities of tryptophan absorption in shadows at λ = 228 nm, respectively, without addition of hormone and with hormone. Concentration of membrane protein C = 0.124 mg/ml. The emission wavelength λ = 332 nm

Due to high intrinsic fluorescence of the hormones, their effect on tryptophan fluorescence quenching cannot be revealed. Nevertheless, the absorption measurements allowed us to calculate K_b, B_{max} and ΔG for these hormones. Overall, they were of the same order of magnitude as those for cortisol. Adrenaline showed a higher value of K_b, a lower amount of bound hormone (B_{max}), and more pronounced changes of ΔG (– 40.2 kJ/mol). The latter fact testifies that adrenaline increases the orderliness of structural components in erythrocyte membranes to a greater extent than other hormones. It provided also a stronger deformation of erythrocyte membranes, which agrees with the data of atomic force microscopy (see figs. 24, 25). It was vital to relate structural transitions in erythrocyte membranes under the action of stress hormones with changes in their elasticity and viscosity.

3.2.3 Changes in microviscosity of erythrocyte membranes under the action of cortisol and catecholamines

The addition of cortisol, adrenaline or noradrenaline to erythrocyte shadows increased microviscosity of erythrocyte membranes (figs. 16, 17). The effect was more pronounced in the presence of adrenaline (an increase by 40%) and less pronounced in the presence of noradrenaline (by 24%) and cortisol (by 25%). Adrenaline was able to increase the membrane microviscosity at a much lower concentration of hormone in the incubation medium. With adrenaline, a plateau was reached at the hormone concentration of $1.7 \cdot 10^{-9}$ M, whereas with noradrenaline and cortisol at $7 \cdot 10^{-8}$ M.

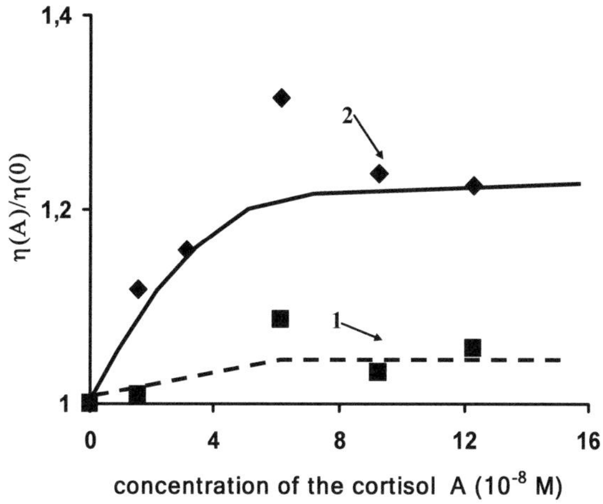

Fig. 16. Changes in the relative microviscosity of membranes *η (A)/η(0)* of erythrocyte shadows at the concentration A of hormone cortisol, where *η(A)* and *η(0* are microviscosities of the membranes, respectively, with cortisol added to the shadows suspension and without hormone. Concentration of shadows C = 0.128 mg protein/ml. Line 1 - changes of relative microviscosity in the region of lipid-lipid interaction, line 2 - changes of relative microviscosity in the region of protein-lipid interaction. Concentration of pyrene in the suspension is $7.7 \cdot 10^{-6}$ M, temperature of the specimens 309.1±0.1 K (36 °C), pH of the suspension 7.35. The measured value of *η (A)/η(0)* exhibits error of 6%

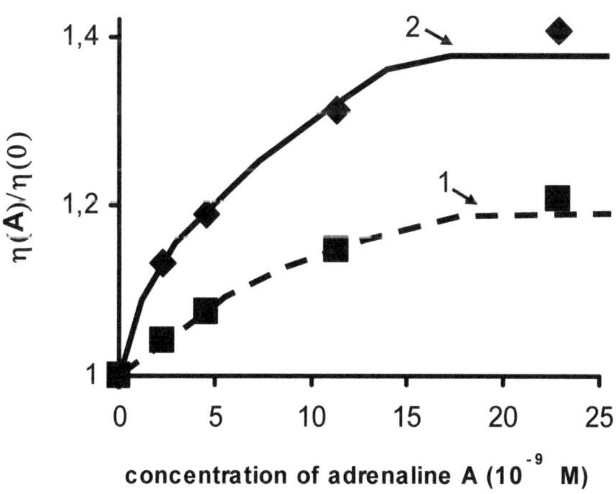

Fig. 17. Changes in the relative microviscosity of membranes *η(A)/η(0))* of erythrocyte shadows at the concentration A of hormone adrenaline, where *η(A)* and *η(0)* - are microviscosities of the membranes, respectively, with adrenaline added to the shadows suspension and without hormone. Concentration of shadows C = 0.128 mg protein/ml. Line 1 - changes of relative microviscosity in the region of lipid-lipid interaction, line 2 - changes of relative microviscosity in the region of protein-lipid interaction. Concentration of pyrene in the suspension is $7.7 \cdot 10^{-6}$ M, temperature of the specimens 309.1±0.1 K (36 °C), pH of the suspension 7.35. The measured value of *η(A)/η(0)* exhibits an error of 6%

For all the hormones, in the region of lipid-protein interactions microviscosity increased at lower concentrations and was more pronounced than in the region of lipid-lipid interactions (fig. 16, 17). An increase in microviscosity of erythrocyte membranes correlated with a decrease of tryptophan absorption in membrane proteins (fig. 13, 14, 15).
Presumably, structural transitions in erythrocyte membranes under the action of stress hormones are initiated to a greater extent in proteins and to a less extent in lipids.

3.2.4 Atomic force microscopy

Under atomic force microscope, erythrocytes of healthy animals looked as large biconcave discs ca. 6 μm in diameter, which agrees with the results obtained by other authors (Wu et al., 2009). At a higher magnification, their surface showed a slight nonuniformity caused most likely by the presence of membrane proteins. When the erythrocyte suspension was supplemented with DMS and ethanol (0.25% of the mixture volume), the surface nonuniformity increased, probably due to denaturating effect of solvent on the surface structural proteins (figs. 18, 19). The pattern changed drastically upon addition of cortisol to erythrocyte suspension with the final concentration 10^{-6} M. On a smooth surface, there appeared numerous meso-stripes loosening the cell membrane structure (figs. 20, 21). These meso-stripes are ca. 7 nm hollows in erythrocyte membrane. They divide the erythrocyte membrane into flat domains.

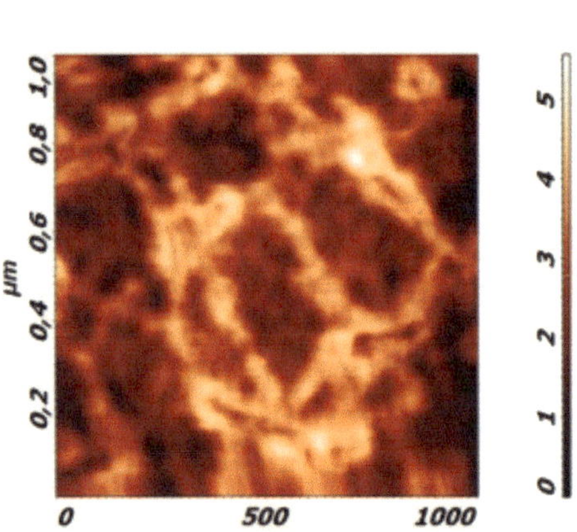

Fig. 18. The control surface of rat erythrocyte. The erythrocyte suspension was supplemented with DMS and ethanol (0.25% of the mixture volume), the surface nonuniformity increased, probably due to denaturating effect of solvent on the surface structural proteins. Scan size 1x1 μm^2

Adrenaline produced more abrupt changes in the structure of erythrocyte membranes (figs. 22, 23). The membrane surface lost its flatness, there appeared convex domains with quasi-staggered arrangement alternating with considerable hollows on the surface. The addition of adrenaline led to domains of size 250 x 250 nm, with smaller domains on the surface of large ones: size 50 x 50 μm^2 and height 5 nm. Noradrenaline caused the formation of domains with the size 100 x 100 μm and height 2 nm (figs. 24, 25).
IR spectroscopy allowed us to reveal the nature of these structural transformations.

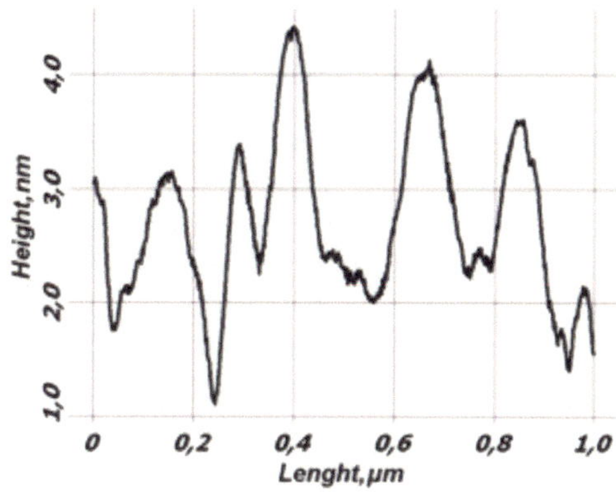

Fig. 19. Center section of the control surface of rat erythrocyte. Section is made from left to right through the center of scan in Fig. 18. Domains with the length 200-250 nm and height 2 nm are seen

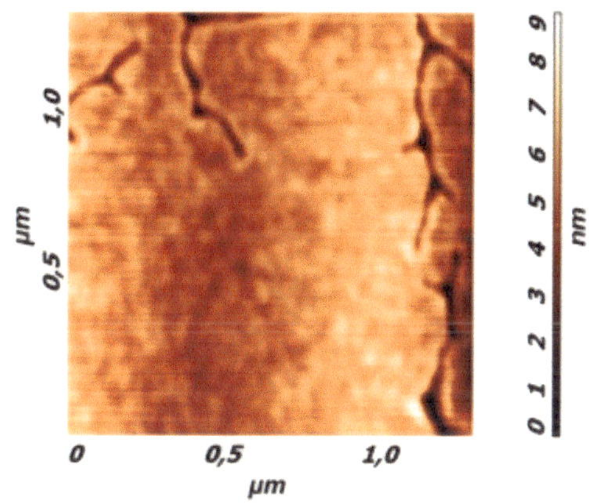

Fig. 20. The surface of rat erythrocyte after adsorption of cortisol. Concentration of the hormone is 10^{-6} M. Scan size 1.3x1.3 μm^2. Deep meso-stripes with bifurcations are seen.

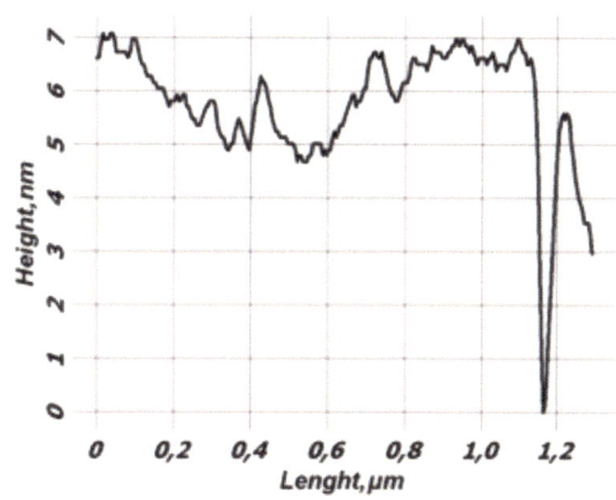

Fig. 21. Center section of the surface of rat erythrocyte after adsorption of cortisol. Section is made from left to right through the center of scan in Fig. 20. The surface is flat, there are hollows of 7 nm depth that divide membrane into domains

3.2.5 Conclusion

Nonspecific mechanisms of the stress hormones interaction with erythrocyte membranes were studied by means of atomic force microscopy, fluorescence analysis, and IR spectroscopy. It was shown that stress hormones (cortisol, adrenaline, noradrenaline) can bind to erythrocyte membranes with high affinity ($K_b \sim 10^6$ M^{-1}). The binding mechanism involves hydrogen bonds and hydrophobic and electrostatic interactions.

Active groups of the hormones (NH_2, $NHCH_3$, keto, and hydroxy groups) interact simultaneously with CO and NH groups both of proteins and phospholipids. This leads to the formation of complex protein-lipid domains that distort the surface of the erythrocyte membrane. Water dipoles are displaced from the domains to adjacent regions and facilitate membrane loosening. The interaction of hormones with the membrane is accompanied by structural transitions of disorder - order (tangle - α-helix, tangle -β-structure) in membrane proteins and structural transitions of order - order in phospholipids. Formation of large domains (clusters) of the lipid-protein and lipid-lipid nature leads to distortion of membranes and deteriorates their elasticity and rheological properties.

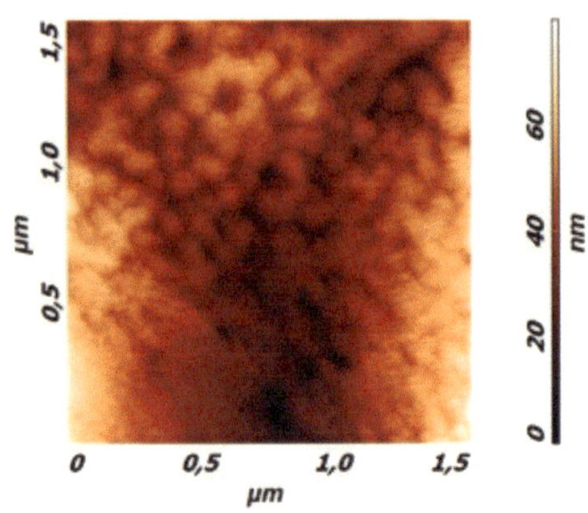

Fig. 22. The surface of rat erythrocyte after adsorption of adrenaline. Concentration of the hormone is 10^{-6} M. Scan size 1.5x1.5 μm^2

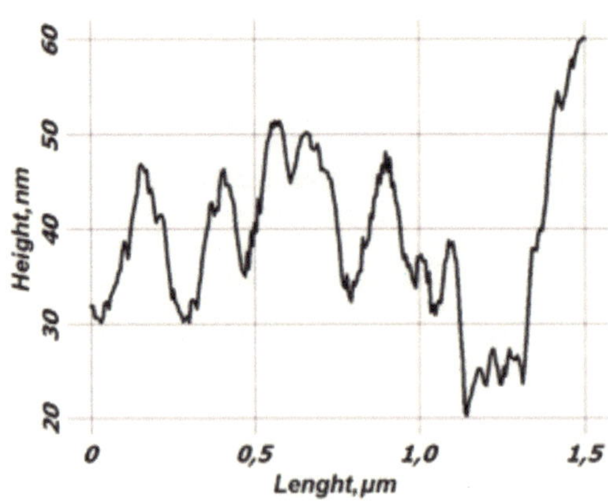

Fig. 23. Center section of the surface of rat erythrocyte after adsorption of adrenaline. Section is made from left to right through the center of scan in Fig. 22. The surface is tuberous, there are domains of size 250 x 250 nm, with smaller domains on the surface of large ones: size 50 x 50 μm^2 and height 5 nm

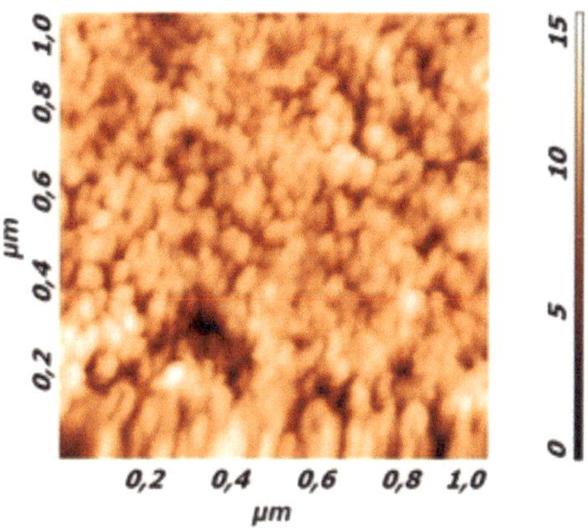

Fig. 24. The surface of rat erythrocyte after adsorption of noradrenaline. Concentration of the hormone is $4 \cdot 10^{-7}$ M. Scan size 1x1 μm^2

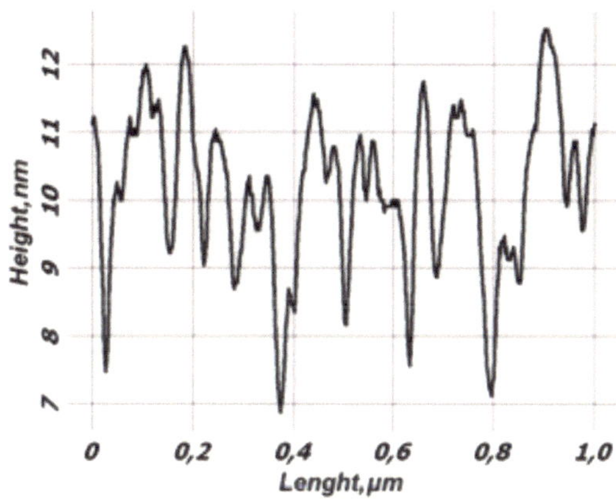

Fig. 25. Center section of the surface of rat erythrocyte after adsorption of noradrenaline. Section is made from left to right through the center of scan in fig. 24. The surface is flat, there are domains of size 100 x 100 µm and height 2 nm

4. Mesomechanics of nanostructural transitions in erythrocyte membranes under the action of metal oxide nanoparticles

The use of nanotechnologies in modern society strongly increased the risk of nanoparticles interaction with the cell membranes of human organism. We studied the effect of metal oxide nanoparticles (SiO_2, ZrO_2) on intact erythrocytes. Such particles possess hydrophobic properties. Nanoparticles of size 5-10 nm penetrate deep into the phospholipid bilayer, which leads to formation of localized deformation meso-stripes (microcracks) (fig. 26b, 26c.). Their formation mechanism also implies the displacement of water dipoles to adjacent regions and membrane loosening due to hydrostatic forces.

Similar fragmentation and fracture is observed in solid crystals in the fields of strong external action. Fig. 26d shows the formation of a domain structure in the bulk of highly nonequilibrium nanostructural aluminium and chains of micropores at its surface layer under severe plastic deformation. Increasing the strain degree causes fracture of the nanostructural material along the porous bands of the local plastic strain (Panin & Egorushkin, 2008).

As shown by experiments, the addition of Al_2O_3 nanocrystals was not accompanied by reliable changes in microviscosity. However, the addition of SiO_2 nanocrystal initiated on the average a 25% decrease in microviscosity both in the region of lipid-lipid and in the region of protein-lipid interaction. A minor rise of microviscosity at the end of the curve is related with strong damage of the membrane and fluorescence of pyrene that "fell out" to the buffer solution and emits light there. In the buffer, the formation of pyrene dimers is more difficult, so the intensity of fluorescence at a wavelength of 468 nm decreases, whereas relative microviscosity increases. Thus, the rise of microviscosity can be attributed to details of the experimental procedure.

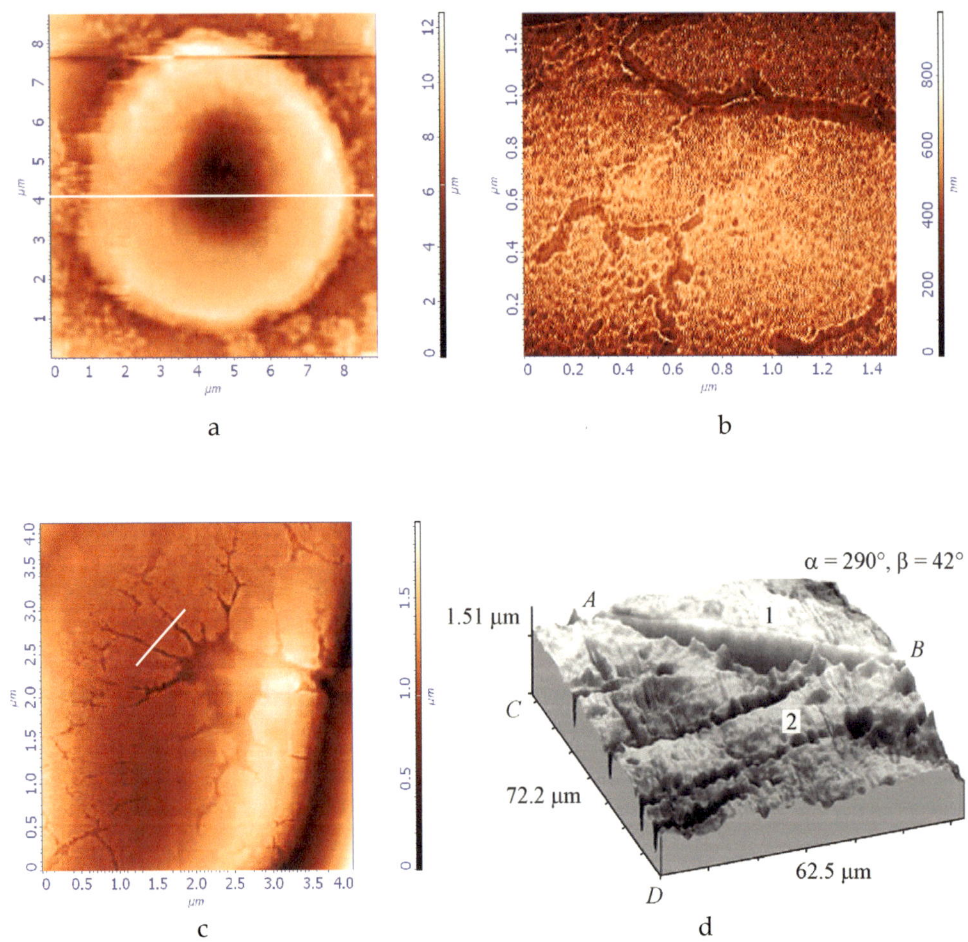

Fig. 26. Structural levels of deformation in liquid and solid crystals
a. intact erythrocyte; formation of localized deformation meso-stripes in erythrocyte under the action of SiO_2 (b) and ZrO_2 (c); formation of micropores and their integration into localized deformation meso-stripes in a sample of commercial Al (d), alternating bending, $17.5 \cdot 10^6$ cycles (Panin & Egorushkin, 2008)

5. Thermodynamics of structural transitions in liquid and solid crystals in the fields of external action

Liquid and solid crystals are homeostatic determinate systems (Panin, 2008).

"Classical thermodynamics leads to the notion of system in equilibrium, such as, e.g., crystal" (Nicolis & Prigogin, 1979). In solid crystals residing in the fields of weak external action, the state of homeostasis is determined by the nature of interatomic interactions. This is the metallic bonding, which has a quantum-mechanical nature. It is responsible for the high shear strength of crystals. According to the literature (Panin & Egorushkin, 2008; Shaniavski, 2007), solid crystal in the fields of strong external action behaves as an open thermodynamically nonequilibrium system. "Pumping" of external energy results in a local loss of shear strength, which is accompanied by local structural-phase transformations in the crystal lattice. The possibility of such transformations is determined by electron energy spectrum of the crystal. This is confirmed by a relationship between packing-defect energy (γ) of the crystal and its electronic structure, i.e., memory elements of the crystal (Panin et al., 1971; Panin, 2008). In solid crystals, which are a structurally inhomogeneous medium, structural-phase transitions occur mainly on the external surface and at the internal interfaces. This is the place where a chessboard distribution of stresses and deformations is formed: the cells under compressive normal stress alternate chequerwise with the cells under tensile normal stress (Panin & Panin, 2007). Tangential stresses also have a chessboard distribution, which is spatially displaced in phase by $\pi/2$. This can explain the physical nature of plastic flow localization at different scale-structural levels under different conditions of loading. The excess molar volume and virtual nodes of a higher energy structure in the interstitial space that are present in the zones of tensile stress allow the occurrence of local structural transformations caused by nonequilibrium state in this zone.

As it was mentioned above, a similar situation is observed for biological membranes under the action of various external factors (metal oxide nanoparticles, stress hormones). In biological membranes considered as liquid crystals, the system-forming bonds are represented by low-energy forces: covalent and hydrogen bonds, hydrophobic and weak electrostatic interactions. They serve as the memory elements and determine also a low shear strength of biological membranes and such notion as membrane "liquidness". It can be stated that homeostatic mechanisms of physical and biological systems in dissipative state are universal.

The concept of homeostatic determinate systems is fundamental for synergetics. It provides deep insight into physical meaning and genesis of the hierarchy of instabilities in self-organizing homeostatic systems, into the nature of interrelations between instabilities and order parameters (Panin, 2008; Panin & Panin, 2004).

Under homeostatic determinate systems we understand such systems where the eventual (actual) result of an action is predicted (determined) via the interaction of signals specific for the given system with its memory elements. Structurally such systems include determinate synthesis, choice of an adequate action program, its outcome, and feedback closed on results-of-action acceptor (fig. 27).

The most complex element of any homeostatic determinate system is the determinate synthesis. Here the dominant motivation, conditional (environmental) and causal (triggering) afferentation interact with the memory elements of the system. This results in decision making and choosing a termodinamic determined action program. Information on the achieved result goes to results-of-action acceptor (feedback), where the anticipated and

actual results are compared. If the goal is not reached, i.e., the actual result does not fit the anticipated one, homeostatic determinate systems switch over to another programs due to changes in the decision made.
Thus, in the expression "homeostatic determinant system" the term "homeostatic" imlies a basic property of the system, its stability, resistance, and the term "determinant" defines the mechanism for achieving this aim and the determinant type of behavior of the system.

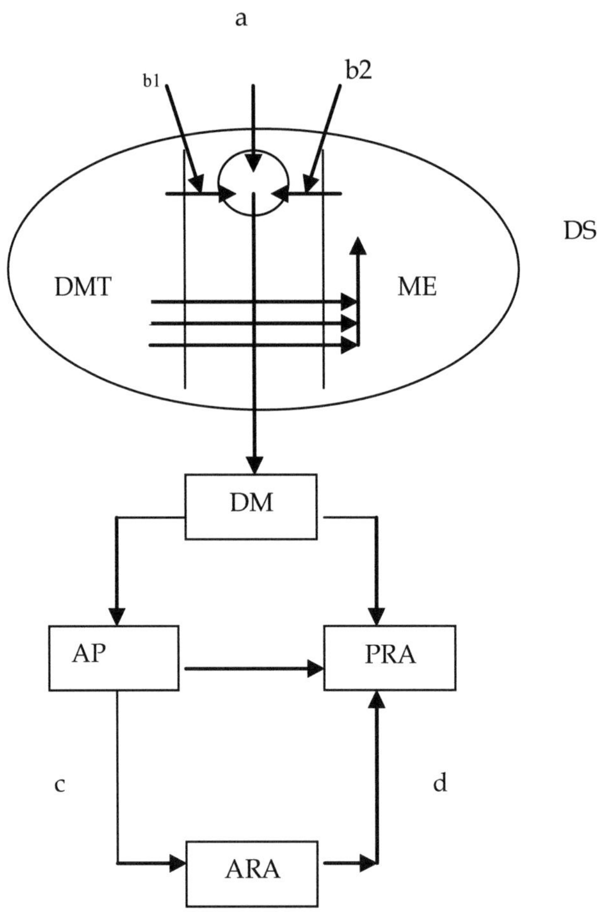

Fig. 27. A schematic diagram of homeostatic determinate system
DMT - dominant motivation: system in a condition of thermodynamic balance; ME - memory elements: metal, covalent and hydrogen bonds, electrostatic and hydrophobic interactions; a — trigger signaling: specific signals able to interact with ME (mechanical load, temperature, pressure, electric discharges, etc.; b_1 and b_2 — circumstantial signaling: additional signaling accompanying the main one, for example, an excited environment; DM — decision making: transition of crystal to a new level of systemic organization; AP — action program: formation of new bonds, which change negenthropy (information) or enthropy; ARA - actual results of action: emergence of a new system in one of energetically allowed state; PRA — predicted result of action: emergence of a new system in the energetically stable state; DS — determinate synthesis, decision-making process; c and d — straight line and feedback

Destruction of solid and liquid crystals increases the molar volume.
A dependence of the Gibbs thermodynamic potential $F(\upsilon)$ on the molar volume υ taking into account local zones of different scale stress concentrators is described by the equation:

$$F(\upsilon) = U - TS + p\upsilon - \Sigma\mu_i C_i, \tag{19}$$

where μ_i - chemical potential, C_i - concentration (fig. 28, Panin & Egorushkin, 2008).
At critical values of molar volume υ_i = (1,2...6), the thermodynamic potential $F(\upsilon)$ has local minima. They reflect local nonequilibrium potentials in the zones of different scale hydrostatic tension. Critical values of υ_i correspond to different levels of homeostasis in a deformable solid:

υ_0 is an equilibrium crystal; the initial level of homeostasis;

υ_1 are the zones of stress microconcentrators where dislocation cores are generated; the next level of homeostasis;

υ_2, υ_3 are the zones of stress meso- and macroconcentrators where local structural-phase transitions with the formation of meso- and macrostripes of local plastic deformation take place; the next levels of homeostasis;

υ_4 corresponds to intersection of curve $F(\upsilon)$ with the abscissa. At a further increase of the local molar volume, changes of the Gibbs thermodynamic potential proceed under the conditions of $F(\upsilon) > 0$, and the system becomes unstable. Various forms of material failure appear; solid crystal starts to behave as a liquid one.

$\upsilon > \upsilon_6$ - the existence of two phases is possible: at $\upsilon = \upsilon_5$ - the vacancy phase atom, at $\upsilon > \upsilon_6$ - different thermodynamic levels of the crystal lattice in a deformable solid, different levels of its homeostasis.

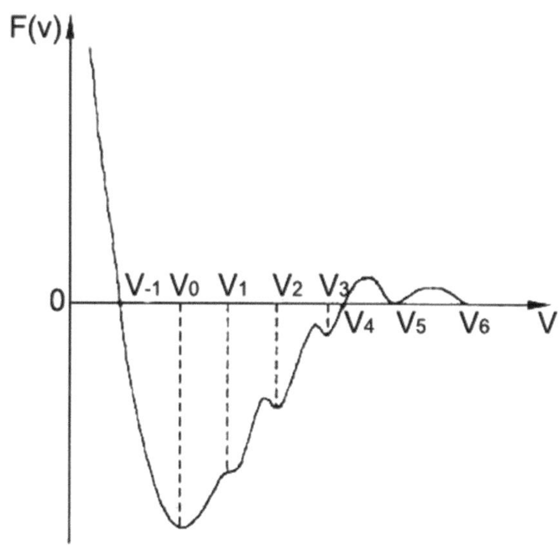

Fig. 28. The dependence of the Gibbs thermodynamic potential F (v) from the molar volume v in the light of local zones of stress concentrators of different scale (Panin & Egorushkin, 2008)

Thus, plastic deformation of solid and liquid heterocrystals in the fields of external action is a multilevel process of their destruction, with the corresponding levels of crystal lattice self-

organization and levels of its homeostasis, i.e., the destruction via different phases of strengthening (self-organization).
Here we see the universal nature of homeostasis of biological and physical systems in dissipative state.
From a physical standpoint, the evolution of material word is based on increasing negentropy, i.e., the degree of orderliness. E. Schrodinger defined it as

$$-S = k \lg (1/D), \quad (20)$$

where -S is the negative entropy, or negentropy; k is the Boltzmann's constant equal to $3.2983 \cdot 10^{-24}$ cal/deg; D is the quantitative measure of disorderliness of atoms in the system, lg (1/D) is the negative logarithm of D, and 1/D is the measure of orderliness. This is the way to estimate negenthropy in liquid and solid crystals.
However, of prime importance for us is that increasing negentropy is always supported by increasing amount of structural information. This can be expressed by the following equation:

$$-S = k \lg(1/D) + \Sigma p_i \cdot \log p_i \quad (21)$$

where p_i is the probability of individual events in the system. Thus, the informational component in this equation determines an increase of negentropy in the system and is related with acquisition of new properties.
Developing the concept about a correspondence between negentropy and structural information, we can present the following equality:

$$-S = k \lg(1/D) = \Sigma p_i \cdot \log p_i \quad (22)$$

Hence,

$$\lg (1/D) = (\Sigma p_i \cdot \log p_i)/k, \text{ and} \quad (23)$$

$$1/D = 10^{(\Sigma pi \cdot \log pi)/k}, \text{ then} \quad (24)$$

$$D = 10^{-(\Sigma pi \cdot \log pi)/k} \text{ or } 1/10^{\Sigma pi \cdot \log pi)/k} \quad (25)$$

Then dependence of Gibbs thermodynamic potential F(υ) on the molar volume υ taking into account local zones of stress concentratotrs is determined by the expression:

$$F(\upsilon) = U - T/10^{(\Sigma pi \cdot \log pi)/k} + p\upsilon - \Sigma \mu_i C_i, \quad (26)$$

These quantitative interrelations underlie the emergence of new homeostatic determinate systems with new properties in the nature.
The principles of systemic organization of matter based on homeostatic determinate systems, where memory elements are represented by metal, covalent and hydrogen bonds, electrostatic and hydrophobic interactions, have gained wide acceptance in physics and chemistry. The formation of hydrogen molecules from hydrogen atom is a simple example of mechanism self-organization.
The structure of hydrogen atom is very simple: a proton with an electron revolving around it. In the atomic state, hydrogen is a gas with a high level of entropy (low level of negenthropy). Its atoms, colliding with each other, excite electrons, thus making them to

occupy different energy levels and sublevels. Such electrons have a large number of degrees of freedom. However, the most probable is the state when electron occupies the lowest energy level (Is). For such a gas, temperature elevation serves as a specific signal interacting with the memory elements (electromagnetic forces). As kinetic energy of the gas increases to a certain value, atoms having electrons at the 1s-level will interact with each other to form hydrogen molecules. Therewith, electrons form a new hybrid wave.
Different programs used for the formation of covalent bonds in hydrogen molecules (fig. 29).

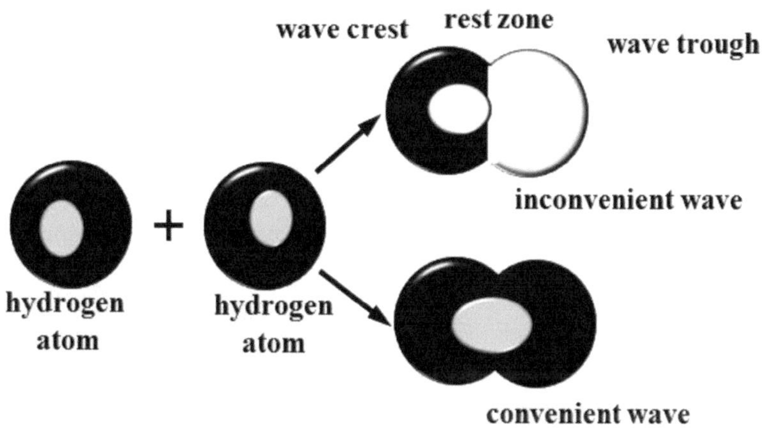

Fig. 29. Programs used for the formation of covalent bonds in hydrogen molecules

At the most stable state of such molecules, the gain in energy attains 4 eV. Another essential aspect of this directional process is increasing the amount of information related to electrons. It is described by the Shannon equation:

$$- H = \Sigma p_i \log p_i \tag{27}$$

In atomic hydrogen, the probability that electron occupies one of allowed energy levels is rather high. A minimum of information is connected with such electrons. When each pair of electrons forms a new hybrid wave, this abruptly decreases the number of degrees of freedom, and their behavior becomes determinate. In a new wave, each electron may stay only in one of two allowed states, which differ in their spin (+1/2 or –1/2). The amount of information connected with such electron increases and depends on the quantity 1/2 · log 1/2. For each pair of electrons this will be equal to log 1/2. Multiplying this quantity with the number of hydrogen molecules in the given gas volume gives the total amount of information connected with electrons. This value corresponds to an increase of negentropy upon transition from atomic to molecular hydrogen.
Decomposition of H_2 molecules yielding atomic hydrogen will give rise to an inverse process: decreasing negentropy (structural information) and growing entropy. Similar mechanisms show up also upon destruction of heterocrystals in the fields of external action.

6. Conclusion

Thus, thermodynamics and mesomechanics of nanostructural transitions in biological membranes as liquid crystals resemble structural transitions that occur in solid crystals in the fields of external action. The distinction is that system-forming binding in biological membranes is represented by covalent and hydrogen bonds, hydrophobic and electrostatic interactions. These are the low-energy bonds. They determine a weak shear strength and such feature as "liquidness" of the membranes. This is very important for functioning of, e.g., erythrocytes that move in the capillary network. Capillary and erythrocyte have comparable sizes, so the movement of erythrocyte is hampered by the frictional force. However, high fluidity of erythrocyte cell membrane makes this process feasible under normal conditions. In extreme states or at pathology related with changes in temperature, blood pH and its salt composition, structural phase transitions are observed in the membranes. There are the smectic A $\rightarrow$ smectic C, smectic $\rightarrow$ cholesteric and nematic $\rightarrow$ isotropic state transitions. In membrane bound proteins, the tangle $\rightarrow$ α-helix and tangle $\rightarrow$ β-structure transitions are essential. They alter the membrane properties, in particular their "fluidity".

Structural phase transitions in erythrocyte membranes occur also under the action of stress hormones (cortisol, adrenaline). The active CO and OH groups of the hormones interact with CO and NH groups of the cell membrane proteins and phospholipids to form hydrogen bonds. Large domains form on the cell surface, which comprise structural proteins and local phospholipid environment. Hydrophobic rings of the hormones enhance hydrophobic interactions in the domains. Water dipoles are displaced to adjacent regions and facilitate there the action of stretching hydrostatic forces. This leads to appearance of nanostructural phase boundaries, along which the membrane damage occurs. The plastic deformation meso-stripes with bifurcations are formed. Viscosity of such membranes increases in the region of lipid-lipid and protein-lipid interactions. Such erythrocytes are able to occlude capillaries. In the heart, this may cause diffusion hypoxia and sudden cardiac arrest. Such cases were observed recently in sportsmen during the high-level competitions (Courson, 2007; Montagnana et al., 2008).

Damage of erythrocyte membranes upon interaction with metal oxide nanoparticles follows the same way. Hydrophobic particles of size 5-10 nm penetrate deep into a lipid bilayer. Due to enhancement of hydrophobic interactions, water dipoles are displaced to adjacent regions. Here, the stretching hydrostatic forces lead to the formation of microcracks and meso-stripes of plastic deformation.

In solid crystals, system-forming bonds are represented by strong interactions. This is the metallic bonding, which has a quantum-mechanical nature. Such crystals possess a high shear strength. However, "pumping" of additional energy in the fields of external action causes a local loss of shear strength, which is accompanied by local structural-phase transformations in the crystal lattice.

Biological membranes and solid crystals are the self-organizing systems. In the process of destruction, they pass several levels of self-organization, with the corresponding hierarchy of instabilities. In terms of synergetics, they are the homeostatic determinate systems. In such systems, memory elements are represented by the system-forming bonds: metallic, covalent and hydrogen bonds, hydrophobic and electrostatic interactions. Exerting some effect on these bonds, external factors induce transition of the system to a new level of homeostasis or self-organization, i.e., to thermodynamically steady states. Ultimately, such transitions may result in complete demolition of the system. All that indicates a mechanistic similarity in the behavior of biological membranes and solid crystals.

Thus, the introduction of principles and regularities of physical mesomechanics in biology provides a deep insight into the mechanism interrelating structure and function of biological

membranes, both in the norm and at systemic membrane pathology (upon variation of hormone concentration, temperature, pH, chemical potencial etc.).
It's may to agree with Henri Poincare, who believed that "the true and only aim of science is to discover unity, but not mechanism".

7. References

Attallah, N.A. & Lata, G.F. (1968). Steroid-protein interactions studied by fluorescence quenching. *Biochim Biophys Acta,* Vol. 168. Issue 2, (October 1968) pp. 321-333, ISSN 0005-2795

Courson, R. (2007). Preventing sudden death on the athletic field; the emergency action plan. *Curr. Sports Med.Rep.*, Vol. 6, No. 2, (Apr. 2007) pp. 93-100, ISSN 1537-890X

Dawson, R.M.C.; Elliot D.C.; Elliot W.H. & Jones K.M. (1986). *Data for biochemical research.* Clarendon Press, ISBN is absent, Oxford, United Kingdom.

Kunitsyn, V.G.; Panin, L.E. & Polyakov, L.M. (2001). Anomalous change of viscosity and conductivity in blood plasma lipoproteins in the physiological temperature range. *Int. J. Quantum Chem.*, Vol. 81, (Feb. 2001) pp. 348-369. ISSN 0020-7608

Leto, T.L. & Marchesi V.T., (1984) A structural model of human erythrocyte protein 4.1. *J. Biol. Chem.*, Vol. 259, N. 7. (Apr. 1984) pp. 4603-4608, ISSN 0021-9258

Montagnana, M.; Lippi, G.; Franchini M.; Banfi, G. & Guidi, G.C. (2008). Sudden cardiac death in young athletes. *Intern. Med.*, Vol. 47, No. 15, (Aug. 2008) pp. 1373-1378, ISSN: 0918-2918

Nicolis, G. & Prigogin I. (1979). Self-organization in nonequilibrium systems. *From dissipative structures to orderliness through fluctuation*, Mir, ISBN, Moscow, Russia

Observation on the Soviet/Canada Transpolar Ski Trek (1992). R.J.Shepard, A. Rode (Ed.), Karger, ISBN 3-8055-5410-9, Basel, Switzerland

Ooi, T.; Itsuka, A.; Onari S. et al. (1988). *Biopolymers,* Imanisi Y. (Ed.), Mir, ISBN is absent, Moscow, Russia

Palek, J. & Sahr, K.E. (1992). Mutations of the red blood cell membrane proteins: from clinical evaluation to detection of the underlying genetic defect. *Blood,* Vol. 80, No. 2, (Jul. 1992) pp. 308-330, ISSN 0006-4971

Panin, L.E. (2008). *Determinate systems in physics, chemistry and biology*. Siberian University publishing house, ISBN 978-5-379-00495-8, Novosibirsk, Russia

Panin, V.E.; Dudarev, E.F. & Bushnev, L.S. (1971). Structure and mechanical properties of substitutional solid solutions, *Metallurgy,* ISBN is absent, Moscow.

Panin, V.E. & Egorushkin, V.E. (2008). Nonequilibrium thermodynamics of a deformed solid as a multilevel system. Corpuscular-wave dualism of plastic shear. *Physical Mesomechanics,* Vol. 11, N. 2. (Mar.-Apr. 2008) pp. 9-30, ISSN 1029-9599

Panin, V.E. & Panin, L.E. (2004). Scale levels of homeostasis in a deformable solid. *Physical Mesomechanics,* Vol. 7, No. 4, (Jul.-Aug. 2004) pp. 5-23, ISSN 1029-9599

Panin, L.E. & Panin, V.E. (2007). Effect of the "chessboard" and mass transfer in interfacial media of organic and inorganic nature. *Physical Mesomechanics,* Vol. 10, No. 6, (Dec. 2007) pp. 5-20, ISSN 1029-9599

Sergeev, P.V.; Galenko-Yaroshevskiy, P.A. & Shymanowski, N.L. (1996). *Studies of biochemical pharmacology,* Gogotova, M.V. (Ed.), R.TS. Farmedinfo, ISBN 5-85556-018-X, Moscow, Russia

Shaniavski, A.A. (2007). *Modeling of fatigue cracking of metals. Synergetics for aviation,* Publishing Hause of Scientific-and-Technical Literature, "Monography", ISBN 978-5-94920-058-2, Ufa, Russia

Wu, G.; Hu, G; Cai, J.; Ma, S.; Wang. X.; Chen, G. & Pan, G. (2009). Time-dependent surface adhesive force and morphology of RBC measured by AFM. *Micron*,Vol. 40, Issue 3., (Apr. 2009) pp. 359-364. ISSN 0968-4328

13

Adsorption Profiles and Solvation of Ions at Liquid-Liquid Interfaces and Membranes

William Kung[1], Francisco J. Solis[2] and Monica Olvera de la Cruz[3]
[1]*Department of Materials Science and Engineering, Northwestern University, Evanston, IL 60208-3108*
[2]*Division of Mathematical and Natural Sciences, Arizona State University, Glendale, AZ 85306*
[3]*Department of Materials Science and Engineering, Department of Chemistry, Department of Chemical and Biological Engineering, Northwestern University, Evanston, IL 60208-3108*
USA

1. Introduction

1.1 Motivations

Electro-chemical and physical processes at liquid-liquid interfaces and across membranes are broadly important in many systems. These processes are relevant for cellular and physiological systems (34; 18), biotechnology (12), everyday applications such as portable batteries, and even in speculative scenarios for the origin of life (31). Biologically, ion transport is essential for the creation of transmembranous and transcellular electric potential, fluid transport, and maintaining cellular volume (1); the failure in the regulation of the above would lead to such conditions as septicemia induced pulmonary edema, electrolyte abnormalities in pyelonephritis of early infancy, hypovolemia and hyponatremia (8). In the chemistry community, there has been a long interest in the rich electrochemistry associated with interfacial charge transfer that is important in, for example, hydometallurgy and two-phase electrolysis (36).

The interaction of ionic species, particularly proteins and nucleic acids, with interfaces and membranes has become exceedingly important in our understanding of biological systems (34) and in processes such as genetic transformation, since the latter specifically involves DNA crossing cellular membranes. Moreover, important biomolecular processes take place more efficiently at heterogeneous media than in homogeneous systems, as demonstrated in the example of DNA renaturation at water-phenol interfaces (12). Therefore, the study of ionic profiles at interfaces also leads to a better understanding of these various biomolecular processes. In general, *in situ* experimental techniques are required to evaluate interface phenomena. For example, x-ray-standing waves measurements provide the profile of ions (2; 3) and also more recently of nucleic acids (4; 21), both along hard surfaces in aqueous solutions (29; 38; 33; 20). Recent experimental techniques have detected ion density profiles at liquid-liquid interfaces with the use of x-ray reflectivity measurements (23; 24). These experiments provide direct knowledge of electrostatic interactions among ionic species

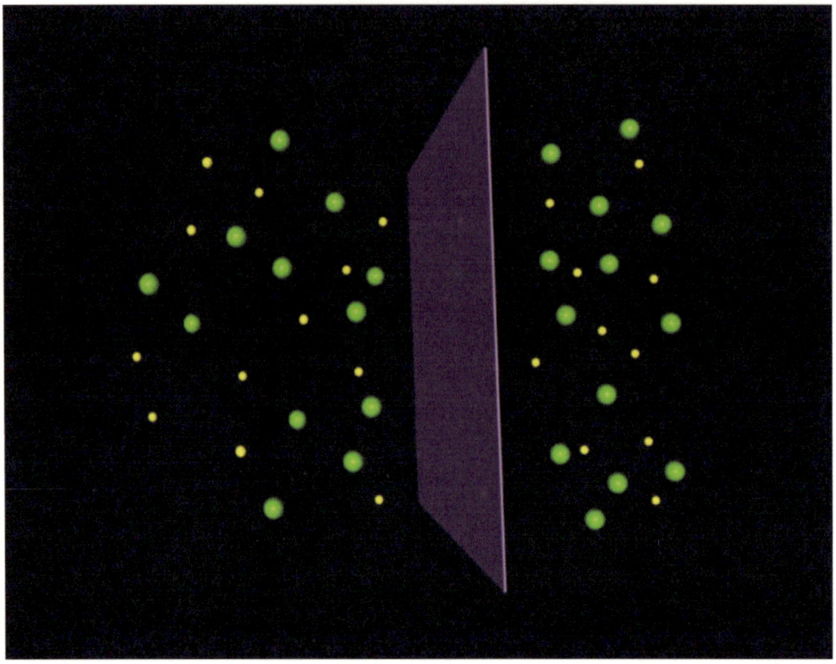

Fig. 1. A single charged plate in salt solution. The system is symmetric with respect to $z = 0$. Without loss of generality, the charged plate is assumed to have a positive surface-charge density σ, with counterions (yellow) and coions (green) present on both sides of the plate.

at interfaces and can be easily extended to analyze the adsorption of proteins and nucleic acids at liquid-liquid interfaces.

In their experimental study (23; 24), Luo *et al.* demonstrated that the predictions from Gouy-Chapman theory did not match well with x-ray reflectivity measurements, presumably due to its neglect of molecular-scale structure in the liquid solution. To better fit their experimental data and the free-energy profile of the ions, the authors instead formally introduce an *ad hoc* term in the electrostatic energy, which presumably includes effects from short-range correlations and which can be theoretically approximated by the potential of mean force, experienced by single ions near the interface, obtainable from molecular dynamics (MD) simulations. Notwithstanding the absence of a fundamental theoretical framework for the aforementioned system of a planar liquid-liquid interface, progress has been made with phenomenological approaches, including the example of a Ginzburg-Landau theory of ions solvation and their distribution around an interface (32).

1.2 Statement of problem

In this work, we will present a top-down thermodynamic formulation for the system of dissolved ions in immiscible polar-binary mixtures. In our formulation, we can naturally incorporate details such as the interaction between ions and background solvents, at the mean-field level, via the Born approximation (25; 26). We will then relate the various hitherto phenomenological parameters, namely, the bulk ion densities in each phase as well as the distribution potential across the interface, to the fundamental parameters of our system that include the dielectric constant of each phase and the size and charge of dissolved ions. In addition, the full nonlinear Poisson-Boltzmann equation follows from a variational principle and can be analytically solved for the electrostatic potential and for the ion density profiles without further approximations. Electro-neutrality, in our formulation, simply manifests from consistency requirement with thermodynamics.

We emphasize that to ensure full analytical tractability, we have not included the full effects of such features as finite ion-size interaction or correlation effects beyond mean-field. While research works that investigate the aforementioned effects as well as other specific types of component interactions are present elsewhere (43; 28; 11), the exact results contained in this work enable a concise quantitative assessment of the importance of finite-ion size interaction and correlation effects in these systems predicted by more complex models.

1.3 Phenomenology

As an important consequence of thermodynamics and electrostatics, we will demonstrate the physics of *differential adsorption* of ions near an interface, which is a general phenomenon occurring between two phases of different dielectric medium wherein ions are selectively driven into confinement near, or expulsion from, the interface based on their charge polarity. In particular, we will consider examples involving both neutral and charged membranes partitioning two chemically different environments. Given the long-standing interest in this topic across diverse disciplines and the many still unresolved issues, it is inevitable that there exists many prior works on the subject with very different emphases and approaches. For example, chemists have traditionally focussed on detailing the microscopic molecular interactions between ions, solvents, and the different types of membranes (36; 19; 7). For the same reason, there exists considerable varied usage of terminology (27) regarding similar concepts. For clarity, we will henceforth explicitly consider in this work only permeable membranes. In addition, to establish notations and conventions, we will refer to the potential gradient across the interface, due to the different existing bulk ions densities in the two phases, as the distribution potential Φ_D. The cases of neutral and charged membranes (regardless of polarity) would correspond to the terminology of nonpolarizable and polarizable liquid-liquid interfaces, respectively, as discussed in chemistry literature. While partial results regarding these systems have been derived in various contexts and guises, our present work represents a unifying framework in a complete, mean-field thermodynamic formulation for the system of dissolved ions in immiscible polar binary mixtures, within which we can derive rigorous relations between the different phenomenological parameters of the system that would otherwise be unrelated when treated in the context of electrostatics alone.

1.4 Chapter organization

In what follows, we will present, in section II, our thermodynamic formalism in terms of a mean-field Ω-potential and derive, via variational principle, the relevant electrostatics and thermodynamic constraints in our system. We will then proceed in section III to consider full solutions to the nonlinear Poisson-Boltzmann equation in this new context and demonstrate the general mechanism of differential adsorption of ions across a liquid-liquid interface. Section IV concludes our paper.

2. Thermodynamics of ion solvation

2.1 Variational approach

In this section, we construct a thermodynamic potential to describe the equilibrium properties of salt ions dissolved in a pair of immiscible fluids. These fluids can be homogeneous or of two respective phases of a segregated mixture. For each phase $\alpha = \{A, B\}$, we indicate

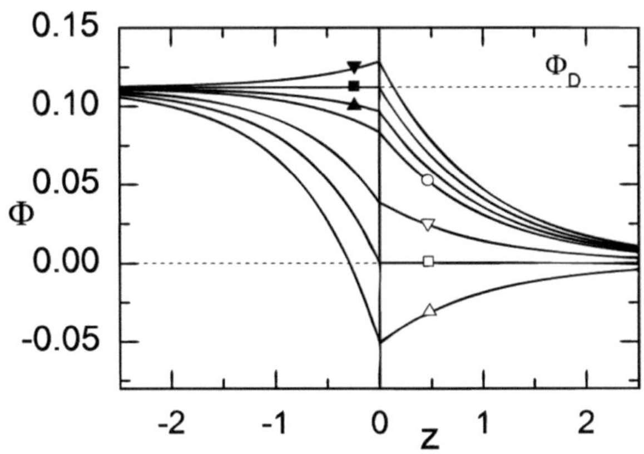

Fig. 2. Comparison of the electrostatic potential across a positively and negatively charged membrane in nitrobenzene/water solution at 68F. The surface charge densities are quoted in units of $\epsilon_A \kappa_A$ and $\Phi'_D \approx 0.112$. The potential is plotted for $\sigma = -1.5(\triangle), \sigma = -0.93(\square)$, $\sigma = -0.5(\triangledown), \sigma = 0(\circ), \sigma = 0.15(\blacktriangle), \sigma = 0.32(\blacksquare), \sigma = 0.5(\blacktriangledown)$.

the dielectric constant as ε_α and denote the ion number bulk densities by $\rho^\alpha_{0,i}$ for each of the ion species i. Far away from the interface, there is a relative potential difference between the phases, known as the distribution potential Φ_D. As we demonstrate below, a number of relations between these bulk quantities arise from the condition of equal chemical potential for each of the species: $\mu_i^A = \mu_i^B$.

2.2 Ω-potential

The thermodynamics with this equilibrium condition are formulated by constructing the Ω-potential (equals to $-PV$ where P is the pressure and V is the volume) for each phase α as follows:

$$\Omega^\alpha = \Omega^\alpha_{el} + \Omega^\alpha_{th} + \Omega^\alpha_{sol}, \tag{1}$$

where we have divided the potential into a contribution due to electrostatics Ω_{el}, to thermodynamics Ω_{th}, and to solvation interactions Ω_{sol}. In what follows, all energy quantities will be measured in units of $k_B T$, all length scales in units of the Bjerrum length $\ell_B = e^2/k_B T$, and all ionic charges quoted as multiples of the fundamental electron charge e. The electrostatic part delineates the free-energy dependence on the dielectric constant ε^α and imposes Gauss' law via a Lagrange-multiplier constraint using the electrostatic potential Φ:

$$\Omega^\alpha_{el} = \int \frac{\epsilon_\alpha}{8\pi} |\mathbf{E}^\alpha(\boldsymbol{x})|^2 - \int \Phi^\alpha \left[\nabla \cdot \frac{\epsilon_\alpha \mathbf{E}^\alpha(\boldsymbol{x})}{4\pi} - \sum_i Z_i \rho_i^\alpha - \rho_F \right]. \tag{2}$$

The total charge density is obtained from the charges of free ions and a contribution ρ_F from fixed charges. We will only consider the case where these charges are present at the interface between media. As these charges are not present in the bulk, they will only contribute to terms determining the boundary conditions for the fields and concentrations.

On the other hand, the thermodynamic contribution consists of the entropy associated with the ions, treated as point particles, in solution as well as their corresponding chemical potentials

$$\Omega^{\alpha}_{th} = \int \left(\sum_i \rho^{\alpha}_i(\boldsymbol{x}) \log \left[\frac{\rho^{\alpha}_i(\boldsymbol{x})}{e} \right] - \sum_i \mu^{\alpha}_i \rho^{\alpha}_i(\boldsymbol{x}) \right) \tag{3}$$

where μ^{α}_i are the chemical potentials of the species within each phase, and they will be specified by the equilibrium constraint that they should be equal across the interface.
Lastly, the solvation interaction between the ions and the respective polar solvent in each phase is modeled by the mean-field Born approximation of polarziation;

$$\Omega^{\alpha}_{sol} = \int \sum_i g^{\alpha}_i \rho^{\alpha}_i, \tag{4}$$

where the Born polarization energy g^{α}_i is given by

$$g^{\alpha}_i = \frac{Z_i^2 e^2}{8\pi\varepsilon^{\alpha} R_i}, \tag{5}$$

for ions with valency Z_i and radius R_i. As expected, we note that an increase in dielectric constant ε^{α} or in the ionic radius R_i for a given valence Z_i facilitates solvation of ions in polar mixtures and leads to an overall lower Born energy. Consequently, Born's model predicts a larger solubility for anions (which are generally larger in size due to reduction in redox chemistry) than for the corresponding cations. As we will see in subsequent sections, both charge properties and the size factor play an important role in the physics of ion solvation near a boundary interface.

2.3 Dimensional analysis and charge neutrality

To examine the consequence of our thermodynamic formulation of ion solvation in immiscible polar binary mixtures, we first perform dimensional analysis on the electrostatic term. We note that the Ω-potential is an extensive quantity by construction: $\Omega^{\alpha}_{el} \propto V$, where V is the volume of the overall system. For the imminscible binary mixtures consisting of one positive and one negative ionic species, respectively, we define the total charge density $\rho^{\alpha}_c(\boldsymbol{x}) = |Z_+|\rho^{\alpha}_+(\boldsymbol{x}) - |Z_-|\rho^{\alpha}_-(\boldsymbol{x})$ and assume that it has a nonvanishing zero mode and observe that its contribution to the energy density is $\left(\rho^{\alpha}_{\mathbf{0}c}\right)^2 \int d\boldsymbol{x} \int d\boldsymbol{y} \frac{1}{|\boldsymbol{x}-\boldsymbol{y}|}$. Based on the dimensionality of the various factors within the integral, it is straightforward to see that $\Omega^{\alpha}_{el}|_{\boldsymbol{k}=\mathbf{0}} \propto V^{5/3}$, for $\rho_{\mathbf{0}c} \neq 0$, which is inconsistent with the thermodynamic construction of the extensive Ω-potential. Therefore, the requirement of consistency with equilibrium thermodynamics automatically imposes charge neutrality in the overall system; namely, $\rho^{\alpha}_{\mathbf{0}} = 0$ (we note that this electroneutrality condition in linearized approaches appears as a restriction to avoid the divergence in the potential (9)).

2.4 Variational principle to system equations and constraints

From Eq. (1), variation with respect to the electric field $\boldsymbol{E}^{\alpha}$, the Lagrange multiplier (electrostatic potential) Φ^{α}, and the number density $\rho^{\alpha}_{\pm}$ readily yields the relations

$$\delta \boldsymbol{E}^{\alpha} : \quad \boldsymbol{E}^{\alpha}(\boldsymbol{x}) = -\nabla\Phi^{\alpha}(\boldsymbol{x}), \tag{6}$$

$$\delta\Phi : \quad \nabla \cdot \left[\frac{\epsilon^{\alpha} \boldsymbol{E}^{\alpha}(\boldsymbol{x})}{4\pi} \right] = \sum_i Z_i \rho^{\alpha}_i, \tag{7}$$

$$\delta\rho^{\alpha}_i : \quad \mu^{\alpha}_i = \log \rho^{\alpha}_i(\boldsymbol{x}) + Z_i \Phi^{\alpha}(\boldsymbol{x}) + g^{\alpha}_i. \tag{8}$$

We note that Eqs. (6) and (7) simply reproduce, respectively, the expressibility of the electric field $\boldsymbol{E}$ in terms of a scalar potential Φ^{α} and the Gauss' law in electrostatics. The last relation in Eq. (8) provides an explicit expression of the chemical potential for each phase in terms of its respective *input* parameters such as the dielectric constants and ion sizes. It is now obvious that the thermodynamic condition of equal chemical potentials across the different phases α would provide further constraints on the respective bulk ionic densities.

To use these relations, we first introduce notation for the limiting values of our quantities away from the interface, i.e. in the bulk. For the potential, we define $\Phi_{\infty}^{A} = \Phi_D$ and $\Phi_{\infty}^{B} = 0$. The difference $\Phi_{\infty}^{A} - \Phi_{\infty}^{B} = \Phi_D$ is the Donnan or distribution potential. The particle densities achieve a uniform value away from the interface, that is their bulk concentration, denoted by $\rho_{0,i}^{\alpha}$. The minimum of the potential occurs when the electric field away from the interface approaches zero, and this in turn is equivalent to the condition of overall electroneutrality:

$$\sum_i Z_i \rho_i^{\alpha} = 0. \tag{9}$$

When we consider the equilibrium of the joint system, minimization of the total Ω potential requires the equality of the chemical potential of each species:

$$\mu_i^A = \mu_i^B. \tag{10}$$

The equations (6-8) are equivalent to the Poisson-Boltzmann equation. In addition, a useful form of the Boltzmann distribution for the equilibrium ionic profiles follows from Eq.(8):

$$\rho_i^{\alpha}(\boldsymbol{x}) \quad = \quad \rho_0^{\alpha} \exp\left[-Z_i(\Phi^{\alpha}(\boldsymbol{x}) - \Phi_{\infty}^{\alpha})\right]. \tag{11}$$

The Poisson-Boltzmann equation and the Boltzmann distribution are often taken as the starting points for most analysis presented in literature on these systems. We note that they follow naturally from our fundamental Ω-potential [Eq. (1)]. Many other properties of electrolyte systems can furthermore be derived by suitable analysis of this functional. The properties of osmotic pressure, for example, have been extensively discussed in this framework (6).

We now obtain explicit solutions for a single symmetric salt dispersed in two media. In this case, the ions have equal charge magnitudes which we take as $Z_{\pm} = \pm 1$. We use the indices $i = +, -$ for these species and note that, due to the electroneutrality condition, their bulk densities are equal. This bulk density will be denoted as $\rho_0^{\alpha} = \rho_{0,+}^{\alpha} = \rho_{0,-}^{\alpha}$.

Taking the logarithm of Eq. (11) readily yields

$$\log \rho_{\pm}^{A}(\boldsymbol{x}) \pm \Phi^{A}(\boldsymbol{x}) \quad = \quad \log \rho_0^{A} \pm \Phi_D \tag{12}$$

Again, a similar expression to Eq. (12), modulus the term containing the distribution potential, applies to the less polar phase B. Now upon substituting Eq. (12) and the corresponding expression for phase B into Eq. (8), as well as imposing equal chemical potentials across boundary, we obtain the following relations

$$\log \rho_0^{A} + \Phi_D - g_{+}^{A} \quad = \quad \log \rho_0^{B} - g_{+}^{B}, \tag{13}$$

$$\log \rho_0^{A} - \Phi_D - g_{-}^{A} \quad = \quad \log \rho_0^{B} - g_{-}^{B}. \tag{14}$$

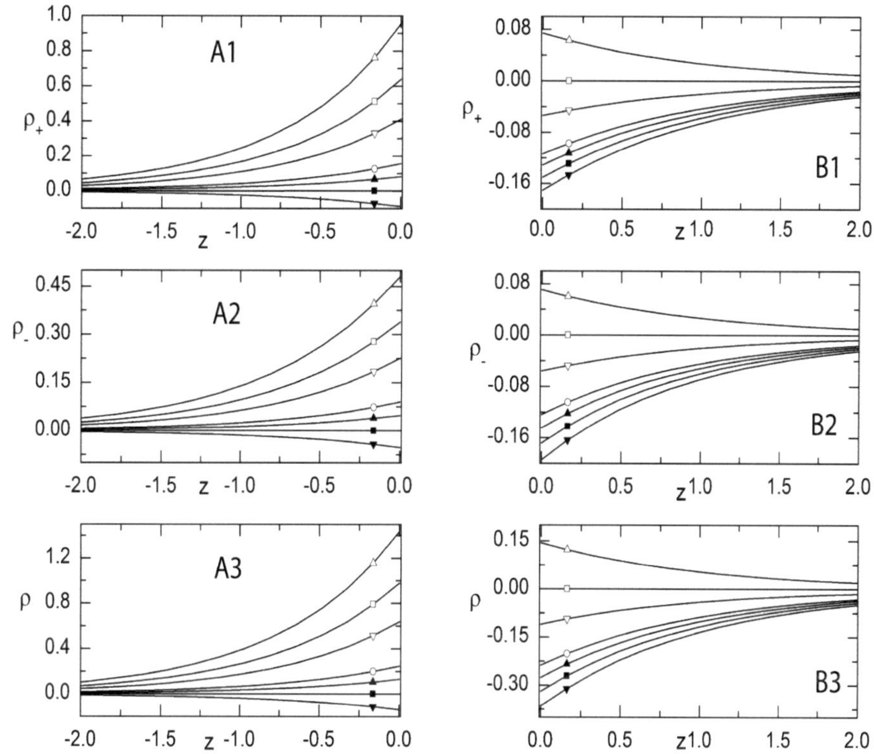

Fig. 3. Comparison of the ion distributions of ions in the two polar phases, $A\,(z<0)$ and $B\,(z>0)$ characterized by ε_A and ε_B, respectively, when the surface density of the interface is $\sigma=-1.5(\triangle)$, $\sigma=-0.93(\square)$, $\sigma=-0.5(\triangledown)$, $\sigma=0(\circ)$, $\sigma=0.15(\blacktriangle)$, $\sigma=0.32(\blacksquare)$, $\sigma=0.5(\blacktriangledown)$. (1) Cations. (2) Anions. (3) Total charge density $\rho=\rho_++\rho_-$. The curves are presented in separate plots because of difference in concentrations. For this particular plot, we use a range of surface charge densities σ in units of $\epsilon_A\,\kappa_A$ and $\Phi'_d=0.112$ for the nitrobenzene/water solution at 68F.

We can now take the sum and difference of the two equations to obtain the relations

$$\log\rho_0^A+\bar{g}_A = \log\rho_0^B+\bar{g}_B, \tag{15}$$

$$2\Phi_D+\Delta g_A = \Delta g_B, \tag{16}$$

where we have defined the average Born polarization energy over ions of both charge polarity, $\bar{g}_\alpha=g_{\alpha+}+g_{\alpha-}$, and the corresponding differential Born energy, $\Delta g_\alpha=g_{\alpha+}-g_{\alpha-}$ between the cations and anions. In terms of these two quantities, we can relate the bulk ion densities and the distribution potential to the dielectric constants ε_α in each phase via

$$\frac{\rho_0^A}{\rho_0^B} = e^{\bar{g}_B-\bar{g}_A}=e^{-\Delta g_\alpha} \tag{17}$$

$$\Phi_D = -\frac{1}{2}\Delta_\alpha\Delta g \tag{18}$$

where $\Delta_\alpha\Delta g=\Delta g_A-\Delta g_B$.

2.5 Summary

In summary, we have thus shown in Eq. (17) that the ratio between the bulk densities $\rho^{\alpha}_{0,\pm}$ is related to the difference of the average Born solvation energy between the two phases, while in Eq. (18) the distribution potential Φ_D is related to the difference in the differential Born energy between ion types in the two phases. We note that even though the above formulation has been done for a neutral liquid-liquid interface wherein the charge distribution of the system resides completely amongst the dissolved ions in the bulk, it is straightforward to generalize to the case where a surface charge exists on the interface itself.
To determine the equilibrium ion density profiles $\rho^{\alpha}_{\pm}(\boldsymbol{x})$ as well as the electrostatic potential $\Phi(\boldsymbol{x})$, it remains for us to solve the Poisson equation [Eq. (7)] after applying the integrability condition [Eq. (6)] as well as the thermodynamic ansatz of the Boltzmann distribution for the ion density [Eq. (11)]. We will illustrate with two classes of examples in the subsequent section involving non-polarizable (neutral) but permeable membranes as well as polarizable (charged) membranes.

3. Electrostatic phenomenology of differential adsorption

3.1 Case study defined

As mentioned in the Introduction, it is now feasible to make precise experimental detection of the ion distributions near liquid-liquid interfaces (23; 24). It has been found that while the mean-field Poisson-Boltzmann framework provides a generally good description for the ion profiles that matches well with experimental results obtained by x-ray structural measurements, the linearized Guoy-Chapman approximation is insufficient for most general cases (24). For completeness, we will now illustrate the solution to the Poisson-Boltzmann equation for the case when the membrane across the two dielectric media is charged (polarizable), including the particular interesting case of the neutral (non-polarizable) interface. Furthermore, our liquid-liquid interface is treated as an infinite two-dimensional plane so that we do not need to consider surface terms. We thus essentially reduce our system to a one-dimensional problem. Denoting the lateral distance from the membrane by z, we will now formulate the full non-linear Poisson-Boltzmann equation with the proper boundary conditions and obtain the corresponding electrostatic potential $\Phi^{\alpha}(z)$ and ion density profiles $\rho^{\alpha}_{\pm}(z)$ in each phase α.

3.2 Analytic mean-field formulation

To begin, we analytically formulate our system in which the two phases A and B are in equilibrium, where each phase is characterized by its respective constant dielectric permittivity ε_A and ε_B, as follows:

$$\varepsilon_A \nabla \cdot \boldsymbol{E}^{\boldsymbol{A}} = \rho_c^A, \quad z > 0, \tag{19}$$

$$\varepsilon_B \nabla \cdot \boldsymbol{E}^{\boldsymbol{B}} = \rho_c^B, \quad z < 0, \tag{20}$$

subject to the following boundary conditions at $z = 0$:

$$\varepsilon_B E_z^B \Big|_{z\to 0^-} \varepsilon_A E_z^A \Big|_{z\to 0^+} = 4\pi\sigma, \tag{21}$$

$$E_x^A|_{z\to 0^+} = E_x^B|_{z\to 0^-}, \tag{22}$$

$$E_y^A|_{z\to 0^+} = E_y^B|_{z\to 0^-}. \tag{23}$$

Making use of Eq. (6) and substituting the Boltzmann distribution for ion density profile [Eq. (11)] , we now take advantage of the one-dimensional nature of our system and collapse the above formulation into a single equation, written in terms of the electrostatic potential $\Phi^{\alpha}(z)$, to obtain the Poisson Boltzmann-equation representation of the Gauss' law:

$$\frac{d^2\Phi^A}{dz^2} = \left(\kappa^A\right)^2 \sinh\left(\Phi^A - \Phi_D\right), \quad z<0, \tag{24}$$

$$\frac{d^2\Phi^B}{dz^2} = \left(\kappa^B\right)^2 \sinh\Phi^B, \quad z>0, \tag{25}$$

where, respectively, $\kappa^{\alpha} = \sqrt{4\pi e^2 \rho_0^{\alpha}/\epsilon^{\alpha} kT}$ is the inverse Debye screening length, and $\rho_0^{\alpha} = Z_+^2 \rho_{0,+}^{\alpha} + Z_-^2 \rho_{0,-}^{\alpha}$ is the weighted sum of the bulk densities of the ions in the two phases, $\alpha = A, B$. The inverse Debye screening length has been so defined as to scale the electrostatic potential Φ^{α} dimensionless. The corresponding boundary conditions for the potential are as follows:

$$\Phi^A\Big|_{z\to 0^-} = \Phi^B\Big|_{z\to 0^+}, \tag{26}$$

$$\varepsilon^A \frac{d\Phi^A}{dz}\Big|_{z\to 0^-} - \varepsilon^B \frac{d\Phi^B}{dz}\Big|_{z\to 0^+} = 4\pi\sigma, \tag{27}$$

$$\frac{d\Phi^A}{dz}\Big|_{z\to -\infty} = 0, \tag{28}$$

$$\frac{d\Phi^B}{dz}\Big|_{z\to +\infty} = 0. \tag{29}$$

3.3 Full nonlinear solution

The general solution to the full nonlinear Poisson-Boltzmann equation can be readily found (41). At present, we simply furnish the particular solution for our system of nonpolarizable liquid-liquid interface specified by the above boundary conditions:

$$\Phi^A(z) = \Phi_D - 4\tanh^{-1}\left(C_A e^{\kappa_A z}\right), \quad z<0, \tag{30}$$

$$\Phi^B(z) = 4\tanh^{-1}\left(C_B e^{-\kappa_B z}\right), \quad z>0. \tag{31}$$

As shown, Eqs. (30) and (31) are expressed in terms of the two integration constants, C_A and C_B, and as such, physically meaningful solutions would correspond to the range of $|C_\alpha| \leq 1$. Alternatively, we can rewrite $C_\alpha = \exp(-\kappa_\alpha z_\alpha)$, and the solutions now take the form of $\Phi^A(z) = \Phi_D - 4\,\mathrm{sgn}(C_A)\tanh^{-1} e^{\kappa_A(z-z_A)}$, and $\Phi^B(z) = 4\,\mathrm{sgn}(C_B)\tanh^{-1} e^{-\kappa_B(z-z_B)}$, where $\mathrm{sgn}[\cdot]$ is the sign function. The choice of signs in the above solutions is such that the integration constants are positive when we have a neutral interface, $\sigma = 0$.
In terms of the integration constants C_A and C_B, the continuity equation in Eq. (26) reads:

$$\frac{(1+C_A)(1+C_B)}{(1-C_A)(1-C_B)} = e^{\frac{\Phi_D}{2}}. \tag{32}$$

while the Gauss' law in Eq. (27) takes the form of

$$\frac{4\epsilon_A \kappa A C_A}{C_A^2 - 1} - \frac{4\epsilon_B \kappa B C_B}{C_B^2 - 1} = 4\pi\sigma. \tag{33}$$

This system of equations admits the following simple algebraic solutions for C_A and C_B:

$$C_A = \frac{\epsilon_A \kappa_A + \epsilon_B \kappa_B \cosh \Phi_D/2 - R}{2\pi\sigma + \epsilon_B \kappa_B \sinh \Phi_D/2}. \tag{34}$$

$$C_B = \frac{\epsilon_B \kappa_B + \epsilon_A \kappa_A \cosh \Phi_D/2 - R}{-2\pi\sigma + \epsilon_A \kappa_A \sinh \Phi_D/2}, \tag{35}$$

where

$$R^2 = (2\pi\sigma)^2 + (\epsilon_A \kappa_A)^2 + (\epsilon_B \kappa_B)^2 + 2\epsilon_A \kappa_A \epsilon_B \kappa_B \cosh \Phi_D/2. \tag{36}$$

Thus, we have now determined the constants of integration explicitly in terms of the input parameters ϵ_A, ϵ_B, and Φ_D.

3.4 Equilibrium ion profiles

We can now write down the positive and negative ion densities in a straightforward manner:

$$\rho_\pm^A(z) = \rho_0^A \left(\frac{1 \pm C_A e^{\kappa_A z}}{1 \mp C_A e^{\kappa_A z}} \right)^2, \tag{37}$$

$$\rho_\pm^B(z) = \rho_0^B \left(\frac{1 \mp C_B e^{-\kappa_B z}}{1 \pm C_B e^{-\kappa_B z}} \right)^2. \tag{38}$$

To illustrate the profiles of the electrostatic potential $\Phi(z)$ and of the ion distributions $\rho_\pm^\alpha(z)$ across an interface, we will now consider, in particular, the system of a polarizable membrane between a nitrobenzene solution [$\varepsilon^B = 35.7\varepsilon_0$ at 68F] of sodium chloride and a water solution [$\varepsilon^A = 80.4\varepsilon_0$ at 68F] of the same salt. Given the size ratio between the ions of $R_+/R_- \approx 0.695$, we obtain the following normalized relations between the various Born solvation energies: $g_+^A \approx 0.662$, $g_-^A \approx 0.446$, $g_+^B \approx 1.44$, and $g_-^B \equiv 1$. Thus, it follows readily that $\rho_0^A/\rho_0^B = e^{1.33} \approx 3.79$ and $\Phi_D \approx 0.112$, which implies that $\kappa^A/\kappa^B \approx 1.30$. We display the resulting plots in Figs. 2 and 3.

3.5 Excess interfacial charge

The excess interfacial charge, $Q_\pm^\alpha$, defined as

$$Q_\pm^\alpha = \int_0^\infty dz \left[\rho_\pm^\alpha(z) - \rho_0^\alpha \right], \tag{39}$$

and due physically to the accumulation or depletion of charges near the interface as the ions migrate across the boundary, can also be obtained in analytical closed-form for the present case of planar interface. We note that the excess charge $Q_\pm^\alpha$ has the dimensions of charge density per unit of transverse area. Upon direct evaluation, we obtain

$$Q_\pm^A = \frac{1}{2\pi} \frac{\epsilon_A \kappa_A C_A}{1 \mp C_A}, \tag{40}$$

$$Q_\pm^B = -\frac{1}{2\pi} \frac{\epsilon_B \kappa_B C_B}{1 \pm C_B}. \tag{41}$$

The total excess charge in each phase α can be simply evaluated by $Q^\alpha = Q_+^\alpha + Q_-^\alpha$. In addition, the net charge flux across the interface can be defined as $\Delta Q = \frac{1}{2}(Q^B - Q^A)$. The

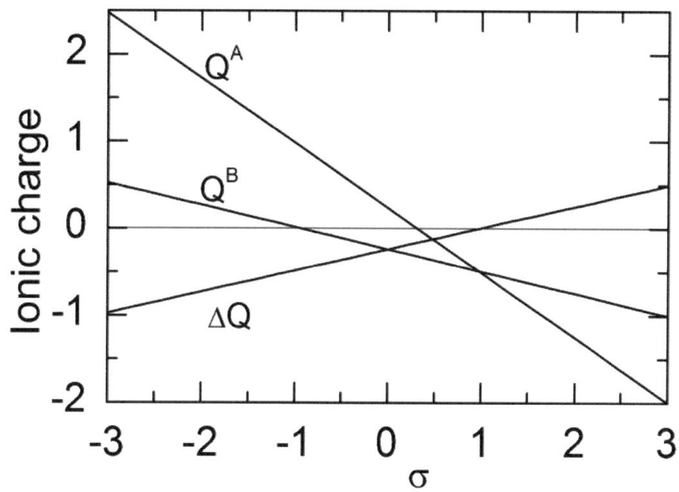

Fig. 4. Plots of the total excess charge Q^A, Q^B and the net charge flux across the interface ΔQ. For this particular plot, we use a range of surface charge densities σ in units of $\epsilon_A \kappa_A$ and $\Phi'_d = 0.112$ for the nitrobenzene/water solution at 68F.

total net charge accumulated in each phase compensates for the charge introduced by the polarizable membrane to the overall system such that the relation $\sigma + Q^A + Q^B = 0$ holds at all times. The plot of Q^A, Q^B and ΔQ for various values of the surface charge density σ is shown in Fig. 4.

3.6 Extension to polarizable membranes

In the case of polarizable membranes, we see that the additional surface charge density can have dramatic effects on the electrostatic potential and charge distributions of the two phases. As shown in Fig. 2, there exists a sufficiently large surface charge density on the interface, $\sigma^{*\alpha}$, at each phase α such that the electric field vanishes as manifested by the corresponding constant electrostatic potential. The further addition of surface charges, either positive or negative, reverses the sign of the potential gradient and of the excess charge in the corresponding phase. The critical surface charge densities are found to be

$$\sigma^{*A} = \frac{1}{2\pi}\epsilon_B \kappa_B \sinh \frac{\Phi_D}{2}, \quad (42)$$

$$\sigma^{*B} = -\frac{1}{2\pi}\epsilon_A \kappa_A \sinh \frac{\Phi_D}{2}. \quad (43)$$

It can be also checked that at these critical surface charge densities, our expressions for the excess charge Q^α and the net charge flux across the interface ΔQ also vanish in the respective phase (Fig. 4). Notwithstanding the seeming linearity that both Q^α and ΔQ follow with respect to the surface density σ, we do observe a slight jump across $\sigma = 0$ thereby establishing a small deviation to perfect linearity.

3.7 Properties of the solutions

We now consider a few general properties of our solutions. Firstly, we note that the distribution potential Φ_D across the interface, itself generated by the different dielectric constants of each phase, is only dependent upon ion types and not of their concentration. Thus, in the absence of any ions in the system, we must then recover a uniform potential. Indeed, we observe exactly that in approaching the limit of zero ionic concentration where the gradient of the potential across the interface gradually decreases; the required net jump Φ_D is being pushed outwardly towards infinity.

Another noteworthy feature is the asymptotic behavior of the potential at large distances from the interface. In the limit of $z \to \infty$, we have $\Phi^A \approx \Phi_D - 4C_A \exp(\kappa_A z)$, and $\Phi^B \approx 4C_B \exp(-\kappa_B z)$. Thus, we see that the integration constants now act as potential sources in the solution form for the linearized version of the Poisson-Boltzmann equation. Given that each integration constant satisfies the constraint of $|C_\alpha| \leq 1$, the asymptotics always arises from a linear source of maximum potential $\Phi_{max} \leq 4$.

Lastly, we remark that our results are consistent with the Born description of solvation energy: since cations are in general smaller in size due to their oxidized state, the solvation free energy is always more negative when compared with their anionic counterpart, provided that they are both subjected to the same solvent of the same constant dielectric permittivity. Therefore, it is always thermodynamically more favorable to dissolve more of the cations in the more polar phase and redistribute the density profiles of the remaining ions accordingly in both phases. The validity of our argument remains even when the membrane becomes charged in the same polarity as the cations. In general, we do not expect that the multivalency of the dissolved ions would qualitatively change the behavior of the differential adsorption phenomenon described herein, at the mean-field level, other than to intensify its effect as evident in the prefactor of our Born expression of solvation energy.

4. Conclusions

4.1 Summary of results

In this work, we have constructed a self-contained thermodynamic formulation of ion solvation in binary immiscible polar mixtures and described the general phenomenon of differential ion adsorption at liquid-liquid interfaces via the full nonlinear Poisson-Boltzmann framework. Assuming the Born model of solvation and using the mechanics of the Ω-potential, our formulation presents a fundamental, mean-field description that relates the experimentally detectable bulk-ion densities in each phase, as well as the distribution potential across the membrane, to the respective dielectric constants and sizes of each ion species present in the two phases. We note that electrostatics alone is insufficient in capturing the physics of ion solvation and interfacial adsorption.

4.2 Future research

Our work can be generalized in several ways. Our formulation models the dissolved ions specifically as point particles and the liquid-liquid interface as sharp. Immiscible multicomponent liquids in some cases can lead to the broadening of interfaces (16; 17). In such instances, adsorption of ionic component in slabs of immiscible liquids has been recently considered using liquid-liquid weak segregation approximations (32). Thus, it is of interest to consider other morphologies and systems than those presented in this work. As a refinement to our model, it is also possible to incorporate further molecular details in the construction of our thermodynamic potential. This can be done, for example, with the explicit inclusion of short-range interactions between the ions by replacing our expression of Ω^{α}_{sol} by

$$\Omega'^{\alpha}_{sol} = \int\int d\boldsymbol{r}d\boldsymbol{r}' G_{sr}\left(\boldsymbol{r}-\boldsymbol{r}'\right)\rho(\boldsymbol{r}')\rho(\boldsymbol{r}) \quad (44)$$

One could then consider components with more complicated structures than point ions, such as polyelectrolytes, and consider the additional effects introduced by these new degrees of freedom (10; 28). Though protein adsorption on hard surfaces have been analyzed (11), a general description of macroion adsorption to liquid interfaces is lacking. Such a general description would be highly useful, as, for example, the phenomenon of DNA adsorption to the interface among liquid surfaces is crucial in cell biology (34; 18). The inclusion of proteins and/or DNA molecules in the system and the study of their corresponding adsorption phenomenon along liquid-liquid interfaces will be presented elsewhere. We point out in passing that the long-range interaction of electrostatics would not contribute at the mean-field level due to overall electro-neutrality.

4.3 Outlook

It is our hope to have demonstrated in this work that the differential adsorption of ions is a general phenomenon in electrostatics that occurs along interfaces between different dielectric media. Such chemical environments are ubiquitous in many biological systems at the cellular and physiological levels (1). In particular, it was pointed out by Onsager (31) regarding the potential importance of such a "primodial oil slick" in the early development of life. The phenomenon of thermodynamically driven, selective confinement of ions and the consequent breaking of charge-conjugation symmetry near a liquid-liquid interface would provide just the physical mechanisms necessary for the essential organic molecules to aggregate in close, two-dimensional proximity to each other, where the process of diffusion would have worked much more efficiently in bringing together these molecules and starting the chain reaction of life than in the otherwise three-dimensional scenario in the bulk. It is our further hope to generalize our results in this work to other interfacial geometries such as the cylindrical and spherical cases.

5. Acknowledgments

This work was supported primarily by the Nanoscale Science and Engineering Initiative of the National Science Foundation under NSF Award Number EEC-0647560. Any opinions, findings and conclusions or recommendations expressed in this material are those of the author(s) and do not necessarily reflect those of the National Science Foundation.

6. References

[1] Alberts, B; *et. al.*, (2002), *Molecular Biology of the Cell*, 4th ed., Garland Publishing, Inc., New York, ISBN-10: 0815332181.

[2] Bedzyk, M. J.; Bilderback, D. H.; Bommarito, G. M.; Caffrey, M. and Schildkraut, J. S., (1988) X-Ray standing waves - a molecular yardstick for biological-membranes, *Science*, Vol.241, No. 4874,1788-1791, ISSN: 0036-8075.

[3] Bedzyk, M. J.; Bommarito,G. M.; Caffrey, M. and Penner, T. L., (1990), Diffuse-double layer at a membrane-aqueous interface measured with x-ray standing waves, *Science* Vol. 248, No. 4951, 52-56, ISSN: 0036-8075.

[4] Cheng, H; Zhang, K; Libera, J. A.; Olvera de la Cruz, M.; and Bedzyk, M. J., (2006), Polynucleotide adsorption to negatively charged surfaces in divalent salt solutions, *Biophysical J.*, Vol. 90, No. 4, 1164-1174, ISSN: 0006-3495;

[5] *CRC Handbook of Chemistry and Physics*, (2007), 88th ed. CRC, London, ISBN-10: 0849304881.

[6] Deserno, M. and von Grunberg, H. H. (2002), Osmotic pressure of charged colloidal suspensions: A unified approach to linearized Poisson-Boltzmann theory, *Phys. Rev. E*, Vol. 66, No. 1, 011401, ISSN: 1063-651X.

[7] Donnan, V. F. G., (1911), *Z. Elektrochemie*, Vol. 17, 572-581.

[8] Eisenhut, M.; (2006), Changes in ion transport in inflammatory disease, *J. Inflamm* (London) Vol. 3, No. 5, 5-15, ISSN .

[9] Ermoshkin A. V. and Olvera de la Cruz, M, (2003), A Modified Random Phase Approximation of Polyelectrolyte Solutions, *Macromolecules*, Vol. 36, 7824-7832, ISSN: 0024-9297.

[10] Evans, R (1979), Nature of the liquid-vapor interface and other topics in the statistical-mechanics of nonuniform, classical fluids, *Adv. Phys.*, Vol. 28, No. 2, 143-200, ISSN: 0001-8732.

[11] Fang, F. and Szleifer, I., (2003), Competitive adsorption in model charged protein mixtures: Equilibrium isotherms and kinetics behavior, *J. Chem. Phys.* Vol. 119, 1053-1065.

[12] Goldar, A. and Sikorav, J. L. (2004),DNA renaturation at the water-phenol interface, *Eur. Phys. J. E*, Vol. 14, No. 3, 211-239 ISSN: 1292-8941.

[13] González-Mozuelos, P.; and Olvera de la Cruz, M.; (2008), Solvent and nonlinear effects on the charge renormalization of nanoparticles within a molecular electrolyte model, *Physica A*, Vol. 387, No. 22, 5362-5370, ISSN: 0378-437.

[14] Gonzalez-Mozuelos, P.; and Olvera de la Cruz, M.; (1995), Ion Condensation in Salt-Free Polyelectrolyte Dilute Solutions, *J. Chem. Phys.*, Vol. 103, 3145, ISSN: 0021-9606.

[15] Hsiao, P. Y.; and Luijten, E.; (2006), Salt-Induced Collapse and Re-expansion of Highly Charged Flexible Polyelectrolytes, *Phys. Rev. Lett*, Vol. 97, No. 14, 148301, ISSN: 0031-9007.

[16] Huang, C; Olvera de la Cruz, M. and Swift, B. W., (1995), Phase separation of ternary mixtures - symmetrical polymer blends, *Macromolecules*, Vol.28, No. 24, 7996-8005 ISSN: 0024-9297.

[17] Huang, C. and Olvera de la Cruz, M., (1996), Adsorption of a minority component in polymer bend interfaces, *Phys. Rev. E.*, Vol. 53, No. 1, 812-819, ISSN: 1063-651X.

[18] Kahn M. E. and Smith H. O. (1984), Transformation in hemophilus - A problem in membrane biology, *J. Membr. Biol.*, Vol. 81, No. 2, 89-103, ISSN 0022-2631.

[19] Karpfen, F. M. and Randles, J. E. B., (1953). Ionic Equilibria and Phase-boundary Potentials in Oil-water Systems. *Trans. Faraday Soc.* Vol. 49, No. 7, 823-831 IDS: UG043.

[20] Lau, A. W. C.; Pincus, P.; Levine, D; Fertig, H.A., (2001), Electrostatic attraction of coupled Wigner crystals: Finite temperature effects, *Phys. Rev. E* Vol.63, No. 5, 051604, ISSN: 1063-651X.

[21] Libera, J. A; Cheng, H.; Olvera de la Cruz, M. and Bedzyk, M. J., (2005), Direct observation of cations and polynucleotides explains polyion adsorption to like-charged surfaces, *J. Phys. Chem. B*, Vol. 109, No. 48, 23001-23007, ISSN: 1520-6106.

[22] Liu, S; and Muthukumar, M.; (2002), Langevin dynamics simulation of counterion distribution around isolated flexible polyelectrolyte chains, *J. Chem. Phys.*, Vol. 116, No.

22, 9975, ISSN: 0021-9606.
[23] Luo. G.; Malkova, S.; Yoo, J.; Schultz, D. G.; Lin, B.; Meron, M.; Benjamin, I.; Vanysek, P.; and Schlossman, M. L., (2006), Ion distributions near a liquid-liquid interface, *Science*, Vol. 311, No.5758, 216-218, ISSN: 0036-8075.
[24] Luo G.; Malkova S.; Yoo J.; Schultz D. G.; Lin B.; Meron, M.; Benjamin, I.; Vanysek, P.; and Schlossman, M. L., (2006), Ion distributions at the nitrobenzene-water interface electrified by a common ion, *J. Electroanalytical Chem.* Vol. 593, No. 1-2, 142-158, ISSN: 0022-0728.
[25] Lynden-Bell, R. M.; Rasaiah, J. C. and Noworyta, J. P. (2001), Using simulation to study solvation in water , *Pure Appl. Chem.*, Vol. 73, No. 11, 1721-1731ISSN: 0033-4545.
[26] Lynden-Bell R. M. and Rasiah J. C. (1997), From hydrophobic to hydrophilic behaviour: A simulation study of solvation entropy and free energy of simple solutes, *J. Chem. Phys.*, Vol. 107, No. 6, 1981-1992, ISSN 0021-9606.
[27] Markin, V. S. and Volkov, A. G., (1989) Interfacial Potentials at the Interface between 2 Immiscible Electrolyte-solutions - Some Problems in Definitions and Interpretation, J. Col. Int. Sci., Vol. 131, No. 2, 382-392.
[28] Nap, R.; Gong, P. and Szleifer, I.; Weak polyelectrolytes tethered to surfaces: effect of geometry, acid-base equilibrium and electrical permittivity, *J. Poly. Sci. B*, Vol. 44, No. 18, 2638-2662, ISBN: 0887-6266.
[29] Netz, R. R.; Andelman, D (2003), Neutral and charged polymers at interfaces, *Physics Reports* Vol. 380, No. 1-2, 1-95 ISSN: 0370-1573.
[30] Olvera de la Cruz, M.; Belloni, L.; Delsanti, M.; Dalbiez, J. P.; Spalla, O.; and Drifford, M.; Precipitation of highly charged polyelectrolyte solutions in the presence of multivalent salts, *J. Chem. Phys.* Vol. 103, No. 13, 5781, ISSN: 0021-9606.
[31] Onsager L. (1974) *Quantum Statistical mechanics in the Natural Sciences*, Plenum Press, New York.
[32] Onuki A. (2006) Ginzburg-Landau theory of solvation in polar fluids: Ion distribution around an interface, *Phys. Rev. E*, Vol. 73, No. 2, 021506, ISSN: 1539-3755.
[33] Pincus P. and Safran S. A., Charge fluctuations and membrane attractions, *Europhys. Lett.* Vol. 42, No. 1, 103-108 ISSN: 0295-5075.
[34] Pollack, G. H. (2003). The role of aqueous interfaces in the cell, *Adv. Colloid Interface Sci.*, Vol. 103, No. 2, 173-196 ISSN: 0001-8686.
[35] Raspaud, E.; Olvera de la Cruz, M.; Sikorav, J. L.; and Livolant, F.; (1998), Precipitation of DNA by polyamines: A polyelectrolyte behavior, *Biophys. Journal* Vol. 74, No. 1, 381-393 ISSN: 0006-3495, and references therein.
[36] Samec Z., (1998). Electrical double-layer at the interface between two immiscible electrolyte solutions, *Chem. Rev.* Vol. 73. No. 4, 617-632, ISSN: 0009-2665.
[37] Sayar, M.; and Holm, C.; (2007), Finite-size polyelectrolyte bundles at thermodynamic equilibrium, *Eurphys. Lett.*, Vol. 77, No. 1, 16001, ISSN: 0295-5075.
[38] Shafir, A.; Andelman, D., (2006), Phase behavior of polyelectrolyte-surfactant complexes at planar surfaces, *Phys. Rev. E*, Vol. 74, No. 2, 021803 SSN: 1539-3755.
[39] Solis, F. J.; and Olvera de la Cruz, M.; (2000), Collapse of flexible polyelectrolytes in multivalent salt solutions *J. Chem. Phys.* Vol. 112, No. 4, 2030 ISSN: 0021-9606.
[40] Stevens, M. J.; and Kremer, K.; (1995), The nature of flexible linear polyelectrolytes in salt-free solution - a molecular-dynamics study, *J. Chem. Phys.* Vol. 103, No. 4, 1669-1690, ISSN: 0021-9606.
[41] Verwey, E. J. W. and Overbeek, J. Th. G. (1999), *Theory of Stability of Lyophobic Colloids*,

Dover Publication Inc., New York, ISBN-13: 978-0486409290.

[42] Yoshikawa, M. K.; Takahashi, M.; Vasilevskaya, V. V.; and Khokhlov, A. R.; (1996), Large Discrete Transition in a Single DNA Molecule Appears Continuous in the Ensemble, *Phys. Rev. Lett.* Vol. 76, No. 16, 3029-3031, ISSN: 0031-9007.

[43] Zwanikken, J. and van Roij, R, (2007) Charged colloidal particles and small mobile ions near the oil-water interface: Destruction of colloidal double layer and ionic charge separation, *Phys. Rev. Lett.* Vol. 99, No. 17, 178301 ISSN: 0031-9007.

Part 2

Application of Thermodynamics to Chemistry, Solid State Physics and Materials Science

14

Calorimetric: A Tecnique Useful in Characterization of Porous Solid

Juan Carlos Moreno[1] and Liliana Giraldo[2]
[1]*Universidad de los Andes, Bogotá,*
[2]*Universidad Nacional de Colombia, Bogotá*
Colombia

1. Introduction

The adsorption of gases in porous solids such as zeolites and activated carbons, has been widely applied in cases of separation, purification and bottling of gases (Ruthven et al. 1994; Yang, 1997; Bastos-Neto et al. 2005a; Figueroa et al., 2008; Belmabkhout and Sayari, 2009). This potential is reflected not only in increasing the number of technical and scientific articles and patents, but also in the world market growth in plants for air separation and purification processes of hydrogen and natural gas and many others (Zimmermann and Keller, 2003). Due to the various applications of porous adsorbents, many research groups in various parts of the world have sought to develop and improve these materials to improve performance in these specific applications (Bastos-Neto et al. 2005b; Arou et al., 2008; Prauchner and Rodriguez-Reinoso, 2008; Rivers and Smith, 2009). In the procedures for obtaining porous solids, it is necessary to control the various process variables such as preparation, carbonization temperature and time, type and concentration of activating agents, among others, since these activation parameters determine the chemical and physical properties of adsorbents. The textural characteristics are the most important properties of the adsorbents, since it indicates the implementation and performance of the solid obtained (Giraldo and Moreno, 2005). In addition, chemical properties also determine the adsorption properties of adsorbent and solid-fluid interactions. Nature of surface groups, hydrophobic or hydrophilic character and acidic or basic behavior are some of the relevant chemical properties of the adsorbents in adsorption processes. Since the physical and chemical properties of an adsorbent determine the application and performance of the same, it is necessary to determine precisely the parameters that characterize these materials such as surface area, microporosity, pore size distribution, heats of adsorption, among others. Several experimental techniques are used to characterize porous materials, for example, mercury porosimetry, adsorption of liquid nitrogen, x-ray diffraction, etc.. The technique most commonly used to characterize the texture of carbon adsorbents (ie surface area, properties of molecular sieve, size distribution of pores, etc.). Is the physical adsorption of gases and vapors. However, immersion calorimetry, with molecular probes of various molecular dimensions, and gas adsorption microcalorimetry techniques are also applied to characterize this type of solid (Denoyel et al., 1993, González et al. 1995; Rouquerol et al. 1999, Navarrete et al., 2004; Garcia-Cuello et al., 2009). The aim of this study is to review the calorimetric methods as a technique for characterization of adsorbent materials. The focus of this study is to review available scientific articles in the literature showing the use of calorimetry to physical characterization of adsorbent materials. The aim is also to describe some types of calorimeters commonly used in adsorption processes.

2. Adsorption calorimetry

The determination of heat of adsorption is essential in the description of gas solid interactions. This is particularly useful when measuring the heat of adsorption are combined with simultaneous measurements of adsorption isotherm (Llewellyn and Maurin, 2005; Garcia-Cuello et al., 2008; Garcia-Cuello et al., 2009). There are many factors to determine the heat of adsorption, namely: to characterize the surface energy of the material (Rouquerol et al., 1999); provide basic data for developing new theories for equilibrium and kinetic adsorption (Zimmermann and Keller, 2003); design and improve plants separation processes such as adsorption and desorption, PSA, VSA, TSA and their combinations (Ruthven et al. 1984; Yang, 1997). Calorimetry was never widely used for characterization of carbons. In fact, its use has been limited to a few researchers who have used calorimetric techniques to specific problems (Menéndez, 1998). However, it has been shown that the adsorption calorimetry in combination with other physical techniques or physical-chemical properties can be used to describe the properties of the surface of a solid (Llewellyn and Maurin, 2005; Giraldo and Moreno, 2005; Garcia-Cuello et al., 2008; Garcia-Cuello et al., 2009). Menéndez (1998) presents a brief review of calorimetric methods applied to physical and chemical surface characterization of carbonaceous adsorbents. Their results indicate great potential of using calorimetry to study the physical structure (ie textural) of carbons, especially when used in conjunction with more traditional techniques based on physical adsorption of gases and vapors. Three types of calorimetric methods have been used to characterize the porous solids, namely, immersion, flow-adsorption calorimetry and gas adsorption. Among the three methods, immersion calorimetry and itrogen adsorption are the most used for the characterization of adsorbents. The next section presents the classifications of the calorimetric curves depending on the classification of adsorption isotherms according to IUPAC. In the following sections will show the advantages of using calorimetric techniques as a tool for characterization of adsorbents through their applications.

2.1 Theoretical microcalorimetric curves (Llewellyn and Maurin, 2005)

The differential enthalpy curve obtained from microcalorimetric experiments is a result of various effects that include both adsorbent-adsorbate interactions and adsorbate-adsorbate. Mechanisms of pore filling and phase transition can be demonstrated as well as structural changes of the adsorbent. In general, the calorimetric curve shows three distinct behaviors, as shown in Figure 1. In each system, an increased amount of gas adsorbed by the sample leads to an increase of interactions between adsorbed molecules. Regarding contributions adsorbate-adsorbent, the interaction of the adsorbed molecule with an energetically homogeneous surface leads to a constant signal.

Finally, in most cases, the adsorbent is energetically heterogeneous due to their pore size distribution (amorphous adsorbents) or a variable surface chemistry (defects, cations, etc.).. Initially, one would expect relatively strong interactions between the adsorbed molecules and surface. The intensity of these interactions decreases as the specific sites are occupied.

Therefore, to energetically heterogeneous adsorbents, a gradual decrease in the calorimetric signal is observed. However, each curve of enthalpy differential current varies because the results of the different contributions are summed.

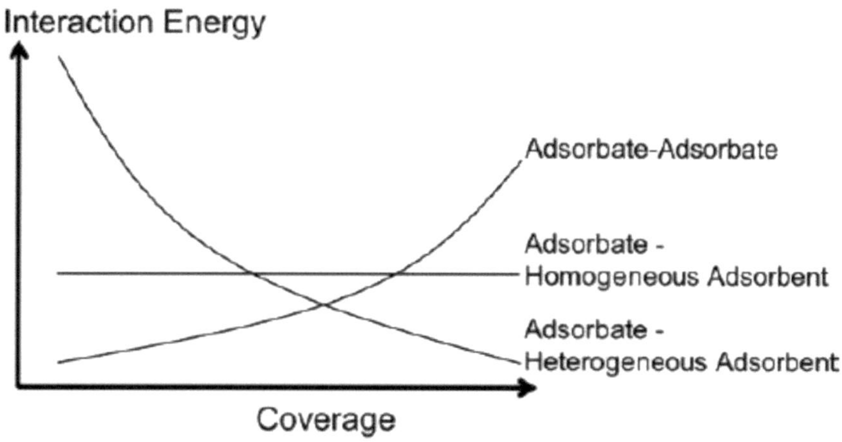

Fig. 1. Hypothetical calorimetric curves illustring the various interactions involved during the adsorption of simple gases at low temperatures . (Llewellyn and Maurin, 2005.)

2.2 Thermodynamic basis of calorimetry

The determination of the heat quantity that is involved in a process, be it physical, chemical, or biological, provides information about the evolution, duration, and intensity of the same, and the experimentation being conducted to determine the magnitude of the heat that is produced or absorbed can be simple or complex according to the required measurement accuracy, the sensitivity of the instruments used, and the amount of energy that is transferred, to name some of the features involved in the determination.

In the development of chemistry, and particularly in the field of thermodynamics, calorimetry has been a factor of undeniable importance. Some authors point to the calorimeter as an instrument that opens up the second part of the pre-classical thermodynamics, which is called calorimetry (Swietoslawski, 1946).

Construction of the first calorimeter by Cavendish (Armstrong, 1964) in 1720, for the determination of the heat of vaporization of water and specific heats of various substances, was the start of the creation of a great variety of designs by leading researchers of that time, among whom names of such importance as Lavoisier, LaPlace, Black and Irvine, Bunsen, Dulong and Petit, and Nernst Euken can be highlighted (Wilhoit, 1967; Rouquerol, 1985).

Due to the large number of systems, phenomena, and conditions of interest, there is no single model of calorimeter, so their diversity is very wide. That is why since the very emergence of calorimeters a variety of equipment has been generated, among which we can examine adiabatic calorimeters, isothermal calorimeters, and isoperibolic calorimeters. In parallel with the development of calorimeters, it was necessary to improve data collection systems, leading to the use of peripheral systems with high sensitivity and precision.

The purposes and applications of calorimeters have greatly expanded the field of study and concepts of calorimetry and therefore of thermodynamics: the heat of solution, combustion, mixing, and vaporization are just some of the determinations that are made using this technique. Because information can be obtained from the measurement of heat, the number

of instrument designs for its determination has increased, taking into account that the energetic effect associated with different interactions may be of short duration or, later in development, hours or even weeks (Swietoslawski, 1946; Hemminger and Hohne, 1984).
The calorimetric determination must then take into account the accuracy required, the temperature, the amount of sample available, the magnitude of the heat involved, the duration of the experiment, and the cost of the instrument.

2.3 General classes and types of signals of calorimeters

The heat transfer can be determined in different conditions, and therefore different kinds of instruments are used as calorimeters, which can be classified into three general groups as described below (Wadso and Goldberg, 2001; Giraldo et al., 1996).

2.3.1 Isoperibolic calorimeter

An isoperibolic calorimeter maintains a constant temperature of the surroundings through the use of a thermostat, while the temperature measurement system may vary over time. There is a thermal resistance R_T, with a magnitude defined between the surroundings and the cell which is being measured, so that the heat exchange depends on the temperature difference between the temperature of the surroundings T_A and the temperature of the cell and measuring system T_C; as T_A is constant then the flow of heat is a function of T_C. If the heat generation within the cell is completed, the temperature T_C approaches the temperature of the surroundings T_A.

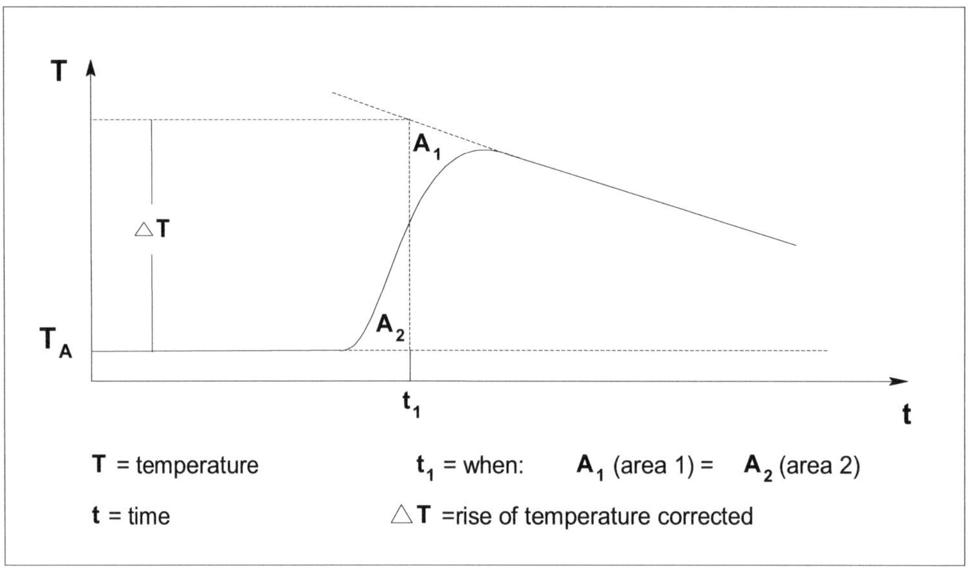

Fig. 2. Typical thermogram determined by an isoperibolic calorimeter. Adapted from Hohne et al. (1996)

Figure 2 illustrates a typical curve of temperature versus time obtained by an isoperibolic calorimeter in the observation of an exothermic event. At the beginning of the experiment the temperature stays close to the temperature of the surroundings T_A; when there is a certain amount of heat in the cell, the temperature increases initially, and then reaches a maximum value to finally begin to decline because T_A is less than T_C, and the magnitude of

the decrease depends on the isolation of the cell, that is, the thermal resistance R_T, which defines the constant heat leak, K_{ft}, a parameter of the device used, which is also a function of the temperature gradient. The amount of heat for the process under consideration is equal to $Q = Cp\Delta T_{corrected}$ where Cp is the heat capacity of the system understudy plus the heat capacity of the cell; ΔTcorrected is the difference in temperature above which gives a correction graphic of small but existing heat leaks, as shown in Figure 1. For accurate measurements it is not absolutely necessary to keep the heat losses as small as possible; however these are reproducible in terms of the temperature difference between the cell and the surrounding area and can be determined by electric calibration (Giraldo et al., 1994).
Taking into account the above considerations in general, isoperibolic calorimeters seek to reduce the heat exchange between the cell where the process is carried out and the surroundings, which is achieved by minimizing the temperature difference between them, decreasing the coefficient of heat transfer, and reducing the duration of heat exchange (Moreno and Giraldo, 2005).

2.3.2 Adiabatic calorimeters

Adiabatic calorimeters further restrict the heat transfer compared with isoperibolic calorimeters; ideally adiabatic calorimeters do not allow heat exchange between the cell and its surroundings. We can consider three ways to achieve this goal:

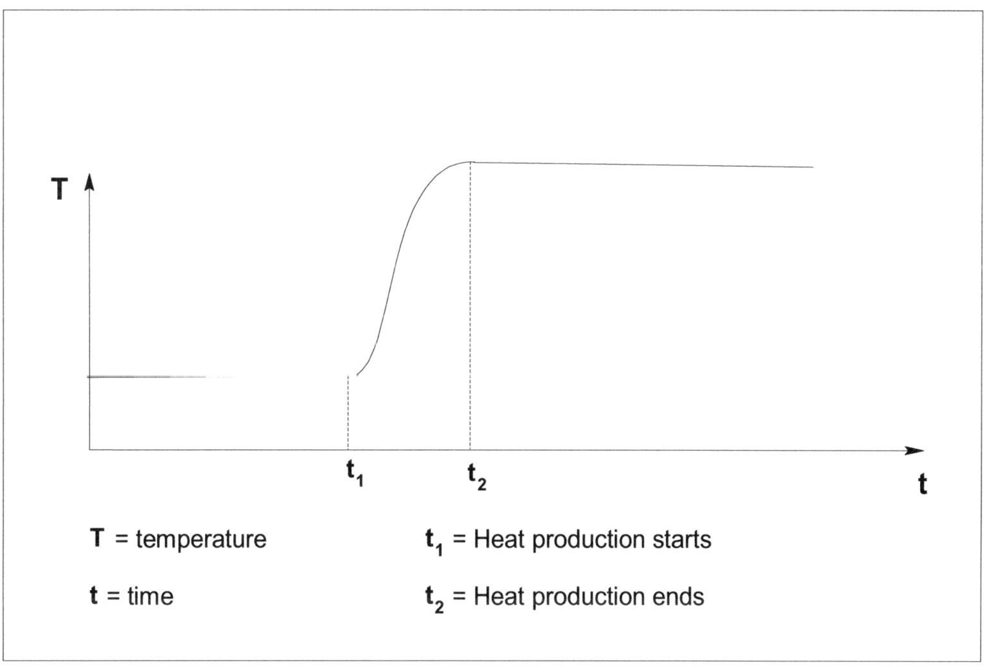

Fig. 3. Typical thermogram of an adiabatic calorimeter. Adapted from Hohne et al. (1996)

1. by carrying out the heat generation so fast that no appreciable amount of heat can enter or leave the cell during the period in which it is carried out;
2. by separating the cell to give a near to infinitely large thermal resistance R_T, so that the system is as isolated as possible;

3. by means of external controls that make the temperature of the surroundings always as close as possible to that of the cell.

During the calorimetric process any heat generated or consumed in the cell leads to a change in temperature. Figure 3 shows the graph of temperature versus time, T vs t. t_1 is generated in a heat effect until t_2. The Δheat can be calculated from measurements of the temperature difference ΔT:

$$Q = Cp\Delta T \tag{1}$$

The heat capacity is easily determined by calibration with the use of electricity. Ideally the slope of the curve of temperature versus time is proportional to heat flow:

$$\frac{dQ}{dt}(t) \approx C\frac{dT}{dt}(t) \tag{2}$$

where the heat flow dQ/dt (t) is obtained directly from curves like that shown in Figure 2.

2.3.3 Isothermal calorimeters

Another way of measuring the energy involved in a process, using a method opposed to the previous two (total isolation in the adiabatic case or allowing a small thermal leak in the isoperibolic case), provides a large exchange of the heat that is produced in the cell. This method is isothermal in nature, and the cell and its surroundings have the same constant temperature ($T_A = T_C$ = constant).

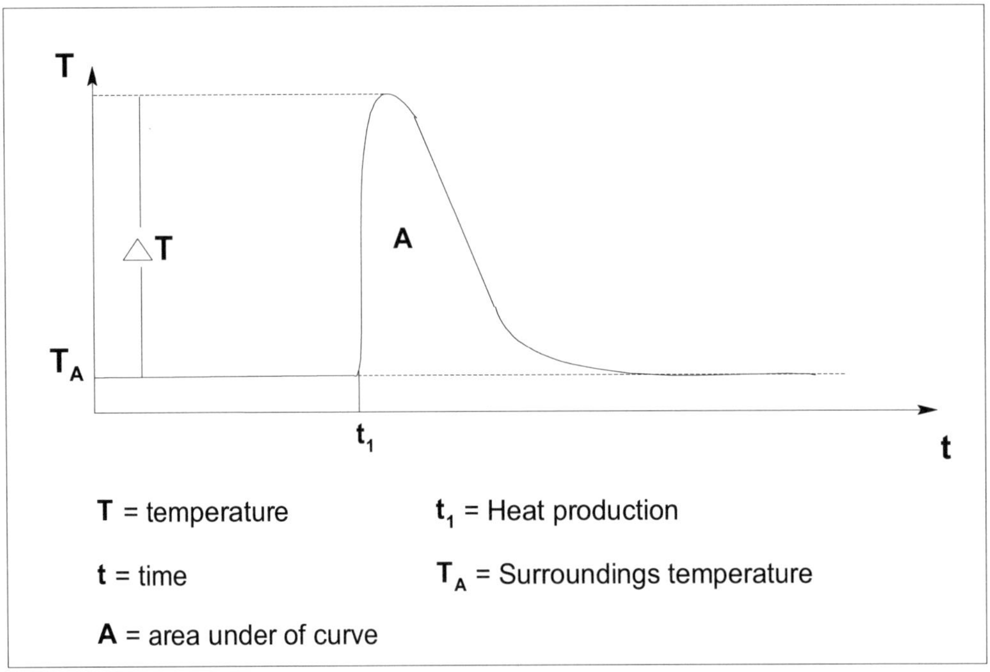

Fig. 4. Typical thermogram of an isothermal calorimeter. Adapted from Hohne et al. (1996)

The isothermal calorimeter has a very small thermal resistance R_T and also the heat capacity of the surrounding area is infinitely large. Taking into account these requirements, under strict conditions isothermal T_A and T_C may remain constant over time and space, but then no heat flow occurs. In real cases, we present a heat flow between the cell, and the surrounding flow is detected by thermal sensors placed between them. The flow is usually due to the small temperature difference between T_A and T_C during the process observed, and the magnitude of this temperature difference depends on the amount of heat released per unit time, the thermal conductivity and geometry of the cell, and the type of insulation of the thermal sensors. Despite these limitations, an isothermal calorimeter design is commonly used where temperatures T_A and T_C may be different from each other but each taken separately is constant throughout the duration of the process that generates the heat flow (Zielenkiewicz, 2000). Figure 4 shows a thermogram (T vs t) obtained with an isothermal calorimeter, where the heat conduction is observed around the drop in temperature after the cell is supplied with a pulse of heat.

2.3.4 Heat conduction calorimetry

Heat conduction calorimetry studies can be undertaken to provide kinetic and analytical thermodynamic data. The potential of the technique is such that Buckton (1995) states "The isothermal microcalorimetry has the ability to record all physical and chemical processes. The range of application of this technique is limited only by the imagination of the researcher and the ability to control the experiment." Some other interesting examples of applications of the technique are presented below.

Heat conduction calorimetry is classified as an isothermal technique (Wadso, 1986) because the variable that is constant in the experiment is the temperature of different parts of the computer. However, there is a local temperature difference which occurs whenever there is heat exchange between the cell and the surroundings (Wadso, 2001), as the heart of the device is composed of the cell, sensors, and a solid body, and this is expected to be the primary mechanism driving the exchange of heat.

In isothermal calorimetry, the surroundings and the cell have the same constant temperature:

$$T_{surrounding} = T_{cell} = \text{constant} \quad (3)$$

It can be seen that the calorimeter has a heat resistance, R_T, which is very small and also that the heat capacity of the surrounding area is infinitely large. Taking into account these requirements, under strictly isothermal conditions $T_{surrounding}$ and T_{cell} can remain constant over time and space, but then no heat flow occurs. In real cases, we present a heat flow between the cell and the surrounding flow is detected by thermal sensors placed between them. The flow is usually due to the small temperature difference between Tcell and Tsurrounding, and during the process observed the magnitude of this temperature difference depends on the amount of heat released per unit time, the thermal conductivity and geometry of the cell, and the type of insulation of the thermal sensors (Hemminger and Hohne, 1984). Despite these limitations, isothermal calorimeters commonly have a design where the temperatures Tcell and Tsurrounding may differ from each other, but each taken separately is constant throughout the duration of the process that generates the heat flow.

Figure 5 shows a diagram of the arrangement of the measuring cell and the surroundings in a heat conduction calorimeter. The cell is connected with them through a thermal resistance R_T, which is an interesting parameter because it relates the flow of heat dQ/dt to the difference in temperature.

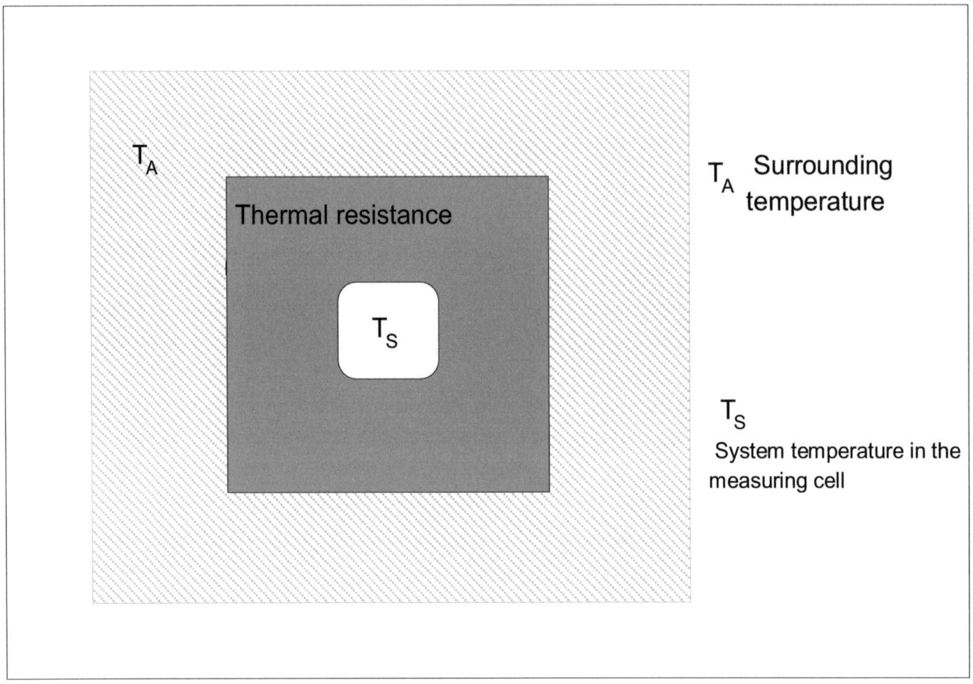

Fig. 5. Diagram of the heat conduction calorimeter

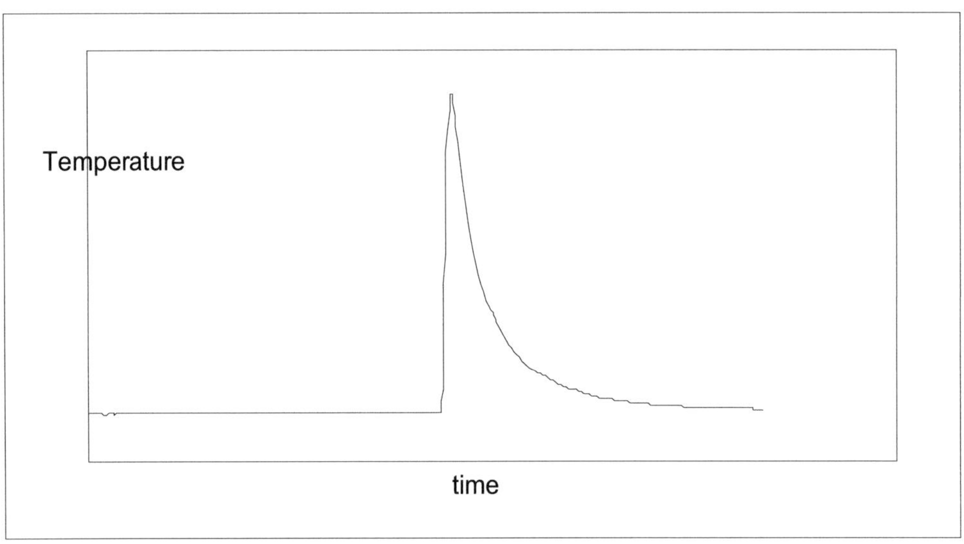

Fig. 6. Thermogram obtained by a heat conduction calorimeter

The temperature difference of the thermal resistance is

$$\Delta T = T_C - T_A \tag{4}$$

and at steady state we obtain the relationship

$$\frac{dQ}{dt} = \frac{\Delta T}{R_T} \tag{5}$$

Integration gives

$$Q = \frac{1}{R_T} \int \Delta T(t) dt \tag{6}$$

Figure 6 shows the form of a thermogram determined by a heat conduction calorimeter in which the thermal effect generated in the cell appears as a pulse in the curve of temperature versus time. As in the case of isothermal calorimetry, the heat flows to the surroundings, and the change in system conditions can be detected by means of a property that is proportional to heat flow, producing a continuous variation in this case. Measuring the sensor system properties as a function of time and through calibration experiments can determine the amount of heat generated by that system.

3. Adsorption calorimetry

Heats of adsorption of pure gases, which are usually obtained from isotherms using the Clapeyron equation, are unreliable unless extra precautions are taken to ensure reversibility and reproducibility. The calculation of heats of adsorption for mixtures from extensions of the Clapeyron equation is impractical. However, we have recently demonstrated that both adsorption isotherms and multicomponent heats of adsorption can be measured accurately and quickly in a single, inexpensive (Sharma et al., 1994; Spiewak et al., 1994; Dunne et al., 1997; García et al,2008). This paper summarizes the design criteria and construction of our combined calorimetric-volumetric apparatus in sufficient detail to reproduce our instrument, with numerous helpful suggestions to avoid some of the pitfalls associated with adsorption calorimetry. The known technology reported in the literature about the calibration systems (Handy et al., 93; Huertemendía et al., 2005) has been used here to evaluate the results obtained in this work.

3.1 Theory

The isosteric heat of adsorption is defined as the difference between the partial molar enthalpies in the gas and adsorbed phases:

$$Q_{st} = \bar{h}^{g} - \bar{h}^{\mathbf{a}} \tag{7}$$

Thus, Q_{st} is the heat of desorption. Even though it is not a heat but the difference of two state functions, the name is well established. The actual heat measured in a particular calorimeter must be related to the thermodynamic definition of isosteric heat in eq 7.

3.2 Idealized calorimeter

An idealized batch calorimeter consists of a dosing cell, a sample cell and a valve between the two, completely enclosed in an isothermal calorimeter at temperature T_0. At the start, the valve is closed, both cells are at temperature T_0, the pressure in the dosing loop is Pd, and the pressure in the sample cell is Pc, with $P\text{d} > P\text{c}$.

When the value is opened, an amount of gas expands from the dosing cell into the sample cell and a portion of this amount adsorbs. The total energy is:

$$U = U^g + U^a = u^g n^g + u^a n^a \tag{8}$$

The total energy U *should include* contributions from the adsorbent, the walls of the sample cell and dosing cell, and the valve.
However, since the temperature is fixed at T_0, these energies are omitted from eq 8 because they are constant and do not contribute to the change in energy. The total amount of gas in both cells is ng. The differential of the total energy is:

$$dU = u^g dn^g + n^g du^g + u^a dn^a + n^a du^a \tag{9}$$

where dU refers to the differential energy change after attainment of adsorption equilibrium. Because the temperature is T_0 before and after adsorption, $du^g = 0$ and

$$dU = u^g dn^g + u^a dn^a + n^a du^a \tag{10}$$

The mass balance is:

$$n^g + n^a = constant \tag{11}$$

so

$$dn^g = -dn^a \tag{12}$$

Substituting eq 12 into eq 10:

$$dU = -u^g dn^a + u^a dn^a + n^a du^a \tag{13}$$

The first law for the combined closed system consisting of the dosing cell, the sample cell and the valve is

$$dU = dQ \tag{14}$$

where dQ is the heat adsorbed by the combined system. For adsorption, dQ is a negative quantity. Combining eqs 13 and 14 we obtain

$$-dQ = u^g dn^a - u^a dn^a - n^a du^a \tag{15}$$

or

$$-\frac{dQ}{dn^a} = u^g - \left[u^a + n^a \frac{du^a}{dn^a} \right] \tag{16}$$

Because $h^a \approx u^a$ and $h^g = u^g + zRT_0$, comparison of eqs 7 and 16 gives

$$Q_{st} = -\frac{dQ}{dn^a} + zRT_0 \tag{17}$$

This result was derived by Hill. The first term is the differential heat measured by the idealized calorimeter, and the second term is the difference between the enthalpy and the internal energy in the equilibrium gas phase. $z=PV/RT$, the compressibility factor in the gas phase, is close to unity for sub atmospheric measurements of isosteric heat. The RT_0 term at 25 °C is 2.5 kJ/mol, and typical isosteric heats of adsorption are in the range 10-50 kJ/mol (O´Neil et al., 85).

3.2.1 Thermopiles theory

The Seebeck effect

If two semiconductors a and b are joined together at the hot point and a temperature difference ΔT is maintained between this point and the cold point, see Figure 6(a), then an open circuit voltage ΔV is developed between the leads at the cold point. This effect, called the Seebeck effect after its discoverer T. J. Seebeck (1770 - 1831), can be mathematically expressed by

$$\Delta V = \alpha_s \Delta T \tag{18}$$

here a_s is the Seebeck coefficient expressed in V/K (or more commonly in µV/K). It was found that only a combination of two different materials, a so-called thermocouple, exhibits the Seebeck effect. For two leads of the same material no Seebeck effect is shown, for reasons of symmetry. It is, however, somehow present because the Seebeck effect is a bulk property and does not depend on a specific arrangement of the leads or the material, nor on a specific way of joining them. This bulk property can be expressed as

$$VE_F / q = \alpha_s VT \tag{19}$$

where E_F is the Fermi energy (and E_F/q = ΦF is the electrochemical potential), and where the Seebeck coefficient αs depends, among other things, upon the chemical composition of the material and upon the temperature.

The Seebeck coefficient of, for example, silicon, can be derived by setting αs as (see Figure 6(b))

$$\alpha_s = \frac{d}{q\,dT}(E_F) \tag{20}$$

For non-degenerate silicon the Seebeck coefficient may be approximated by using simple Maxwell- Boltzmann statistics. Three main effects are present.

First, with increasing temperature the silicon becomes more intrinsic:

$$\frac{d}{q\,dT}(E_F)\;{}_{E_C-E_F} = -\frac{\kappa}{q}\left(\ln(N_c / n) + \frac{3}{2}\right) \tag{21}$$

where E_c is the conduction-band edge energy, N_c the conduction-band density of states, n the electron density (fixed by the doping concentration) n_d and k the Boltzmann constant.

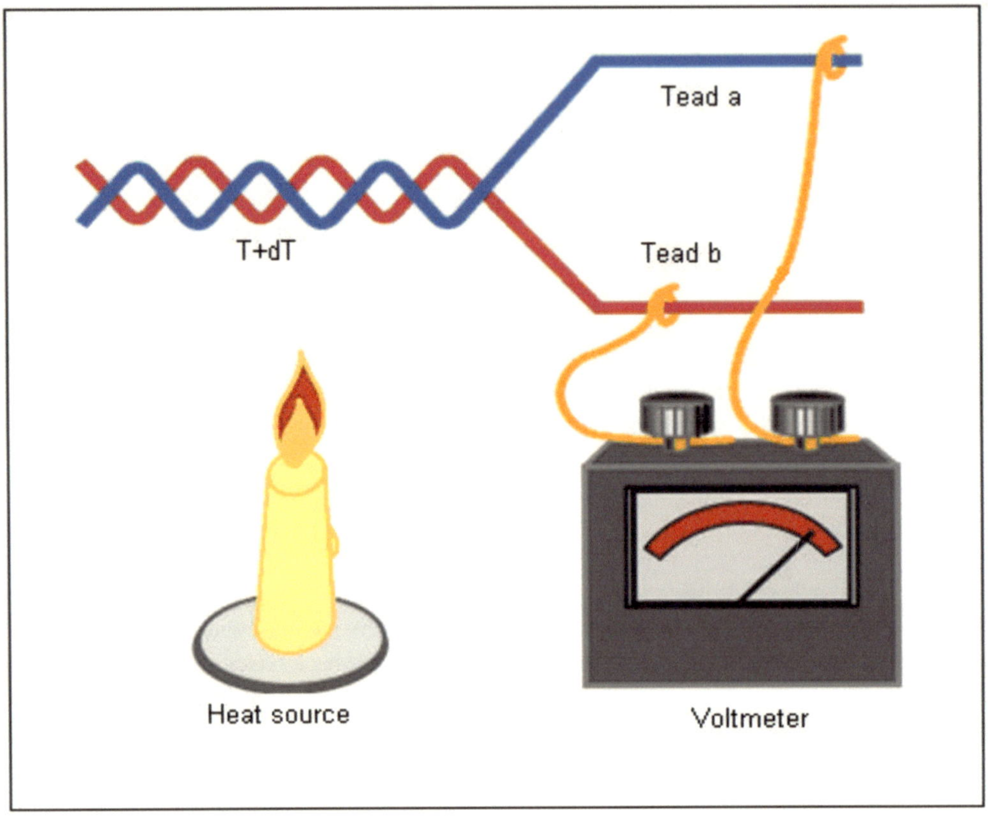

Fig. 6a. Seebeck Efect

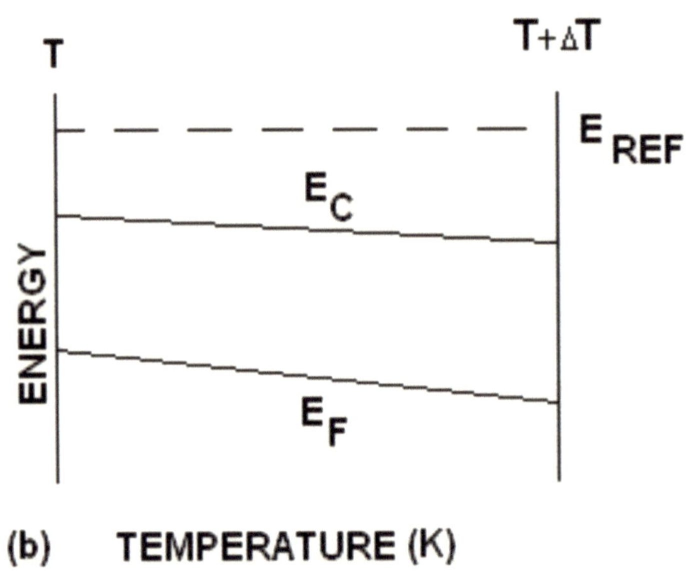

Fig. 6b. Variation of E_F due to ΔT

Secondly, with increasing temperature the charge carriers have a higher average velocity, leading to charge build-up on the cold side of the silicon.
Moreover, the scattering of charge carriers is usually energy (and thus temperature) dependent, likewise leading to charge build-up on the cold or hot side of the silicon, depending on whether the hot carriers can move more freely than the cold carriers or are 'trapped' by increased scattering:

$$\frac{d}{q\,dT}\left(E_F\right)_\tau = -\frac{\kappa}{q}(1+s) \tag{22}$$

where τ is the relaxation time (mean free time between collisions) and s is the exponent describing the relation between τ and the charge-carrier energy.
Finally, the temperature difference in the silicon causes a net flow of phonons from hot to cold. In a certain temperature region (-10 - 500 K) and for non-degenerate silicon, a transfer of momentum from acoustic phonons to the charge carriers can occur. As there is a net phonon momentum directed from hot to cold, this will drag the charge carriers towards the cold side of the silicon. This effect may be represented by:

$$\frac{d}{q\,dT}\left(E_F\right)_{\phi_n} = -\frac{\kappa}{q}\phi_n \tag{23}$$

in which Φ_n denotes the phonon drag effect. In sum, the total Seebeck coefficient in non-degenerate silicon becomes:

$$\alpha_s = -\frac{\kappa}{q}\left\{\ln\left(N_c / n\right) + \frac{5}{2} + s_n + \phi_n\right\} \quad n-type \tag{24}$$

$$\alpha_s = +\frac{\kappa}{q}\left\{\ln\left(N_v / p\right) + \frac{5}{2} + s_p + \phi_p\right\} \quad p-type \tag{25}$$

where s is of the order -1 to 2, and where he phonon-drag contribution Φ ranges from 0, for highly-doped silicon, to approximately 5, for low-doped silicon, at 300 K, while Φ ranges from 0, for highly-doped silicon, to 100, for low-doped silicon, at low temperature (100K).
In practice, the Seebeck coefficient may be approximated, for the range of interest for use in sensors and at room temperature, as a function of electrical resistivity:

$$\alpha_s = \frac{m\kappa}{q}\ln\left(p / p_o\right) \tag{26}$$

3.3 Experimental

3.3.1 Design criteria

The desired equilibrium information for adsorbed mixtures is the pressure and composition of the gas phase above the adsorbent for a given loading, as well as the heat evolved for differential increases in the loading. Because we considered direct calorimetric measurements of differential heats to be more reliable than differentiation of isotherms at various temperatures, the instrument was built around a Tian-Calvet calorimeter. Practical limitations on the ability to integrate the heat flow in the calorimeter as a function of time required that equilibrium be established in 15 min or less. The necessity of establishing equilibrium within 15 min of changing the sample loading placed a stringent limitation on the design. The major limitation for the attainment of adsorption equilibrium is gas-phase mixing in the region above the sample. On the basis of a typical gas-phase diffusion coefficient of 0.1 cm^2s^{-1}, a tube length of even 10 cm will result in mixing times of 1000 s. This imposes significant challenges on the instrument design. While imposed circulation would alleviate this problem, forced flow would also complicate the design of the calorimeter because of convective heat losses. The maximum distance within our equipment (from the bottom of the sample cell to the diaphragm of the pressure transducer) was approximately 10 cm. The pressure transducer was chosen for its small dead volume. The leak valve for the composition measurements was welded directly on the top of the cell to minimize the dimensions of the apparatus. These design criteria could only be met by a custom-made calorimeter. In general this calorimeter is based in literature design and experience of our laboratory (Giraldo et al., 98)

3.3.2 Practical calorimeter

In the idealized calorimeter, the temperature of the gas in the sample loop decreases upon expansion while the temperature of the gas in the sample cell increases as it is compressed by the incoming gas. In the absence of adsorption, heat is absorbed by the dosing loop and heat is liberated by the sample cell until the pressures equalize and the temperature returns to T_0. For a perfect gas, the two effects cancel because the enthalpy of a perfect gas is a function only of temperature.

Our design is a modification of the idealized calorimeter in which only the sample cell is placed in the calorimeter. Because the dosing loop and valve are external to the calorimeter, adding a dose of gas to the sample cell generates an exothermic heat of compression in the sample cell which is not cancelled by absorption of heat in the dosing loop. The spurious heat of compression calculated from eq 21 is subtracted from the total heat registered by the calorimeter in order to obtain the heat of adsorption.

3.3.3 Description of the Instrument

A diagram of the calorimeter apparatus is shown in Figure 7, the components are described in Table 1. A picture of the sample cell and its connections is shown in Figure 8. The stainless steel cube is the sample cell for the adsorbent and adsorbate. The use of stainless steel to maximize heat conduction through the top of the cell is a crucial element of the design. The stainless cube is surrounded on all four sides and on the bottom by square thermal flow meters (show in the picture) obtained from the Melcor Corporation™. Each thermopile is a 40x40x2 mm ceramic plate with about 240 embedded thermocouples for detecting temperature differences across the plate.

The signal from these thermopiles was input to a data acquisition system with a computer.

The sample cell slides into cubical holes cut into an aluminum block (10x13x8 cm, mass 1 kg). A silicone based heat-sink compound was used to ensure good thermal contact between the Al block and the thermopiles and between the thermopiles and the stainless steel.

The cubic stainless steel cell shown in Figure 8 (on the top) was inserted into a Cajon fitting, which provides a vacuum seal by compression of a Viton O-ring. The Cajon fitting connects to a custom-made T connection onto which the leak valve, the pressure head, the connection to vacuum and the 0.01-in. bore tube from the dosing loop are welded. The leak valve is connected through a 1/4-in.-o.d. stainless-steel tube; the pressure head is connected through a 1/4-in. δ NPT fitting; the valve that opens to vacuum is connected through a 1/4-in. VCR fitting. The pressure head was chosen for its small dead space (2.0 cm^3). The total dead space is 17.8 cm^3 for the (empty) sample cell, the dead space inside the pressure head, the lines to vacuum, the dosing loop and the RGA leak valve.

No.	Description	Model No.
1	Gas 1 inlet	
2	Gas 2 inlet	
3	To vacuum pump	
4	Three-way valve	
5	Inlet valve to the dosing loop	
6	Outlet valve from the dosing loop	
7	Pressure transducer for the dosing loop	Teledyne™
8	Liquid nitrogen trap	
9	Valco six-way valve	
10	Calibrated dosing loop (5 cm3)	
11	Variable leak valve	Granville-Phillips 203™
12	Cell outlet valve	
13	Pressure transducer for the cell	Edwards™, 655 and 622
14	Reference cell	
15	Calorimeter cell	
16	K-type thermocouple	
17	Thermopiles	Tellurex Corporation™
18	Heat sink (aluminium block)	
19	Mass spectrometer (RGA)	HP GC/MS 5890 series II 5972 MDS
20	Turbopump	Pfeiffer ™
21	Data acquisition board	
22	Computer	

Table 1. Components of microcalorimeter. Key to figure 1

Gas was introduced to the sample cell from the dosing loop using a six-port Valco sampling valve connected to a small bore (0.01-in.-i.d.) tube. The small diameter of the tube prevents back mixing of the mixture into the dosing loop. This tube enters the T-shaped connector from the back (the welded connection does not appear on Figure 3) and extends downward, with the opening 5 cm above the bottom of the sample cell. Two small metal cylinders with a Viton O-ring between them were inserted in the NPT connection to the pressure head to make a vacuum seal.

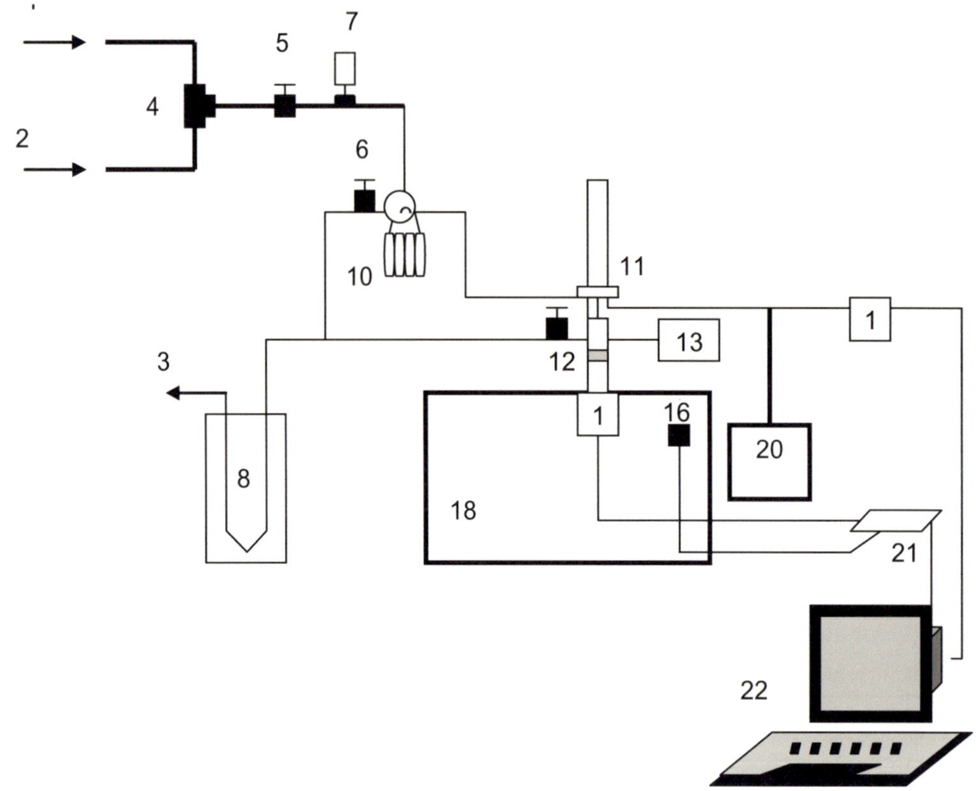

Fig. 7. Schematic diagram of microcalorimeter system and auxiliary equipment

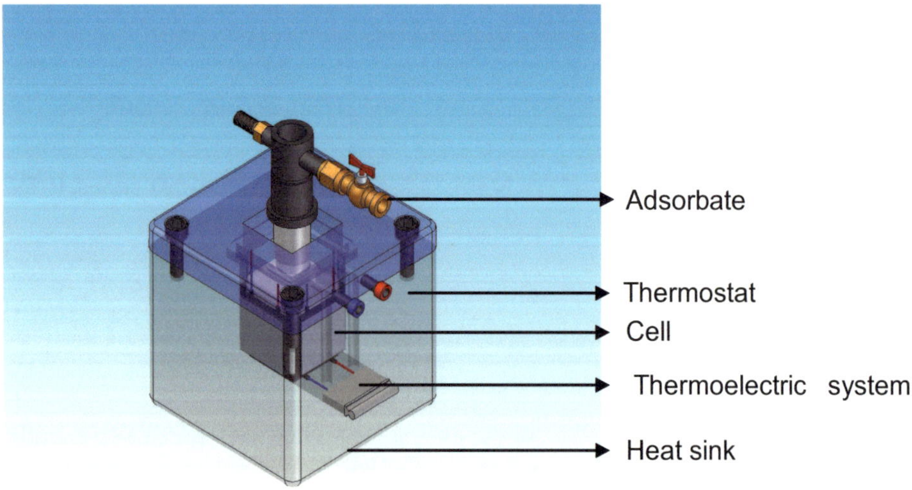

Fig. 8. Picture of the stainless steel sample cell and connections to the pressure head, vacuum line, dosing loop and leak valve. The stainless steel sample cell is surrounded by thermopiles set into an aluminum heat sink

The adsorbent was covered with a 1.5 cm layer of glass chips to minimize heat loss through the top of the cell and regenerated in situ.

3.3.4 Electric calibration of the adsorption micro calorimeter.

To establish the correct operation of the micro calorimeter prior to connecting it to the volumetric adsorption unit, we evaluated its sensitivity by determining the calorimeter constant.

The calibration constant gives the voltage generated by the calorimeter when a given amount of heat is emitted from inside the microcalorimetric cell.

There are two methods to determine the calibration constant (**K**):

Determination of the calibration constant by application of electric power.

This method is based on the dissipation of electric work (**W_e**) by an electric resistor through which an electric current (i) passes for a certain amount of time (**t**). This generates a voltage (**V_t**) in the micro calorimeter and this is measured.

The micro calorimeter calibration constant (**K**) is given by:

$$K = \frac{We}{\int V_t dt} = \frac{V_c it}{\int V_t dt} \tag{27}$$

Where **V_c** is the voltage applied to the resistor, **i** is the current that passes through it, and **t** is the time expressed in seconds.

Determination of the constant by the stationary state method.

This is an alternate method to the one above, which is useful to compare and evaluate whether the constant (**K**) assessed by the above method is correct. The method consists on applying a constant voltage (**V_c**) through the micro calorimeter electric resistor until the voltage generated by the calorimeter (**V_t**) reaches the condition of stationary state. Under these conditions, **K** is given by:

$$K = \left(\frac{V_c i}{V_t}\right)_{stationary} \tag{28}$$

3.3.5 Spurious heat of compression in sample cell

Before taking a measurement, the dosing loop and the sample cell are both at the temperature T_0 of the experiment; the pressure inside the sample cell is P_c, and the pressure in the dosing loop is some higher pressure P_d. Increments of gas are added to the sample cell by opening the valve between the dosing loop and the cell.

The temperature of the gas inside the dosing loop falls because of the expansion, while the temperature of the gas inside the sample cell rises as it is compressed by the incoming gas. The calorimeter measures both the latent heat of adsorption and the sensible heat liberated by the compressed gas as it cools to the temperature of the calorimeter. This sensible heat must be subtracted from the heat registered by the thermopiles to obtain the heat of adsorption.

The spurious heat term generated by compression of the gas inside the cell was determined by expanding gas from the dosing loop into a sample cell containing no adsorbent. For a 10 cm^3 dosing loop and for a dead space of 18 cm^3 in the sample cell, the linear correlation

$$Q_{sp} = a\ \Delta \mathbf{P} \tag{29}$$

for the experimental data shown in Figure 9 yields a = 4.42 x10-4 J/Torr. ΔP is the driving force for the irreversible expansion: the pressure difference between the dosing loop and the sample cell.

The correlation ignores the effect of adsorption as gas enters the sample cell. For the case of weak adsorption, when only a small fraction of the gas entering the sample cell actually adsorbs, the approximation is justified.

For the case of strong adsorption, when most of the gas entering the sample cell adsorbs, the spurious heat of compression is negligible compared to the heat of adsorption.

Thus, for strong adsorption (95% of gas dose adsorbs) or weak adsorption (5% of gas dose adsorbs), the approximation of considering that the heat of compression is independent of adsorption is acceptable. We have no proof that the correction for the spurious heat of compression is negligible in the intermediate case when about 50% of the gas dose adsorbs, but the excellent agreement of both strong and weakly adsorbing gases with the Clapeyron equation is indirect evidence that eq 29 is adequate for both strongly and weakly adsorbing gases.

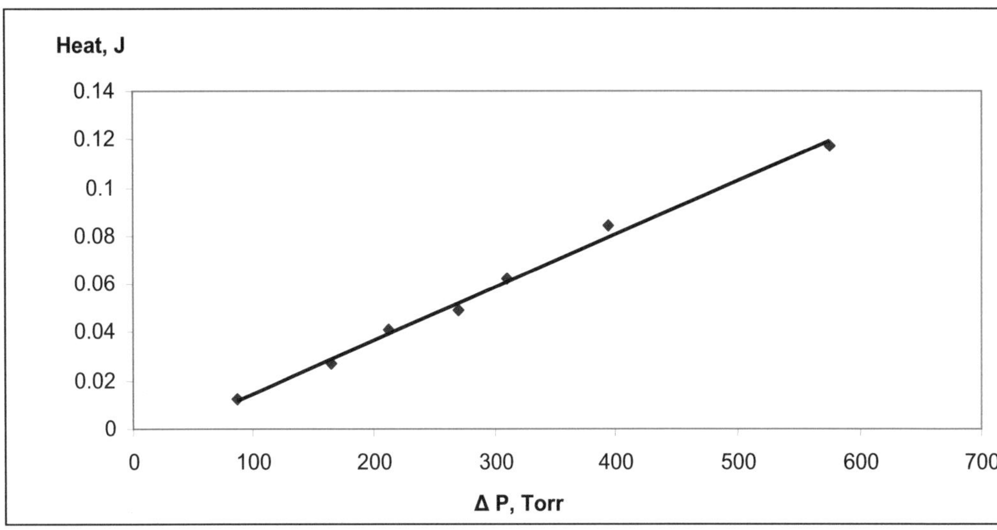

Fig. 9. Linear correlation for a spurious sensible heat term in adding a dose of gas. The difference is the pressure in the dosing loop minus the pressure in the sample cell before opening the valve

Other calorimeters are designed for isothermal introduction of gas to the sample cell. This is accomplished by adding increments of gas slowly through a needle valve so that the temperature of the gas in the dosing loop is equal to the temperature in the sample cell (T_0). In the absence of adsorption, the reversible isothermal introduction of a gas sample generates an exothermic heat inside the sample cell equal to RT_0 per mole of gas added; the signal for this spurious heat term can be eliminated by adding the same amount of gas to a reference cell wired in reverse polarity. Isothermal dosing is effective for the measurement of heats of adsorption of pure gases. For mixtures, the fast, irreversible addition of increments of gas shortens the time required for mixing and equilibration.

3.3.6 Verification of adsorption equilibrium

The mixing time required when a new dose of gas is added to the sample cell containing a gaseous mixture but no adsorbent is about 15 min. Sampling the gas phase continuously to check for equilibrium is impracticable because the amount of gas sampled over 30 min would affect the mass balance used to calculate the amount adsorbed.

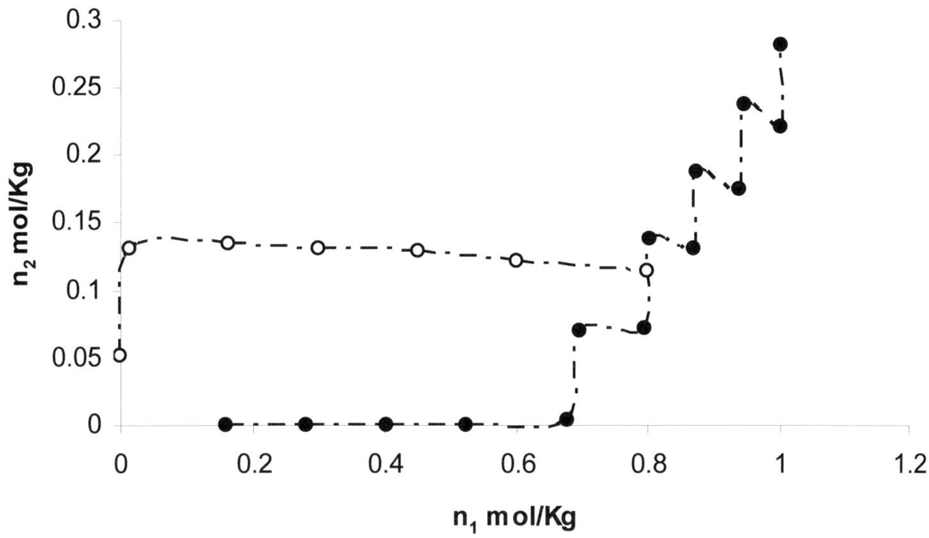

Fig. 10. Adsorption of mixtures of SF_6 and CH_4

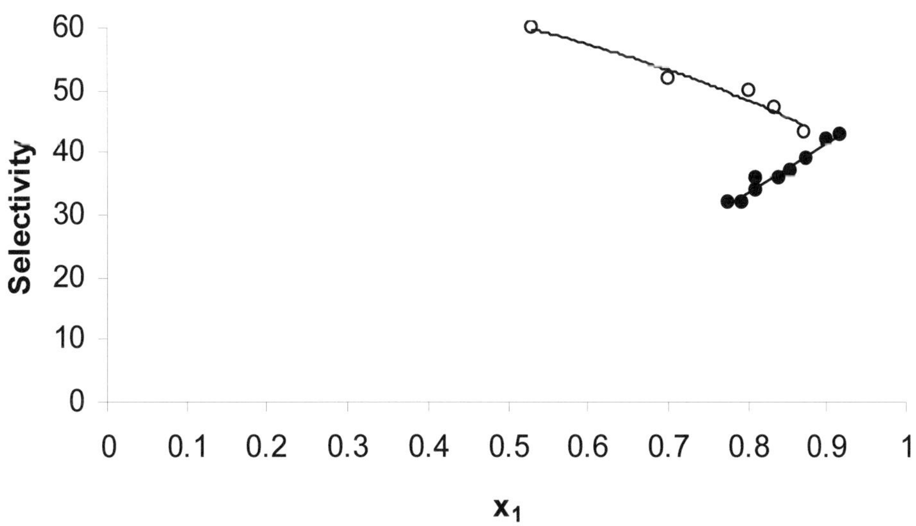

Fig. 11. Selectivity of SF_6 relative to CH_4 at 24,5 °C. symbols are the same as those in figure 5

Two methods were used to verify the attainment of equilibrium for mixture adsorption. The first method is to fit the experimental data to a model which is thermodynamically consistent; agreement of the model with the experimental data is an indirect but robust method of verifying equilibration. A second, direct method is to verify that a particular point is independent of the path to reach that point. Figure 10 shows an example for the adsorption of mixtures of SF_6 (component 1) and CH_4 (component 2). The closed and open circles indicate two paths from zero loading to point A; the arrows show the direction of the paths. These two paths intersect at $n_1 = 0.78$ and $n_2 = 0.12$, or a mole fraction $x_1 = 0.87$. Figure 11 shows the selectivity for the same two paths; the selectivity curves intersect at $x_1 = 0.89$. Therefore, within an uncertainty of about 1%, the selectivity is independent of the path followed by the system. There is excellent agreement among them and with values reported in literature (Siperstein et al., 99).

3.3.7 Determination of differential heats from finite doses

The amount dosed Δn must be small enough to measure the differential heat but large enough to generate an accurate heat signal Q. Because the differential heat is defined as the ratio of $Q/\Delta n$ in the limit as Δn goes to zero, the error associated with finite increments needs to be examined.

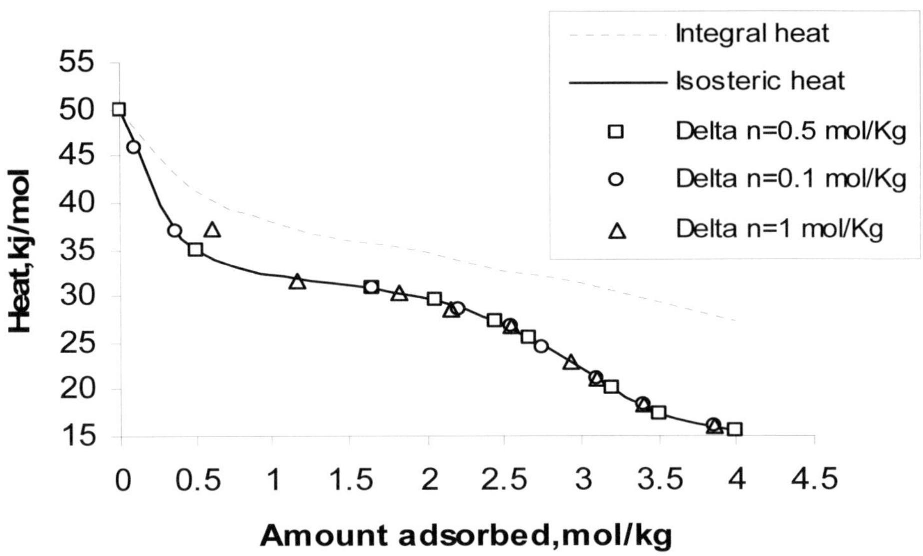

Fig. 12. Comparison of the differential heat of adsorption (solid line) with experimental heats determined with finites doses of gas. The dashed line is the integral heat of adsorption. Heats determined experimentally with small doses of the order 0.12 mol/Kg agree very well with the exact differential heat

Assume that the differential heat $q_d(n)$ is given exactly by the polynomial:

$$q_d(n) = q_0 + d_1 n + d_2 n^2 + d_3 n^3 + \dots \tag{30}$$

For a finite amount of gas adsorbed ($\Delta n = n_2 - n_1$), the approximate differential heat $q\delta$ measured experimentally is

$$q_\delta = \frac{\int_{n_1}^{n_2} q_d(n)dn}{n_2 - n_1} \tag{31}$$

representing the average value of the differential heat measured at the average loading ($n_1 + n_2$)/2. Comparison of q_δ with the exact differential heat at the same average loading gives the error:

$$q_\delta - q_d = \frac{d_2}{12}(n_1 - n_2)^2 + \frac{d_3}{8}(n_1 + n_2)(n_1 - n_2)^2 + ... \tag{32}$$

The error is of order of (n_1 - n_2) ^{2}. Because the leading term of the error is also proportional to the second derivative of the heat curve, $q_\delta = q_d$ for linear heat curves.
Figure 12 shows hypothetical differential (solid line) and integral (dashed line) heats of adsorption. The points show approximate heats qδ calculated from eq 25 for finite doses n_2 - n_1 = 0.1, 0.5, and 1.0 mol/kg. Only for finite doses as large as 1 mol/kg can the difference between the exact differential q_d and the approximate q_δ be appreciated. Typical experimental values of Δn are of order 0.1 mol/kg. Except for abrupt changes of heat with coverage associated with phase transitions, the error associated with the use of finite doses of gases to measure the differential heat is negligible. There is agreement among them and with values reported in literature (Siperstein et al., 99).
It is convenient to report differential heats of adsorption at the loading n2 instead of the average loading ($n_1 + n_2$)/2. This introduces errors larger than that predicted by eq 24, especially when the slope of the heat curve is large.

3.3.8 Alternating dosings of each component

Two independent dosings (A and B) are required to measure the individual differential heats of adsorption (q_1 and q_2) from a binary mixture (Siperstein et al., 99) :

$$Q^A = \Delta n_1^A q_1 + \Delta n_2^A q_2 \tag{33}$$

$$Q^B = \Delta n_1^B q_1 + \Delta n_2^B q_2 \tag{34}$$

where Q^A and Q^B are the heats registered by the calorimeter and Δn_1 and Δn_2 are the amounts adsorbed, or desorbed, of components 1 and 2, respectively. When the system of equations (33) and (34) is solved, the individual heats of adsorption are

$$q_1 = \frac{Q^A \Delta n_2^B - Q^B \Delta n_2^A}{\Delta n_1^A \Delta n_2^B - \Delta n_1^B \Delta n_2^A} \tag{35}$$

$$q_2 = \frac{Q^B \Delta n_1^A - Q^A \Delta n_1^B}{\Delta n_1^A \Delta n_2^B - \Delta n_1^B \Delta n_2^A} \tag{36}$$

Dosing of one component generates a positive incremental adsorption of that component which is normally 1 or 2 orders of magnitude larger than the accompanying desorption of the other component. The solution of eqs 27 and 28 requires that the dosing of the components be alternated; successive dosings of the same component generate an indeterminate solution.

4. Experimental results

4.1 Electric calibration of the adsorption micro calorimeter: (by applying electric power)

Table 2 shows the calibration constants obtained for the micro calorimeter operation conditions. Additionally, it presents the values reported at different voltage levels, which range between 12.34 ± 0.12 W V^{-1} and 16.67 ± 0.32 W V^{-1}. These values show the sensitivity of the micro calorimeter built here, which is higher than that of equipments reported in literature and even of those built in our laboratory previously. This constitutes a considerable contribution to the construction of this type of instruments.

Figure 13 shows a typical thermogram obtained with the calorimeter, which corresponds to an electric energy of 1 joule inside the cell, which contains only air. From this type of thermograms the calibration constant (K) can be obtained. It is necessary to point out the great stability of the baseline before and after the thermal effect.

Total electrical energy (J)	Electrical power (mW)	Calibration constant, K ($W.V^{-1}$)*
4.500	25.00	22.21 ± 0.31
3.000	16.00	21.45 ± 0.11
2.000	9.000	23.68 ± 0.06
0.700	4.000	24.25 ± 0.21
0.180	1.000	25.41 ± 0.23
0.045	0.250	22.64± 0.14

Table 2. Calibration constants obtained for the micro calorimeter obtained applying a known electrical power

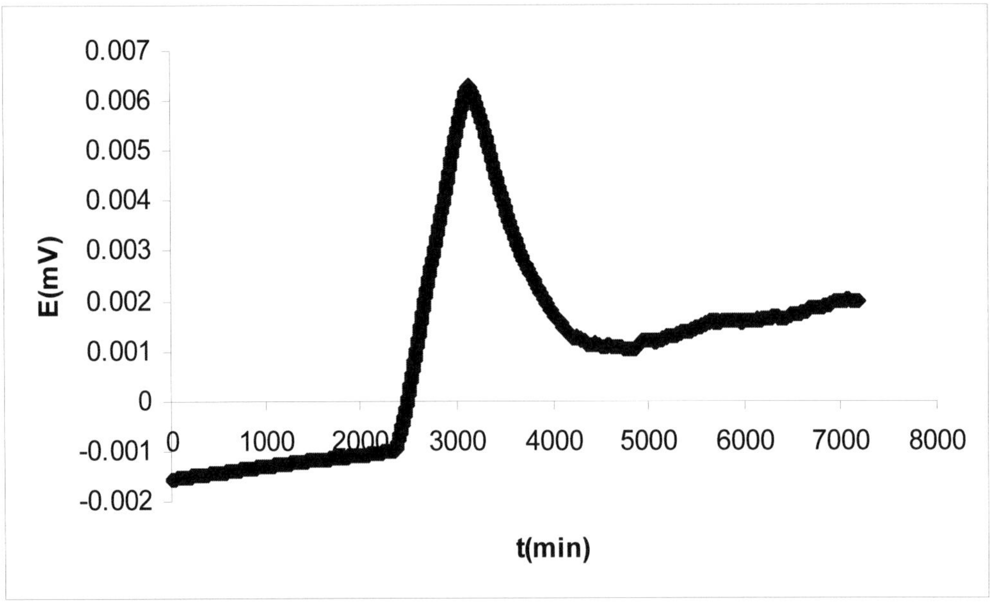

Fig. 13. A typical thermogram of electric calibration by electrical input

4.2 Calibration by stationary state method

Table 3 shows the calibration constants obtained for the equipment built here. These results strongly agree with the previous methodology and with the laboratory previous works.
The voltage signal from the calorimeter was determined as a function of the rate of heat dissipation (dQ/dt) I^2R in a platinum resistance wire wrapped around the outside of the cell in thermal contact with the cell wall and the thermopiles. Similar difficulties were encountered by Handy (Handy et al., 93) the voltage-to-power ratio for a resistor inside the cell was 9% lower than that for an externally wrapped resistance wire. The difference was attributed to heat losses. We chose the Clapeyron equation as the more reliable method of calibration.

Applied voltaje (V)	Electrical power (mW)	Calibration constant, K ($W.V^{-1}$)*
0.254	0.060	24.12 ± 0.31
0.567	0.230	24.15± 0.15
1.009	0.980	23.45 ± 0.64
2.084	3.920	13.23 ± 0.22
3.096	8.860	22.62 ± 0.41
4.084	15.92	23.84 ± 0.45
5.096	24.75	22.22± 0.2

Table 3. Steady state calibration constants obtained for the microcalorimeter

Figure 14 illustrates a thermogram obtained when an electric power of approximately 10mW disperses inside the micro calorimetric cell.
A secondary calibration of the calorimeter (0.045W/V) is based upon the Clapeyron equation applied to a series of adsorption isotherms measured in a separate, high-precision volumetric apparatus for ethane on silicalite (MFI) synthesized in our laboratory. The calibration constant for ethane was confirmed by excellent agreement of calorimetric data with the Clapeyron equation for SF_6, CO_2, and CH_4. The calibration constant is independent of the amount of adsorbent in the cell.
For the presentation of experimental results, it would be helpful that one of the variables, such as the total pressure or fugacity of one of the components, could be held constant.
However, the necessity of alternating doses generates a locus similar to the closed circles shown in Figure 6. The inability to obtain data along some locus, such as an isobar, is annoying but does not affect the analysis of the experimental data for activity coefficients and excess functions. After covering the entire phase diagram for a binary mixture by varying the preloading of the pure components, a model fitting the experimental data can be used to generate loci such as isobars or constant loading of one component.
Figure 7. Comparison of the differential heat of adsorption (solid line) with experimental heats determined with finites doses of gas. The dashed line is the integral heat of adsorption. Heats determined experimentally with small doses of the order 0.12 mol/Kg agree very well with the exact differential heat.
Since our first measurements of heats of adsorption from binary mixtures reported in 1997 for CH_4 and C_2H_6 in silicalite and for CO_2 and C_2H_6 in NaX, we have completed experiments for four other binary mixtures; one of them is reported in this paper (SF_6 and CH_4 on NaX). Presently, we are computing thermodynamic excess properties for these mixtures, especially adsorbed-phase activity coefficients, excess free energy, excess entropy, and heat of mixing in

the adsorbed phase. It is interesting that all of the excess functions are negative: activity coefficients are less than unity, and the heat of mixing is exothermic in every case.

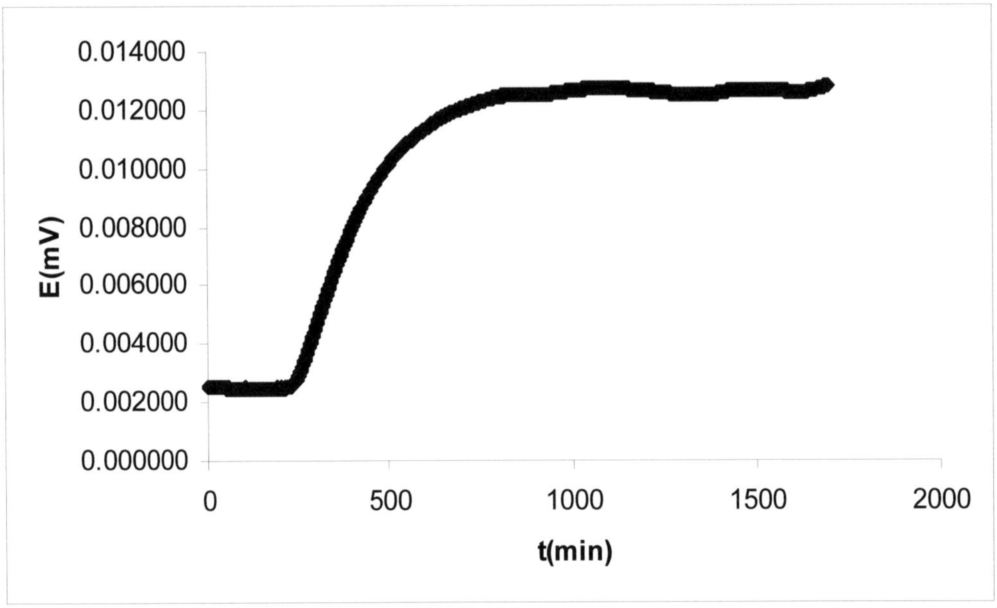

Fig. 14. A typical thermogram of electric calibration by stationary state

5. Variation of the noise levels in the baseline of an adsorption microcalorimeter

To measure precisely the heat, various elements can be used. The thermometer can be very stable, but only by a period of time that exceeds the experiment. However, other sensors can be used, such as termistors that are particularly useful for several reasons: high thermometric sensitivity; built with simple and inexpensive components; and at last, they have a small mass then, the time constant can be also small. The principal disadvantage is that the termistor heats itself. Some electronic circuits are sometimes used when termistors are employed to measure the temperature (Hansen and Hart, 2004), like the modified Wheatstone bridge, one of the most used because of its simplicity, easy operation and high sensitivity. However, when a high sensitive system is required, like in calorimetric studies for the gas-solid interphase, a Calvet-type adsorption calorimeter is the chosen. In these equipments the measure system consists in a group of thermo couples in series that form the thermo battery, which is the measure thermo element. These thermo elements use the Seebeck effect, where a difference in the temperature generates a difference in the potential (Martínez et al., 2004). A thermo battery works very well in theory, but errors are common in practice, which are attributed to small electrical signals generated by the unions, connections and imperfections in the electrical circuit employed in the laboratories to obtain the thermo electrical signal. Within them it can be mentioned: a) the length, thickness, polarity of the terminal wires in the thermo batteries; b) the number of connections in the system; c) the magnitude of the thermal gradients along the wires in the thermal elements. These factors can generate noise levels that could invalidate the calorimetric measures if taken into account that the magnitude of the thermal effect in adsorption calorimetry, in some cases, are about μW. In this context, it is called noise to every undesired signal that overlaps the genuine signal and it is not directly related to the thermal measure and could distort it. There are three kinds of fundamental noises in every electronic component, two of which are: a) thermal noise or Johnson; b) shot noise.

5.1 Thermal noise or Johnson

Every kind of resistances generates, by themselves and by their terminals, a voltage with random fluctuations like thermal noise or Johnson, which is caused by the charge carrier random movement in conductor materials. It is always produced at temperatures above absolute zero (-273°C or 0 K).
This kind of noise is related to the no continuous nature of the electrical current, formed by a discrete charges flow that causes statistical fluctuations in the current. The shot noise, as well as the thermal noise, is a kind of white noise.
In this work, we studied the noise signal behavior generated in an adsorption micro calorimeter built in our laboratory, based in others built previously, respect to the applied potent and the temperature, to establish if the noise level can eventually affect the measures in the gas-solid interface (García et al., 2008).

5.2 Experimental

The measures are realized in a micro calorimeter designed in our laboratory, which basic scheme is shown in Figure 15. I consist basically in two parts: a) adsorption part and b) the micro calorimeter itself. Each one of these parts is detailed in the figure.

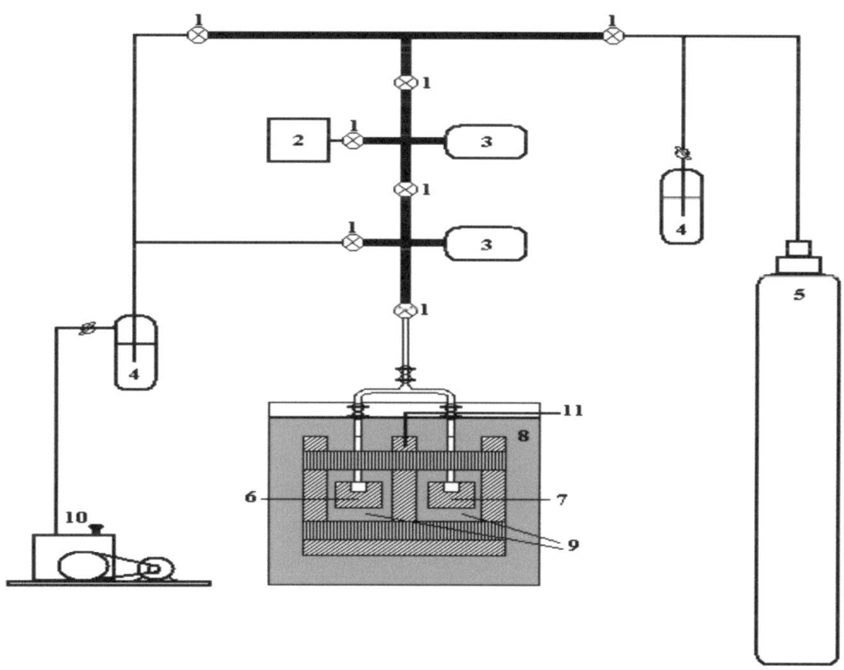

Fig. 15. Adsorption Micro calorimeter
1. Precision Valves 2. Calibration volume or storage. 3. Pressure Transducers. 4. Nitrogenous Traps 5. Adsorbate. 6. Reference Cell. 7. Reaction Cell. 8. Heat Storage. 9. 3D-Type Heat Sensors. 10. Vacuum System. 11. Temperature Control Sensor

A detail of the calorimetric cell is shown in Figure 16. It consists in two calorimetric cells; one of them acts as the reference cell and the other as the measure cell. The 3D-Type heat sensors used are zoomed in the same figure.
The equipment is insolated from the surroundings to control the temperature by a special material with low thermal conductivity. The calorimeter has two resistances with the same magnitude, which value is about 698.32 KΩ. The measurements scheme is shown in Figure 17; it is highlighted with an interrupted line, the external electrical connections between the adsorption micro calorimeter and the data entry system, which contribute to the noise in the signal measurements.

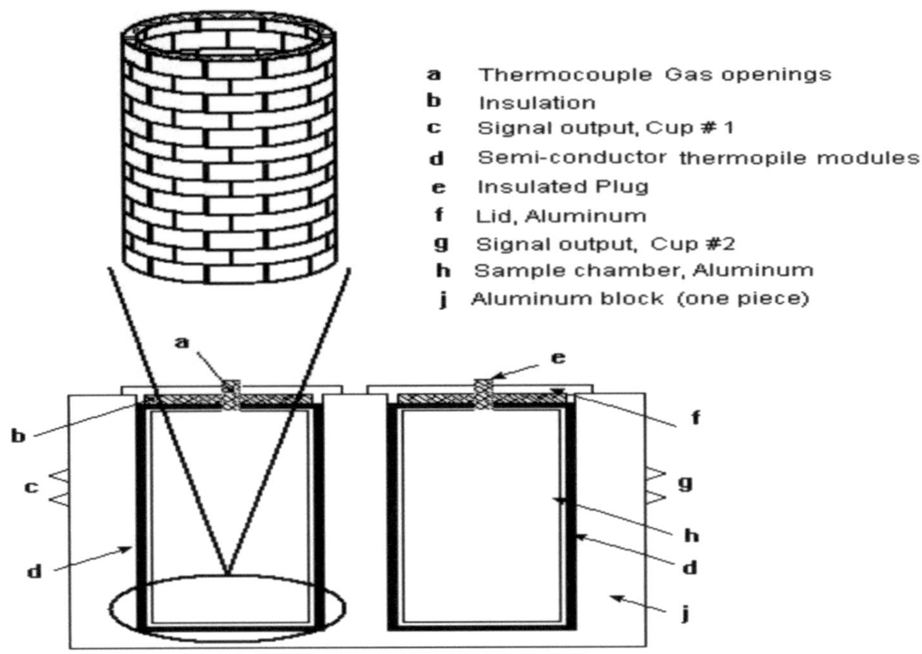

Fig. 16. Calorimetric Cell in the adsorption micro calorimeter

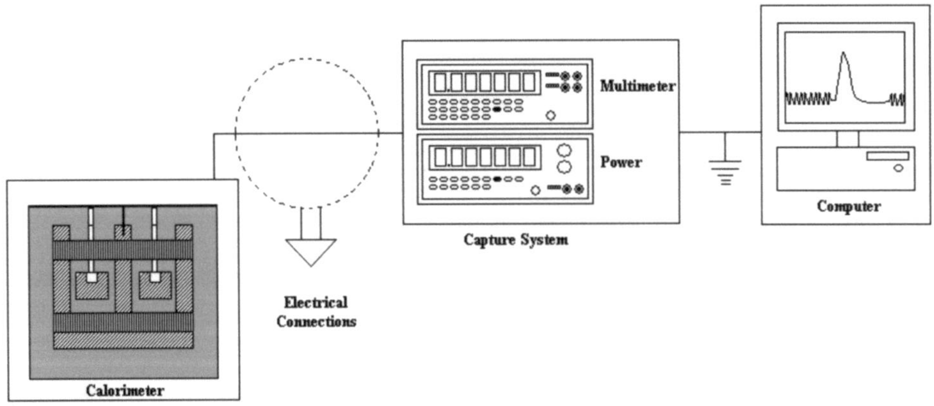

Fig. 17. Sistema de captura de datos

To determine the noise level, different signal potencies are applied by means of the electrical resistance with a fix electrical work at different temperatures and under experimental work conditions in the surroundings. It was also realized some essays at a fix temperature and varying the electrical work level. A highly stabilized source Agilent™ E3649A Al model applies the potential and a multimeter Agilent™ 34401 with 6½ numbers enters the data. This multimeter is connected to a PC by a GPIB interface, where the signals are evaluated.

5.3 Results

The results at a constant temperature of 18°C, where obtained by putting the system in an air thermostat and varying the applied potential and the electrical work levels. The applied potential varied from 0.40 to 1.5 volts that corresponds to electrical works from 0.19 to 2.70 Jules. Figure 18 shows that increasing the electrical work magnitude, increases the noise level about 10 µV and repeated tests for the lowest values of electrical work increase the noises values about 15 µV. These results are interesting under an experimental point of view in fine calorimetry like adsorption because, independently of this values magnitude, it is important to compare them with the thermal effect magnitude and then, quantify the possible error introduced in the measures. When dissipating a small electrical work, there is a tendency to about 1 µV noise in the baseline; when shooting bigger electrical works, it increases to magnitude orders no too large compared to the measurements.

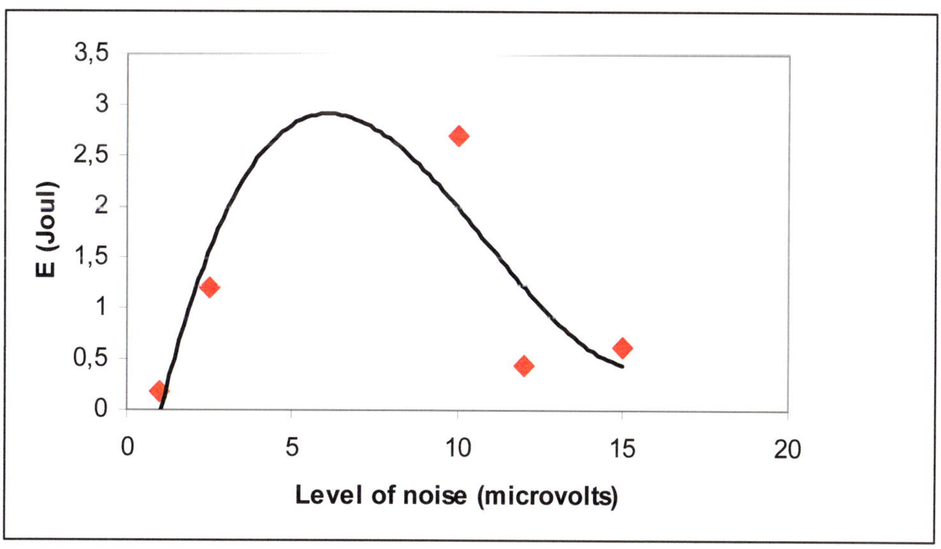

Fig. 18. Peak to Peak Noise Level at 18 °C varying the Voltage Level

Although in specialized bibliography there are some studies about the noise level in calorimeters and/or in calorimetric measurements and different magnitude orders have been reported in different kind of calorimeters (Degroote and García, 2005), this work focuses in the importance of determining in a precise way, the signals that are not associated with what is under measure, like noises, which are due to the connections between the different parts of the equipment.

Figure 19 shows a calibration potenciogram for the constructed calorimeter with a zoom in the baseline signal part; it is clear that the noise level is insignificant respect to the signal, with a noise value in this case of 0.5 μV for a dissipated electrical work of 0.42 Jules.

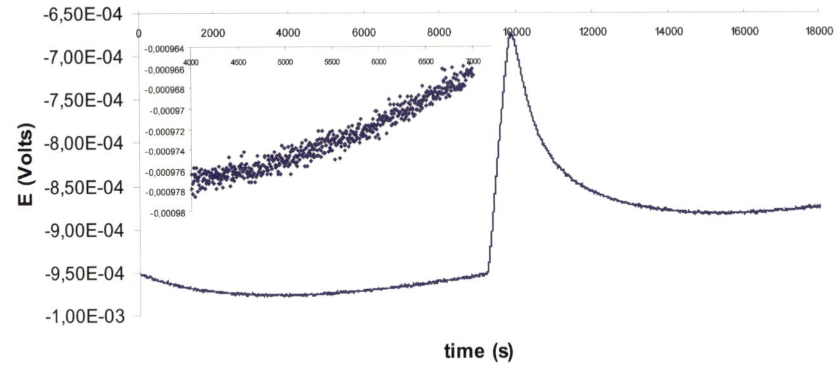

Fig. 19. Potenciogram: Applied Work 0.42 Jules; Peak to Peak Noise 0.5

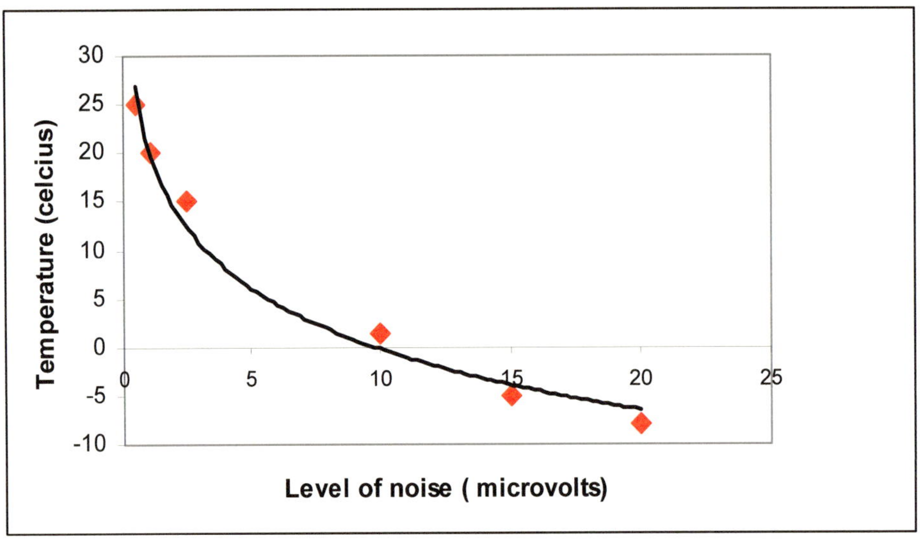

Fig. 20. Peak to Peak Noise Level applying a 0.42 Jules Work

In this work, it was studied if the time constant value (τ) of the equipment could be affected by the noise level magnitude. It was established that it is not true, obtaining repetitive values about 120 seconds, quite similar to commercial equipment like SETARAM™.

Figure 20 shows the behavior of the calorimetric system when is varied the temperature at a fix level of electrical work (0.42 Jules). This essay is realized taking into account that the equipment was designed to work in a temperature range from -196°C to 600°C. The figure shows that near room temperature the signal noise level is about 0.5 μV, which permits without any doubt to carry out experiments with magnitudes about 5 mV with a good precision. At low temperatures the noise level increases up to about 20 μV. It must be taking into account that when the temperature decreases in the calorimeter, a thermal gradient is

generated between the heat sensor terminals inside the calorimeter and the ones connected to its external part, causing a higher noise level in the signal. Certainly, it is observed here the thermal noise or Johnson that is present at temperatures higher than -273°C, where the connections can be considered resistances that generate by means of their terminals a voltage with random fluctuations. This generates a random movement in the charge carrier in conducting materials. The resistance has a conduction band with electrons free that tends to move freely in any direction, the thermal energy of the surrounding provokes this random movement that at the same time increases the temperature.
In Figure 21 it is shown the noise level monitoring keeping constant the temperature at 18°C by more than 36 hours.

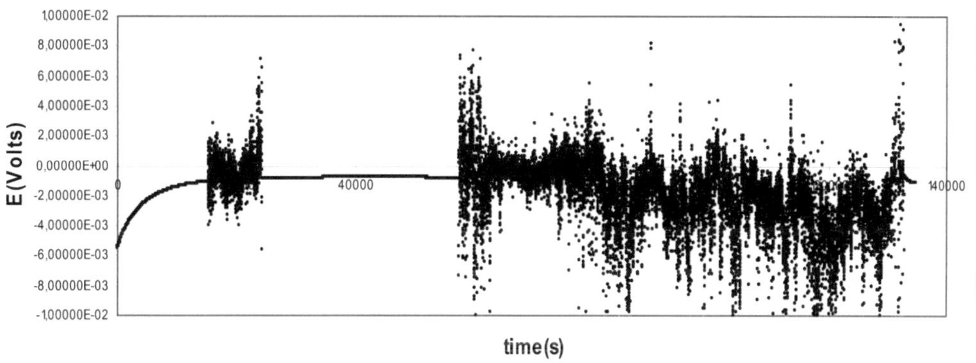

Fig. 21. Noise Signal in the Calorimeter during 36 hours

It is noticed that the stability in the baseline is very important, where the noise level is under 0.5 μV; the scale is zoomed to a better visualization of the noise signal. It is important to recall that the electrical connections generate noises that can also be associated to the shot-type noise.

6. Conclusions

It was built an adsorption micro calorimeter at an affordable price, which is useful to measure adsorption heats and solid surfaces reactions. The equipment works in a temperature range from 77 K to 500 K. It was demonstrated experimentally in this work that it works at 298 K. Calorimetric cells in glass and stainless steel were prepared for the treatment of samples in static form. These cells enable the study of various catalysts in general. The time constants in vacuum and at atmospheric pressure were determined. These constants show that this equipment is useful to study processes with slow kinetic. The sensitiveness of the equipment is high. Finally the noise in the signal is very small and do not affect the measures.

7. References

Armstrong, G.T. (1964). The calorimeter and its influence on the development of chemistry. *J. Chem. Ed.*, 44 (6), 297-307, ISSN 00219584.

Aroua, M.K.; Daud, W.M.A.W.; Yin, C.Y.; Adinata, D. (2008). Adsorption capacities of carbon dioxide, oxygen, nitrogen and methane on carbon molecular basket derived from polyethyleneimine impregnation on microporous palm shell activated carbon. *Separation and Purification Technology*, 62, 609-613, ISSN 13835866

Bastos Neto, M ; Torres, A. E. B. ; Azevedo, D. C. S. ; Cavalcante JR, C. L. (2005a). A Theoretical and Experimental Study of Charge and Discharge Cycles in a Storage Vessel for Adsorbed Natural Gas. *Adsorption-Journal of the International Adsorption Society*, 11, 147-157, ISSN 09295607

Bastos Neto, M ; Torres, A. E. B. ; Azevedo, D. C. S. ; Cavalcante JR, C. L. (2005b). Methane Adsorption Storage Using Microporous Carbons Obtained From Coconut Shells. *Adsorption-Journal of The International Adsorption Society*, 11, 911-915, ISSN 09295607

Belmabkhout, Y.; Sayari, A. (2009). Adsorption of CO_2 from dry gases on MCM-41 silica at ambient temperature and high pressure. 2: Adsorption of CO2/N2, CO2/CH4 and CO2/H2 binary mixtures, *Chemical Engineering Science*, 64, 3729 - 3735, ISSN 00092509

Buckton G. (1995). Application of isothermal microcalorimetry in the pharmaceutical sciences. *Thermochim. Acta*, , 117-125, ISSN 00406031

Degroote, E.; García Ybarra, P.L. (2005). Flame propagation over liquid alcohols Part I. Experimental results *J. Therm. Anal. Cal.*, 80, 541-548, ISSN 14182874

Denoyel, R; Fernandez-Colinas, J.; Grillet, Y.; Rouquerol J. (1993). Assessment of the surface area and microporosity of activated charcoals from immersion calorimetry and nitrogen adsorption data *Langmuir*, 9, 515-518, ISSN 07437463.

Dunne, J.A.; Rao, M.; Sircar, S.; Gorte, R.J.; Myers, A.L. (1997). Calorimetric Heats of Adsorption and Adsorption Isotherms. 3. Mixtures of CH_4 and C_2H_6 in Silicalite and Mixtures of CO_2 and C_2H_6 in NaX. *Langmuir*, 13, 4333-4339, ISSN 07437463.

Figueroa, J. D.; Fout, T.; Plasynski, S.; Mcilvried, H.; Srivastava, R. D. (2008). Advances in CO2 capture technology - The U.S. Department of Energy's Carbon Sequestration Program, *International Journal of Greenhouse Gas Control*, 2, 9-20, ISSN 1750-5836

Garcia-Cuello, V; Moreno-Piraján, J. C.; Giraldo, L.; Sapag, K.; Zgrablich, G. (2009). A new microcalorimeter of adsorption for the determination of differential enthalpies. *Microporous and Mesoporous Materials*, 120, 239–245, ISSN 13871811.

Garcia-Cuello, V; Moreno-Piraján, J. C.; Giraldo, L.; Sapag, K.; Zgrablich, G. (2008). Determination of Differential Enthalpy and Isotherm by Adsorption Calorimetry. *Research Letters in Physical Chemistry*, 5, 4 pages, ISSN 16876873

Giraldo, L.; Moreno, J.C.; Gómez, A. (1994) Construcción de un calorímetro isoperibólico de inmersión de precisión. *Rev. Col. Quim.*, 23, 1-14, 01202804

Giraldo, L.; Moreno, J.C.; Gómez, A. (1996) Instrumentación calorimétrica: calorimetros isoperibolicos y de conduccion de calor. *Revista Química Actualidad y Futuro*, 5, 1996, 90-95, ISSN 01213644

Giraldo, L.; Moreno, J.C.; Gómez, A. (1996) A heat-conduction flow microcalorimeter for solute transfer enthalpy determinations: design and. calibration. *Instrumentation Science & Technology*, 26, 521-530, ISSN 15256030

González, M.T.; Sepúlveda-Escribano, A.; Molina-Sabio, M.; Rodríguez-Reinoso, F. (1995) Correlation between Surface Areas and Micropore Volumes of Activated Carbons Obtained from Physical Adsorption and Immersion Calorimetry. *Langmuir*, 11 , 2151-2155, ISSN 07437463

Gravelle, P.C. (1972) Heat-flow microcalorimetry and its application to heterogeneous Catalysis, *Adv. Catal.*, 22, 191–263, 16154150

Gravelle, P.C. (1985) Aplication of adsorption Calorimetry to the study of heterogeneous catalysis reactions. *Thermochimica Acta*, 96, 365-376, ISSN 00406031

Handy, T.L.; Sharma, S.B.; Spiewak, B.e.; Dumesic, J.A. (1993) A Tian-Calvet heat-flux microcalorimeter for measurement of differential heats of adsorption. *Meas Sci. Technol.*, 4, 1350-1357, ISSN 09570233

Hansen, L.D.; Hart, R.M. (2004) The art of calorimetry. *Thermochim. Acta*, 417, 257-262, ISSN 00406031 .

Hemminger W. and Hohne G. (1984) Calorimetry fundamentals and practice. Ed. Verlag Chemie. Florida.

Hohne, G.W.H.; Hemminger, W.; Flammersheim, H.J. (1996) Differencial Scanning Calorimetry. *An Introduction for practitioners.*1st Edition; Springer; Berlin

Huertemendía, M.; Giraldo, L.; Parra, D.; Moreno, J.C. (2005) Adsorption microcalorimeter and its software: design for the establishment of parameters corresponding to different models of adsorption isotherms. *Instrumentation Science & Technology*, 33, 645-660, ISSN 15256030

Kjems, J.K.; Passell, L.; Taub, H.; Dash, J.G.; Novaco, A.D. (1976) Neutron scattering study of nitrogen adsorbed on basal-plane-oriented graphite. *Phys. Rev. B*, 13, 1446-1462, 01631829.

Llewellyn, P.L.; Grillet, Y.; Rouquerol, J.; Martin, C.; Coulomb, J.P. (1996) Thermodynamic and structural properties of physisorbed phases within the model mesoporous adsorbent M41S (pore diameter 2.5 nm). *Surf. Sci.*, 352–354, 468-474, ISSN 00396028.

Llewellyn, P. L.; Maurin, G. (2005) Gas adsorption microcalorimetry and modelling to characterize zeolites and related materials. *C. R. Chimie*, 8, 283–302, ISSN 16310748.

Martinez, P.; L. Giraldo, E. Vargas, J.C. Moreno. (2005) Isoperibolic calorimetric cell uit electronic integrator circuit for temperature measurement. *Instrumentation Science & Technology*, 33, 415-420, ISSN 15256030.

Menéndez, J. A. (1998) On the use of calorimetric techniques for the characterization of carbons: A brief review. *Thermochimica Acta*, 312, 79–86, ISSN 00406031 .

Moreno, J.C., Giraldo, L. (2005) Influence of thermal insulation of the surroundings on the response of the outpun electrical signal in a heat conduction calorimetric unit. *Instrumentation Science & Technology*, 33, 415-425, ISSN 15256030.

Moreno, J. C.; Giraldo, L. (2005) Setups for simultaneous measurement of isotherms and adsorption heats. *Review of Scientific Instruments*, 76, 1-8, ISSN 00346748

Moreno-Piraján, J. C.; Giraldo, L.; Garcia-Cuello, V.; Sapag, K.; Zgrablich G. (2008) Design, calibration and test of a new Tian-Calvet Heat flow Microcalorimeter for Measurement of differential heats of adsorption. *Instrumentation Science & Technology*. 36, 455-475. ISSN 15256030.

Navarrete, R; Llewellyn, P.; Rouquerol, F.; Denoyel, R.; Rouquerol, J. (2004) Calorimetry by immersion into liquid nitrogen and liquid argon: a better way to determine the internal surface area of micropores. *Journal of Colloid and Interface Science*, 277, 383–386, ISSN 00219797

O'Neil, M.; Louvrien, R.; Phillips, J. (1985) New microcalorimeter for the measurement of differential heats of adsorption of gases on high surface area solids. *Rev. Sci. instrum.*, 56, 2312-2317, ISSN 00346748

O'Neil, M.; Phillips, J. (1987) Differential microcalorimetric study of chemical adsorption processes on a microporous solid. *J. Phys. Chem.*, 91, 2867- 2874, ISSN 0022-3654.

Prauchner, M. J.; Rodriguez-Reinoso, F. (2008) Preparation of granular activated carbons for adsorption of natural gas. *Microporous and Mesoporous Materials,* 109, 581-584, ISSN 13871811.

Rios, R. B.; Silva, F. W. M.; Torres, A. E. B.; Azevedo, D. C. S. ; Cavalcante Jr., C. L. (2009) Adsorption of methane in activated carbons obtained from coconut shells using H_3PO_4 chemical activation, *Adsorption*, 15, 271-277, ISSN 0929-5607.

Rouquerol, F.; Rouquerol, J.; Sing, K. Adsorption by Powders & Porous Solids. Academic Press, San Diego, 1999.

Rouquérol, J. (1985) The contribution of microcalorimetry to the solution of problems involving a liquid/solid or a gas/solid interface, especially in physisorption. *Thermochim Acta*, 96, 377-390, ISSN 00406031

Ruthven, D. M.; Shamsuzzaman, F.; Knaebel, K.S. (1994) Pressure Swing Adsorption, VCH Publishers, New York/Weinheim,

Sharma, S.B.; Miller, J.T.; Dumesic, J.A. (1994) Microcalorimetric Study of Silica- and Zeolite-Supported Platinum Catalysts. *Journal of Catalysis.*, 148, 198-204, ISSN: 0021-9517 .

Sing, K.S.W.; Everett, D.H.; Haul, R.A.W.; Moscou, L.; Pierotti, R.A.; Rouquérol, J.; Siemieniewska, T., (1985) Reporting physisorption data for gas/solid systems with special reference to the determination of surface area and porosity. *Pure Appl. Chem.* 57, 603, ISSN 00334545.

Siperstein, F.; Gorte, R.J.; Myers, A.L. (1999) A New Calorimeter for Simultaneous Measurements of Loading and Heats of Adsorption from Gaseous Mixtures. *Langmuir*, 15, 1570-1576, ISSN 07437463.

Spiewak, B. E.; Handy, B. E.; Sharma, B.; Dumesic, J.A. (1994) Microcalorimetric studies of ammonia adsorption on γ-Al_2O_3, HNa-Y zeolite, and H-mordenite. *Catalysis Letters.*, 23, 207-213, ISSN 1011372X.

Swietoslawski, W. (1948) Microcalorimetry. Ed. Reinhold. New York.

Wadso I., (1986) Bio-calorimetry. *Trends Biotechnol.*, 4, 45-51, ISSN 01677799

Wadso, I.; Goldberg, R.N. (2001) Standards in isothermal microcalorimetry. *Pure Appl. Chem.*, 73, 1625-1639 ISSN 00334545

Wilhoit, R.C. (1967) Recent developments in calorimetry: Part two. Some associated measurements. *J. Chem Ed.*, 44, A571, ISSN 00219584.

Wirawan, S. K.; Creaser, D. (2006) CO_2 adsorption on silicalite-1 and cation exchanged ZSM-5 zeolites using a step changeresponse method. *Microporous and Mesoporous Materials*, 91, 196–205, ISSN 13871811

Yang, R. T. (1997) Gas Separation by Adsorption Processes, Butterworths, London, Imperial College Press, London, UK

Zielenkiewicz, W. (2000) Comparative measurements in isoperibol calorimetry: uses and misuses. *Thermochim. Acta*, 347, 15-20, ISSN 00406031

Zimmermann, W.; Keller, J. U. (2003) A new calorimeter for simultaneous measurement of isotherms and heats of adsorption. *Thermochimica Acta*, 403, 31 – 41, ISSN 00406031

15

Dissociation Energies of O–H Bonds of Phenols and Hydroperoxides

Denisov Evgeny and Denisova Taisa
Institute of Problems of Chemical Physics RAS
Russia

1. Introduction

The bond dissociation energy is very important characteristic of molecule that usually refers to standard thermodynamic state in gas phase (298 K, pressure 1 atm). When these thermodynamic quantities are known they become an invaluable tool for calculation of activation energies and rate constants of homolytic reactions.

Phenols are widely used as antioxidants for stabilization of organic compounds and materials. Many organic compounds are oxidized by oxygen due to contact with air. Phenols are used for stabilization of variety organic products such as polyolefins, polystyrene, and rubbers (Hamid, 2000; Scott, 1980; Pospisil & Klemchuk, 1990; Scott, 1993), monomers (Mogilevich & Pliss, 1990), hydrocarbon fuels (Denisov & Kovalev, 1990), lubricants (Kuliev, 1972), edible fats and oils (Emanuel & Lyaskovskaya, 1961), cosmetics, drugs *etc.*

Autooxidation of substrate RH proceeds as free radical chain process with participation of free radicals $R^\bullet$ and $RO_2^\bullet$ and formation of hydroperoxide ROOH (Emanuel et al., 1967; Mill & Hendry, 1980; Denisov & Khudyakov, 1987; Denisov & Afanas'ev, 2005). The kinetic scheme of chain oxidation of a hydrocarbon is presented below.

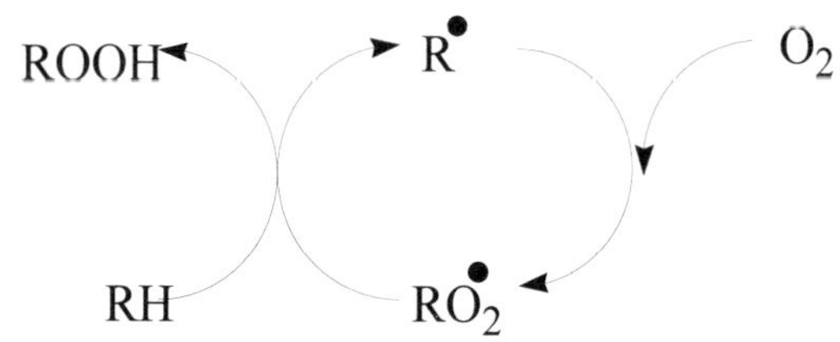

Phenols (ArOH) act as chain breaking inhibitors of this process. They stop developing of chain oxidation by reacting with peroxy radicals (Emanuel et al., 1967; Mill & Hendry, 1980; Denisov & Khudyakov, 1987; Roginsky, 1988; Denisov & Azatyan, 2000; Denisov & Afanas'ev, 2005; Lucarini & Pedulli, 2010).

$$RO_2^\bullet + ArOH \rightarrow ROOH + ArO^\bullet$$

$$RO_2^\bullet + ArO^\bullet \rightarrow \text{Molecular products}$$

The first reaction is the limiting step of chain termination. The rate and activation energy of this reaction depends on the dissociation energy of O–H bond of reacting phenol and formed hydroperoxide. So, these characteristics of phenols and hydroperoxides are the key values in thermodynamics of antioxidative action of phenols.

Natural phenols play very important role in preventing free radical oxidation in living bodies. Vitamin E (tocopherols) is present in cellular membranes and edible oils and functions as efficient inhibitor of lipid peroxidation in biomembranes (Burton & Ingold, 1986; Denisov & Afanas'ev, 2005). Polyfunctional phenols (flavonoids, flavones *etc.*) play important role in the biology of plants (regulation of gene expression, gene silencing, organization of metabolic pathways) (Grotewold, 2006). The bond dissociation energies for such biologically important phenols were estimated recently (Denisov & Denisova, 2009).

2. Experimental methods of estimation of dissociation energy of O−H bonds of phenols

2.1 Calorimetric method

The most of phenoxyl radicals are unstable and rapid disappear by reactions of recombination and disproportionation (Denisov & Khudyakov, 1987; Roginsky, 1988; Denisov & Afanas'ev, 2005). However, there are a few stable phenoxyl radicals and one of them is 2,4,6-tri-*tert*-butylphenoxyl. In contrast to most free radicals, solutions of the 2,4,6-tri-*tert*-butylphenoxyl radical may be prepared and these solutions are stable in the absence of oxygen and other reactive compounds. This unique property was used by L. Mahoney *et al.* for to perform the direct calorimetric determination of the heat of its reaction in systems were the heats of formation of the other reactants and products are known. The following reaction was chosen for such study (Mahoney et al., 1969).

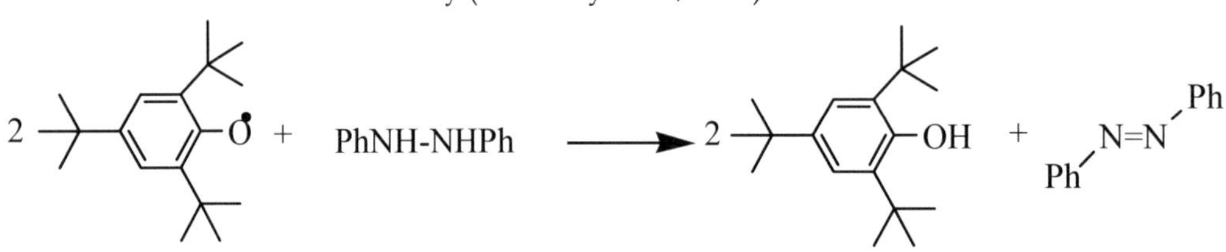

In the result of this reaction all molecules of hydrazobenzene are transformed into *trans*-azobenzene when phenoxyl is taken into excess.

The calorimeter with base line compensator and sample injection assembly was used for this study. Plots of the calories absorbed *vs.* moles of compound dissolved were linear with essentially zero intercepts. The values of the partial molar enthalpies of solution at infinite dilution for phenol and for azobenzene were not altered by the presence of the 2,4,6-tri-*tert*-butylphenoxyl radical in the solvent. For the determination of the heats of reaction of the 2,4,6-tri-*tert*-butylphenoxyl with hydrazobenzene a concentrated solution of the radical (0.2 − 0.5 M) was used. The enthalpy of reaction was measured in three solvents: CCl_4, C_6H_6 and PhCl. The enthalpies of solid reactants solution were estimated and were found to be equal (in benzene): 13.2 kJ/mol for ArOH, 11.8 kJ/mol for analogue of $ArO^•$, 19.7 kJ/mol for PhNHNHPh, and 21.1 kJ/mol for *trans*-PhN=NPh. They were taken into account at calculation of enthalpy of reaction (ΔH). From the last the difference of enthalpies was calculated:

$$\Delta H_f^0(ArO^•) - \Delta H_f^0(ArOH) = 121.9 \pm 0.4 \text{ kJ/mol (298 K, 1 atm., benzene)} \quad (1)$$

As well as this difference is equal to $D_{O\text{-}H}$ + $\Delta H_f^0(H^\bullet)$ and $\Delta H_f^0(H^\bullet)$ = 218 kJ/mol (Lide, 2004) the $D_{O\text{-}H}$ = 339.9 kJ/mol. The following values of reactant and product were used at this calculation: ΔH_f^0(PhNHNHPh) = 221.3 kJ/mol and ΔH_f^0(*trans*-PhN=NPh) = 320.0 kJ/mol. Recently the following correction was proposed with using the new value of ΔH_f^0 (*trans*-PhN=NPh) = 310.4 kJ/mol according that $D_{O\text{-}H}$ = 335.1 kJ/mol (Mulder et al., 2005).

2.2 Chemical Equilibrium (CE)

The value of $D_{O\text{-}H}$ of 2,4,6-tri-*tert*-butylphenol appeared to be very claiming in the method of chemical equilibrium due to stability of formed phenoxyl radical. Chemical equilibrium

$$Ar_1O^\bullet + Ar_iOH \rightleftharpoons Ar_1OH + Ar_iO^\bullet$$

where $Ar_1O^\bullet$ and $Ar_iO^\bullet$ are reactive free phenoxyl radicals, cannot be achieved in solution owing to very fast recombination or disproportionation of these species. Such an equilibrium can be attained only when both radicals are stable and do not enter the recombination reaction. In this case the equilibrium concentrations of both radicals can be determined by electron paramagnetic resonance spectroscopy or spectrophotometrically (Mahoney & DaRooge, 1975; Belyakov at al., 1975, Lucarini at al., 1994). The equilibrium constant (*K*) is calculated from the ratio of the equilibrium concentrations of the molecules and radicals:

$$K = \frac{[Ar_1OH]_\infty [Ar_iO^\bullet]_\infty}{[Ar_iOH]_\infty [Ar_1O^\bullet]_\infty} \tag{2}$$

The equilibrium enthalpy (ΔH) is calculated from the temperature dependence of equilibrium constant *K*. On the other hand, the equilibrium enthalpy of this reaction is equal to the difference between the dissociation energies ($D_{O\text{-}H}$) of the bonds involved in the reaction

$$\Delta H = D_{O\text{-}H}(Ar_iOH) - D_{O\text{-}H}(Ar_1OH) \tag{3}$$

provided that solvation of reactants makes an insignificant contribution to the equilibrium. This can be attained by carrying out experiments in no polar solvents. As the reference phenoxyl radical $Ar_1O^\bullet$ were used 2,4,6-tri-*tert*-butylphenoxyl ((Mahoney & DaRooge, 1975; Lucarini at al., 1994), galvinoxyl (Belyakov at al., 1975), and ionol (Lucarini et al., 2002). Calculations of the equilibrium constant *K* from the reactant concentration ratio were followed by calculations of the Gibbs free energy of equilibrium:

$$\Delta G = -RT\ln K \tag{4}$$

The equilibrium enthalpy was determined using the temperature dependence of the equilibrium constant. Experience showed that in such reactions one has $\Delta H \cong \Delta G$ within the limits of error in measurements. The dissociation energies of the O–H bonds in phenols thus measured are listed in Table 1.

It was found that this approach can also be used in studies of the systems characterized by recombination of phenoxyl radicals formed. Here, the reaction conditions are chosen in such a fashion that the equilibrium between reactants is attained rapidly and the phenols are consumed slowly, so it is possible to monitor each reactant and calculate the equilibrium constant. This allowed the range of phenols with known O−H bond dissociation energies to be extended.

Having determined the equilibrium constant, it is possible to estimate the O−H bond strength difference between two phenols. To calculate the absolute value of $D_{O\text{-}H}$, one must know the $D_{O\text{-}H}$ value for one phenol. For 2,4,6-tri-*tert*-butylphenol, $D_{O\text{-}H}$ = 339.9 kJ/mol (see **2.1**), which is in good agreement with the $D_{O\text{-}H}$ value for unsubstituted phenol ($D_{O\text{-}H}$ = 369.0 kJ/mol, see (Denisov & Denisova, 2000). For galvinol, $D_{O\text{-}H}$ = 329.1 kJ/mol. The error in estimating $D_{O\text{-}H}$ values is only 1.1 kJ/mol (Lucarini at al., 1994). When equilibrium between $Ar_iO^•$ and Ar_1OH was studied in the solvent (S) that forms hydrogen bond with phenol, the difference in hydrogen bond enthalpies was taken into account. The point of difference is that sterically no hindered phenols form hydrogen bond Ar_1OH...S and 2,4,6-tri-*tert*-butylphenol (Ar_1OH) practically does not. Therefore the enthalpy of hydrogen bond Ar_1OH...S should be abstracted from enthalpy of equilibrium. For example, hydrogen bond enthalpy $\Delta H(C_6H_5OH...C_6H_6) = -4$ kJ/mol (Mulder et al., 2005).

In order to evaluate the dissociation energies of O−H bonds in various phenols, Mahoney and DaRooge (Mahoney & DaRooge, 1975) measured the equilibrium constant $K = k_1/k_{-1}$ for the reactions

$$RO_2^• + Ar_iOH \underset{k_{-1}}{\overset{k_1}{\rightleftharpoons}} ROOH + Ar_iO^•$$

To this end, 9,10-dihydroantracene (RH) was oxidized with oxygen in the presence of corresponding hydroperoxide (ROOH) and phenol (Ar_iOH) at 333 K with azoisobutyronitryl as initiator. The experimental conditions were chosen in such a manner that the equilibrium was established in the system and the chain termination step was limited by the recombination of $Ar_iO^•$ and $ROO^•$ radicals. In this case the rate (v), of the chain oxidation process is satisfactorily described by the following equation:

$$v = k_p[\mathrm{RH}]\sqrt{\frac{k_{-1}[\mathrm{ROOH}]v_i}{2k_tk_1[\mathrm{Ar}_i\mathrm{OH}]}}, \qquad (5)$$

where k_p and k_t are the rate constants for the reactions of $RO_2^•$ with RH and $RO_2^•$ with $ArO^•$, respectively, and v_i is the initiation rate. The dependence of the chain oxidation rate v on the ROOH and ArOH concentrations was used to determine the $k_p/(2k_tK)^{1/2}$ ratio and then the equilibrium constant $K = k_1/k_{-1}$ was calculated using known values of ratio $k_p(2k_t)^{-1/2}$. The $\Delta D = D(\mathrm{ArO{-}H}) - D(\mathrm{ROO{-}H})$ values were determined assuming that $\Delta H = \Delta G = -RT\ln K$. The values of D(ArO−H) calculated relative to the O−H bond dissociation energy of *sec*-ROOH (365.5 kJ/mol) are listed in Table 1. The values of $D_{O\text{-}H}$ measured for the same phenol in different papers are in good agreement. For example, for 4-methylphenol (*para*-cresol) $D_{O\text{-}H} = 360.7 \pm 1.0$ kJ/mol, for 4-*tert*-butylphenol $D_{O\text{-}H} = 359.1 \pm 1.6$ kJ/mol, for 4-methoxyphenol $D_{O\text{-}H} = 348.2 \pm 1.1$ kJ/mol.

Substituents of Phenol, Phenol	Ar_1OH or ROOH	ΔD, kJ/mol	D, kJ/mol	Ref
3-Me	ROOH	−2.5	363.0	Howard & Ingold, 1965
4-Me	ROOH	−3.7	361.8	Howard & Ingold, 1965
4-Me	ROOH	−6.1	359.4	Mahoney & DaRooge, 1975
4-Me	2,4,6-tri-*tert*-butylphenol	20.9	360.8	Lucarini & Pedulli, 2010
4-CMe_3	ROOH	−4.6	360.9	Howard & Ingold, 1965
4-CMe_3	ROOH	−6.0	359.5	Mahoney & DaRooge, 1975
4-CMe_3	2,4,6-tri-*tert*-butylphenol	17.1	357.0	Lucarini & Pedulli, 2010
4-Ph	ROOH	−12,5	353.0	Mahoney & DaRooge, 1975
2-OMe	2,4,6-tri-*tert*-butylphenol	19.2	359.1	Lucarini & Pedulli, 2010
3-OMe	ROOH	0.7	364.8	Howard & Ingold, 1965
4-OMe	ROOH	−16.5	349.0	Howard & Ingold, 1965
4-OMe	ROOH	−16.6	348.9	Mahoney & DaRooge, 1975
4-OMe	2,4,6-tri-*tert*-butylphenol	6.7	346.6	Lucarini & Pedulli, 2010
3-COOEt	ROOH	7.8	373.3	Mahoney & DaRooge, 1975
4-NH_2	2,4,6-tri-*tert*-butylphenol	−12.1	327.8	Lucarini & Pedulli, 2010
4-Cl	ROOH	1.9	367.4	Howard & Ingold, 1965
2-Me, 6-Me	ROOH	−15.3	359.2	Howard & Ingold, 1965
2-Me, 6-Me	2,4,6-tri-*tert*-butylphenol	13.8	353.7	Lucarini & Pedulli, 2010
3-Me, 5-Me	2,4,6-tri-*tert*-butylphenol	22.6	362.5	Lucarini & Pedulli, 2010
2-Me, 6-CMe_3	ROOH	−20.1	345.4	Howard & Ingold, 1965
2-CMe_3, 6-CMe_3	ROOH	−11.1	354.4	Howard & Ingold, 1965
2-CMe_3, 6-CMe_3	ROOH	−12.2	353.3	Mahoney & DaRooge, 1975
2-CMe_3, 6-CMe_3	2,4,6-tri-*tert*-butylphenol	6.7	346.6	Lucarini & Pedulli, 2010
3-CMe_3, 5-CMe_3	ROOH	−4.5	361.0	Mahoney & DaRooge, 1975
3-CMe_3, 5-CMe_3	2,4,6-tri-*tert*-butylphenol	22.6	362.5	Lucarini & Pedulli, 2010
2-OMe, 4-Me	2,4,6-tri-*tert*-butylphenol	11.3	351.2	Lucarini & Pedulli, 2010

2-OMe, 4-OMe	2,4,6-tri-*tret*-butylphenol	4.2	344.1	Lucarini & Pedulli, 2010
3-OMe, 5-OMe	2,4,6-tri-*tret*-butylphenol	23.0	362.9	Lucarini & Pedulli, 2010
2-OMe, 6-OMe	2,4,6-tri-*tert*-butylphenol	8.4	348.3	Lucarini & Pedulli, 2010
2-Me, 4-Me, 6-Me	2,4,6-tri-*tert*-butylphenol	6.3	346.2	Lucarini & Pedulli, 2010
2-Me, 4-CN, 6-Me	ROOH	2.5	368.0	Howard & Ingold, 1965
2-CMe_3, 4-Me, 6-CMe_3	Galvinol	14.6	343.7	Belyakov at al., 1975
2-CMe_3, 4-Me, 6-CMe_3	2,4,6-tri-*tert*-butylphenol	0.3	340.2	Jackson & Hosseini, 1992
2-CMe_3, 4-Me, 6-CMe_3	2,4,6-tri-*tert*-butylphenol	−0.8	339.1	Lucarini & Pedulli, 2010
2-CMe_3, 4-CMe_3, 6-CMe_3	Galvinol	11.5	340.6	Belyakov at al., 1975
2-CMe_3, 4-CH_2Ph, 6-CMe_3	Galvinol	6.9	336.0	Belyakov at al., 1975
2-CMe_3, 4-Ph, 6-CMe_3	Galvinol	9.7	338.8	Belyakov at al., 1975
2-CMe_3, 4-Ph, 6-CMe_3	2,4,6-tri-*tert*-butylphenol	0.0	339.9	Lucarini & Pedulli, 2010
2-CMe_3, 4-CH=CHPh, 6-CMe_3	2,4,6-tri-*tert*-butylphenol	−9.6	330.3	Lucarini & Pedulli, 2010
2-CMe_3, 4-OMe, 6-CMe_3	Galvinol	−2.7	325.7	Belyakov at al., 1975
2-CMe_3, 4-OMe, 6-CMe_3	2,4,6-tri-*tert*-butylphenol	−13.5	326.4	Jackson & Hosseini, 1992
2-CMe_3, 4-OMe, 6-CMe_3	2,4,6-tri-*tert*-butylphenol	−12.1	327.8	Lucarini & Pedulli, 2010
2-CMe_3, 4-$OCMe_3$, 6-CMe_3	2,4,6-tri-*tert*-butylphenol	−5.9	334.0	Howard & Ingold, 1965
2-CMe_3, 4-$OCMe_3$, 6-CMe_3	Galvinol	2.0	331.1	Belyakov at al., 1975
2-CMe_3, 4-CHO, 6-CMe_3	2,4,6-tri-*tert*-butylphenol	15.2	355.1	Jackson & Hosseini, 1992
2-CMe_3, 4-CHO, 6-CMe_3	2,4,6-tri-*tert*-butylphenol	12.5	352.4	Lucarini & Pedulli, 2010
2-CMe_3, 4-C(O)Me, 6-CMe_3	Galvinol	24.1	353.2	Belyakov at al., 1975
2-CMe_3, 4-CH=NOH, 6-CMe_3	2,4,6-tri-*tert*-butylphenol	−4.8	335.1	Jackson & Hosseini, 1992
2-OH, 4-CMe_3, 6-CMe_3,	Ionol	−7.1	331.5	Lucarini et al., 2002
2- $CEtMe_2$, 4-OH, $CEtMe_2$,	Ionol	−0.8	337.8	Lucarini et al., 2002
2-CMe_3, 4-CH_2NMe_2, 6-CMe_3	2,4,6-tri-*tert*-butylphenol	0.7	340.6	Jackson & Hosseini, 1992
2-CMe_3, 4-NO_2, 6-CMe_3	2,4,6-tri-*tert*-butylphenol	15.5	355.4	Jackson & Hosseini, 1992
2-CMe_3, 4-NO_2, 6-CMe_3	2,4,6-tri-*tert*-butylphenol	15,5	355.4	Lucarini & Pedulli, 2010
2-CMe_3, 4-Cl, 6-CMe_3	Galvinol	16.3	345.4	Belyakov at al., 1975

2-CMe_3, 4-Cl, 6-CMe_3	2,4,6-tri-*tert*-butylphenol	5.0	344.9	Lucarini & Pedulli, 2010
2-CMe_3, 4-COOMe, 6-CMe_3	2,4,6-tri-*tert*-butylphenol	12.1	352.0	Lucarini & Pedulli, 2010
2-CMe_3, 4-COOH, 6-CMe_3	2,4,6-tri-*tert*-butylphenol	13.0	352.9	Lucarini & Pedulli, 2010
2-CMe_3, 4-CN, 6-CMe_3	ROOH	−7.7	357.8	Howard & Ingold, 1965
2-CMe_3, 4-CN, 6-CMe_3	2,4,6-tri-*tert*-butylphenol	12.5	352.4	Lucarini & Pedulli, 2010
2-CMe_3, 4-SMe, 6-CMe_3	2,4,6-tri-*tert*-butylphenol	−7.5	332.4	Lucarini & Pedulli, 2010
2-CMe_3, 4-S(O)Me, 6-CMe_3	2,4,6-tri-*tert*-butylphenol	6.3	346.2	Lucarini & Pedulli, 2010
2-CMe_3, 4-SO_2Me, 6-CMe_3	2,4,6-tri-*tert*-butylphenol	6.7	346.6	Lucarini & Pedulli, 2010
2-CMe_3, 4-CMe_3, 6-SMe	2,4,6-tri-*tert*-butylphenol	8.8	348.7	Lucarini & Pedulli, 2010
2-OMe, 4-OMe, 6-OMe	2,4,6-tri-*tert*-butylphenol	−5.0	334.9	Lucarini & Pedulli, 2010
2-Me, 3-Me, 4-OMe, 6-Me	2,4,6-tri-*tert*-butylphenol	−8.4	331.5	Lucarini & Pedulli, 2010
2-Me, 3-Me, 4-OMe, 5-Me, 6-Me	2,4,6-tri-*tert*-butylphenol	2.9	342.8	Lucarini & Pedulli, 2010
2-Naphthol	ROOH	−6.4	359.1	Mahoney & DaRooge, 1975
Galvinol	2,4,6-tri-*tert*-butylphenol	−11.5	328.4	Howard & Ingold, 1965
Indol	2,4,6-tri-*tert*-butylphenol	−14.8	325.1	Howard & Ingold, 1965

Table 1. The values of D_{O-H} of substituted phenols estimated by **CE** method (D_{O-H} = 339.9 kJ/mol for 2,4,6-tri-*tert*- butylphenol D_{O-H} = 329.1 kJ/mol for galvinol, and D_{O-H} = 365.5 kJ/mol for *sec*-ROOH (Denisov & Denisova, 2000))

2.3 Low pressure pyrolysis of substituted anisoles (VLPP)

The direct approach to estimation of dissociation energy of O-H bond of phenol via reaction

$$C_6H_5OH \rightarrow C_6H_5O^\bullet + H^\bullet$$

cannot be successful due to the presence of a competing tautomerization of phenol to the cyclohexa-2,4-dienon (Zhu & Borzzelli, 2003.). An indirect way to assess the phenolic O−H bond dissociation energy is by studying the temperature dependence of the rate constant for O−C bond dissociation in phenyl ethers, such as anisoles (Suryan et al., 1989a; Suryan et al., 1989b) and benzylphenyl ether (Pratt at al., 2001). In these studies, dissociation rates of substituted anisoles were determined in the gas phase by a method of very-low-pressure pyrolysis (**VLPP**). This method provides a straightforward means for determining decomposition rates in the absence of bimolecular reactions. Anisoles are especially suited for study by this method, since their low O−Me bond strengths (ca. 268 kJ/mol) cause them to homolyze under relatively mild (for **VLPP**) conditions (800-1200 K). The decomposition of substituted anisoles was found to proceed exclusively by simple homolysis (Suryan et al., 1989a).

General operating principles of the **VLPP** technique have been described by Golden et al (Golden et al., 1973). Anisole decomposition were performed at T = 800 ÷ 1200 K and pressure $p = 10^{-2} \div 10^{-4}$ Torr monitored periodically and percentage dissociation was reproducible to ±1% (Suryan et al., 1989b). An electron impact quadruple mass spectrometer, tuned to 70-eV ionization energy, was used to monitor reactant decomposition. Unimolecular reaction rate constants, k_{uni}, under **VLPP** conditions were calculated from the equation (Suryan et al., 1989a).

$$k_{uni}/k_e = x/(1-x) = (I_0 - I)/I. \tag{6}$$

where, k_e, the escape rate constant is $3.965(T/M)^{1/2}$ s^{-1} for the 3-mm aperture reactor, M is the molecular weight, T is the temperature (K), and x represents the fraction of reactant decomposed. The latter value was derived from mass spectrometer signal intensity of the parent molecular ion before reaction (I_0) and after reaction (I) at an ionization energy of 70 eV (Suryan et al., 1989a).

Y	2-Y		3-Y		4-Y		Ref
	ΔD, kJ/mol	D_{O-H}, kJ/mol	ΔD, kJ/mol	D_{O-H}, kJ/mol	ΔD, kJ/mol	D_{O-H}, kJ/mol	
Me	−10.9	358.1	−2.1	366.9	−7.9	361.1	Suryan et al., 1989b
$CH=CH_2$	−10.5	358.5					Suryan et al., 1989b
OMe	−17.6	351.4	−4.6	264.4	−16.3	352,7	Suryan et al., 1989a
OMe	−17.1	351.9			−13.0	356.0	Pratt at al., 2001
C(O)Me	−6.0	333.9	0.8	369.8	2.5	371.5	Suryan et al., 1989b
OH	−29.9	310.0	1.2	370.2	−10.5	358.5	Suryan et al., 1989a
OH	−30.1	338.9			−11.3	357.7	Pratt at al., 2001
NH_2	−30.9	309.8	−1.7	367.3	−12.5	356.5	Suryan et al., 1989b
CN	−0.9	339.0	−0.4	368.6	1.2	370.2	Suryan et al., 1989b
NO_2	−5.6	334.3	−2.1	366.9	4.6	373.6	Suryan et al., 1989b
F	−8.0	361.0	3.8	372.8	−4.6	264.4	Suryan et al., 1989b
Cl	−9.2	359.8	0.8	369.8	−4.6	264.4	Suryan et al., 1989a
Br	−7.1	361.9					Suryan et al., 1989a
CF_3					9.2	378.2	Pratt at al., 2001

Table 2. Dissociation energies of O−H bonds (in kJ/mol) in substituted phenols estimated by **VLPP** technique, $D_{O-H}(C_6H_5OH)$ = 369.0 kJ/mol

For a meaningful comparison of rate constants for the different reactions, they were converted to their high-pressure (collision-frequency independent) values. This was done with RRKM theory (Robinson & Holbrook, 1972). The pre-exponential A factor of $10^{15.5}$ s^{-1} was assumed for all reactions. Activation energies were used for to calculate D_{O-C} at standard conditions using the equation (Mulder et al., 2005):

$$D_{O-C} = E_{uni} + RT_m - \Delta C_p(T_m - 298), \tag{7}$$

where T_m is the average temperature of experiment, ΔC_p is the average change in heat capacity between T_m and T = 298 K. The errors in rate constants, measured at 50% decomposition, were approximately 10%.
As well as the enthalpy of exchange reaction

$$PhOH + YC_6H_4OMe \rightleftharpoons YC_6H_4OH + PhOMe$$

was proved to be very small (Suryan et al., 1989a) one can use the differences in dissociation energy of C–O bonds of substituted anisoles for evaluation of dissociation energy of O–H bonds of phenols. The results of calculation of dissociation energy for O–H bonds in substituted phenols taking into account $D_{O-H}(C_6H_5OH)$ = 369.0 kJ/mol (Denisov & Denisova, 2000) are presented in Table 2.

2.4 Estimation of bond dissociation energy from kinetic measurements

In the framework of the intersecting parabolas model (**MIP**) the transition state of a radical reaction, for example

$$RO_2^{\bullet} + ArOH \rightarrow ROOH + ArO^{\bullet}$$

is treated as a result of intersection of two potential energy curves (Denisov, 1997; Denisov, 1999; Denisov & Denisova, 2000; Denisov & Afanas'ev, 2005; Denisov et al., 2003). One of them characterises the potential energy of the stretching vibration of the attacked ArO–H bond as a function of the vibration amplitude while the other characterises the potential energy of the vibration of the ROO–H bond being formed. The stretching vibrations of the bonds are considered harmonic. A free radical abstraction reaction is characterised by the following parameters:

1. classical enthalpy ΔH_e. which includes the zero-point vibrational energy difference between the bond being cleaved and the bond being formed;
2. classical potential activation barrier E_e which includes the zero-point vibrational energy of the bond being cleaved;
3. parameter r_e equal to the total elongation of the bond being cleaved and the bond being formed in the transition state;
4. coefficient b ($2b^2$ is the force constant of the bond being cleaved);
5. parameter α (α^2 is the force constant ratio of the reacting bonds);
6. pre-exponential factor A_0 per equally reactive cleaved bond (in the molecule) involved in the reaction. The rate constant is expressed via Arrhenius equation: $k = n_{OH}A_0\exp(-E/RT)$ where n_{OH} is a number of OH groups with equireactivity. All these values are connected via equation:

$$br_e = \alpha\sqrt{E_e + \Delta H_e} + \sqrt{E_e}\,. \tag{8}$$

The **MIP** method allows the variety of radical reactions to be divided into classes using experimental data. Each class is characterised by the same set of parameters mentioned above. An individual reaction belonging to a certain class is characterised by the classical enthalpy H_e and classical activation energy E_e that for the written above reaction is expressed by the equation (at $\Delta H_e(1 - \alpha^2) << (br_e)^2$):

$$\sqrt{E_e} = 0.496\, br_e + 0.507\, (br_e)^{-1}\Delta H_e, \quad (9)$$

were parameter br_e = 13.16 (kJ/mol)$^{1/2}$ for reaction $RO_2^{\bullet}$ + phenol and br_e = 14.30 (kJ/mol)$^{1/2}$ for reaction $RO_2^{\bullet}$ + sterically hindered phenol (Denisov & Denisova, 2000). A method for estimating the bond strengths from the kinetic data developed in the framework of the **MIP** model, involves a number of versions (Denisov, 1995a; Denisov, 1995b; Denisov, 1997; Denisov, 1999; Denisov & Tumanov, 2005).

1. With the measured rate constants k_i or the k_i/k_1 ratios for a reaction class of the type $RO_2^{\bullet}$ + ArOH it is possible to estimate the difference in bond strengths between the Ar_iOH and Ar_1OH compounds. The rate constant ratio k_i/k_1 allows to calculate the activation energies difference:

$$\Delta E_i = E_i - E_1 = RT\ln(n_i k_1 / n_1 k_i) \quad (10)$$

where n_1 and n_i are respectively the numbers of equally reactive bonds in the Ar_1OH and Ar_iOH molecules attacked by the $RO_2^{\bullet}$ radical. From Eqn. (8) it follows that the relation between the ΔE_i value and the enthalpy difference $\Delta(\Delta H_i)$ between these reactions has the form:

$$\Delta(\Delta H_i) = 1.945\, br_e \left(\sqrt{E_{e1} + \Delta E_i} - \sqrt{E_{e1}}\right) + 0.0274\, \Delta E_i \quad (11)$$

On the other hand, the enthalpy difference $\Delta(\Delta H_i) = D_i - D_1$ and the strength of the $Ar_iO{-}H$ bond can be calculated using the equation:

$$D_i = D_1 + 1.945\, br_e \left(\sqrt{E_{e1} + \Delta E_i} - \sqrt{E_{e1}}\right) + 0.0274\, \Delta E_i \quad (12)$$

Thus, calculations of the D_i values require knowledge of the D_1, E_{e1}, k_1/k_i ratio, and the parameters characterising a given reaction class (parameter br_e and coefficient α). Experience showed that the error in ΔD measurements is 1.5 ÷ 2.5 kJ/mol.

2. Using different radicals (e.g. $R_iOO^{\bullet}$) in a series of radical reactions $R_iOO^{\bullet}$ + RH belonging to the same class allows the approach to be employed for evaluating the $R_iOO{-}H$ bond strengths. In this case Eqn. (12) takes the form:

$$D_i = D_1 + 2\alpha^{-2} br_e \left(\sqrt{E_{e1}} - \sqrt{E_{e1} + \Delta E_i}\right) + (\alpha^{-2} - 1)\Delta E_i \quad (13)$$

where E_{e1} refers to reaction $R_1O_2^{\bullet}$ + RH, coefficients α and b_{re} see in Handbook (Denisov & Denisova, 2000). This version of the technique was used in estimating the D_{O-H} values in hydroperoxides (see **4.1**).

3. The D_i value for a free radical abstraction reaction can also be evaluated from the absolute value of the rate constant k_i. (Denisov & Tumanov, 2005).

   Limitation imposed on the method is associated with the enthalpy of reaction (Denisov, 1997; Denisov & Denisova, 2000). The case in point is that Eqn (8) is valid for reactions belonging to the same reaction class and characterised by enthalpies lying in the interval $\Delta H_{emin} < \Delta H_e < \Delta H_{emax}$. The activation energy for a highly exothermic reaction ($\Delta H_e < \Delta H_{emin}$) is nearly zero (more correctly, $E = 0.5RT$) and independent of ΔH. Therefore the enthalpies ΔH_e of these reactions cannot be estimated from the E_e values. The ΔH_{emin} value depends on the br_e and α parameters and on the zero-point vibrational energy of the bond being cleaved. The activation energies of highly endothermic reactions ($H_e > \Delta H_{emax}$) is approximately equal to ΔH:

$$E = \Delta H + 0.5\,RT. \tag{14}$$

The bond dissociation energies estimated from experimental data on reactions of peroxyl radicals with phenols in hydrocarbon solutions are given in Table 3.

Substituents of Phenol, Phenol	Phenol, YC_6H_4OH		2,6-Di-*tert*-butylphenol	
	ΔD_{O-H}, kJ/mol	D_{O-H}, kJ/mol	ΔD_{O-H}, kJ/mol	D_{O-H}, kJ/mol
H	0.0	369	7.7	347.6
2-Me	−9.1	359.9		
3-Me	−2.3	366.7		
4-Me	−6.8	362.2	0.0	339.9
2-CMe$_3$	−15.2	353.8		
4-CMe$_3$	−8.9	360.1	0.0	339.9
4-Ph			−1.3	338.6
4-CH$_2$Ph			0.7	340.6
4-CMe$_2$Ph			1.1	341.0
4-CHPh$_2$			3.3	343.2
4-OMe			−8.3	331.6
4-OCMe$_3$	−21.2	347.8	−7.9	332.0
2-OH	−29.4	339.6		
3-OH	0.1	369.1		
4-OH	−17.0	352.0		
4-COOH	2.7	371.7	9.8	349.7
4-C(O)H			8.8	348.7
4-C(O)Me			6.5	346.4
4-COOCMe$_3$			8.8	348.7
4-CH$_2$COOH			−2.3	337.6
4-CH$_2$COOMe			3.8	343.7
4-(CH$_2$)$_2$COOC$_{18}$H$_{37}$			0.6	340.5
4-CH$_2$NH$_2$			−4.2	335.7
4-NHAc			−12.3	327.6
4-NO			7.0	346.9
4-NO$_2$	3.8	372.8	19.0	358.9
4-CN			13.4	353.3
4-Cl			5.5	345.4
4-SPh			7.4	347.3
2-Me, 3-Me	−13.5	355.5		
2-Me, 4-Me	−8.5	360.5		

2-Me, 6-Me	−14.4	354.6		
3-Me, 4-Me	−14.4	354.6		
3-Me, 5-Me	−4.3	364.7		
2-Me, 4-CMe_3	−9.1	359.9		
2-CMe_3, 4-CMe_3	−9.5	359.5		
3-CMe_3, 5-CMe_3	−6.6	362.4		
2-CMe_3, 4-OMe	−25.6	343.4		
2-OH, 4-CMe_3	−26.6	342.4		
2-Me, 4-OH	−19.2	349.8		
2-Me, 4-Me, 6-Me	−21.5	347.5		
2-Me, 4-Me, 5-Me	−12.2	356.8		
2-CMe_3, 4-Me, 6-Me	−13.2	355.8		
2-CMe_3, 4-CMe_3, 6-CMe_3	−13.1	355.9		
2-Me, 4-CH_2NH_2, 5-Me	−20.9	348.1		
2-Me, 4-Cl, 5-Me	−17.6	351.4		
2-$S(CH_2)_2CN$, 4-Me, 6-CHMePh	−21.5	347.5		
2-Me, 4-CH_2NH_2, 6-CMe_3	−18.4	350.6		
2-Me, 3-Me, 4-Me, 6-Me	−21.4	347.6		
2-Me, 3-Me, 5-Me, 6-Me	−17.8	351.2		
2-Me, 3-Me, 4-Me, 5-Me, 6-Me	−28.5	340.5		
2-OH, 3-CMe_3, 5-CMe_3	−28.7	340.3		
2-OH, 3-CMe_3, 6-CMe_3	−29.5	339.5		
3-Me, 4-CH_2COOH, 5-Me	−19.0	350.0		
2-Me, 4-CH_2CN, 6-Me	−15.0	354.0		
2-Me, 3-Me, 4-OH, 5-Me	−24.3	344.7		
2-OMe, 3-OMe, 4-OH, 6-Me	−25.2	343.8		
1-Naphthol	−25.6	343.4		
2-Naphthol	−15.2	353.8		
1-Hydroxyfluorene	−30.7	338.3		
1-Hydroxyphenanthrene	−14.3	354.7		
2-Hydroxyphenanthrene	−2.0	367.0		
3-Hydroxyphenanthrene	−6.5	362.5		
4-Hydroxyphenanthrene	−12.8	356.2		
3,8-Pyrendiol	−53.1	315.9		
3,10-Pyrendiol	−51.3	317.7		

Table 3. The values of D_{O-H} for phenols estimated by **MIP** (Denisov, 1995a; Denisov & Denisova, 2000)

2.5 Photoacoustic calorimetry (PAC)

Photoacoustic calorimetry is a thermodynamic method of estimation of bond dissociation energies in solution. The physical principle of **PAC** is the following (Rothberg et al., 1983; Simon & Peters, 1983; Grabowski et al., 1984). Very rapid heat release from a photoinitiated process in liquid generates a pressure wave. This wave propagates through the solution at the speed of sound. Detection and quantification of this pressure wave is the bases of the technique.

The values of dissociation energies of O−H bonds in phenols were estimated by (Wayner et al., 1995; Wayner et al., 1996; Laarhoven et al., 1999). Di-*tert*-butyl peroxide was used as photoinitiator. Pulses from a nitrogen laser (λ = 337.1 nm; pulse width 10 ns; power, 10 mJ per pulse; repetition rate, 5 Hz) were used to photolyse di-*tert*-butyl peroxide in solution.

$$Me_3COOCMe_3 + h\nu \rightarrow 2\,Me_3CO^\bullet$$

The formed *tert*-butoxy radicals react very rapidly with phenol with evolution of heat.

$$Me_3CO^\bullet + ArOH \rightarrow Me_3COH + ArO^\bullet$$

Substituents of Phenol	ΔD, kJ/mol	D, kJ/mol
4-H	0.0	369.0
4-CMe_3	−7.9	361.1
4-OMe	−24.3	344.7
4-CF_3	13.4	382.4
4-CN	23.4	392.4
4-Cl	1.7	370.7
2-Me, 4-Me, 6-Me-	−23.0	346.0
2-CMe_3, 4-Me, 6-CMe_3	−32.2	336.8
2-CMe_3, 4-CMe_3	−21.8	347.2
2-Me, 4-OMe, 6-Me-	−42.3	326.7

Table 4. Dissociation energies of O−H bonds of phenols measured by **PAC**, benzene, T = 298 K, (Wayner et al., 1996; Laarhoven et al., 1999)

An iris ensured that only a very small part of the light passed as a fine beam through the centre of the cell, and a low powered lens was used to correct for the slight divergence of the beam. The heat evolved as a result of the photoinitiated reactions caused a shock wave in the solution, which was transmitted at the speed of sound to the cell wall. Here, the primary wave and its many reflections were detected in a time-resolved mode by a piezoelectric transducer. The transducer signals were amplified and were recorded on a storage oscilloscope. A quartz plate was used to reflect part of the laser beam to a reference device, so that corrections could be made for variations in the laser power. On prolonged irradiation some drift in this device occurred, presumably because small convection currents were set up in the solution. However, the problem was easily overcome by slowly flowing the solution through the cell with a peristaltic pump. Signals from the reference transducer were amplified and were stored in a second channel of the oscilloscope. The time profiles of the photoacoustic waves were quite reproducible so long as the geometry of the apparatus remained unchanged. The measurements from a line of laser shots were averaged to give the amplitudes of the photoacoustic waves due to sample and reference. In the photoacoustic experiment, an important condition was that the heat evolved in a given reaction was released in a time that was short with respect to the response of the transducer. This was tested by different ways. The samples used in the photoacoustic experiments were always carefully deoxygenated by nitrogen or argon purging and were flowed through the cuvette so as to avoid problems associated with sample depletion and product formation. Samples that were oxygen-sensitive were always prepared in an inert atmosphere since oxidation generally gave rise to colored impurities which affected the optical properties of the solutions. The results of $D_{O\text{-}H}$ estimation of a line substituted phenols are listed in Table 4. The results on $D_{O\text{-}H}$ estimation in C_6H_5OH see in paragraph 3.

2.6 Acidity – oxidation potential method (AOP)

The theoretical basis for acidity – oxidation – potential method (**AOP**) (Bordwell & Bausch, 1986; Bordwell et al., 1991) lies in thermochemical cycle:

$$\mathrm{ArOH} \rightleftharpoons \mathrm{ArO^-} + \mathrm{H^+} \qquad (pK_a)$$

$$\mathrm{ArO^-} \rightleftharpoons \mathrm{ArO^\bullet} + e^- \qquad (E_{ox}(\mathrm{ArO^-}))$$

$$\mathrm{H^+} + e^- \rightleftharpoons \mathrm{H^\bullet} \qquad (E_r(\mathrm{H^+}))$$

$$\mathrm{ArOH} \rightleftharpoons \mathrm{ArO^\bullet} + \mathrm{H^\bullet} \qquad (D_{O\text{-}H})$$

Equilibrium acidity measurements and oxidation potentials, both measured in Me_2SO solution and can be combined to obtain relative homolytic dissociation energy of O−H bond of phenol. Since E_{red} for the proton is constant, differences in the sum of the oxidation potentials of the anions and the acidity constants for their conjugate acids (pK) can be taken as measures of relative bond dissociation energy:

$$\Delta D_{O\text{-}H}(\mathrm{kJ/mol}) = 5.73\Delta pK_{ArOH} + 96.48\Delta E_{ox}(\mathrm{ArO^-}) \qquad (15)$$

This approach is limited in practice by the irreversibility of the oxidation potentials for most anions. Nevertheless, there was observed that when families of anions wherein basicities have been changed by remote substitution are used, good correlations between E_{ox} and pK_{ArOH}, or between E_{ox} and log k for electron-transfer reactions, are often obtained. In these instances, the extent of irreversibility of E_{ox} throughout the family appears to be similar enough to permit estimates of relative bond dissociation energies by this method.
Lind et al. measured the constant of equilibrium in aqueous solution:

$$\mathrm{ArO^-} + \mathrm{ClO_2^\bullet} \rightleftharpoons \mathrm{ArO^\bullet} + \mathrm{ClO_2^-}$$

using pulse radiolysis technique (Lind et al., 1990). The values of $\Delta D_{O\text{-}H}$ estimated by two methods are listed in Table 5.

Substituents of Phenol, Phenol	ΔD_{O-H}, kJ/mol, (Bordwell & Cheng, 1991)	ΔD_{O-H}, kJ/mol, (Arnett et al., 1990)	ΔD_{O-H}, kJ/mol, (Lind et al., 1990)
H	0.0	0.0	0.0
2-Me	−6.9		
3-Me	−1.9		
4-Me	−4.8	−18.7	−0.9
3-Me, 5-Me	−3.1		
2-Me, 6-Me	−18.2		
4-CMe_3	−4.8	−7.3	
2-CMe_3, 6-CMe_3	−32.4		
2-CMe_3, 4-CMe_3, 6-CMe_3	−32.0		
4-Ph	−9.4		
2-OMe	−16.1		
3-OMe	1.5		
4-OMe	−22.0	−22.2	−23.4
4-O^-	−70.5		−66.1
4-OH	−34.9		−33.5
3-NH_2	−7.7		
3-NMe_2	−8.2		
4-NH_2	−52.5		−53.1
4-NMe_2	−40.0		−58.6
4-F		−1.9	−3.3
2-Cl	0.6		
3-Cl	8.2		
4-Cl	1.9	7.5	−2.5
3-Cl, 5-Cl	16.9		
3-Cl, 4-Cl , 5-Cl	13.6		
4-Br	3.6		−0.4
4-I			−1.2
3-CF_3	16.5		
4-CF_3	22.8		
3-SO_2Me	10.2		
4-SO_2Me	21.5		
3-C(O)Me	8.2		
4-C(O)Me	12.3		2.1
4-C(O)Ph	11.1		
4-OC(O)Me		−11.5	
4-CO_2^-			7.1
3-CN	16.9		
4-CN	18.2	6.1	19.7
3-NO_2	18.6		
4-NO_2	20.3	9.1	25.1
1-Naphthol	−24.5		
2-Naphthol	−7.7		

Table 5. The values of ΔD_{O-H} of substituted phenols measured by electrochemical techniques

3. Dissociation energies of O−H bonds of phenols

3.1 Dissociation energy of O−H bond of C_6H_5OH

As was shown earlier, the O−H bond of simplest phenol C_6H_5OH plays a key role among a line of different substituted phenols. The absolute value of $D_{O-H}(ArOH)$ can be calculated via difference of bond dissociation energies: $\Delta D_{O-H} = D_{O-H}(ArOH) - D_{O-H}(C_6H_5OH)$ and the last can be estimated by a line of methods. The values of $D_{O-H}(C_6H_5OH)$ were measured during last 20 years and are collected in Table 6.

Year	$D_{O-H}(C_6H_5OH)$, kJ/mol	Method	Ref.
1989	374.5	**VLPP** (gas)	Suryan et al., 1989a
1990	369.5	Shock tube (gas)	Walker & Tsang, 1990
1991	375.9	**AOP** (Me_2SO)	Bordwell & Cheng, 1991
1995	365.3	**PAC** (C_6H_6)	Wayner et al., 1995
1996	369.4	**CE** (C_6H_6)	Lucarini et al.,1996
1998	376.6	Negative ion cycle (gas)	DeTuri & Ervin, 1998
2000	369.0	Recommended	Denisov & Denisova, 2000
2003	368.2	Recommended	Luo, 2003
2004	359.0	Negative ion cycle (gas)	Angel & Ervin, 2005
2005	362.3	Recommended	Mulder et al., 2005

Table 6. The values of $D_{O-H}(C_6H_5OH)$ estimated by different techniques

It is seen from Table 6 that experimental values of $D_{O-H}(C_6H_5OH)$ vary from 359 to 377 kJ/mol, the mean value of $D_{O-H}(C_6H_5OH) = 369.0 \pm 5.7$ kJ/mol. This value coincides with that recommended in Handbook (Denisov & Denisova, 2000) and the last is in good agreement with D_{O-H} of hydroperoxides (see paragraph 4).

In recent years, quantum chemical methods, particularly density functional theory, are often used for the assessment of the dissociation energy of O−H bond in phenols. Let us know that the results of calculation as a rule, differ substantially from the experimental values. As we see experimental $D_{O-H}(C_6H_5OH) = 369 \pm 6$ kJ/mol and calculated values are sufficiently lower (see Table 7).

Method	D_{O-H}, kJ/mol	Method	D_{O-H}, kJ/mol
6-31G	332.8	6-31+G(d,p)	344.2
6-31G(d)	327.7	6-311+G(d,p)	347.5
6-31G(d,p)	346.2	6-311+G(2d,p)	348.0
6-31G(d,p)	341.7	6-311+G(2d,2p)	350.7
6-31G(d,p′)	361.3	6-311+G(3d,p)	350.1
6-31+G(d)	328.9	G-3	369.0

Table 7. Values of $D_{O-H}(C_6H_5OH)$ calculated by density functional theory (Wright et al., 1997; Luzhkov, 2005; Mulder et al., 2005)

3.2 Dissociation energies of O−H bond of substituted phenols

As was shown earlier, the influence of substituent of aromatic ring on D_{O-H} is very important. As well as each technique has its specific peculiarities that may influence on the measured value ΔD_{O-H} it would be useful to compare them. This comparison is performed in Table 8.

Substituents of Phenol, Phenol	CE	VLPP	MIP	AOP	PAC	D, kJ/mol
	ΔD, kJ/mol	ΔD, kJ/mol	ΔD, kJ/mol	ΔD, kJ/mol	ΔD, kJ/mol	
2-Me		−10.9	−9.1	−6.9		360.0 ± 1.6
3-Me	−6.0	−2.1	−2.3	−1.9		365.9 ± 1.7
4-Me	−8.3 ± 1.0	−7.9	−6.8	−4.8		361.6 ± 1.3
4-CMe_3	−10.0 ± 1.6		−8.9	−4.8	−7.9	361.1 ± 0.5
4-Ph	−16.0			−9.4		356.3 ± 3.3
3-OH		1.2	0.1			368.4 ± 0.7
4-OH		−10.9 ± 0.4	−17.0	−34.9 (?)		355.1 ± 3.0
2-OMe	−9.9	−17.6		−16.1		354.5 ± 3.3
3-OMe	−4.2	−4.6		1.5 (?)		364.6 ± 0.2
4-OMe	−20.8 ± 1.1	−14.6 ± 1.7		−22.0	−24.3	346.6 ± 1.4
3-NH_2		−1.7		−7.7		364.3 ± 3.0
4-NH_2	−41.2	−12.5 (?)		−52.5		322.2 ± 5.6
3-CN		−9.4		16.9 (?)		359.6
4-CN		1.2		18.2	23.4	389.8 ± 2.6
3-NO_2		−2.1		18.6 (?)		366.9
4-NO_2		4.6	3.8	20.3		373.2
4-F		−4.6		−2.6		365.4
2-Cl		−9.2		0.6 (?)		359.8
3-Cl		0.8		8.2 (?)		369.8
4-Cl	−1.6	−4.6		1.9	1.7	368.4 ± 2.7
4-CF_3		9.2		22.8 (?)		378.2
2-Me, 6-Me	−12.5 ± 2.7			−18.2		353.7 ± 2.8
3-Me, 5-Me	−6.5			−3.1		364.2 ± 1.7
3-CMe_3, 5-CMe_3	−8.0		−6.6			361.7 ± 0.7
2-CMe_3, 6-CMe_3	−19.0 ± 3.4		−22.6	−32.4 (?)		349.1 ± 1.6
2-CMe_3, 4-Me, 6-CMe_3	−27.0 ± 1.5		−13.1(?)	−32.0	−32.2	338.6 ± 2.4
1-Naphthol			−25.6	−24.5		344.0 ± 0.6
2-Naphthol	−9.9		−15.2	−7.7		358.1 ± 3.1
Indol	−43.9		−41.8			326.2 ± 1.0

Table 8. Comparison of ΔD (kJ/mol) = $D(Ar_iOH) - D(C_6H_5OH)$ for substituted phenols estimated by various methods, $D(C_6H_5OH)$ = 369.0 kJ/mol (see Tables 1−5); data marked by (?) were not included in the calculation of mean values of ΔD

We observe a good agreement between ΔD measured by different methods for phenols with alkyl, alkoxyl, and amino substituents. However, for phenols with polar electronegative substituents such as Cl, CF_3, CN, and NO_2 method **AOP** gives much higher absolute values of ΔD in comparison with another methods. The possible explanation of these discrepancies lies in strong additional solvation of polar groups in such strong polar solvent as dimethylsulfoxide used in **AOP** method.

3.3 Dissociation energies of O−H bonds of natural phenols

Bioantioxidants play very important role in biological processes and are under intensive investigation during last a few decades. Among them only for tocopherols and ubiquinols were measured their O−H bonds dissociation energies (Denisov & Denisova, 2000; Luo, 2003). These important characteristics for many natural phenolic compounds (flavonoids et al.) were estimated only recently (Denisov & Denisova, 2008). The list of these data is given in Table 9.

Phenol	Site of O−H bond	D_{O-H}, kJ/mol	Method	Ref.
α-Tocopherol	6	328.9	**CE**	Jackson & Hosseini, 1992
α-Tocopherol	6	330.1	**CE**	Lucarini et al., 1996
α-Tocopherol	6	330.0	**MIP**	Denisov, 1995
α-Tocopherol	6	327.3	**CE**	Lucarini et al., 1994
α-Tocopherol	6	323.4 ± 8.0	**PAC**	Wayner et al., 1996
α-Tocopherol	6	338.5	**AOP**	Bordwell & Liu, 1996
β-Tocopherol	6	335.2	**MIP**	Denisov & Denisova, 2009
β-Tocopherol	6	335.6	**MIP**	Denisov, 1995
β-Tocopherol	6	335.3 ± 2.0	**MIP**	Denisova & Denisov, 2008
γ- Tocopherol	6	334.8	**MIP**	Denisov & Denisova, 2009
γ-Tocopherol	6	335.1	**MIP**	Denisov 1995
γ-Tocopherol	6	334.9 ± 2.0	**MIP**	Denisova & Denisov, 2008
δ- Tocopherol	6	341.4	**MIP**	Denisov & Denisova, 2009
δ- Tocopherol	6	342.8	**MIP**	Denisov, 1995
δ- Tocopherol	6	335.6	**PAC**	Wayner et al., 1996
δ- Tocopherol	6	341.5 ± 2.0	**MIP**	Denisova & Denisov, 2008

Ubiquinol-0	1,4	343.8	**MIP**	Denisov, 1995
Ubiquinol-0	1,4	346.0	**MIP**	Denisova & Denisov, 2008
Ubiquinol-2	1,4	344.3	**MIP**	Denisov, 1995
Ubiquinol-2	1,4	345.7	**MIP**	Denisova & Denisov, 2008
Ubiquinol-6	1,4	344.3	**MIP**	Denisov, 1995
Ubiquinol-6	1,4	345.7	**MIP**	Denisova & Denisov, 2008
Ubiquinol-9	1,4	344.8	**MIP**	Denisova & Denisov, 2008
Ubiquinol-10	1,4	345.6	**MIP**	Denisova & Denisov, 2008
5-Hydroxy-2,4,6,7-tetramethyl-2,3-dihydrobenzo[b]furan	5	326.7	**MIP**	Denisov & Denisova, 2009
5-Hydroxy-2,2,4,6,7-pentamethyl-2,3-dihydrobenzo[b]furan	5	326.4	**MIP**	Denisov & Denisova, 2009
5-Hydroxy-2,2,4,6,7-pentamethyl-2,3-dihydrobenzo[b]furan	5	323.4	**CE**	Jackson & Hosseini, 1992
5-Hydroxy-2-carboxy-2,4,6,7-tetra-methyl-2,3-dihydrobenzo[b]furan	6	334.0	**MIP**	Denisov & Denisova, 2009
6-Hydroxy-5,7,8-trimethylchromane	6	330.9	**MIP**	Denisov & Denisova, 2009
6-Hydroxy-2-hydroxymethyl-2,5,7,8-tetramethylchromane	6	330.9	**MIP**	Denisov & Denisova, 2009
6-Hydroxy-2-methoxy-2,5,7,8-tetramethylchromane	6	334.4	**MIP**	Denisov & Denisova, 2009
6-Hydroxy-2-carboxy-2,5,7,8-tetramethylchromane	6	336.5	**MIP**	Denisov & Denisova, 2009
6-Hydroxy-2-methylcarboxy-2,5,7,8-tetramethylchromane	6	333.3	**MIP**	Denisov & Denisova, 2009
6-Hydroxy-2-carboxymethyl-2,5,7,8-tetramethylchromane	6	333.0	**MIP**	Denisov & Denisova, 2009
6-Hydroxy-2-methylcarboxymethyl-2,5,7,8-tetramethylchromane	6	330.9	**MIP**	Denisov & Denisova, 2009
6-Hydroxy-2,2,5,7,8-pentamethylchromane	6	328.9	**MIP**	Denisov & Denisova, 2009

6-Hydroxy-2,2,5,7,8-pentamethylchromane	6	328.4	**MIP**	Denisova & Denisov, 2008
6-Hydroxy-2,2,8-trimethyl-5,7-diethylchromane	6	331.9	**MIP**	Denisova & Denisov, 2008
6-Hydroxy-2,2-dimethyl-5,7-diethylchromane	6	333.6	**MIP**	Denisova & Denisov, 2008
6-Hydroxy-2,2,8-trimethyl-5,7-diisopropylchromane	6	337.0	**MIP**	Denisova & Denisov, 2008
6-Hydroxy-2,2-dimethyl-5,7-diisopropylchromane	6	335.7	**MIP**	Denisova & Denisov, 2008
6-Hydroxy-5-methyl-7-*tert*-butylchromane	6	332.5	**MIP**	Denisova & Denisov, 2008
6-Hydroxy-5-isopropyl-7-*tert*-butylchromane	6	340.7	**MIP**	Denisova & Denisov, 2008
6-Hydroxytocol	6	340.1	**MIP**	Denisova & Denisov, 2008
6-Hydroxy-5,7-diethyltocol	6	333.9	**MIP**	Denisova & Denisov, 2008
6-Hydroxy-5,7-diethyl-8-methyltocol	6	331.9	**MIP**	Denisova & Denisov, 2008
6-Hydroxy-5,7-diisopropyltocol	6	335.7	**MIP**	Denisova & Denisov, 2008
6-Hydroxy-5,7-diisopropyl-8-methyltocol	6	336.3	**MIP**	Denisova & Denisov, 2008
6-Hydroxy-5-methyl-7-*tert*-butyltocol	6	333.2	**MIP**	Denisova & Denisov, 2008
6-Hydroxy-5-isopropyl-7-*tert*-butyltocol	6	339.3	**MIP**	Denisova & Denisov, 2008
6-Hydroxy-5,7,8-trimethyl-3,4-dihydro-2*H*-1-benzothiopyran	6	332.5	**MIP**	Denisova & Denisov, 2008
6-Hydroxy-2,5,7,8-tetramethyl-3,4-dihydro-2*H*-1-benzothiopyran	6	333.9	**MIP**	Denisova & Denisov, 2008
6-Hydroxy-2,2,5,7,8-pentamethyl-3,4-dihydro-2*H*-1-benzothiopyran	6	333.4	**MIP**	Denisova & Denisov, 2008
6-Hydroxy-2-phytyl-2,5,7,8-tetramethyl-3,4-dihydro-2*H*-1-benzothiopyran	6	333.3	**MIP**	Denisov, 1995
6-Hydroxy-2-phytyl-2,5,7,8-tetramethyl-3,4-dihydro-2*H*-1-benzothiopyran	6	334.7	**MIP**	Denisova & Denisov, 2008

6-Hydroxy-2-methylcarboxy-2,5,7,8-tetramethyl-3,4-dihydro-2*H*-1-benzothiopyran	6	337.8	**MIP**	Denisova & Denisov, 2008
6-Hydroxy-2,5-dimethyl-2-phytyl-7,8-benzochromane	6	321.4	**MIP**	Denisova & Denisov, 2008
6-Hydroxy-2,5-dimethyl-2-phytyl-7,8-benzochromene	6	322.1	**MIP**	Denisova & Denisov, 2008
6-Hydroxy-4,4,5,7,8-pentamethyl-3,4-dihydro-2*H*-1-benzothiopyran	6	329.8	**MIP**	Denisova & Denisov, 2008
5,7,8-Trimethylselenotocol	6	335.7	**MIP**	Denisova & Denisov, 2008
5-Hydroxy-2,4-dimethyl-2,3-dihydro-benzo[b]selenophene	5	334.5	**MIP**	Denisova & Denisov, 2008
5-Hydroxy-2-methyl-2,3-dihydro-benzo[b]selenophene	5	342.3	**MIP**	Denisova & Denisov, 2008
Kaempferol	7,4′	348.9	**MIP**	Denisova & Denisov, 2008
Morin	7,4′	363.6	**MIP**	Denisova & Denisov, 2008
Ubichromenol	1,4	350.2	**MIP**	Denisova & Denisov, 2008
Quercetin	4′	343.0	**MIP**	Denisova & Denisov, 2008
(−)-Epicatechin	4′	346.2 ± 1.8	**MIP**	Denisova & Denisov, 2008
(−)-Epicatechin	4′	339.7	**CE**	Lucarini et al., 2002
6,7-Dihydroxyflavone	6	332.3	**MIP**	Denisova & Denisov, 2008
7,8-Dihydroxyflavone	8	333.0	**MIP**	Denisova & Denisov, 2008
Chrysin	7	357.1	**MIP**	Denisova & Denisov, 2008
Galangin	7	363.1	**MIP**	Denisova & Denisov, 2008
Dihydroquercetin	4′	343.6	**MIP**	Denisova & Denisov, 2008

Catechin	4′	348.1 ± 1.3	**MIP**	Denisova & Denisov, 2008
Hesperitin	3′	353.8	**MIP**	Denisova & Denisov, 2008
Fisetin	4′	348.0 ± 1.7	**MIP**	Denisova & Denisov, 2008
Gallic acid	4	347.4	**MIP**	Denisova & Denisov, 2008
Propyl gallate	4	334.6	**MIP**	Denisova & Denisov, 2008
Propyl gallate	4	339.7	**CE**	Lucarini et al., 2002
Myricetin	4′	340.9	**MIP**	Denisova & Denisov, 2008
(−)-Epigallocatechin	4′	344.6	**MIP**	Denisova & Denisov, 2008
Rutin	4′	343.2 ± 0.6	**MIP**	Denisova & Denisov, 2008
Hesperidin	3′	345.8	**MIP**	Denisova & Denisov, 2008
Luteolin	4′	342.7	**MIP**	Denisova & Denisov, 2008
Nordihydroguaiaretic acid	4,4′	351.3	**MIP**	Denisova & Denisov, 2008
Caffeic acid	4	339.8	**MIP**	Denisova & Denisov, 2008
(−)-Epigallocatechin gallate	4′,4″	338.7 ± 0.3	**MIP**	Denisova & Denisov, 2008
β-Glucogallin	4	335.0	**MIP**	Denisova & Denisov, 2008
(−)-Epicatechin gallate	4′,4″	339.6 ± 1.3	**MIP**	Denisova & Denisov, 2008
Tannic acid	4	341.6	**MIP**	Denisova & Denisov, 2008
Pentagalloylglucose	4	338.1	**MIP**	Denisova & Denisov, 2008

Table 9. The values of D_{O-H} of natural phenols (D_{O-H}(α-tocopherol) = 330.0 kJ/mol), the second column contains the sites of O−H groups with equireactivity

As seen from the data in Tab1e 9, all ubiquinols possess virtually the same dissociation energy of the O−H bond, which is independent of the length of a 2-substituent. An average value of $D_{O\text{-}H}$ is 345.2 ± 0.8 kJ/mol (8 measurements). It should be borne in mind that both OH groups in ubiquinols are involved in hydrogen bonding with the *ortho*-methoxy groups. Phenolic groups form very strong intramolecular hydrogen bonds. For instance, in a phenolic crown ether (for the structure, see Ref. (Pozdeeva et al., 1989) the enthalpy of formation of such a bond is 21 kJ/mol, whereas the concentration of free OH groups is only 0.1% at 323 K. According to a theoretical calculation performed by density functional theory (Heer et al., 2000) the difference $\Delta\Delta G^{\neq}$ (298 K) for two transition states for the abstraction of an H atom by the methoxyl radica1 from a hydrogen-bonded OH group of ubiquinol-0 and from a free OH group is 7.5 kJ/mol. Assuming that $\Delta E(RO_2^{\bullet}$ + ubiquinol) = $\Delta\Delta G^{\neq}(MeO^{\bullet}$ + ubiquinol) and using Eqn. (12), we obtain the value of $D_{O\text{-}H}$ = 329.0 kJ/mol for a free OH group of ubiquinols, *i. e.,* almost the same as that for α-tocopherol.

In tocopherols, the value $D_{O\text{-}H}$ depends on the number and arrangement of methyl groups in the aromatic ring (see Table 9). At the same time, replacement of methyl groups by ethyl, isopropyl, and *tert*-butyl ones in positions 5 and 7 affects slightly the $D_{O\text{-}H}$ values of tocols:

Compound	R^1	R^2	$D_{O\text{-}H}$, kJ/mol
6-Hydroxytocol	H	H	340.1
6-Hydroxy-5,7-diethyltocol	Et	Et	333.9
6-Hydroxy-5,7-diisopropyltocol	Me_2CH	Me_2CH	335.7
6-Hydroxy-5-isopropyl-7-*tert*-butyltocol	Me_2CH	Me_3C	339.3

Substitution of phytyl (Pht) for the methyl substituent in position 2 exerts virtually no effect on the value of $D_{O\text{-}H}$ of 6-hydroxychromanes: $D_{O\text{-}H}$(2-Me, 2-Me) − $D_{O\text{-}H}$(2-Me, 2-Pht) = 0.4 ± 0.7 kJ/mol (7 pairs of compounds from Table 9 were compared). In 6-hydroxy-5,7,8-trimethylchromanes, the nature of substituents in position 2 virtually has no impact on the dissociation energy of the O−H bond: for seven phenols, the average value is 332.8 ± 2.0 kJ/mol. However, the nature of 5,7,8-substituents in 6-hydroxy-2,2-dimethyl-chromane appreciab1y influences the value of $D_{O\text{-}H}$ altering it in the range 328 to 341 kJ/mol. Substitution of a naphthalene ring for the benzene ring on going from γ-tocopherol (3) to 6-hydroxy-2,5-dimethyl-2-phytyl-7,8-benzochromane reduces $D_{O\text{-}H}$ by 13 kJ/mol (see Table 9). Substitution of S and Se for O atoms in α-tocopherol results in a sma1l decrease in dissociation energy: $D_{O\text{-}H}$ are 330.0, 334.0 and 335.7 kJ/mol for α-tocopherol, S-, and Se-analogs, respectively. The presence or the absence of methyl groups in position 2 in 6-hydroxy- 5,7,8-trimethyl-3,4-dihydrobenzothiopyrans does not affect the dissociation energy of the O−H bond: $D_{O\text{-}H}$ (in kJ/mol) = 332.5 (2-H, 2-H), 333.9 (2-H, 2-Me), 333.4 (2-Me, 2-Me) (see Tab1e 9).

The molecules of chrysin and galangin contain two phenolic groups. One of them, *viz.,* the OH group in position 5 is linked to the adjacent carbonyl group by a hydrogen bond, hence it is the 7-OH group that is the most reactive. As a result, for chrysin and galangin the number of equireactive O−H bonds ($n_{O\text{-}H}$ is 1), and $D_{O\text{-}H}$ differ little: 357.1 (chrysin) and 363.1 (galangin) kJ/mol, respectively. For morin, $D_{O\text{-}H}$ of the OH groups in positions 7 and 4′ are apparently roughly the same and equal to 363.6 kJ/mol ($n_{O\text{-}H}$ = 2).

In catechol (1,2-dihydroxybenzene), one hydroxyl group weakens the adjacent O−H bond (thus, in pyrogallol $D_{O\text{-}H}$ = 339.9 kJ/mol, while in phenol $D_{O\text{-}H}$ = 369.0 kJ/mol). A hydrogen bond increases effective strength of O−H bond roughly by 10 kJ/mol (for comparison, in quercetin $D_{O\text{-}H}$ = 343.0 kJ/mol and in hesperitin $D_{O\text{-}H}$ = 353.8 kJ/mol); in the latter compound, the peroxyl radical attacks the OH group linked to a methoxy group by a hydrogen bond. Thus, in catechol only one OH group reacts with $RO_2^{\bullet}$, the second is inactive due to the formation of a strong hydrogen bond.

An average value of $D_{O\text{-}H}$ in flavanones and flavones is 344.7 ± 2.6 kJ/mol. In 1,2,3-trihydroxybenzenes only one O−H bond actively reacts with $RO_2^{\bullet}$, whereas the other two are involved in hydrogen bonding. For such phenols a fairly wide range of $D_{O\text{-}H}$ values is observed: 347.4 (gallic acid), 337.2 ± 2.5 (propyl gallate), 340.9 (myricetin), 344.6 ((−)-epigallocatechin), 343.2 (rutin), 345.8 (hesperidin) and 338.7 kJ/mol ((−)-epigallocatechin gallate).

Thus, dissociation energies of the O−H bond in natural antioxidants ranges from 330 (for α-tocopherol) to 364 kJ/mol (for morin). These compounds compose a group with very close values of of $D_{O\text{-}H}$; it includes tocopherols, ubiquinols, flavones, flavanones and gallates. The diversity of their structures seems to be associated with the peculiarities of the media where they manifest their antioxidant activity.

3.4 Influence of structure on $D_{O\text{-}H}$ of phenols

The most important factor affecting the strength of the O−H bond is the stabilization of a phenoxyl radical due to the overlap of the unpaired electron orbital of the oxygen atom with the π-electrons of the benzene ring. The stabilization energy can be judged by a comparison of the dissociation energy of the O−H bond in phenol (PhOH) ($D_{O\text{-}H}$ = 369 kJ/mol) and in an aliphatic alcohol ROH ($D_{O\text{-}H}$ = 432 kJ/mol) (Luo 2003). The difference is 63 kJ/mol therefore reactions of peroxyl radicals ($D_{O\text{-}H}$ = 365.5 kJ/mol) with most of phenols, which are essential for the inhibition of oxidation, are exothermic.

The second factor that influences the dissociation energy of the O−H bonds of substituted phenols is the inductive effect of alkyl, in particular, methyl, groups. Below the data are given that illustrate the role of the inductive effect of methyl groups on the dissociation energy of the O−H bond in phenols: $\Delta D = D(MeC_6H_4OH) - D(C_6H_5OH)$ kJ/mol (see Tables 2-4, 8).

Substituent	ΔD, kJ/mol	Substituent	ΔD, kJ/mol	Substituent	ΔD, kJ/mol
2-Me	−9.0	2-Me, 3-Me	−13.5	3-Me, 4-Me	−14.4
3-Me	−3.1	2-Me, 4-Me	−8.5	3-Me, 5-Me	−4.8
4-Me	−7.4	2-Me, 6-Me	−15.3	2-Me, 4-Me, 6-Me	−13.2

We observe stronger effect on the $D_{O\text{-}H}$ for *ortho-* and *para*-methyl groups than that of *meta*-methyl group. There is no additivity in action of two or more methyl groups on $D_{O\text{-}H}$ of substituted phenol. Analogous effect is also observed in tocopherols: the more methyl substituents are present in the benzene ring of a tocopherol the weaker is its O−H bond.

Phenol	n_{Me}	$D_{O\text{-}H}$, kJ/mol
α-Tocopherol	3	330.0
β-Tocopherol	2	335.5
γ-Tocopherol	2	335.5
δ-Tocopherol	1	361.5

The third factor is the enhancement of the stabilization of a phenoxyl radical due to interaction of the *p*-electrons of the N or O atom of amino- and alkoxy- substituent or a hydroxyl group with the π-electrons of the benzene ring (mesomeric effect). The magnitude of this effect is clearly seen from the following comparison:

Substituent of Phenol	Δ*D*, kJ/mol	Substituent of Phenol	Δ*D*, kJ/mol	Substituent of Phenol	Δ*D*, kJ/mol
3-OH	−0.6	3-MeO	−4.4	3-NH_2	−4.7
4-OH	−13.9	4-MeO	−19.1	4-NH_2	−46.8

All these substituents reveal a weak effect on D_{O-H} in *meta*-position and very strong in *para*-position.

The fourth factor is influence of electronegative substituents that attract the π-electron density of benzene ring and often increase the dissociation energy of O−H bond. Down are given examples of such influence.

Substituent of Phenol	Δ*D*, kJ/mol	Substituent of Phenol	Δ*D*, kJ/mol	Substituent of Phenol	Δ*D*, kJ/mol
3-CN	−9.4	3-NO_2	−2.1	4-COOH	2.7
4-CN	1.2	4-NO_2	4.2	4-CF_3	9.2

It is seen that influence of substituent depends on position: in *meta*-position these substituents decrease D_{O-H} and in *para*-position increase D_{O-H}. Quite another effect have haloid substituents (F, Cl, Br): in *meta*-position they increase D_{O-H} of phenols and in *ortho*- and *para*-position diminish it (see Table 2).

The fifth factor is the stereoelectronic one, which has been discussed in detail by Burton et al. (Burton et al., 1985). The point is that for bicyclical phenols, such as hydroxychromanes and hydroxybenzofurans, an important parameter is the angle θ between the C−O bond of the annulated oxygen-containing ring and the plane of the benzene ring. The smaller this angle the larger the overlap of the *p*-electron orbitals of the O atom of the pyran or furan ring with the π-electcons of the benzene ring and the higher the stabilization energy of the phenoxyl radical. This is exemplified by the data given below:

Phenol

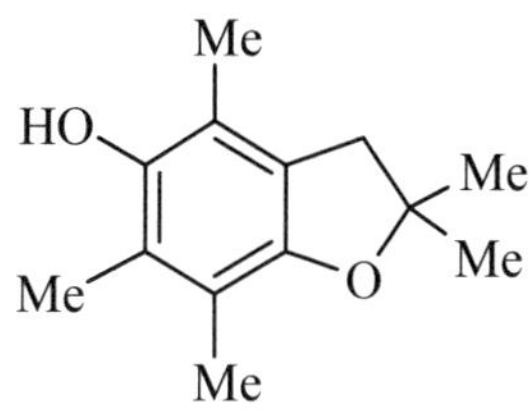

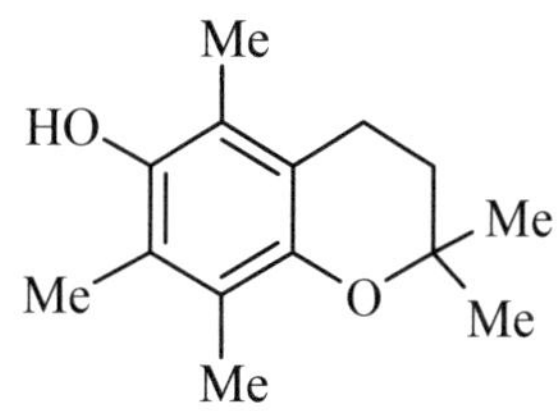

Angle θ, deg	6	17
D_{O-H}, kJ/mol	326.4	328.9

The sixth factor is the intramolecular hydrogen bonding. The value of input of the intramolecular hydrogen bond into D_{O-H} of *ortho*-methoxyphenol can be evaluated from comparison of ΔD_{O-H} estimated by **CE** and **VLPP** methods. In the **CE** method, the equilibrium

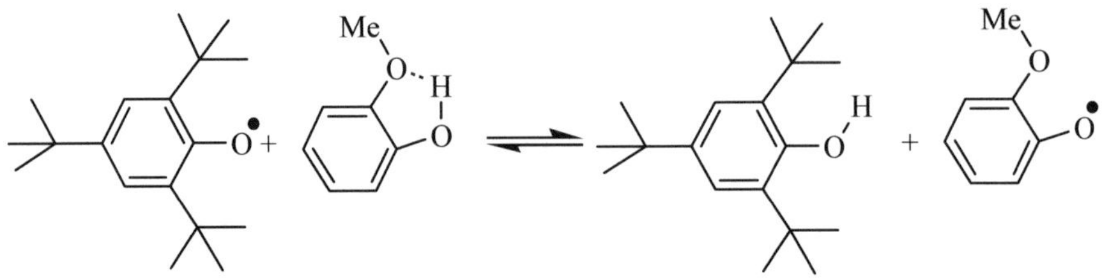

proceeds with *ortho*-methoxyphenol with intramolecular hydrogen bond and the value of $\Delta D_{O\text{-}H}$ = −9.9 kJ/mol (see Table 8) included the input of hydrogen bond into energy of *ortho*-methoxyphenoxyl radical stabilization. In the **VLPP** method $\Delta D_{O\text{-}H}$ = −17.6 kJ/mol (see Table 8) was calculated from comparison of activation energies of decomposition of anisole and *ortho*-methoxyanisole and evidently does not include the energy of the hydrogen bond. So, we can evaluate the difference in O−H bond dissociation energies in *ortho*-methoxyphenol with and without hydrogen bond as $\Delta D_{O\text{-}H...O}$ = −9.9 − (−17.6) = 7.7 kJ/mol. A very close value ($\Delta D_{O\text{-}H...O}$ = 7.5 kJ/mol) gives a quantum chemical calculation of the Gibbs energy of the transition state for the reaction of the methoxyl radical with *ortho*-methoxyphenol in two distinct states, viz., with a free OH group and with that bound by a hydrogen bond (Heer et al., 2000).
The influence of remote hydrogen bond on $D_{O\text{-}H}$ value of phenol was found recently by Foti et al. (Foti et al., 2010). The comparison of reactivity of 3 substituted phenols in their reactions with peroxyl ($k(RO_2^\bullet)$) and diphenylpycrylhydrazil (DPPH$^\bullet$) radicals demonstrated the diversity as the result of formation of remote intermolecular hydrogen bond.

Phenol	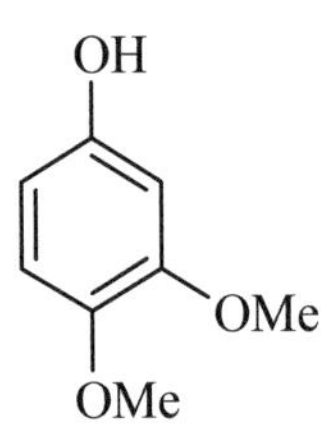	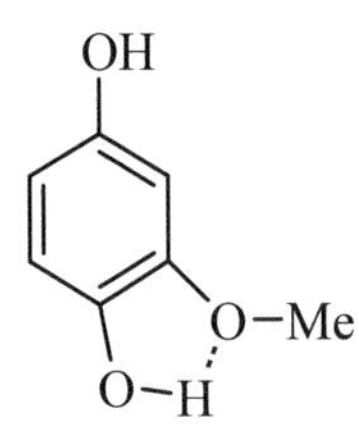	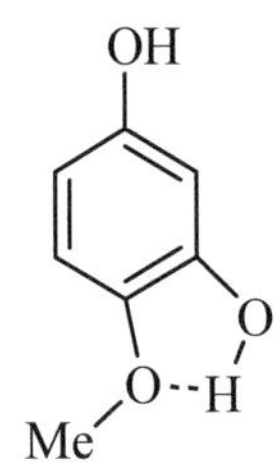
$k(RO_2^\bullet)$, l/mol s (303 K)	4.7×10^5	7.2×10^5	2.5×10^5
k(DPPH$^\bullet$), l/mol s (303 K)	1.8×10^3	4.4×10^3	2.0×10^2
$\Delta D_{O\text{-}H}$, kJ/mol	0.0	-2.5	1.6

One sees that remote hydrogen bonds have appreciable effect on the phenolic bond dissociation energy. Intermolecular *para*-OH...*meta*-OMe hydrogen bond weaken, while *meta*-OH...*para*-OMe hydrogen bond strengthen O−H bond dissociation energy compared with similarly substituted 3,4-dimethoxyphenol.

4. Thermochemistry of hydroperoxides

4.1 Dissociation energies of O−H bonds of hydroperoxides

The O−H bond dissociation energy in *tert*-butyl hydroperoxide was measured by Holmes et al. using masspectrometry technique and appearance energy measurements (Holmes et al., 1991) and was found be equal to 258.6 kJ/mol. All tertiary alkylperoxy radicals has the same activity in reactions of hydrogen atom abstraction (Howard, 1972) and hence all tertiary alkylhydroperoxides have the same D_{O-H}. Secondary alkylperoxy radicals are more active than tertiary as hydrogen atom acceptors due to the last have lower D_{O-H} (Mill & Hendry, 1980; Denisov & Afanas'ev, 2005). The difference in activation energies of reactions:

$$R'O_2^\bullet + RH \rightarrow R'OOH + R^\bullet$$

between *sec*-$R'O_2^\bullet$ and *tert*-$R'O_2^\bullet$ was found be equal to 3.1 ± 1.6 kJ/mol (Denisov & Denisova, 1993). The calculated via **MIP** method (Eqn. 13) value of $\Delta D_{O-H} = D_{O-H}(sec\text{-}R'O_2^\bullet) - D_{O-H}(tert\text{-}R'O_2^\bullet)$ is equal to 6.9 kJ/mol and hence $D_{O-H}(sec\text{-ROOH}) = D_{O-H}(tert\text{-ROOH}) + 6.9 = 365.5$ kJ/mol. All primary and secondary alkylhydroperoxide have practically the same D_{O-H} (Howard, 1972). In accordance with these data is equilibrium constant between secondary and tertiary hydroperoxides and peroxy radicals (Howard et al., 1968).

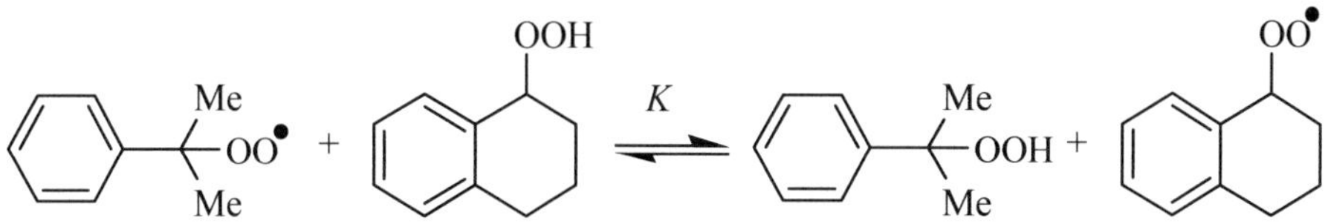

This equilibrium was found be moved to the left and equilibrium constant $K = 0.24$ (303 K). As well as enthalpy of equilibrium $\Delta H \cong \Delta G = -RT\ln K$, so $\Delta H = \Delta D_{O-H} = 3.6$ kJ/mol. Mahoney and DaRooge studied the equilibrium between *sec*-peroxyl radical and 2,4,6-tri-*tert*-butylphenol (Mahoney & DaRooge, 1975).

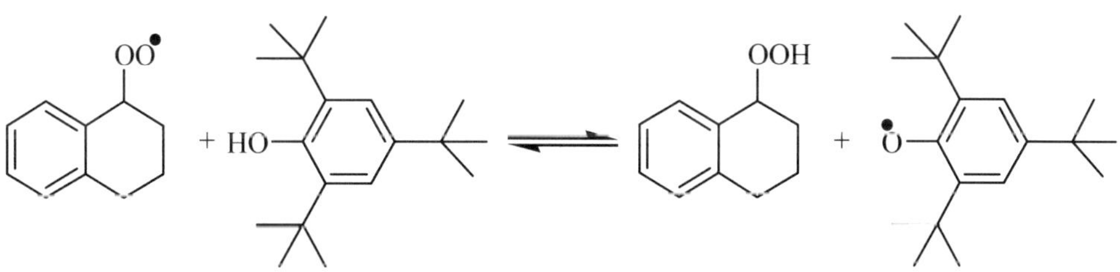

The value of D_{O-H}(ROOH) calculated from enthalpy of this equilibrium was found to cover the interval 362 ÷ 369 kJ/mol. All data mentioned above are in agreement with the recommended value D_{O-H}(*sec*-ROOH) = 365.5 kJ/mol (Denisov & Denisova, 2000, Denisov et al., 2003). For the O–H bond dissociation energy of hydrogen peroxide is recommended the value D_{O-H}(HOOH) = 369.0 kJ/mol (Luo, 2003, Lide, 2004).

Functional groups (Y = OH, >C(O) *etc.*) in hydroperoxides influence on their O–H bond dissociation energies. The problem of estimation of D_{O-H}(YROOH) for such hydroperoxides was solved recently by using **MIP** in application to kinetic experimental data on co-oxidation of hydrocarbons with compounds YRH including functional groups (Denisova & Denisov, 2004). From kinetics of co-oxidation of YRH with hydrocarbon (RH) ratios of rate constants $k_Y(YROO^\bullet + RH)/k(ROO^\bullet + RH)$ were calculated and then estimated the differences in activation energies: $\Delta E = E(YROO^\bullet + RH) - E(ROO^\bullet + RH) = RT\ln(k/k_Y)$. These values of ΔE opened the way to estimate the O–H bond dissociation energies in hydroperoxides with functional groups (see Eqn. 13 and Table 10).

ROO–H	*D*, kJ/mol	ROO–H	*D*, kJ/mol
$R^1R^2CHOO–H$	365.5	$R^1R^2R^3COO–H$	358.6
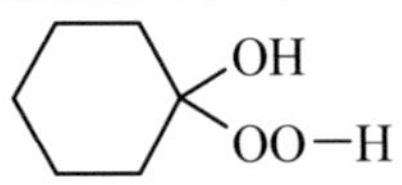	362.1	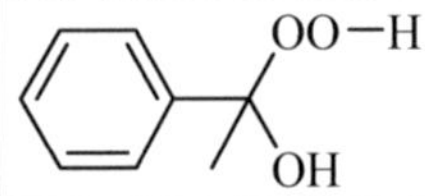	358.7
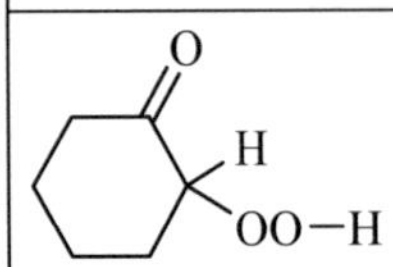	369.8	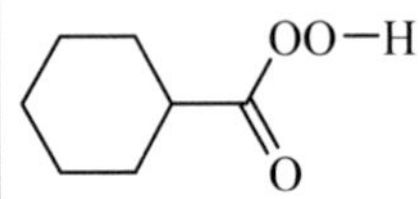	376.9
$R^1C(O)OO–H$	387.1	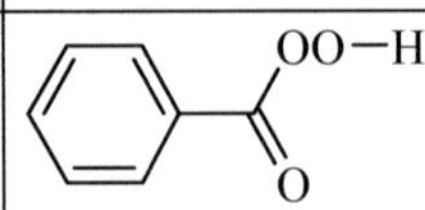	403.9
$R^1OCH(OO–H)R$	367.3	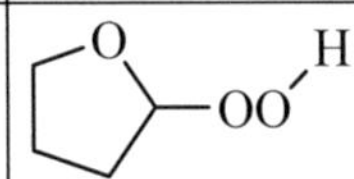	367.6
$R^1R^2CHOC(OO–H)R^1R^2$	358.4	$R^1OCH(OOH)Ph$	374.8
$R^1C(O)CH(OO–H)R$	369.8	AcOCH(OOH)Ph	378.0
$R^1R^2NC(OO–H)CHMe$	364.1	$CCl_3CCl_2OO–H$	413.1

Table 10. The O−H bond dissociation energies in hydroperoxides with functional groups (Denisova & Denisov, 2004)

There were measured the rate constants for reactions of haloid substituted methyl and ethyl peroxyl radicals with nonsaturated fatty acids (Huie & Neta, 1997). These data can be used for evaluation of D_{O-H} of substituted hydroperoxides in the scope of **MIP**. The ratios of rate constants $k_i(R_iO_2^{\bullet} + RH)/k_1(R_1O_2^{\bullet} + RH)$ at $T = 298$ K and values of ΔD and D of O−H bonds of ROOH calculated by Eqns. 10 and 13 are presented in Table 11. The following parameters were used for calculation: $\alpha = 0.814$, $br_e = 15.21$ (kJ/mol)$^{1/2}$, $A_0 = 1.0 \times 10^7$ l/mol s (Denisov & Denisova, 2000), $E_{e1}(HO_2^{\bullet}$ + linoleic acid) = 36.3 kJ/mol and $E_{e1}(CCl_3O_2^{\bullet}$ + linolenic acid) = 25.1 kJ/mol.

$R_1O_2^{\bullet}$	$R_iO_2^{\bullet}$	RH	k_1/k_i	ΔD, kJ/mol	*D*, kJ/mol
$HO_2^{\bullet}$	$CCl_3O_2^{\bullet}$	Linoleic acid	0.012	40.1	409.1
$CCl_3O_2^{\bullet}$	$CF_3O_2^{\bullet}$	Linolenic acid	0.16	19.5	428.6
$CCl_3O_2^{\bullet}$	$CBr_3O_2^{\bullet}$	Linolenic acid	0.92	0.9	410.0
$CCl_3O_2^{\bullet}$	$CF_3CHClO_2^{\bullet}$	Linolenic acid	3.67	−12.7	396.4

Table 11. Dissociation energies of O−H bonds in haloid substituted hydroperoxides calculated by **MIP**

Recently these values were used for calculation of enthalpy of exchange equilibrium reaction (Denisova, 2007):

$$RO_2^{\bullet} + YROOH \rightleftharpoons ROOH + YROO^{\bullet}$$

where $\Delta H = D_{O-H}$(YROOH) − D_{O-H}(ROOH) in nonpolar solvent. The reactions and values of ΔH for reactions of exchange between $Me_3CO_2^{\bullet}$ and YROOH are listed in Table 12.

YROOH	D_{O-H}, kJ/mol	ΔH, kJ/mol	K (T = 300 K)	K (T = 350 K)
HOOH	369.0	10.4	1.55×10^{-2}	2.80×10^{-2}
sec-ROOH	365.5	6.9	6.29×10^{-2}	9.34×10^{-2}
tert-ROOH	358.6	0.0	1.00	1.00
cyclo-C_6H_{10}(OH)OOH	362.1	3.5	2.46×10^{-1}	3.00×10^{-1}
RPhC(OH)OOH	359.8	1.2	0.62	0.66
ROC(OOH)R	367.3	8.7	3.06×10^{-2}	5.03×10^{-2}
R_2CHOC(OOH)R_2	358.4	−0.2	1.08	1.07×10^{-2}
ROCH(OOH)Ph	374.8	16.2	1.51×10^{-3}	3.82×10^{-3}
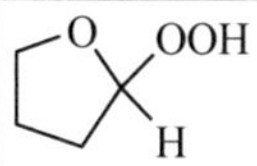	367.6	9.0	2.71×10^{-2}	4.54×10^{-2}
RC(O)OOH	387.1	28.5	1.09×10^{-5}	5.58×10^{-5}
R_3CC(O)OOH	376.9	18.3	6.51×10^{-4}	1.86×10^{-3}
cyclo-C_6H_{11}C(O)OOH	376.9	18.3	6.51×10^{-4}	1.86×10^{-3}
PhC(O)OOH	403.9	45.3	1.30×10^{-8}	1.73×10^{-7}
RC(O)CH(OOH)R	369.8	11.2	1.12×10^{-2}	2.13×10^{-2}
RC(O)CH(OOH)Ph	376.4	17.8	7.06×10^{-4}	2.21×10^{-3}
CCl_3CCl_2OOH	413.1	54.5	3.51×10^{-10}	7.35×10^{-8}
$CHCl_2CCl_2OOH$	411.6	53.0	5.92×10^{-10}	1.23×10^{-8}

Table 12. The values of ΔH for reactions of exchange: $Me_3CO_2^{\bullet}$ + YROOH $\rightleftharpoons$ Me_3CO_2H + $YROO^{\bullet}$ (Denisova, 2007)

4.2 Decomposition of α-hydroxyhydroperoxides and α-hydroxyperoxyl radicals

The oxidation of alcohols, their co-oxidation with other organic compounds, and deep steps of hydrocarbon oxidation yield α-hydroxyperoxyl radicals (Denisov & Afanas'ev, 2005). The last participate in the following reactions:

$$R^1R^2C(OH)OO^{\bullet} + R^1R^2CH(OH) \rightarrow R^1R^2C(OH)OOH + R^1R^2C^{\bullet}(OH)$$

$$R^1R^2C(OH)OO^{\bullet} \rightarrow R^1R^2C(O) + HO_2^{\bullet}$$

The formed α-hydroxyhydroperoxide decomposes into carbonyl compound and hydrogen peroxide.

$$R^1R^2C(OH)OOH \rightarrow R^1R^2C(O) + H_2O_2$$

The thermodynamics and kinetic of these reactions were analysed recently in paper (Denisov & Denisova, 2006). Hydroxyhydroperoxides are labile compounds and are not amenable to thermochemical measurements of their enthalpy of formation (Denisov et al.,

2003). Earlier, comparing the enthalpies of formation of hydroperoxides ROOH with ΔH(RH) and ΔH(ROH), we observed a symbatic change in these values, namely, the correlations: $\Delta H(\text{ROOH}) = a + \Delta H(\text{RH})$ and $\Delta H(\text{ROOH}) = b + \Delta H(\text{ROH})$ (Denisov & Denisova, 1988). This regularity was extended to the enthalpy of formation of α-hydroxyhydroperoxides. The following procedure was used to estimate $\Delta H(R^1R^2C(OH)OOH)$. First, the difference of the formation enthalpies of R_2CHOOH and R_2CH_2 was determined, and then the enthalpy of formation of $R^1R^2C(OH)OOH$ was calculated as the algebraic sum: ΔH_f^0

$$\Delta H_f^0(R^1R^2C(OH)OOH) = \Delta H_f^0(R^1R^2CHOH) + \{\Delta H_f^0(R^1R^2CHOOH) - \Delta H_f^0(R^1R^2CH_2)\} \quad (16)$$

The results of calculation of the enthalpies of formation for fourteen α-hydroxyhydroperoxides by Eqn. (16) are given in Table 13. This calculation implies that the replacement of H by OOH varies the enthalpy of formation of the corresponding alcohol by the same value as in the case of this substitution in RH. The validity of this approach was qualitatively verified by comparing the enthalpies of two equilibrium reactions of cyclohexanone with ROOH and H_2O_2 (Denisov & Denisova, 2006). The values of decomposition enthalpies of α-hydroxyhydroperoxides are listed in Table 13.

α-Hydroxyhydroperoxide	$-\Delta H_f^0(R^1R^2C(O))$, kJ/mol	$-\Delta H_f^0(R^1R^2C(OH)OOH)$, kJ/mol	ΔH, kJ/mol
$CH_2(OH)OOH$	108.8	261.1	15.9
MeCH(OH)OOH	165.7	320.0	17.9
EtCH(OH)OOH	187.4	336.8	13.0
$Me_2C(OH)OOH$	217.1	366.1	12.6
PrCH(OH)OOH	207.5	353.5	9.6
EtMeC(OH)OOH	240.6	385.8	8.8
MePrC(OH)OOH	259.0	401.7	6.3
PhCH(OH)OOH	37.7	190.0	15.9
PhMeC(OH)OOH	86.6	231.8	8.8
cyclo-C_6H_{11}CH(OH)OOH	235.1	389.5	18.0
cyclo-C_5H_8(OH)OOH	192.5	349.2	20.3
cyclo-C_6H_{10}(OH)OOH	225.9	377.7	22.4
cyclo-$C_{12}H_{22}$(OH)OOH	351.5	495.7	7.8
α-Tetralyl-(OH)OOH	90.8	246.0	18.8

Table 13. Enthalpies (ΔH) of decomposition of α-hydroxyhydroperoxides in nonpolar solvents

According data presented in Table 10 the strength of the OO–H bond in α-hydroxycyclohexyl hydroperoxide is equal to 362.1 kJ/mol. Therefore, using the expression for the strength of the OO–H bond:

$$D_{O-H} = \Delta H_f^0(R_2C(OH)OO^\bullet) + \Delta H_f^0(H^\bullet) - \Delta H_f^0(R_2C(OH)OOH) \quad (17)$$

and supposing D_{O-H} = 362.1 kJ/mol for all hydroxyhydroperoxides we obtain the following thermochemical equation for the enthalpy of formation of the α-hydroxyperoxyl radical ($\Delta H_f^0(H^\bullet)$ = 218.0 kJ/mol (Luo, 2005)):

$$\Delta H_f^0(R_2C(OH)OO^\bullet),\ kJ/mol\ = \Delta H_f^0(R_2C(OH)OOH) + 144.1 \qquad (18)$$

The ΔH_f^0 (>C(OH)OO$^\bullet$) values calculated for decomposition of fifteen radicals by Eq. (18) are given in Table 14. The enthalpy (ΔH) of decomposition of the α-hydroxyperoxyl radical was calculated by the Eqn. 19:

$$\Delta H(decay),\ kJ/mol = \Delta H_f^0(R_2C(O)) + \Delta H_f^0(HO_2^\bullet) - \Delta H_f^0(R_2C(OH)OO^\bullet) \qquad (19)$$

where $\Delta H_f^0(HO_2^\bullet)$ = 14.6 kJ/mol (Lide, 2004). The results of calculation of the enthalpies of degradation of the α-hydroxyperoxyl radicals are presented in Table 14.

α-Hydroxyperoxyl radical	$-\Delta H_f^0$(>C(OH)OOH)), kJ/mol	$-\Delta H_f^0$(>C(OH)OO$^\bullet$), kJ/mol	ΔH, kJ/mol
$CH_2(OH)OO^\bullet$	261.1	117.0	22.8
MeCH(OH)OO$^\bullet$	320.0	175.9	24.8
EtCH(OH)OO$^\bullet$	336.8	192.7	20.0
$Me_2C(OH)OO^\bullet$	366.1	222.0	19.5
PrCH(OH)OO$^\bullet$	353.5	209.4	16.5
EtMeC(OH)OO$^\bullet$	385.8	241.7	15.7
MePrC(OH)OO$^\bullet$	401.7	257.6	13.2
$Me_2CHMeC(OH)OO^\bullet$	401.7	257.6	9.9
PhCH(OH)OO$^\bullet$	190.0	45.9	22.8
PhMeC(OH)OO$^\bullet$	231.8	87.7	15.7
cyclo-$C_6H_{11}CH(OH)OO^\bullet$	389.5	245.4	25.0
cyclo-$C_5H_8(OH)OO^\bullet$	349.2	205.1	27.2
cyclo-$C_6H_{10}(OH)OO^\bullet$	377.7	233.4	22.4
cyclo-$C_{12}H_{22}(OH)OO^\bullet$	495.7	351.6	15.0
α-Tetralyl-(OH)OO$^\bullet$	246.0	101.9	25.7

Table 14. Enthalpies of decomposition of α-hydroxyperoxyl radicals

4.3 Enthalpies of reactions of peroxyl radicals with phenols

As was written earlier, the reaction of peroxyl radicals with phenols is the main reaction of action of phenols as antioxidants. The enthalpy of this reaction is equal to difference of bond dissociation energies of two O–H bonds:

$$\Delta H = D_{O-H}(ArOH) - D_{O-H}(ROOH)\,. \qquad (20)$$

The values of ΔH calculated for 108 reactions of different $R_iO_2^\bullet$ with phenols (Ar_jOH) are presented in Table 15. It is seen from Table 15 that enthalpy of reaction strongly depends on $D_{O-H}(ArOH)$ as well as on $D_{O-H}(ROOH)$.

$Ar_jOH/R_iO_2^•$	ΔH, kJ/mol					
	tert-ROO•	*sec*-ROO•	$HO_2^•$	AcCH(OO•)Me	AcOO•	PhC(O)OO•
PhOH	10.4	3.5	0.0	1.7	−18.1	−34.9
4-Me-C_6H_4OH	3.0	−3.9	−7.4	−5.7	−25.5	−42.3
4-OMe-C_6H_4OH	−8.7	−15.6	−19.1	−17.4	−37.2	−54.0
4-OH-C_6H_4OH	−3.5	−10.4	−13.9	−12.2	−32.0	−48.8
4-NH_2-C_6H_4OH	−36.4	−43.3	−46.8	−45.1	−64.9	−81.7
4-COMe-C_6H_4OH	12.9	6.0	2.5	4.2	−15.6	−32.4
2,4-$(Me)_2$-C_6H_3OH	1.9	−5.0	−8.5	−6.8	−26.6	−43.4
2,6 -$(Me)_2$-C_6H_3OH	−4.0	−10.9	−14.4	−12.7	−32.5	−49.3
Ionol	−15.6	−22.5	−26.0	−24.3	−44.1	−60.9
α-Tocopherol	−28.6	−35.5	−39.0	−37.3	−57.1	−73.9
β-Tocopherol	−23.3	−30.2	−33.7	−32.0	−51.8	−68.6
δ-Tocopherol	−17.1	−24.0	−27.5	−25.8	−45.6	−62.4
Ubiquinol-0	−13.3	−20.2	−23.7	−22.0	−41.8	−58.6
Quercetin	−15.6	−22.5	−26.0	−24.3	−44.1	−60.9
Chrysin	−1.5	−8.4	−11.9	−10.2	−30.0	−46.8
Morin	5.0	−1.9	−5.4	−3.7	−23.5	−40.3
Campherol	−9.7	−16.6	−20.1	−18.2	−38.2	−55.0
Myricetin	−17.7	−24.6	−28.1	−26.4	−46.2	−63.0

Table 15. Enthalpies (kJ/mol) of reactions $R_iO_2^• + Ar_jOH \rightarrow R_iOOH + Ar_jO^•$ (Eqn. 20)

5. References

Angel, L. & Ervin, K. (2004). Competitive Threshold Collision-Induced Dissociation: Gas-Phase Acidity and O−H Bond Dissociation Enthalpy of Phenol. *J. Phys. Chem. A*, Vol. 108, No. 40, 8346-8352, ISSN 1089-5639

Arnett, E., Amarnath, K., Harvey, N. & Venimadhavan, S. (1990). Heterolysis and Homolysis Energies of Some Carbon-Oxygen Bonds. *J. Am. Chem. Soc.*, Vol. 112, No. 20, 7346-7353, ISSN 0002-7863

Belyakov, V., Shanina, E., Roginsky, V. & Miller, V. (1975). O−H Bond Energies and the Inhibiting Action of Sterically Hindered Phenols. *Izv. Akad. Nauk SSSR, Ser. Khim. (Russian)*, No. 12, 2685-2691, ISSN 0002-3353

Bordwell, F. & Bausch, M. (1986). Acidity-Oxidation-Potential (AOP) Values as Estimates of Relative Bond Dissociation Energies and Radical Stabilities in Dimethyl Sulfoxide Solution. *J. Am. Chem. Soc.*, Vol. 108, No. 8, 1979-1985, ISSN 0002-7863

Bordwell, F. & Cheng, J.-P. (1991). Substituent Effects on the Stabilities of Phenoxyl Radicals and the Acidities of Phenoxyl Radical Cations. *J. Am. Chem. Soc.*, Vol. 113, No. 5, 1736-1743, ISSN 0002-7863

Bordwell, F., Cheng, J.-P., Ji, G.-Z., Satish, A. & Zhang, X. (1991). Bond Dissociation energies in DMSO Related to the Gas Phase. *J. Am. Chem. Soc.*, Vol. 113, No. 26, 9790-9795, ISSN 0002-7863

Bordwell, F. & Liu, W.-Z. (1996). Equilibrium Acidities and Homolytic Bond Dissociation Energies of N−H and/or O−H Bonds in N-Phenylhydroxylamine and Its Derivatives. *J. Am. Chem. Soc.*, Vol. 118, No. 37, 8777-8781, ISSN 0002-7863

Burton, G., Doba, T., Gabe, E., Hughes, L., Lee, F., Prasad, L. & Ingold, K. (1985). Autoxidation of Biological Molecules. 4. Maximizing the Antioxidant Activity of Phenols. *J. Am. Chem. Soc.*, Vol. 107, No. 24, 7053-7065, ISSN 0002-7863

Burton, G. & Ingold, K. (1986). Vitamin E: Application of the Principles of Physical Organic Chemistry to the Exploration of Its Structure and Function. *Acc. Chem. Res.*, Vol. 19, No. 7, 194-201, ISSN 0001-4842

Denisov, E. (1995a). A new semiempirical method of estimation of activity and bond dissociation energies of antioxidants. *Polymer Degradation and Stability*, Vol. 49, No. 1, 71-75, ISSN 0141-3910

Denisov, E. (1995b). Evaluation of the dissociation energies of the O–H bonds of phenols on the basis of kinetic measurements. *Russ. J. Phys. Chem.*, Vol. 69, No. 4, 563-574, ISSN 0036-0244

Denisov, E. (1997). New empirical models of radical abstraction reactions. *Russ. Chem. Rev.*, Vol. 66, No. 10, 859-876, ISSN 0036-021X

Denisov, E. (1999). Models for Abstraction and Addition Reactions of Free Radicals, In: *General Aspects of the Chemistry of Radicals.* Alfassi, Z. (Ed.), 79-137, Wiley and Sons, ISBN 0-471-98760-3, Chichester, UK

Denisov, E. & Afanas'ev, I. (2005). *Oxidation and Antioxidants in Organic Chemistry and Biology,* Taylor and Francis, ISBN 0-8247-5356-9, Boca Raton, FL

Denisov, E. & Azatyan, V. (2000). *Inhibition of Chain Reaction,* Gordon and Breach Science Publishers, ISBN 90-6994-002-7, London

Denisov, E. & Denisova, T. (1988). The Enthalpies of Formation of Peroxy Radicals and Dissociation Energies, of C–O and O–O Bonds in Peroxy Radicals and Hydroperoxides. *Zh. Fiz. Khim. (Russian)*, Vol. 62, No. 2, 304-309, ISSN 0044-4537

Denisov, E. & Denisova, T. (1993). Kinetic Parameters of the Reactions $RO_2^{\bullet}$ + RH in the Framework of the Parabolic Model of Transition State. *Kinet. Catal.*, Vol. 34, No. 2, 173-179, ISSN 0023-1584

Denisov, E. & Denisova, T. (2000). *Handbook of Antioxidants.* CRC Press, ISBN 0-8493-9004-4, Boca Raton, FL

Denisov, E. & Denisova, T. (2006). Kinetics and Thermodynamics of Formation and Degradation of α-Hydroperoxy Radicals. *Pet. Chem.*, Vol. 46, No. 6, 373-383, ISSN 0965-5441

Denisov, E. & Denisova, T. (2009). The Reactivity of Natural Phenols. *Russ. Chem. Rev.*, Vol. 78, No. 11, 1047-1073, ISSN 0036-021X

Denisov, E., Denisova, T. & Pokidova, T. (2003). *Handbook of Free Radical Initiators.* John Wiley & Sons, ISBN 0-471-20753-5, Hoboken, NJ

Denisov, E. & Khudyakov, I. (1987). Mechanisms of action and reactivities of the free radicals of inhibitors. *Chem. Rev.*, Vol. 87, No. 6, 1313-1357, ISSN 0009-2665

Denisov, E. & Kovalev, G. (1983). *Okislenie i Stabilizatsiya Reactivnykh Topliv (Oxidation and Stabilization of Jet Fuels),* Khimiya Publishers, Moscow (Russian)

Denisov, E. & Tumanov, V. (2005). Estimation of the bond dissociation energies from the kinetic characteristics of liquid-phase radical reactions. *Russ. Chem. Rev.*, Vol. 74, No. 9, 825-858, ISSN 0036-021X

Denisova, T. (2007). Kinetics and Thermodynamics of the Exchange Reactions of Peroxy radicals with Hydroperoxides. *Kinet. Catal. (Engl. Transl.)*, Vol. 48, No. 5, 609-614, ISSN 0023-1584

Denisova, T. & Denisov, E. (2004). Estimation of the O–H Bond Dissociation Energy from Kinetic Data for Hydroperoxides with Functional Group. *Pet. Chem.*, Vol. 44, No. 4, 250-255, ISSN 0965-5441

Denisova, T. & Denisov, E. (2008). Dissociation energies of O–H bonds in natural antioxidants. *Russ. Chem. Bull., Int. Ed.*, Vol. 57, No. 9, 1858-1866, ISSN 1066-5285

Emanuel, N., Denisov, E. & Maizus, Z. (1967). *Liquid-Phase Oxidation of Hydrocarbons*, Plenum Press, Number 66-12888, New York

Emanuel, N. & Lyaskovskaya, Yu. (1961). *Tormozhenie Protsessov Okisleniya Zhirov. (Inhibition of Fat Oxidation)*, Pische-promizdat, Moscow (Russian)

Foti, M., Amorati, R., Pedulli, G. F., Daquino, C., Pratt, D. & Ingold, K. (2010). Influence of "Remote" Intramolecular Hydrogen Bonds on the Stabilities of Phenoxyl Radicals and Benzyl Cations. *J. Org. Chem.*, Vol. 75, No. 13, 4434-4440, ISSN 0022-3263

Golden, D., Spokes, G. & Benson, S. (1973). Very Low-Pressure Pyrolysis (VLPP); A Versatile Kinetic Tool. *Angew. Chem. Int. Ed. Egl.*, Vol. 12, No. 7, 534-546, ISSN 0570-0833

Grabowski, J., Simon, J. & Peters, K. (1984). Heat of Formation of Diphenylcyclopropenone by Photoacoustic Calorimetry. *J. Am. Chem. Soc.*, Vol. 106, No. 16, 4615-4616, ISSN 0002-7863

Hamid, S. (Ed.), (2000). *Handbook of Polymer Degradation*, Marcel Dekker, ISBN 0-8247-0324-3, New York

de Heer, M., Mulder, P., Korth, H.-G., Ingold, K. & Lusztyk, J. (2000). Hydrogen Atom Abstraction Kinetics from Intramolecularly Hydrogen Bonded Ubiquinol-0 and other (Poly)methoxy Phenols. *J. Am. Chem. Soc.*, Vol. 122, No. 10, 2355-2360, ISSN 0002-7863

Holmes, J., Lossing, F. & Mayer, P. (1991). Heats of Formation of Oxygen Containing Organic Free Radicals from Apperange Energy Measurements. *J. Am. Chem. Soc.*, Vol. 113, No. 26, 9723-9728, ISSN 0002-7863

Howard, J. (1972). Absolute Rate Constants for Reactions of Oxyl Radicals, In: *Advances in Free Radical Chemistry*. Williams, G. (Ed.), Vol. 4, 49-173, Logos Press, ISBN 0236176625, London

Howard, J. & Ingold, K. (1965). The kinetics of inhibited oxidation of tetralin. *Can J. Chem.*, Vol. 43, No. 10, 2724-2728, ISSN 0008-4042

Howard, J., Schwalm, W. & Ingold, K. (1968). Absolute Rate Constants for Hydrocarbon Autoxidation. VII. Reactivities of Peroxy Radicals Toward Hydrocarbons and Hydroperoxides. In: *Oxidation of Organic Compounds-1; Adv. Chem. Ser. 75,* Gould, R. (Ed), *Proceedings of the International Oxidation Symposium*, Vol. 75, pp. 6-23, LCCC 967-7520, Stanford Research Institute, San Francisco, Calif. Aug. 28 - Sept. 1, 1967, American Chemical Society, Washinton

Huie, R. E. & Neta, P. (1997). Kinetic Studies of Organic Peroxyl Radicals in Aqueous Solutions and Mixed Solvents, In: *Peroxyl Radicals*. Alfassi, Z. B. (Ed.), 235-281, Wiley and Sons, ISBN-10: 0471970654, ISBN-13: 9780471970651, Chichester, UK

Jackson, R. & Hosseini, K. (1992). Phenol-Phenoxyl Radical Equilibria by Electron Spin Resonance: are Radicals Derived from Tocopherol and Analogues Exceptionally Stabilized? *J. Chem. Soc., Chem. Commun.*, No. 13, 967-968, ISSN 0022-4936

Kuliev, A. (1972). *Khimiya i Tekhnologiya Prisadok k Maslam i Toplivam (Chemistry and Technology of Additives to Oils and Fuels)*, Khimiya Publishers, 3-14-7-69-72, Moscow (Russian)

Laarhoven, L., Mulder, P. & Wayner, D. (1999). Determination of Bond Dissociation Enthalpies in Solution by Photoacoustic Calorimetry. *Acc. Chem. Res.*, Vol. 32, No. 4, 342-349, ISSN 0001-4842

Lide, D. (Ed.), (2004-2005). *Handbook of Chemistry and Physics*. 85 Edition. CRC Press, ISBN 0-8493-0485-7, Boca Raton, FL

Lind, J., Shen, X. Eriksen, T. & Merenyi, G. (1990). The One Electron Reduction Potencial of 4-Substituted Phenoxyl Radicals in Water. *J. Am. Chem. Soc.*, Vol. 112, No. 2, 479-482, ISSN 0002-7863

Lucarini, M., Mugnaini, V. & Pedulli, G. F. (2002). Bond Dissociation Enthalpies of Polyphenols: the Importance of Cooperative Effects. *J. Org. Chem.*, Vol. 67, No. 3, 928-931, ISSN 0022-3263

Lucarini, M. & Pedulli, G. F. (2010). Free radical intermediates in the inhibition of the autoxidation reaction. *Chem. Soc. Rev.*, Vol. 39, 2106-2119, ISSN 0306-0012

Lucarini, M., Pedulli, G. F. & Cipollone, M. (1994). Bond Dissociation Enthalpy of α-Tocopherol and Other Phenolic Antioxidants. *J. Org. Chem.*, Vol. 59, No. 17, 5063-5070, ISSN 0022-3263

Lucarini, M., Pedrielli, P., Pedulli, G. F., Cabiddu, S. & Fattuoni, C. (1996). Bond Dissociation Energies of O–H Bonds in Substituted Phenols from Equilibration Studies. *J. Org. Chem.*, Vol. 61, No. 26, 9259-9263, ISSN 0022-3263

Luo, Y-R. (2003). *Handbook of Bond Dissociation Energies in Organic Compounds,* CRC Press, ISBN 0-8493-1589-1, Boca Raton, FL

Luzhkov, V. (2005). Mechanisms of antioxidant activity: The DFT study of hydrogen abstraction from phenol and toluene by the hydroperoxyl radical. *Chem. Phys.*, Vol. 314, No. 1-3, 211-217, ISSN 0301-0104

Mahoney, L. & DaRooge, M. (1975). The Kinetic Behavior and Thermochemical Properties of Phenoxy Radicals. *J. Am. Chem. Soc.*, Vol. 97, No. 16, 4722-4731, ISSN 0002-7863

Mahoney, L. Ferris, F. & DaRooge, M. (1969). Calorimetric Study of the 2,4,6-Tri-t-butylphenoxy Radical in Solution. *J. Am. Chem. Soc.*, Vol. 91, No. 14, 3883-3889, ISSN 0002-7863

Mill, T. & Hendry, D. (1980). Kinetics and Mechanisms of Free Radical Oxidation of Alkanes and Olefins in the Liquid Phase, In: *Comprehensive Chemical Kinetics. Liquid-Phase Oxidation,* Bamford, C. & Tipper, C. (Eds.), Vol. 16, 1-87, Elsevier, ISBN 0-444-41860-1, Amsterdam

Mogilevich, M. & Pliss, E. (1990). *Okislenie i Okislitel'naya Polimerizatsiya Nepredel'nykh Soedinenii (Oxidation and Oxidative Polymerisation of Unsaturated Compounds),* Khimiya Publishers, ISBN 5-7245-0564-9, Moscow (Russian)

Mulder, P., Korth, H-G., Pratt, D., DiLabio, G., Valgimigly, L., Pedulli, G.F. & Ingold, K. (2005). Critical Re-evaluation of the O–H Bond Dissociation Enthalpy in Phenol. *J. Phys. Chem. A*, Vol. 109, No. 11, 2647-2655, ISSN 1089-5639

Pratt, D., de Heer, M., Mulder, P. & Ingold, K. (2001). Oxygen-Carbon Bond Dissociation Enthalpies of Benzyl Phenyl Ethers and Anisoles. An Example of Temperature Dependent Substituent Effects. *J. Am. Chem. Soc.*, Vol. 123, No. 23, 5518-5526, ISSN 0002-7863

Pospisil, J. & Klemchuk, P. (Eds.), (1990). *Oxidation Inhibition in Organic Materials,* Vol. 1-2, CRC Press, ISBN-10: 08493-4767X ISBN-13: 9780849347672, Boca Raton, FL

Pozdeeva, N., Yakushchenko, I., Aleksandrov, A., Luzhkov, V. & Denisov, E. (1989). Reactivity of Crown-phenol and Its Phenolate in the Reaction with Styrene Peroxy Radicals, Role of the Hydrogen Bond. *Kinet. Catal.* Vol. 30, No. 1, 22-28, ISSN 0023-1584

Robinson, P. & Holbrook, K. (1972). *Unimolecular Reactions*, Wiley-Interscience, ISBN-10: 0471728144, ISBN-13: 978-04-71728146, London

Roginsky, V. (1988). *Phenol'nye Antioksidanty (Phenolic Antioxidants), Nauka*, ISBN 5-02-001416-8, Moscow (Russian).

Rothberg, L., Simon, J., Bernstein, M. & Peters, K. (1983). Pulsed Laser Photoacoustic Calorimetry of Metastable Species. *J. Am. Chem. Soc.*, Vol. 105, No. 11, 3464-3468, ISSN 0002-7863

Scott, G. (Ed.) (1980). *Developments in Polymer Stabilisation-3*, Vol. 3, Applied Science Publishers, ISBN 0-85334-890-1, London

Scott, G. (Ed.) (1993). *Mechanisms of Polymer Degradation and Stabilisation*, Elsevier Applied Science, ISBN 1-85166-505-6, London

Simon, J. & Peters, K. (1983). Determination of the Heat of Reaction for the Formation Diphenylcarbene from Diphenyl-diazomethane using Photoacoustic Calorimetry. *J. Am. Chem. Soc.*, Vol. 105, No.15, 5156-5158, ISSN 0002-7863

Suryan, M., Kafafi, S. & Stein, S. (1989a). The Thermal Decomposition of Hydroxyl- and Methoxy-substituted Anisoles. *J. Am. Chem. Soc.*, Vol. 111, No. 4, 1423-1436, ISSN 0002-7863

Suryan, M., Kafafi, S. & Stein, S. (1989b). Dissociation of Substituted Anisoles: Substituent Effects on Bond Strengths. *J. Am. Chem. Soc.*, Vol. 111, No. 13, 4594-4600, ISSN 0002-7863

DeTuri, V. & Ervin, K. (1998). Proton transfer between Cl^- and C_6H_5O–H. O–H bond energy of phenol. *Int. J. Mass Spec-trom. Ion Proc.*, Vol. 175, 123-132, ISSN 0168-1176

Walker, J. & Tsang, W. (1990). Single-pulse Shock Tube Studies on the Thermal Decomposition of *n*-Butyl Phenyl Ether, *n*-Pentylbenzene and Phenotole and the Heat of Formation of Phenoxy and Benzyl Radicals. *J. Phys. Chem.*, Vol. 94, No. 8, 3324-3327, ISSN 0022-3654

Wayner, D., Lusztyk, E., Ingold, K. & Mulder, P. (1996). Application of Photoacoustic Calorimetry to the Measurement of the O–H Bond Strength in Vitamin E (α- and δ-tocopherol) and Related Phenolic Antioxidants. *J. Org. Chem.*, Vol. 61, No. 18, 6430-6433, ISSN 0022-3263

Wayner, D., Lusztyk, E., Page, D., Ingold, K., Mulder, P., Laarhoven, L. & Aldrich, H. (1995). Effects of Solvation on the Enthalpies of Reaction of *tert*-Butyl Radicals with Phenol and on the Calculated O–H Bond Strength in Phenol. *J. Am. Chem. Soc.*, Vol. 117, No. 34, 8737-8744, ISSN 0002-7863

Wright, J., Carpenter, D., McKay, D. & Ingold, K. (1997). Theoretical Calculation of Substituent Effects on the O–H Bond Strength of Phenolic Antioxidant Related to Vitamin E. *J. Am. Chem. Soc.*, Vol. 119, No. 18, 4245-4252, ISSN 0002-7863

Zhu, L. & Bozzelli, J. (2003). Kinetic and Thermochemistry for the Gas-Phase Keto-Enol Tautomerism of Phenol <=> 2,4- Cyclohexadienone. *J. Phys. Chem. A*, Vol. 107, No. 19, 3696-3703, ISSN 1089-5639

16

Determination of the Constants of Formation of Complexes of Iron(III) and Acetohydroxamic Acid

Fabrice PL Andrieux[1], Colin Boxall[1] and Robin J Taylor[2]
[1]*Lancaster University*
[2]*National Nuclear Laboratory*
UK

1. Introduction

Simple hydroxamic acids (XHA) are hydrophilic organic compounds with the general formula RCONHOH that can act as O,O donor ligands with high affinities for hard cations such as Fe^{3+}, Np^{4+} and Pu^{4+} with which they form 5-membered chelate rings (Muri et al.,2002; Desaraju & Winston,1986; Carrott et al.,2007).

They have a wide range of applications in many fields, including as enzyme inhibitors, soil enhancers, fungicides, mutagens, carcinogens, DNA cleavers, drug delivery systems (Ghosh,1997), ion exchange (Vernon,1982) and materials research (Ghosh,1997). Siderophores such as the desferrioxamines contain multiple hydroxamate functionalities and are naturally occurring ligands specifically used by fungi for the sequestration of iron from the environment (Raymond et al.,1984; Renshaw et al.,2002; Monzyk & Crumbliss,1979). The potential role of such hydroxamate siderophores in the mobilisation of actinide ions within the environment has also been considered (John et al.,2001).

More recently, XHAs have also been identified as useful reagents for the control of Pu and Np in advanced PUREX and UREX processes proposed for the processing of spent nuclear fuel (Birkett et al.,2005). In such separation processes, U, Np and Pu are separated according to oxidation state specific aqueous/non aqueous solvent extraction in the presence of acetohydroxamic acid (AHA) (Carrott et al.,2007). In this context, Fe(III) is a useful non-radioactive analogue for the interrogation of the complexation behaviour of actinides.

It is also well known that hydroxamic acids are susceptible to hydrolysis, particularly in acidic solutions (Ghosh,1997). Whilst there have been many studies of hydroxamic acid hydrolysis and their complexation reactions with metal ions, there have been relatively few studies of the stability of the metal-hydroxamate complexes towards hydrolysis. We have previously reported on the kinetics of the hydrolysis of hydroxamic acids both in free solution and when bound to metal ions and developed a kinetic model describing this process at room temperature (Andrieux et al.,2007; Andrieux et al.,2008; Carrott et al.,2008). However, the applications of hydroxamic acids in biological-related fields requires an understanding of the behaviour of these systems at temperatures of biological interest, for example in vivo (310K) or temperatures at which proteins start to denature ($\sim$ 325K) (Roos,2006). Additionally, the temperature dependencies of the equilibrium constants are useful in gaining a better understanding of the complexation processes and the role of enthalpy and entropy in complex formation. As a precursor to any temperature-dependent kinetic study of Fe(III)-AHA

hydrolysis, comprehensive and reliable thermodynamic data are required on the Fe(III)-AHA complexation reactions, which is the focus of this study.
Furthermore, we are applying the kinetic models developed for the Fe(III)-AHA system to the kinetics of the hydrolysis of hydroxamic acids when bound to actinide ions such as Pu^{4+} and Np^{4+} in support of advanced PUREX and UREX processes (Andrieux et al.,2008).
Thus, a study of the temperature dependence of the speciation of the Fe^{3+}-AHA system will both provide a useful addition to the thermodynamic database of Fe(III) in its own right and support advanced PUREX/UREX process development through its role as a non-active analogue of e.g. the Np^{4+}-AHA system.
As mentioned above, hydroxamic acids readily form stable complexes with Fe^{3+} ions according to reactions described in eqs. (1)-(3):

$$Fe^{3+} + HL \rightleftharpoons FeL^{2+} + H^+ \qquad K_1 = \frac{[H^+][FeL^{2+}]}{[Fe^{3+}][HL]} \tag{1}$$

$$FeL^{2+} + HL \rightleftharpoons FeL_2^+ + H^+ \qquad K_2 = \frac{[H^+][FeL_2^+]}{[FeL^{2+}][HL]} \tag{2}$$

$$FeL_2^+ + HL \rightleftharpoons FeL_3 + H^+ \qquad K_3 = \frac{[H^+][FeL_3]}{[FeL_2^+][HL]} \tag{3}$$

where L represents the deprotonated acetohydroxamato ligand and all equilibrium constants are calculated from concentrations rather than activities.
Values of the equilibrium constant for the formation of the monoacetohydroxamatoiron(III) complex were reported in the literature with Monzyk et al. suggesting K_1 = 109 (Monzyk & Crumbliss,1979) without specifying an equilibrium temperature, while Schwarzenbach et al. report a value for the log K_1 = 2.28 at 293 K (Schwarzenbach & Schwarzenbach,1963), corresponding to K_1 = 190.5. The inconsistency of these values together with a lack of data pertaining to their temperature dependence has prompted us to conduct our own determination of values for K_1, K_2 and K_3. This, we have done through measurement of the electronic absorption spectra of a range of mixtures of the three Fe(III)-AHA complexes and the development of a simple analytical method for the extraction of values of equilibrium constants from those spectra. We have also deployed this method for the measurement of the temperature dependence of these equilibrium constants prior to the construction of useful temperature-dependent speciation diagrams.

2. Materials and methods

2.1 Materials

All reagents, including HNO_3 (70%, AnalaR, BDH Chemicals Ltd., Poole, Dorset, UK) and AHA (Sigma-Aldrich Ltd, UK) were obtained from reputable suppliers (Aldrich, Fluka) at the highest available purity and used as received. AHA was stored in a conventional refrigerator at 4°C in order to prevent its decomposition. Solutions were prepared using doubly distilled water, produced by a home-made still and further purified by a deionisation system (E pure model 04642, Barnstead/Thermodyne, Dubuque, Iowa, USA) to a resistivity of 1.8×10^5 Ω m.

2.2 Methods

UV-visible absorbance spectra of complex and complex-precursor solutions were measured by spectrophotometry (Diode Array model 8452A, Hewlett Packard, USA) fitted with a

HP89090A Peltier temperature controller. Solutions for spectroscopic measurement were prepared from stock (Fe(III)) or freshly made solutions (AHA, prepared immediately prior to experiment) and pipetted into an optical cuvette (pathlength 1 cm). The complexant (AHA) was added last. The volume of complexant required was always small and had no effect on the temperature of the receiving solution when added. The spectra were then measured as soon as possible after mixing. All experiments were performed at pH < 1 in order to minimise hydrolysis of free ferric ions and the formation of iron hydroxides and oxyhydroxides, as discussed in the following section.

3. Results and discussion

3.1 Iron hydrolysis

It is well known that Fe^{3+} readily hydrolyses in water in accordance with:

$$Fe^{3+} + H_2O \rightleftharpoons FeOH^{2+} + H^+ \tag{4}$$

the first step in a series of deprotonation / polymerisation reactions that ultimately result in the formation of ferric hydroxides and oxyhydroxides. Using the most recently published, critically assessed Gibbs energies of formation of the various species involved (Cornell & Schwertmann,1996; Heusler & Lorenz,1985), the equilibrium constant for this first deprotonation can be calculated as being equal to 6.78 x 10^{-3} at 298 K, corresponding to a $pK_a(Fe^{3+} / FeOH^{2+})$ of 2.17.
Figure 1 shows the speciation of the two iron-containing species of equation (4) as a function of pH, calculated at T=298K for a total iron(III) concentration of 0.5 mol dm^{-3} and an ionic strength of 0.1 mol dm^{-3}. From this, it may be seen that $FeOH^{2+}$ will become the dominant species above pH=2.17, but that it will also be present in significant amounts at lower pH values. In fact, 10% of all iron(III) is present as the hydroxide at pH values as low as 1.2. In order to facilitate subsequent calculations, it is therefore advisable to carry out spectroscopic measurements at pH values up to 1 (where no more than 6.3% of all iron(III) is present as the hydroxide).

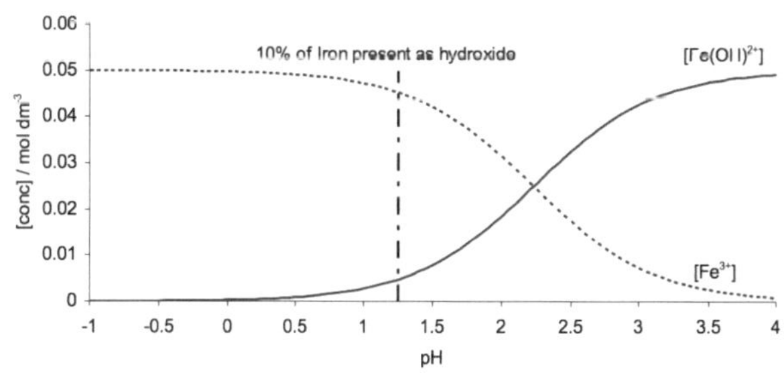

Fig. 1. Speciation diagram for Fe^{3+} and $FeOH^{2+}$ as a function of pH for a total iron(III) concentration of 50x10^{-3} mol dm^{-3} calculated at an ionic strength of 0.1 mol dm^{-3}

The equilibrium constant for the reaction shown in Eq. 4 has been studied as a function of temperature (T = 298-353 K) and ionic strength (I = 0.1-2.67 mol dm^{-3}) in perchlorate media by Sapieszko et al. (Sapieszko et al.,1977). Using their data, we have calculated that, in order to prevent significant formation of hydrolysed iron(III) under the conditions of our experiments (T = 293-323 K, and I $\sim$ 0.1 mol dm^{-3}), the value of the pH to be used in the determination of the equilibrium constants of formation of the acetohydroxamatoiron(III) complexes should not exceed 1 at T $>$ 300 K. Any data presented at pH $>$ 1 at T $>$ 300 K is shown for the purposes of context only.

3.2 General methodology

The general approach used in the determination of the thermodynamic constants of formation of the various iron(III)-AHA complexes may be described by the following. In a solution containing any amount of iron(III) and acetohydroxamic acid, the concentrations of the various species in solution as a result of the establishment of the equilibria described in equation (1) to (3) may be expressed as follows:

$$[FeL^{2+}] = \xi_1 \tag{5}$$

$$[FeL_2^+] = \xi_2 \tag{6}$$

$$[FeL_3] = \xi_3 \tag{7}$$

$$[Fe^{3+}] = [Fe^{3+}]_i - \xi_1 - \xi_2 - \xi_3 \tag{8}$$

$$[HL] = [HL]_i - \xi_1 - 2\xi_2 - 3\xi_3 \tag{9}$$

$$[H^+] = [H^+]_i + \xi_1 + 2\xi_2 + 3\xi_3 \tag{10}$$

where $[Fe^{3+}]_i$, $[HL]_i$ and $[H^+]_i$ represent the initial concentration of iron(III), acetohydroxamic acid and nitric acid, respectively.

These may then be used to express the equilibrium constants for the formation of the various complexes in terms of ξ_1, ξ_2 and ξ_3 as the only unknown parameters, and the resulting equations may be solved as simple polynomial expressions. The following section describes the hypotheses made and the specific methodology used in the determination of the equilibrium constant of formation of the monoacetohydroxamatoiron (III) complex.

3.3 Thermodynamics of monoacetohydroxamatoiron (III) complex formation

The determination of K_1 is best carried out under conditions where the monoacetohydroxamatoiron(III) complex is present in a vast excess compared to both the bis- and trisacetohydroxamatoiron(III) complexes, i.e. in the presence of a very large excess of iron compared to the hydroxamate. Under such conditions, equation (1) may be expressed in terms of the parameters defined in equations (5)-(10) as:

$$K_1 = \frac{\xi_1([H^+]_i + \xi_1)}{([Fe^{3+}]_i - \xi_1)([HL]_i - \xi_1)} \tag{11}$$

which contains two unknown parameters K_1 and ξ_1. In order to solve equation (11) it is then necessary to use two data sets for each temperature considered. This may be achieved by varying the pH of the solution, thereby altering the concentration of the complex. It is then possible, from the spectroscopic data to express the complex concentration of one of the samples as a function of the complex concentration of the other, *via*:

$$\frac{\xi_{1_1}}{\xi_{1_2}} = \frac{Abs_1}{Abs_2} = x \tag{12}$$

where the very last digit of the subscript in each term refers to the arbitrarily assigned experiment number and x is the ratio of the absorbance due to the mono-complex in each of the samples considered. For each of the two experiments, equation (11) may then be re-written as:

$$K_1 = \frac{\xi_{1_1}([H^+]_{i_1} + \xi_{1_1})}{([Fe^{3+}]_{i_1} - \xi_{1_1})([HL]_{i_1} - \xi_{1_1})} \quad \text{for experiment 1} \tag{13}$$

and

$$K_1 = \frac{x\xi_{1_1}([H^+]_{i_2} + x\xi_{1_1})}{([Fe^{3+}]_{i_2} - x\xi_{1_1})([HL]_{i_2} - x\xi_{1_1})} \quad \text{for experiment 2} \tag{14}$$

These two equations may then be equated and re-arranged to give the polynomial expression of the second order shown in equation (15).

$$\begin{aligned} &\xi_{1_1}\left([H^+]_{i_1} + \xi_{1_1}\right)\left([Fe^{3+}]_{i_2} - x\xi_{1_1}\right)\left([HL]_{i_2} - x\xi_{1_1}\right) - \\ &x\xi_{1_1}\left([H^+]_{i_2} + x\xi_{1_1}\right)\left([Fe^{3+}]_{i_1} - \xi_{1_1}\right)\left([HL]_{i_1} - \xi_{1_1}\right) = 0 \end{aligned} \tag{15}$$

Solution of equation (15) yields a value of ξ_{1_1} and enables the calculation of K_1. Three data sets are sufficient to calculate three values of K_1 allowing for an average to be taken. Figure 2 shows the UV-visible spectra recorded from three solutions at different pH values at 313K. From that, it may be seen that the λ_{max} value is independent of pH.

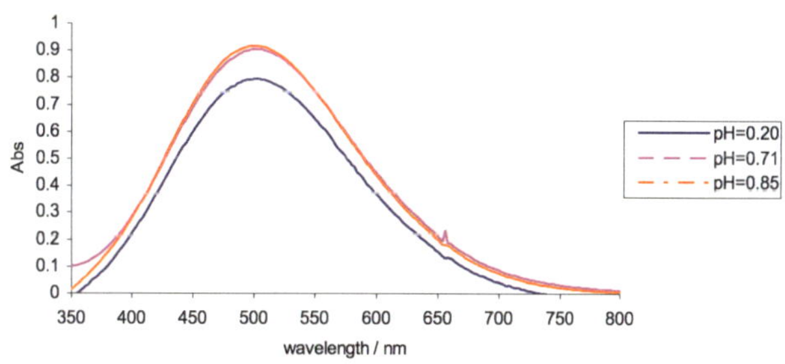

Fig. 2. Typical UV-visible spectra of a solution containing the monoacetohydroxamatoiron (III) complex (initial solution composition: [Fe^{3+} = 50x10^{-3} mol dm^{-3}, [HL] = 10^{-3} mol dm^{-3}) recoded at T = 313K and pH = 0.20, 0.71 and 0.85 respectively. λ_{max} = 510nm.

Table 1 shows the absorbance of three solutions containing the monoacetohydroxamatoiron (III) complex at three different pH values at the range of temperature considered, together with the calculated average and standard deviation values of K_1 as a function of temperature. One notation of the Van't Hoff isochore, equation (16), predicts that Ln(K_1) should be linearly dependent on T^{-1}. Figure 3 shows such a plot for the data of Table 1.

$$Ln(K) = -\frac{\Delta H^o}{RT} + \frac{\Delta S^o}{R} \tag{16}$$

T/K	Abs @ 510nm, pH = 0.20	Abs @ 510nm, pH = 0.71	Abs @ 510nm, pH=0.85	Average K_1	Standard deviation
293	0.803	0.921	0.912	65.25	0.68
298	0.793	0.916	0.902	40.65	5.79
303	0.793	0.912	0.889	28.22	11.4
313	0.779	0.894	0.877	14.58	4.17
323	0.667	0.858	0.842	4.24	0.54

Table 1. Determination of K_1 in accordance with equation (15) as a function of temperature, calculated from solutions with $[Fe^{3+}]_i$ = 50x10^{-3} mol dm^{-3} and $[HL]_i = 10^{-3}$ mol dm^{-3}

where R is the universal gas constant and ΔH^o and ΔS^o are the standard enthalpy and standard entropy of reaction, respectively, assumed to be constant over the temperature range of our experiments.

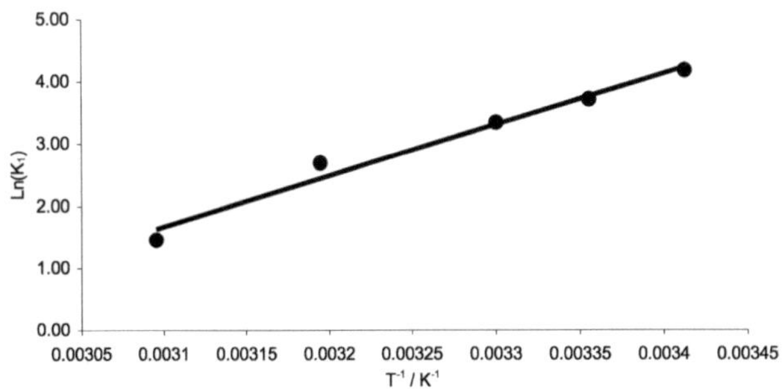

Fig. 3. Ln(K_1) vs. T^{-1} for the formation of the monoacetohydroxamatoiron(III) complex. Regression line : $Ln(K_1) = 8253T^{-1} - 23.934; R^2 = 0.9783$

The good linearity of the plot in figure 3 supports our earlier assumption that ΔH^o and ΔS^o are invariant with regards to temperature in the range 293-323K.
From the regression line of figure 3, and according to equation (16), the standard enthaply of reaction is found to be -68.6 kJ mol^{-1} and the standard entropy of reaction is found to be -199.0 J mol^{-1} K^{-1}. The negative value of ΔH^o indicates that the complexation process is exothermic in nature, while the negative value of ΔS^o suggests that the process is accompanied by a loss of entropy and is therefore enthalpically driven.

3.3.1 Extinction coefficient determination

In addition to having determined a reliable value of K_1 over the range of temperature considered, the determination of the various concentrations of the monoacetohydroxamatoiron(III) complex at a range of pH values and temperatures affords the possibility of calculating the extinction coefficient of the former, ϵ_1, by use of Beer's Law:

$$A = \epsilon l \xi_1 \tag{17}$$

where A is the absorbance of the sample at the wavelength of interest (here λ_{max} = 510nm) and l the pathlength of the cell (1 cm). The extinction coefficient of the monoacetohydroxamatoiron(III) complex is then found to be $\epsilon_1 = 1005 \pm 13$ dm^3 mol $^{-1}$ cm^{-1}, with all calculated values being within 1.5% of the calculated average, an acceptable level of experimental error.

3.4 Thermodynamics of bisacetohydroxamatoiron (III) complex formation

The determination of the value of K_2 necessitates the preparation of solutions where the bisacetohydroxamatoiron(III) complex (FeL_2^+) is in equilibrium with either the monoacetohydroxamatoiron(III) or the trisacetohydroxamatoiron(III) (FeL_3) complex. This presents an additional difficulty compared to the determination of K_1 as it is not possible to consider both ξ_1 and ξ_3 to be negligible compared to ξ_2. However, knowledge of the individual absorption spectra and the extinction coefficient of the monoacetohydroxamatoiron(III) complex provide us with a simple means of deconvoluting spectrometric data relating to solutions containing a mixture of the mono- and bis- complexes and alleviating this issue. The description of the method used forms the basis of the next section.

3.4.1 Deconvolution of spectrometric data

The spectrum of a solution containing a mixture may be deconvoluted *via* the exploitation of the law of additivity of absorbance, which states that as long as two or more species do not interact chemically with one another the absorbance of the mixture (A_T) is equal to the sum of the absorbances of each individual species, according to:

$$A_T = \sum_i A_i = \sum_i \epsilon_i \, c_i \, l \tag{18}$$

where c is the concentration of each individual species i. Using this principle, the spectrum of the FeL_2^+ may be obtained by subtracting the spectrum of FeL^{2+}, multiplied by a scaling factor, from the measured spectrum of the mixture. The scaling factor is adjusted until the λ_{max} of the deconvoluted spectrum of the bisacetohydroxamatoiron(III) complex matches the literature values of 470nm (Desaraju & Winston,1986). This technique assumes that the concentration of FeL_3 is negligible compared to that of the two other complexes. Figure 4 shows the typical deconvolution of a spectrum recorded for a solution containing 5x10^{-4} mol dm^{-3} iron(III) and 0.2 mol dm^{-3} HL at pH=0.61, T=298K.

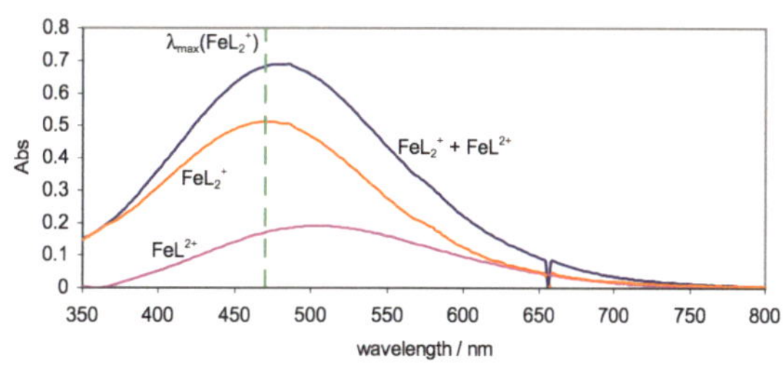

Fig. 4. Typical deconvolution of a spectrum recorded for a solution containing 5x10^{-4} mol dm^{-3} iron(III) and 0.2 mol dm^{-3} HL at pH=0.61, T=298K

3.4.2 Determination of K_2

The formation of the bisacetohydroxamatoiron(III) complex may be envisaged as either the stepwise formation of FeL^{2+} followed by the formation of FeL_2^+ or the simultaneous formation of a mixture of both complexes, the latter resulting in the definition of a new equilibrium constant for the formation of the bisacetohydroxamatoiron(III) complex from acetohydroxamic acid and iron(III), K_2'. Although this leads to a more complex mathematical expression, it is the preferred option in the determination of ξ_2 as it explicitly takes into account the presence of free iron(III) as well as that of the monoacetohydroxamatoiron(III) complex .

$$K_2' = K_1 K_2 = \frac{[FeL_2^+][H^+]^2}{[Fe^{3+}][HL]^2} \tag{19}$$

equation (19) may be expressed in terms of the parameters defined in equations (5)-(10), under conditions where ξ_3 is negligible as:

$$K_2' = \frac{\xi_2\left([H^+]_i + \xi_1 + 2\xi_2\right)^2}{\left([Fe^{3+}]_i - \xi_1 - \xi_2\right)\left([HL]_i - \xi_1 - 2\xi_2\right)^2} \tag{20}$$

Using the extinction coefficient for the monoacetohydroxamatoiron(III) complex and the deconvoluted spectra for the mixture of mono- and bisacetohydroxamatoiron(III), it is possible to extract values of ξ_1 for all data recorded, leaving equation (20) with only two unknown parameters (K_2' and ξ_2). Using an approach similar to that used in the determination of K_1, it is possible to use the spectra recorded for two samples at the same temperature but different pH values to extract values of ξ_2 and enable the determination of K_2. Here the ratio of concentration is defined as y:

$$\frac{\xi_{2_1}}{\xi_{2_2}} = \frac{Abs_1}{Abs_2} = y \tag{21}$$

where the very last digit of the subscript in each term refers to the arbitrarily attributed experiment number. This results in the establishment of the soluble system of two equations with two unknowns:

$$K_2' = \frac{\xi_{2_1}\left([H^+]_{i_1} + \xi_{1_1} + 2\xi_{2_1}\right)^2}{\left([Fe^{3+}]_{i_1} - \xi_{1_1} - \xi_{2_1}\right)\left([HL]_{i_1} - \xi_{1_1} - 2\xi_{2_1}\right)^2} \tag{22}$$

$$K_2' = \frac{y\xi_{2_1}\left([H^+]_{i_2} + \xi_{1_2} + 2y\xi_{2_1}\right)^2}{\left([Fe^{3+}]_{i_2} - \xi_{1_2} - y\xi_{2_1}\right)\left([HL]_{i_2} - \xi_{1_2} - 2y\xi_{2_1}\right)^2} \tag{23}$$

which may be equated and re-arranged to yield the 6th order polynomial expression:

$$\xi_{2_1}\left([H^+]_{i_1} + \xi_{1_1} + 2\xi_{2_1}\right)^2 \left([Fe^{3+}]_{i_2} - \xi_{1_2} - y\xi_{2_1}\right)\left([HL]_{i_2} - \xi_{1_2} - 2y\xi_{2_1}\right)^2 -$$
$$y\xi_{2_1}\left([H^+]_{i_2} + \xi_{1_2} + 2y\xi_{2_1}\right)^2 \left([Fe^{3+}]_{i_1} - \xi_{1_1} - \xi_{2_2}\right)\left([HL]_{i_1} - \xi_{1_1} - 2\xi_{2_1}\right)^2 = 0 \tag{24}$$

Equation (24) may not be trivially solved analytically but may be solved by a straightforward computation. A spreadsheet calculating the result of the lefthand side of equation (24) as a function of ξ_{2_1} is generated and the roots are inspected to identify the single root corresponding to a meaningful value for the concentration of FeL_2^+ (i.e. $0 < \xi_{2_1} \leq [Fe^{3+}]_{i_1}$). This value is then used to calculate the equilibrium concentrations of all other species in solution according to equations (5)-(10) enabling the determination of K_2 according to equation (2). Three data sets allow for three values of K_2 to be calculated and an average to be taken. Table 2 shows the values of the absorbance of the mono- and bisacetohydroxamatoiron(III) complex from the deconvoluted data, together with the calculated average and standard deviation values of K_2.

T/K	Abs_{mono} @ 510nm			Abs_{bis} @ 470nm			Average K_2	Standard Deviation
	pH=0.40	pH=0.61	pH=0.74	pH=0.40	pH=0.61	pH=0.74		
293	0.419	0.199	0.080	0.338	0.505	0.696	2.638	2.055
298	0.449	0.190	0.085	0.317	0.512	0.691	2.561	1.934
303	0.449	0.199	0.090	0.314	0.502	0.686	2.340	1.782
313	0.469	0.239	0.120	0.285	0.464	0.656	1.759	1.175
323	0.489	0.259	0.140	0.241	0.431	0.629	1.438	0.9488

Table 2. Determination of K_2 in accordance with equation (24) and (2) as a function of temperature calculated from solution of initial composition: $[Fe^{3+}] = 5x10^{-4}$ mol dm^{-3} and [HL] = 0.2 mol dm^{-3}

Figure 5 shows the data of table 2 plotted in accordance with equation (16) assuming that the values of ΔH^o and ΔS^o are temperature-independent over the range of temperatures considered. This then enables the determination of the standard enthalpy and entropy of formation of the bisacetohydroxamatoiron(III) complex from the monoacetohydroxamatoiron(III) complex and acetohydroxamic acid; ΔH^o = -17.09 kJ mol^{-1} and ΔS^o = -49.7 J mol^{-1} K^{-1}. The negative values obtained for both parameters indicates that the formation of the bisacetohydroxamatoiron(III) complex is exothermic and enthalpically driven, as was the case for the monoacetohydroxamatoiron(III) complex.

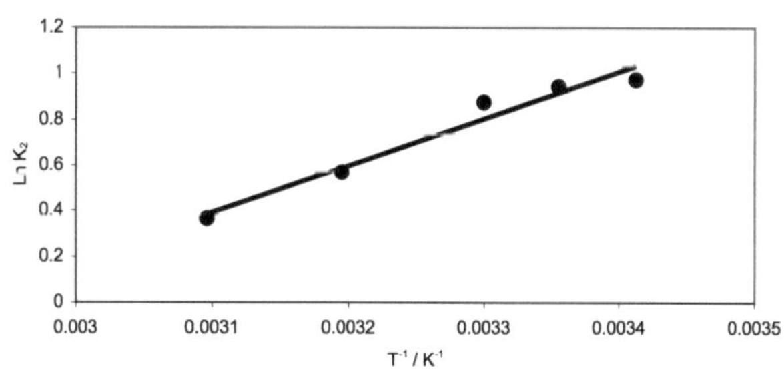

Fig. 5. LnK_2 vs. T^{-1} for the formation of the bisacetohydroxamatoiron(III) complex. Regression line : $Ln(K_2) = 2055T^{-1} - 5.982; R^2 = 0.9619$

3.4.3 Extinction coefficient determination

The deconvolution of the spectra recorded in this study, together with the determination of the concentration of the bishydroxamatoiron(III) complex in each solution, allows us to determine the extinction coefficient for this complex at λ_{max} = 470nm. This is done through the use of the ξ_2 equivalent of Beer's law shown in equation (17). The extinction coefficient for the bisacetohydroxamatoiron(III) complex was found to be $\epsilon_2 = 1650 \pm 150$ mol dm^{-3} cm^{-1}.

3.5 Thermodynamics of trisacetohydroxamatoiron (III) complex formation

Using a strategy analogous to that used in the determination of K_2, it is possible to determine a value of K_3. Again, to simplify the mathematics, solutions of acetohydroxamic acid and iron(III) are prepared that contain a mixture of the bis- and trisacetohydroxamatoiron(III) complex only (i.e. ξ_1 is negligible compared to ξ_2 and ξ_3). It is worth noting here that solutions containing only the trisacetohydroxamatoiron(III) complex, which would further simplify the mathematical treatment of the data, would require the dissolution of a quantity of acetohydroxamic acid at a level above its water solubility making this option unviable. Figure 6 shows a typical spectra recorded for this experiment, where the signals from the bis- and trisacetohydroxamatoiron(III) complexes have been deconvoluted using the methodology described in section 3.4.1, i.e. by subtracting the normalised averaged spectra of FeL_2^+ from the real solution spectra in order to isolate that of the trisacetohydroxamatoiron(III) complex. The literature value of the wavelength of maximum absorption for the latter (λ_{max} = 420nm) (Renshaw et al.,2002) was used to determine the value of the scaling factor.

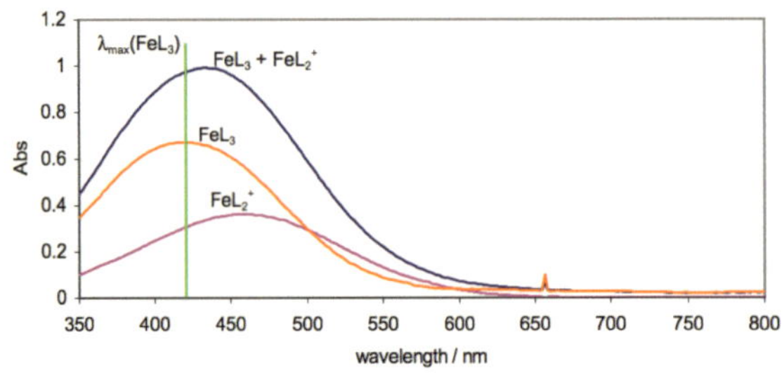

Fig. 6. Typical deconvolution of a spectrum recorded for a solution containing 4x10^{-4} mol dm^{-3} iron(III) and 6.3 mol dm^{-3} HL at pH = 0.84, T = 298K

Again, the formation of the trisacetohydroxamatoiron(III) complex may be envisaged as a series of reactions taking place sequentially and resulting in the formation of complexes of increasing order or as a one step process resulting in the formation of a mixture of complexes. Whilst the former would result in a simpler mathematical expression, the latter explicitly takes into account the presence of all species. It does however necessitate the definition of a thermodynamic equilibrium constant for the formation of the trisacetohydroxamatoiron(III) complex from iron(III) and acetohydroxamic acid, K_3':

$$K_3' = K_1 K_2 K_3 = \frac{[FeL_3][H^+]^3}{[Fe^{3+}][HL]^3} \tag{25}$$

which may be expressed as a function of the equilibrium concentrations defined in equations (5)-(10) and under conditions where ξ_1 is negligible as:

$$K_3' = \frac{\xi_3 \left([H^+]_i + 2\xi_2 + 3\xi_3\right)^3}{\left([Fe^{3+}]_i - \xi_2 - \xi_3\right)\left([HL]_i - 2\xi_2 - 3\xi_3\right)^3} \tag{26}$$

Using the extinction coefficient for the bisacetohydroxamatoiron(III) complex and the deconvoluted spectra for the mixture of bis- and trisacetohydroxamatoiron(III), it is possible to extract values of ξ_2 for all data recorded, leaving equation (26) with only two unknown parameters (K_3' and ξ_3). Using the approach used in the determination of K_2 and K_1, it is possible to use the spectra recorded for two samples at the same temperature but different pH values to extract values of ξ_3 and enable the determination of K_3. Here the ratio of concentration is defined as z:

$$\frac{\xi_{3_1}}{\xi_{3_2}} = \frac{Abs_1}{Abs_2} = z \tag{27}$$

where the very last digit of the subscript in each term refers to the arbitrarily attributed experiment number. This results in the establishment of the soluble system of two equations with two unknowns:

$$K_3' = \frac{\xi_{3_1}\left([H^+]_{i_1} + 2\xi_{2_1} + 3\xi_{3_1}\right)^3}{\left([Fe^{3+}]_{i_1} - \xi_{2_1} - \xi_{3_1}\right)\left([HL]_{i_1} - 2\xi_{2_1} - 3\xi_{3_1}\right)^3} \tag{28}$$

$$K_3' = \frac{z\xi_{3_1}\left([H^+]_{i_2} + 2\xi_{2_2} + 3z\xi_{3_1}\right)^3}{\left([Fe^{3+}]_{i_2} - \xi_{2_2} - z\xi_{3_1}\right)\left([HL]_{i_2} - 2\xi_{2_2} - 3z\xi_{3_1}\right)^3} \tag{29}$$

which may be equated and re-arranged to yield the 8th order polynomial expression:

$$\begin{aligned}&\xi_{3_1}\left([H^+]_{i_1} + 2\xi_{2_1} + 3\xi_{3_1}\right)^3\left([Fe^{3+}]_{i_2} - \xi_{2_2} - z\xi_{3_1}\right)\left([HL]_{i_2} - 2\xi_{2_2} - 3z\xi_{3_1}\right)^3 - \\ &z\xi_{3_1}\left([H^+]_{i_2} + 2\xi_{2_2} + 3z\xi_{3_1}\right)^3\left([Fe^{3+}]_{i_1} - \xi_{2_1} - \xi_{3_1}\right)\left([HL]_{i_1} - 2\xi_{2_1} - 3\xi_{3_1}\right)^3 = 0\end{aligned} \tag{30}$$

Once again, an analytical solution to equation (30) may not be found trivially, but it may be computed by generating a spreadsheet calculating the value of the expression at the lefthand side of equation (30) as a function of ξ_{3_1}. The roots of equation (30) are then inspected and the single answer corresponding to a meaningful value of ξ_{3_1} (i.e. $0 < \xi_{3_1} \leq [Fe^{3+}]_{i_1}$) taken as the equilibrium concentration of the trisacetohydroxamatoiron(III) complex. The values of ξ_2 are calculated from the deconvoluted spectra and the extinction coefficient of the bisacetohydroxamatoiron(III) complex.
Table 3 shows the values of the absorbance of the bis- and trisacetohydroxamatoiron(III) complex from the deconvoluted data, together with the calculated average and standard deviation values of K_3.

T/K	Abs_{bis} @ 470nm			Abs_{tris} @ 420nm			Average	Standard
	pH=0.44	pH=0.84	pH=0.93	pH=0.44	pH=0.84	pH=0.93	K_3	Deviation
293	0.540	0.370	0.086	0.341	0.656	0.200	0.0151	0.0015
298	0.510	0.360	0.085	0.434	0.670	0.198	0.0179	0.0038
303	0.490	0.330	0.080	0.473	0.802	0.200	0.0222	0.0057
313	0.450	0.290	0.075	0.543	0.842	0.206	0.0274	0.0080
323	0.380	0.270	0.070	0.553	0.856	0.210	0.0323	0.0091

Table 3. Determination of K_3 in accordance with equation (30) as a function of temperature calculated from solution of initial composition: $[Fe^{3+}]$ = 4x10^{-4} mol dm^{-3} and [HL]=6.3 mol dm^{-3}

Figure 7 shows the data of table 3 plotted in accordance with equation (16) assuming that the values of ΔH^o and ΔS^o are temperature-independent over the range of temperatures considered. This then enables the determination of the standard enthalpy and entropy of formation of the trisacetohydroxamatoiron(III) complex from the bisacetohydroxamatoiron(III) complex and acetohydroxamic acid, respectively ΔH^o = +19.74

kJ mol^{-1} and ΔS^o = +32.9 J mol^{-1} K^{-1}. These values are both positive suggesting that the process is endothermic in nature and entropically driven, in contrast to the processes associated with the formation of the mono- and bisacetohydroxamatoiron(III) complexes. This has implication for the temperature-dependent speciation of the iron(III)-acetohydroxamic acid system which we shall return to shortly.

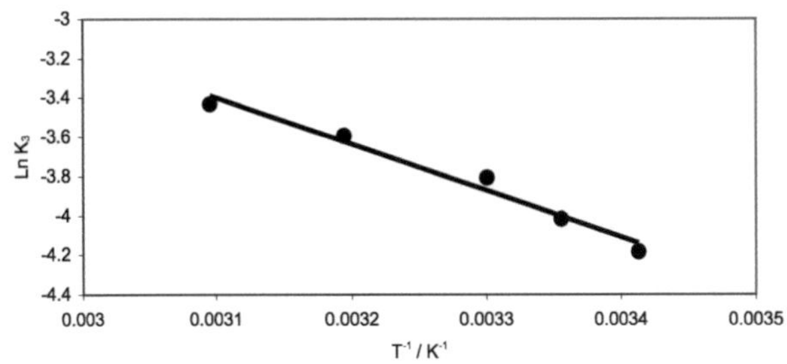

Fig. 7. LnK_3 vs. T^{-1} for the formation of the trisacetohydroxamatoiron(III) complex. Regression line : $Ln(K_3) = -2374T^{-1} + 3.961; R^2 = 0.9752$

3.5.1 Extinction coefficient determination

The deconvolution of the spectra recorded in this study, together with the determination of the concentration of the trishydroxamatoiron(III) complex in each solution, allows us to determine the extinction coefficient for this complex at λ_{max} = 420nm. This is done through the use of the ξ_3 equivalent of Beer's law shown in equation (17). The extinction coefficient for the trisacetohydroxamatoiron(III) complex was found to be ϵ_3 = 3630 $\pm$ 290 mol dm^{-3} cm^{-1}.

3.6 Temperature-dependent speciation of the iron (III) - acetohydroxamic acid system

With the temperature-dependent value of K_1, K_2 and K_3 determined, it is possible to generate a series of speciation diagrams for the Fe-AHA system as a function of pH, HL concentration (expressed as pHL=$-\log_{10}[HL]$) and temperature. These diagrams are generated by calculating the equilibrium concentrations of each of the iron(III)-containing species (Fe^{3+}, FeL^{2+}, FeL_2^+ and FeL_3) present in solution for a set of known starting conditions (i.e. $[Fe^{3+}]_i$, $[HL]_i$, $[H^+]_i$, and T) through the use of equations (1)-(3) and (5)-(10).

3.6.1 Speciation as a function of acidity

Figure 8 shows the speciation of the Fe-AHA system as a function of [HL] expressed as pHL at 293K for a range of pH values. As is to be expected from equations (1)-(3) an increase in pH (decrease in $[H^+]$) favours the formation of higher order complexes a direct result of LeChatelier's principle. It is important to stress that these speciation diagrams represent meaningful solution compositions up to a pH value of 2 as above this value iron(III) undergoes hydrolysis to form iron oxyhydroxides which are purposely overlooked in the calculation of these diagrams.

3.6.2 Speciation as a function of acetohydroxamic acid concentration

Figure 9 shows the speciation of the Fe-AHA system as a function of pH at 293K for a range of acetohydroxamic acid concentrations. Again as expected from equations (1)-(3) an increase in acetohydroxamic acid concentration results in the production of higher complexes due to LeChateliers principle.

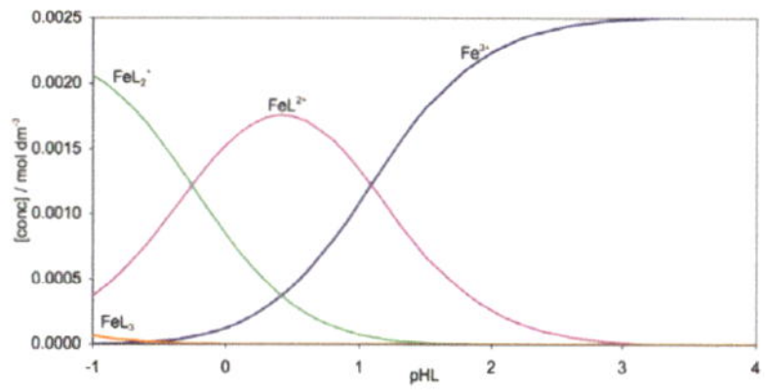

(a) pH = - 0.77

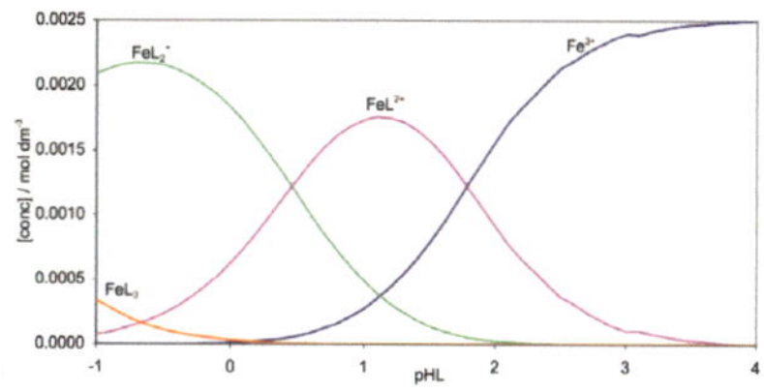

(b) pH = 0

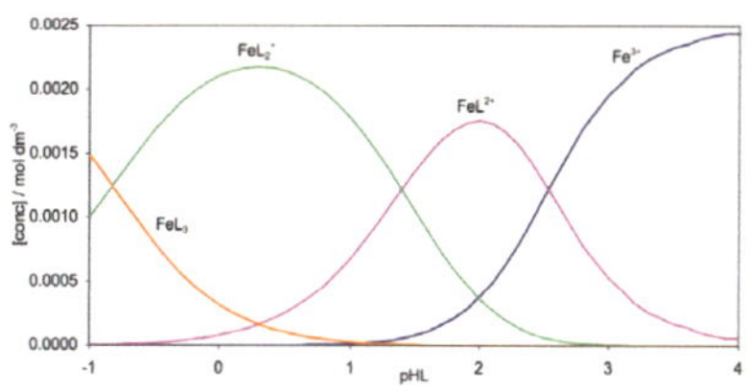

(c) pH = 1

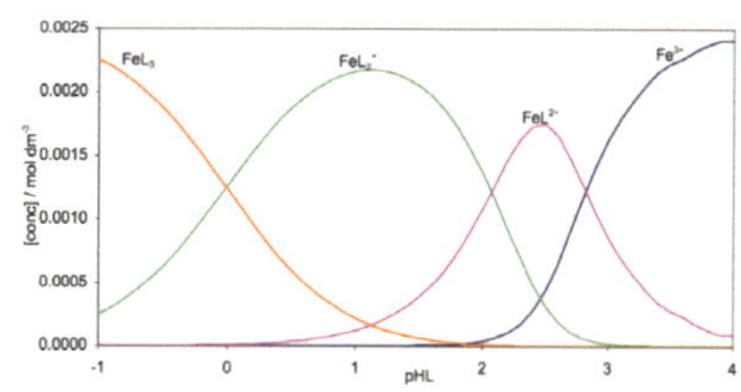

(d) pH = 2

Fig. 8. Speciation diagrams showing the concentration of Fe^{3+}, FeL^{2+}, FeL_2^+ and FeL_3 as a function of initial acetohydroxamic acid concentration (expressed as pHL) calculated for a total iron(III) concentration of $2.5x10^{-3}$ mol dm^{-3} at a temparature of 293K and pH values of -0.77, 0, 1 and 2 as per legend.

3.6.3 Speciation as a function of temperature

Figure 10 shows the speciation of the Fe-AHA system as a function of acetohydroxamic acid concentration expressed as pHL at pH = 1 for a range of temperatures. From this set of figures, it is apparent that an increase in temperature results in the production of free iron(III) and the trisacetohydroxamatoiron(III) complex at the expense of both the mono- and bisacetohydroxamatoiron(III) complexes. This is to be expected from the exothermic nature of the reaction producing the latter two compounds and the endothermic nature of the reaction resulting in the formation of the trisacetohydroxamatoiron(III) complex.

3.7 Verification of assumptions

During the course of this study a number of assumption have been made in order to simplify the mathematical treatment of the data collected. It is now necessary to return to these assumptions and verify that they are justified.

3.7.1 Determination of K_1

In the determination of the equilibrium constant for the formation of the monoacetohydroxamatoiron(III) complex an assumption was made that the concentrations of the bis- and trisacetohydroxamatoiron(III) complex were negligible when compared to the concentration of the monoacetohydroxamatoiron(III) complex. Figure 11 shows the speciation diagrams recorded in the conditions of the experiment at 293 and 323K, the temperature range boundaries. Under the conditions of the experiment, a maximum of 0.3% of all complexed iron(III) is present in the form of the bis- or trisacetohydroxamatoiron(III) complex, which strongly support our assumption that FeL^{2+} is the predominant Fe-AHA species in solution.

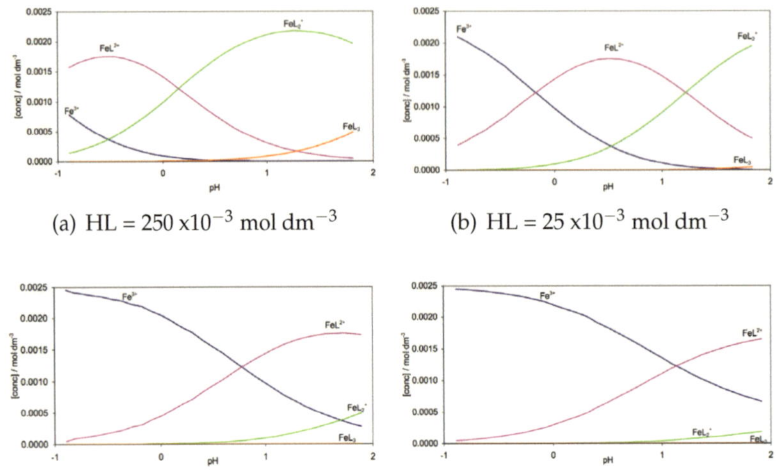

(a) HL = 250 x10^{-3} mol dm^{-3} (b) HL = 25 x10^{-3} mol dm^{-3}

(c) HL = 4 x10^{-3} mol dm^{-3} (d) HL = 2.5 x10^{-3} mol dm^{-3}

Fig. 9. Speciation diagrams showing the concentration of Fe^{3+}, FeL^{2+}, FeL_2^+ and FeL_3 as a function of pH calculated for a total iron(III) concentration of 2.5x10^{-3} mol dm^{-3} at a temparature of 293K and acetohydroxamic acid concentration of 250 x10^{-3}, 25 x10^{-3}, 4 x10^{-3}, 2.5 x10^{-3} mol dm^{-3} as per legend, corresponding to AHA:Fe ratios of 100:1, 10:, 1.6:1 and 1:1.

3.7.2 Determination of K_2

Figure 12 shows the speciation diagram for the Fe-AHA system calculated for the specific experimental conditions under which K_2 was determined. These confirm that a maximum of 2.7% of all complexed iron is present as the trisacetohydroxamatoiron(III) complex, validating the assumption that the mono- and bisacetohydroxamatoiron(III) complexes are the predominant species.

3.7.3 Determination of K_3

Figure 13 shows the speciation diagram for the Fe-AHA system calculated for the specific experimental conditions under which K_3 was determined. These confirm that a maximum of 2.5% of all complexed iron is present as the monoacetohydroxamatoiron(III) complex, confirming, as we assumed, that the bis- and trisacetohydroxamatoiron(III) complexes are the predominant species.

4. Conclusion

The thermodynamic constants for all three equilibria that obtain in the iron(III)-AHA system have been calculated and the associated enthalpies and entropies for the formation of these complexes have been determined. These are summarised in Table 4 together with the molar absorptivity for each of the three complexes. These new values have been used to calculate speciation diagrams for the Fe-AHA system as a function of pH, AHA concentration and temperature.

Inspection of the molar absorptivities measured for all three complexes reveal that it increases as the number of ligands bound to the iron increases. This is thought to be in line with the

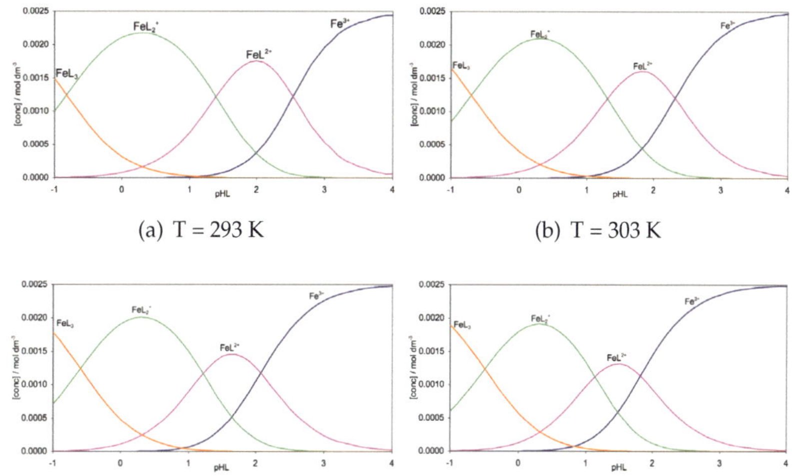

(a) T = 293 K

(b) T = 303 K

(c) T = 313 K

(d) T = 323 K

Fig. 10. Speciation diagrams showing the concentration of Fe^{3+}, FeL^{2+}, FeL_2^+ and FeL_3 as a function of acetohydroxamic acid concentration (expressed as pHL) calculated for a total iron(III) concentration of 2.5x10^{-3} mol dm^{-3} at pH = 1 and T = 293K, 303K, 313K, 323K as per legend.

complex formed	K @ 298K unitless	ΔH^o $kJmol^{-1}$	ΔS^o $Jmol^{-1}K^{-1}$	ϵ $dm^{-3}mol^{-1}cm^{-1}$
FeL^{2+}	42.64	-68.6	-199.0	1005
FeL_2^+	2.51	-17.09	-49.7	1650
FeL_3	1.18x10^{-2}	19.74	32.9	3630

Table 4. Thermodynamic data pertaining to the formation of the various iron(III)-AHA complexes

greater availability of ligands to contribute to the ligand-to-metal charge-transfer bands that dominate the absorption spectra of the Fe(III)-AHA system.

It is worth noting that both the entropy and enthalpy become progressively more positive as the number of bound ligands increases, i.e., complexation becomes enthalpically less favourable but entropically more favourable as the L:M ratio increases.

The change in entropy as individual reactions progress reflects a balance of two effects:

(i) an increase due to the chelate effect, as a result of the establishment of bonds in a 5-membered ring arrangement as the ligand binds to the metal centre while two water molecules from the inner solvation shell of iron(III) and a proton from the hydroxamic acid are released. This is accompanied by a re-ordering of the outer solvation sphere around the complex and a decrease in the overall charge on the complex.

(ii) a decrease due to the overall transfer of charge from the bulky Fe^{3+} ion to the much smaller proton leading to an increase in charge density and simultaneous solvent ordering around the charge.

In the formation of FeL^{2+}, process (ii) is presumably dominant but becomes less so as the ligand number increases due to the decrease in charge of the formed complex and associated

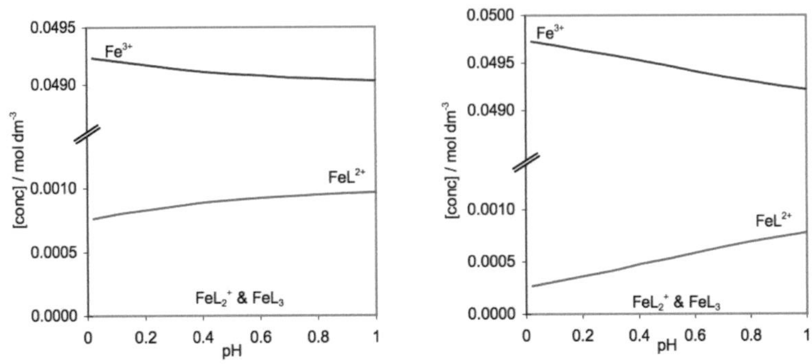

(a) T = 293 K (b) T = 303 K

Fig. 11. Speciation diagrams showing the concentration of Fe^{3+}, FeL^{2+}, FeL_2^+ and FeL_3 as a function of acetohydroxamic acid concentration (expressed as pHL) calculated for the conditions under which K_1 was determined.

decreases in its influence on the ordering of the solvent. Thus as the influence of (ii) decreases, the impact of (i) becomes proportionately greater until it dominates and the complexation process exhibits an overall favourable formation entropy with the neutral FeL_3 complex.

The progression in enthalpy suggests that, as the degree of complexation increases, the energy balance between Fe-O bonds breaking (E_{bb}) and bonds forming (E_{bf}) changes from $|E_{bb}| < |E_{bf}|$ in the FeL^{2+} to $|E_{bb}| > |E_{bf}|$ in the FeL_3 complex, i.e., the replacement of two H_2O molecules by an AHA ligand becomes less favourable. This is presumably related to steric considerations as the hydroxamate ligand is much bulkier than the two water molecules it displaces. Thus, as the number of hydroxamato ligands around the metal increases, the resultant ligand-ligand repulsion forces increase, acting against the enthalpy gained by complexation.

The work carried out in this study supports the wider understanding of the behaviour of hydroxamic acids as chelating ligands for hard cations such as Fe(III), Pu(IV) and Np(IV), in particular the kinetics of hydrolysis of these ligands in the presence of such cations in the context of nuclear reprocessing and specifically the development of advanced reprocessing flowsheets. As such the determination of the thermodynamic equilibrium constants for the formation of the various Fe-AHA complexes, allows for our model of the kinetics of hydrolysis of hydroxamic acids in the presence of non-oxidising metal cations (Andrieux et al.,2007) to be extended to temperatures away from room temperature and determine the activation energy and pre-exponential factor associated with the hydrolysis processes, giving access to valuable mechanistic information. This work is currently ongoing in our laboratories and will be the subject of a forthcoming publication.

5. Acknowledgements

The authors thank the Nuclear Decommissioning Authority (NDA) for financial support and for a post-doctoral research fellowship for FA. CB is supported by the Lloyds Register Educational Trust (LRET) an independent charity working to achieve advances in transportation, science, engineering and technology education, training and research worldwide for the benefit of all.

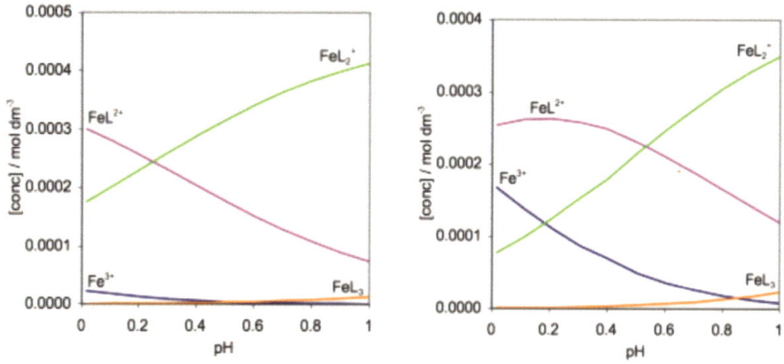

(a) T = 293 K (b) T = 303 K

Fig. 12. Speciation diagrams showing the concentration of Fe^{3+}, FeL^{2+}, FeL_2^+ and FeL_3 as a function of acetohydroxamic acid concentration (expressed as pHL) calculated for the conditions under which K_2 was determined.

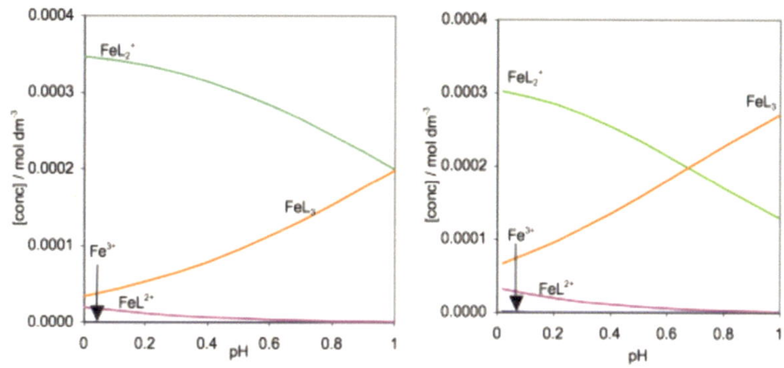

(a) T = 293 K (b) T = 303 K

Fig. 13. Speciation diagrams showing the concentration of Fe^{3+}, FeL^{2+}, FeL_2^+ and FeL_3 as a function of acetohydroxamic acid concentration (expressed as pHL) calculated for the conditions under which K_3 was determined.

6. References

[Andrieux et al.,2008] Andrieux, F.P.L.; Boxall, C.; Mason, C. & Taylor, R.J. (2008). The hydrolysis of hydroxamic acid complexants in the presence of non-oxidising metal ions 2: Neptunium(IV) ions. *J. Sol. Chem.*, Vol.37, (September 2008) p. 215-232, ISSN 1572-8927

[Andrieux et al.,2007] Andrieux, F.P.L.; Boxall, C. & Taylor, R.J. (2007). The hydrolysis of hydroxamic acid complexants in the presence of non-oxidising metal ions 1: Ferric ions. *J. Sol. Chem.*, Vol.36, (September 2008) p. 1201-1217, ISSN 1572-8927

[Birkett et al.,2005] Birkett, J.E.; Carrott, M.J.; Fox, O.D.; Jones, C.J.; Maher, C.J.; Roube, C.V.; Taylor, R.J. & Woodhead, D.A. (2005). Recent developments in the Purex process for nuclear fuel reprocessing: Complexant based stripping for uranium-plutonium separation. *Chimia* Vol.59, (2005) p. 898-904, ISSN 0009-4293

[Carrott et al.,2008] Carrott, M.J.; Fox, O.D.; LeGurun, G.; Jones, C.J.; Mason, C.; Taylor, R.J.; Andrieux, F.P.L. & Boxall, C. (2008) Oxidation-reduction reactions of simple hydroxamic acids and plutonium(IV) ions in nitric acid. *Radiochim. Acta* Vol.96, (2008) p.333-344, ISSN 0033-8230

[Carrott et al.,2007] Carrott, M.J.; Fox, O.D.; Maher, C.J.; Mason, C.; Taylor, R.J.; Sinkov, S.I. & Choppin, G.R. (2007) Solvent extraction behaviour of plutonium(IV) ions in the presence of simple hydroxamic acids. *Solv. Extr. Ion Exch.* Vol.25 (2007), p.723-746 ISSN 0736-6299

[Cornell & Schwertmann,1996] Cornell, R.M. & Schwertmann, U. (1996) *The Iron Oxides: Structure, Properties, Reactions, Occurrences and Uses*, VCH, ISBN 3527302743, New York

[Desaraju & Winston,1986] Desaraju, P. & Winston, A. (1986) Synthesis and iron complexation studies of bis-hydroxamic acids. *J. Coord. Chem.* Vol.14, (1986) p.241-248 ISSN 0095-8972

[Ghosh,1997] Ghosh, K.K. (1997) Kinetic and mechanistic aspects of acid catalysed hydrolysis of hydroxamic acids. *Indian J. Chem.* Vol.36B, (1997) p.1089-1102 ISSN

[Heusler & Lorenz,1985] Heusler, K.E.& Lorenz, W.J.: Iron, Ruthenium and Osmium In: *Standard Potentials in Aqueous Solution* Bard, A.J.; Parsons, R.; Jordan, J. (Ed.), p. 391. Dekker, ISBN:0824772911, New York

[John et al.,2001] John, S.G.; Ruggiero, C.E.; Hersman, L.E.; Tung, C.-S. & Neu, M.P. Siderophore mediated plutonium accumulation by microbacterium flavescens (JG-9). *Environ. Sci. Technol.* Vol.35, (2001) p.2942-2948 ISSN 1520-5851

[Monzyk & Crumbliss,1979] Monzyk, B. & Crumbliss, A.L. Mechanism of ligand substitution on high-spin iron(III) by hydroxamic acid chelators. Thermodynamic and kinetic studies on the formation and dissociation of a series of monohydroxamatoiron(III) complexes. *J. Am. Chem. Soc.* Vol.101, (1979) p.6203-6213 ISSN 0002-7863

[Muri et al.,2002] Muri, E.M.F.; Nieto, M.J.; Sindelar, R.D. & Williamson, J.S. Hydroxamic acids as pharmacological agents. *Current Med. Chem.* Vol.9, (2002) p.1631-1653 ISSN: 0929-8673

[Raymond et al.,1984] Raymond, K.N.; Freeman, G.E. & Kappel, M.J. Actinide-specific complexing agents: Their structural and solution chemistry. *Inorg. Chim. Acta* Vol.84, (1984) p.193-204 ISSN: 0020-1693

[Renshaw et al.,2002] Renshaw, J.C.; Robson, G.D.; Trinci, A.P.J.; Wiebe, M.G.; Livens, F.R.; Collinson, D. & Taylor, R.J. Fungal siderophores: Structures, functions and applications. *Mycol. Res.* Vol.106, (2002) p.1123-1142 ISSN: 0953-7562

[Roos,2006] Roos, Y.H. Phase transitions and transformations in food systems. In: *Handbook of Food Engineering, 2nd edition* Heldmann, D.R., Lund, D.B. (Eds.), p. 328. CRC, ISBN: 0824753313, Boca Raton

[Sapieszko et al.,1977] Sapieszko, R.S.; Patel, R.C. & Matijevic, E. Ferric hydrous oxide sols. 2. Thermodynamics of aqueous hydroxo and sulfato ferric complexes. *J. Phys. Chem.* Vol.81, (1977) p.1061-1068 ISSN: 1089-5639

[Schwarzenbach & Schwarzenbach,1963] Schwarzenbach, G. & Schwarzenbach, K. Hydroxamatkomplexe, I: Die Stabilität der Eisen(III)-Komplexe einfacher Hydroxamsäuren und des Ferrioxamins B. *Helv. Chim. Acta* Vol.46, (1963) p.1390-1400 ISSN: 0018-019X

[Vernon,1982] Vernon, F. Chelating ion exchangers. The synthesis and use of poly(hydroxamic acid) resins. *Pure Appl. Chem.* Vol.54, (1982) p.2151-2158 ISSN: 1365-3075

17

Obtaining Thermodynamic Properties and Fluid Phase Equilibria without Experimental Measurements

Lin, Shiang-Tai and Hsieh, Chieh-Ming
Department of Chemical Engineering, National Taiwan University
Taiwan

1. Introduction

The knowledge of thermodynamic properties and phase equilibria of pure and mixture fluids is crucial for the design, development and optimization of chemical, biochemical, and environmental engineering (Sandler, 2006). Conventional thermodynamic models, containing empirical or semi-empirical parameters, are very useful for the correlation of experimental data and have limited predictive capability for conditions beyond the experimental measurements. While a large amount of thermodynamic experiments have been conducted over the past decades, it has been estimated that about 2000 new chemicals are being synthesized everyday (Campbell, 2008). It would be a daunting and impractical task attempting to obtain the thermodynamic properties of these new chemicals and their mixtures with existing chemicals from experiment. Experimental determination of thermodynamic properties can also be quite challenging because of the toxic nature of some chemicals and the conditions of interest may reach the detection limit of modern apparatus (low vapor pressure, infinite dilution, etc.). It is therefore highly desirable to have a reliable means to estimate these important data prior to measurements, or even before the synthesis of the chemicals.

In this chapter, we illustrate how the thermodynamic properties and phase behavior of fluids can be obtained from the combination of advances in computational chemistry and the theories of statistical and classical thermodynamics. Instead of regression to experimental data, the model parameters in an equation of state, such as the Peng-Robinson equation of state (Peng & Robinson, 1976), can be determined from the results of first principle solvation calculations. The consequence is a completely predictive approach for almost any thermodynamic properties of all types of fluids (pure and mixture) at all conditions (above and below the critical point) without the need of any experimental data. We show that this is a practical approach for the prediction of pure fluid properties (Hsieh & Lin, 2008) such as the vapor pressure, liquid density, the critical temperature and pressure, as well as the vapor-liquid (Hsieh & Lin, 2009b), liquid-liquid (Hsieh & Lin, 2010), and solid-liquid phase equilibria of mixture fluids. This method also allows for accurate predictions of the distribution of a trace amount of pollutant between two partially miscible liquids, such as the octanol-water partition coefficient (Hsieh & Lin, 2009a). Conventionally, models were developed for a certain specific property of interest.

For example, a model that can be used for the vapor pressure cannot be used for the Henry's law constant, and vice versa. The method we have developed is general and can be used to determine all aspects of properties of a chemical under any conditions. Furthermore, the method developed here does not suffer from the problem of missing parameters that is commonly seen in the group contribution method and applicable to many more problems than the quantitative structure-activity relationship for property estimation. The accuracy of property predictions from our method is comparable or superior to most existing methods.

2. Equations of state

A thermodynamic equation of state is the mathematical equation that describes the interrelationship between thermodynamic variables (Prausnitz et al., 2004). One of such equations is the pressure(P)-volume(V)-temperature(T)-composition($\underline{x}$) equation of state, which is commonly expressed as

$$z=f(T,\underline{V},\underline{x}) \tag{1}$$

where $z=P\underline{V}/RT$ is the compressibility factor of a fluid. (Note that $\underline{V}=V/N$ is the molar volume, $\underline{x}=(x_1,x_2,\ldots,x_c)$ denotes the mole fractions of all the components.) In the case of an ideal gas, the function $f(T,\underline{V},\underline{x})$ is unity regardless of temperature and concentration of the fluid, i.e., $z=1$. In general, the compressibility factor of a fluid is a function of its temperature, pressure, and composition. For example, a widely used Peng-Robinson equation of state (Peng & Robinson, 1976) describes the compressibility of a fluid as follows

$$z(T,\underline{V},\underline{x})=\frac{\underline{V}}{\underline{V}-b(\underline{x})}-\frac{a(T,\underline{x})}{RT}\frac{\underline{V}}{\underline{V}(\underline{V}+b)+b(\underline{V}-b)} \tag{2}$$

where $a(T,\underline{x})$ and $b(\underline{x})$ are the energy and volume parameters, respectively. (For a review of other types of equations of state, please refer to (Kontogeorgis & Folas, 2010, Poling et al., 2001, Prausnitz et al., 2004, Sandler, 2006)). Conventionally, the two interaction parameters in the Peng-Robinson equation of state are determined from the critical properties of the fluid. For example, for pure fluids

$$a_i(T)=0.45724\frac{R^2T_{c,i}^2}{P_{c,i}}\left[1+\kappa_i\left(1-\sqrt{\frac{T}{T_{c,i}}}\right)\right]^2 \tag{3}$$

and

$$b_i=0.07780\frac{RT_{c,i}}{P_{c,i}} \tag{4}$$

where $T_{c,i}$, $P_{c,i}$ are the critical temperature and pressure of substance i, and the parameter k is

$$\kappa_i=0.37464+1.54226\omega_i-0.26992\omega_i^2 \tag{5}$$

with ω_i being the acentric factor. For mixtures, it is necessary to determine the composition dependence of the interaction parameters $a(T,\underline{x})$ and $b(\underline{x})$. One common approach is to assume the a quadratic composition dependence, such as in the van der Waals one-fluid mixing rule

$$a(T,\underline{x}) = \sum_{i=1}^{c}\sum_{j=1}^{c} x_i x_j a_{ij} \tag{6}$$

and

$$b(\underline{x}) = \sum_{i=1}^{c}\sum_{j=1}^{c} x_i x_j b_{ij} \tag{7}$$

where a_{ij} and b_{ij} are determined from the combining rule

$$\begin{aligned} a_{ij} &= \sqrt{a_i(T)a_j(T)}(1-k_{ij}) \\ b_{ij} &= (b_i + b_j)/2 \end{aligned} \tag{8}$$

with k_{ij} being the binary interaction parameter whose value must be determined from fitting to experimental data. There are other more advanced mixing rules, where $a(T,\underline{x})$ and $b(\underline{x})$ are obtained by matching the equation of state to other physical properties (such as the excess Gibbs free energy). Interested readers are refer to the book (Kontogeorgis & Folas, 2010) for further details.

2.1 Thermodynamic properties and phase equilibria from an equation of state

Equations of state are widely adopted in process simulators, such as the AspenPlus (AspenTech, 2007), for chemical processes. All thermodynamic properties of a fluid can be obtained from an accurate equation of state, together with the ideal gas heat capacities of all the fluid components. For example, the Gibbs free energy of a mixture fluid at certain temperature and pressure can be calculated from

$$\underline{G}(T,P,\underline{x}) = \underline{G}^{IGM}(T,P,\underline{x}) + RT(z-1) - RT\ln z + \int_{T,\underline{V}=\infty}^{T,\underline{V}} \left[\frac{RT}{\underline{V}} - P\right] d\underline{V} \tag{9}$$

where $\underline{G}^{IGM}(T,P,\underline{x})$ is the Gibbs free energy of an ideal gas mixture at the same condition as the fluid of interest and can be determined from

$$\underline{G}^{IGM}(T,P,\underline{x}) = \sum_{i=1}^{c} x_i \underline{G}_i^{IG}(T,P) + RT\ln x_i \tag{10}$$

where $\underline{G}_i^{IG}(T,P)$ is the Gibbs free energy of pure species i in an ideal gas state at T and P. The ideal gas contribution can be obtained with high accuracy with modern computational chemistry (Foresman & Frisch, 1996, Ochterski, 2000). As seen in equation 9, the property deviation from ideal gas contributions can be obtained from an equation of state. When the Peng-Robinson equation of state (eqn. 2) is used, equation 9 becomes

$$\underline{G}(T,P,\underline{x}) = \underline{G}^{IGM}(T,P,\underline{x}) + RT(z-1) - RT\ln(z-B) - \frac{A}{2\sqrt{2}B}\ln\left[\frac{z+(1+\sqrt{2})B}{z+(1-\sqrt{2})B}\right] \tag{11}$$

where $A=Pa/(RT)^2$ and $B=Pb/RT$ are dimension less quantities. Other thermodynamic properties, such as the internal energy, enthalpy, entropy, Helmholtz free energy, ect., can be determined in a similar fashion (Sandler, 2006).
When there are multiple phases coexisting in a system, the equilibrium composition of each phase can be determined from the equivalence of fugacity of each chemical in all phases

$$\overline{f}_i^I(T,P,\underline{x}^I)=\overline{f}_i^{II}(T,P,\underline{x}^{II})=\overline{f}_i^{III}(T,P,\underline{x}^{III})=\cdots \tag{12}$$

where the superscripts I, II, III, etc indicate the phase of interest. Therefore, determination of fugacity is the key in phase equilibrium calculations. This quantity can be calculated if an equation of state of the fluid is known

$$\ln\frac{\overline{f}_i(T,P,\underline{x})}{x_iP}=-\ln z+\frac{1}{RT}\int_{T,\underline{V}=\infty}^{T,\underline{V}}\left[\frac{RT}{\underline{V}}-N\left(\frac{\partial P}{\partial N_i}\right)_{T,V,N_{j\neq i}}\right]d\underline{V} \tag{13}$$

and when the Peng-Robinson equation of state and van der Waals mixing rule is assumed for the fluid, equation 13 becomes

$$\ln\frac{\overline{f}_i(T,P,\underline{x})}{x_iP}=\frac{B_i}{B}(z-1)-\ln(z-B)-\frac{A}{2\sqrt{2}B}\left[\frac{2\sum_{j=1}^{c}x_iA_{ij}}{A}-\frac{B_i}{B}\right]\ln\left[\frac{z+(1+\sqrt{2})B}{z+(1-\sqrt{2})B}\right] \tag{14}$$

where $A_{ij}=Pa_{ij}/(RT)^2$.
From the above examples, it is clear that an accurate equation of state is very important for obtaining the thermodynamic properties and phase behaviours of a fluid. However, in conventional approaches the necessary parameters in an equation of state ($a(T,\underline{x})$ and $b(\underline{x})$) require input of several experimental data, including the critical properties and acentric factor of pure substances, and some mixture data (such as the vapor-liquid equilibrium data) for determining k_{ij}. The need of experimental input severely restricts the application of an equation of state. For example, the critical properties of high molecular weight compounds (e.g., polymers, proteins, heavy organics, etc.) are not experimentally available as they would decompose before reaching the critical point. More importantly, the need for binary interaction parameter k_{ij} posses issues for processes involving new chemicals (e.g., drug discovery) where scarce or even no data is available. We will illustrate how the use of computational chemistry, in particular, quantum mechanical solvation calculations, to resolve these difficulties.

2.2 Solvation properties from an equation of state

The solvation free energy $\Delta\underline{G}^{*sol}$, as defined by Ben-Naim (Ben-Naim, 1987), is the work needed for transferring of a molecule (solute) from an ideal gas phase to a solution (solvent) under constant temperature T and pressure P. Such a free energy is commonly computed from a hypothetical two-step process, as illustrated in Figure 1: (1) creating a cavity to size of the solute (the corresponding work is referred to as the cavity formation free energy $\Delta\underline{G}^{*cav}$) and (2) placing the solute into the cavity (the corresponding work referred to as the solvation charging free energy $\Delta\underline{G}^{*chg}$), that is

$$\Delta\underline{G}^{*sol} = \Delta\underline{G}^{*cav} + \Delta\underline{G}^{*chg} \tag{15}$$

The solvation free energy can also be obtained from an equation of state (Lin et al., 2007)

$$\frac{\Delta\underline{G}^{*sol}(T,P,\underline{x})}{RT} = (z-1) + \int_{\underline{V}=\infty}^{\underline{V}} \frac{(1-z)}{\underline{V}} d\underline{V} \tag{16}$$

If the fluid obeys the Peng-Robinson equation of state, equation 16 becomes

$$\frac{\Delta\underline{G}^{*sol}(T,\underline{V})}{RT} = (z-1) + \ln\frac{\underline{V}}{\underline{V}-b} + \frac{a}{bRT}\left[\frac{1}{2\sqrt{2}}\ln\frac{\underline{V}+b(1-\sqrt{2})}{\underline{V}+b(1+\sqrt{2})}\right] \tag{17}$$

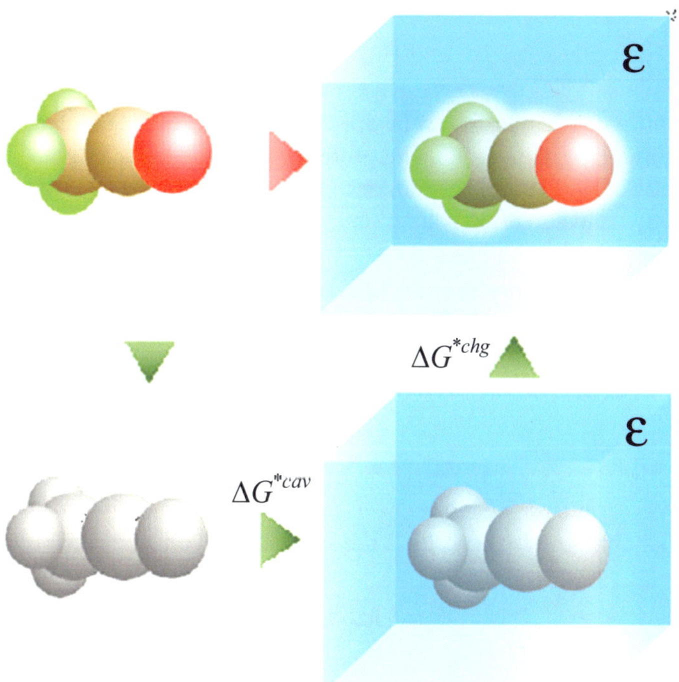

Fig. 1. The solvation process can be separated into two steps: (1) creating a cavity to size of the solute ($\Delta\underline{G}^{*cav}$) and (2) placing the solute into the cavity ($\Delta\underline{G}^{*chg}$)

Since the cavity formation free energy accounts for mostly repulsive interactions between the solute and the solvent, it can be estimated from an equation of state by setting the interaction parameter a to zero.

$$\frac{\Delta\underline{G}^{*cav}}{RT} = \left(\ln\frac{\underline{V}}{\underline{V}-b} + \frac{b}{\underline{V}-b}\right) \tag{18}$$

Therefore, the changing free energy must be

$$\frac{\Delta \underline{G}^{*chg}}{RT}=\frac{a}{bRT}\left[\frac{1}{2\sqrt{2}}\ln\frac{\underline{V}+b(1-\sqrt{2})}{\underline{V}+b(1+\sqrt{2})}-\frac{b\underline{V}}{\underline{V}^2+2b\underline{V}-b^2}\right]=\frac{aC}{bRT} \tag{19}$$

where the variable C represents the terms in the square brackets, and its value depends on the density of the fluid. Equations 19 provides a new way of obtaining the equation of state parameters from solvation calculations

$$a(T,\underline{x})=\frac{b(\underline{x})}{C(\underline{V})}\Delta \underline{G}^{*chg}(T,\underline{V},\underline{x}) \tag{20}$$

The choice of density for evaluation of C must be consistent with that using in the solvation calculations. Our previous studies (Hsieh & Lin, 2009b) showed that C=-0.623 is a good choice for the Peng-Robinson equation of state.
The size parameter $b(\underline{x})$ can be approximated in a way similar to the van der Waals mixing rule

$$b(\underline{x})=\sum_i x_i b_i \tag{21}$$

and the size of pure component b_i can be approximated using the volume of the cavity (see Fig. 1) used in solvation calculations. Therefore, equations 20 and 21 provide a new route to the temperature and composition dependence of parameters in the Peng-Robinson equation of state completely from solvation properties.

2.3 Equation of state parameters from first principle solvation calculations

A solvation model is necessary for the charging free energy in equations 20. Here we choose to use the method proposed by Lin et al. (Lin et al., 2007), in which the charging free energy is the summation of contributions from all components in the mixture

$$\Delta \underline{G}^{*chg}=\sum_{i=1}^{c}\Delta \underline{G}_{i/j}^{*chg}(T,\underline{x}) \tag{22}$$

and the molecular charging free energy is the sum of four contributions (Lin et al., 2004),

$$\Delta \underline{G}_{i/j}^{*chg}(T,\underline{x})=\Delta \underline{G}_i^{*is}+\Delta \underline{G}_i^{*cc}+\Delta \underline{G}_i^{*dsp}(T)+\Delta \underline{G}_{i/j}^{*res}(T,\underline{x}) \tag{23}$$

where the superscripts *is*, *cc*, *res*, and *dsp* denote ideal solvation, charge-averaging correction, restoring, and dispersion contribution to the solvation charging free energy. In this solvation model, the solute molecule is first dissolved into a fluid of infinite dielectric constant (or a conductor), where the solute molecule is perfectly screened. The first three terms on the right and side of equation 23 account for the free energy change in this process. Finally, the chemical nature of the solvent is restored (reducing the dielectric constant from infinity to its physical value at temperature T), and the corresponding free energy is referred to as the restoring free energy (the last term of eqn. 23).
The ideal solvation term is the difference in energy when the solute is in the ideal gas and in the conductor state

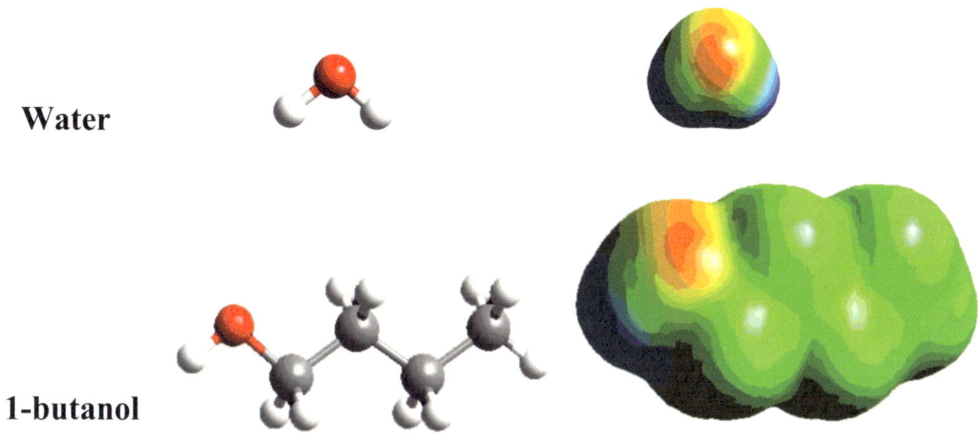

Fig. 2. Molecular structure and screening charge distribution on molecular cavity (intensity of charge density: -0.025 to 0.025) obtained from quantum mechanical geometry optimization and COSMO calculation, respectively, for water and 1-butanol

$$\Delta \underline{G}_i^{*is} = E_i^{COSMO} - E_i^{IG} \tag{24}$$

where E_i^{IG} and E_i^{COSMO} are the total energy of molecule i in the ideal gas and ideal conductor states determined from first-principles quantum mechanical (QM) and COSMO solvation calculations (Klamt & Schuurmann, 1993, Lin & Hsieh, 2006). Also obtained from the COSMO calculations are the screening charges on the cavity surface. These charges are useful for the evaluation of other free energy terms in equation 23. Figures 2 illustrates the optimal conformations from QM calculations and the screening charge distribution on molecular cavity from COSMO calculations for two selected compounds. For the subsequent restoring free energy calculation, the raw screening charge data (σ^*) are smoothed according to the following equation

$$\sigma_m = \frac{\sum_n \sigma_n^* \frac{r_n^2 r_{eff}^2}{r_n^2 + r_{eff}^2} \exp\left(-f_{decay} \frac{d_{mn}^2}{r_n^2 + r_{eff}^2}\right)}{\sum_n \frac{r_n^2 r_{eff}^2}{r_n^2 + r_{eff}^2} \exp\left(-f_{decay} \frac{d_{mn}^2}{r_n^2 + r_{eff}^2}\right)} \tag{25}$$

where d_{mn} is the distance between surface segments m and n, the parameter f_{decay} (= 3.57) controls the correlation distance between any two segments, r_n is the radius of segment n, and $r_{eff}=(a_{eff}/\pi)^{1/2}$ (with a_{eff} = 7.50 Å^2) is the radius of a standard surface segment. As a result, it is necessary to consider the energy shift associated with the charge averaging process

$$\Delta \underline{G}_i^{*cc} = f_{pol}^{1/2}\left[E_{diel}(\underline{q}) - E_{diel}(\underline{q}^*)\right] \tag{26}$$

where f_{pol} is the polarization factor (determined to be 0.6916 (Lin et al., 2004)); the dielectric energy is defined as

$$E_{diel}(\underline{q}) = \frac{1}{2}\sum_{v}\phi_v q_v \tag{27}$$

here q_v is the screening charge at some position v on the cavity surface and ϕ_v is the electrostatic potential due to the solute at position v. Both q_v and ϕ_v are obtained from the COSMO solvation calculation.

The screening charges distributed on the cavity surface are quantified by the sigma-profile, $p(\sigma)$, based on the fraction of surfaces possessing a certain charge density value σ

$$p_i(\sigma) = \frac{A_i(\sigma)}{A_i} \tag{28}$$

where A_i is the total surface area of species i. To better account for the interactions between species that may form a hydrogen bond, the surfaces may be further categorized to four types: surfaces that do not involve in a hydrogen bond A_i^{nhb}, surfaces from hydroxyl groups A_i^{hydro}, surfaces from amino groups of primary and secondary amines A_i^{amino}, and all other surfaces that may be involved in a hydrogen bond A_i^{other}, such as -NO_2. Therefore, the sigma-profile is refined as

$$\begin{aligned} p(\sigma) &= \frac{A_i^{nhb}(\sigma) + A_i^{hydro}(\sigma) + A_i^{amino}(\sigma) + A_i^{other}(\sigma)}{A_i} \\ &= p^{nhb}(\sigma) + p^{hydro}(\sigma) + p^{amino}(\sigma) + p^{other}(\sigma) \end{aligned} \tag{29}$$

Figure 3 illustrates the sigma-profile of a few common chemicals.

The sigma-profile is in fact the probability distribution of screening charges. In a mixture, such probability distribution can be calculated as the mole-fraction average of contributions from the pure components

$$p_S(\sigma) = \frac{\sum_i x_i A_i p_i(\sigma)}{\sum_i x_i A_i} \tag{30}$$

The sigma profile reflects the electronic nature of a chemical species and is unique to each chemical. Therefore, it was used to determine the restoring free energy (Lin & Sandler, 2002, Hsieh & Lin, 2010)

$$\frac{\Delta \underline{G}_{i/j}^{*res}}{RT} = \frac{A_i}{a_{eff}} \sum_{s}^{\text{nhb, hydro, amino, other}} \left[\sum_{\sigma_m} p_i^s(\sigma_m^s) \ln \Gamma_j^s(\sigma_m^s) \right] \tag{31}$$

where Γ is the segment activity coefficient

$$\ln \Gamma_j^t(\sigma_m^t) = -\ln \left\{ \sum_{s}^{\text{nhb, hydro, amino, other}} \sum_{\sigma_n} p_j^s(\sigma_n^s) \exp\left[\frac{-\Delta W(\sigma_m^t, \sigma_n^s)}{RT} + \ln \Gamma_j^s(\sigma_n^s) \right] \right\} \tag{32}$$

The segment interaction energy ΔW is determined by considering the electrostatic and hydrogen bonding interactions between two contacting surfaces

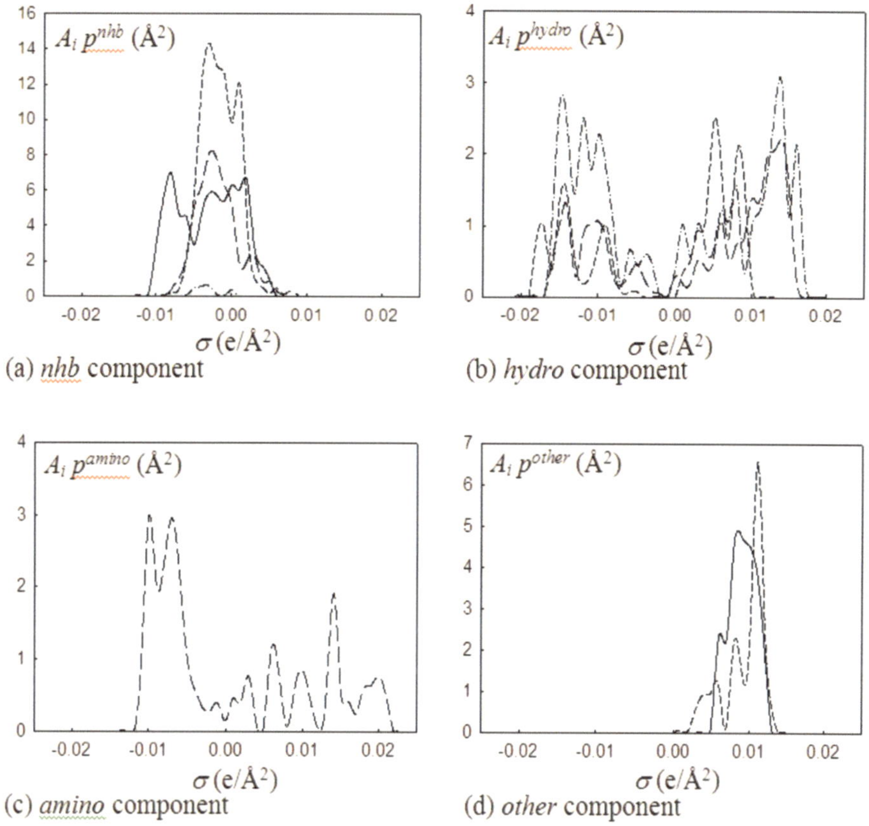

Fig. 3. The sigma profiles for four representative compounds: acrylonitrile (solid line), monoethanolamine (long dashed line), isobutyric acid (short dashed line), and water (dashed-dotted line). A_i is the molecular surface area of species i.

$$\Delta W(\sigma_m^t, \sigma_n^s) = f_{pol}\frac{0.3 a_{eff}{}^{3/2}}{2\varepsilon_0}\left(\sigma_m^t + \sigma_n^s\right)^2 - c_{hb}(\sigma_m^t, \sigma_n^s)(\sigma_m^t - \sigma_n^s)^2 \tag{33}$$

where ε_0 is the permittivity of vacuum; coefficient $c_{hb}(\sigma_m^t, \sigma_n^s)$ is temperature-independent and its value is determined according to the type of the interacting segments

$$c_{\text{hb}}(\sigma_m^t, \sigma_n^s) = \begin{cases} c_{\text{HH}} & \text{if s = t = hydro and } \sigma_m^t \cdot \sigma_n^s < 0 \\ c_{\text{AA}} & \text{if s = t = amino and } \sigma_m^t \cdot \sigma_n^s < 0 \\ c_{\text{OO}} & \text{if s = t = other and } \sigma_m^t \cdot \sigma_n^s < 0 \\ c_{\text{HA}} & \text{if s = hydro, t = amino, and } \sigma_m^t \cdot \sigma_n^s < 0 \\ c_{\text{HO}} & \text{if s = hydro, t = other, and } \sigma_m^t \cdot \sigma_n^s < 0 \\ c_{\text{AO}} & \text{if s = amino, t = other, and } \sigma_m^t \cdot \sigma_n^s < 0 \\ 0 & \text{otherwise} \end{cases} \tag{34}$$

The six parameters (c_{HH}, c_{AA}, c_{OO}, c_{HA}, c_{HO}, and c_{AO}) in equation 34 are universal constants. Their values have been determined using experimental vapor pressure data of selected pure substances and selected liquid-liquid equilibria, and do not need to be changed.

Universal Parameters			
Parameter	Value	Parameter	Value
C_{PR}	-0.623	c_{HH} (kcal/mol)(Å^4/e^2)	1757.9468
a_{eff} (Å^2)	7.50	c_{AA} (kcal/mol)(Å^4/e^2)	1121.4047
f_{pol}	0.6916	c_{OO} (kcal/mol)(Å^4/e^2)	1757.9468
f_{decay}	3.57	c_{HA} (kcal/mol)(Å^4/e^2)	2462.3206
$A_{dsp,HB}$ (J/mol/K/Å^2)	-465876.8150	c_{HO} (kcal/mol)(Å^4/e^2)	933.4108
$B_{dsp,HB}$ (J/mol/Å^2)	-429.5556	c_{AO} (kcal/mol)(Å^4/e^2)	2057.9712
$C_{dsp,HB}$ (J/mol/K^2/Å^2)	-141.8436		
$A_{dsp,RING}$ (J/mol/K/Å^2)	-0.9181		
$B_{dsp,RING}$ (J/mol/Å^2)	-365.0667		

Atom Specific Parameters			
atom type	radius R_i (Å)	$A_{dsp,i}$ (J/mol/K/Å^2)	$B_{dsp,i}$ (J/mol/Å^2)
H	1.30	0.1694	-191.4602
C	2.00	0.1694	-191.4602
N	1.83	0.4045	-207.9411
O	1.72	0.2701	-178.0767
F	1.72	0.1806	-125.7842
Cl	2.05	0.1566	-201.7754

Table 1. Parameters and their values in the PR+COSMOSAC EOS

The dispersion solvation free energy is considered to be proportional to the exposed surface area of the atom comprising the molecule (Cramer & Truhlar, 1991, Cramer & Truhlar, 2008, Klamt, 1995, Klamt et al., 1998, Lin et al., 2004, Lin & Sandler, 2002, Still et al., 1990, Tomasi & Persico, 1994). Here we propose a slightly modified form to better describe ring containing and hydrogen bonding chemicals.

$$\Delta \underline{G}_i^{*dsp}(T) = \sum_j s_j (A_{dsp,j} T + B_{dsp,j}) + \underline{G}_{HB}^{*dsp}(T) + \underline{G}_{RING}^{*dsp}(T) \quad (35)$$

where s_j is the total exposed surface area of atom type j, $A_{dsp,j}$ and $B_{dsp,j}$ are the dispersion parameters of atom type j, and $\underline{G}_{HB}^{*dsp}$ and $\underline{G}_{RING}^{*dsp}$ are the empirical corrections for hydrogen-bonding and cyclic (or aromatic) containing molecules. The expressions of these two terms are

$$\underline{G}_{HB}^{*dsp}(T) = \frac{1}{N_{HBH}} \left\{ \frac{A_{dsp,HB}}{1 + \exp\left[-\left(T - B_{dsp,HB}\right) / C_{dsp,HB}\right]} \right\} \quad (36)$$

and

$$\underline{G}_{RING}^{*dsp}(T) = N_{AR}(A_{dsp,RING} + B_{dsp,RING}) \tag{36}$$

where N_{HBH} and N_{AR} are the number of hydrogen-bonding donors (the hydrogen atoms connecting to either nitrogen, oxygen, or fluorine atom) and the number of atoms involved in any ring-structure, respectively. All the parameters in the proposed model have been determined by regression to experimental vapor pressure of selected compounds or experimental data of binary liquid-liquid equilibrium. They are summarized in Table 1. Once the solvation charging free energy at some given temperature and solution composition is determined from equations 22 to 36, the energetic parameter $a(T,\underline{x})$ in the PR EOS can be calculated from equation 20. This approach is referred to as PR+COSMOSAC hereafter. It is useful to note that the fugacity coefficient can be determined analytically from the PR+COSMOSAC equation of state as follows (Hsieh and Lin, 2010)

$$\ln \frac{\overline{f}_i(T,P,\underline{x})}{x_i P} = \frac{\Delta G_{i/S}^{*chg}}{RT} - \ln\left(1 - \frac{Pb}{zRT}\right) + \frac{Pb_i}{zRT - Pb} - \ln z \tag{37}$$

3. Prediction of thermodynamic properties and phase equilibria

The PR+COSMOSAC equation of state contains 15 universal constants and 3 element specific coefficients, as shown in Table 1. This set of parameters is all that is needed to determine all kinds of thermodynamic properties and phase equilibria of any system. In the following, we illustrate some of the properties predicted from PR+COSMOSAC. The procedure of calculation of thermodynamic properties can be found in most text books such as (Prausnitz et al., 2004, Sandler, 2006).

3.1 Thermodynamic properties of pure fluids

As a first example, the PR+COSMOSAC is used to predict the vapor-liquid coexisting curve of pure fluids all the way to the critical point. Figure 4 shows the pressure-volume and pressure-temperature diagrams for two chemical species having very different critical pressures: 1,3-propylene glycol (high P_c= 9.5 MPa), cyclooctane (low P_c= 3.57 MPa). The solid and grey lines are the results from the proposed method and the original PR EOS. The temperature dependence of the vapor pressure is well described for these compounds (Fig. 4a). The deviations of liquid molar volume and the gas molar volume from proposed model are similar to those from original PR EOS (Fig. 4b). Although the agreement is not perfect, the present model shows how effectively a theoretically-based statistical mechanical model can describe the temperature-dependent parameter $a(T)$ in the PR EOS.

Another important application of the PR+COSMOSAC equation of state is the prediction of the normal boiling point of environmentally important chemicals. The normal boiling point is an indication of the volatility of a chemical, and therefore is an important quantity needed in many chemical, biochemical, and environmental studies. These hazardous chemicals are often complex in their chemical structure and have a low vapor pressure at ambient conditions, making direct experimental measurement quite challenging. Figure 5 shows the comparison of predicted and measured normal boiling point of 63 environmentally significant compounds (including chemical such as dichloro acetic aced, D-limonene, isoprene, perchlorocyclopentadiene, acridine, 2-chlorobiphenyl (PCB-1), etc.). As can be seen that there is a very good agreement between the predicted and experimental values (most of the data points lay on the diagonal line). The overall average absolute error from the PR+COSMOSAC model is 3.8% for the normal boiling temperature.

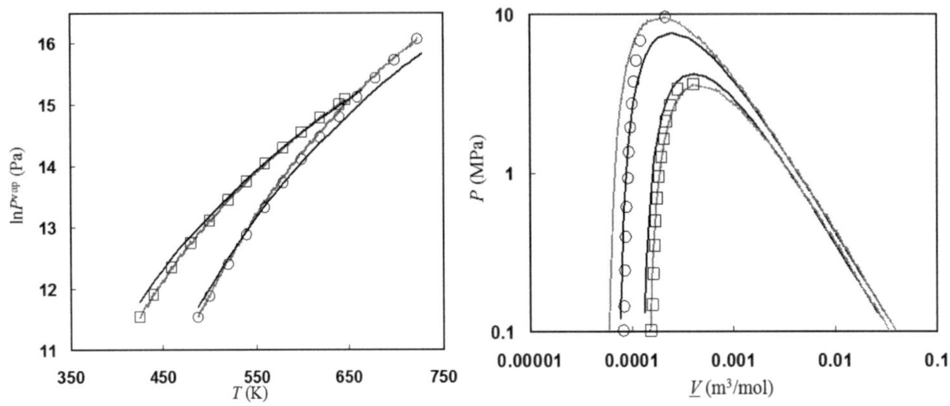

Fig. 4. The pressure-temperature (a) and pressure-volume (b) diagrams from the original PR EOS (gray lines) and the PR+COSMOSAC with (solid lines) for cyclooctane and 1,3-propylene glycol. The experimental data (DIPPR, 2008) are shown in squares and circles for the two compounds, respectively

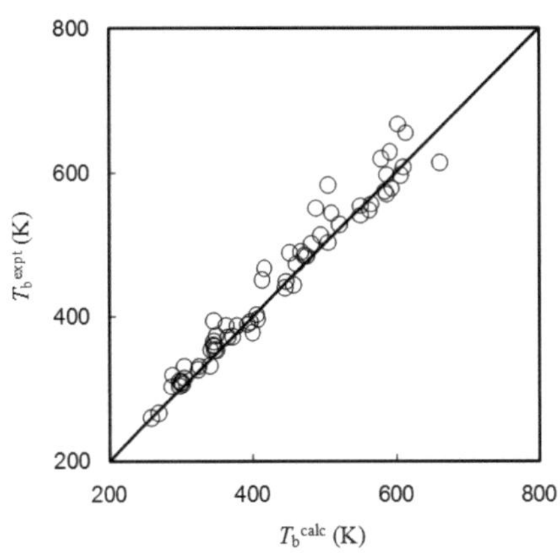

Fig. 5. Comparison of normal boiling temperatures for environmentally important chemicals from experiments and predictions from PR+COSMOSAC equation of state

Unlike the conventional approach where the critical properties are used to determine the interaction parameters in the Peng-Robinson equation of state (eqn. 3 to 5), the PR+COSMOSAC determines these parameters from solvation calculations (eqn. 20 and 21). Therefore, critical properties of pure fluids can be now calculated. Figure 6 illustrates the comparison of experimental and predicted critical properties for compounds whose experimental critical properties are available in the DIPPR database (346, 431, and 270 compounds in critical pressure P_c, critical temperature T_c, and critical volume $\underline{V}_c$, respectively). The critical volume is estimated from the volume of solvation cavity in COSMO calculation. The predicted $\underline{V}_c$ are in good agreement with experiment for small compounds (e.g., $\underline{V}_c < 0.6$ m³/kmol). It has been observed previously that the critical volume is highly correlated with the molecular size (Kontogeorgis et al., 1997). We also found a good linear correlation between the value of parameter b determined from the solvation cavity and that from T_c and P_c in the PR EOS. These results show that the atomic radii (Table 1) used in establishing the solvation cavity are adequate for describing the volume parameter b. The larger deviations (underestimation) found for larger compounds could be attributed to the ignorance of conformation flexibility in current calculations (e.g., long chain alkanes are modeled as linear but they could be folded in reality) and/or the ignorance of molecular shape effects in the PR EOS (the cavity term in PR EOS is valid for spherical molecules (Lin, 2006)).

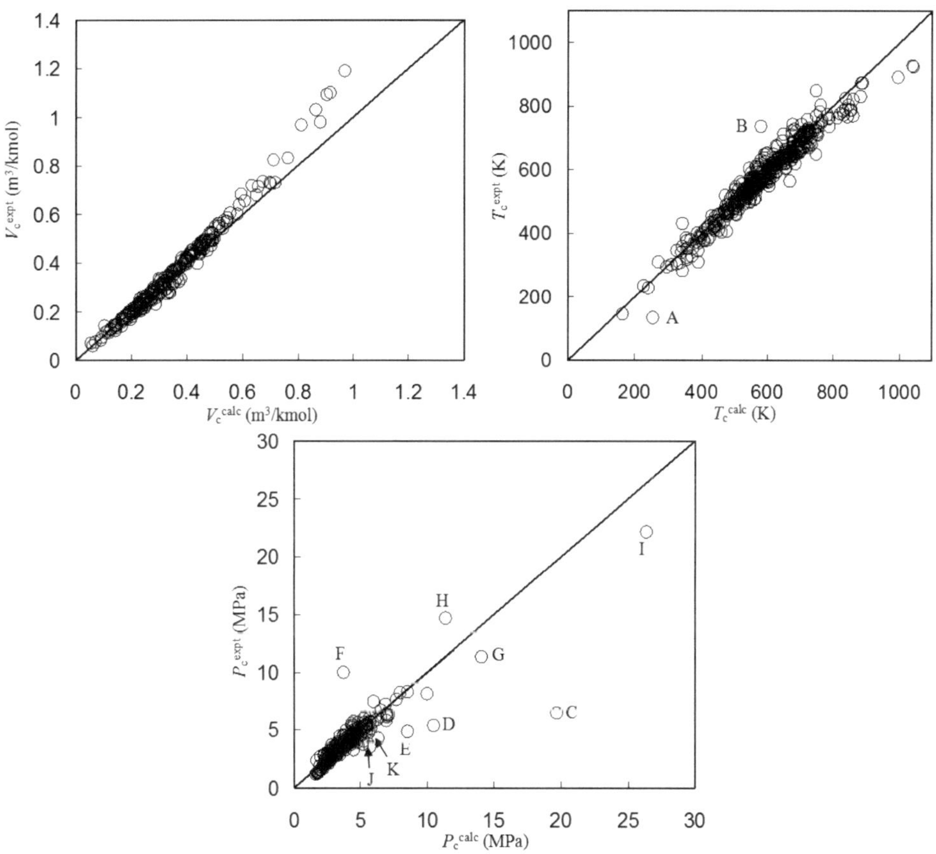

Fig. 6. Comparison of critical properties from experiments and predictions. The marked species are (A) carbon monoxide, (B) methacrylic acid, (C) hydrogen fluoride, (D) hydrogen cyanide, (E) acetonitrile, (F) nitrogen tetroxide, (G) ammonia, (H) hydrazine, (I) water, (J) carbon monoxide, and (K) propionitrile

The deviations in T_c are less than 120 K, except for carbon monoxide and methacrylic acid, marked by A and B, respectively, in Figure 6. Once $\underline{V}_c$ and T_c are known, the critical pressure can be calculated from $P_c = z_c RT_c/\underline{V}_c$. While the PR EOS has a fixed value of critical compressibility factor ($z_c = 0.307$) for all compounds, experimental values of z_c range from 0.2 to 0.3 for most chemicals (DIPPR, 2008, Poling et al., 2001). Therefore, we have rescaled the calculated P_c by 0.26/0.307 in Figure 6 for better accuracy. [Note that the use of $z_c = 0.26$ is recommended for calculation of P_c only and is not used in any other property calculations (e.g. the vapor pressure).] The poorly predicted P_c found in Figure 6 (marked as C to K) are caused either by the constant critical compressibility factor ($z_c = P_cV_c/RT_c$) in the PR EOS or the error in the predicted T_c. For example, hydrogen fluoride (marked C) has $z_c = 0.117$ and ammonia (marked G) has $z_c = 0.242$ but has a deviation of 73 K in predicted T_c.

3.2 Vapor-liquid, liquid-liquid, and vapor-liquid-liquid equilibria

Figure 7 shows the *P-x-y* diagram for morpholine/n-octane binary mixtures at two temperatures, 330 K and 340 K. While the predictions from PR+COSMOSAC (dashed lines) are in reasonable agreement with experiments (open circles), the predicted total pressure are slightly too high in the low morpholine concentration regime. This is a result of the over prediction of the vapor pressure of n-octane from PR+COSMOSAC. The predictions can be further improved if some experimental data is used to correct for the inaccuracy found for the vapor pressure.

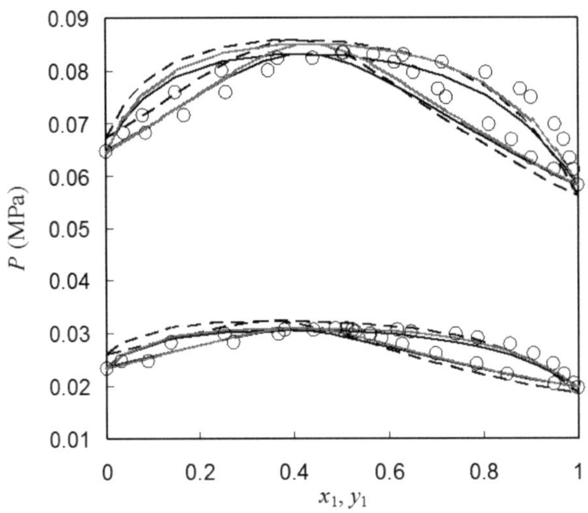

Fig. 7. *P-x-y* phase diagram of vapor-liquid equilibrium for morpholine (1) + n-octane (2) (a type II system). The dashed, solid, and gray lines are predicted results from PR+COSMOSAC, PR+COSMOSAC+$T_cP_c\omega$, and PR+COSMOSAC+P^{vap}, respectively. The experimental data, taken from (Wu et al., 1991), are shown as open circles

$$\Delta \underline{G}_{i/S}^{*chg} = \Delta \underline{G}_{i/S}^{*chg}[\text{eqn. } 23] + \Delta \underline{G}_{i,\text{Corr}}^{*chg}(T) \tag{38}$$

One possible correction is to adjust $\Delta \underline{G}_{i,\text{Corr}}^{*chg}(T)$ so that correct vapor pressure of species *i* is obtained at temperature *T*. This approach, denoted as PR+COSMOSAC+P^{vap}, ensures correct pressures in the VLE phase diagram in the pure fluid limits. Figure 7 also shows the predictions from PR+COSMOSAC+P^{vap} (gray). It can be seen that the predictions are significantly improved.

One limitation of PR+COSMOSAC+P^{vap} is that it is not applicable for problems at temperatures above the critical point of its component (e.g. determination the solubility of a gas in liquid). Figure 8 shows the vapor-liquid equilibrium diagram of 2-propanol/water binary mixture at five temperatures. While the PR+COSMOSAC model captures the general phase behavior of this mixture, it is quantitatively inaccurate due to the under prediction of the vapor pressure of water. At temperatures above the critical point of 2-propanol (~508 K) the correction of eqn. 38 cannot be applied. Another possible correction method is to use the critical properties (T_c and P_c) and acentric factor (ω),

$$\Delta \underline{G}_{i,\mathrm{Corr}}^{*chg}(T) = -\Delta \underline{G}_{i/i}^{*chg}[\text{eqn. 23}] + \frac{a_i C}{b_i} \tag{39}$$

Equation 39 ensures the same vapor pressure as that from the Peng-Robinson EOS is obtained. This approach is denoted as PR+COSMOSAC+$T_cP_c\omega$. Using this correction, the vapor pressure of water is correctly obtained and the phase diagram of 2-propanol/water can be accurately predicted (see the solid curves in Figure 8).

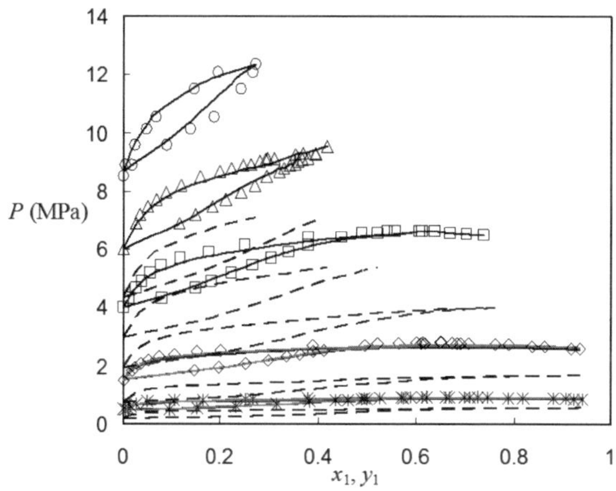

Fig. 8. P-x-y phase diagram of vapor-liquid equilibrium for 2-propanol (1) + water (2) (a type III system). The lines have the same meanings as in Figure 7. The experimental data, taken from Barr-David and Dodge (Barr-David & Dodge, 1959) and Sada and Morisue (Sada & Morisue, 1975), are shown as open symbols

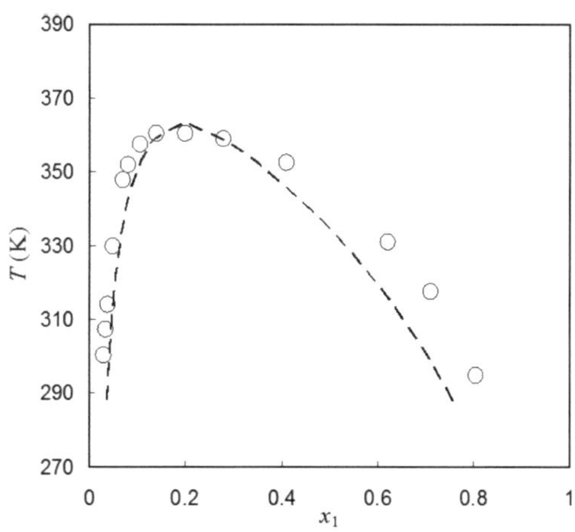

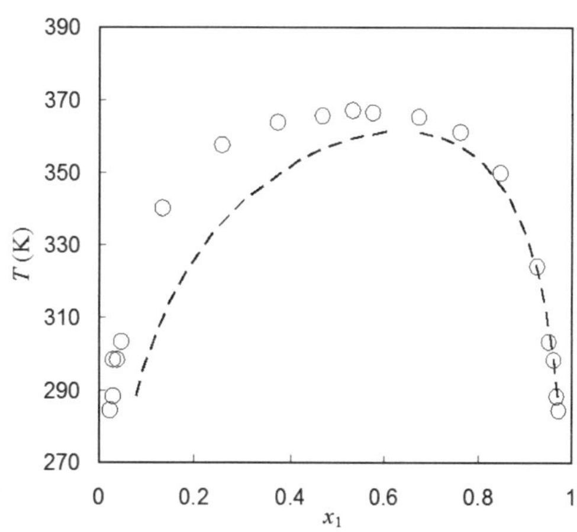

Fig. 9. Comparison of predicted (dashed lines) and experimental (open circles)(Sørensen & Arlt, 1979) liquid-liquid equilibrium for furfural (1) + 2,2,5-trimethylhexane (2) (left panel) and 2,4-pentanedione (1) + water (2) (right panel)

Figure 9 illustrates two examples of liquid-liquid equilibrium predictions from PR+COSMOSAC. In the mixture of furfural/2,2,5-trimethylhexane, no hydrogen bond is present. It can be seen that the PR+COSMOSAC describes the miscibility gap accurately all the way to the upper critical solution temperature. In the second example of 2,4-pentanedione/water, hydrogen bonds may be present between like species (water) and unlike species (water-2,4-pentanedione). The predictions from PR+COSMOSAC are again in quantitative agreement with the experiments.

As one last example of fluid phase equilibria predictions from PR+COSMOSAC, Figure 10 shows the vapor-liquid-liquid equilibrium of water/ethyl acetate at one atmospheric pressure. The PR+COSMOSAC model, without use of any experimental data, is able to capture the general features of VLLE. Because of the under estimation of the vapor pressure of water, the predicted normal boiling point is too high, leading to an over estimation of the three phase coexisting temperature. However, when the critical properties are introduced, i.e., the PR+COSMOSAC+$T_cP_c\omega$, the predictions are in excellent agreement with experiment both in the VLE, LLE, and the coexisting temperature. In general, accurate VLLE predictions can be achieved if a model is accurate on both VLE and LLE. Furthermore, good description on LLE helps in getting better liquid phase compositions; whereas good description on VLE helps in getting better equilibrium temperature and gas phase composition. From the previous examples, the PR+COSMOSAC model usually provides good descriptions for the liquid phase non-ideality; however, its performance in the phase vapor is less accurate, mainly because of its deficiency in describing the vapor pressure of pure fluids. The PR+COSMOSAC+$T_cP_c\omega$ model thus serves as an effective compromise for both vapor and liquid phase equilibrium predictions.

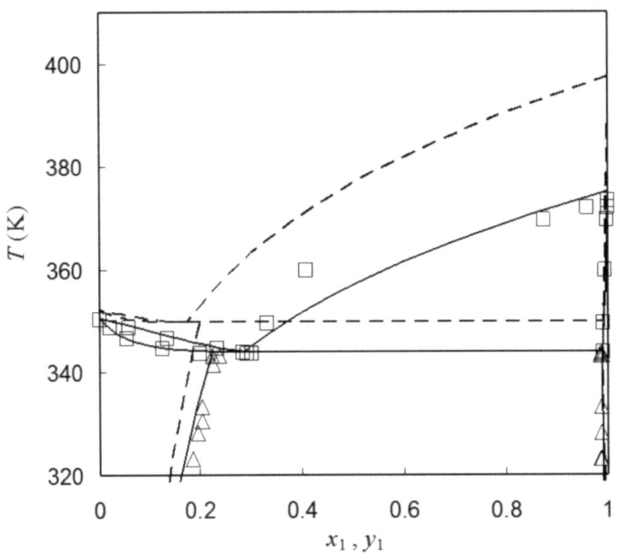

Fig. 10. Comparison of VLLE from experiments and predictions for water(1) + ethyl acetate(2). The open squares and triangles are experimental VLE and LLE(Gmehling et al., 1977, Sørensen & Arlt, 1979). The dashed lines and solid lines are results from PR+COSMOSAC and PR+COSMOSAC+$T_cP_c\,\omega$, respectively

3.3 Solubility of drugs

The knowledge of drug solubility is important in the design of drug manufacturing processes. Solvent screening is often necessary to identify optimal operating conditions for synthesis, reaction, and purification (Constable et al., 2007, Modarresi et al., 2008). However, experimental approaches are both time-consuming and costly since there essentially are an infinite number of combinations of possible solvent mixtures. Furthermore, as the drug discovery techniques continue to improve, the number of potential drug candidates increases significantly. It is impractical to measure the solubility data for all drug candidates in all possible solvent combinations. Thus, a predictive thermodynamic model can help to overcome this obstacle. Numerous thermodynamic methods are used to predict the solubility of organic compounds in the literature (Chen & Crafts, 2006, Chen et al., 2008, Klamt et al., 2002, Mullins et al., 2008, Shu & Lin, 2010, Tung et al., 2008). Here, we illustrate the prediction of drug solubility from PR+COSMOSAC.

The solubility of 52 drug compounds [from the smallest iodine (2 atoms) to the largest testosterone (49 atoms)] in 37 different pure solvents are considered. There are a total of 171 drug-solvent pairs for drug solubility in pure solvent and 156 mixture solvent combinations (298 systems), including 3 ternary solvent mixtures (10 systems) and 1 quaternary solvent mixture (1 system). The solubility data cover a wide range of values, from 10^{-1} to 10^{-6} (by mole fractions) over a temperature of 273.15 K to 323.15 K. In the calculation of solid solubility, the following relations (derived from considering solid-liquid equilibrium) is used

$$\ln x_i = \frac{\Delta H_i^{fus}}{R}\left(\frac{1}{T} - \frac{1}{T_{m,i}}\right) - \ln \frac{\overline{f}_i^S(T,P,\underline{x})}{x_i f_i(T,P)} \tag{40}$$

where $\overline{f}_i^S$ is the fugacity of the drug in the solvent, f_i is the fugacity of a hypothetical pure liquidus drug, $T_{m,i}$ and ΔH_i^{fus} are the melting temperature and the enthalpy of fusion at melting temperature of the drug. Experimental (Marrero & Abildskov, 2003, Linstrom & Mallard, 2010) values of $T_{m,i}$ and ΔH_i^{fus} are used, while the PR+COSMOSAC model is used for the fugacities.

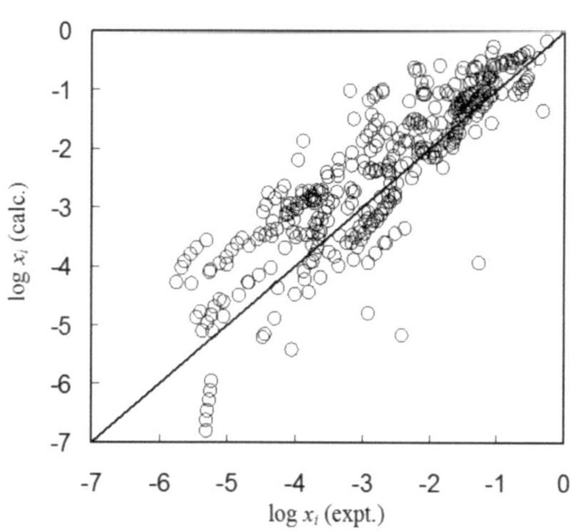

Fig. 11. Comparison of solubility (in logarithm) of 52 drugs in 37 pure solvents (362 data points) from experiments and predictions from PR+COSMOSAC

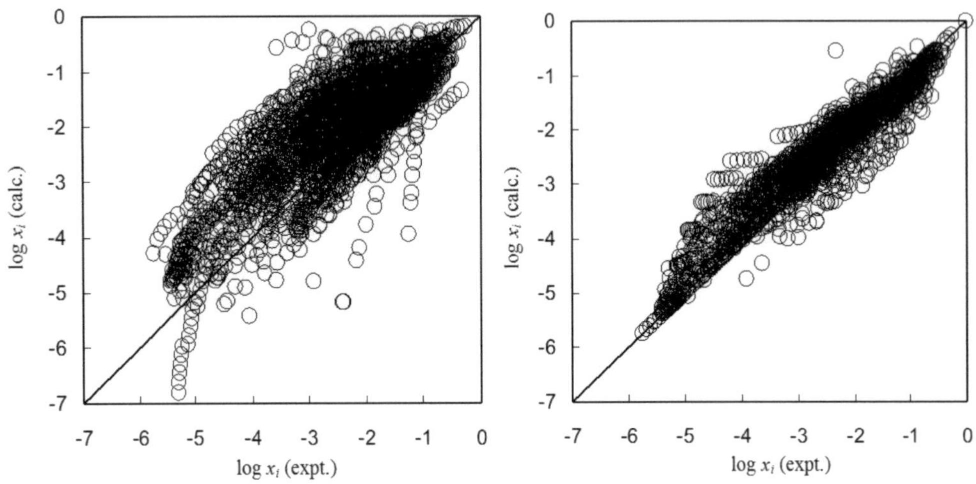

Fig. 12. Comparison of solubility (in logarithm) of 52 drugs in mixed solvents (2894 data points) from experiments and predictions from PR+COSMOSAC (left panel) and PR+COSMOSAC+G^{dsporr} (right panel)

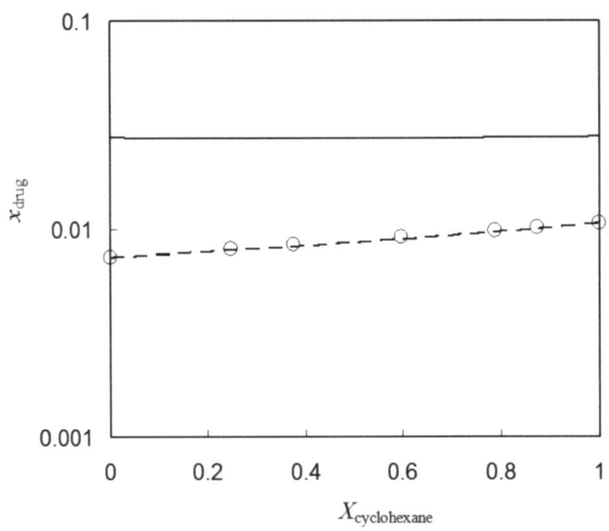

Fig. 13. Solubility of benzil in the solvent of cyclohexane and n-octane at 298.15 K from PR+COSMOSAC (solid line) and PR+COSMOSAC+G^{dsporr} (dashed line). Experimental data(Acree & Rytting, 1982) are shown in open circles. The x-axis X is the solvent fraction in the solute free solvent mixtures

Figure 11 shows the comparison between the prediction and experimental drug solubility of a drug in pure solvent (362 data points). The overall root-mean-square error (RMSE) in the logarithm of solubility from PR+COSMOSAC are 1.78 (corresponding to 495% in percentage error). Figure 12 shows the predictions in the case of mixed solvents. The overall RMSE is 1.40 (304%), slightly lower than that in pure solvents. Also can be seen from Figures 11 and 12 is that the PR+COSMOSAC model usually over predicts the solubility. Figure 13 illustrates the solubility of benzyl in the solvent mixture of cyclohexane and n-octane at

298.15 K. It can be seen that PR+COSMOSAC captures the concentration dependence of the solubility on the solvent composition (in this case, a nearly linear dependence), however, the predicted solubilities are in general too high. Figure 14 illustrates the solubility of acetanilide in the solvent of water and dioxane. In this case, the solubility can vary by about two hundreds folds (from 0.0011 in pure water to 0.21 in pure dioxane) as the composition of the solvent changes. The PR+COSMOSAC model is able to capture such large variations in solubility due to the change in solvent compositions.

The inaccuracy in PR+COSMOSAC can be attributed to its lack of accuracy in describing the drug-solvent interactions. The predictions in mixed solvent can thus be improved if the experimental solubility data of the drug in pure solvents are introduced to correct any error in the PR+COSMOSAC model, i.e.,

$$\Delta \underline{G}_{i,\mathrm{Corr}}^{*chg}(T) = \sum_k G_{ik}^{\mathrm{dsp,corr}} X_k \tag{41}$$

where $G_{ik}^{\mathrm{dsp,corr}}$ is the dispersion free energy correction coefficient for solute i in solvent k; X_k is the solvent fraction in the solute free solvent mixtures. In the case of a single solvent, the correction term becomes

$$\Delta \underline{G}_{i,\mathrm{Corr}}^{*chg}(T) = G_{ik}^{\mathrm{dsp,corr}} \tag{42}$$

and the coefficient $G_{ik}^{\mathrm{dsp,corr}}$ can be determined from fitting to experimental solubility data of the drug in pure solvent i. This approach is denoted as PR+COSMOSAC+G^{dsporr}. By using the solubility in pure solvents, the RMSE of the predicted solubility in mixed solvents (see also Figure 12) significantly reduces to 0.65 (or 91%). In Figures 13 and 14, it can be seen that the results from PR+COSMOSAC+G^{dsporr} (dashed line) are in excellent agreement with the experiments.

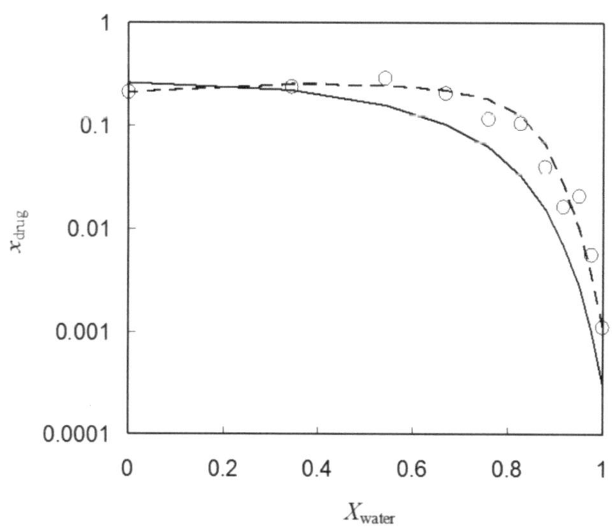

Fig. 14. Solubility of acetanilide in the solvent of water and dioxane at 293.15 K from PR+COSMOSAC (solid line) and PR+COSMOSAC+G^{dsporr} (dashed line). Experimental data(Bustamante et al., 1998) are shown in open circles. The x-axis X is the solvent fraction in the solute free solvent mixtures

Figure 15 shows an interesting example where a maximum value of solubility is observed at some intermediate concentrations of the solvent combinations. At 298.15 K, the solubility of sulphisomidine in pure water is 0.0009 and in pure dioxane is 0.0025. The solubility reaches a maximum value of 0.011 when the composition of the water is around 0.54. While the PR+COSMOSAC model (solid line) captures the general solvent composition dependence, the PR+COSMOSAC+G^{dsporr} model (dashed line) is able to predict the existence of such solubility maximum.

Although the accuracy of PR+COSMOSAC in the prediction of drug solubility is still far from accurate, it is capable of providing a priori predictions for compounds without binary interaction parameters and experimental data (the enthalpy of fusion and normal melting temperature of drug are needed). This is very useful for in the early stage of drug discovery and the design of purification processes in the pharmaceutical industry.

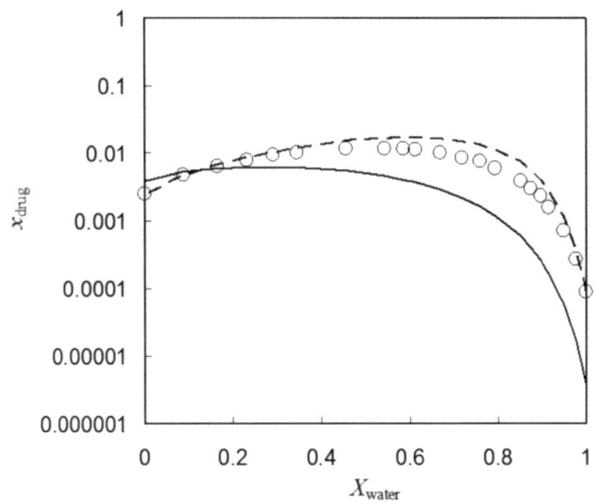

Fig. 15. Solubility of sulphisomidine in the solvent of water and dioxane at 298.15 K from PR+COSMOSAC (solid line) and PR+COSMOSAC+G^{dsporr} (dashed line). Experimental data (Martin et al., 1985) are shown in open circles. The x-axis X is the solvent fraction in the solute free solvent mixtures.

4. Conclusion

The employment of *ab initio* solvation calculation in determination of cubic equation of state parameters for pure and mixture fluids, denoted as PR+COSMOSAC, has led to a new way for describing fluid phase equilibria without input of experimental data such as critical properties. The solvation calculation presented in this work is capable of capturing the correct composition and temperature dependence of the interaction parameter $a(T,\underline{x})$, while the solvation cavity and mole-fraction weighted summation is a good estimate for volume parameter $b(\underline{x})$. The PR+COSMOSAC EOS is able to provide reasonable predictions on vapor pressures, liquid densities and critical properties for pure fluids and vapor-liquid equilibrium (VLE), liquid-liquid equilibrium (LLE), vapor-liquid-liquid equilibrium (VLLE) for mixtures, and solid-liquid equilibrium (SLE) with a single model and a single set of parameters. The use of this method in the predictions of solubility of drugs in pure and mixture solvents is also validated. Although not shown here, this method can be used to determine other properties such as heat of vaporization, excess properties, distribution coefficient, infinite dilution activity coefficient, etc.

Limited by the accuracy in vapor pressure predictions, this approach presently provides only qualitative results for VLE predictions; however, in the case of mixtures, the predicted accuracy can be improved significantly if the critical properties and acentric factor (PR+COSMOSAC+$T_cP_c\omega$) or the vapor pressure (PR+COSMOSAC+P^{vap}) are used. The accuracy from PR+COSMOSAC may, in some cases, be inferior to existing group contribution methods, e.g., PSRK or modified UNIFAC. However, because of the proximity effects, methods based on concept of group contributions (e.g., PSRK) may sometimes fail badly if used for compounds that do not belong to the family of compounds used in the parameterization. Furthermore, unlike the group contribution methods (PSRK or modified UNIFAC) whose parameter matrix was optimized against a large set of experimental data, the PR+COSMOSAC contains only a few (about 33) non-species dependent, universal parameters. There is no issue of missing parameters if a new chemical species is involved. The computation-demanding QM calculations have to be done only once for each chemical species and can be stored in a database. Once the database is established (e.g. the VT COSMO database (Mullins et al., 2008)), the time need for phase equilibrium calculations using PR+COSMOSAC is similar to that using group contribution methods on a modern personal computer. We consider the PR+COSMOSAC as an ideal complementary method when the existing models are not applicable or no experimental data are available.

5. References

Acree, W.E. & Rytting, J.H. (1982). Solubility in binary solvent systems. 1. Specific versus nonspecific interactions, *Journal of Pharmaceutical Sciences*, 71, 2, pp. 201-205,0022-3549.

Aspentech (2007), *Aspen Plus Unit Operation Models*, Aspen Technology Inc: Burlington, MA

Barr-David, F. & Dodge, B.F. (1959). Vapor-Liquid Equilibrium at High Pressures. The Systems Ethanol-Water and 2-Propanol-Water., *J Chem Eng Data*, 4, 2, pp. 107 - 121

Ben-Naim, A. (1987). Solvation Thermodynamics, Plenum Press, New York.

Bustamante, P.; Romero, S.; Pena, A.; Escalera, B. & Reillo, A. (1998). Enthalpy-entropy compensation for the solubility of drugs in solvent mixtures: Paracetamol, acetanilide, and nalidixic acid in dioxane-water, *Journal of Pharmaceutical Sciences*, 87, 12, pp. 1590-1596,0022-3549.

Campbell, J.A. (2008), Chemical Reaction, in *Microsoft Encarta Online Encyclopedia*, Microsoft Corporation.,http://encarta.msn.com

Chen, C.C. & Crafts, P.A. (2006). Correlation and prediction of drug molecule solubility in mixed solvent systems with the Nonrandom Two-Liquid Segment Activity Coefficient (NRTL-SAC) model, *Industrial & Engineering Chemistry Research*, 45, 13, pp. 4816-4824,0888-5885.

Chen, C.C.; Simoni, L.D.; Brennecke, J.F. & Stadtherr, M.A. (2008). Correlation and prediction of phase behavior of organic compounds in ionic liquids using the nonrandom two-liquid segment activity coefficient model, *Industrial & Engineering Chemistry Research*, 47, 18, pp. 7081-7093,0888-5885.

Constable, D.J.C.; Jimenez-Gonzalez, C. & Henderson, R.K. (2007). Perspective on solvent use in the pharmaceutical industry, *Organic Process Research & Development*, 11, 1, pp. 133-137,1083-6160.

Cramer, C.J. & Truhlar, D.G. (1991). General parameterized scf model for free-energies of solvation in aqueous-solution, *Journal of the American Chemical Society*, 113, 22, pp. 8305-8311,0002-7863.

Cramer, C.J. & Truhlar, D.G. (2008). A universal approach to solvation modeling, *Accounts of Chemical Research*, 41, 6, pp. 760-768,0001-4842.

Dippr (2008), DIPPR801 *Thermodynamic Properties Database*, Brigham Young University: Provo

Foresman, J.B. & Frisch, A. (1996). Exploring Chemistry With Electronic Structure Methods: *A Guide to Using Gaussian*, Gaussian Inc., 0-9636769-4-6,

Gmehling, J.; Onken, U. & Arlt, W. (1977). *Vapor-Liquid Equilibrium Data Collection Dechema*, Frankfurt.

Hsieh, C.M. & Lin, S.T. (2008). Determination of cubic equation of state parameters for pure fluids from first principle solvation calculations, *AIChE Journal*, 54, 8, pp. 2174-2181,0001-1541.

Hsieh, C.M. & Lin, S.T. (2009a). Prediction of 1-octanol-water partition coefficient and infinite dilution activity coefficient in water from the PR plus COSMOSAC model, *Fluid Phase Equilibria*, 285, 1-2, pp. 8-14,0378-3812.

Hsieh, C.M. & Lin, S.T. (2009b). First-Principles Predictions of Vapor-Liquid Equilibria for Pure and Mixture Fluids from the Combined Use of Cubic Equations of State and Solvation Calculations, *Industrial & Engineering Chemistry Research*, 48, 6, pp. 3197-3205,0888-5885.

Hsieh, C.M. & Lin, S.T. (2010). Prediction of liquid-liquid equilibrium from the Peng-Robinson+COSMOSAC equation of state, *Chemical Engineering Science*, 65, 6, pp. 1955-1963

Hsieh, C.T.; Lee, M.J. & Lin, H.M. (2008). Vapor-Liquid -Liquid Equilibria for Aqueous Systems with Methyl Acetate, Methyl Propionate, and Methanol, *Industrial & Engineering Chemistry Research*, 47, 20, pp. 7927-7933,0888-5885.

Klamt, A. & Schuurmann, G. (1993). COSMO - A new approach to dielectric screening in solvents with explicit expressions for the screening energy and its gradient, *Journal Of The Chemical Society-Perkin Transactions* 2, 5, pp. 799-805

Klamt, A. (1995). Conductor-like screening model for real solvents - A new approach to the quantitative calculation of solvation phenomena, *Journal of Physical Chemistry*, 99, 7, pp. 2224-2235

Klamt, A.; Jonas, V.; Burger, T. & Lohrenz, J.C.W. (1998). Refinement and parametrization of COSMO-RS, *Journal of Physical Chemistry A*, 102, 26, pp. 5074-5085

Klamt, A.; Eckert, F.; Hornig, M.; Beck, M.E. & Burger, T. (2002). Prediction of aqueous solubility of drugs and pesticides with COSMO-RS, *Journal of Computational Chemistry*, 23, 2, pp. 275-281,0192-8651.

Kontogeorgis, G.M.; Smirlis, I.; Yakoumis, I.V.; Harismiadis, V. & Tassios, D.P. (1997). Method for estimating critical properties of heavy compounds suitable for cubic equations of state and its application to the prediction of vapor pressures, I*ndustrial & Engineering Chemistry Research*, 36, 9, pp. 4008-4012,0888-5885.

Kontogeorgis, G.M. & Folas, G.K. (2010). Thermodynamic Models for Industrial Applications: From Classical and Advanced Mixing Rules to Association Theories John Wiley & Sons Ltd, 978-0-470-69726-9, West Sussex.

Lin, S.T. & Sandler, S.I. (2002). A priori phase equilibrium prediction from a segment contribution solvation model, *Industrial & Engineering Chemistry Research*, 41, 5, pp. 899-913

Lin, S.T.; Chang, J.; Wang, S.; Goddard, W.A. & Sandler, S.I. (2004). Prediction of vapor pressures and enthalpies of vaporization using a COSMO solvation model, *Journal of Physical Chemistry A*, 108, 36, pp. 7429-7439

Lin, S.T. (2006). Thermodynamic equations of state from molecular solvation, *Fluid Phase Equilib.*, 245, 2, pp. 185-192

Lin, S.T. & Hsieh, C.M. (2006). Efficient and accurate solvation energy calculation from polarizable continuum models, *Journal of Chemical Physics*, 125, 12, pp. 124103(124101-124110),0021-9606.

Lin, S.T.; Hsieh, C.M. & Lee, M.T. (2007). Solvation and chemical engineering thermodynamics, *Journal of the Chinese Institute of Chemical Engineers*, 38, 5-6, pp. 467-476,0368-1653.

Marrero, J. & Abildskov, J. (2003). Solubility and Related Properties of Large Complex Chemicals, *DECHEMA*, Frankfurt am Main.

Martin, A.; Wu, P.L. & Velasquez, T. (1985). Eetended Hildebrand solubility approach - Sulfonamides in binary and ternary solvents, *Journal of Pharmaceutical Sciences*, 74, 3, pp. 277-282,0022-3549.

Modarresi, H.; Conte, E.; Abildskov, J.; Gani, R. & Crafts, P. (2008). Model-based calculation of solid solubility for solvent selection - A review, *Industrial & Engineering Chemistry Research*, 47, 15, pp. 5234-5242,0888-5885.

Mullins, E.; Liu, Y.A.; Ghaderi, A. & Fast, S.D. (2008). Sigma profile database for predicting solid solubility in pure and mixed solvent mixtures for organic pharmacological compounds with COSMO-based thermodynamic methods, *Industrial & Engineering Chemistry Research*, 47, 5, pp. 1707-1725,0888-5885.

Linstrom, P.J. & Mallard, W.G. (2010). NIST Chemistry WebBook, *NIST Standard Reference Database Number 69*, Gaithersburg. (http://webbook.nist.gov/chemistry/)

Ochterski, J.W. (2000), Thermochemistry in Gaussian, Gaussian, Inc.,http://www.gaussian.com/g_whitepap/thermo.htm

Peng, D. & Robinson, D.B. (1976). New 2-constant equation of state, *Industrial & Engineering Chemistry Fundamentals*, 15, 1, pp. 59-64,0196-4313.

Poling, B.E.; Prausnitz, J.M. & O'connell, J.P. (2001). *The properties of gases and liquids* McGraw-Hill, New York.

Prausnitz, J.M.; Lichtenthaler, R.N. & De Azevedo, E.G. (2004). *Molecular Thermodynamics of Fluid-Phase Equilibria*, Pearson Education Taiwan Ltd., Taipei.

Sørensen, J.M. & Arlt, W. (1979). Liquid-Liquid Equilibrium Data Collection, *Dechema*, Frankfurt.

Sada, E. & Morisue, T. (1975). Isothermal Vapor-Liquid Equilibrium Data of Isopropanol-Water System, *Journal of Chemical Engineering of Japan*, 8, 3, pp. 191-195

Sandler, S.I. (2006). *Chemical, Biochemical, and Engineering Thermodynamics*, John Weily & Sons, 978-0471661740, New York.

Shu, C.C. & Lin, S.T. (2010). Prediction of drug solubility in mixed solvent systems using the COSMO-SAC activity coefficient model, *Industrial & Engineering Chemistry Research*, in press,

Still, W.C.; Tempczyk, A.; Hawley, R.C. & Hendrickson, T. (1990). Semianalytical treatment of solvation for molecular mechanics and dynamics, *Journal of the American Chemical Society*, 112, 16, pp. 6127-6129,0002-7863.

Tomasi, J. & Persico, M. (1994). Molecular-interactions in solution - An overview of methods based on continuous distributions of the solvent, *Chemical Reviews*, 94, 7, pp. 2027-2094

Tung, H.H.; Tabora, J.; Variankaval, N.; Bakken, D. & Chen, C.C. (2008). Prediction of pharmaceutical solubility via NRTL-SAC and COSMO-SAC, *Journal of Pharmaceutical Sciences*, 97, 5, pp. 1813-1820,0022-3549.

Wu, H.S.; Locke, W.E. & Sandler, S.I. (1991). Isothermal Vapor-Liquid-Equilibrium of Binary-Mixtures Containing Morpholine, *Journal of Chemical & Engineering Data*, 36, 1, pp. 127-130,0021-9568.

18

Complex Fluid Phase Equilibrium Modeling and Calculations

Gholamreza Vakili-Nezhaad
Department of Petroleum & Chemical Engineering, College of Engineering, Sultan Qaboos University, Muscat
Oman &
Department of Chemical Engineering, Faculty of Engineering, University of Kashan, Kashan
Iran

1. Introduction

Thermodynamic modeling and calculations of different complex fluid mixtures are presented in this chapter. This study includes four parts. In the first part phase equilibrium calculations of the system polystyrene-methylcyclohexane with the Sanchez-Lacombe equation of state is presented using continuous thermodynamics. In this part the Sanchez-Lacombe equation of state is used to model the stability and cloud-point curves of polystyrene (PS) in methylcyclohexane (MCH) systems. An algorithm based on the work of Browarzik and Kowalewski is applied. Three different polymers are studied. Two of them are monodisperse, and the third is polydisperse. To describe the polydispersity of polystyrene, Schulz-Flory distribution function is considered. One of the monodisperse systems shows lower critical point (LCP) and upper critical point (UCP) curves at a certain temperature region, which turn into hour-glass shaped cloud-point curves by lowering the temperature. Excellent agreement with the experimental data is observed. Polymer parameters are fitted to the experimental data. These parameters are used in modeling the other systems and the results are compared with experimental data. The second part of this work presents the application of continuous thermodynamics to investigate the limited miscibility of methanol–gasoline blends. To predict the liquid–liquid equilibrium of these systems, the Gaussian distribution function was used to represent the composition of paraffins in the gasoline. The naphthenes and aromatics were represented by model compounds. A model has been developed using three different continuous versions of the UNIFAC model. Methanol is an associating component, and association affects phase equilibria. Therefore, the CONTAS (continuous thermodynamics of associating systems) model based on the Flory–Huggins equation, for multicomponent methanol–gasoline blends has also been investigated. The predicted results including the cloud point curve, shadow curve and phase separation data have been compared with experimental data and good agreement was found for the two UNIFAC and CONTAS models. In part three a method based on continuous thermodynamics has been presented for calculating the vapor pressure of undefined composition mixtures. In order to verify the proposed method the experimental data of two real samples have been used. In this work by using the combination of the Trouton's rule and the Clausius–Clapeyron equation in the well-known equation of Antonie, the continuous version of the Raoult's law has been developed and the results of the modeling and calculations have been compared with the experimental data and good agreement was found. Moreover, for improving the results of the ideal continuous mixture, modeling of non-ideal behavior of such mixtures has been carried out by introducing the continuous version of the UNIFAC activity coefficient in the formulation and a better result has been obtained. In the last section a new approach based on the genetic algorithm has been proposed for solving three phase flash calculations containing two liquid phases and one vapor phase. Based on this approach choosing the initial guesses for the compositions of the involved phases is not an important step, which in the ordinary three phase algorithms affects directly the convergence of the calculations. A real problem has been solved with this approach and very good results have been obtained in comparison with the experimental data. φ-φ approach has been adopted for vapor-liquid-liquid equilibrium (VLLE) calculations based on the Peng-Robinson equation of state.

2. Phase equilibrium calculations of polymer solution systems using continuous thermodynamics

The Sanchez-Lacombe equation of state is used to model the stability and cloud-point curves of polystyrene (PS) in methylcyclohexane (MCH) systems. An algorithm based on the work of Browarzik and Kowalewski (2002) is applied. Three different polymers are studied. Two monodisperse systems and one polydisperse system have been studied in this section. To describe the polydispersity of polystyrene, Schulz-Flory distribution function is considered. One of the monodisperse systems shows lower critical point (LCP) and upper critical point (UCP) curves at a certain temperature region, which turn into hour-glass shaped cloud-point curves by lowering the temperature. Excellent agreement with the experimental data is observed. Polymer parameters are fitted to the experimental data. These parameters are used in modeling the other systems and the results are compared with experimental data (Behnam, 2008).

2.1 Introduction

The system polystyrene (PS) in methylcyclohexane (MCH) has been studied by some researchers (Xiong & Kiran, 2000; Wachnic & Van Hook, 2004) and a lot of experimental data exist for this system. Narita et al. (2003) determined a semiempirical equation of Gibbs free energy of mixing (ΔG) for the system based on the Flory-Huggins equation. The equation of ΔG was used in preparation of controlled size distribution microcapsules. Vanhee et al. (1994) used a nearly monodisperse PS. They applied a pressure-pulse -induced scattering (PPIS) technique and obtained spinodal and cloud-point curves. They observed the pressure-concentration diagram shows upper critical point (UCP) and lower critical point (LCP) behavior for some temperatures. These curves come closer and turn into hour-glass shaped curves by lowering the temperature. Browarzik and Kowalewski (2004) studied the mentioned system by using SWP (Sako-Wu-Prausnitz) equation of state which is able to predict polymer solutions. They also discussed the transition of the UCP/LCP behavior into the hour-glass diagrams for polydisperse polymer and studied the polydispersity effect on stability and phase equilibria. Koak et al. (1998) measured phase equilibria for PS-MCH systems in the vicinity of the upper critical solution temperature (UCST) in the glass tube Cailletet apparatus for two samples of polystyrene. One was nearly monodisperse and the other was polydisperse. Here, the Sanchez-Lacombe equation of state has been chosen. Three different PS solutions based on the measurements of Vanhee et al. (1994) and Koak et al. (1998) were studied (Behnam, 2008). The method developed by Browarzik and Kowalewski (2002, 2004) was applied to obtain the stability and phase equilibria equations. To describe the molecular weight polydispersity of the polymer, Schulz-Flory distribution function which is appropriate to characterize PS samples (Browarzik & Kowalewski, 2002; Enders & de Loos, 1997) were considered and the results were compared with experimental findings.

2.2 Spinodal, cloud-point and critical conditions

Continuous thermodynamics is an appropriate method in treatment of multi-component mixtures such as polydisperse polymer solutions. A continuous distribution function is used to represent the polymer (B). The solvent is denoted as A. Considering the polymer described by the distribution function $W_S(r)$, where r shows the segment numbers of the molecules of the polymer species, $W_S(r)\,dr$ is the segment fraction of all polymer species with segment numbers between r and $r + dr$. So the integral of $W_S(r)\,dr$ over the entire range of r is 1. The limits of the integral are usually 0 and ∞. Browarzik and Kowalewski (1999) developed a method based on an EOS to calculate spinodal curves and critical points of a system containing a polydisperse polymer, a solvent and a gas. In 2002, they studied a system containing a polymer and a solvent and determined the segment-molar Helmholtz energy of a mixture whose non-linear parts with respect to segment mole fractions can be presented as follows (Browarzik & Kowalewski, 2002) :

$$\Delta A_s = RT\left\{\int \frac{\psi}{r} W_s(r)\ln[\psi W_s(r)]\,dr + \frac{1-\psi}{r_A}\ln(1-\psi)\right\} - \frac{RT}{\bar{r}^M}\ln V_s + J_s \tag{1}$$

The linear parts do not influence the stability. The first term on the right-hand side of Eq. (1) is the Flory-Huggins contribution. R is the universal gas constant and T is the temperature. $W_s(r)$ is the segment –molar distribution function of the polymer. r_A is the segment number of the solvent. ψ is the total segment-mole fraction of the polymer. Since it was assumed the segment number is proportional to the molecular weight, ψ is the total weight fraction of the polymer as well. $\bar{r}^M$ is the number-averaged segment number of the mixture which reads:

$$\frac{1}{\bar{r}_M} = \frac{1-\psi}{r_A} + \frac{\psi}{\bar{r}_B};\quad \frac{1}{\bar{r}_B} = \int \frac{W_s(r)}{r}\,dr \tag{2}$$

$\bar{r}_B$ is the number-averaged segment number of the polymer B and is proportional to number average molar mass of the polymer ($\bar{M}_n$). By division of molar volume through $\bar{r}^M$ the segment-molar volume, Vs, is obtained. The last term on the right hand side of Eq. (1), J_s, is related to the equation of state and is obtained by:

$$J_s = \int_{V_S}^{\infty}\left(p - \frac{RT}{\bar{r}_M V_s}\right) dV_s \tag{3}$$

In order to find spinodal points, the following equation should be solved:

$$K_0 = \alpha_{VV}\alpha_{\psi\psi} - (\alpha_{V\psi})^2 = 0 \tag{4}$$

$$\alpha_{VV} = J_{VV} + \frac{RT}{(V_s)^2}\left\{\frac{1-\psi}{r_A} + \frac{\psi}{\overline{r}_B(1+U)}\right\} \tag{4a}$$

$$\alpha_{\psi\psi} = J_{\psi\psi} + RT\left\{\frac{1}{r_A(1-\psi)} + \frac{1}{\overline{r}_B(1+U)\ \psi}\right\} \tag{4b}$$

$$\alpha_{V\psi} = J_{V\psi} + \frac{RT}{V_s}\left\{\frac{1}{r_A} - \frac{1}{\overline{r}_B(1+U)}\right\} \tag{4c}$$

The parameters J_{VV}, $J_{V\psi}$ and $J_{\psi\psi}$ are the second derivatives of J_s with respect to Vs and ψ. U is non-uniformity and shows polydispersity of the polymer and is defined by:

$$U = \frac{\overline{r}_B^{(1)}}{\overline{r}_B} - 1 = \frac{\overline{M}_W}{\overline{M}_n} - 1\ ;\quad \overline{r}_B^{(1)} = \int r W_s(r)\ dr\ ;\quad \overline{r}_B^{(2)} = \int r^2\ W_s(r) dr \tag{5}$$

$\overline{M}_W$ and $\overline{M}_n$ are mass average molar mass and number average molar mass of the polymer respectively. Browarzik and Kowalewski (2004) developed two equations for critical points. The first is from Eq. (4) and the second is obtained by solving Eq. (6).

$$K_1 = L_V\alpha_{\psi\psi} - L_\psi\alpha_{V\psi} + L_r\overline{r}_B^{(1)}\left\{\frac{\psi}{V_s}\left[1 - \frac{\overline{r}_B^{(2)}}{\left(\overline{r}_B^{(1)}\right)^2}\right]\alpha_{\psi\psi} - \frac{\overline{r}_B^{(2)}}{\left(\overline{r}_B^{(1)}\right)^2}\alpha_{V\psi}\right\} = 0 \tag{6}$$

$$L_V = \alpha_{\psi\psi}\left\{J_{VVV} - \frac{2RT}{(V_s)^3}\left[\frac{1-\psi}{r_A} + \frac{\psi}{\overline{r}_B^{(1)}}\right]\right\} + \alpha_{VV}J_{\psi\psi V} - 2\alpha_{V\psi}\left\{J_{V\psi V} - \frac{RT}{(V_s)^2}\left[\frac{1}{r_A} - \frac{1}{\overline{r}_B^{(1)}}\right]\right\} \tag{6a}$$

$$L_\psi = \alpha_{\psi\psi}\left\{J_{VV\psi} + \frac{RT}{(V_s)^2}\left[-\frac{1}{r_A} + \frac{2}{\overline{r}_B^{(1)}}\right]\right\}$$

$$+\alpha_{VV}\left\{J_{\psi\psi\psi} + \frac{RT}{r_A(1-\psi)^2}\right\} - 2\alpha_{V\psi}\left\{J_{V\psi\psi} - \frac{RT}{V_s}\frac{1}{\psi\,\overline{r}_B^{(1)}}\right\} \tag{6b}$$

$$L_r = -\frac{RT}{\left[\overline{r}_B^{(1)}\right]^2}\left\{\frac{\alpha_{\psi\psi}}{(V_s)^2} + \frac{\alpha_{VV}}{\psi^2} + \frac{2\alpha_{V\psi}}{\psi\ V_s}\right\} \tag{6c}$$

To calculate phase equilibria, the fugacity coefficients are used. The equilibrium condition for the phases I and II reads (Browarzik & Kowalewski, 2002):

$$X_i^I \phi_i^I = X_i^{II} \phi_i^{II} \tag{7}$$

$$\psi_i = X_i \frac{r_i}{\bar{r}^M} \tag{8}$$

X_i is the mole fraction of the component i , ψ_i is the segment-mole fraction of i and r_i shows the segment-number of i. Combining Eq. (7) with Eq. (8) for solvent and the polymer species yields :

$$\left(1-\psi^I\right)\ \phi_A^I\ \bar{r}^{MI} = \left(1-\psi^{II}\right)\ \phi_A^{II}\ \bar{r}^{MII} \tag{9}$$

of the component i, Φi, may be obtained by:

$$\ln\phi_i = \frac{1}{RT}\int_{\bar{V}}^{\infty}\left(\frac{\partial p}{\partial n_i} - \frac{RT}{\bar{V}}\right) d\bar{V} - \ln\left(\frac{p\bar{V}}{nRT}\right) \quad 10)$$

$\bar{V}$ is the total volume ($\bar{V}=nV$) and V is the molar volume.

The fugacity coefficients of the solvent A and the species of the polymer B may be expressed by (Browarzik & Kowalewski, 2002):

$$\ln\phi_A = r_A B_A - \ln\bar{r}^M - \ln\left(\frac{pV_s}{RT}\right) \tag{11}$$

$$\ln\phi_B(r) = rB_B - \ln\bar{r}^M - \ln\left(\frac{pV_s}{RT}\right) \tag{11a}$$

B_A and B_B depend on the equation of state which is used for modeling. Combining Eq. (9) with Eq. (11) yields:

$$\frac{1}{r_A}\ln\left(\frac{\left(1-\psi^I\right)\ V_s^{II}}{\left(1-\psi^{II}\right)\ V_s^I}\right) + B_A^I - B_A^{II} = 0 \tag{12}$$

To represent the polymer, Schulz-Flory distribution function is chosen:

$$W_s^I(r) = \frac{k^k}{\Gamma(k)\ \bar{r}_B^I}\left(\frac{r}{\bar{r}_B^I}\right)^k \exp\left[-k\frac{r}{\bar{r}_B^I}\right] \tag{13}$$

Γ is the Γ-function and $k=1/U$. The distribution function of the phase II is also a Schulz-Flory function with the parameter $\bar{r}_B^{II}$ and the same value of k. Introducing distribution functions of both phases into Eq. (9a) and combining with Eq. (11a) considering the resultant equation is valid for all amounts of r between 0 and ∞, following equations are obtained :

$$\overline{r}_B^{II} = \overline{r}_B^{I}\left(\frac{\psi^{II}V_s^{I}}{\psi^{I}V_s^{II}}\right)^{U/U+1} \tag{14}$$

$$\left(\frac{\psi^{I}V_s^{II}}{\psi^{II}V_s^{I}}\right)^{U/U+1} - 1 + \overline{r}_B^{I}U\left(B_B^{I} - B_B^{II}\right) = 0 \tag{15}$$

For given temperature, the parameters of pressure p, the mass fraction of the polymer in phase II ψ^{II}, Vs^{I} and Vs^{II} are unknown. For given p and ψ^{II}, the values of Vs^{I} and Vs^{II} can be calculated by using the EOS. To find p and ψ^{II}, Eqs .(12) and (15) are solved simultaneously.

2.3 Sanchez-Lacombe equation of state

The Sanchez-Lacombe EOS (Sanchez & Lacombe, 1978; Sanchez & Lacombe, 1976) is used for calculation of stability and phase equilibria of the system PS-MCH. Gauter and Heidemann (2001) used this EOS for modeling polyethylene-solvent mixtures. The Sanchez-Lacombe equation has proven to be suitable for polymer systems. This EOS is used in the form reported by Koak and Heidemann (1996):

$$\frac{p}{RT} = \frac{1-d}{V} - \frac{d^2}{b}\ln\left(\frac{V - b/d}{V}\right) - \frac{a/RT}{V^2} \tag{16}$$

$$b = \sum_{i=1}^{n_C}\sum_{j=1}^{n_C} X_i X_j b_{ij}\ ; \qquad b_{ij} = d_i d_j v_{ij} = d_i d_j\left(v_{ii} + v_{jj}\right)/2 \tag{16a}$$

$$a = \sum_{i=1}^{n_C}\sum_{j=1}^{n_C} X_i X_j a_{ij}\ ; \qquad a_{ij} = d_i d_j \varepsilon_{ij} v_{ij} \qquad \varepsilon_{ij} = \left(1 - k_{ij}\right)\sqrt{\varepsilon_{ii}\ \varepsilon_{jj}} \tag{16b}$$

$$d = \sum_{i=1}^{n_C} X_i d_i \tag{16c}$$

In order to apply continuous thermodynamics method, the Sanchez-Lacombe EOS is expressed in segment-molar notation. The segment-molar parameters a_s and b_s are obtained by division of a and b through $(\overline{r}^{M})^2$ and the parameter d_s is obtained by division of d through $\overline{r}^{M}$. Mole fractions change into weight fractions through Eq. (8).The parameters d_A and d_B are:

$$d_A = d_{s,A}\, r_A\ ; \quad d_B = d_{s,B}\, \overline{r}_B \tag{17}$$

The Sanchez-Lacombe EOS in segment-molar notation is obtained by:

$$p = \frac{RT}{\overline{r}^{M}V_s} - \frac{RTd_S}{V_s} - \frac{RTd_S^2}{b_S}\ln\left(\frac{V_s - b_S/d_S}{V_s}\right) - \frac{a_S}{V_s^2} \qquad 18)$$

$$a_S = \psi_A^2\, d_{s,A}^2 \varepsilon_{AA} v_{AA} + \psi_B^{\,2} d_{s,B}^2\, \varepsilon_{BB} v_{BB} + 2\psi_A \psi_B d_{S,A} d_{S,B} \varepsilon_{AB} v_{AB} \qquad 18a)$$

$$b_S = \psi_A^2\, d_{s,A}^2 v_{AA} + \psi_B^{\,2} d_{s,B}^2\, v_{BB} + 2\psi_A \psi_B d_{S,A} d_{S,B} v_{AB} \tag{18b}$$

$$d_S = \psi_A d_{s,A} + \psi_B d_{s,B} \quad 18c)$$

By introducing the EOS into Eq. (3) J_s is found:

$$J_S = RTd_S\left[1+\ln\left(\frac{V_s - b_S/d_s}{V_s}\right)\left(\frac{V_s d_s}{b_S} - 1\right)\right] - \frac{a_S}{V_s} \quad (19)$$

J_{vv}, $J_{\psi\psi}$ and $J_{V\psi}$ are needed for calculation of spinodal curves. To calculate the cloud-point curves, Eqs. (10) and (11-11a) are used to find B_i (i=A, B).

$$B_i = d_{S,i}\left[Fi\frac{b_S}{d_S} + \ln\left(\frac{V_s - b_S/d_S}{V_s}\right)(F_i V_s - 1) - \frac{H_i}{RTV_s}\right] \quad (20)$$

$$F_A = \frac{2d_S}{b_S}\left(1 - \frac{d_S\tilde{b}_A}{b_S}\right); \qquad \tilde{b}_A = \psi_A d_{s,A}v_{AA} + \psi_B d_{S,B}v_{AB} \quad (20a)$$

$$F_B = \frac{2d_S}{b_S}\left(1 - \frac{d_S\tilde{b}_B}{b_S}\right); \qquad \tilde{b}_B = \psi_B d_{s,B}v_{BB} + \psi_A d_{S,A}v_{AB} \quad (20b)$$

$$H_A = 2\ \tilde{a}_A; \qquad \tilde{a}_A = \psi_A d_{s,A}\varepsilon_{AA}v_{AA} + \psi_B d_{S,B}\varepsilon_{AB}v_{AB} \quad (20c)$$

$$H_B = 2\ \tilde{a}_B; \qquad \tilde{a}_B = \psi_B d_{s,B}\varepsilon_{BB}v_{BB} + \psi_A d_{S,A}\varepsilon_{AB}v_{AB} \quad (20d)$$

2.4 Pure-component parameters

To calculate the segment numbers of the solvent and the polymer, the MCH molecules are chosen as reference segment with r_A=1. Since it was assumed the segment number is proportional to the molecular weight M, the segment number r of the polymer species is (Browarzik & Kowalewski, 2002):

$$r = 10.1852\frac{M}{1000} \quad (21)$$

Three pure-component parameters are needed for both the solvent and the polymer. Koak (1997) determined these parameters of Sanchez-Lacombe EOS for MCH. Gauter and Heidemann (2001) determined equations to calculate three pure-component parameters of Sanchez-Lacombe EOS from critical temperature, critical pressure and acentric factor:

$$d = 5.1178 + 13.5698w + 5.9404w^2 - 1.2952w^3 \quad (22)$$

$$\varepsilon = \frac{RT_C}{2}\frac{\left(1+\sqrt{d}\right)^2}{d} \quad (23)$$

$$v = \frac{RT_C}{P_C}\left[\ln\left(\frac{1+\sqrt{d}}{\sqrt{d}}\right) - \frac{\sqrt{d}-1/2}{d}\right] \tag{24}$$

For MCH, the resultant parameters and those of Koak (1997) are as follows:

$$v_{AA} = 0.01443 \ \frac{m^3}{kgmol}; \ \varepsilon_{AA} = 4.2734 \ \frac{Mpa.\ m^3}{kgmol} \ ; \ d_{S,A} = 8.6222\ ; \tag{25}$$

three polystyrene samples are used for modeling. They differ in their average molar mass and polydispersity. The first sample (PS1) used by Vanhee et al. (1994) had weight-average molar mass $\bar{M}_w$=17500 (g/mol) and the ratio $\frac{\bar{M}_w}{\bar{M}_n}$ =1.06. To find three pure-component parameters $d_{s,B}$, υ_{BB}, ε_{BB} ,and the interaction parameter k_{AB} ,experimental data of PS1 is used. Since the interaction parameter has an important role and the results are extremely sensitive to this parameter, a temperature dependence is necessary to get good results. So, at the first step, the parameters were fitted to the experimental spinodal pressures at 20.8 ° C. Two extreme points on both UCP/LCP curves and an arbitrary point on the UCP curve are used. The parameter υ_{BB} was fixed and the remaining parameters were obtained. The best set of these parameters is:

$$v_{BB} = 0.007 \ \frac{m^3}{kgmol}; \ \varepsilon_{BB} = 4.5742379 \ \frac{Mpa.\ m^3}{kgmol} \ ; \ d_{S,B} = 7.93475659\ ; \tag{26}$$

$$k_{AB}=0.0014877 \tag{26a}$$

To find the temperature dependence of k_{AB}, a linear temperature function is assumed. For PS1, experimental spinodal points and cloud-points at 20.6 and 20.7 ° C are used for fitting:

$$k_{AB} = 0.0014877 - 0.00035\ (T - 293.95) \tag{27}$$

There are transitions of UCP/LCP curves into hour-glass shaped two phase region curves for PS1 at 21.2 °C. Lowering the temperature causes the two miscibility gaps to come closer together. They merge and form an hour-glass shaped two-phase region at 20.6 °C. After formation of the hour-glass shaped curves, at temperatures below the merging of LCP and UCP, the critical points disappear and spinodal curves are not tangent to cloud-point curves anymore. It was shown that the hour-glass shape for the spinodal curves and for the cloud-point curves are formed at same temperatures. These are not the case in polydisperse samples (Browarzik & Kowalewski, 2002). Koak et al. (1998) reported cloud-point data for two samples of PS in MCH. They measured phase equilibria in the glass tube Cailletet apparatus. Temperatures and compositions were in the region of UCST behavior and pressures ranged from 1 to 14 MPa. One sample (PS2) was nearly monodisperse with $\bar{M}_w$=31600 and $\bar{M}_n = 29100$ (g/mol). Another PS3 had $\bar{M}_w$=250000 and $\bar{M}_n = 64000$

(g/mol). Experimental temperatures ranged from 297-306 K for PS2 and 323-331 K for PS3. The mass fractions of the polymer ranged from 0.03 to 0.27 for PS2 and between 0.02 and 0.14 for PS3. To calculate cloud-point curves of PS2 and PS3 systems, pure-component parameters of MCH and PS of the PS1 are used. Since k_{AB} can depend on the molecular weight, we fitted it again to experimental cloud-point data. Given data, at ψ_B=0.2634 for PS2 and at ψ_B=0.136, for PS3 are used to fit the function k_{AB} (T) given by Eqs. (28) and (29).

$$\text{For PS2:} \quad k_{AB} = -0.00039749270543\ T + 0.120703166 \tag{28}$$

$$\text{For PS3:} \quad k_{AB} = -0.00039740070\ T + 0.128231617 \tag{29}$$

Calculated cloud-point data for other concentrations for both PS2 and PS3 systems deviate from experimental data to some extent. The reason might be the effect of temperature on the function k_{AB} (T).

2.5 Discussion

The Sanchez-Lacombe equation of state may be able to model the stability and cloud-point curves of polystyrene (PS) in methylcyclohexane (MCH) systems. Polydispersity of polystyrene is described by using continuous thermodynamics. PS/MCH systems may show lower critical point (LCP) and upper critical point (UCP) curves at a certain temperature region. These curves turn into hour-glass shaped curves by lowering the temperature. Sanchez-Lacombe equation of state can predict this transition. By formation of hour-glass shaped curves, the critical points no longer exist, and spinodal curves are not tangent to cloud-point curves anymore. The hour-glass shape temperatures for the spinodal curves and for the cloud-point curves are the same. However, for polydisperse case the transition temperatures for spinodal and cloud-point curve may be different. The treatment may be able to predict other monodisperse and polydisperse systems if the temperature is not in a wide range.

3. Liquid-liquid equilibrium calculations for methanol-gasoline blends using continuous thermodynamics

In this section we present the application of continuous thermodynamics to investigate the limited miscibility of methanol-gasoline blends (Nasrollahi et al., 2009). To predict the liquid–liquid equilibrium of these systems, the Gaussian distribution function was used to represent the composition of paraffins in the gasoline. The naphthenes and aromatics were represented by model compounds. A model has been developed using three different continuous versions of the UNIFAC model. Methanol is an associating component, and association affects phase equilibria. Therefore, the CONTAS (continuous thermodynamics of associating systems) model based on the Flory–Huggins equation, for multicomponent methanol-gasoline blends has also been investigated. The predicted results including the cloud-pint curve, shadow curve and phase separation data have been compared with experimental data and good agreement was found for the two UNIFAC and CONTAS models (Nasrollahi et al., 2009).

3.1 Introduction

Ethanol, methanol and their blends with gasoline are known as important alternative fuels for motor engines (Agarwal, 2007; Yuksel & Yuksel, 2004). A serious problem encountered in using gasoline-alcohol blends as motor fuel at low temperatures is the separation of the mixture into two liquid phases (French & Malone, 2005; Cox, 1979). Therefore, it is important to investigate the limited miscibility of alcohol-gasoline blends. In alcohol-gasoline systems phase separation is strongly influenced by the water content, temperature and composition of aromatic components in the gasoline. Gasoline is a multicomponent mixture of paraffins, naphthenes and aromatics. In order to apply conventional thermodynamics to phase equilibrium calculations of a complex mixture, a complete chemical analysis of the mixture is required. Even if such analysis were available, inclusion of all components would lead to massive calculations. The pseudo-component (Chorn & Mansoori, 1989; Neau et al., 1993) and continuous thermodynamics (Cotterman et al., 1985; Du & Mansoori, 1986) methods are the main approaches presented in the literature for phase equilibrium calculations of complex mixtures. In the pseudo–component approach the multicomponent mixture is represented by key components or model compounds. Whereas, in the continuous thermodynamics approach the composition of these complex mixtures is described by a continuous distribution function of a measurable variable like molecular weight, carbon number or boiling point temperature. This model is suitable for phase equilibrium calculation of systems containing many similar components like petroleum fluids, natural gases, vegetable oils and polymer solutions. In a majority of the published works, gasoline has been considered as a mixture of naphthenic, aromatic and paraffinic representative components and the pseudo-component approach has been applied to predict liquid-liquid equilibrium of gasoline-methanol blends (Leeper & Wankat, 1982). Ruzicka et al. (1983) developed a method based on the group-contribution concept to describe the fossil fuels in terms of model compounds. They generated LLE data over the temperature range of - 20 °C to 20 °C and subsequently developed LLE models to predict their experimental results (Ruzicka et al., 1986). They characterized gasoline by a set of naphthenic, aromatic and paraffinic model compounds to predict limited miscibility of methanol-gasoline blends applying the UNIFAC model. It was concluded that gasoline can be characterized as a mixture of three arbitrary model compounds. Kehlen et al. (1988) studied liquid-liquid equilibrium of systems containing petroleum fractions like gasoline-methanol blends applying the continuous thermodynamics method and a continuous version of the modified UNIFAC model. Strong H-bonds in methanol molecules lead to the formation of methanol chain associates in the gasoline mixture which strongly affects phase equilibria. Different methods for phase equilibria calculations of associating systems are proposed in the literature. Recently the highly accurate statistical association fluid theory (SAFT) based on physical theory has been applied extensively ([Huang & Radosz, 1991). Since the SAFT equation of state requires a high numerical expense, a simpler G^E-model which is called the CONTAS (continuous thermodynamics of associating systems) model has recently been introduced by Browarzik (2004). The model, which is suitable in the moderate pressure regions, is based on the Flory–Huggins theory and continuous thermodynamics approach. In this model the composition of chains of associates in the mixture is described by a continuous distribution function derived from the mass action law. Browarzik applied the CONTAS model to different binary associating systems (Browarzik, 2005), and presented the equations to calculate excess properties like the excess Gibbs energy and the excess enthalpy (Browarzik, 2007).

In this section both the pseudo component and the continuous thermodynamics approaches have been employed. The naphthenic and aromatic components in the gasoline were represented by two model compounds. The structure of these model compounds for the gasoline mixture is presented by Ruzicka et al. (1983). The compositions of paraffinic components were described by the Gaussian distribution function. The CONTAS model is applied to the multicomponent gasoline-methanol blends and the results obtained from the CONTAS model and a continuous version of the UNIFAC model, the Dortmund and Lyngby modified UNIFAC models were compared with experimental data (Ruzicka et al., 1986).

3.2 Theoretical background

A mixture of continuous paraffinic components (p), methanol (B) and two discrete naphthenic (n) and aromatic (a) model compounds were considered. The TBP distillation curve and properties of the investigated gasoline are presented by Ruzicka et al. (1986). The following PNA analysis is given: 64.6% paraffins, 27.6% naphthenes and 7.8% aromatics. They determined the structures of naphthenic and aromatic model compounds of this gasoline mixture. The symmetrical Gaussian distribution function which is suitable for petroleum fluids is employed to show the distribution of paraffinic components.

$$F_p(N) = \frac{1}{\sigma\sqrt{2\pi}} \exp\{\frac{-(N-\overline{N})^2}{2\sigma^2}\} \tag{30}$$

The independent variable N is the carbon number of paraffins. The Gaussian distribution function is characterized by mean carbon number $\overline{N}$ and standard deviation σ. For paraffinic components in the gasoline $\overline{N}$ =7.8 and $\sigma^2 = 0.5$.

3.2.1 The CONTAS model

The distribution function of associates is a function of association degree or the number of methanol monomers forming the associates. It indicates the mole fraction of chain associates consisting of r monomers. This function is normalized such that:

$$\int_1^\infty W(r)dr = 1 \tag{31}$$

The average number of the association degree is given by:

$$\overline{r}_B = \int_1^\infty rW(r)dr \tag{32}$$

Assuming methanol monomer as the standard segment, the segment number of naphthenes, aromatics and paraffins is defined as follows (Browarzik, 2004b):

$$r_i = V_{Wi} / V_{WB}, \qquad i = n, a \tag{33}$$

$$r_p(N) = V_{Wp}(N) / V_{WB} \tag{34}$$

Where V_{Wi} is the Van der Waals volume of component *i*. For the discrete components, the segment number takes constant values while the segment number of continuous paraffinic components is a function of carbon number. V_{Wp} in Eq. (34) is calculated by:

$$V_{Wp}(N) = \sum_{m} \upsilon_{p,m}(N) V_{Wm} \tag{35}$$

Here, V_{Wm} denotes the Van der Waals volume of the group of type m and $\upsilon_{p,m}$ is the number of functional groups of type m in a paraffinic component consisting of *N* carbon atoms. The number of different functional groups in the paraffinic, naphthenic and aromatic components of a petroleum fraction has been approximated by a second order polynomial function of carbon number (Kehlen et al., 1988). The average segment number of paraffins and the mixture are defined by:

$$\bar{r}_p = \int_{N0}^{Nt} r_p(N) F_p(N) dN \tag{36}$$

$$\bar{r}^M = (1 - x_B)(x_p^g \bar{r}_p + x_n^g r_n + x_a^g r_a) + x_B \bar{r}_B \tag{37}$$

The quantity x_B is the mole fraction of associates in the gasoline-methanol mixture and x_i^g is the mole fraction of component *i* in the gasoline. The following equation can be written for the association equilibrium:

$$B_{r-r'} + B_{r''} \Leftrightarrow B_r \tag{38}$$

By applying the mass action law to the association equilibrium, the association constant is obtained (Browarzik, 2004).

$$K^L = \frac{x_B\, W(r) \gamma_B(r)}{x_B\, W(r - r') \gamma_B(r - r') x_B\, W(r') \gamma_B(r')} \tag{39}$$

Due to the similarity of associating systems and polymer solutions, the activity coefficient in this equation is calculated by using the Flory-Huggins equation of polymer solutions:

$$\ln \gamma_B(r) = \ln(r / \bar{r}^M) + 1 - r / \bar{r}^M + r \ln \gamma_{s,B} \tag{40}$$

The activity coefficients of naphthenes, aromatics and paraffins as well as the methanol activity coefficient are obtained by applying the Flory-Huggins theory. The activity coefficients of discrete naphthenic and aromatic components are:

$$\ln \gamma_i = \ln(r_i / \bar{r}^M) + 1 - r_i / \bar{r}^M + r_i \ln \gamma_{s,i}, \qquad i = n, a \tag{41}$$

The activity coefficient of each paraffinic component as a function of carbon number is:

$$\ln \gamma_p(N) = \ln(r_p(N) / \bar{r}^M) + 1 - r_p(N) / \bar{r}^M + r_p(N) \ln \gamma_{s,p} \tag{42}$$

In the Flory-Huggins equation the intermolecular interactions are described by the quantity $\gamma_{s,i}$ which is called the segment molar activity coefficient.

$$\ln \gamma_{s,i} = \frac{1}{T}(\chi_i - \sum_i \psi_i \chi_i)^2 \tag{43}$$

In this equation the χ_i-prameter is related to the solubility parameter of *i*-th component as follows:

$$\chi_i = (\frac{V_s}{R})^{0.5} \sigma_i \tag{44}$$

Where V_s is the molar volume of standard segment and σ_i is the solubility parameter of i-th component. The values of χ_i-parameters for the paraffins, naphthenes, aromatics and methanol should be fitted to experimental data. The parameter χ_i for the paraffinic components and methanol is assumed to be independent of carbon number and association degree. The summation is written over the paraffinic, naphthenic, aromatic and methanol components. ψ_i is the segment mole fraction of component *i* and it is given by:

$$\psi_i = x_i \bar{r}_i / \bar{r}^M \tag{45}$$

The quantity x_i is a function of true mole fraction of methanol and average number of association degree (Browarzik, 2004). Using this expression and Eqs. (37) and (45), the following equation for the segment mole fraction of methanol is derived:

$$\psi_B = \frac{x_B^t}{x_B^t + (1 - x_B^t)(x_p^g \bar{r}_p + x_n^g r_n + x_a^g r_a)} \tag{46}$$

Combining Eqs. (39) and (40) and after some rearrangement, the molar distribution function of associates is expressed by the following equation:

$$W(r) = -\lambda \exp[\lambda(r-1)] \bar{r}_B / r \tag{47}$$

Where λ takes negative values and it can be obtained by solving the following equation and with the aid of the expression for the temperature dependence of K^L (Browarzik, 2004).

$$\exp(\lambda) + K^L e \psi_B \lambda = 0, \qquad \lambda < 0 \tag{48}$$

Substitution of Eq. (47) into Eq. (32) and integration of this equation with respect to *r* results in:

$$1/\bar{r}_B = \lambda \exp(-\lambda) \times \{0.57721567 + \ln|\lambda| + \frac{\lambda}{1\times 1!} + \frac{\lambda^2}{2\times 2!} + \frac{\lambda^3}{3\times 3!} + \cdots + \frac{\lambda^n}{n \times n!}\} \tag{49}$$

In order to perform phase equilibrium calculations, phase I is assumed to be at equilibrium with phase II and the quantity K_i is defined by the following equation:

$$K_i = \gamma_i^{I} / \gamma_i^{II} \tag{50}$$

With the aid of Eq. (50) K_i is obtained for the paraffinic, naphthenic, aromatic and methanol components.

$$K_i = \exp(r_i \rho_i) \bar{r}_m^{II} / \bar{r}_m^{I} \qquad i = n, a \tag{51}$$

$$K_p(N) = \exp(r_p(N) \rho_p) \bar{r}_m^{II} / \bar{r}_m^{I} \tag{52}$$

$$K_B(r) = \exp(r \rho_B) \bar{r}_m^{II} / \bar{r}_m^{I} \tag{53}$$

Where

$$\rho_i = 1 / \bar{r}_m^{II} - 1 / \bar{r}_m^{I} + \ln \gamma_{s,i}^{I} - \ln \gamma_{s,i}^{II} \tag{54}$$

3.2.2 The semi-continuous UNIFAC models

In the UNIFAC model the methanol component is considered as a discrete component like naphthenes and aromatics. This is in contrast with the CONTAS model where methanol is considered as a continuous mixture of chain associates with a different association degree. The UNIFAC model is a suitable model for the prediction of activity coefficients of a mixture including different principal types of hydrocarbons. Using the UNIFAC model, the quantity K_i can be separated into two combinatorial and residual parts.

$$\ln K_i = \ln K_i^C + \ln K_i^R \tag{55}$$

To estimate the activity coefficient of paraffins, the continuous version of the UNIFAC model is employed (Kehlen et al., 1988). The K_p^C (N) and K_p^R (N) are calculated according to Eq. (50).

$$\ln K_p^R(N) = \sum_k v_{p,k}(N) Q_k \xi_k \tag{56}$$

$$\xi_k = \ln \sum_m \frac{Q_m \bar{v}_m^{II} \psi_{mk}}{\sum_n Q_n \bar{v}_n^{II}} - \ln \sum_m \frac{Q_m \bar{v}_m^{I} \psi_{mk}}{\sum_n Q_n \bar{v}_n^{I}} + \sum_m \frac{Q_m \bar{v}_m^{II} \psi_{km}}{\sum_n Q_n \bar{v}_n^{II} \psi_{nm}} - \sum_m \frac{Q_m \bar{v}_m^{I} \psi_{km}}{\sum_n Q_n \bar{v}_n^{I} \psi_{nm}} \tag{57}$$

$$\bar{v}_m = v_{B,m} x_B + (1 - x_B)[v_{n,m} x_n^g + v_{a,m} x_a^g + x_p^g \int_{N0}^{Nt} F_p(N) v_{p,m}(N) dN] \tag{58}$$

$$\psi_{nm} = \exp(-a_{nm} / T) \tag{59}$$

$$\ln K_p^C(N) = \ln\frac{\bar{r}^{II}}{\bar{r}^{I}} + r_p(N)\left(\frac{1}{\bar{r}^{II}} - \frac{1}{\bar{r}^{I}}\right) + \frac{z}{2}q_p(N)\left[\left(\frac{r_p(N)}{q_p(N)}\left(\frac{\bar{q}^{II}}{\bar{r}^{II}} - \frac{\bar{q}^{I}}{\bar{r}^{I}}\right)\right) + \ln\left(\frac{\bar{r}^{I}}{\bar{r}^{II}} \times \frac{\bar{q}^{II}}{\bar{q}^{I}}\right)\right] \tag{60}$$

$$\bar{r} = r_B x_B + (1 - x_B)[r_n x_n^g + r_a x_a^g + x_p^g \int_{N0}^{Nt} F_p(N) r_p(N) dN] \tag{61}$$

$$\bar{q} = q_B x_B + (1 - x_B)[q_n x_n^g + q_a x_a^g + x_p^g \int_{N0}^{Nt} F_p(N) q_p(N) dN] \tag{62}$$

$$r_p(N) = \sum_k v_k^p(N) R_k \tag{63}$$

$$q_p(N) = \sum_k v_k^p(N) Q_k \tag{64}$$

Here, the parameters R_k and Q_k are the volume and surface area parameters for group *k*, respectively, and *z* is the coordination number (*z* = 10). a_{nm} is the group interaction parameter between groups *n* and *m*. The UNIFAC group interaction parameter table, group volume and group surface parameters at temperatures between 10ºC and 40ºC have been presented by Magnussen et al. (1981). The average number of groups of type k in the mixture and the average volume and surface parameters are given by Eqs. (58), (61) and (62). In these equations the summations are written over the discrete naphthenic, aromatic and methanol components and the continuous paraffinic components. The activity coefficients are also calculated by employing the two continuous versions of modified Lyngby and Dortmund UNIFAC models. In contrast to the UNIFAC model, in the Lyngby and Dortmund modified UNIFAC models the group interaction parameters are described as function of temperature and the combinatorial terms are slightly modified. The group interaction parameters, group volume and group surface area parameters in the UNIFAC models are extracted from literatures (Magnussen et al., 1981; Larsen et al., 1987; Gmehling & Li, 1993). In the modified UNIFAC models the residual parts are the same as in the UNIFAC model. Inserting the expression for the combinatorial part of the activity coefficient of the continuous Lyngby modified UNIFAC model into Eq. (50) leads to the following equations (Kehlen et al., 1988).

$$\ln K_p^C(N) = \ln\frac{\tilde{r}^{II}}{\tilde{r}^{I}} + r_p^{2/3}(N)\left(\frac{1}{\tilde{r}^{II}} - \frac{1}{\tilde{r}^{I}}\right) \tag{65}$$

$$\tilde{r} = r_B^{2/3} x_B + (1 - x_B)[r_n^{2/3} x_n^A + r_a^{2/3} x_a^A + x_p^A \int_{N0}^{Nt} F_p(N) r_p^{2/3}(N) dN] \tag{66}$$

Using the Dortmund modified UNIFAC model, the quantity K_i^C may be calculated according to:

$$\ln K_p^C(N) = \ln\frac{\hat{r}^{II}}{\hat{r}^{I}} + r_p^{\frac{3}{4}}(N)\left(\frac{1}{\hat{r}^{II}} - \frac{1}{\hat{r}^{I}}\right) + \frac{z}{2}q_p(N)\left[\left(\frac{r_p(N)}{q_p(N)}\left(\frac{\bar{q}^{I}}{\bar{r}^{I}} - \frac{\bar{q}^{II}}{\bar{r}^{II}}\right)\right) + \ln\left(\frac{\bar{r}^{I}\bar{q}^{II}}{\bar{r}^{II}\bar{q}^{I}}\right)\right] \tag{67}$$

$$\hat{r} = r_B^{3/4}x_B + (1-x_B)\left[r_n^{3/4}x_n^A + r_a^{3/4}x_a^A + x_p^A \int_{N0}^{Nt} F_p(N) r_p^{3/4}(N)dN\right] \tag{68}$$

Eqs. (55) to (68) can also be applied to the discrete naphthenic, aromatic and methanol components. For these components the quantities which are functions of carbon number are the constant quantities and they are not related to carbon number.

3.2.3 Liquid-liquid equilibrium

By applying the equality of chemical potentials in the UNIFAC models and equality of segment-molar chemical potentials in the CONTAS model, the following phase equilibrium equations can be derived for continuous and discrete components of the mixture.

$$x_i^{II} = x_i^{I} K_i \tag{69}$$

$$x_p^{II} F_p^{II}(N) = x_p^{I} F_p^{I}(N) K_p(N) \tag{70}$$

Where:

$$x_i = (1-x_B)x_i^g \qquad i = n, a, p \tag{71}$$

To predict the quantities K_i and $K_p(N)$, the activity coefficients are calculated by using the CONTAS and three UNIFAC models. The x_B quantity in the CONTAS model is the mole fraction of methanol chain associates and in the UNIFAC models it is the true mole fraction of methanol component.

The integration of Eq. (70) with respect to N results in:

$$x_p^{II} - x_p^{I} \int_{N0}^{Nt} F_p^{I}(N) K_p(N) dN = 0 \tag{72}$$

Multiplying Eq. (70) by N and integrating this expression, leads to the following equation for the mean carbon number of paraffinic components in phase II.

$$\overline{N}^{II} = \frac{\int_{N0}^{Nt} N F_p^{I}(N) K_p(N) dN}{\int_{N0}^{Nt} F_p^{I}(N) K_p(N) dN} \tag{73}$$

In the CONTAS model the phase equilibrium equation for the methanol component can be written as follows:

$$x_B^{II} W^{II}(r) = x_B^{I} W^{I}(r) K_B(r) \tag{74}$$

Using Eqs. (45) and (53), Eq (74) may be written as:

$$\psi_B^{II} W^{II}(r) / \overline{r}_B^{II} = \psi_B^{I} W^{I}(r) \exp(r\rho_B) / \overline{r}_B^{I} \tag{75}$$

Inserting the distribution function of associates from Eq (47) into Eq. (75) and integrating this equation leads to the final equilibrium equation for methanol component.

$$\lambda^{I} - \lambda^{II} + \rho_B = 0 \tag{76}$$

To investigate the phase split calculations a set of mass balance and equilibrium equations are needed. In this process the gasoline-methanol mixture in the feed phase is separated into two liquid equilibrium phases. The mass balance equation for naphthenes, aromatics, paraffins and methanol is given by

$$x_i^f = x_i^I + \phi(x_i^{II} - x_i^I) \tag{77}$$

Where parameter ϕ is the number of moles of phase II based on 1 mole feed. The mass balance equation for the paraffinic components consisting of N carbon atoms can be written as:

$$x_p^f F_p^f(N) = (1-\phi) x_p^I F_p^I(N) + \phi x_p^{II} F_p^{II}(N) \tag{78}$$

The relation between the mean carbon number of the feed and the equilibrium phases may be obtained by multiplying Eq. (78) by N and integrating this equation.

$$\overline{N}^I = \frac{\overline{N}^f x_p^f - \phi \overline{N}^{II} x_p^{II}}{(1-\phi) x_p^I} \tag{79}$$

In Eqs. (77) to (79), the quantities x_i^f, x_i^I, x_i^{II} are the true mole fractions of i-th component in the feed, phase I and phase II respectively. Therefore, to use these equations in the CONTAS model, x_i quantities should be replaced by x_i^t. For the cloud-point and shadow curves calculations all quantities of the first phase including the distribution function of paraffins and composition of paraffins, naphthenes, aromatics and methanol are known. Numerical solution of four equilibrium equations leads to the estimation of equilibrium temperature and mole fraction of components in the second phase. Equilibrium equations in the UNIFAC models are Eq. (70) for paraffins and three other equations which are written based on Eq. (69) for the discrete naphthenic, aromatic and methanol components. However, in the CONTAS model for the associating methanol component, Eq. (76) is used instead of Eq. (69). In the first step the calculations are made based on the guessed values of unknown quantities. By using equilibrium equations and the Newton-Raphson method the new quantities are calculated and compared with the previous ones. To estimate these quantities, It is necessary to know the distribution function of paraffins in the second phase which is related to the mean carbon number of paraffins in the second phase. $\overline{N}^{II}$ may be obtained with the aid of Eq. (73). This iterative procedure is repeated to obtain the final results. To apply the UNIFAC models, quantities $K_p(N), K_n, K_a$ and K_B has to be determined from Eqs.

(55) to (68). In the CONTAS model, the Eqs. (51) to (53) for the K_i quantities are related to the $\bar{r}_m$ and ψ_i. To calculate $\bar{r}_m$ and ψ_i from Eqs. (37) and (45), we need to know $\bar{r}_B$ and mole fraction of associates. Mole fraction of associates can be obtained from the true mole fraction of methanol (Browarzik, 2004a). Based on Eq. (49) $\bar{r}_B$ is a function of λ which is determined using Eqs. (46) and (48). For the phase separation calculations, the feed quantities like feed mole fraction of methanol and temperature are known and the quantities of the two equilibrium phases are determined by solving the equilibrium and mass balance equations. The guessed quantities in the first step of calculations are true mole fraction of paraffins, naphthenes and methanol in the first phase and true mole fraction of methanol in the second phase. True mole fraction of paraffins, naphthenes and aromatics in the second phase are obtained from Eq. 48. Additionally, according to Eq. (79) $\overline{N}^I$ is calculated from the mean carbon number of paraffins in the second and feed phase.

3.3 Results and discussion

The LLE results of the gasoline-methanol blends based on the four different models UNIFAC, Lyngby modified UNIFAC, Dortmund modified UNIFAC and CONTAS at temperatures between -20 °C to 20 °C (Ruzicka et al., 1986) were obtained and compared with experimental data. The results of the CONTAS model are obtained using four adjustable parameters in Eq. (43) which are fitted to the LLE experimental data. The adjusted parameters are $\chi_p = 15.7$ K$^{0.5}$, $\chi_n = 12.0179$ K$^{0.5}$, $\chi_a = 3.5819$ K$^{0.5}$, $\chi_B = 23.8539$ K$^{0.5}$.

Unlike in binary systems, in polydisperse systems the solubility diagrams consist of a cloud-point curve and a shadow curve. The cloud-point curve indicates the composition of the first phase as a function of temperature at the onset of the second phase formation. The composition of the nearly formed phase is given by the shadow curve. The intersection point of the cloud-point curve and shadow curve is the critical point. The solubility diagrams indicate the phase separation boundaries. The methanol mole fraction in the methanol-lean part of the cloud-point curve gives the maximum amount of methanol that can be added to gasoline to avoid phase separation. The mutual miscibility of methanol and gasoline is related to temperature so that decreasing the temperature leads to a decrease in miscibility. The effect of the aromatics concentration on the liquid-liquid equilibrium of the methanol-gasoline blend was studied and showed that the two-liquid phase regions will be larger for the gasoline-methanol mixtures with lower amounts of aromatics. Furthermore we have a decrease in the upper critical solution temperature with increasing the concentration of aromatic components of gasoline. Thus the solubility of methanol and gasoline improves by increasing the aromatic content of the mixture. This effect can be explained based on the solubility of aromatics, naphthenes and paraffins with methanol. In contrast to paraffins and naphthenes which show a limited miscibility with methanol, aromatics are completely miscible with methanol. Therefore, adding aromatics to a system containing paraffins and naphthenes results in lowering of the critical solution temperature. Moreover, the effect of polydispersity or the variance of the paraffinic components on the solubility diagrams was studied. This study shows that by increasing the variance of the distribution function of paraffins (σ^2), the critical solution temperature and the distance between cloud-point curve and shadow curve increases. As mentioned above methanol and gasoline are partially miscible and a limited amount of methanol can be blended with gasoline. Adding more methanol leads to the separation of the mixture into two equilibrium liquid phases. To predict the composition of paraffins, naphthenes, aromatics and methanol in the two formed liquid phases, phase separation calculations which are based on the mass balance and phase equilibrium equations, are necessary. The results of the phase separation calculations in comparison with experimental data based on the four different models CONTAS, UNIFAC, Lyngby and Dortmund modified UNIFAC were obtained. These results include the mole fraction of methanol and different hydrocarbon families in the lean and rich methanol phases at five temperatures over the range of -20 °C to 20 °C. It was found that in comparison with the experimental data, the results obtained from the CONTAS and UNIFAC models are more accurate than the results of the Dortmund and Lyngby modified UNIFAC models. The calculated compositions using the UNIFAC and CONTAS models are in reasonable agreement with the experimental ones. It was shown that the Dortmond and Lyngby modified UNIFAC models have not properly predicted experimental data. The results of the phase separation of methanol-gasoline blends predicted by the UNIFAC and CONTAS models, at temperatures 20 °C and -20 °C, for different feed compositions of methanol were obtained. These calculations show that the UNIFAC model, especially at temperatures between 0 °C to 20 °C, gives better agreement with the experimental data than the CONTAS model. This fact can be explained by the group interaction parameters of the UNIFAC model which have been developed for LLE calculations at temperatures between 10 °C to 40 °C (Magnussen et al., 1981). The upper critical solution temperature of the systems containing paraffins, naphthenes and methanol is about 300 K to 360 K (Kehlen et al., 1988). By adding aromatics to this system, the critical temperature decreases so that the critical temperature for the gasoline-methanol mixture cannot be higher than 360 K. However, the critical temperature predicted by the UNIFAC model is about 405 K and this temperature for the CONTAS model is about 329 K. Therefore, the CONTAS model predicts the critical solution temperature better than the UNIFAC model and it is an appropriate model for the liquid-liquid equilibrium prediction at high temperatures (Nasrollahi et al., 2009).

3.4 Conclusions

To predict the cloud-point curve, shadow curve and phase separation results of methanol-gasoline blends, the continuous thermodynamics procedure was applied and the four different models CONTAS, UNIFAC, Lyngby and Dortmund modified UNIFAC were examined. The CONTAS model includes four fitting parameters which are related to the solubility parameters of paraffins, naphthenes, aromatics and methanol. The common group interaction parameters were applied to predict limited miscibility of methanol-gasoline blends using the UNIFAC and modified UNIFAC models. The highest possible composition of methanol that can be added to gasoline decreases by decreasing the temperature and its value is estimated at the lean-methanol part of the cloud-point curve. The influence of the aromatics content and polydispersity of paraffins on the liquid-liquid equilibrium results were investigated. Due to the complete miscibility of aromatics and methanol, a larger amount of aromatics in the mixture leads to a lowering of the critical solution temperature. Furthermore by increasing the variance of the distribution function of paraffins, the critical temperature increases and the cloud-point and shadow curves get closer. The results calculated using the UNIFAC and CONTAS models are in good agreement with the experimental data. The prediction of the UNIFAC model in the temperature range of -20 °C to 20 °C is more accurate than the CONTAS model. However, the CONTAS model is more suitable for liquid-liquid equilibrium calculations at high temperatures.

4. Continuous thermodynamics of petroleum fluids fractions

A method based on continuous thermodynamics is introduced here for calculating the vapor pressure of undefined composition mixtures (Vakili-Nezhaad et al., 2001). In order to verify the proposed method the experimental data of an Indonesian petroleum fluid as well as a detailed true boiling point data on the lube-oil cut SAE 10 of Tehran Refinery have been used. In this work by using the combination of the Trouton's rule and the Clausius–Clapeyron equation in the well-known equation of Antonie, the continuous version of the Raoult's law has been developed and the results of the modeling and calculations have been compared with the experimental data and good agreement was found. Moreover, for improving the results of the ideal continuous mixture, modeling of non-ideal behaviour of such mixtures has been carried out by introducing the continuous version of the UNIFAC activity coefficient in the formulation and a better result has been obtained.

4.1 Modeling and calculations

Here we have carried out the boiling point calculations for two cases (i) ideal mixtures; and (ii) non-ideal mixtures. In the first case the continuous version of the Raoult's law has been presented and in the second case the continuous version of the UNIFAC model was used for developing the results of the ideal mixtures to the non-ideal mixtures.

4.1.1 Case one: ideal mixture

In general, the equilibrium vapor pressure of a multicomponent mixture with N distinct species can be calculated by the following equation (Smith et al., 2005):

$$P = \sum_{i=1}^{N} x_i P_i^{sat} \gamma_i \tag{80}$$

where, P is the total pressure, x_i is the mole fraction of the ith component, P_i^{sat} is the saturation pressure and γ_i is the activity coefficient of the ith component in the mixture. For the ideal mixtures because of the unity of all activity coefficients, this equation can be reduced to the following simple form:

$$P = \sum_{i=1}^{N} x_i P_i^{sat} \tag{81}$$

This equation is known as the Raoult's law and indicates that the partial pressure of the ith component in the mixture is equal to the product of its mole fraction by the saturated vapor pressure (at the mixture temperature). If the number of components in the mixture goes to a very large number, in such a way that the composition of the mixture cannot be presented in usual forms of mole or mass fractions, a continuous distribution function can be applied for defining the composition of the mixture. In such cases the continuous version of the Eq. (81) may be written as follows:

$$P = \int F(I) P^{sat}(I) \mathrm{d}I \tag{82}$$

where, $F(I)$ is a suitable distribution function of a proper characterization index 'I' such as the molecular weight or boiling point. In application of Eq. (82), an equation for expressing P^{sat} is required. In this work the well-known equation of Antonie has been applied (Smith et al., 2005),

$$\log_{10} P^{sat} = A_A - \frac{B_A}{t + C_A} \tag{83}$$

where, P^{sat} is the vapor pressure of a pure component at temperature t and A, B and C are the special constants for each component. It is obvious that for the application of Eq. (82) the continuous form of the Antonie's equation is needed. If we combine the Trouton's rule with the Clapeyron– Clausius equation, the following equation can be obtained:

$$P^{sat}(T) = P_{atm} \exp\left[10.58\left(1 - \frac{T}{T_S}\right)\right] \tag{84}$$

where T is the boiling point in distribution functions (in Kelvin) and Ts, is the temperature at which distillation occurs. Therefore, by combining Eq. (82) and Eq. (84) we have,

$$P = \int_{IBP}^{FBP} P_{atm} \exp\left[10.58\left(1 - \frac{T}{T_S}\right)\right] F(t)\,\mathrm{d}T \tag{85}$$

The lower and upper limits of the integral (IBP and FBP) are the initial and final boiling points of the petroleum fluids fractions, respectively. Therefore, if we have the distribution function of the mixture, we will be able to calculate the equilibrium vapor pressures of such complex mixtures. To verify the proposed method we have examined various distribution functions. These equations were used in the Eq. (85) for calculating the normal boiling point of the lube-oil cut SAE 10 of the Tehran Refinery. The error analysis based on these distribution functions showed that Eq. (86) has the minimum absolute average deviation (AAD) compared with the other distribution functions. The probability density function which has been used for our calculations can be expressed in the following form (Riazi, 1997):

$$F(T) = \frac{1}{T_0} \frac{B^2}{A} \left(\frac{T - T_0}{T_0}\right)^{B-1} \exp\left[-\frac{B}{A}\left(\frac{T - T_0}{T_0}\right)^{B}\right] \tag{86}$$

All the parameters of this equation which have been calculated by the regression analysis method can be given as:

$$A = 0.01862 \tag{87}$$

$$B = 3.5298 \tag{88}$$

$$To = 554.45 \text{ K} \tag{89}$$

$F(T)$ from Eq. (86) was introduced in Eq. (85) and the integral was calculated by numerical methods. The results of the calculations are:

$$T = 674.45 \text{ K} \tag{90}$$

$$P = 1.014 \times 105\ Pa \tag{91}$$

4.1.2 Case two: non-ideal mixtures

It is evident that the proposed formulation for ideal mixture could not handle the related calculations for the non-ideal mixtures, therefore, we have to generalize the results of the non-ideal mixtures. In this case the equilibrium vapor pressure of the continuous mixture M can be written as:

$$P = \int F(I) P^{sat}(I) \gamma(I,T) \mathrm{d} I \tag{92}$$

where $\gamma(I,T)$ is the continuous version of the activity coefficient of the mixture components. The composition of the complex mixtures such as petroleum fractions are not well defined therefore, in this work we have used the predictive model of UNIFAC. For applying this model for the petroleum fluids fractions its continuous version is required, but when using the continuous form of the UNIFAC model for phase equilibrium calculations, it is necessary to know the analytical expressions for the number of the various functional groups in the mixture. A polynomial of order 2 has been used by some researchers (Kehlen et al., 1988) for calculating the number of different functional groups in the light petroleum fractions but these equations are not suitable for heavier petroleum fractions such as lube-oil cut which is our interest in the present work. Therefore, we have proposed higher order polynomials (up to 8) for this purpose (Vakili-Nezhaad, 1999). The coefficients of these equations have been obtained using the detailed experimental data on lube-oil cut SAE 10 of Tehran Refinery and an Indonesian oil (Jaubert et al., 1995) by using the powerful numerical software of TABLE CURVE (TC). Since a petroleum fraction have been made mainly from three different homologue series as paraffinic, naphthenic and aromatic hydrocarbons, we have divided these hydrocarbon homologues to two parts, i.e. aromatic hydrocarbons and paraffinic+naphtenic hydrocarbons. Therefore, by using this approach the continuous version of UNIFAC model for each continuous parts of the mixture including aromatic or paraffinic+naphthenic parts were obtained. In the UNIFAC model the activity coefficient comprises of two parts, i.e. combinatorial and residual, therefore, the activity coefficient of the continuous ensembles of aromatic or naphthenic+paraffinic parts were obtained. For the ideal mixtures it has been demonstrated that the best distribution function for the examined mixture was Eq. (86), therefore for the non-ideal mixture too we used this function and the following results were obtained (Vakili-Nezhaad et al., 2001):

$$T = 390.94 \text{ K} \tag{93}$$

$$P = 1.502 \times 10^7 \text{ Pa} \tag{94}$$

In order to verify the accuracy of the proposed method, these theoretical results were compared with the experimental data of an Indonesian petroleum fluid (Jaubert et al., 1995). This comparison indicated a good agreement between theoretical results and experimental data. Thus it can be said that the proposed formulation based on the application of the continuous form of the activity coefficient model can improve the accuracy of the calculations (Vakili-Nezhaad et al., 2001).

4.2 Conclusion

Based on the concepts of continuous thermodynamics a method has been proposed for the vapor pressure calculation of the complex petroleum fluids fractions. At first the continuous version of the Raoult's law has been developed for applying in those mixtures that comprise of similar species and their behaviour can be considered as ideal mixtures. Comparison between modeling and experimental data on the lube-oil cut SAE 10 of the Tehran Refinery showed that for these types of petroleum fractions the proposed modeling and calculations give the best results with the Riazi distribution function. The above mentioned method has been developed to the case of non-ideal mixtures using the continuous version of the UNIFAC-model and the results have been compared with the experimental data of an Indonesian oil, and better accuracy has been obtained (Vakili-Nezhaad et al., 2001).

5. Three phase flash calculations using genetic algorithm approach

In this section a new approach based on the genetic algorithm is introduced for solving three phase flash calculations containing two liquid phases and one vapor phase (Vakili-Nezhaad et al., 2010). Based on this approach choosing the initial guesses for the compositions of the involved phases is not an important step, which in the ordinary three phase algorithms affects directly the convergence of the calculations. A real problem has been solved with this approach and very good results have been obtained in comparison with the experimental data. Phi-Phi approach has been adopted for vapor-liquid-liquid equilibrium (VLLE) calculations based on the Peng-Robinson equation of state (Vakili-Nezhaad et al., 2010).

5.1 Introduction

Information about vapor-liquid-liquid equilibrium (VLLE) is essential for many chemical processes and the separation operations (Christov & Dohrn, 2002). Among these industrial processes the recovery of organic acids from dilute solution resulting from fermentation processes and the extraction of aromatic compounds from petroleum fluid fraction are only two examples of the potential of VLLE operations in the chemical and petrochemical industries (Taghikhani et al., 2001; Vakili-Nezhaad et al., 2001; Vakili-Nezhaad et al., 2002; Vakili-Nezhaad et al., 2004). For designing and optimization of the liquid extraction processes as well as various separation equipments one has to know the equilibrium conditions due to complex nature of the VLLE calculations it can be said that for obtaining reliable results one has to examine every specific case to check its convergence. To confirm this point one may refer to the works of Fang-Zhi et al (1993) and Yokozeki (2004). Therefore as an important result it may be mentioned that presentation of a unique algorithm for VLLE calculation is a crucial step in the subject. In this work a simple and efficient approach based on the genetic algorithm has been proposed for VLLE calculations and good results have been obtained for a ternary system (Vakili-Nezhaad et al., 2010).

5.2 Problem statement

The compositions of the liquids and vapor phases are given by the following equations:

$$X1(I) = \frac{\mathrm{s(I)}}{\mathrm{x(1)*(1 - R1(I)) + x(2)*(R1(I) / R2(I)) + R1(I)}} \tag{95}$$

$$X2(I) = \frac{\mathrm{(s(I)*R1(I) / R2(I))}}{\mathrm{x(1)*(1 - R1(I)) + x(2)*(R1(I) / R2(I) - R1(I)) + R1(I)}} \tag{96}$$

$$Y(I) = \frac{(\mathrm{s(I)} * \mathrm{R1(I)})}{\mathrm{x(1)} * (1 - \mathrm{R1(I)}) + \mathrm{x(2)} * (\mathrm{R1(I)} / \mathrm{R2(I)}) + \mathrm{R1(I)}} \tag{97}$$

x(1) and x(2), equals to the number of moles of the liquid phases No. 1 and No. 2 respectively. Considering the above relations, it is obvious that in a three-phase system the value of x(1) and x(2) must be known. These values will be the answers of a nonlinear system of equations which will be obtained from the following equations:

$$\sum_{i=1}^{3} X2(i) = 1 \tag{98}$$

$$\sum_{i-1}^{3} (X1(i) - Y(i)) = 0 \tag{99}$$

Using these two relations, the system of equation will be obtained:

$$f_1 = \sum_{k=1}^{3} \frac{\mathrm{s(k)} * \mathrm{R1(k)} / \mathrm{R2(k)}}{\mathrm{x(1)} * (1 - \mathrm{R1(k)}) + \mathrm{x(2)} * (\mathrm{R1(k)} / \mathrm{R2(k)} - \mathrm{R1(k)}) + \mathrm{R1(k)}} - 1 = 0 \tag{100}$$

$$f_2 = \sum_{k=1}^{3} \frac{\mathrm{s(k)} * (1 - \mathrm{R1(k)})}{\mathrm{x(1)} * (1 - \mathrm{R1(k)}) + \mathrm{x(2)} * (\mathrm{R1(k)} / \mathrm{R2(k)} - \mathrm{R1(k)}) + \mathrm{R1(k)}} = 0 \tag{101}$$

where,

$$\mathrm{R1(I)} = [1.4529 \quad 0.2103 \quad 0.0001] \tag{102}$$

$$\mathrm{R2(I)} = [0.1518 \quad 0.2876 \quad 0.0009] \tag{103}$$

R1 and R2 are the K-values of different components in the mixture. These values have been calculated by the Peng-Robinson equation of state. Now by considering the above equations we can carry out the VLLE calculations for ternary system. The feed consists of Methane (1) - Ethane(2) - n-Octane (3). In this work, we have chosen the composition of feed as s(1)=0.8054, s(2)=0.1385, s(3)=0.0560.

5.3 Genetic algorithm

The genetic algorithm (GA) is a stochastic search technique based on the mechanism of natural selection and natural genetics to imitate living beings for solving difficult optimization problems with high complexity and an undesirable structure. The GA approach represents a powerful, general-purpose optimization paradigm in which the computational process mimics the theory of biological evolution (Holland, 1992; Storn & Price, 1997). It has been successfully used in job-shop scheduling, production planning, line balancing and process optimization. Goldberg (1989) proposed the most common and useful form of GA. Different from traditional point-to-point descending and ascending search techniques, a GA starts from one set of random solutions called a population. Each individual solution in the population is called a chromosome. At each generation, the GA performs genetic operations such as crossover and mutation on the selected chromosomes to yield offspring to produce the next generation. During each generation, these chromosomes evolve into better fitness by applying an evolution operation, called the selection. From generation to generation, eventually, the chromosomes in the population will converge. In this case, the best chromosome is found. Generally, the basic steps of a GA approach in solving an optimization problem can be summarized in Table 1.

1. Represent the problem variable as a chromosome of a fixed length, and choose the size of a chromosome population, the crossover probability, and the mutation probability.
2. Define a fitness function to measure the fitness of an individual chromosome in the problem domain.
3. Randomly generate an initial population of chromosomes.
4. Calculate the fitness of each individual chromosome.
5. Select a pair of chromosomes for mating from the current population. Parent chromosomes are selected with a probability related to their fitness. Highly fit chromosomes have a higher probability of being selected formatting. The roulette wheel method is usually applied in chromosome selection.
6. Create a pair of offspring chromosomes by applying the genetic operators: crossover and mutation.
7. Place the created offspring chromosomes in the new population.
8. Repeat step 5 until the size of the new chromosome population is equal to the size of the initial population.
9. Replace the initial (parent) chromosome population with the new (offspring) population.
10. Repeat steps 4–9 until the termination criterion is satisfied.

Table 1. The Genetic Algorithm Steps

5.4 Discussion

In this work, The Genetic toolbox in MATLAB® has been used (Chipperfield, 2007). Equations (100) and (101) are the objective functions (f_1 and f_2) which have been combined to form a unique objective function or fitness function. This function may be read as:

$$F = \frac{w_1 f_1 + w_2 f_2}{w_1 + w_2} \tag{104}$$

As X1, X2 and Y are mole fractions they must be in the interval [0, 1] the nonlinear constraints have been considered. The genetic algorithm toolbox uses the Augmented Lagrangian Genetic Algorithm (ALGA) to solve nonlinear constraint problems. The Genetic Algorithm in MATLAB was run with the parameters reported in Table 2.

The best solution in last generation is x(1)=0.4132 and x(2)=0.2095 which by these value the composition of two liquid phases and one vapor phase are as follows:

$$X1 = [0.657123 \quad 0.214483 \quad 0.128288] \tag{105}$$

$$X2 = [0.828907 \quad 0.156835 \quad 0.014254] \tag{106}$$

$$Y = [\,0.954734 \quad 0.045106 \quad 0.000013] \tag{107}$$

The experimental data on the liquid phases are as follows:

$$X1 = [0.682\ \ 0.187\ \ 0.131] \quad (108)$$

$$X2 = [0.806\ \ 0.158\ \ 0.036] \quad (109)$$

In comparison with these data we find that the method handled the relevant calculations very well.

Population size	20
Parameters	2 $0 \leq x(1) \leq 1$, $0 \leq x(2) \leq 1$
Generation	30
Nonlinear constraints	$X_2(i)-1\leq 0$ $X_1(i)-1\leq 0$ $Y_1(i)-1 \leq 0$, for i=1:3
Hybrid with Local Optimization	Pattern search
Encoding	Real value

Table 2. Parameters Used in Genetic Algorithm Calculation

6. Concluding remarks

Thermodynamic modeling and calculations of different complex fluid mixtures were presented in this chapter. Phase equilibrium calculations for some complex fluid systems were presented. Continuous thermodynamics was applied for modeling and phase equilibrium calculations of polymer solutions and petroleum fluids. This approach (continuous thermodynamics) was shown as a powerful tool for these kinds of complex phase equilibrium calculations. In the last part of the present work a new approach of genetic algorithm (GA) for doing phase equilibrium calculations was introduced, and it was shown that GA can be applied for some complex vapor-liquid-liquid equilibrium systems. Application of GA in fluid phase equilibria is going to be increased and developed as a reliable tool.

7. Acknowledgments

This work was partially supported by the internal grant no. IG/ENG/PCED/10/02 of Postgraduate Studies & Research, College of Engineering of Sulatn Qaboos University in Muscat, Oman.

8. References

Agarwal, A. K. (2007). Biofuels (alcohols and biodiesel) applications as fuels for internal combustion engines, *Progress in Energy and Combustion Science*, Vol. 33, No. 3, pp. 233–271.

Behnam, S. (2008). Continuous Thermodynamics of Polymer Solutions, M.Sc. Thesis, Isfahan University of Technology, Isfahan, Iran.

Browarzik, D. (2004). Phase-equilibrium calculations for *n*-alkane + alkanol systems using continuous thermodynamics, *Fluid Phase Equilibria*, Vol. 217, No. 2, pp 125–135.

Browarzik, D. (2005). Phase-equilibrium calculations for non-aqueous and aqueous associating systems using continuous thermodynamics, *Fluid Phase Equilibria*, Vol. 230, No. 1-2, pp. 143–152.

Browarzik, D. (2007). Continuous thermodynamics of binary associating systems, *Fluid Phase Equilibria*, Vol. 254, No. 1-2, pp. 174–187.

Browarzik, D. & Kowalewski, M. (2002). Calculation of the cloud-point and the spinodal curve for the system methylcyclohexane/polystyrene at high pressures, *Fluid Phase Equilibria*, Vol. 194-197, No. 30 March 2002, pp. 451-467.

Browarzik, D. & Kowalewski, M. (2004). Calculation of stability and of the phase equilibrium on the system methylcyclohexane + polystyrene based on an equation of state, In: *Thermodynamic properties of complex fluid mixtures*, Maurer, G. & Forschungsgemeinschaft, D. (Eds.), pp. 488-508, Wiley-VCH.

Browarzik, D. & Kowalewski, M. (1999). Calculation of the stability and of the phase equilibrium in the system polystyrene+cyclohexane+carbon dioxide based on equations of state , *Fluid Phase Equilibria*, Vol. 163, No. 1, pp. 43–60

Chipperfield, A. (2007). *Genetic Algorithm Toolbox for Use with Matlab*, University of Sheffield.

Chorn, L. G. & Mansoori, G. A. (1989). Multi-component fractions characterization: principles and theories, In: C7+ *Fraction Characterization*, Chorn, L. G. & Mansoori, G. A. (Eds.), pp. 1-10, Advances in Thermodynamics, Taylor & Francis Press., New York.

Christov, M. & Dohrn, R. (2002). High-pressure fluid phase equilibrium experimental method and system investigated, *Fluid Phase Equilibria*, Vol. 202, pp. 153-218.

Cotterman, R. L.; Bender, R. & Prausnitz, J. M. (1985). Phase equilibria for mixtures containing very many components, *Industrial & Engineering Chemistry Process Design & Development*, Vol. 24, pp. 194–203.

Cox, F. W. (1979). Physical properties of gasoline-alcohol blends, BETC/RI-79/4, U.S. Department of Energy, Bartlesville, OK, September 1979.

Du, P. C. & Mansoori, G. A. (1986). Phase equilibrium computational algorithms of continuous mixtures, *Fluid Phase Equilibria*, Vol. 30, No. 1, pp. 57–64.

Enders, S. & de Loos, Th. W. (1997). Pressure dependence of the phase behaviour of polystyrene in methylcyclohexane, *Fluid Phase Equilibria*, Vol. 139, No. 1-2, pp. 335–347.

Fang-Zhi, L.; Mori, H.; Hiraoka, S. & Yamada, I. (1993). Phase Equilibrium and Simulation Method for Heterogeneous Azeotropic Distillation, *Journal of Chemical Engineering of Japan*, Vol. 26, No.1, pp. 41-47.

French, R. & Malone, P. (2005). Phase equilibria of ethanol fuel blends, *Fluid Phase Equilibria*, Vol. 228-229, February 2005, pp. 27-40.

Gauter, K. & Heidemann, R. A. (2001). Modeling polyethylene-solvent mixtures with the Sanchez–Lacombe equation, *Fluid Phase Equilibria*, Vol. 183-184, 1 July 2001, pp. 87-97.

Gauter, K. & Heidemann, R. A. (2000). A Proposal for Parametrizing the Sanchez−Lacombe Equation of State, *Industrial & Engineering Chemistry Research*, Vol. 39, No. 4, pp. 1115-1117.

Gmehling, J.; Li., J. & Schiller, M. (1993). A modified UNIFAC model. 2. Present parameter matrix and results for different thermodynamic properties, *Industrial & Engineering Chemistry Research*, Vol. 32, No. 1, pp. 178–193.

Goldberg, D.E. (1989). *Genetic algorithms in search, Optimization and Machine Learning*, Addison-Wesley.

Holland, J. H. (1992). Genetic Algorithms, *Scientific American*, Vol. 267, No. 1, pp. 66–72.

Huang, S. H. & Radosz, M. (1991). Equation of state for small, large, polydisperse, and associating molecules: extension to fluid mixtures, *Industrial & Engineering Chemistry Research*, Vol. 30, No. 8, pp. 1994–2005.

Jaubert, J. N.; Neau, E.; Penelox, A.; Fressigne, C. & A. Fuchs, A. (1995). Phase Equilibrium Calculations on an Indonesian Crude Oil Using Detailed NMR Analysis or a Predictive or a Predictive Method to Assess the Properties of the Heavy Fractions, *Industrial & Engineering Chemistry Research*, Vol. 34, pp. 640–655.

Kehlen, H.; Ratzsch, M. T.; Ruzicka, V. &, Sadowski, G. (1988). Continuous Thermodynamics of the Liquid-Liquid Equilibrium for Systems Containing Petroleum Fractions, *Z. Phys. Chemie.* Lipzig, Vol. 269, pp. 908–916.

Koak, N. (1997). Polymer Solution Phase Behavior, PhD Thesis, The University of Calgary, Calgary, Canada.

Koak, N. & Heidemann, R. A. (1996). Polymer−Solvent Phase Behavior near the Solvent Vapor Pressure, *Industrial & Engineering Chemistry Research*, Vol. 35, No. 11, pp. 4301–4309.

Koak, N.; de Loos, Th. W. & Heidemann, R. A. (1998). Upper critical solution temperature behavior of the system polystyrene + methylcyclohexane. Influence of CO_2 on the liquid-liquid equilibrium, *Fluid Phase Equilibria*, Vol. 145, No. 2, pp. 311–323.

Larsen, B. L.; Rasmussen, P. & Fredenslund, A. (1987). A modified UNIFAC group-contribution model for prediction of phase equilibria and heats of mixing, *Industrial & Engineering Chemistry Research*, Vol. 26, No. 11, pp. 2274–2286.

Leeper, S. A. & Wankat, P. C. (1982). Gasohol production by extraction of ethanol from water using gasoline as solvent, *Industrial & Engineering Chemistry Process Design and Development*, Vol. 21, No. 2, pp. 331-334.

Magnussen, T.; Rasmussen, P. & Fredenslund, A. (1981). UNIFAC parameter table for prediction of liquid-liquid equilibriums, *Industrial & Engineering Chemistry Process Design and Development*, Vol. 20, No. 2, pp. 331–339.

Narita T., Yamamoto T., Hosoya E. & Dobashi, T. (2003). Gibbs Free Energy Expression for the System Polystyrene in Methylcyclohexane and Its Application to Microencapsulation, *Langmuir*, Vol. 19, No. 13, pp. 5240-5245.

Nasrollahi, F; Roodpeyma, Sh. & Vakili-Nezhaad, G. R. (2009). Liquid–liquid equilibrium calculations for methanol–gasoline blends using continuous thermodynamics, *Fluid Phase Equilibria*, Vol. 284, No. 1, pp. 1-9.

Neau, E.; Jaubert, J. N. & Rogalski, M. (1993). Characterization of heavy oils, *Industrial & Engineering Chemistry Research*, Vol. 32, No. 6, pp. 1196–1203.

Riazi, M. R. (1997). A Continuous Model for C, Plus Fraction Characterization of Petroleum Fluids, *Industrial & Engineering Chemistry Research*, Vol. 36, pp. 4299–4307.

Ruzicka, V.; Fredenslund, A. & Rasmussen, P. (1983). Representation of petroleum fractions by group contribution, *Industrial & Engineering Chemistry Process Design and Development*, Vol. 22, No. 1, pp. 49–53.

Ruzicka, V.; Frydova, R. & Novak, J. (1986). Liquid—liquid equilibrium in methanol + gasoline blends *Fluid Phase Equilibria*, Vol. 32, No. 1, pp. 27-47.

Sanchez, I. C. & Lacombe, R. H. (1978). Statistical Thermodynamics of Polymer Solutions, *Macromolecules*, Vol. 11, pp. 1145–1156.

Sanchez, I. C. & Lacombe, R. H. (1976). An elementary molecular theory of classical fluids: pure fluids, *The Journal of Physical Chemistry*, Vol. 80, pp. 2352–2362.

Smith, J. M.; Van Ness, H. C. & Abbott, M. M. (2005). *Introduction to Chemical Engineering Thermodynamics*, 5th Edition, McGraw-Hill Co., New York.

Storn, R. & Price, K. (1997). Differential evolution—a simple and efficient heuristic for global optimization over continuous spaces, *Journal of Global Optimization*, Vol. 11, pp. 341–359.

Taghikhani, V.; Vakili-Nezhaad, G. R.; Khoshkbarchi, M. K. & Shariaty-Niassar, M. (2001). Liquid-Liquid Equilibrium of Water + Propionic Acid + Methyl Butyl Ketone and of Water + Propionic Acid +Methyl Isopropyl Ketone, *Journal of Chemical & Engineering Data*, Vol. 46, pp. 1107-1109.

Torkelson, J. M.; M. Tirrell, M. & Frank, C. W. (1984). Polystyrene-methylcyclohexane solutions undergoing phase separation: a study by fluorescence spectroscopy, *Macromolecules*, Vol. 17, No. 8, pp. 1505-1512.

Vanhee, S.; Kiepen, F.; Brinkmann, D.; Borchard, W.; Koningsveld, R. & Berghmans, H. (1994). The system methylcyclohexane/polystyrene. Experimental critical curves, cloud-point and spinodal isopleths, and their description with a semi-phenomenological treatment, *Macromol. Chem. Phys.*, Vol. 195, No. 2, pp. 759–780.

Vakili-Nezhaad, G. R. (1999). Equilibrium Behavoir of the System NMP + Lube Oil Cut, Ph. D. Thesis, Amir-Kabir University of Technology, Tehran, Iran.

Vakili-Nezhaad, G. R.; Modarress, H. & Mansoori, G. A. (2001). Continuous thermodynamics of petroleum fluids fractions, *Chemical Engineering & Processing*, Vol. 40, pp.431–435.

Vakili-Nezhaad, G. R., Mohsen-Nia, M.; Taghikhani, V.; Behpoor, M. & Aghahosseini, M. (2004). Salting out effect of NaCl and KCL and the Ternary LLE data for the system of Water + Propionic Acid + isobutyl methyl ketone and of Water + Propionic Acid + isopropyl methyl ketone, *Journal of Chemical Thermodynamics*, Vol. 36, pp. 341-348.

Vakili-Nezhaad, G. R.; Taghikhani, V. & Karmand, S. (2002). Comparison of Different Activity coefficient Models by Using PRO/II program, *International Mathematical Journal*, Vol. 2, No.11, pp. 1067-1079.

Vakili-Nezhaad, G. R.; Vahidipour, S. M. & Dargahi, M. (2010). Three Phase Flash Calculations Using Genetic Algorithm Approach, *Asian Journal of Chemistry*, Accepted for Publication.

Wilczura-Wachnic, H. & Alexander Van Hook, W. (2004). Liquid–liquid phase equilibria for some polystyrene–methylcyclohexane mixtures, *European Polymer Journal*, Vol. 40, No. 2, pp. 251- 257.

Xiong, Y. & Kiran, E. (2000). Kinetics of pressure-induced phase separation (PIPS) in polystyrene+methylcyclohexane solutions at high pressure, *Polymer*, Vol. 41, No. 10, pp. 3759-3777.

Yokozeki, A. (2004). Refrigerants of ammonia and n-Butane mixtures, *Equilibrium*™, Feb.2004, pp. 20-24.

Yuksel, F. & B. Yuksel, B. (2004). The use of ethanol–gasoline blend as a fuel in an SI engine, *Renew Energy*, Vol. 29, No. 7, pp. 1181–1191.

19

Thermodynamics of Viscodielectric Materials

R. Díaz-Calleja[1] and E. Riande[2]
[1]*I.T.E. Universidad Politécnica de Valencia, Valencia*
[2]*Instituto de Ciencia y Tecnología de Polímeros, CSIC, Madrid*
Spain

1. Introduction

Extended Irreversible Thermodynamics (EIT) [1,2] has proved to be a useful tool to analyze the non-equilibrium behaviour of complex materials close to the linear regime. The axiomatic formulation of the EIT ([1], p.54) furnishes a formal structure that can be used as the basis of the study of different type of systems, i.e., dielectric/magnetic[3-5] and in particular viscoelastic materials [6-8]. Basically EIT extends the scope of the Classical Irreversible Thermodynamics (CIT) as developed by Prigogine, Onsager and more recently by De Groot and Mazur [9] among others. The basic idea is to formulate a generalized entropy function that depends not only on the conserved classical variables, but also on the dissipative fluxes. This implies abandoning the local equilibrium hypothesis. In the particular case of viscoelastic materials, the dissipative flux is the viscous component of the stress tensor.

It is important to note that owing to the memory exhibited by viscoelastic materials the mechanical and dielectric response of these substances to mechanical and electric perturbation fields applied at time t not only depends on the actual fields at t but also on the history of the materials in the range $-\infty < \vartheta \leq t$. In fact the phenomenological theory of mechanical relaxations is nearly the counterpart of dielectric relaxations [10] in such a way that in the general case both phenomena can be considered coupled processes. To account for the interactions between mechanical and dielectric compliance in the continuum it is convenient to describe adequately such a coupling in terms of the polarization vector and the stress tensor. However, since the symmetric part of the stress tensor only accounts for the translational hydrodynamic and taking into account that polarization arises from rotation of dipoles, in what follows that part will not be considered because does not contribute to the dielectric relaxation. It will be also assumed that the relative velocity of the medium with respect to dipoles rotation is negligible. In this way only the antisymmetric part of the stress tensor produces an effective polarization under an electric field. This is the specific way to locally relate the dipole rotation with the mechanical friction.

2. Balance equations

The linear momentum equation in local form can be written

$$div\boldsymbol{\sigma} + \rho\mathbf{b} = \rho\ddot{\mathbf{u}} \tag{1}$$

where **b** is the volume force and **u** is the displacement; bold symbols are used for tensors of order ≥ 1.
The flux or current equation for polarization charges can be written as

$$\frac{d\rho}{dt} + div\mathbf{J} = 0, \quad \mathbf{J} = \frac{d\mathbf{P}}{dt} \tag{2}$$

where **P** is the polarization vector.
In local form the more convenient expression for the energy balance, excluding radiation and other thermal effects is

$$\mathbf{E} \cdot \dot{\mathbf{P}} + \mathbf{Q} grad\mathbf{v} = \rho \dot{u} \tag{3}$$

where **E** is the electric field, u is the internal energy per unit of mass, ρ is the mass density and **Q** is the sum of the mechanical stress tensor $\boldsymbol{\sigma}$ and the Maxwell stress tensor **T.**
In absence of magnetic fields or induced magnetization the Maxwell stress tensor is expressed by the following equation

$$\mathbf{T} = \mathbf{DE} - \frac{1}{2}\varepsilon_0 E^2 \mathbf{I} \tag{4}$$

where **D** is the dielectric displacement, ε_0 the vacuum permittivity and **I** represents the unit tensor.
The tensor **Q** can be decomposed in the corresponding symmetric and antisymmetric parts as follows

$$\mathbf{Q}^{s} = \frac{1}{2}\left(\mathbf{Q} + \mathbf{Q}^{T}\right), \quad \mathbf{Q}^{a} = \frac{1}{2}\left(\mathbf{Q} - \mathbf{Q}^{T}\right) \tag{5}$$

where super index T indicates transposition operation.
The gradient of velocity tensor can be decomposed in a similar way, *i.e.*,

$$grad\mathbf{v} = \mathbf{L} = \mathbf{D} + \mathbf{W} \tag{6}$$

On account that the inner product of a symmetric tensor by an antisymmetric tensor is nil, the following expression holds

$$\mathbf{QL} = \mathbf{Q}^{s}\mathbf{D} + \mathbf{Q}^{a}\mathbf{W} \tag{7}$$

However, owing to the fact that the symmetric part of the total stress tensor is not coupled with dipoles rotation and **W** is associated to the angular rotation vector

$$\boldsymbol{\omega} = \frac{1}{2}\nabla \times \mathbf{v} \tag{8}$$

the two vectors $\boldsymbol{\omega}$ and **W** are related in the usual way, *i.e.*,

$$\varepsilon_{pqi}\omega_i = \frac{1}{2}\varepsilon_{pqi}\varepsilon_{ijk}v_{k,j} = \frac{1}{2}\left(\delta_{pj}\delta_{qk} - \delta_{pk}\delta_{qj}\right)v_{k,j} = \frac{1}{2}\left(v_{q,p} - v_{p,q}\right) = W_{qp} \tag{9}$$

where the index notation was used, comma indicates derivation and ε_{ijk} is the alternating tensor. As a consequence, one can write

$$\mathbf{Q}^a\mathbf{W} = \mathbf{Q}^a\boldsymbol{\varepsilon\omega} \tag{10}$$

Then eq. (3) can be rewritten in the following way

$$\rho\frac{du}{dt} = \mathbf{E}\dot{\mathbf{P}} + \mathbf{Q}^a\boldsymbol{\varepsilon\omega} \tag{11}$$

The conservation of the angular moment has to be considered, *i.e.*,

$$I\dot{\boldsymbol{\omega}} = \boldsymbol{\varepsilon}\mathbf{Q}^a \tag{12}$$

where I is the mean inertia moment of the rotating dipoles.
In general, the entropy production per unit of time is the sum of the entropy flux from the exterior plus the internally generated entropy. Accordingly, one can write

$$\rho\frac{ds}{dt} = -div\mathbf{J}_S + \sigma_s \tag{13}$$

where $\mathbf{J}_s$ is the entropy flux and σ_s the entropy production per unit of volume and time.

3. Entropy equation

According to the usual methodology of the EIT, it will assumed that there exists a regular enough function, called generalized entropy defined over a set of variables

$$\eta = \hat{\eta}(u, \rho_p, \mathbf{J}, \mathbf{Q}) \tag{14}$$

This function is such that the corresponding generalized Gibbs equation can be written as

$$\rho\frac{d\eta}{dt} = \frac{\rho}{T}\frac{du}{dt} + \frac{\phi}{T}\frac{d\rho_p}{dt} + \alpha_1\mathbf{J}\frac{d\mathbf{J}}{dt} + \alpha_2\mathbf{Q}^a\frac{d\mathbf{Q}^a}{dt} \tag{15}$$

where ρ_p is the polarization density and ϕ is the electrical potential. In principle ϕ and the coefficients α_i should be functions of the conserved variables, u, ρ_p, but here for simplicity will be considered constant quantities.
Then, the following equation will be assumed for the extended entropy flux,

$$\mathbf{J}_\eta = \mu_1\mathbf{J} + \mu_2'\mathbf{J}\mathbf{Q}^a \tag{16}$$

where μ_i is subject to the same restrictions as α_i.
By using eqs. (14) and (15) the entropy production is given by

$$\begin{aligned}\sigma_\eta &= \rho\frac{d\eta}{dt} + div\mathbf{J}_\eta = \\ &= \frac{\rho}{T}\frac{du}{dt} + \frac{\phi}{T}\frac{d\rho_p}{dt} + \alpha_1\mathbf{J}\frac{d\mathbf{J}}{dt} + \alpha_2\mathbf{Q}^a + \mu_1 div\mathbf{J} + \mathbf{J}grad\mu_1 + \mu_2'\mathbf{J}div\mathbf{Q}^a + \mu_2'\mathbf{Q}^a(grad\mathbf{J})^a\end{aligned} \tag{17}$$

According to eqs. (1), (2) and (11) and after grouping terms, eq. (17) becomes

$$\sigma_\eta = \dot{\mathbf{P}}\left(\frac{\mathbf{E}}{T} + \alpha_1 \ddot{\mathbf{P}} - \frac{1}{\varepsilon_0 \chi T}\mathbf{P} + \mu_2' div\mathbf{Q}^a\right) + \mathbf{Q}^a\left(\frac{1}{T}\boldsymbol{\varepsilon\omega} + \alpha_2 \dot{\mathbf{Q}}^a + \mu_2'(grad\mathbf{J})^a\right) \tag{18}$$

Notice that in the development of eq. (18) the following expressions were used

$$\mu_1 = \frac{\phi}{T}, \quad grad\phi = -\frac{\mathbf{P}}{\chi\varepsilon_0} \tag{19}$$

where χ is the dielectric susceptibility. Eq. (18) has the form of a linear combination of products of two forces (terms between brackets) by the corresponding fluxes, $\dot{\mathbf{P}}$ and $\mathbf{Q}^a$. According to the methodology of CIT the forces can be expressed as linear functions of current fluxes and their spatial and temporal derivatives (entire or fractional) in order to account for the polarization inhomogeneity. In this way,

$$\begin{aligned}
&\frac{\mathbf{E}}{T} + \alpha_1 \ddot{\mathbf{P}} - \frac{1}{\varepsilon_0 \chi T}\mathbf{P} + \mu_2' div\mathbf{Q}^a = \\
&= \xi_1\left(\frac{d^\alpha \mathbf{P}}{dt^\alpha} + \frac{d^\beta \mathbf{P}}{dt^\beta}\right) + \xi_2 div\,(grad\mathbf{P})^a + \xi_3 div\left(grad\left(\frac{d^\alpha \mathbf{P}}{dt^\alpha} + \frac{d^\beta \mathbf{P}}{dt^\beta}\right)\right)^a + \xi_3' div\mathbf{Q}^a \\
&\frac{1}{T}\boldsymbol{\varepsilon\omega} + \alpha_2 \dot{\mathbf{Q}}^a + \mu_2'(grad\dot{\mathbf{P}})^a = \\
&= \xi_4 \mathbf{Q}^a + \xi_5 (grad\mathbf{P})^a + \xi_6\left(grad\left(\frac{d^\alpha \mathbf{P}}{dt^\alpha} + \frac{d^\beta \mathbf{P}}{dt^\beta}\right)\right)^a
\end{aligned} \tag{20}$$

where fractional derivatives have been introduced. By grouping terms eq. (20) can be rewritten as

$$\begin{aligned}
&\frac{\mathbf{E}}{T} + \alpha_1 \ddot{\mathbf{P}} - \frac{1}{\varepsilon_0 \chi T}\mathbf{P} + \mu_2 div\mathbf{Q}^a = \\
&= \xi_1\left(\frac{d^\alpha \mathbf{P}}{dt^\alpha} + \frac{d^\beta \mathbf{P}}{dt^\beta}\right) + \xi_2 div(grad\mathbf{P})^a + \xi_3 div\left(grad\left(\frac{d^\alpha \mathbf{P}}{dt^\alpha} + \frac{d^\beta \mathbf{P}}{dt^\beta}\right)\right)^a \\
&\frac{1}{T}\boldsymbol{\varepsilon\omega} + \alpha_2 \dot{\mathbf{Q}}^a + \mu_2'(grad\dot{\mathbf{P}})^a = \\
&= \xi_4 \mathbf{Q}^a + \xi_5 (grad\mathbf{P})^a + \xi_6\left(grad\left(\frac{d^\alpha \mathbf{P}}{dt^\alpha} + \frac{d^\beta \mathbf{P}}{dt^\beta}\right)\right)^a
\end{aligned} \tag{21}$$

where $\mu_2 = \mu_2' - \xi_3'$.

Eq. (21b) contains the angular velocity $\boldsymbol{\omega}$ which can be eliminated in favour of $\mathbf{Q}^a$ by using (12). According to eq. (12) and premultiplying by the alternating tensor $\boldsymbol{\varepsilon}$, one obtains

$$\boldsymbol{\varepsilon}\dot{\boldsymbol{\omega}} = \frac{\boldsymbol{\varepsilon\varepsilon}\mathbf{Q}^a}{I} \tag{22}$$

According to eq. (9), and taking into account the antisymmetric character of $\mathbf{Q}^a$, the following relationship holds

$$\varepsilon_{pqi}\varepsilon_{ijk}Q_{hl}^{a} = \left(\delta_{ih}\delta_{jl} - \delta_{il}\delta_{hj}\right)Q_{hl}^{a} = Q_{ij}^{a} - Q_{ji}^{a} = 2Q_{ij}^{a} \tag{23}$$

and eq. (22) can be written as

$$\boldsymbol{\varepsilon}\dot{\boldsymbol{\omega}} = \frac{2\mathbf{Q}^{a}}{I} \tag{24}$$

In principle eq. (21) and (24) should provide the solution to our problem. However, in order to interpret the results, it is more convenient to consider some particular cases. To start with let we assume $\alpha = 1$; after omission of the fractional derivatives of order β, eq. (21) can be written as

$$\begin{aligned} &\frac{\mathbf{E}}{T} + \alpha_1\ddot{\mathbf{P}} - \frac{1}{\varepsilon_0\chi T}\mathbf{P} + \mu_2 div\mathbf{Q}^a = \xi_1\dot{\mathbf{P}} + \xi_2 div(grad\mathbf{P})^a + \xi_3 div(grad\dot{\mathbf{P}})^a \\ &\frac{1}{T}\boldsymbol{\varepsilon\omega} + \alpha_2\dot{\mathbf{Q}}^a + \mu_3(grad\dot{\mathbf{P}})^a = \xi_4\mathbf{Q}^a + \xi_5(grad\mathbf{P})^a \end{aligned} \tag{25}$$

where $\mu_3 = \mu_2' - \mu_6$.

From eq. (24) and after taking derivatives, eq. (25b) becomes

$$\frac{2}{IT}\mathbf{Q}^a + \alpha_2\ddot{\mathbf{Q}}^a + \mu_3(grad\ddot{\mathbf{P}})^a = \xi_4\dot{\mathbf{Q}}^a + \xi_5(grad\dot{\mathbf{P}})^a \tag{26}$$

The second term of the right hand side of eq. (25a) can be written as

$$div(grad\mathbf{P})^a = \frac{1}{2}\left(\Delta\mathbf{P} - graddiv\mathbf{P}\right) \tag{27}$$

For a unidirectional propagation wave with vector wave $\mathbf{k} = (k,0,0)$ under an electrical field transversal to the propagation, i.e., $\mathbf{E} = (0,E_2,E_3)$, the components for the dielectric susceptibility χ_{ij} tensor are

$$\chi_{ij} = 0, \quad k_i k_j = 0, \quad \forall i \neq j; \quad \chi_{11} = 0 \tag{28}$$

Notice that the constitutive relationship between polarization and electric field is given by

$$P_i = \varepsilon_0\chi_{ij}E_j \tag{29}$$

Accordingly, eq. (27) can be written as

$$\mathbf{k}\cdot\left(\mathbf{k}\mathbf{P}^a\right) = \frac{1}{2}\left(k^2\mathbf{P} - \mathbf{k}\left(\mathbf{k}\cdot\mathbf{P}^a\right)\right) \tag{30}$$

This expression in conjunction with $(graddiv\mathbf{P})_i = k_j k_i P_i = 0$, leads eq. (27) to

$$div(grad\mathbf{P})^a = \frac{1}{2}\Delta\mathbf{P} \tag{31}$$

Elimination of $\mathbf{Q}^a$ between eq. (25b) and (26), and considering eq. (21), the following expression is obtained after multiplying by $\varepsilon_0\chi IT^2\mu_2$,

$$\begin{aligned}
&\mathbf{P}+\dot{\mathbf{P}}T(2\xi_1\varepsilon_0\chi-\xi_4 I)+\ddot{\mathbf{P}}T(\alpha_2 I-\xi_1\xi_4\varepsilon_0\chi IT-2\alpha_1\varepsilon_0\chi)+\dddot{\mathbf{P}}\varepsilon_0\chi IT^2(\alpha_2\xi_1+\alpha_1\xi_4)-\\
&-\ddddot{\mathbf{P}}\alpha_1\alpha_2\varepsilon_0\chi IT^2+\frac{1}{2}\Delta\mathbf{P}(2\xi_2\varepsilon_0\chi T)+\frac{1}{2}\Delta\dot{\mathbf{P}}\varepsilon_0\chi T(2\xi_3-\xi_5\mu_2 IT-\xi_2\xi_4 IT)+\\
&+\frac{1}{2}\Delta\ddot{\mathbf{P}}\varepsilon_0\chi T^2 I(\alpha_1\xi_2+\mu_2\mu_3-\xi_3\xi_4)+\frac{1}{2}\Delta\dddot{\mathbf{P}}(\alpha_2\xi_3\varepsilon_0\chi IT^2)-\\
&-\varepsilon_0\chi(2\mathbf{E}-\xi_4 IT\dot{\mathbf{E}}+\alpha_2 IT\ddot{\mathbf{E}})=0
\end{aligned} \tag{32}$$

If electromechanical coupling is absent, *i.e.*, $\mu_2=0$ one obtains

$$\frac{\mathbf{E}}{T}+\alpha_1\ddot{\mathbf{P}}-\frac{1}{\varepsilon_0\chi T}\mathbf{P}=\xi_1\dot{\mathbf{P}}+\xi_2 div(grad\mathbf{P})^a+\xi_3 div(grad\dot{\mathbf{P}})^a \tag{33}$$

Rearrangement of eq. (33) in conjunction with eq. (31) leads to

$$\mathbf{P}+\tau\dot{\mathbf{P}}+\lambda_1\ddot{\mathbf{P}}=\varepsilon_0\chi\mathbf{E}+\frac{1}{2}D_1\Delta\mathbf{P}+\frac{1}{2}D_2\Delta\dot{\mathbf{P}} \tag{34}$$

where

$$\tau=\varepsilon_0\chi T\xi_1,\quad \lambda_1=-\alpha_1\varepsilon_0\chi T,\quad D_1=-\varepsilon_0\chi T\xi_2,\quad D_2=-\varepsilon_0\chi T\xi_3 \tag{35}$$

Equation (34) is a telegrapher type equation for the polarization propagation. The velocity of the propagation c is given by

$$c^2=\frac{D_1}{2\lambda_1} \tag{36}$$

If $D_1=D_2=0$, eq. (34), becomes a Debye type equation with inertial effects, *i.e.*, the so-called Rocard equation is obtained. Finally, if $\lambda_1=0$, the classical Debye equation is recovered.

4. Complex dielectric permittivity

In this section, the linear response to alternating electrical fields will be considered. For this purpose the Laplace transform will be used. By substituting eq. (24) in the temporal derivative of eq. (21b) one obtains

$$\frac{2\mathbf{Q}^a}{IT}+\alpha_2\ddot{\mathbf{Q}}^a+\mu_2(grad\ddot{\mathbf{P}})^a=\xi_4\dot{\mathbf{Q}}^a+\xi_5(grad\dot{\mathbf{P}})^a+\xi_6 grad\left(\frac{d^{\alpha+1}\mathbf{P}}{dt^\alpha}+\frac{d^{\beta+1}\mathbf{P}}{dt^\beta}\right)^a \tag{37}$$

In order to simplify the calculations and without losing significant generality, it will be assumed $\xi_6=0$. In an analogous way, taking $\xi_3=0$, eqs. (33) and (37) can be written as

$$\begin{aligned}
&\frac{\mathbf{E}}{T}+\alpha_1\ddot{\mathbf{P}}-\frac{1}{\varepsilon_0\chi T}\mathbf{P}+\mu_2 div\mathbf{Q}^a=\xi_1\left(\frac{d^\alpha\mathbf{P}}{dt^\alpha}+\frac{d^\beta\mathbf{P}}{dt^\beta}\right)+\xi_2 div(grad\mathbf{P})^a\\
&\frac{2}{IT}\mathbf{Q}^a+\alpha_2\ddot{\mathbf{Q}}^a+\mu_2(grad\ddot{\mathbf{P}})^a=\xi_4\dot{\mathbf{Q}}^a+\xi_5(grad\dot{\mathbf{P}})^a
\end{aligned} \tag{38}$$

The two first derivatives of eq. (38a) and the divergence of (38b) are used to eliminate $\mathbf{Q}^{a}$. Then the Laplace transform of the resulting expression, in conjunction with eqs. (28) and (31), lead to

$$\chi^* = \chi\left[\left(1+\lambda_1 s^2+\tau_D(s^{\alpha}+s^{\beta})\right)-\frac{1}{2}\frac{k^2}{\xi_1}\left(\xi_2+\frac{\mu_2\xi_5 s(1+\lambda_2 s)}{\xi_4 s(1+\lambda_3 s)-\dfrac{2}{IT}}\right)\right]^{-1} \tag{39}$$

where

$$\tau_D=\xi_1\varepsilon_0\chi T,\quad \lambda_1=-\frac{\alpha_1}{\xi_1},\quad \lambda_2=-\frac{\mu_3}{\xi_5},\quad \lambda_3=-\frac{\alpha_2}{\xi_4} \tag{40}$$

Taking $\chi=\chi_0-\chi_\infty$, adding the instantaneous component of the polarization χ_∞, and using, $s=i\omega$, eq. (39) becomes

$$\chi^* = \chi_\infty+\frac{\chi_0-\chi_\infty}{\left(1-\lambda_1\omega^2+\tau_D((i\omega)^{\alpha}+(i\omega)^{\beta})\right)-\dfrac{1}{2}\dfrac{k^2}{\xi_1}\left(\xi_2+\dfrac{i\mu_2\xi_5\omega(1+i\lambda_2\omega)}{i\xi_4\omega(1+i\lambda_3\omega)-\dfrac{2}{IT}}\right)} \tag{41}$$

Taking into account that

$$\chi^*=\varepsilon_r^*-1,\ \chi_0=\varepsilon_0-1,\ \chi_\infty=\varepsilon_\infty-1 \tag{42}$$

Eq. (41) can be expressed in terms of the permittivity by

$$\varepsilon^* = \varepsilon_\infty+\frac{\varepsilon_0-\varepsilon_\infty}{\left(1-\lambda_1\omega^2+\tau_D((i\omega)^{\alpha}+(i\omega)^{\beta})\right)-\dfrac{1}{2}\dfrac{k^2}{\xi_1}\left(\xi_2+\dfrac{i\mu_2\xi_5\omega(1+i\lambda_2\omega)}{i\xi_4\omega(1+i\lambda_3\omega)-\dfrac{2}{IT}}\right)} \tag{43}$$

Equation (43) is the equation for the asymmetric absorption and dispersion of a viscodielectric material in a heterogeneous medium, with inertial viscoelastic coupling effects. The heterogeneity of the medium is taken into account through the wave vector k and the viscoelastic coupling is accounted for by means of the coefficient μ_2. The asymmetry is present through the different behaviour at low and high frequencies of both sides of the corresponding transition peak. This asymmetry has been introduced in the model by means of the degrees α and β of the fractional derivatives. Dielectric and mechanical inertial effects are expressed through λ_1 and I, respectively.

It should be noted that in the present formulation it is compulsory to take into account inertial effects because if $I=0$ in eq. (43), the viscoelastic-dielectric coupling disappears. To avoid this problem it should be necessary to introduce in eq. (24) a term corresponding to the spin transport.

It is noteworthy that, strictly speaking, the molecule "sees" the local field, i.e., the field screened by the dipolar cloud surrounding the molecule. However in the preceding analysis the distinction between the applied and local field has not been considered, thus making equivalent macroscopic and microscopic responses. This a common practice in the analysis of the dielectric relaxation.

5. References

Jou, D.,Casas-Vázquez, J.,Lebon, G.,Extended Irreversible Thermodynamics, Springer, Berlin, 1993.

Müller, I.,Ruggeri, T.,Extended Thermodynamics, Springer, Berlin,N. York, 1993.

Kluitenberg, G.A., On dielectric and magnetic relaxation phenomena and non-equilibrium thermodynamics, Physica,68,(1973),75-92.

Kluitenberg, G.A., On dielectric and magnetic relaxation phenomena and vectorial internal degrees of freedom in thermodynamics, Physica 87A,(1977),302-330.

Kluitenberg, G.A., On vectorial internal variables and dielectric and magnetic relaxation phenomena, Physica 109A,(1981),91-122.

Tschoelgl, N., The Phenomenological Theory of Linear Viscoelasticity. Chapters 3 and 5. Springer, Berlin, 1989.

Riande, E,Díaz-Calleja, R, Prolongo, M.G., Masegosa, R.M.,Salom, C.,Polymer Viscoelasticity:Stress and Strain in Practice, Chapter10. M.Dekker, N. York, 2000.

Kuiken, G.D.C.,Thermodynamics of Irreversible Processes, Wiley, Chichester, England, 1994.

De Groot, S.R.,Mazur, P.,Non-Equilibrium Thermodynamics, North Holland Publ. Co. Amsterdam, 1962.

E. Riande, R. Díaz-Calleja, Electrical Properties of Polymers, M. Dekker, 2004. Chap.3.

20

Volatile Precursors for Films Deposition: Vapor Pressure, Structure and Thermodynamics

Igor K. Igumenov, Tamara V. Basova and Vladimir R. Belosludov
Nikolaev Institute of Inorganic Chemistry SB RAS, Novosibirsk
Russia

1. Introduction

At present the processes of chemical vapour deposition (MO CVD) and physical vapour deposition (PVD) are widely used to obtain the layers of different materials. A possibility to obtain materials of the required composition (elementary, complex, composite, etc.) is determined by the chemical nature of a precursor. By now a set of criteria for the precursors to meet the MO CVD and PVD standards have already been established (Maury, 1995). Among the parameters from which the choice of a precursor is made, the most important one is its ability to transfer to the gaseous phase at temperatures not exceeding the temperature of molecules decomposition, i.e., its volatility. The most important quantitative characteristic of volatility is the pressure of saturated vapour of the compound at defined temperature. A number of experimental data showing volatility of different classes of metal complexes with organic ligands are known. Practically all precursors crystallize in lattices of a molecular type. The relation to molecular crystals suggests that energy of intermolecular interaction is much less than the energy of intramolecular bonding. The structure of metal complexes with organic ligands is determined by different factors depending on the nature of metal and coordinated ligands.

The current review summarizes the data of saturated vapor pressure and thermodynamic parameters of the sublimation process of some complexes with organic ligands as reported in the literature as well as in our works. Dependence of volatility on type of substituents in the ligand, on central atom, and on donor atom in the ligand is discussed using metal β-diketonate (Chart 1) and phthalocyanine (MPc) derivatives as examples. The volatility of metal β-diketonates and phthalocyanines is analysed from the standpoint of their molecular and crystal structure. One important aspect of chemistry of precursors for MO CVD and PVD is a possibility to predict volatility from the data of the compound structure. A description of the microscopic model created in the framework of the statistical mechanics to describe the volatility of metal complexes with organic ligands is given. It follows from the structural studies that interatomic spacing in a molecule is comparable to interatomic spacing in different adjacent molecules. With such location of the molecules of complexes in the crystal lattice, it is necessary to take into account both van der Waals and electrostatic interactions. The correlation between these interactions determines the specific features of the behavior of this class of molecular crystals.

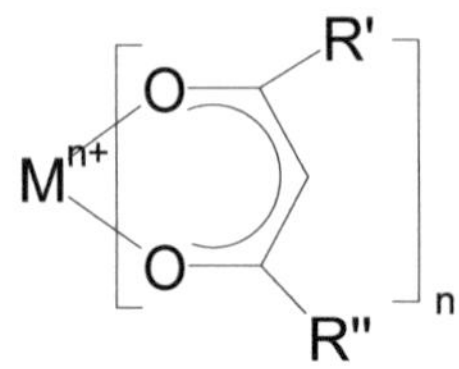

R′	R″	Name	Abbreviation
CH_3	CH_3	pentane-2,4-dionate (acetylacetonate)	aa
$C(CH_3)_3$	$C(CH_3)_3$	2,2,6,6-tetramethylheptane-3,5-dionate (dipivaloylmethanate)	dpm
CF_3	$C(CH_3)_3$	1,1,1-trifluoro-5,5-dimethylhexane-2,4-dionate (pivaloyltrifluoroacetonate)	pta
CH_3	$C(CH_3)_3$	2,2-dimethylhexane-3,5-dionate (pivaloylacetonate)	pva
CF_3	CF_3	1,1,1,6,6,6-hexafluoropentane-2,4-dionate (hexafluoroacetylacetate)	hfa
C_6F_{11}	C_6F_{11}	1,5-di(undecafluorocyclohexyl)-pentane-2,4-dionate	dhfa
CF_3	CH_3	1,1,1-trifluoropentane-2,4-dionate (trifluoroacetylacetate)	tfa
C_6H_5	C_6H_5	1,3-diphenylpropane-1,3-dionate (dibenzoylmethanate)	dbm
C_6H_5	CH_3	1-phenylbutane-1,3-dionate (benzoylacetonate)	ba
C_6H_5	CF_3	1-phenyl-4,4,4-trifluorobutane-1,3-dionate (benzoyltrifluoroacetonate)	btfa
CH_3	CH_3	pentane-2-on-4-iminate (ketoiminate) / N atom in place of one O atom	ki
CH_3	CH_3	pentane-2,4-diiminate (diiminate) / N atoms in place of O atoms	diim
CH_3	CF_3	1,1,1-trifluorpentane-2-on-4-iminate / N atom in place of one O atom	ktfa

Chart 1. Structure of metal β-diketonate molecules and notation conventions of the names of some metal β-diketonates

2. Experimental data on measuring temperature dependence of vapor pressure for metal complexes with organic ligands

2.1 General considerations, the macroscopic approach

The process of sublimation and evaporation of different compounds including complexes of metals with organic ligands is related to the phase transitions of the first order which are described in thermodynamic approach by the Clausius-Clapeyron (CC) equation (Callen, 1985). The temperature dependence of equilibrium pressure in the gaseous phase *P(T)* is derived from CC equation by integration after introduction of a number of assumptions (Callen, 1985):

$$\ln P(T) = q / RT + B \tag{1}$$

where heat of transition is chosen as $q = -\Delta_e h$, where $\Delta_e h$ is enthalpy of transition at a certain pressure and temperature T_o, while the constant of integration B is found in different models and is commonly chosen as $B=\Delta_e s/R$, where $\Delta_e s$ is entropy of crystal-vapor transition at $P(T)$=1 atm and $R=N_o k$, where k is the Boltzmann constant and N_o is the Avogadro constant. Formula (1) makes it possible to find $\Delta_e h$ and $\Delta_e s$ from the experimentally obtained relationship $lnP(T)=f(T)$. However, for unambiguous description of a thermodynamic system additional information on the composition of vapor is needed since different processes associated with changes of the molecular mass, eg. oligomerization or dissociation, may proceed in the gaseous phase.

2.2 Methods of vapor pressure measurements, limitations and specific features of compounds

The absolute value of measured pressure in the first place and thermal stability of the investigated compound in the second place are of crucial importance when the method for investigation of the saturated vapor pressure is chosen. There are well know methods for the study of heterogeneous equilibrium in which gaseous and condensed phases take part; those methods may be classified into static, quasistatic, dynamic (flow method and Langmuir method) and kinetic (Knudsen effusion) methods (Margrave, 1967; Ingram & Drauart, 1962). Different methods may be used to study different ranges of vapor pressure and temperatures. Most of the results of measurement of saturated vapor pressure for volatile compounds have been obtained by different variants of the static method, flow method and Knudsen method as well. All these methods in different modifications have been used for measurements of the temperature dependence of the saturated vapor pressure of precursors.

2.2.1 Static methods of vapor pressure measurements

In the static methods, measurement of vapor pressure is carried out in a closed vessel (Beech & Lintonbon, 1971; Margrave, 1967). The temperature of the vessel is maintained constant to establish equilibrium between the condensed phase and vapor. The equilibrium pressure is determined either directly with a measuring instrument - mercury, ionization, spiral, membrane or other type of manometer (direct methods), or by determination of the substance amount in the known volume (indirect methods). The method allows approaching the equilibrium both from the "bottom" and from the "top", which is one of its main advantages. This method allows determining the overpressure arising in the system due to thermal dissociation of the precursor at a given temperature. The most acceptable range of pressure measured by this method is 1-760 Torr.

2.2.2 Flow method

In this method, a flow of inert gas is passed over a heated substance and carries the substance vapors into the condenser (Beech & Lintonbon, 1971; Margrave, 1967). The quantity is determined by the weight, chemical analytical or radiation method. The vapor pressure of the substance p is determined by the Henry law: $p=P^*n/(n+N)$, where P is the total pressure; n is the number of moles of the transferred substance and N is the number of moles of the inert gas.

The method is applicable for measurement of low vapor pressures of precursors. The dynamic range of the measured pressures is from 10^{-3} to 1 Torr. The main advantage of the flow method is the possibility of measurement of the precursor partial pressure against a background of its decomposition. If the compound exists in several molecular forms in the vapor phase, the measured pressure is averaged. In this case for calculation of the thermodynamic parameters of vaporization it is necessary to know the ratio of these molecular forms in the vapor phase.

2.2.3 Knudsen method

Knudsen method is based on the measurement of the rate of effusion of the equilibrium vapor through a small orifice. It can be used for the measurement of extremely small values of vapor pressures - 10^{-7}–10^{-1} Torr. Knudsen method has different modifications; the most powerful one is its combination with mass spectrometer. This variant allows controlling the composition of vapor phase during measurements.

2.2.4 Indirect methods of vapor pressure measurements

In a limited number of works, different variants of the thermal analysis (Logvinenko et al., 1998; Bublik et al., 1978) as well as gas-liquid chromatography (Irving & Ribeiro da Silva, 1976) are used for the calculation of vapor pressure values. Different calorimetric methods were also used to determine the value of sublimation enthalpy for different metal complexes with organic ligands (Ribeiro da Silva & Goncalves, 1998; Berg & Truempe, 1960).

2.3 Analysis of adequacy of the experimental data on precursors vapor pressure

The measurement of vapor pressure for the same compounds carried out by different methods or sometimes even by the same method but performed by different investigators were not always in good agreement. The temperature dependences of vapor pressure $P(T)$ for chromium tris-acetylacetonate are presented in Fig. 1 and Table 1 for comparison.

The difference in $P(T)$ leads to rather important divergence in the values of enthalpies and entropies of sublimation. Moreover close values of the precursor vapor pressure are described by sets of thermodynamic parameters with quite different values (Table 1) in some independent measurements. In most published work neither structural modifications of the investigated complexes nor their purity were specified, while both factors have significant influence on the value of vapor pressure as well as on the value of sublimation enthalpy of the compounds. In the majority of published work the enthalpy of evaporation is used as the sole criterion for comparison of volatility of different compounds.

This approach is found to be incorrect because the differences in the volatility of various molecular complexes are described by different sets of thermodynamic parameters and the results of evaporation enthalpy change as well as entropy change should be considered. To analyze the reasons of these divergences the following parameters should be taken into account: 1 - characterizations the investigated compound (method of synthesis, purity, thermal properties); 2 - methods of vapor pressure measurement (technique details, pressure and temperature measurement errors, and limitations of the methods); 3 - analysis of observations. Hereafter, where applicable, data collection of independent measurements will be used.

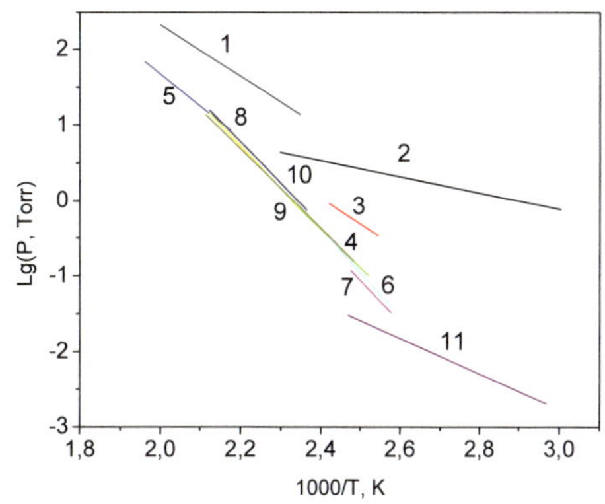

Fig. 1. LgP(1/T) dependence for $Al(aa)_3$ measured by different methods (Table 1)

№	Метод	ΔH, kJ/mol	ΔS°, J/K·mol	Range T, K	Ref
1	St (isoten.)	65.1	119.5	426-500	Berg & Truempe, 1960
2	St (isoten.)	20.5	4.16	333-435	Berg & Truempe, 1965
3	St (isoten.)	65.7	103.1	393-413	Fontain et. al., 1972
4	St (membr.)	96.9	170.0	397-430	Wolf et. al., 1972
5	St (membr.)	80.8±2.9	138±4	460-510	
6	St (membr.)	111±4	204	388-413	Sachinidis & Hill, 1980
7	St (bulb.)	105±2	187	388-404	Sachinidis & Hill, 1980
8	St (membr.)	108.4±1.8	197.1±3.8	426-471	Igumenov et. al., 1979
9	GLC	100.6	172.9	403-473	Bublik et. al., 1978
10	GLC	105.2	191.1	423-471	Mazurenko & Gerasimchuk, 1993
11	Knudsen	45.0±0.4	26.8±1.2	337-405	Techil et. al., 1981

Table 1. Enthalpy and entropy of sublimation of $Al(aa)_3$ obtained by different methods

2.4 Vapor pressure of metal β-diketonates

At present metal β-diketonate derivatives are one of the most widely used class of precursors for MO CVD. Metal β-diketonates are typical members of metal chelates. The metal β-diketonates are very good model objects because it is always possible to choose such series of compounds which have essentially different distribution of intramolecular electronic density and, accordingly, different energies of intermolecular interaction, but for all that they retain similar crystal and molecular structures.

The structure of homoligand metal β-diketonates (Chart 1) allows to change the peripheral surrounding of the molecule in the crystal lattice by variation of the type of substituents in ligand (R^1, R^2=CH_3, CF_3, $C(CH_3)_3$, C_6H_5 etc.). It is also possible to change both the donor atom (X, Y =O, N, S) and the central metal while keeping the molecule geometric parameters and types of crystal packing very close.

2.4.1 Dependence of metal β-diketonate vapor pressure on type of substituents in ligand

Experimental data for metal β-diketonates demonstrate a very strong dependence of their saturated vapor pressure on the type of substitutes in ligand (Irving & Ribeiro da Silva, 1976;

Ribeiro da Silva & Goncalves, 1998). Typical temperature dependences of vapour pressure for Cu (Berg & Truempe, 1960), and Pb (Fontain et. al., 1972) β-diketonates with different substitutes in ligand are given in Fig. 2. Table 2 provides the respective values of enthalpy and entropy of sublimation for these compounds. The data analysis reveals some empirical regularity in the change of saturated vapor pressure as a function of temperature depending on the type of substituents in ligand (CH_3, CF_3, $C(CH_3)_3$, C_3F_7, C_6H_5 and their combinations).
Substitution of both trifluoromethyl groups in ligant for methyl radicals leads to increase of the vapor pressure of monomeric β-diketonates by three or four orders of magnitude. The same order of pressure change is observed in bis-β-diketonates of Pt (Wolf et. al., 1972), Pd and tris-complexes of Al (Berg & Truemper, 1965), Ir and Rh (Sachinidis & Hill, 1980).

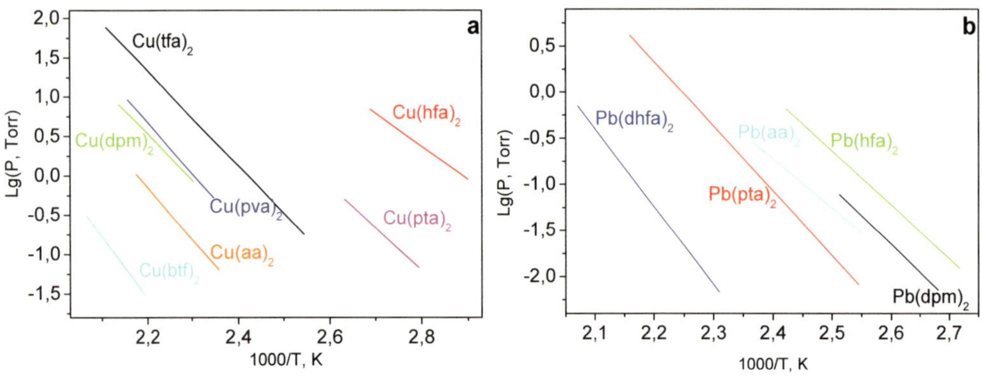

Fig. 2. LgP(1/T) dependence for different Cu(II) (**a**) and for Pb(II) (**b**) β-diketonates

Compound	Method	ΔH, kJ/mol	ΔS°, J/K·mol	Range T, K
$Cu(aa)_2$	membr.	125.5±9.2	218.0±20.1	424-460
$Cu(tfa)_2$	membr.	115.9±0.8	225.5±2.5	393-474
$Cu(hfa)_2$	membr.	80.8±1.8	178.2±5.4	345-372
$Cu(dpm)_2$	membr.	111.7±4.6	200.8±8.8	434-468
$Cu(pva)_2$	membr.	123.8±2.7	230.1±6.3	426-464
$Cu(btf)_2$	flow	151.5±6.7	248.1±14.2	456-484
$Cu(pta)_2$	flow	102.2±3.0	208.0±8.0	358-380
$Pb(hfa)_2$	flow	111.7±1.3	211.8±3.2	368-413
$Pb(dpm)_2$	flow	117,5.±2.8	218.8±7.3	373-398
$Pb(aa)_2$	flow	102.4±5.0	176.7±12.4	392-423
$Pb(pta)_2$	flow	134.4±1.4	246.9±3.2	393-463
$Pb(dhfa)_2$	flow	161.4±1.5	276.1±3.3	433-483

Table 2. Enthalpy and entropy of sublimation of different Cu(II) and Pb(II) β-diketonates

For zirconium tetrakis-β-diketonates (Igumenov et. al., 1979), the substitution of both trifluoromethyl groups for methyl ones leads to pressure increase by 4-4.5 orders of magnitude. Quite different situation is observed for lead bis-β-diketonates (Fig. 2b). The observed change is no more than 0.5 order at 100 °C. Moreover, the lead complex with fluorinated ligand has the lower vapor pressure than non-fluorinated one. This quite rare situation seems to be caused by pseudo-polymeric structure of these lead complexes in the condensed phase (Mazurenko & Gerasimchuk, 1993). In the gas phase only monomeric molecules of these lead complexes have been detected. Decrease of vapor pressure with introducing fluorinated substituents was also observed for copper ethylene-diamine-bis-acetylacetonate derivatives (Techil et. al., 1981) and for dimethylgold(III) fluorinated ketoiminate derivatives (Wolf et. al., 1972). β-diketonate complexes with aromatic substitutes, eg. phenyl, have substantially lower vapor pressure. Thus, substitution of one trifluoromethyl group for phenyl group in copper hexafluoroacetylacetonate leads to decrease of the vapor pressure 5.5 orders of magnitude (Fig. 2a).

Therefore, analyzing all data presented above, we can formulate the following regularity. The vapor pressure increases with consequential introduction of fluorinated substituents into the ligand if strong specific interaction is absent in the condensed phase for bis-, tris- and tetrakis- metal β-diketonates. Introduction of bulky tert-butyl substituents into the ligand also increases the volatility insignificantly in most cases.

2.4.2 Dependence of metal β-diketonates vapor pressure on central atom

The central atom has a strong influence on the value of vapor pressure of monomeric β-dikentonates. Temperature dependences of vapor pressure of different metal β-diketonates with identical ligands are shown in Fig. 3. Enthalpy and entropy of sublimation of acetylacetonates and dipivaloylmethanates of different metals are given in Table 3. The pressure change at 100 °C is 1.5 orders of magnitute for acetylacetone derivatives (Fig. 3a) and up to 3 orders for metal hexafluoroacetylacetonates (Fig. 3c).

The aluminum and zirconium complexes are the most volatile while the platinum and iridium complexes are characterized by the lowest volatility. Lead β-diketonates demonstrate different behaviour. $Pb(aa)_2$ is one of the most volatile, while $Pb(hfa)_2$ has the lowest volatility among the respective derivatives.

As a whole, it is possible to claim that complexes having a planar structure possess the lower vapor pressure than the tris-complexes. This conclusion is also confirmed by the measurements of temperature dependence of vapor pressure for noble metal diethyldithiocarbamates (Igumenov et. al., 1978).

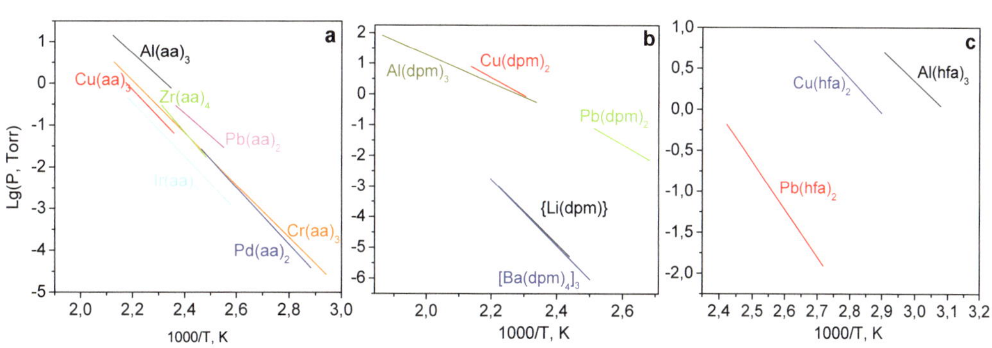

Fig. 3. LgP(1/T) dependence for acetylacetonates (a), dipivaloylmethanates (b) and hexafluoroacetylacetonates (c) of different metals

Complex	Method	ΔH, kJ/mol	ΔS°, J/K·mol	Range T, K	Ref.
$Al(aa)_3$	membr.	108.4±1.8	197.1±3.8	426-471	Igumenov et al., 1979
$Cr(aa)_3$	5 methods	119.7±	209.3±	340-470	Fedotova et al., 1999
$Cu(aa)_2$	flow	125.5±9.2	218.0±20.1	424-460	Igumenov et al., 1978
$Zr(aa)_4$	flow	138.8±4.0	255.4±9.1	403-433	Morozova et al., 1996
$Pd(aa)_2$	Knudsen	130.1±2.8	235.2±7.1	347-406	Zharkova et al., 2000
$Ir(aa)_3$	MCV	124.0±2.1	208.9±4.9	387-459	Igumenov et al., 1985
$Pb(aa)_2$	flow	102.4±5.0	176.7±12.4	392-423	Krisyuk et al., 1998
$\{Li(dpm)\}_4$	Knudsen	192.2±7.2	312.2±7.3	410-450	Filatov et al., 2006
$Al(dpm)_3$	membr.	87.3±0.7	143.7±1.4	360-385	Igumenov et al., 1979
$Cu(dpm)_2$	membr.	111.7±4.6	200.8±8.8	434-468	Igumenov et al., 1978
$Pb(dpm)_2$	flow	117,5.±2.8	218.8±7.3	373-398	Krisyuk et al., 1998
$[Ba(dpm)_2]_3$	Knudsen	208.7±5.4	350.9±2.5	400-455	Alikhanyan et al., 1994

Table 3. Enthalpy and entropy of sublimation of acetylacetonates and dipivaloylmethanates of different metals

2.5 Vapor pressure of metal phthalocyanines

The literature devoted to the investigation of thermal behaviour in the gaseous phase and vapor pressure data is available only for a limited number of phthalocyanines. The detailed review of vapor pressure of some phthalocyanines is given in (Basova et. al., 2010). It was shown that the crystal structure of the investigated compound plays a significant role during these measurements. The results are expected to be largely distorted if α-MPc or the mixture of different phthalocyanine crystalline modifications are used for the measurement of saturated vapor pressure, since the α→β phase transition is expected to proceed during these experimental conditions. In order to avoid such highly possible distortion of results, thermally stable crystal modifications were used during all our investigations; namely β-ZnPc, β-CuPc (Brown, 1968) and triclinic modification (phase φ_{II}) of VOPc (Ziolo et. al., 1980).

The temperature dependence of the saturated vapor pressure of the unsubstituted phthalocyanines without additional ligands in axial positions as well as those of vanadyl and titanyl phthalocyanines are presented in Fig. 4. The calculated thermodynamic parameters of vaporization are listed in Table 4. The phthalocyanines with additional ligands in axial positions $M(IV)Cl_2Pc$, where M=Si, Sn, or Ge, have lower vapor pressure in comparison with Sn(II)Pc which doesn't have axial ligands (Bonderman et. al., 1970; Shaulov et. al., 1972; Shaulov et. al., 1975). This is not only due to their larger molecular weights but also due to features of their crystal structure (Basova et. al., 2010).

While saturated vapor pressure data of some unsubstituted phthalocyanines are already available in the literature, data on substituted phthalocyanines in gaseous phase are scarce. The vapor pressure of only two halogen-substituted phthalocyanines, hexadecachloro- and hexadecabromophthalocyanines (H_2PcCl_{16} and H_2PcBr_{16}) has however been studied (Bonderman et. al., 1970). Data on vapor pressure for hexadecafluorinated phthalocyanines $MPcF_{16}$ have been obtained for the first time by our research group and were published recently (Semyannikov et. al., 2006; Kol'tsov et. al., 2004; Semyannikov et. al., 2008). Measurements on vapor pressure for $MPcF_{16}$, M=Cu, Zn, VO and their unsubstituted analogues are presented in Fig. 4. The order of volatility for some unsubstituted phthalocyanines is as follows: CuPc > ZnPc > VOPc. However, the volatility order for the fluorinated analogues of the same compounds is in the following order; $CuPcF_{16}$ ~ $VOPcF_{16}$ > $ZnPcF_{16}$. The ΔH_T and ΔS^o_T of these phthalocyanines increase from $CuPcF_{16}$ to $ZnPcF_{16}$. The introduction of 16 electron-acceptor fluorine atoms into the phthalocyanine ring leads to the decrease of the volatility of $CuPcF_{16}$ and $ZnPcF_{16}$ in comparison with their unsubstituted analogues while the volatility of $VOPcF_{16}$ becomes higher than that of unsubstituted VOPc.

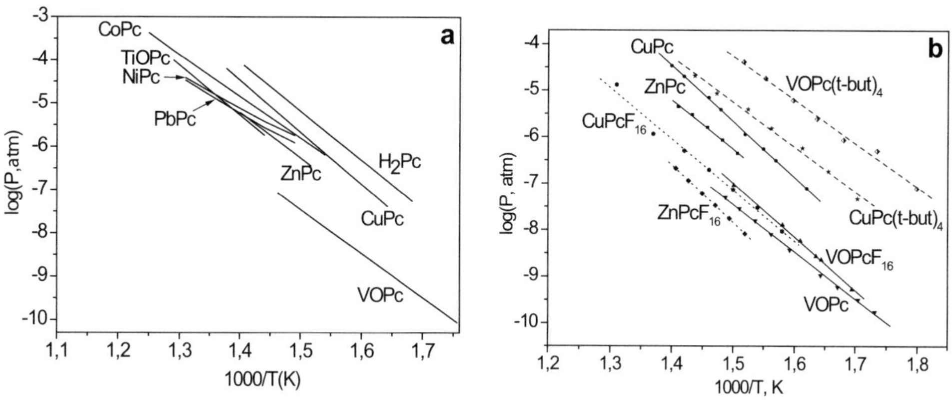

Fig. 4. Temperature dependence of saturated vapor pressure of some unsubstituted (**a**) and substituted (**b**) phthalocyanines

Compound	ΔH_T, kJ/mol	ΔS^o_T, J/mol·K	T, °C	Ref.
β-H_2Pc	223.8±1.3	236.4±2.1	325-425	Semyannikov et. al., 2000
β-CuPc	231.8±2.1	238.9±2.9	345-440	Semyannikov et. al., 2000
β-ZnPc	201.3±7.9	182.4±11.3	390-438	Semyannikov et. al., 2006
CoPc	183.7±13.8	164.8±18.8	375-527	Bonderman et. al., 1970
NiPc	144.6	-	400-490	Yase et. al., 1995
PbPc	156.3	-	400-490	Yase et. al., 1995
VOPc	194.6±3.3	149.4±5.0	305-400	Semyannikov et. al., 2006
TiOPc	220.9	-	420-500	Yase et. al., 1995
AlClPc	236.4±1.7	252.7±2.5	315-430	Semyannikov et. al., 2000
AlFPc	266.9±2.5	253.9±3.7	385-495	Semyannikov et. al., 2000
$(AlPc)_2O$	509.6±2.9	527.6±3.7	485-565	Semyannikov et. al., 2000
$SiCl_2Pc$	151.0±11.7	-	341-410	Shaulov et. al., 1972
$GeCl_2Pc$	148.1±12.6	-	327-438	Shaulov et. al., 1972
$SnCl_2Pc$	218.4±17.6	196.2±25.1	427-496	Bonderman et. al., 1970
$CuPcF_{16}$	200.8±7.1	163.2±9.2	360-490	Kol'tsov et. al., 2004
$ZnPcF_{16}$	236.4±1.7	204.6±2.5	385-440	Semyannikov et. al., 2008
$VOPcF_{16}$	220.5±4.2	197.1±7.1	320-400	Semyannikov et. al., 2008
H_2PcCl_{16}	141.0±17.6	227.6±41.4	127-182	Bonderman et. al., 1970
H_2PcBr_{16}	109.2±16.3	131.8±36.4	153-230	Bonderman et. al., 1970
$CuPc(t\text{-}but)_4$	185.7±6.3	177.8±9.6	310-440	Plyashkevich et. al., 2010
$VOPc(t\text{-}but)_4$	179.5±7.5	187.4±12.6	315-440	Plyashkevich et. al., 2010

Table 4. Thermodynamic parameters of sublimation of phthalocyanines

The difference in MPc and $MPcF_{16}$ (M=Cu, Zn) volatility may be explained by different crystal packing and therefore different intermolecular bonding in the solids. While the crystal structures of unsubstituted phthalocyanines are well described in the literature (Engel, 2002), the crystal structure of their fluorinated derivatives is still unknown due to some difficulties in growing their single crystals.

The suggestion that the intermolecular interaction in the unit cell of $ZnPcF_{16}$ is stronger than that for unsubstituted ZnPc derivative (Schlettwein et. al., 2000) was made on the basis of analysis of the electron absorption spectra using exciton model described by Kasha (Kasha, 1965) for molecular crystals of aromatic compounds. This was further confirmed by measurements and calculations found in the literature (Schlettwein et. al., 1999). Another behavior was observed for VOPc and its fluorinated derivative $VOPcF_{16}$. While VOPc is a nonplanar polar molecule with the vanadyl group located perpendicular to the macrocycle, the outer phenyl rings making an angle of about 7° with respect to the C-N inner ring (Ziolo et. al., 1980). Both the nonplanarity and the dipolar character of VOPc molecule result in a specific polymorphism that differs significantly from that of planar phthalocyanines (Mizugushi et. al., 1995).

Measurements of vapor pressure for Cu(II) and vanadyl tetra-tertbutyl phthalocyanines ($MPc(t\text{-}but)_4$) are also presented in Fig. 4 (Plyashkevich et. al., 2010). The introduction of four branched tert-butyl substituents leads to the increase of vapor pressure and the decrease of sublimation enthalpy (Table 4) in comparison with their unsubstituted analogues.

2.6 Effect of donor atoms in ligand on vapor pressure

Only the limited number of reliable data is available in literature which allows to demonstrate the influence of donor atoms on the saturated vapor pressure of the complexes. These are copper β-diketonates and their imino-derivatives (Fig. 5) as well as β-diketonates complexes of dimethylgold(III) and their imino- and thio-analogues (Semyannikov, 1988). In these examples, substitution of oxygen for another donor atom does not lead to change of the molecular structure of the complexes but influences on their vapor pressure.

In Cu(II) complexes, substitution of one of the oxygen atoms in ligand for nitrogen leads to increase of vapor pressure in the case of aliphatic ligands (Fig. 5, curves $Cu(aa)_2$ and $Cu(ki)_2$). For complexes with chelate node (O_4) and (N_4) (curves $Cu(aa)_2$ and $Cu(NH,NH)_2$, respectively), vapor pressure is practically the same. A considerable narrowing of the range of volatility change is observed for ketoiminate complexes with different type of substituents in ligand.

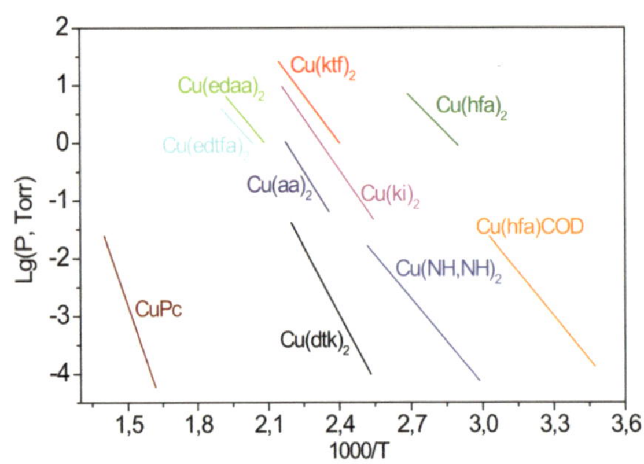

Fig. 5. Temperature dependence of vapor pressure of different Cu(II) complexes

Even more considerable changes of the volatility are observed for tetradentate ligands of 02N2 type. The structure of the copper complex with ethylendiamine-bis-acetylacetone differs from that of the ketiminate complex only by ethylene bridge CH_2-CH_2- binding two ketoimine ligands. However, even this insignificant change in the molecular structure leads to decrease of vapor pressure by one order of magnitude. Substitution of one methyl group in this ligand for trifluoromethyl one (Fig. 5, curves $Cu(edaa)_2$ and $Cu(edtfa)_2$, respectively) decreases the vapor pressure significantly.
It is also interesting to compare the vapour pressure of copper(II) β-diketonates with that of the complexes belonging to different classes. The temperature dependence of saturated vapor pressure for copper(II) β-diketonates, and their imino-derivatives, as well as copper(II) dithiocarbamate, and copper phthalocyanine are also presented in Fig. 5. Thermodynamic parameters of their sublimation are presented in Table 5.

Compound	Method	ΔH, kJ/mol	ΔS, J/K·mol	Range T, K	Refs.
$Cu(aa)_2$	membr.	125.5±9.2	218.0±20.1	424-460	Igumenov et. al., 1978
$Cu(ki)_2$	membr.	114.2±1.3	210.2±3.0	393-463	Chumachenko et. al., 1979
$Cu(ktf)_2$	membr.	104.6±1.7	196.2±4.2	416-466	Chumachenko et. al., 1979
$Cu(edaa)_2$*	membr.	94.6±2.0	142.2±5.0	480-520	Chumachenko et. al., 1979
$Cu(hfa)_2$	membr.	80.8±1.8	178.2±5.4	345-372	Igumenov et. al., 1978
$Cu(edtfa)_2$*				490-525	Chumachenko et. al., 1979
Cu(hfa)cod	Knudsen	97.1±1.4	207.9±18.4	288-330	Liskovskaya et. al., 2001
$Cu(NH,NH)_2$	Knudsen	95.8±13.4	152.0±5.0	335-397	Unpublished results
$Cu(dtk)_2$	flow	149.9±3.1	247.7±7.5	395-455	Zemskova et. al., 1998
CuPc	Knudsen	231.8±2.1	238.9±2.9	618-713	Semyannikov et. al., 2000

Table 5. Thermodynamic parameters of sublimation of different volatile Cu (II) complexes

Phthalocyanines are characterized by the lowest volatility and maximum value of sublimation enthalpy (Table 5) among the above mentioned compounds. The difference between the vapor pressure of $Cu(hfac)_2$ and CuPc is more than 7 orders of magnitude.
In the case of dimethylgold(III) β-diketonates (Semyannikov, 1988), the introduction of trifluoromethyl group increases the vapor pressure while for ketoiminate complexes the introduction of trifluoromethyl radical leads to its decrease. In general the ketoiminate complexes possess the lower vapor pressure than the corresponding β-diketonate derivatives. The substitution of oxygen in ketoimenates for sulfur atom leads to further decrease of vapor pressure. Therefore, the order of volatility for dimethylgold complexes depending on the type of donor atoms of ligand is as follows: (O,O) > (O,N) > (N,S).
It is possible to conclude on the basis of the available data on vapor pressure of different precursors that substitution of oxygen for sulfur leads to considerable decrease of volatility. The substitution of oxygen for nitrogen may lead both to insignificant increase of vapor pressure and to its decrease.

3. Structural aspects of metal complexes with organic ligands

3.1 Features of the interaction of molecules in crystal lattice

The volatile metal complexes with organic ligands are crystallized in lattices of a molecular type. The approaches of the classic organic crystal chemistry are applicable for their description, namely, the van der Waals radiuses, theory of closest packing of molecules and atom-atomic approximation (Kitaigorodsky, 1971).

The structure of metal complexes with organic ligands depends on the nature of metal and coordinated ligands. The main difference between metal volatile complexes with organic ligands and organic compounds is a specific intramolecular distribution of electronic density with a positive charge localized on the central metal atom and a negative charge localized mostly on donor atoms and partially on atoms of the carbon moiety of the ligand. Therefore it is necessary to take into account not only van de Waals interaction but also electrostatic one. Obviously, the specificity and anisotropy of intermolecular contacts are more distinct for metal complexes with organic ligands than for organic molecules. However, no systematic crystal chemical analysis of the regularities of precursor molecules packing in crystals and their nearest intermolecular contacts has been done even for single class of the compounds. The analysis of some interaction features will be done using metal β-diketonates as model compounds.

The main contribution to the energy of crystal lattice is made by van de Waals (nonspecific) contacts of peripheral atoms of adjacent molecules. The number and type of these contacts are determined by molecules packing in the crystal. Structure of complexes depends on electronic and dimensional factors and primarily on the central atom radius and the radical size in ligand. Bulky substitutes in the ligand side chain cause in a number of cases the suppression of polymerization and formation of monomeric complexes sometimes with the lower coordination number. For example, the substitution of methyl radicals to tert-butyl ones in the acetylacetonates of the metals of the first transition series leads to formation of the monomeric complexes. The latter have a higher vapor pressure which is caused by the following reasons: decrease of the coordination number from 6 to 4; decrease of total intermolecular contacts; decrease of precursor molecular weight. However, different situation is observed in the case of metals with big ionic radiuses and high coordination numbers, eg. alkaline-earth metals. For example, barium dipivaloylmethanate, which is widely used as a precursor in MO CVD, is a tetramer (Otvay & Rees, 2000), while introducing more bulky substitutes (R^1=-C(n-C_6H_{13})(CH_3)(OCH_3), R^2=-CH(n-C_4H_9)(CH_3)) into the ligand leads to decrease of the oligomerization extent to 2 (Igumenov et. al., 1996). Although the molecular weight of the latter barium complex decreases substantially, its volatility remains lower than that of barium dipivaloylmethanate.

In Ref. (Baidina, 2001) the crystal chemical analysis of over than 30 complexes of Cu(II) with β-diketones is performed. The symmetry of crystals is various – from triclinic *PI* to tetragonal *I4₁/a*. Most of the investigated complexes have a trans-configuration. Cys-configuration is observed only for five complexes. In all considered complexes the central copper atom is bidentately coordinated by two oxygen atoms of β-diketonate ligand (L) with formation of the six-membered chelate metallocycle. Cu^{2+} ion tends, as a rule, to complete its coordination environment to pyramidal (4+1) and bipyramidal (4+1+1) or (4+2), adding oxygen, carbon or fluorine atoms of the neighboring molecules to axial positions. As a result various types of molecules association arise, eg. dimers, stacks and chains. The formation of dimers is typical for heteroligand and cys-complexes. As a rule, atoms of metal cycle participate in their formation, and the molecules are bound in dimers via their center of symmetry. Cu…Cu distance is in the range 3.39-5.46 Å.

The formation of one-dimensional infinite stacks of identical molecules is typical for homoligand trans-isomers. The planes of coordination squares of copper atoms are parallel in the stack. The angle formed by these planes with the stack axis is in the range of 41.3–

60.4° and Cu...Cu distance varies from 4.4 to 8.3 Å. Carbon atoms (C_γ) of metallocycles often take part in the stacks formation. In this case Cu...C_γ distance is within 3.01-3.44 Å. The chains of complexes are formed as a rule due to additional coordination of the copper atom with atoms of the end groups of adjacent molecules. Comparing to the stacks, a wider range of atoms belonging to end groups takes part in the formation of the chains and the complexes themselves can be rotated relative one another. Among the uniform chains (with the same Cu...Cu distances) it is possible to identify linear and zigzag ones. Cu...Cu distance is in the range from 6.9 to 9.1 Å in all chain structures.

Compound	Structure	Interaction with closest molecule			E_{sh}, energy of shortest contacts	$E_{rep.}$, repulsion of all atoms	E, energy of Van der Waals interaction
		E_M, kJ/mol	shortest contact				
			r, Å	E, kJ/mol	kJ/mol		
$Cu(aa)_2$	$P2_1/n$, Z=2	-19,1	3.017...C_γ	2,9	5,8	92,1	-101,1
$Cu(ba)_2$	$P2_1/n$, Z=2	30,9	3.013...C_{Ph}	3,0	5,9	84,1	-138,9
$Cu(dbm)_2$	$I2_1/c$, Z=4	-37,3	3.145...C_{Ph} 3.367...C_{Ph}	2,1 0,3	3,4	114,2	-193,3
$Cu(thd)_2$	$P2_1/b$, Z=2	-18,4	-	-	-	64,0	-147,3
$Cu(btfa)_2$	$P2_1/n$, Z=2	-37,2	3.251...C_{Ph} 3.389...C_{Ph}	0,3 0,2	1,8	62,6	-153,1
Cu(aa)(hfa)	P(-1), Z=2	-15,4 -24,9	2.704...O 3.186...C_γ	5,4 1,1	6,5	66,2	-100,8
$Cu(aa)_2$	$P2_1/n$, Z=2	-24,8	3.124...C_γ	1,7	3,3	80,7	-120,5
$Cu(3\text{-}Meaa)_2$	P(-1), Z=1	-26,0	3.434...C_{Me}	1,2	4,3	103,2	-120,1
$Cu(3\text{-}Phaa)_2$	$P2_1/c$, Z=2	-26,0	-	-	-	79,3	-168,4

Table 6. Energy characteristics of copper(II) β-diketonates

The presence of shortened contacts in the copper β-diketonate crystals causes mutual repulsion between the neighboring molecules. The contribution of this type of interaction to the total energy of crystal lattice was calculated for a number of complexes (Table 6). The copper complexes considered above have different vapor pressure. The differences in the saturated vapour pressure value may reach some orders of magnitude and are determined by Van der Waals and electrostatic interaction of peripheral atoms of adjacent molecules, as well as by specific interactions.

3.2 Molecular structure in the gaseous phase

Only a few works devoted to investigation of the structure of metal complexes with organic ligands in the gaseous phase are known (Shibata et. al., 1986; Giricheva et. al., 1999). These works are devoted to β-diketonate derivatives. Unfortunately the analysis of these studies does not allow us to make unambiguous conclusions. However it is possible to mention that upon transition of the compound from the crystal to gaseous phase the M...O distance slightly increases within the measurement error if the molecule symmetry and coordination of the central atom do not change. M...O distance may decrease sufficiently if the molecule symmetry and coordination number change. The structural parameters of ligands change only slightly for the most structures. In some cases the angle of chelate cycle bending in the O…O line may also change.

In general the obtained electronographic data show that the structure of precursor molecules does not practically change in the gaseous phase. It means that the data on spatial arrangement of atoms in the molecules obtained from the data of X-ray structural analysis may be used for description of the molecule in the gaseous phase.

3.3 Dependence of thermodynamic parameters of sublimation processes on the structure of complexes

The analysis of the total thermodynamic data for sublimation processes of metal complexes with organic ligands shows that the values of sublimation enthalpy change within 80-150 kJ/mol and the enthropy changes to 250 J/K·mol for most of the investigated compounds. These values are quite comparable with the respective thermodynamic parameters for organic crystals having close molecular weight (300-500 a.u.).

Substitution of one -C_6H_5 group for -CF_3 in copper complexes leads to a difference in vapor pressure by 5.5 orders of magnitude. The sublimation enthalpies for $Cu(hfa)_2$ and $Cu(btf)_2$ are equal to 80.8 and 151.5 kJ/mol (Table 2), respectively. Their sublimation entropies are 178.2 and 248.1 J/K·mol. That is, the differences of 70.7 kJ/mol and 69.9 J/K·mol in thermodynamic parameters determine the above mentioned difference in the vapor pressure of these complexes.

It is possible to estimate the influence of enthalpy and entropy on the saturated vapor pressure of metal β-diketonates, assuming averaged values of thermodynamic parameters typical for this class of compounds. Let ΔH_T=105 kJ/mol, ΔS_T =188 J/K·mol and T=423 K, then at S_t=const the change of ΔH_T by 8 kJ/mol leads to a change of the saturated vapor pressure by 1 order of magnitude. At H_T=const the respective pressure changes are caused by the difference in entropy by 19 J/K·mol. Therefore, the enthalpy could change the vapor pressure of the considered copper complexes by 8.6 orders of magnitude and the entropy – by 3.4 orders of magnitude. This estimation is quite close to the observed difference in the saturated vapor pressure of $Cu(hfa)_2$ and $Cu(btf)_2$ (5.5 orders of magnitude). A regular decrease of the thermodynamic parameters values is observed for the following order: $M(aa)_n$ – $M(tfa)_n$ – $M(hfa)_n$ which is parallel to change of the volatility of these chelates.

It is interesting to mention that similar differences of the volatility of acetylacetonate and hexafluroacetylacetonate complexes are described by different sets of thermodynamic parameters. In the majority of published work the enthalpy is used as the sole criterion for comparison of volatility of different compounds. These data indicate that such approach is found to be incorrect because the differences in the volatility are described by different sets of thermodynamic parameters and the results of evaporation enthalpy change as well as entropy change should be considered (Zharkova et. al., 2000; Igumenov et. al., 1985).

4. Theoretical calculation of enthalpy and entropy of sublimation, and correlation with experiment

4.1 Microscopic theory of the volatility of metal complexes with organic ligands

For characterization of the dependence of equilibrium pressure of the gaseous phase over the crystal it is possible to use microscopic description of both gaseous and crystal phases in the framework of statistical thermodynamics. The equilibrium pressure *P(T)* in the gaseous phase is found from the condition of equality of chemical potentials of the gaseous μ^{gas} and crystal μ^{solid} phases. In order to find the chemical potentials of gaseous and crystal phases, the rigid molecules approximation in which the contributions of the intramolecular freedom degrees of molecules are neglected should be used. It has already been shown that the molecule structure in gaseous and crystal phase changes insignificantly. Therefore, this change can be neglected. In consideration of the binary equilibrium, only contributions of rotational and translatory motion of molecules into thermodynamic functions of the gaseous phase are taken into account. In the case of thermodynamic functions of crystal phase the intermolecular interaction should also be considered.

In the method of lattice dynamics in harmonic (quasi-harmonic) approximation, the analytical expressions describing dependence of the thermodynamic function of solid on parameters of intermolecular interaction is well known. The chemical potential of the crystal phase can be written as follows:

$$\mu^{solid} = \frac{1}{N}\left[F(V,T) + PV\right], \tag{2}$$

where N is the number of molecules in crystal; V is the crystal volume and F(V,T) is free energy of the crystal presented as:

$$F(V,T) = U + E_{vib}(V,T) - kTs_{vib}^{solid}(V,T) \tag{3}$$

where U is the potential energy of the crystal; $E_{vib}(V, T)$ is the crystal vibrational energy and $s_{vib}^{solid}(V,T)$ is the crystal entropy that can be found from the well known formula (Leifried & Ludwig, 1961, Belosludov et. al., 1994):

$$E_{vib}(V,T) = 6\mathrm{N}\hbar\int d\omega \coth(\hbar\omega / 2kT)\omega g(\omega) \tag{4}$$

$$\mathrm{s}_{\mathrm{vib}}^{solid}(V,T) - 6\mathrm{N}\int d\omega\{\frac{\hbar\omega}{2kT}\coth(\frac{\hbar\omega}{2kT}) \ \ln[2\sinh(\frac{\hbar\omega}{2kT})]\}g(\omega) \tag{5}$$

In (4), (5) $g(\omega)$ is the density of the phonon state of the crystal presented as:

$$g(\omega) = \frac{1}{6N}\sum_{\vec{q}j}\delta(\omega - \omega_j(\vec{q})) \tag{6}$$

where $\omega_j(\vec{q})$ are the frequencies of crystal vibrations which are expressed by the following system of equations in rigid molecules approximation:

$$\begin{aligned} m_k\omega^2(q)U_\alpha{}^t(k,q) &= \sum_{k'\beta} D_{\alpha\beta}^t(q,k,k')U_\beta^t(k',q) + D_{\alpha\beta}^{tr}(q,k,k')U_\beta^r(k',q) \\ \sum_\beta I_{\alpha\beta}(k)\omega^2(q)U_\beta^r(k,q) &= \sum_{k'\beta} D_{\alpha\beta}^{rt}(q,k,k')U_\beta^t(k',q) + D_{\alpha\beta}^{rr}(q,k,k')U_\beta^r(k',q) \end{aligned} \tag{7}$$

where $D_{\alpha\beta}^{ii'}(\boldsymbol{q},k,k')$, $(a,\beta=x,y,z)$ are translational $(i,i'=t)$, rotational $(i,i'=r)$ and mixed ($i=t$, $i'=r$ or $i=r,i'=t$) elements of the molecular crystal dynamic matrix, the expressions for which are presented in Refs. (Belosludov et. al., 1988; Belosludov et. al., 1994). $U^{i'}_a(k,\boldsymbol{q})$, $(a,\beta=x,y,z)$ is the vibration amplitude, m_k and $I_{a\beta}(k)$ are the mass and inertia tensor of the k-th molecule in the unit cell.

In the quasiharmonic approximation, the free energy of crystal has the same form as in the harmonic approximation but the structural parameters at fixed volume depend on temperature. If the vapor phase is seen as the ideal gas, it is possible to use the known expressions for enthalpy h^{gas} and entropy s^{gas} of the ideal gas in rigid molecules approximation to find the chemical potential μ^{gas} (Stull et al., 1969):

$$\mu^{gas} = kT\ln P + h^{gas}(T) - kTs^{gas}(T) \tag{8}$$

$$h^{gas}(T) = \frac{5}{2}kT \tag{9}$$

$$s^{gas}(T) = 3 + \ln\left[kT\left(\frac{mkT}{2\pi\hbar}\right)^{3/2}\right] + \ln\frac{(kT)^{3/2}[8\pi I_1 I_2 I_3]^{1/2}}{\sigma\hbar^3} + \ln(8\pi^3 kI_r T)^{1/2} \tag{10}$$

where m is molecule mass, I_1, I_2, I_3 are the molecule inertia moments; σ is the symmetry number; I_r is the moment of fragment rotation around its axis.
Using (3)-(5) and (8), the equation (1) can be write as follows:

$$\ln P(T) = -\Delta_e h / kT + \Delta_e s \tag{11}$$

where enthalpy of phase transition $\Delta_e h$ and entropy of phase transition $\Delta_e s$ are:

$$\Delta_e h = \frac{5}{2}kT - \frac{U}{N} - 6\hbar\int d\omega \coth(\hbar\omega / 2kT)\omega g(\omega) - \frac{PV_{solid}}{N} \tag{12}$$

$$\begin{aligned}\Delta s_e = 3 + \ln\left[kT\left(\frac{mkT}{2\pi\hbar}\right)^{3/2}\right] + \ln\frac{(kT)^{3/2}[8\pi I_1 I_2 I_3]^{1/2}}{\sigma\hbar^3} + \ln(8\pi^3 kI_r T)^{1/2} - \\ -6\int d\omega\left\{\frac{\hbar\omega}{2kT}\coth\frac{\hbar\omega}{2kt} - \ln\left[2\sinh\frac{\hbar\omega}{2kT}\right]\right\}g(\omega)\end{aligned} \tag{13}$$

Therefore, on the basis of the parameters of molecular and crystal structure and atom-atom potential functions, formulas (12) and (13) allow us to estimate enthalpy and entropy of the crystal-gas transitions and equilibrium pressure of the gaseous phase of precursor.

4.2 Calculation of Van der Waals contribution to enthalpy of sublimation

Analysis of equation (12) shows that the main contribution to the enthalpy $\Delta_e h$ is made by the potential energy of crystal U. Other members of the equation (12) are connected with the kinetic energy of the gas phase and vibration energy of the crystal. In the first approximation these members can be neglected. $\Delta_e h$ is approximately equal to the potential energy of crystal and can be presented as:

$$\Delta_e h = -U / N = -(U_{VVI} + U_{EI}^{mol}) / N \tag{14}$$

where U_{VVI} and U_{EI}^{mol} are van der Waals and electrostatic interactions of the molecular crystal, respectively. The electrostatic interaction is determined by the interaction of unlike charges non-uniformly distributed over precursor molecules.

To find the van der Waals portion of intermolecular interaction energy for molecular crystals, the atom-atom method of potential functions (AAP) is commonly used. This method is based on the assumption that energy of molecules interaction is equal to sum interactions of atoms in the molecule (additivity). It is also assumed that interaction potentials are of a universal character, depending only on the sort of atom and not depending either on molecule composition or on oxidation degree of the central atom.

The method of atom-atom potential (AAP) functions and its application have been described in different reviews and monographs (Kitaigorodsky et. al., 1971; Dashevskii et.al., 1982; Timofeeva et. al., 1980). Thus, eg. calculation of the lattice energy of molecular crystals comes to determination of all possible distances between pairs of nonvalency bound atoms and to calculation of the lattice sums of attraction and repulsion. Obviously such a procedure presupposes the use of positional parameters of atoms, i.e., it is necessary to know molecular and crystal structure of the investigated compound.

Van der Waals interaction was calculated with use of interatomic potential $V_{mn}(r_{ij})$ between i and j atoms of different n and m molecules.

$$U_{VVI} = \frac{1}{2}\sum_{nm}\sum_{ij} V_{mn}(r_{ij}) \tag{15}$$

Calculations of the intermolecular interaction energy were carried out in the framework of AAP method. Semiempirical Buckingham potential $V_{mn}(\vec{r}_{ij})$ was used (abbreviated as "6-exp" potential):

$$V_{mn}(r_{ij}) = B_{mn}\exp(-\alpha_{mn} r_{ij}) - \frac{A_{mn}}{r_{ij}^6} \tag{16}$$

where the first term characterizes repulsion and the second one characterizes attraction; indexes m and n indicate the sort of atoms; A, B and α are empirical parameters; r_{ij} is the interatomic space between i and j atoms of different molecules. The choice of this potential is determined by ease of calculations of energetic and dynamic properties of crystal.

Let's consider some features of the calculation of van der Waals interaction contribution to the molecular crystal lattice energy. The "6-exp" potentials were calculated by equalities of Kitaigorodsky (Kitaigorodsky & Mirsky, 1961).

$$\mathrm{A} = 1.79\varepsilon_0 r_0^6, \qquad \mathrm{B} = 6.364\cdot 10^2\varepsilon_0, \qquad \alpha = 13.6/r_0 \tag{17}$$

where ε_0 and r_0 are the depth of the potential well and equilibrium distance between the atoms. Interaction potentials of light atoms were taken from (Gordon & Kim, 1972). Parameters for metal-metal interactions were estimated from $\varepsilon_0 = f(N)$ dependence (where N is the number of metal atoms), that was plotted using the known parameters for noble gases (Timofeeva et. al., 1980; Crowell & Chang, 1963).

4.3 Contribution of electrostatic interaction

Historically two approaches are used for the estimation of electrostatic energy of crystal: calculation taking into account the interactions of charges on atoms (ion crystals) and calculations on the basis of multipoles of molecules (molecular crystals). However, the specific structural features of the molecules of metals complexes with large organic ligands

and their arrangement in the crystal lattice violate a condition of applicability of multipoles–multipole approximation at $a<<r$, where a is the largest space between charges in molecule; r is the least distance between the charges of different molecules.
The long-range Coloumb potential must be taken into account along with the van der Waals interaction. Calculation of electrostatic interaction can be performed by direct summation of interactions of all pairs of electric charges of different molecules since the multipole expansion is impossible for metal β-diketonates. The summation of interactions of all pairs of electric charges *<ij>* of different molecules can be written as follows:

$$U_{El}^{mol} = \frac{1}{2}\sum_{\langle ij \rangle} \frac{Z_i Z_j}{r_{ij}} \tag{18}$$

where Z_i is the effective charge of *i*-atom.
The formula allowing to find the electrostatic energy of molecular crystal (18) from specified distribution of molecule charges have been obtained in Ref. (Belosludov et.al., 1988).

$$U_{El}^{mol} = \frac{1}{2}(\vec{E}\vec{P}) + Z_{\min}^2 \frac{\alpha(r)}{r} \tag{19}$$

where $\vec{E}$ is the macroscopic field, $\vec{P}$ is the dipole moment of elementary cell, r is the least distance between centers of inertia of molecules, $Z_{\min}$ is the least in absolute value charge in the molecule, $\alpha(r)$ is the Madelung constant for molecular crystal which can be found from the equation (20):

$$\alpha(r) = \frac{c}{2}\sum_{l',k,k',m,m'} \xi_m \xi_{m'} H(c|\vec{x}(l'k'm' - \vec{x}(km)| / r) + O(c^3) \tag{20}$$

where $H(x)$ is integral of errors (21):

$$H(x) = \frac{2}{\sqrt{\pi}}\frac{1}{x}\int_x^{\infty} e^{-x^2} dx\,, \tag{21}$$

$Z_{\min} = Z_{\min}\xi_m$ is the *m*-th point charge of molecule; $\vec{x}(lkm)$ is the radius-vector of the point charge of the *k*-th molecule found in the *1st* cell; index 1 is the set of three indices (l_1, l_2, l_3) of the cell (l_i=0, ± 1, ± 2,...), $\vec{x}(km) = \vec{x}(0km)$, $\vec{x}(m) = \vec{x}(0m)$, c is an arbitrary small parameter chosen so as to satisfy the following inequality:

$$c << r|\vec{x}(m') - \vec{x}(m)|^{-1} < 1,\ \ r / a_{\max} \le 1\,, \tag{21}$$

In calculations of the contribution of electrostatic interaction to the energy of crystals lattice of metal β-diketonates the following assumptions were used:

1. the charge of the central atom was calculated from $qMe^{n+} = nqH^+$, where q is the charge of enol hydrogen atom in uncoordinated ligand;
2. in variating the central atom charge based on the principle of electrical neutrality, the compensating charges were distributed only within the chelate cycle, and the charges of the peripheral atoms were not changed;
3. calculations were carried out using one geometry for the complexes of the same type. Only complexes with known positional parameters of all atoms were chosen for calculations.

The charge distribution in metal β-diketonate molecules was calculated by quantum chemical methods (Schaaf et. al., 1976). The calculation results of the electrostatic energy of some metal β-diketonates are given in Table 7.

Chelate	Z_e	-U. kJ/mol
$Fe(aa)_3$	0.51	11.97
$Fe(aa)_3$	1.17	12.02
$Fe(aa)_3$	2.10	9.88
$Fe(tfa)_3$	0.40	9.17
$Fe(tfa)_3$	1.20	7.95
$Fe(tfa)_3$	2.04	6.68
$Fe(hfa)_3$	1.23	11.39
$Fe(dbm)_3$	1.08	8.08
$Cu(dbm)_2$	0.72	7.30

Table 7. Electrostatic energy of some β-diketonate crystals (Z_e is the charge on metal atom)

The results show that the variation of charge inside the chelate ring has not substantial influence on the intermolecular interaction energy of metal β-diketonates. Therefore the main contribution to the electrostatic energy is made by interaction between peripheral parts of molecules, and the electrostatic energy is not practically dependent on the type of ligand substituents and the nature of central metal. For this reason, the contribution of electrostatic interaction to sublimation enthalpy was taken equal to 10 kJ/mol for all compounds.

4.4 Entropy of sublimation of molecular crystals

4.4.1 Entropy of crystal

The entropy of crystal phase can be estimated if the density of phonon states $g(\omega)$ which can be calculated using the method of lattice dynamics (Belosludov et. al., 1994) is known. In this method, frequencies of crystal vibrations $\omega_j(q)$ are found from equations of motion (7). These equations describe the motion of mass centers of molecule and molecule rotation. Interaction between molecules was described by the potential in atom-atom approximation. As in the case of calculation of sublimation enthalpy, van der Waals interaction including both repulsion of molecules at a close distance and long-range dispersive attraction as well as electrostatic interaction were considered. The same parameters of the potentials as in the case of calculation of sublimation enthalpy were also used.

4.4.2 Entropy of gaseous phase

The entropy of vapor phase was calculated using eq. 10. For calculation of the moments of inertia and the moments of molecule rotation the positional atom parameters (x,y,z) were used. The values of rotation barriers of molecule fragments were chosen according to the recommendations presented in Ref. (Stull et. al., 1969).

4.5 Comparison between experimental and calculated data on vapor pressure

Structural data for the investigated acetylacetonates were taken from Cambridge Structural Data Base (CSDB). All known structural data for Al(III), Cr(III), Fe(III) and Ir(III) tris-acetylacetonates are presented in (Prokuda et. al., 2006). The structures determined by X-

Ray diffraction method with reasonably low R-factors and all defined atoms including hydrogen were chosen for calculations.

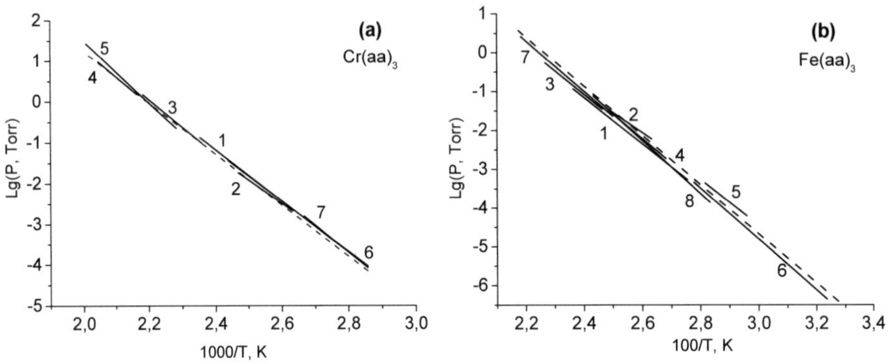

Fig. 6. Comparison of the experimental (—) and calculated (---) LgP(1/T) dependence for $Cr(aa)_3$ (a): 1 – flow method (Götze et. al., 1970); 2 – (Melia & Merrifield, 1970); 3 – membrane method (Wolf et. al., 1972); 4 – membrane method (Techil et.al., 1981); 5 – chromatography (Bublik et al., 1978); 6 – Knudsen method (Malkerova et. al., 1990); 7 - Knudsen method (Semyannikov et.al., 2001) and for $Fe(aa)_3$ (b): 1 - flow method (Götze et. al., 1970); 2 - (Melia & Merrifield, 1970); 3 – Knudsen method; 4 – (Sachinidis et. al., 1980); 5 – Langmuir method (Fedotova et. al., 1993); 6 - Knudsen method (Malkerova et. al., 1990); 7 - Knudsen method (Stabnikov et.al., 2001); 8 - flow method (Stabnikov et.al., 2001)

Dynamic and thermodynamic properties for different metal acetylacetonate crystals were calculated. A comparison of the experimental and calculated data on temperature dependence of saturated vapor pressure for $Cr(aa)_3$ and $Fe(aa)_3$ is given as an example in Figs. 6(a) and 6(b), respectively. The experimental data obtained by different methods are shown in these graphs. Good agreement between the experimental and calculated dependencies confirms the reliability of the obtained data.

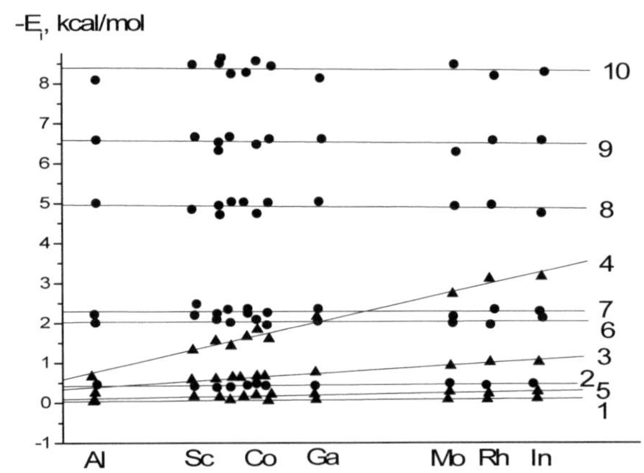

Fig. 7. Contribution of different types of interaction into the energy of Van der Waals interaction for metal tris-acetylacetonates: 1 - M–M; 2 - M–O; 3 - M–C; 4 - M–CH_3; 5 - O–O; 6 - O–C; 7 - C–C; 8 - O–CH_3; 9 - CH_3–CH_3; 10 - CH_3–C

Contribution of different types of interaction into the energy of Van der Waals interaction for metal tris-acetylacetonates is presented in Fig. 7. The existing evaluation of parameters of metal-metal interactions using known *e* and *r* values for inert gases is a suitable technique but it should be supported by experimental data.
Table 8 presents experimental and calculated values of enthalpy and entropy of sublimation for some metal acetylacetonates.

Complex		-ΔH, kJ/mol	ΔS°, J/(K·mol)
$Cu(aa)_2$	experiment	125.5±9.2	218.0±20.1
	calculation	107.9	202.3
$Al(aa)_3$	experiment	104.5	188.4
	calculation	112.9	
$Cr(aa)_3$	experiment	119.7	203.9
	calculation	120,4	222.4
$Ir(aa)_3$	experiment	127.9	215
	calculation	139	
$Fe(aa)_3$	experiment	124.4	224.3
	calculation	121.4	220.9

Table 8. Experimental and calculated values of enthalpy and entropy of sublimation of metal acetylacetonates

The values of enthalpy and entropy were obtained by averaging of experimental data on enthalpy of the investigated metal acetylacetonates presented in literature. The calculations of sublimation enthalpy were carried out using "6-exp" potential (eq. 15). The calculations of entropy of sublimation for $Cu(aa)_2$ and $Cr(aa)_3$ were carried out according to equation 13. Contribution of electrostatic interaction to lattice energy was taken as a constant value 10 kJ/mol in the calculations.

5. Conclusions

A systematic analysis of experimental data on temperature dependencies of vapor pressure of molecular complexes with organic ligands using metal β-diketonates and phthalocyanines as examples was performed. The volatility of β-diketonates and phthalocyanines was analysed from the standpoint of their molecular and crystal structure. The complexes considered in this review are characterized by different vapor pressure. The differences in the saturated vapor pressure value may reach some orders of magnitude and are determined by the Van der Waals and electrostatic interaction of the peripheral atoms of adjacent molecules, as well as by a specific interaction, the type and number of which depend on the type of molecules packing in the crystal.
Along with other volatile metal complexes with organic ligands the phthalocyanines crystallize in the molecular type lattices. Approaches of classical organic crystal chemistry such as the Van der Waals radii algorithm, theory of closest packing of molecules and atom-atom approximation are usually used for their description. The structure of complexes depends on electron and size factors, most of all, on the type of central metal atom and on the type of substituents in the phthalocyanine ring.
The vapor pressure is described by enthalpy and entropy of sublimation; in the majority of published work the enthalpy of evaporation is used as the sole criterion for comparison of

volatility of different compounds. This approach is found to be incorrect because the differences in the volatility of various phthalocyanines are described by different sets of thermodynamic parameters and the results of evaporation enthalpy change as well as entropy change should be considered.

The general theoretical approach to the quantitative description of the volatility of precursors for MO CVD has been developed on the basis of statistical thermodynamics. Numerical procedures for calculating temperature dependencies of vapor pressure in the atom-atom approximation on the basis of the known crystal chemical data have been proposed and realized for a number of precursors. Good agreement between the experimental and calculated dependencies confirms the reliability of the suggested approach.

6. Acknowledgements

The authors thank Dr. A. Turgambaeva and Dr. P. Stabnikov for useful discussions and assistance in the graphs preparation.

7. References

Alikhanyan, A.; Malkerova, I.; Kuzmina, N.; Ivanov, V. & Kaul A. (1994) Thermodynamic investigation of barium dipivaloylmethanate $Ba(dpm)_2$, *Zh. Neorg. Khim.*, Vol. 39, No 9, pp. 1534-1538, ISSN 0044-457X.

Baidina, I.A. (2001). *Crystal chemical investigation of the molecular crystals of copper(II) with β-diketones*, Ph.D. Thesis, Novosibirsk, Russia.

Basova, T.; Semyannikov, P.; Plyashkevich, V.; Hassan, A. & Igumenov, I. (2009). Volatile Phthalocyanines: Vapor Pressure and Thermodynamics, *Critical Reviews in Solid State and Materials Sciences*, Vol. 34, No 3-4, pp. 180-189, ISSN 1040-8436.

Beech, G. & Lintonbon, R. (1971). Thermal and kinetic studies of some complexes of 2,4-pentanedione, *Thermochim. Acta.*, Vol. 3, No 2, pp. 97-105, ISSN 0040-6031.

Belosludov, V.; Lavrentiev, M. & Syskin, S. (1988). Dynamical properties of the molecular-crystals with electrostatic interaction taken into account—low-pressure ice phases (Ih and Ic). *Phys. Status Solidi (b)*, Vol. 149, No 1, (September 1988) pp. 133–142, ISSN 0370-1972.

Belosludov, R. V.; Igumenov, I. K.; Belosludov, V. R. & Shpakov, V. P. (1994). Dinamical and thermodinamical properties of the acetylacetones of copper, aluminium and rhodium. *Mol. Phys.* Vol. 82, No 1, pp. 51-66, ISSN 0026-8976.

Berg, E. & Truempe, J. (1960). A study of the volatile characteristics of various metal β-diketone chelates, *J. Phys. Chem.*, Vol. 64, No 4, (April 1960) pp. 487-490, ISSN 0022-3654.

Berg, E. & Truemper, J. (1965). Vapor pressure-temperature data for various metal β-diketone chelates, *Analyt. Chim. Acta.*, Vol. 32, pp. 245-252, ISSN 0003-2670.

Bonderman, D.; Cater E. & Bennett W. (1970). Vapor pressure, mass spectra, magnetic susceptibilities and thermodynamics of some phthalocyanine compounds, *J. Chem. Eng. Data*, Vol. 15, No 3, (July 1970) pp. 396-400, ISSN 0021-9568.

Brown, C. (1968). Crystal structure of beta-copper phthalocyanine, *J. Chem. Soc. A.*, Vol. 10, pp. 2488-2493, ISSN 0368-1769.

Bublik, G.N.; Mazurenko, E.A. & Volkov, S.V. (1978). Chromatographic method of determination of vapour pressure of metal coordination compounds. *Ukr. Khim. Zhurn.*, Vol. 44, No 11, pp. 12-14-1217, ISSN 0041-6045.

Callen, H. (1985). *Thermodynamics and an Introduction to Thermostatistics*, John Wiley & Sons, ISBN 0471862568, USA.

Chumachenko, U.I.; Igumenov, I.K. & Zemskov, S.V. (1979). Saturated vapour pressure of bidentate and tetradentate copper(II) β-ketoiminates, *Russian J. Coord. Chem.*, Vol. 5, No 11, pp. 1625-1628, ISSN 0132-344X.

Crowell, A. & Chang, C. (1963). Constants for a (6-exp) Potential between Simple Molecules and Graphite, *J. Chem. Phys.*, Vol. 38, No 10, pp. 2584-2586, ISSN 0021-9606.

Dashevski, B.G. (1982) *Conformational analysis of organic molecules*, Khimiya, Moscow.

Engel, M. (2002). Single-Crystal Structures of Phthalocyanine Complexes and Related Macrocycles, In: *The Porphyrin Handbook*, Kadish, K.; Smith K.; Guilard, R. (Eds.) Vol. 20, pp. 1-242. Academic Press, ISBN-10 0123932009, ISBN-13 978-0123932006.

Fedotova, N.; Morozova, N.; Igumenov, I.; Gerasimov, P. & Gerasimova, A. (1993). Thermodynamic investigation of iron(III) tris-beta-diketonates, *Russ. J. Coord. Chem.* Vol. 19, No 8, pp. 622-629, ISSN 0132-344X.

Fedotova, N.; Mikheev, A.; Gelfond, N.; Igumenov, I.; Morozova, N. & Tuffias, R. (1999). Modeling of mass-transportation of tris-(acetylacetonato)chromium(III) at atmospheric pressure, *J.Phys. IV, France.* Vol. 09, No PR8, (September 1999) pp. 251-258, ISSN 1155-4339.

Filatov, E.S.; Sysoev, S.V.; Zelenina, L.N.; Chusova, T.P.; Logvinenko, V.A.; Semyannikov, P.P. & Igumenov, I.K. (2006). Thermodynamic study of a series lithium β-diketonates, *J. Thermal Anal. Cal.* Vol. 86, No 2, (November, 2006) pp. 537-539, ISSN 1388-6150.

Fontain, R.; Pommier, C. & Guiotchon, G. (1972). Ètude de la pression de vapeur et de la stabilité thermique de chélates de chrome et d'aluminium. *Bull. Soc. Chim. France,* Vol. 8, pp. 3011-3015, ISSN 0037-8968.

Götze, H.-J.; Bloss, K. & Molketin H. (1970). Dampfdruckbestimmung von Acetylacetonaten, *Z. Phys. Chem.* Vol. 73, No 4-6, (September 1970) pp. 314-320.

Giricheva, N.; Girichev, G.; Belova, N.; Kuz'mina, N.; Kaul, A. & Petrov, V. (1999). Investigation of the structure and energetics of β-diketonates. VIII. Molecular structure of the $Ba(dpm)_2$ monomer according to gas phase electron diffraction data, *J. Struct. Chem.* Vol. 40, No 4, (July 1999) pp. 527-532, ISSN 0022-4766.

Gordon, R. & Kim Y. (1972). Theory for the Forces between Closed Shell Atoms and Molecules, *J. Chem. Phys.* Vol. 56, No 6, pp. 3122-3133, ISSN 0021-9606.

Igumenov, I.K.; Chumachenko, U.I. & Zemskov, S.V. (1978). Study of the volatility of some copper(II) β-diketonates, *Russian J. Coord. Chem.* Vol. 4, No 2, pp. 163-168, ISSN 0132-344X.

Igumenov, I.K.; Chumachenko, U.I. & Zemskov, S.V. (1979). Study of the volatility of some aluminium(III) β-diketonates, *Russian J. Coord. Chem.* Vol. 5, No 1, pp. 34-38, ISSN 0132-344X.

Igumenov, I., Isakova, V. & Zemskov, S. (1985). Saturated vapour pressure of iridium(III) and rhodium(III) β-diketonates, *Russ. J. Coord. Chem.* Vol. 11, No 10, pp. 1377-1380, ISSN 0132-344X.

Igumenov, I.; Semyannikov, P.; Belaya, S.; Zanina, A.; Shergina, S. & Sokolov, I. (1996). New volatile β-diketonate complexes of barium with sterically hindered methoxy-β-diketones as precursors for CVD, *Polyhedron*, Vol. 15, No 24, (September 1996) pp. 4521-4530, ISSN 0277-5387.

Ingram, M. & Drauart, J. (1962). *Investigations at high temperatures*, Mir, Moscow, USSR.

Irving, R. & Ribeiro da Silva, M.A.V. (1976). Thermochemistry of tris(l,1,1-trifluoropentane-2,4-dionato)- and tris(2,2,6,6-tetramethylheptane-3,5-dionato)-aluminium(III), *J. Chem. Soc., Dalton Trans.*, Vol. 19, pp. 1940-1942, ISSN 0300-9246.

Kasha, M.; Rawls H. & El-Bayoumi, A. (1965). The exciton model in molecular spectroscopy, *Pure Appl. Chem.* Vol. 11, No. 3-4, pp. 371-392, ISSN 0033-4545.

Kitaigorodsky, A. & Mirsky, K. (1961). Calculation of the lattice energy of molecular crystals, *Kristalografia,* Vol. 6, No 4, pp. 507-514, ISSN 0023-4761.

Kitaigorodsky, A.I. (1971). *Molecular Crystals*, Nauka, Moscow.

Kol'tsov, E.; Basova, T.; Semyannikov P. & Igumenov, I. (2004). Synthesis and investigations of copper hexadecafluorophthalocyanine $CuPcF_{16}$, *Mat. Chem. Phys.* Vol. 86, No 1, (July 2004) pp. 222-227, ISSN 0254-0584.

Krisyuk, V.; Turgambaeva, A. & Igumenov, I. (1998). Volatile Lead β-Diketonates as CVD Precursors, *Chemical Vapor Deposition*, Vol. 4, No 2, (March 1998) pp. 43-46, ISSN 0948-1907.

Leifried, G. and Ludwig, W. 1961. *Theory of Anharmonic Eff ects in Crystal*, Academic Press, New York.

Liskovskaya, T.; Bulusheva, L.; Okotrub, A.; Krupoder, S.; Semyannikov, P.; Asanov, I.; Igumenov, I.; Manaev, A.; Traven, V. & Cherkov, A. (2001). (HFA)Cu. 1,5-COD as the prospective precursor for CVD-technologies: The electronic structure, thermodynamic properties and process of formation of thin copper films, *J. Phys. IV France* Vol. 11, No PR3, (August 2001) pp. 69-76, ISSN 1155-4339.

Logvinenko, V.; Gavrilova, G. & Morozova, N. (1998). CRTA for the Thermoanalytical Screening of Volatile Compounds 2. The volatility of trifluoroacetylacetonates of Al, Cr and Zr studied by the method of quasi-equilibrium thermogravimetry, *J. Therm. Anal.* Vol. 52, No 2, (March 1998) pp. 341-344, ISSN 1388-6150.

Malkerova, I.; Alikhanyan, A.; Sevastyanov, V.; Grinberg, Ya. & Gorgoraki, V. (1990). Features of thermal behaviour of 3d-transition metal acetylacetonates, *Russ. J. Inorg. Chem.* Vol. 35, No 2, pp. 413-418, ISSN 0044-457X.

Margrave, J. (1967). *The Characterization of High-Temperature Vapors*, John Wiley and Sons, New York.

Maury, F. (1995). Recent Trends in the Selection of Metal-Organic Precursors for MOCVD Process, *J. Phys. IV France,* Vol. 05, No C5 (June 1995) pp. 449-463, ISSN 1155-4339.

Mazurenko, E.A. & Gerasimchuk, A.I. (1993). Metal coordination compounds in gaseous phase, *Ukr. Khim. Zhurn.* Vol. 59, No 5, pp. 526-536, ISSN 0041-6045.

Melia T.P. & Merrifield R. (1970). Vapour pressures of the tris(acetylacetonato) complexes of scandium(III), vanadium(III) and chromium(III), *J. Inorg. Nucl. Chem.,* Vol. 32, No 5 (May 1970), pp. 1489-1491, ISSN 0277-5387.

Mizugushi, Z.; Rihs, G. & Karfunkel, H. (1995). Optical and non-linear optical properties of vanadium oxide phthalocyanine films, *J. Phys. Chem.* Vol. 99, No 441, (November 1995) pp. 6217-16227, ISSN 0022-3654.

Morozova, N.; Sysoev, S.; Igumenov, I. & Golubenko, A. (1996). Study of temperature dependence of saturated vapour pressure of zirconium(IV) β-diketonates, *J. Therm. Anal.* Vol. 46, No 5, (May 1996) pp. 1367-1373, ISSN 1388-6150.

Otvay, D. & Rees, W. (2000) Group 2 element β-diketonate complexes: synthetic and structural investigations, *Coord. Chem. Rev.* Vol. 210, No 1, (December 2000) pp. 279-328, ISSN 0010-8545.

Plyashkevich, V.; Basova, T.; Semyannikov, P. & Hassan, A. (2010) Vapour pressure of tetra-tert-butyl substituted phthalocyanines, *Thermochim. Acta*, Vol. 501, No 1-2 (March 2010) pp. 108-111, ISSN: 0040-6031.

Prokuda, O.; Belosludov, V. ; Igumenov, I. & Stabnikov P. (2006). Calculations of van der Waals interaction energies for Al, Cr, Fe, and Ir acetylacetonate crystals, *J. Struct. Chem.*, Vol. 47, No 6 (November 2006) pp. 1032-1041, ISSN 0022-4766.

Ribeiro da Silva, M.A.V. & Gonçalves, J.M. (1998). Standard molar enthalpies of formation of eight copper(II) fluorinated β-diketonates. Mean (Cu−O) bond dissociation enthalpies, *J. Chem. Thermodyn.* Vol. 30, No 12 (December 1998) pp. 1465-1631, ISSN: 0021-9614.

Sachinidis, J. & Hill, J. (1980). A re-evaluation of the enthalpy of sublimation of some metal acetylacetonate complexes, *Thermochim. Acta.* Vol. 35, No 1, (January 1980) pp. 59-66, ISSN 0040-6031.

Schaaf, T.F.; Avanzino, S.C.; Jolly, W.L. & Sievers, R.E. (1976). Core binding energies of some metal β-diketonates and β-diketones in the vapour phase, *J. Coord. Chem.* Vol. 5, No 3, pp. 157-161, ISSN 0095-8972.

Schlettwein, D.; Graaf, H.; Meyer, J.-P.; Oekermann, T. & Jaeger, N. (1999). Molecular Interactions in thin films of hexadecafluorophthalocyaninatozink as compared to islands of N,N′-dimethylperylene-3,4,9,10-biscarboximide, *J. Phys. Chem. B.* Vol. 103, No 16 (April 1999) pp. 3078-3086, ISSN 1520-6106.

Schlettwein, D.; Tada, H. & Mashiko, S. (2000). Substrate-induced order and multilayer epitaxial growth of substituted phthalocyanine thin films, *Langmuir*, Vol. 16, No 6, (January 2000) pp. 2872–2881, ISSN 0743-7463.

Semyannikov, P.P. (1988). *Mass spectrometric investigation of thermal stability of β-diketonates of noble metals in gaseous phase*, Ph.D. Thesis, Novosibirsk.

Semyannikov, P.; Basova, T.; Grankin V. & Igumenov, I. (2000). Vapor Pressure of some metal phthalocyanines, *J. Porphyrins Phthalocyanines* Vol. 4, No 3, (Apr & May 2000) pp. 271- 277, ISSN 1088-4246.

Semyannikov, P.; Igumenov, I.; Trubin, S. & Asanov, I. (2001). In situ mass spectrometry during thermal CVD of the tris-acetylacetonates of 3-d transition metals, *J. Phys. IV France*, Vol. 11, No PR3, (August 2001) P. 995-1003, ISSN 1155-4339.

Semyannikov, P.; Basova, T.; Trubin, S.; Kol'tsov, E. & Igumenov, I. (2006). Synthesis and investigations of hexadecafluorophthalocyanine $CuPcF_{16}$, *J. Porphyrins Phthalocyanines,* Vol. 10, No. 8 (August 2006) pp. 1034-1039, ISSN 1088-4246.

Semyannikov, P.; Basova, T.; Trubin, S.; Kol'tsov, E.; Plyashkevich V. & Igumenov, I. (2008). Vapor pressure of some metal phthalocyanines, *Russian J. Phys. Chem. A*, Vol. 82, No 2, pp. 159-163, ISSN 0036-0244.

Shaulov, Y.; Priselkov, Y.; Lopatkina, I. & Markova, I. (1972). Saturated vapor pressures of silicon and germanium phthalocyanines, *Russian J. Phys. Chem.* Vol. 46, No 4, pp. 857-859, ISSN 0036-0244.

Shaulov, Y.; Lopatkina, I.; Kirjukhin I. & Krasulin G. (1975). Determination of vapor pressure of phthalocyanines, Russian J. Phys. Chem. Vol. 49, pp. 252-253, ISSN 0036-0244.

Shibata, S. Iijima, K.; Inuzuka, T. & Sato, T. (1986). Molecular structures of dipivaloylmethane complexes of lanthanides of samarium to holmiun as determined by gas electron diffraction, *J. Mol. Struct.*, Vol. 144, No 3-4, (May 1986) pp. 351-357, ISSN 0022-2860.

Stabnikov, P.; Sysoev, S.; Vanina, N.; Trubin, S.; Semyannikov, P. & Igumenov, I. (2001). Vapour pressure of Fe(III) beta-diketonates, *Electronic Journal "Investigated in Russia"*, Vol. 4, pp. 237-245, http://zhurnal.ape.relarn.ru/articles/2001/023.pdf, ISSN 1819-4192

Stull D.R, Westrum E.F. Jr. & Sinke C.G. (1969). *The Chemical Thermodynamics of Organic Compounds*, Wiley, N.Y.-London-Sydney-Toronto.

Techil, R.; Ferro, O.; Benchivenni, L. & Pelino, M. (1981). A thermodynamic study of the sublimation processes of aluminium and copper acetylacetonates, *Thermochim. Acta.* Vol. 44, No 2, (March 1981) pp. 213-222, ISSN 0040-6031.

Timofeeva, T.; Chernikova, N. & Zorkii, P. (1980). Theoretical Calculation of the Spatial Distribution of Molecules in Crystals, *Russ. Chem. Rev.*, Vol. 49, No 6, pp. 509–525, ISSN 0042-1308.

Wolf, W.; Siewers, R. & Brown, G. (1972). Vapor pressure measurements and gas chromatographic studies of the solution thermodynamics of metal beta-diketonates, *Inorg. Chem.* Vol. 11, No 9 (September 1972) pp. 1995-2002, ISSN 0020-1669.

Yase, K.; Takahashi, Y.; Ara-Kato N. & Kawazu, A. (1995). Evaporation rate and saturated vapor pressure of functional organic materials, *Jpn. J. Appl. Phys.*, Vol. 34, Part 1, No. 2A (February 1995) pp. 636-637, ISSN 0021-4922.

Zemskova, S.M.; Stabnikov, P.A.; Sysoev, S.V.; Igumenov, I.K. (1998). MOCVD Precursors: Thermodynamic Properties and Crystal Lattice Energy, *The Electrochem. Soc. Meetings Abstract*, Vol. 2, p. 812, ISSN 1091-8213.

Zharkova, G.; Stabnikov, P.; Grankin, V.; Semyannikov, P.; Igumenov, I. & Zemskov, S. (2000). Palladium(II) β-Diketonates: Volatility and Energy of the Crystal Lattice *Russian J. Coord. Chem.*, Vol. 26, No 8, pp. 576-581, ISSN 1070-3284

Ziolo, R.; Griffiths C. & Troup, J. (1980). Crystal-structure of vanadyl phthalocyanine, phase-II, *J. Chem. Soc. Dalton Trans.*, Vol. 11, pp. 2300-2302, ISSN 1477-9226.

21

Thermochemistry and Kinetics of the Reactions of Apatite Phosphates with Acid Solutions

Mohamed Jemal
Faculty of Science, Chemistry Department, Applied Thermodynamics Laboratory, Tunis El Manar University, 2092 Tunis El Manar, Tunisia

1. Introduction

The main component of phosphate ore derive from fluorpapatite, $Ca_{10}(PO_4)_6F_2$, by a faint substitution of calcium by magnesium and sodium, phosphate ions by carbonate and fluorides by the latter and hydroxyl ions. Mineral component of calcified tissues (bone, dentine and enamel) derives from hydroxyapatie, $Ca_{10}(PO_4)_6(OH)_2$.

Thermochemistry of these synthetic compounds and similar products began around the middle of last century (Gottshall, 1958). During the next 25 year period, few works have been performed in order to determine thermochemical quantities of these compounds, but a lot of work has been undertaken in the beginning of 70s on the synthesis and characterization of such products. Reactions of acid solutions on natural phosphates have also been widely studied, this is because these reactions lead to phosphoric acid which is the basic intermediate for phosphate fertilizer production. Fertilizers such as Mono-Ammonium Phosphate, (MAP), $NH_4(H_2PO_4)$, Di-Ammonium Phosphate (DAP), $(NH_4)_2HPO_4$, are produced by the reaction of ammonia on phosphoric acid solution. The slurry is then granulated at high temperature leading to the fertilizer product. Triple Super Phosphate (TSP) results from the reaction of phosphoric acid on the phosphate ore.

High purity (food grade) phosphoric acid is usually produced in two steps. The first one consists in reducing apatite by carbon in presence of silica at a temperature up to 1500°C, to produce native phosphorus which is then oxidized into P_4O_{10}. The latter reacts with water to produce H_3PO_4. This process is called 'dry' process in opposition to the 'wet' one in which the phosphoric acid results from the reaction of phosphate with sulfuric acid or with a mixture of the latter with diluted phosphoric acid solution. Wet process produces an acid solution containing large amounts of impurities, some of them coming from the phosphate ore and the other from reacting acid. Depending on the reaction temperature and the concentration of the acid, the wet process produces also a by-product which can be calcium sulphate dihydrate (gypsum), $CaSO_4,2H_2O$, or hemihydrate, $CaSO_4,0.5H_2O$, or anhydrous Ca-sulfate, $CaSO_4$, (Becker, 1989). Because of the exothermicity of the reactions occurring in this process, a large amount of thermal energy is generated allowing to heat the reaction vessel at temperature up to 120°C. Due to low cost and simplicity of acid production by wet process, fertilizers are manufactured using phosphoric acid resulting from this process.

This paper deals with the determination of the formation enthalpies of a series of apatites, the general formula of which is $M_{10}(PO_4)_6Y_2$ in which M could be Ca, Sr, Ba, Cd or Pb or a

combination of Ca with one of the other metallic elements. Y could be F, OH or Cl. A variety of products have been synthesized, characterized then dissolved in acid solution using an isoperibol calorimeter in order to determine the enthalpies of solution. Combining these quantities with others determined in the same device or picked from literature allowed to determine the enthalpies of formation. Dissolution of the two metal compounds enables to determine the enthalpies of mixing of the limit products in the solid state. Some other experiments have been recently performed on B-type Ca-carbonate apatites and allowed to characterize the substitution of PO_4 by CO_3 ions.
Dissolution of samples of fluorapatite at different temperatures has also been followed microcalorimetrically using a C 80 SETARAM microcalorimeter and kinetic models have been proposed in accordance with the thermogenesis curves. In addition attack of a phosphate ore sample by phosphoric and phosphoric/sulphuric acid solutions has been performed and revealed various phenomena which have not been previously reported.

2. Synthesis of the apatites

Two main procedures have been used in order to synthesize the products. The 'wet' one consists in adding drop by drop a solution of metal nitrate solution into a boiling diammonium phosphate solution at a pH between 9 and 11. Precipitation is then performed by maintaining ebullition for at least one hour, then the solid is separated and ignited at various temperature, depending on the product in concern. This procedure which is called as 'direct method' allows to synthesize hydroxyapatites with cadmium (Ben Cherifa, 2002) or lead (Ntahomvukiyé, 1997)- It also permitted to obtain (Ca,Cd) and (Ca,Sr) hydroxyapatite solid solutions. The reverse method consisting in droping the metal solution into phosphate solution allows to synthesize hydroxyapatites containg only Ca or Pb, or Ca-Sr (Ben Cherifa and Jemal, 2004) or Ca-Pb (Ntahomvukiyé, 1997) couples..
Fluorapatites containing one or two metals have also been prepared according to one or the other method but the phosphate solution has been previously added with ammonium fluoride. This procedure allowed to synthesize Ca, Pb and Sr fluorapatites or Ca-Sr (Khattech and Jemal, 1997), Ca-Cd (Ben Cherifa et al, 2001a) or Ca-Pb (Ntahomvukiye et al, 1997a and 1997b) fluorapatites solid solutions.
Chlorapatites have been usually obtained by the 'solid state' procedure which consists in igniting during several hours at a temperature up to 1100°C a mixture of trimetallic phosphate $M_3(PO_4)_2$ with MCl_2. This procedure allowed to synthesize $Ca_{10}(PO_4)_6Cl_2$ (Ben Cherifa 2002), $Ba_{10}(PO_4)_6Cl_2$ (Khattech, 1996), $Cd_{10}(PO_4)_6Cl_2$ (Ben Cherifa and Jemal, 2001b) together with Ca-Cd and Ca-Sr chlorapatite solid solutions.

3. Isoperibol calorimeter

The solids have been dissolved at 25°C in a 9% weight nitric acid solution using an isoperibol calorimeter which is composed of a 500mL Dewar flask immersed in a thermostated bath. In this device, the heat effect results in a variation of the temperature of the reactional medium, which is detected by a thermistance probe acting as one of the four arms of a Wheatstone bridge connected to a recorder through a '177 Keitley' current amplifier. The device is provided with an electrical resistance connected to a stabilized DC current supplier. The magnitude of the current was measured by a A.C power-meter 'GPM-8212' and the heat duration was accurately measured using a digital electronic timer with

0.01 s resolution. 350 mL solution are introduced in the flask, and the solid is preliminarily isolated in a Pyrex ball manufactured at the extremity of a 5 mm diameter tube which is supplied with a thin bar aimed at breaking the ball to start the reaction.
Experiments started by searching a quasi-steady state in which the electrical current is practically nil and the baseline deviates slightly from the horizontal line. Injection of electrical or chemical energy in the medium results in a deviation of the signal, which then becomes parallel to the previous baseline. Energy resulting from the reaction is calculated considering the corresponding shift between the baselines and the mean calibration coefficient. Calibration coefficient 'E' is calculated by dividing the calibration energy (J) over the resulting shift 'd' (mm) on the recorded paper. Three calibration experiments are usually performed and the error on enthalpy is determined considering the scatter between the extreme values of 'E'.
Various amounts 'm_i' (mg) of each product were dissolved in the same volume of solution and the molar dissolution enthalpy 'ΔH_i' was determined. Enthalpy per mass unit was calculated according to a mathematical treatment in which the statistical weight of an experimental result depends on the corresponding variance. This procedure leads to express the dissolution enthalpy as: $\Delta_{sol}H = (\sum_i w_i m_i \Delta H_i)/(\sum_i w_i m_i^2)$, in which w_i is the variance, it was here assimilated to the reverse of the square of the error. The theoretical basis of this calculation was developed by Sands, 1974 and Pattengill and Sands, 1979.

4. Thermochemistry of the apatites

4.1 Formation enthalpy of non substituted apatites

In order to determine the enthalpy of formation of any compound from dissolution results, one procedure consists in considering a particular reaction involving the compound to be studied and other solid or liquid reactants and products for which the formation enthalpies are reported in literature. This reaction can be possible, impossible or even hypothetical. Measurements of the heat evolved on dissolution of the whole constituents in the same solvent enable to derive the enthalpy of this reaction, and so, to deduce the formation enthalpy of the product in concern. This way of doing is useful particularly when dissolution enthalpy value depends on the concentration of the resulting solution that is when interaction of ions in the final solution is dependent on their concentrations even when the latter are very low. In that case the formation enthalpy of the compound have to be calculated by considering values corresponding to infinite dilution. This is the case when dissolving whitlockite, $Ca_{18}Mg_2H_2(PO_4)_{14}$. This compound can be involved in the following reaction:

$$16/3\ Ca_3(PO_4)_2 + 2/3\ Mg_3(PO_4)_2 + 2CaHPO_4 \rightarrow Ca_{18}Mg_2H_2(PO_4)_{14} \qquad \text{(R1)}$$

Dissolution of various amounts of withlockite and of stoichiometric mixture of Ca and Mg phosphates have been performed in a 46%weight HNO_3 solution using a conduction calorimeter (Calsol) and the results are reported in figure 1. Least square processing method led to express the standard molar enthalpies 'y' *versus* the final concentration 'x' as $y = ax^2 + bx + c$, from which the enthalpy of (R1) reaction was deduced at infinite dilution as $\Delta H°(298) = -203.3 \pm 1.6$ kJ mol^{-1}, leading to $-27.93.10^3$ kJ mol^{-1} as the standard enthalpy of formation of whittlockite (Ben Abdelkader et al., 2004).
Variation of dissolution enthalpy on final concentration can be neglected when the drawing of the enthalpy resulting from dissolving various amounts 'm' of the same product is linear versus 'm'. This was the case for more simple apatite compounds.

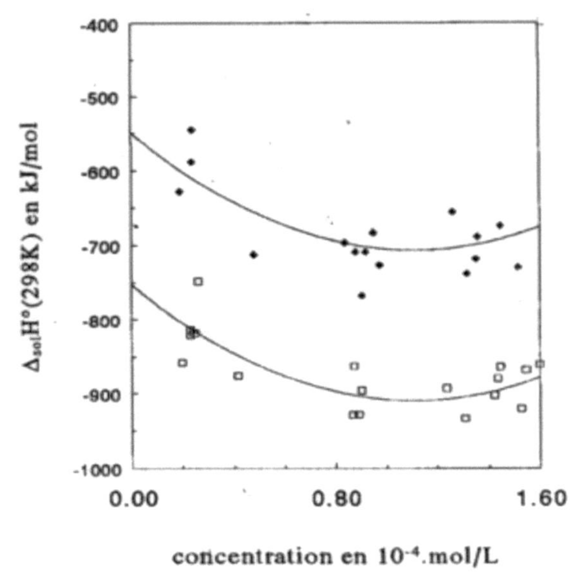

Fig. 1. Variation of molar enthalpy of dissolution of whithlockite (■) and of the stoichiometric mixture (□) as a function of the solid concentration in the final solution. Solvent : 46%weight HNO_3 aqueous solution.

Determination of the enthalpy of formation using dissolution results can also be performed by involving the entities or compounds appearing in the dissolution reaction in other processes (dissolution, dilution or mixing) in order to get a succession of reactions the "summation" of which leads to the formation reaction of the product in concern. This way of doing allowed us to determine the enthalpies of formation of a number of hydroxy, chloro, carbonate and fluor-apatites.

For example, in order to determine the formation enthalpy of Ca-fluorapatite, $Ca_{10}(PO_4)_6F_2$, one can imagine the following succession of processes.

$$\{10Ca(NO_3)_2 + 6\,H_3PO_4 + 2\,HF\}_{sln} \rightarrow Ca_{10}(PO_4)_6F_2\,(sd) + 20\,\{HNO_3\}_{sln} \qquad (1)$$

$$20\,[HNO_3, 35.35\,H_2O] \rightarrow 10\,H_2 + 10\,N_2 + 30\,O_2 + 707\,H_2O \qquad (2)$$

$$10[Ca(NO_3)_2, 4H_2O] + sln \rightarrow 10\,\{Ca(NO_3)_2 + 4\,H_2O\}_{sln} \qquad (3)$$

$$10\,Ca + 10\,N_2 + 50\,O_2 + 40\,H_2 \rightarrow 10\,Ca(NO_3)_2, 4H_2O\,(sd) \qquad (4)$$

$$6\,[H_3PO_4, 0.756\,H_2O] + sln \rightarrow \{6\,H_3PO_4 + 4.536\,H_2O\}_{sln} \qquad (5)$$

$$9\,H_2 + 6\,P + 12\,O_2 + 4.536\,H_2O \rightarrow 6\,[H_3PO_4, 0.756\,H_2O] \qquad (6)$$

$$2\,[HF, 1.708\,H_2O] + sln \rightarrow 2\,\{HF, 1.708\,H_2O\}_{sln} \qquad (7)$$

$$H_2 + F_2 + 3.416\,H_2O \rightarrow 2\,[HF, 1.708\,H_2O] \qquad (8)$$

$$47.952\{H_2O\}_{sln} \rightarrow 47.952\,H_2O + sln \qquad (9)$$

$$40\ H_2O \rightarrow 40\ H_2 + 20\ O_2 \quad (10)$$

$$10\ Ca(sd) + 6\ P(sd) + 12\ O_2(g) + F_2(g) \rightarrow Ca_{10}(PO_4)_6F_2\ (sd) \quad (11)$$

The subscript 'sln' means 'in solution' which is here a 9%weight HNO_3 aqueous solution, or [HNO_3, 35.35 H_2O]

In this succession, step (1) is the reverse of the fluorapatite dissolution. That dissolution is supposed to lead to a solution containing dissolved calcium nitrate, phosphoric and fluorhydric acids. Steps (2), (4), (6), (8) and (10) correspond to the formation of particular products or entities, the corresponding enthalpies are picked from literature (Handbook of Chemistry and Physics, 2006/2007). Steps (3), (5) (7) and (9) correspond to dissolution of Ca-nitrate tetrahydrate in the same solvent and dilution of H_3PO_4 and HF solutions and water in the same solvent, respectively. The corresponding enthalpies are measured in the same device using the same procedure as for the apatite. Step (11) corresponds to formation reaction of $Ca_{10}(PO_4)_6F_2$ according to the definition. Enthalpy of formation of chlorapatite has been determined with a similar succession but HF solution in step (7) has been replaced by a HCl solution.

This way of doing enabled to determine enthalpies of formation of various apatites and the results are gathered in table 1. However, due to lack of uncertainties on literature values, errors on these quantities can not be estimated.

$M_{10}(PO_4)_6Y_2$	Y = F	Y = OH	Y = Cl
$Ca_{10}(PO_4)_6Y_2$	-13548	-13305	-13179
$Sr_{10}(PO_4)_6Y_2$	-13604	-13373	-13233
$Ba_{10}(PO_4)_6Y_2$	-13564	-13309	-13246
$Cd_{10}(PO_4)_6Y_2$	-8795	-8648	-8463
$Pb_{10}(PO_4)_6Y_2$	-8529	-8261	-8204

Table 1. Standard enthalpies of formation (in kJ mol^{-1}) of the apatites at 25°C

More recent determination of the formation enthalpies for Ca-hydroxy and chlor-apatites led to -13399 kJ mol^{-1} and -13231 kJ mol^{-1} (Cruz et al., 2005). These values differ from that of table 1 by 0.7% and 0.3% respectively.

Neglecting the entropy factor, one can notice the following remarks:

i. whatever is the metal, fluor compound is more stable than hydroxy one which is more stable than the chlor- compound. This gives an explanation of what was observed experimentally, that is, as soon as fluor is introduced in the precipitation medium, fluor compound precipitates. On the other hand, except for lead, chlorapatites were prepared only by solid state reactions because in the 'wet' procedure hydroxyapatites precipitate even in presence of choride ions.

ii. for the same 'Y', alkali earth compounds are more stable than cadmium one which is more stable than the lead one.

Values in this table have been used by Flora et al. (2004) to determine the lattice enthalpies of these compounds and to derive consistent values for the enthalpy of formation of (PO_4^{3-}, g) using the cycles reported in figure 2.

Using the correlation of Jenkins and Glasser (2003), Flora et al. calculated standard Gibbs energy of dissolution of a series of apatites including those which are not yet synthesized. A negative value of this quantity indicates a significant solubility for the apatite. This is the case for zinc and magnesium apatites, leading the authors to conclude on 'the uncertain existence of these apatites'. In fact a significant solubility of an apatite suggests impossibility of synthesizing it by 'wet' procedure, but this does not eliminate the possibility of synthesis by solid state reaction.

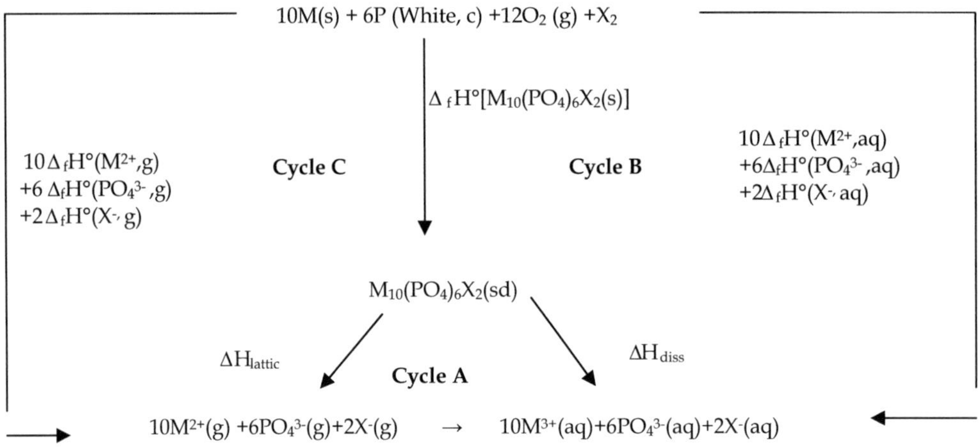

Cycle A : $\Delta H_{lattice}=\Delta H_{diss}-10\Delta_{hyd}H°(M^{2+},g)-6\Delta_{hyd}H°(PO_4^{3-},g)-2\Delta_{hyd}H°(X^{-},g)$

Cycle B : $\Delta H_{diss}=10\Delta_f H°(M^{2+},aq)+6\Delta_f H°(PO_4^{3-},aq)+2\Delta_f H°(X^{-},aq)-\Delta_f H°(M_{10}(PO_4)X_2,s)$

Cycle C : $\Delta_f H°(PO_4^{3-},g)=1/6[\Delta H_{lattice}+\Delta_f H°(M_{10}(PO_4)X_2,s)-10\Delta_f H°(M^{2+},g)-2\Delta_f H°(X^{-},g)]$

Fig. 2. Thermochemical cycles to derive the lattice enthalpies of the apatites and the formation enthalpy of gaseous PO_4^{3-} at 25°C and to calculate again the dissolution enthalpies (Flora et al., 2004).

4.2 Estimation of standard Gibbs energy of formation

Comparing the stability of similar compounds on the basis of the formation enthalpies supposes the entropy factor be non significant compared to enthalpy. This way of doing remains valid when the enthalpies of formation are significantly different. It also supposes the entropy of formation of the compounds be equal zero. Taking into account the discrepancy between the formation enthalpies of alkali earth apatites and those of cadmium or lead one can compare their stability on the basis of this approach. But when the formation enthalpies are close to each other, entropy factor becomes important and can modify the stability order.

Instead of affecting the zero value to the formation entropy of the compound, an other approach can be developed for better estimating this quantity. This approach consists in involving the compound in a particular reaction for which the formation entropies of reactants and products are tabulated. This leads to determine the upper value of the formation entropy of the compound in concern. By assimilating $\Delta_f S$ be equal to this limit, one can derive a new value of the Gibbs energy.

For example Ca-fluorapatite , CaFap, can be involved in the following reaction:

$$9\,CaO + 3/2\,P_4O_{10} + CaF_2 \rightarrow Ca_{10}(PO_4)_6F_2 \qquad \text{(R2)}$$

Taking into account the literature data (Handbook, 2006-2007 and JANAF, 1971 for P_4O_{10}) , the entropy of this reaction is derived as: $\Delta_{R2}S(T_0) = S°(T_0, CaFap) - 904.6$ J mol^{-1} K^{-1}, where

T_0 is the standard temperature (298.15 K) and S°(T_0, CaFap) the standard entropy of CaFap at 298.15 K. Entropy of R2 reaction is negative, consequently S°(T_0,Fap) < 904.6 J mol^{-1} K^{-1}. The formation entropy of Fap is the entropy of the following reaction:

$$10\ Ca(sd) + 6\ P(sd) + 12\ O_2(g) + F_2(g) \rightarrow Ca_{10}(PO_4)_6F_2 \qquad (R3)$$

It can be expressed as Δ_fS°(T_0, CaFap) = S°(T_0, CaFap) - 3327,6 J mol^{-1} K^{-1}. Taking into account the previous inequality one can derive the following inequality as Δ_fS°(T_0, CaFap) < -2423 J mol^{-1} K^{-1}. So according to this demarche, the upper limit of Δ_fS°(T_0, CaFap) is -2423 J mol^{-1} K^{-1}. If we assume the formation entropy to be equal this limit, one can derive the standard formation Gibbs energy at T_0 as -12826 kJ mol^{-1} instead of -13548 kJ mol^{-1} assuming Δ_fS°(T_0, CaFap) = 0

A similar calculation has been made for Ca-hydroxyapatite, CaHap, considering the following reactions:

$$10\ CaO + 3/2\ P_4O_{10} + H_2O_{(lq)} \rightarrow Ca_{10}(PO_4)_6OH)_2 \qquad (R4)$$

$$10\ Ca(sd) + 6\ P(sd) + 13\ O_2(g) + H_2 \rightarrow Ca_{10}(PO_4)_6(OH)_2 \qquad (R5)$$

and calculation led to -12560 kJ mol^{-1} for the standard Gibbs energy of formation instead of -13305 kJ mol^{-1} assuming Δ_fS°(T_0, CaHap) = 0

This approach has been applied to all the apatites listed in table 1 after involving them in reactions having fluorides, oxides, chlorides, H_2O and P_4O_{10} as reactants. Calculation results are reported in table 2 (Jemal, 2004)

$M_{10}(PO_4)_6Y_2$	Y = F	Y = OH	Y = Cl
$Ca_{10}(PO_4)_6Y_2$	-12826	-12560	-12462
$Sr_{10}(PO_4)_6Y_2$	-12845	-12586	-12477
$Ba_{10}(PO_4)_6Y_2$	-12834	-12553	-12539
$Cd_{10}(PO_4)_6Y_2$	-8042	-7872	-7718
$Pb_{10}(PO_4)_6Y_2$	-7781	-7481	-7457

Table 2. Estimated standard Gibbs energies of formation (in kJ mol^{-1}) of the apatites at 25°C

Table 2 shows that Ca-hydroxyapatite is more stable than Ba-hydroxyapatite, whereas neglecting the entropy factor, table 1, shows the reverse and the difference between their Δ_fG°(T_0) is greater than between their Δ_fH°(T_0). However with this approach one can not explain what was observed experimentally during synthesis of Pb-chlorapatite. This compound has been precipitated in aqueous medium whereas all the other chlorapatites have never been obtained by 'wet' method synthesis.

4.3 Thermochemistry of substituted Ca-apatites

Natural apatites are thought to be Ca-apatites in which other metals substitute for Calcium, carbonate or silicate ions for phosphate and carbonate for fluoride or hydroxyl ions. Silicate phosphate apatites are found in high-temperature geological settings such as metamorphic rocks in Oklo (Gabon) and Hoggar (Algeria) or volcanic ejecta. A thermochemical study performed on neodymium silicate Ca-fluorapatites, or britholites, showed a decrease in stability as the amount of couple substitution (Nd,SiO_4) for (Ca,PO_4) increases (Ardhaoui et al., 2006).

X ray diffraction and thermochemical study of substitution of calcium by magnesium in hydroxyl and fluor-apatites showed a limited substitution not exceeding 7% atom in the former and 10 % atom in the latter (Ben Abdelkader et al., 2001). However, substitution by Sr, Cd or Pb in Ca-hydroxy and Ca-fluorapaties is possible over the all range of composition (khattech and Jemal, 1997) (Ben Cherifa and Jemal, 2004) and (Ntahomvukiye et al., 1997) respectively. Thermochemical results showed an increase or decrease in the enthalpy of formation as the substitution increases, depending on the values of the corresponding quantity for the limit apatites. However, heat of dissolution in acid solutions exibited an extremum at a composition close to that at which Rietveld X ray analysis showed the substituting metal moving preferably from one metallic site to the other in the apatite structure.

Substitution by CO_3 anion has been widely studied structurally and spectroscopically (Elliott, 1994 and Legeros, 1991). There are two kinds of carbonate apatites. In 'A' type F- or OH- ions are replaced by CO_3^{2-}. Carbonate ions are centered on the helicoidal senary axis of the network along the tunnels. This substitution is obtained by a solid gas reaction at high temperature, 900°C, under CO_2 gas in very dry conditions. In 'B' type carbonate apatites CO_3^{2-} ions occupy PO_4^{3-} position. There is no ambiguity about the substitution mechanism in A-type carbonate apatites, but for B-type carbonate apatites several mechanisms have been proposed in literature all of them suppose two kinds of CO_3^{2-} in such compounds. These compounds have been subjected to thermal decomposition associated to gas chromatography analysis and the results showed departure of CO_2 in two steps, confirming the presence of two kinds of carbonates in Na free (Khattech and Jemal, 1985 and 1987) and Na bearing (El Feki et al., 1991) carbonate apatites..

De Maeyer and Verbeeck (1993) showed that in Na bearing carbonate apatites the various formulae proposed in literature for B-type hydroxy-compounds derive from the following fundamental mechanisms:

I) $Ca^{2+} + PO_4^{3-} \leftrightarrow V^{Ca} + CO_3^{2-} + V^{OH}$
II) $Ca^{2+} + 2PO_4^{3-} \leftrightarrow V^{Ca} + 2CO_3^{2-}$
III) $Ca^{2+} + PO_4^{3-} \leftrightarrow Na^{+} + V^{OH}$
IV) $Ca^{2+} + OH^{-} \leftrightarrow Na^{+} + V^{OH}$
V) $PO_4^{3-} \leftrightarrow CO_3^{2-} + OH^{-}$

where V^{Ca} and V^{OH-} are vacancies of Ca and OH respectively.

Let a, b, c, d, and e the contributions of mechanisms I to V in the composition of the unit cell respectively, the carbonate apatite can be described by the general formula as:

$$Ca_{10-(a+b+c+d)}(Na_{(c+d)})(PO_4)_{6-(a+2b+c+e)}(CO_3)_{(a+2b+c+e)}(OH)_{2-(a+d-e)} \quad (I)$$

Na bearing carbonate apatites have never been subjected to themochemical study, but Na free hydroxyapatites have been recently thermochemically studied using the isoperibol calorimeter, and their enthalpies of formation were derived (Bel Hadj Yahia and Jemal, 2010). The general formula of the latter compounds derives from the previous one with 'c' and 'd' equal to zero, this gives the following formula:

$$Ca_{10-(a+b)}(PO_4)_{6-(a+2b+e)}(CO_3)_{(a+2b+e)}(OH)_{2-(a-e)} \quad (II)$$

Hydroxyapatites having various compositions have been dissolved in a 9% weight nitric acid solution and their dissolution reactions have been combined with other processes in a succession of reactions similar to that proposed above for Ca-fluorapatite, in order to get the enthalpies of formation. Table 3 gathers the formulae of these products determined by chemical analysis of Ca, PO_4 and carbonate and their enthalpies of solution and formation. One can notice a decrease of $\Delta_f H°(T_0)$ suggesting an increase in stability as CO_3 amount increases until around 1.3 CO_3 and 0.8 OH.

	$\Delta_{sol}H°T_0)$/kJ mol^{-1}	$\Delta_f H°T_0)$/kJ mol^{-1}
$Ca_{9.83}(PO_4)_{5.82}(CO_3)_{0.28}(OH)_{1.64}$	- 387.76	- 13133.3
$Ca_{9.60}(PO_4)_{5.55}(CO_3)_{0.53}(OH)_{1.49}$	- 397.44	- 12763.3
$Ca_{9.35}(PO_4)_{5.34}(CO_3)_{0.67}(OH)_{1.34}$	- 396.05	- 12352.6
$Ca_{9.06}(PO_4)_{4.95}(CO_3)_{0.83}(OH)_{1.61}$	- 440.99	- 11640.0
$Ca_{8.93}(PO_4)_{4.83}(CO_3)_{1.06}(OH)_{1.25}$	- 421.71	- 11583.0
$Ca_{8.82}(PO_4)_{4.74}(CO_3)_{1.31}(OH)_{0.80}$	- 389.58	- 11618.7
$Ca_{8.86}(PO_4)_{4.65}(CO_3)_{1.37}(OH)_{1.03}$	- 420.20	- 11575.4
$Ca_{8.76}(PO_4)_{4.59}(CO_3)_{1.44}(OH)_{0.87}$	- 400.91	- 11474.1
$Ca_{8.68}(PO_4)_{4.52}(CO_3)_{1.58}(OH)_{0.64}$	- 387.48	- 11443.9
$Ca_{8.51}(PO_4)_{4.41}(CO_3)_{1.63}(OH)_{0.53}$	- 401.27	- 11148.3

Table 3. Standard enthalpies of solution and formation for various B-type carbonate Ca-hydroxyapatites.

Taking into account the existence of only two types of carbonates, among 'a', 'b' and 'c' parameters one has to be nil. Thermochemical results have been treated statistically in order to find which couple of parameters has to be considered in the chemical formula in agreement with experimental results. The mathematical model is based on the linear regression on two independent variables. Test of significance of regression coefficients and analysis of variance showed that the most probable substitutions in this kind of products are the couple substitutions (CO_3,OH) for (PO_4) and ($2CO_3$,V_{Ca}) for ($2PO_4$), and so thermochemistry showed that the most probable formula of these compounds is $Ca_{10-(a+b)}(PO_4)_{6-(a+2b)}(CO_3)_{(a+2b)}(OH)_{(2-a)}$.

It was interesting to determine the influence of carbonate introduction on the lattice enthalpy. This can be made by comparing the formation enthalpy of the apatite to that of the stoichiometric mixture of solids containing the same entities as in apatite. Let $Ca_x(PO_4)_{(6-y)}(CO_3)_x(OH)_t$ (with $2x = 18 - y + t$) the general formula of the CO_3-apatite, the effect of carbonate is tightly related to the enthalpy of the following reaction:

$$(1/3)\ (x-y-t/2)\ Ca_3(PO_4)_2 + y\ CaCO_3 + (t/2)\ Ca(OH)_2 \rightarrow Ca_x(PO_4)_{(6-y)}(CO_3)_x(OH)_t \quad (R6)$$

Enthalpy of R6 reaction ($\Delta_R H$) has been determined from literature for the reactants (Handbook 2006/07 and Ben Abdelkader et al., 1999) and from table 3 for the apatites. Figure 3 shows the variation of $\Delta_R H$ over both CO_3 and OH amounts in the apatite lattice. One can notice a maximum around 0.8 CO_3 and 1.6 OH, confirming the presence of two kinds of carbonates and showing the variable effect on the lattice enthalpy depending on the side around the maximum.

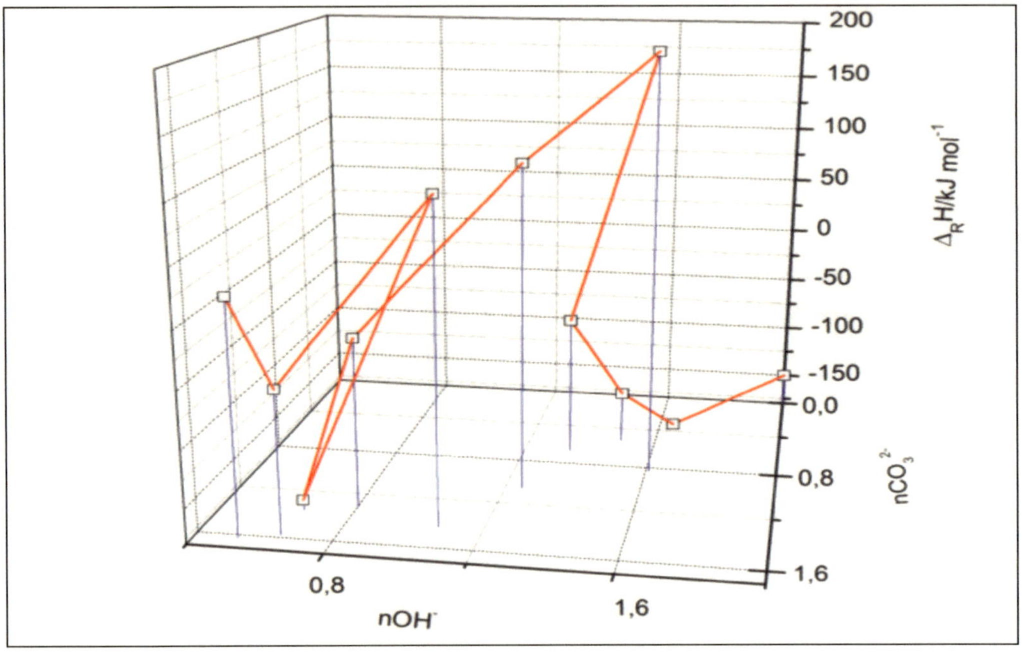

Fig. 3. Variation of enthalpy of reaction R6, $\Delta_R H$, over both CO_3 and OH amounts in B-type CO3 apatite lattice

5. Microcalorimetry for the attack of phosphates by acid solutions.

Attack of phosphates by acid solutions have been widely studied in literature. Due to the importance of this reaction in manufacturing fertilizers, most of the papers deal with phosphate ores and a few with synthetic phosphates such as fluorapatite, hydroxyapatite or tricalcium phosphate (TCP), monocalcium phosphate monohydrate or other Ca-phosphates. Simulation of the hemihydrate process for H_3PO_4 production has been also developed (Gioia et al., 1977) together with dihydrate process used in Rusaifa (Jordan) pilot plant (Abu-Eishah Samir, I. and Abu-Jabal Nizar, M., 2001). A number of papers developed kinetic models for the attack by H_3PO_4 (Huffman et al., 1957; van der Sluis et al., 1987, Dorozhkin, 1996 and Ben Brahim et al., 1999), HNO_3 (Bayramoglu et al., 1995) or H_2SO_4 (Shakourzadeh et al., 1984 and Sevim et al., 2003). In these works analysis of the solution was mainly used as an experimental technique to follow the evolution of the reaction. Conductance measurements, optical polarization microscopy or isoperibol calorimetry has also been used by some authors. Analysis of the solution proceeds discontinuously and the other techniques do not undergo high sensitivity allowing to follow accurately *in situ* the process over time, particularly in the beginning when the reactions are very rapid.

Microcalorimetry overcomes these drawbacks and gives much more precise information on what is happening *in situ*. Consequently it offers the possibility to propose an attack mechanism showing the intermediate steps which appear during the process and leading to the kinetic model for each of them. This information can be used in order to choose the experimental conditions likely to increase the yield of the attack or to lead to a slurry having particular properties, i.e. containing one or two of the calcium sulfates. This technique has been used to study the attack of synthetic fluorapatite and a sample of a Tunisian phosphate ore by phosphoric acid or by a mixture of the latter with sulfuric acid.

5.1 Experimental procedure

C-80 SETARAM microcalorimeter is a differential device operating at temperature up to 300°C. It is composed of a thermally isolated jacket provided with two thermopiles each one containing hundreds of thermocouples surrounding a cavity in which a reaction cell is introduced and tightly fixed. The heat effect evolved in the cell is detected by the thermopile then appears as a heat flow on the recorder. The cell is composed of two compartments separated by a cover and aimed at containing separate reactants. The reaction starts by a reversing mechanism putting alternatively the jacket in the returned position at which the cover falls and the reactants are mixed. In order to minimize the stirring effect one of the cells is provided with reactants and the other with an equivalent amount of an 'inert' compound (here is water), and the difference of the signals coming from the cells is recorded over time. Thermograms begin as a horizontal baseline with a heat flow close to zero. Mixing the reactants by the reversing mechanism results in a peak ending by a horizontal line when the process is over. Integration of the peak leads to the heat effect taking into account the calibration constant.
It could be thought that he heat effect appearing in the reaction cell is detected as soon as possible and so the thermogramm profile gives a real picture of what is happening during the process, allowing to derive kinetics of the phenomena. This true only when the process in low, but when it is fast, the signal produced at the reaction cell, or thermogenesis signal, is delayed and deformed before appearing on the recorder. This difficulty is usually resolved by recording the heat curve, or thermogramm, resulting from a phenomenon having a well-known shape profile. Joule effect energy has this particular property. When providing an electrical resistance with a stabilized current during a certain time 't', the heat profile is nil in the beginning, then increases abruptly when switching on and stays constant during 't' time. It falls vertically to zero value when switching of. Because of the thermal resistance of the device, this rectangular shape, or 'échelon' shape, appears differently on recording. However a mathematical treatment of the recorded signal allows to find back the thermogenesis signal. This treatment consists in expressing the recorded signal as a sum of two exponential functions each one having a particular time constant 'τ', as:

$$s(t) =[\tau_1/(\tau_1- \tau_2)][1-e^{-(t/\tau_1)}] + [\tau_2/(\tau_2- \tau_1)][1-e^{-(t/\tau_2)}] \quad (1)$$

Iteration procedure allows to calculate values of time constants corresponding to tightly superposition of calculated and recorded curves. These values of 'τ' will be then used to calculate back the thermogenesis curve resulting from Joule effect. Entrance signal, or thermogenesis curve, was calculated using the two first derivatives of s(t), ds(t)/dt and $ds(t)^2/dt^2$, according to the following expression (Dubès et al., 1976):

$$e(t) = s(t) + (\tau_1-\tau_2)\, ds(t)/dt + (\tau_1\tau_2).ds(t)^2/dt^2 \quad (2)$$

The 'échelon' shape of the so calculated curve confirms the validity of the time constant values. It should be noticed that these values depend on the nature of couple of cells introduced in the device (Brahim, 2006). SETARAM provides Joule effect calibration cells very different from the reaction cells, consequently it was necessary to provide the latter with electrical resistances in order to get the time constants values in the same conditions as for the chemical process (Brahim et al., 2005). Electrical resistance fixed in the reaction cell was connected to stabilized DC supplier in order to deliver a certain amount of Joule effect energy. Figure 4 shows the recorded and calculated, or deconvoluted, signals obtained for a electrical input in the reaction cell, with τ_1 = 404 seconds and τ_2 = 30 seconds. One can notice the large difference between the two signals and the 'échelon' shape of the calculated one.

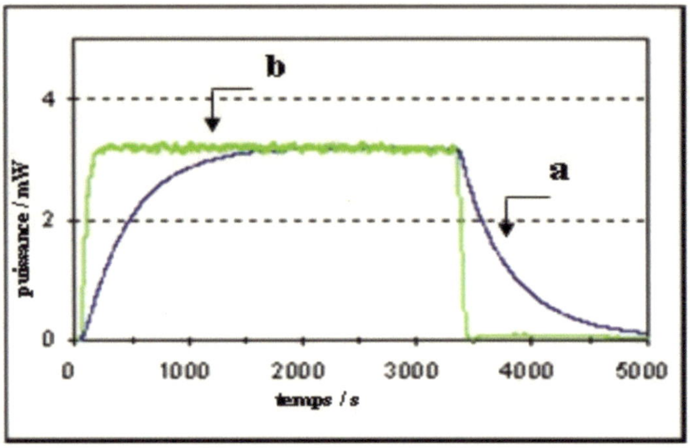

Fig. 4. Recorded (a) and calculated (b) signals for an electrical energy supplied in the reaction cell at 25°C

Time constants determined in that conditions were used to calculate the deconvoluted curves resulting from chemical processes, and the device so provided with electrical resistances has been checked by a key reaction, which is the saponification of ethyl acetate. Thermodynamic and kinetic parameters obtained with this reaction are in good agreement with the published values (Willson et al. 1995).

5.2 Attack of fluorapatite by phosphoric acid solutions

Various amounts of Fap have been dissolved in phosphoric acid solution having 10%, 18% and 30% weight P_2O_5, and the heat effects were calculated by integration of the peaks whereas the kinetic scheme were determined using the deconvoluted curves.

5.2.1 Thermodynamics

The drawing of the heat effect over Fap mass 'm' is composed of 2 to 3 segments, depending on the acid concentration. Figure 5 shows the graph obtained with the 18% P_2O_5 solution. One can notice the presence of 3 segments corresponding to 'a', 'b' and 'c' domains. In the first one no solid was observed, whereas domains 'b' and 'c' correspond to precipitation of Mono-Calcium Phosphate Monohydrate (MCPM), $Ca(H_2PO_4)_2$,H_2O, and of a mixture of the latter with Di-Calcium Phosphate Dihydrate (DCPD), $CaHPO_4$,$2H_2O$, respectively. The presence of these compounds has been checked by X ray diffraction analysis performed on the solid precipitated in these domains.

The slope of line 'a' gives the enthalpy of solution of Fap in 18% weight P_2O_5 solution as $\Delta_{sol}H°(T_0)$= -171.1 kJ mol^{-1}. Lines 'a' and 'b' intercept at m = 53.71 mg for 4.5 mL solution, leading to 0.118 mol L^{-1} for the solubility of MCPD in the acid solution, and the difference between the slopes of lines 'a' and 'b' equals the enthalpy of precipitation of MCPH per unit mass and leads to -5.27 kJ mol^{-1}. This value is in good agreement with literature value -5.0 ± 0.1 kJ mol^{-1} (Elasri et al., 1995). Deconvoluted curves obtained from thermograms recorded in 'b' domain reveals two separate peaks corresponding to dissolution of Fap and precipitation of MCPM respectively, whereas crude thermogramm contains a peak and a faint shoulder, Figure 6. In region 'c' deconvolution curve contains 3 overlapping peaks corresponding to the two previously mentioned phenomena and precipitation of DCPD respectively. Similar phenomena have been observed for 10% and 30% weight P_2O_5 solutions (Brahim et al., 2006).

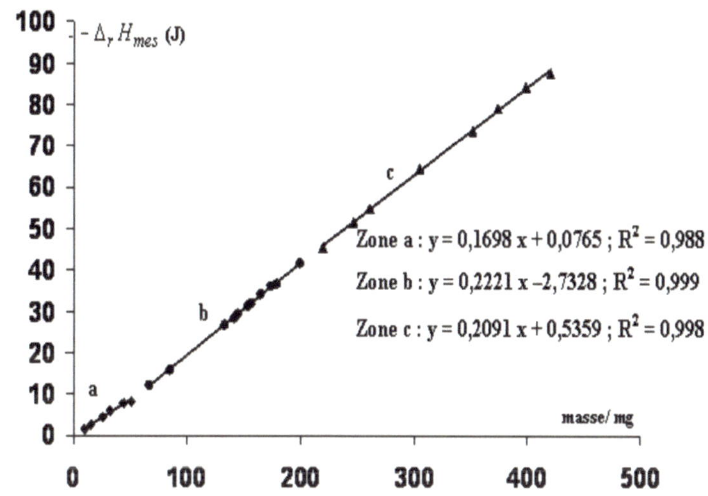

Fig. 5. Heat of solution of Fluorapatite in 18% weight P_2O_5 solution *versus* dissolved mass 'm' at 25°C in 4.5 mL solution

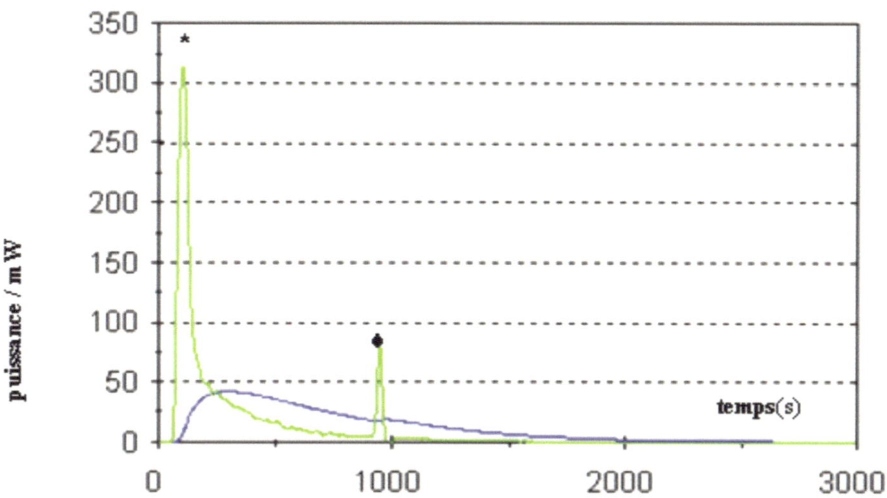

Fig. 6. Crude thermogram (blue) and deconvoluted (green) curves for the attack of Fap with 18% solution

5.2.2 Kinetic schemes

5.2.2.1 Attack by phosphoric acid at 25°C.

Deconvoluted curve gives the more real picture of what is happening during the process, and so it can be used to derive a kinetic model or scheme. The procedure begins by supposing a mathematical model then deriving the corresponding heat flow expression. This expression contains kinetic (order rates and constants) and thermodynamic (enthalpies) parameters. Iteration performed on the heat flow expression allows to find out the best values of parameters leading to the coincidence between calculated and deconvoluted curves. An other criterion of validity of the model consists in comparing the enthalpy deduced from iteration to the experimental value obtained by integrating the thermogram.

As an example let us consider the dissolution of fluorapatite, Fap, (domain 'a' in figure 5). In this domain kinetic models with one step, two opposite or two successive steps have been supposed then the corresponding heat flow expressions were derived, but the coincidence between iterated and deconvoluted curves was obtained only with a two successive reaction model with rate order one for calcium. Here only the latter was considered because of the few amount of Fap compared to that of the acid

In this model Fap (A) reacts with the solution to give an entity (B) which transforms into (C) according to the scheme:

$$A + sol \xrightarrow{k_1} B \quad \text{with} \quad k_1 \text{ as rate constant}$$

$$B \xrightarrow{k_2} C \qquad k_2 : \text{rate constant}$$

The reaction rates are expressed as : $r_1 = k_1\, C(A)$ and $r_2 = k_2\, C(B)$ with C(A) and C(B) are the concentrations of A and B species respectively. If the first reaction rate was independent from that of the second , concentration of 'A' species can be expressed as $C(A) = C_o(A) \exp(-k_1 t)$ in which $C_o(A)$ is the initial concentration of A and equals the calcium concentration corresponding to hypothetical complete dissolution of Fap mass in the 4,5mL solution.

B entity appears through reaction 1 and disappears through 2, so it concentration is expressed as:

$$-dC(B)/dt = k_2 C(B) - k_1 C(A)$$

Integrating this expression leads to: $C(B) = \Omega_1 \exp(-k_1 t) + \Omega_2 \exp(-k_2\, t)$ in which Ω_1 and Ω_2 are integration constants. Taking account the absence of B species at zero time leads to : $\Omega_1 = -\Omega_2$ and so C(B) is derived as:

$$C(B) = C_o(A) \frac{k_1}{k_1 - k_2} [\exp(-k_2 t) - \exp(-k_1 t)]$$

Masse conservation of matter imposes : $C(A) + C(B) + C(C) = C_o(A)$ and so concentration of C is derived as :

$$C(C) = C_o(A)[\,1 - \frac{k_2 \exp(-k_1 t) - k_1 \exp(-k_2 t)}{k_2 - k_1}\,]$$

Let q_1 and q_2 the energy amounts resulting from 1 and 2 reactions respectively, these quantities equal $q_1 = [C_o(A) - C(A)]V\Delta_1H$ and $q_2 = C(C)\ V\Delta_2H$ with V the solution volume and Δ_1H and Δ_2H the molar enthalpies of reactions 1 and 2 respectively. The derivative of the total energy ($q = q_1 + q_2$) is expressed as:

$$\frac{dq}{dt}=\frac{m_A}{M_A}[\left[\frac{\Delta_2H\ k_1k_2}{k_1-k_2}(\exp(-k_2t)-\exp(-k_1t))+\Delta_1H\ k_1\exp(-k_1t)\right]]$$

in which M_A is the molar weight of Fap and m_A mass initially introduced. This expression gives the variation of heat flow over time according to this model and contains the four unknown parameters k_1, k_2, Δ_1H and Δ_2H.

Iteration procedure has been performed on this expression in order to get suitable values of these parameters and the results are shown on figure 7 for 18% and 30% weight P_2O_5 solutions with different mass of dissolved Fap. Dissolution enthalpies of Fap giving the best coincidence for 18% P_2O_5 solution are Δ_1H = -182.5 kJ mol^{-1} and Δ_2H = 10.3 kJ mol^{-1}and the sum -172.2 kJ mol^{-1} is close to that determined experimentally by integrating the recorded peak (-171.0 ± 2.0 kJ mol^{-1}). For 30% P_2O_5 solution iteration enthalpies equal –158.3 and 10.3 kJ mol^{-1} respectively and their sum –148.0 is also close to the experimental value (-147.0 ± 1.7 kJ mol^{-1}).

So dissolution of Fap in phosphoric acid solution likely occurs in two successive steps, the first one could be a simple dissolution and the second the formation of the $[Ca(H_2PO_4]^+$ complex ion in the dissolved state, according to the chemical scheme :

$$Ca_{10}(PO_4)_6F_{2\,(s)}\ +\ 14\,\{H_3PO_4\}_{sol}\ \rightarrow\ \{10\,Ca^{2+}\ +\ 20\,H_2PO_4^-\ +\ 2\,HF\}_{sol}\quad\text{(step 1)}$$

$$10\,\{Ca^{2+}\}_{sol}\ +\ 10\,\{H_2PO_4^-\}_{sol}\ \rightarrow\ \{10Ca(H_2PO4)^+\}_{sol}\quad\text{(step 2)}$$

$$Ca_{10}(PO_4)_6F_{2(s)}+\{14\,H_3PO_4\}_{sol}\rightarrow\{10Ca(H_2PO_4)^+ + 10H_2PO_4^- + 2\,HF\}_{sol}\quad\text{(the global reaction)}$$

$[Ca(H_2PO_4]^+$ complex ion has been reported in literature and is stability constant has been determined at 25°C as 5.1.

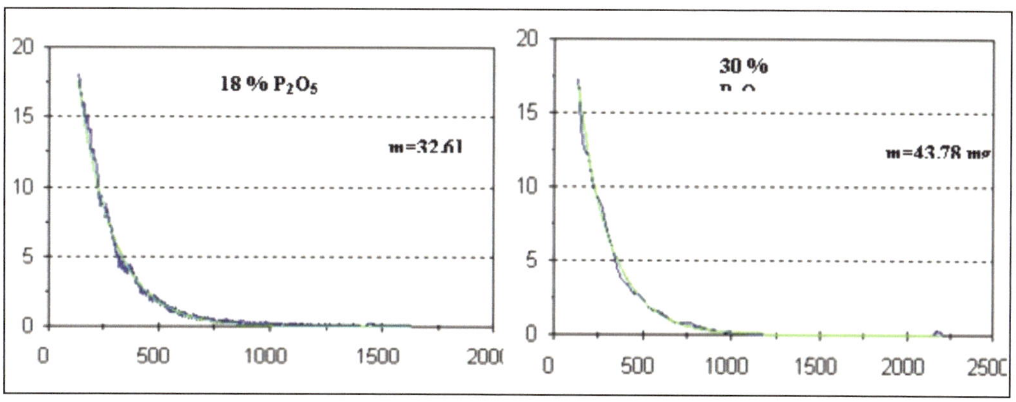

Fig. 7. Examples of coincidence between iterated (green) and deconvoluted (blue) curves for dissolution of Fap in 18% and 30% weight P_2O_5 solutions at 25°C.

In that hypothesis, the second peak corresponds to precipitation of MCPM according to the reaction:

$$\{ Ca(H_2PO_4)^+ + H_2PO_4^- + H_2O \}_{sol} \rightarrow Ca(H_2PO_4)_2,H_2O_{(sd)}$$

A kinetic scheme has been also established for this process by considering the second peak in the deconvoluted curve on figure 6 as corresponding to only one reaction:
Let $C([Ca(H_2PO_4]^+)$ the concentration of $[Ca(H_2PO_4]^+$ in the solution at t time, because of the relatively high amounts of $H_2PO_4^-$ and H_2O, the precipitation rate of MCPM is expressed taking into account only the concentration of calcium, as:

$$r = -dC([Ca(H_2PO_4]^+) / dt = k [C([Ca(H_2PO_4]^+)]^n$$

with k and n the order and constant rate. $[Ca(H_2PO_4]^+$ amount at time t equals the difference between its amount at initial time and that of the reacting quantity and so:

$$C([Ca(H_2PO_4]^+) = C_0([Ca(H_2PO_4]^+) - C_{react}([Ca(H_2PO_4]^+)$$

The heat amount evolved at time t is expressed as : $q = C_{react}([Ca(H_2PO_4]^+)V\Delta_rH$ and the total heat is $Q = C_0([Ca(H_2PO_4]^+)V\Delta_rH$, with V the total volume of solution and Δ_rH the molar enthalpy of the reaction.
The heat flow is then derived as:

$$\frac{dq}{dt} = k(V\Delta_rH)^{1-n}(Q-q)^n$$

Iteration has been performed for various amounts of Fap dissolved in 18 and 30% weight P_2O_5 solutions. An example of coincidence between the calculated and deconvoluted curves of the second peak is shown in figure 8. Iteration values of 'n' order, 'k' constant and 'Δ_rH' enthalpy are in the ranges: [0.48; 0.52]; [3.10^{-3}; 16.10^{-3}] and [-5.14; -5.16]/kJ mol^{-1}, respectively.

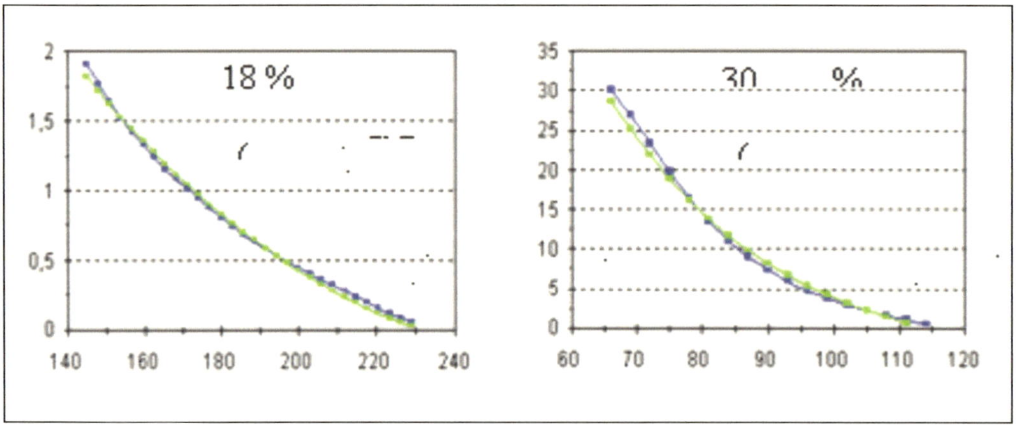

Fig. 8. Examples of coincidence between calculated (blue) and deconvoluted (green) curves for precipitation of Mono-Calcium Phosphate Monohydrate from the solution resulting from dissolution of fluorapatites at 25°C.

'$\Delta_r H$' value agrees with the experimental one obtained by dissolving a commercial sample of MCPM (-5.2 kJ mol^{-1}) and with the values reported in the thermodynamics section § 5.2.1.

5.2.2.2 Attack by phosphoric acid at higher temperatures.

Amounts of Fap close to 30 mg have been attacked by the acid solution at temperatures up to 65°C. Figure 9 shows the molar global enthalpy as a function of temperature. One can notice a particular behavior which can not be explained by the variation of reaction enthalpy according to Kirchhoff relationship, that is:

$$\Delta_r H°(T) = \Delta_r H°(T_0) + \Delta_r C_P°(T - T_0) \; ; \text{ supposing } \Delta_r C_P° \text{ constant.}$$

This relation leads a smooth variation in the [T_0 - T] temperature range. The presence of an extremum in the molar enthalpy curve suggests a mechanism modification around 45°C.

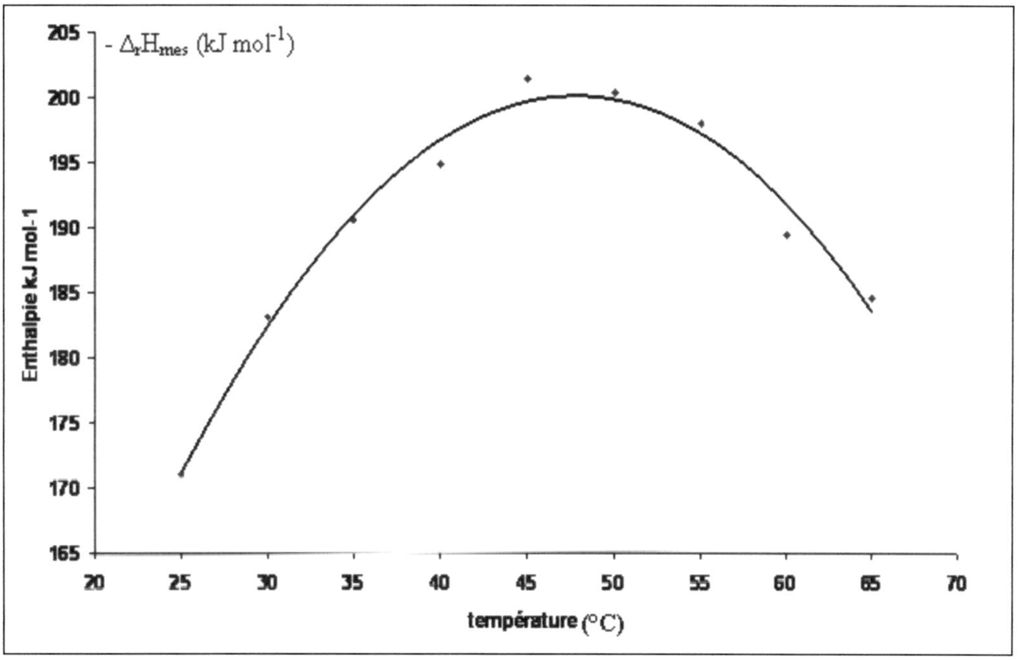

Fig. 9. Molar global enthalpy of the attack of fluorapatite by the acid mixture *versus* temperature.

The kinetic model based on two successive reactions seems to be suitable until 40°C giving a good coincidence between the deconvoluted and calculated curves and enthalpy values not exceeding 2% the experimental ones. For higher temperature the coincidence was very bad and the one step model seems to be much more suitable. This model gave values of the order rate and reaction enthalpy in the ranges [0.98 ; 1.15] and [-185.5 ; - 202.2] kJ mol^{-1}. Experimental values of the latter are in the range [-184.6 ; -200.4] kJ mol^{-1}. An activation energy of the process in that range of temperature was calculated as 101 kJ mol^{-1} (Brahim, K. et al., 2008), whereas in the lower temperature range the activation energy for the first step was around 20 kJ mol-1. The process seems to be governed by a diffusion phenomenon at lower temperature and by a chemical reaction when temperature increases.

5.3 Attack by the mixture of phosphoric and sulphuric acids.

Phosphoric acid is produced industrially by mixing the ore with sulphuric acid or with recycled phosphoric acid solution together with sulphuric acid. Attack by the acid mixture is supposed to begin by the dissolution of the ore in phosphoric acid. Then production of phosphoric acid occurs together with precipitation of phosphogypsum. This scheme has never been evidenced.

In order to model this attack, microcalorimetry has been used to follow at temperature up to 65°C the attack of various amounts of fluorapatite by a solution obtained by mixing 80% volume of a 20% weight P_2O_5 solution with 20% volume of a 98% weight H_2SO_4 solution.

At 25°C the heat effect resulting from the attack of various masses of Fap, 'm' (25 to 115 mg) by 4.5 ml solution is represented as a straight line the slope of which leads to -407.2 kJ mol-1 Fap.

Various kinetic models with one or two steps have been supposed for the process, but no one led to an overlapping between the calculated and deconvoluted curves. The whole phenomenon seems to be more complex. Consequently, Avrami mathematical model has been applied in order to determine the number of steps in the process. In this model the transformed fraction 'x' of a reactant is related to time 't' by the following relation $-\ln(1-x) = kt^n$ with k and n as constants. At the origin this model has been developed theoretically in order to determine the number of directions crystallization occurs (uni, bi or tridimentional crystallization). It has then been extended to chemical transformations and allowed to determine the number of the major reactions involving in a chemical process.

According to this model, $\ln[-\ln(1-x)]$ as a function of lnt has a straight line shape in a domain corresponding to one predominant phenomenon, and so the drawing of that function for the whole process gives the number of successive phenomena it contains. This number equals that of straight lines existing on the graph (Avrami graph).

In the present case, the transformed fraction of Fap at 't' time was calculated by dividing the heat evolved until t time, q, by the total heat amount, Q. This ratio equals that of the corresponding surface areas. Figure 10 shows the Avrami curves for various masses of Fap attacked by 4.5 mL solution during 8 hours under shaking. One can notice two straight lines each one corresponding to a predominant reaction. Supposing the first step be the dissolution of Fap and the second the precipitation of gypsum, one can deduce the time at which the latter begins (≈ 55 seconds).

It should be noticed that because of the low diffusion process across the gypsum shell around the grains, the attack by the acid mixture doesn't finish even after 8 hours maturation time under shaking and the calorimeter signal didn't reach the horizontal line after several hours. However, for comparison this time has been chosen for the experiments above.

Suppose the dissolution of Fap giving $[(CaH_2PO_4)+]$ ions, the second step corresponds to the following reaction:

$$[(CaH_2PO_4)^+]_{sol} + [H^+]_{sol} + 2[H_2O]_{sol} \rightarrow CaSO_4,2H_2O(sd) + [H_3PO_4]_{sol}$$

A kinetic scheme with only one step starting at 55 seconds seemed to be the most suitable for this reaction, with the same mathematical formalism as for precipitation of MCPM § 5.2.2.1. Calculation for various amounts of Fap led to the best coincidence with values of kinetic and thermodynamic parameters in the ranges [1.98 ; 2], [0.1039 ; 0.1062] and [-229.0 ; 231.2] kJ mol^{-1} for the order, the constant and the molar enthalpy respectively (Antar et al., 2006). The order rate of calcium is in good agreement with value 2 reported in literature.

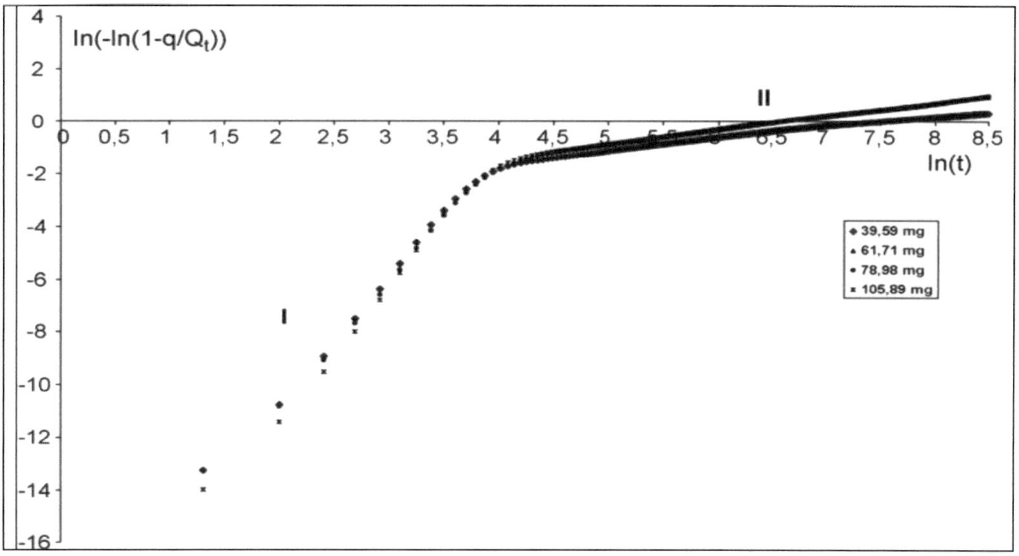

Fig. 10. Avrami graph for various masses of Fluorapatite attacked by 4.5 mL of mixture acid solution at 25°C

Experimental determination of the heat effect of the reaction above has been performed by determining the difference between the energy resulting from the addition of the mixture acid solution to 'pure' H_3PO_4 solution or to a H_3PO_4 solution previously provided with Fap. The results gave -224.5 kJ mol^{-1} Fap, which differs from the iteration value by 2.3% (Antar, 2007), confirming the attribution of the step appearing at 55 seconds to gypsum precipitation. The presence of two steps in the whole process appears on the Avrami curves and not on the crude and deconvoluted curves which contain only one peak at 25°C.
Adding the mean value of the gypsum precipitation enthalpy (-230 kJ mol^{-1}) to that resulting from dissolution of Fap in 18% weight P_2O_5 solution -171.6 kJ mol-1 (mean value) leads to -401.6 kJ mol^{-1} which differs from that resulting from the attack of Fap with the acid mixture (-407.0) by 1.3 %, confirming the two steps in the whole process.
When the attack temperature increases, a second peak appears and overlaps partly the first one till 55°C when they separate. X ray diffraction performed on the solid present between the peaks and at the end of the process reveals the transitory formation of calcium sulphate hemihydrate, $CaSO_4,0.5H_2O$, (HH), which transforms into dihydrate (DH) or gypsum when the time elapses. That transformation lasts about 20 hours at 55°C. Figure 11 shows an example of crude and deconvoluted curves obtained at 55°C, together with a zooming in the region of each peak.

5.4 Transformation of Ca-sulfate hemihydrate into dihydrate at 55°C

Various amounts 'm' of Fap have been attacked by the acid mixture at 55°C, and the graphs representing the heat effects over 'm' gave -27.9 kJ and -13.1 kJ per mol of HH and DH respectively.
A kinetic model with two successive Ca-order one reactions was developed for the first peak and gave the better coincidence of the calculated and deconvoluted curves. The iteration enthalpy mean values are -150.3 and -120.3 kJ per Fap mole. Their sum (-270.6) differs from the peak integration one (-278.7) by about 3%. Complementary experiments allowed to

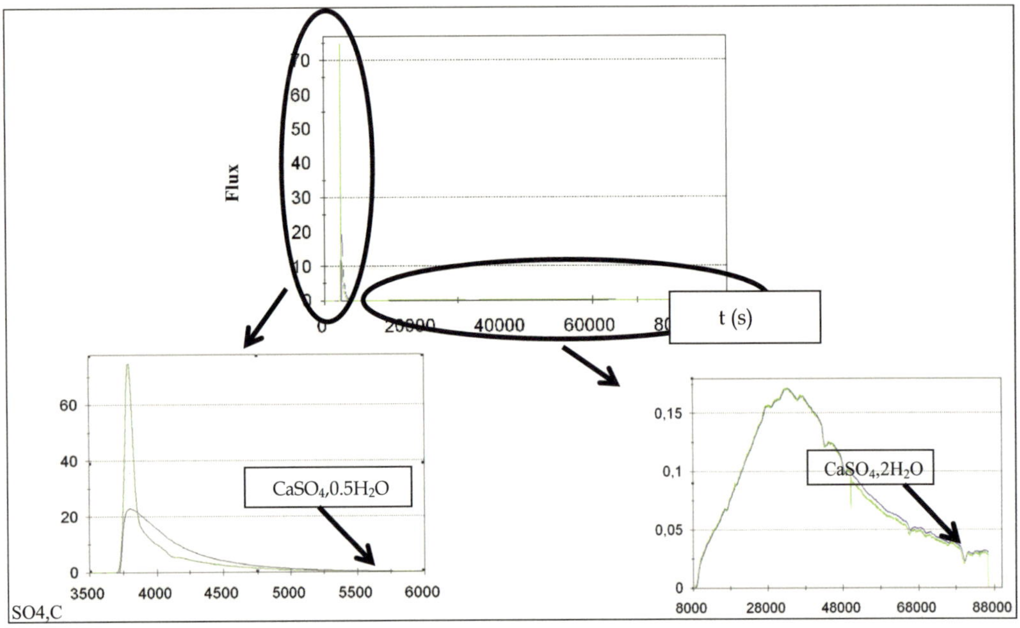

Fig. 11. Deconvoluted (green) and crude (blue) curves for the attack of fluorapatite by phosphoric and sulfuric acid mixture at 55°C

attribute the first step in the first peak to dissolution of Fap and neutralization of $H_2PO_4^-$ into H_3PO_4, and the second step to precipitation of Ca-sulfate hemihydrate, (HH), according to the following reactions:
For the first peak:

$$Ca_{10}(PO_4)_6F_2(s) + 20\{H^+\}_{sol} \rightarrow \{10Ca^{2+} + 6H_3PO_4 + 2HF\}_{sol} \qquad 1^{st}\ step$$

$$\{Ca^{2+}\}_{sol} + \{SO_4^{2-}\}_{sol} + 1/2\,\{H_2O\}_{sol} \rightarrow CaSO_4 \cdot 1/2H_2O(s) \qquad 2^{nd}\ step$$

with mean iteration enthalpy values as -150.3 kJ and -120.3 kJ per mole of Fap. This corresponds to -15.0 and --12.0 kJ per mole of Ca-sulfate, respectively.
and for the second peak:

$$CaSO_4{,}0.5H_2O(s) + 3/2\ H_2O \rightarrow CaSO_4{,}2H_2O(s)$$

A kinetic model based on a dissolution/precipitation process has been developed for the second peak, in which dissolution rate of HH, r_{dis}, is proportional to under saturation (Ce – [Ca]) where Ce is the Ca concentration corresponding to solubility of HH, whereas precipitation rate of DH, r_{pp}, is of order rate 2 for Ca concentration, and so:

$$r_{dis} = k_1(Ce - [Ca^{2+}]) \text{ and } r_{pp} = k_2\,[Ca2+]^2$$

where k_1 and k_2 are constant rate.
The global rate is controlled by the Ca^{2+}concentration, so: $d[Ca^{2+}]/dt = r_{dis} - r_{pp}$

$$d[Ca^{2+}]/dt = k_1(Ce - [Ca^{2+}]) - k_2\,[Ca^{2+}]^2$$

$$\text{giving} \quad d[Ca^{2+}]/dt + k_2[Ca^{2+}]^2 + k_1[Ca^{2+}] = k_1Ce$$

Usual methods for resolution did not make it possible to solve this equation. But Maple 6 software gave the two following solutions:

$$[Ca^{2+}] = [1/2k_2][-k_1 + C_1th\{C_1t/2 + \ln C_2\}]$$

and

$$[Ca^{2+}] = [1/2k_2][-k_1 + C_1th\{C_1t/2 + \ln(-C_2)\}]$$

with C_1 and C_2 two constants depending both on k_1, k_2 and Ce.

$$C_1 = [4k_1Cek_2 + k_1^2]^{1/2} \text{ and } C_2 = [1/2k_2Ce][4k_1C_1 - 2k_1^2 - 8k_1 k_2 Ce - 4C_1Cek_2 - 4Ce^2 k_2^2]^{1/2}$$

The adopted solution is given by the first relation because C_2 constant should have a positive value.

Let Q_1 and Q_2 denote the heat quantities released by hemihydrate dissolution and dihydrate precipitation reactions respectively. Variation of each of these quantities is proportional to the corresponding reaction rate as:

$$dQ_1/dt = V\Delta_{dis}Hk_1(Ce - [Ca^{2+}]) \text{ and } dQ_2/dt = V\Delta_{pp}Hk_2[Ca^{2+}]^2$$

with $\Delta_{dis}H$ and $\Delta_{pp}H$ the molar enthalpies of dissolution of HH and precipitation of DH, respectively.

The total energy is $Q = Q_1 + Q_2$ and is variation is expressed as:

$$dQ/dt = V\Delta_{dis}k_1(Ce - [Ca^{2+}]) + V\Delta_{\pi\pi}Hk_2[Ca^{2+}]^2$$

The final heat flow expression is expressed as:

$$dQ/dt = V\Delta_{dis}Hk_1[Ce - (1/2k_2)(-k_1 + C_1th\{C_1t/2 + \ln C_2\})] + V\Delta_{pp}Hk_2[(1/2k_2)(-k_1 + C_1th\{C_1t/2 + \ln C_2\})]^2$$

Iteration on this expression has been performed in order to get the suitable values of the 6 parameters, $\Delta_{dis}H$, $\Delta_{pp}H$, k_1, k_2, Ce and C_2, giving the better coincidence between calculated and recorded curves corresponding to the second peak. C_2 parameter was included because of the complexity of its expression. Due to the low rate of the transformation deconvoluted and recorded curve overlap as shown on figure 11. Figure 12 gives an example of calculated and recorded curves obtained for 39.41 mg Fap. For the other masses of Fap, coincidence is similar and the whole iteration values of k_1, k_2, $\Delta_{dis}H$, $\Delta_{pp}H$, Ce and C_2 pertain to the intervals:[4.15×10^{-5} ; 5.00×10^{-5} s^{-1}], [1.45×10^{-3} ; 1.99×10^{-3} L mol^{-1} s^{-1}], [11.90 ; 12.30 kJ mol^{-1}], [−25.10 ; −24.90 kJ mol^{-1}], [3.12×10^{-4} ; 3.27×10^{-4} mol L^{-1}] and [31.01 ; 59.33], respectively (Antar and Jemal, 2007). It can be noticed that the mean iteration value of $\Delta_{dis}H$ of HH (+12.10 kJ mol^{-1}) is very close to the opposite of the previous iteration enthalpy of precipitation of hemihydrate (-12.0 kJ per mole of Ca-sulfate).

5.5 Attack of a phosphate ore.

A natural phosphate (NP) ore sample containing 23.6 %P_2O_5, 38.4 % CaO and 15.1 % calcite has been attacked by the 20% weight P_2O_5 solution and by the mixture of acids at different temperatures.

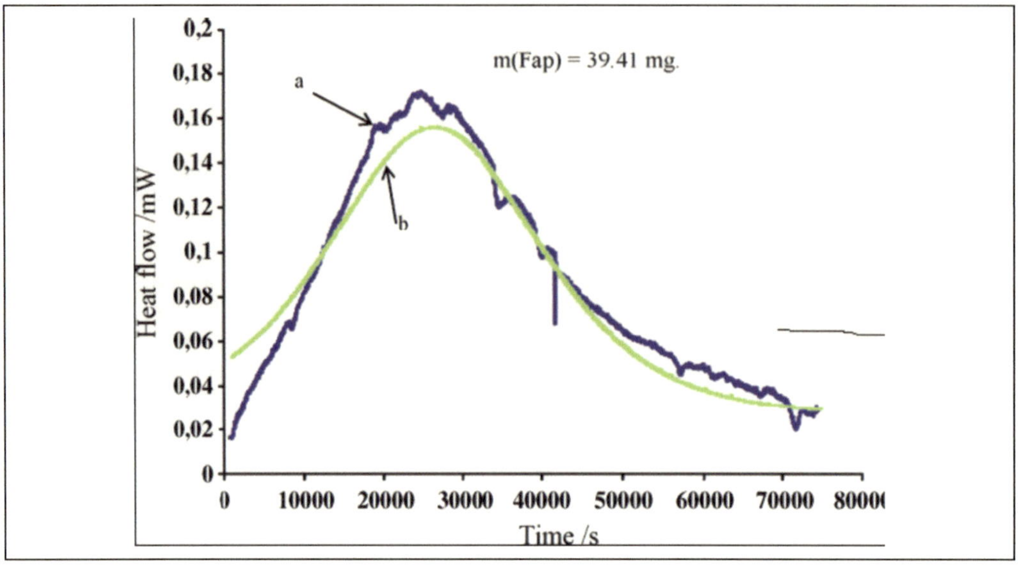

Fig. 12. calculated (b) and recorded (a) curves for the transformation of Ca-sulfate hemihydrate into dehydrate at 55°C (second peak).

The drawing of the heat evolved by attacking at 25°C few amounts of a sample with a granulometry ≤ 500 μm in 4.5 mL phosphoric acid solution showed only dissolution with -233.6 J per gram ore. A very close value has been obtained with a sample obtained by grinding the former in an agate mortar. Attack of greater amounts in the same conditions led to an Avrami curve with two successive phenomena corresponding to dissolution and precipitation of Di-Calcium Phosphate Dihydrate, DCPD, the latter starting at about 2.5 minutes (Brahim, 2006). DCPD has been identified by X ray diffraction performed on the isolated solid.

Attack by the acid mixture was very long and gave a two overlapped peaks signal which didn't become horizontal even after several hours. After 8 hours maturation time under shaking the heat effect was -161.5 J g^{-1} for the ≤ 500 μm sample and -305.5 J g^{-1} for the grinded one, confirming the dependence of the yield of the process on the grain size of the ore. As for Fap, complementary experiments consisting in adding the acid mixture to a H_3PO_4 and to a H_3PO_4/NP solutions allowed to deduce the enthalpy value for complete NP attack as -377.6 J g^{-1}, and so after 8 hours maturation time only 43% and 81% of non grinded and grinded samples reacted. X ray diffraction performed on the solids isolated after 8 hours reaction time showed a mixture of Ca-sulphate hemihydrate and dehydrate. As with Fap, reaction of NP with the acid mixture leads to HH wich transforms into DH. This was observed for NP at 25°C whereas for Fap for higher temperatures. The presence of impurities in the solid results in lowering the temperature of transitory formation of HH

Figure 13 shows what was obtained by attacking various amounts of NP at 25°C by 4.5 mL of the acid mixture solution. It should be noticed that the peaks overlap more when the amount of NP increases. Avrami curves show the second peak starting at around 100 seconds for a solid mass until about 100 mg and started earlier when the solid amount increases. This is in agreement with literature data concerning the decrease of germination time when the over saturation increases.

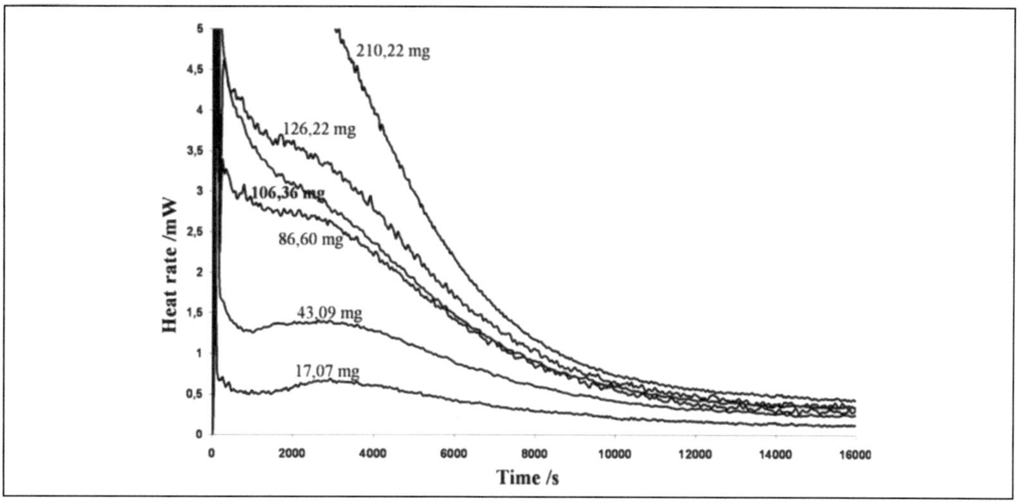

Fig. 13. Thermograms obtained by attacking various amounts of NP by phosphoric and sulfuric acid mixture at 25°C.

When the attack temperature increases, the second peak departs from the former but they didn't separate and the signal remains with only two peaks till 55°C where a third peak appeared at about 18.5 hours and didn't finish even after 45 hours under shaking, Figure 14 shows a crude thermogramm recorded during about 33 hours with 32.63 mg of NP attacked with 4.5 mL of acid mixture solution at 55°C (Antar and Jemal, 2008).

X ray analysis performed on the solid isolated at different times reveals that the first peak corresponds to dissolution of the ore and precipitation of a mixture of HH and AH (anhydrous Ca-sulfate), the second peak to transformation of HH into DH and the third to transformation of AH into DH.

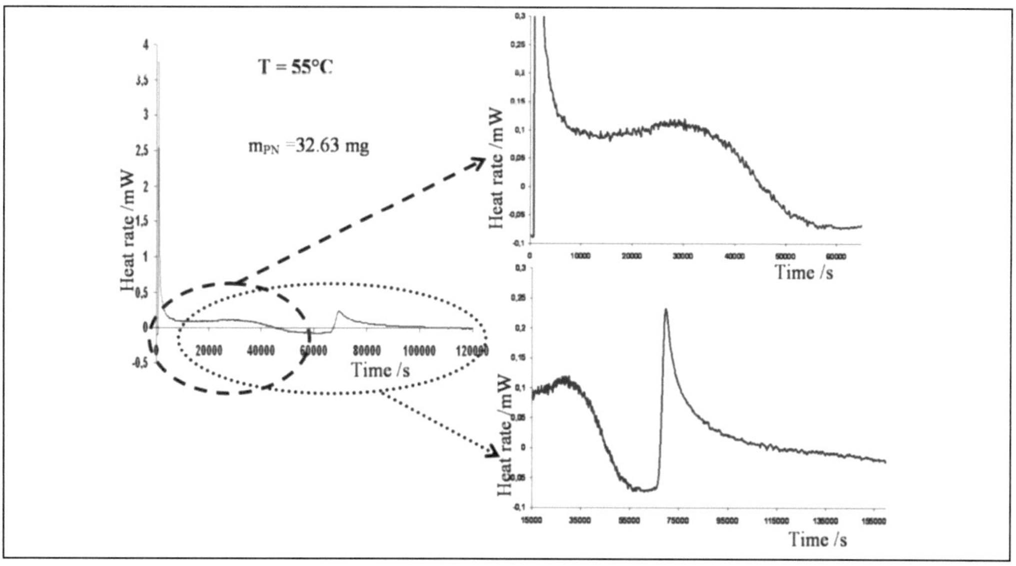

Fig. 14. Thermogram recorded with NP attacked with acid mixture at 55°C

6. Conclusion

Isoperibol dissolution calorimetry allowed to determine not only the solution enthalpies but also other important quantities such as the enthalpies of formation or mixing. These quantities are tightly related to lattice energy of the compounds and can be used to calculate the enthalpies of the reaction in which the compounds are involved. Dissolution enthalpy is also useful to determine the solubility of a compound at any temperature from only one experimental solubility value. Experimentation is very easy but the technique has its own limits.

Microcalometry is more sophisticated and sensitive and can give information on both thermodynamics and kinetics of the processes. Its application to transformations of phosphates reveals phenomena which have not be previously seen with other techniques. But for a rapid process, the recorded curve doesn't give a real image of what is happening in the reacting medium. Consequently, deconvolution procedure is necessary to find out the thermogenesis curve, however for more precise determination, the time constants of the device have to be determined in the same conditions as for the chemical process. This is possible after suitable modifications operated on the device.

As opposed to other techniques, kinetic models deduced from microcalorimetric results can be checked not only mathematically but also by comparing the heat effect resulting from mathematical model to the experimental one. This represents an additional criterion for the validity of the model. However the procedure is applicable for a one step process whatever is the order rate, provided it affects only one reactant concentration, but for a two step process the procedure is limited to low order rate kinetic equations, otherwise integration becomes impossible and the heat flow expression can't be derived.

7. Aknowledgements

The author would like to thank his collaborators K. Antar, K. Ardhaoui, F. Bel Haj Yahia, S. Ben Abdelkader, A. Ben Cherifa, K. Brahim, I. Ntahomvukiyé and I. Khattech.

8. References

Abu-Eishah Samir, I. and Abu-Jabal Nizar, M., Parametric study on the production of phosphoric acid by the dihydrate process, *Chem. Eng.*, 81, (2001), 231-50

Antar, K., Brahim, K. and M. Jemal, Etude cinétique et thermodynamique de l'attaque d'une fluorapatites par des mélanges d'acides sulfurique et phosphorique à 25°C, *Thermochim. Acta*, 449, (2006), 35-41

Antar, K., *Contribution à l'étude de l'attaque d'une fluorapatites par le mélange sulfo-phosphorique. Application au phosphate naturel*, Thèse de Doctorat, (2007), Université de Tunis El Manar, Tunisia.

Antar, k. and Jemal, M., Kinetics and thermodynamics of the attack of fluorapatites by a mixture of sulfuric and phosphoric acids, *Thermochim. Acta*, 452, (2007), 71-75

Antar, K. and Jemal, M., Kinetics and thermodynamics of the attack of a phosphate ore by acid solutions at different temperatures, *Thermochim. Acta*, 474, (2008), 32-35

Ardhaoui, K., Coulet, M.V., Ben Cherifa, A., Carpena, J., Rogez, J. and Jemal, M., Standard enthalpy of formation of neodimium fluorapatites, *Thermochim. Acta*, 444, (2006), 190-194

Bayramoglu, M., Demircilolu, N. and Tekin, T., Dissolution kinetics of Mazidagi phosphate rock in HNO3 solution, *Int. J. Miner. Process*, 43 (1995), 240-54

Becker, P., *Phosphates and Phosphoric Acid*, Second Edition (1989) Marcel Decker Inc., New York, Basel

Bel Hadj Yahia, F. and Jemal, M. Synthesis, structural analysis and thermochemistry of B-type carbonate apatites, *Thermochim. Acta*, 505 (2010) 22-32

Ben Abdelkader, S., Ben Cherifa, A., Khattech, I. and Jemal, M., Synthèse, caractérisation et thermochimie du phosphate trimagnésien et du phosphate tricalcique, *Thermochim. Acta*, (1999), 334, 123-29

Ben Abdelkader, S., Khattech, I. Rey, C. and Jemal, M. Synthèse et thermochimie d'apatites calco-magnésiennes hydroxylées et fluorées, *Thermochim. Acta*, 376, (2001), 25-36

Ben Abdelkader, S., Ben Cherifa, A., Coulet, V.M., Khattech, I. Rogez, J. and Jemal, M. Enthalpie de formation de la whitlockite Ca10Mg2H2(PO4)14, *Thermochim. Acta*, 77, (2004), 863-71

Ben Brahim F., Mgaïdi, A. and Elmaaoui, M.,Kinetics of teaching of Tunisian phosphate ore particles in dilute phosphoric acid solutions, *Can. J. Chem. Eng.* 77 (Febrary 1999), 136-42

Ben Cherifa, A., Nounah, A., Lacout, J.L. and Jemal, M. Synthèse et thermochimie de phosphates au cadmium. Partie II: Enthalpie standard de formation de fluorapatites calco-cadmiées, *Thermochim. Acta*, 366, (2001a), 7-13

Ben Cherifa, A. and Jemal, M. Synthèse et thermochimie de phosphates au cadmium. Partie I: cas du phosphate tricalcique et de la chlorapatite, *Thermochim. Acta*, 366, (2001b), 1-6

Ben Cherifa, A. *Synthèse, caractérisation et détermination de grandeurs thermochimiques de phosphates à base de calcium et/ou de cadmium*, Thèse de Doctorat, Université de Tunis El Manar, (Mai 2002), Tunis, Tunisia

Ben Cherifa, A. and Jemal, Enthalpy of formation and mixing of calcium-cadmium phosphoapatites, *Phos. Res. Bull.*, 15 (2004), 113-118

Brahim, K., Khattech, I., Dubès, J.P. and Jemal, M., Etude cinétique et thermodynamique de la dissolution de la fluorapatites dans l'acide phosphorique, *Thermochim. Acta*, 436 (2005), 43-50

Brahim, K. *Contribution à l'étude thermodynamique et cinétique de l'attaque phosphorique d'une fluorapatite. Application à un phosphate naturel*, Thèse de Doctorat, (2006) Université de Tunis El Manar, Tunisia

Brahim, K., Antar, K., Khattech, I. and Jemal, M. Etude thermodynamique et cinétique de l'attaque de la fluorapatites par l'acide phosphorique, *Ann. Chim. Sci. Mat.*, 31(5), (2006), 611-20

Brahim, K., Antar, K., Khattech, I. and Jemal, M., Effect of temperature on the attack of fluorapatite by a phosphoric acid solution, *Sci. Res. Essay*, (3(1), (2008), 35-39

Cruz Fernando, J.A.I., da Piedade Manuel, E.E., Calado Jorge, C.G., Standard molar enthalpies of formation of hydroxyl-, chlor- and bromapatite, *J. Chem. Thermodyn.*, 37, (2005), 1061-70

De Maeyer , E.A.P. and Verbeeck, R.M.H., Possible substitution mechanisms for sodium and carbonate in calcium hydroxyapatites, *Bull. Soc. Chim. Belg.*, 102, N°9 (1993), 601-9

Dorozhkin, S.V., Dissolution kinetics of single fluorapatite crystals in phosphoric acid solution under the conditions of the wet-process phosphoric acid production, *J. prakt. Chem.*, 338 (1996), 620-26

Dubès, J.P. Barrès, M. and Tachoire, H., Correcteur automatique pour calorimètres à conduction et analyseurs calorimétriques différentiels, *C. R. Acad. Sci.*, Série C, 283 (1976), 163-66

Dubès, J. P. *Déconvolution de la réponse instrumentale par filtrage inverse en calorimétrie à conduction*, Thèse de Doctorat, 1985), Université de Provence, Aix-Marseille I, France

Elasi, S., Ben Cherifa, A., Bouhaouss, A., Ferhat, M. and Jemal, M, Mécanisme de dissolution du p)hosphate tricalcique β et de l'hydroxyapatite dans l'acide phosphorique, *Thermochim. Acta*, 249 (1995) 121-26

El Feki, H., Khattech, I. and Jemal, M. Décomposition thermique de fluorapatites carbonates sodées, *J. Chim. Phys.*, 88, (1991), 1885-92

Elliott, J.C., (1994) *Structure and Chemistry of the Apatites and Other Calcium Orthophosphates*, Elsevier Sciences B.V., Amsterdam

Flora, N.J. and Jenkins, H.D.B., Lattice energies of apatites and the estimation of $\Delta_f H°(PO_4^{3-}$, g), *Inorg. Chem.*, 43,7, (2004), 2040-45

Gottshall, A.J., *J. South Africa*, 11 (1958), 45

Huffman, E.O., Cate, W.E., Deming, M.E. and Elmore, K.L., Rates of solution of calcium phosphates in phosphoric acid solutions, *Agri. Food Chem.*, 5(4), (April 1957), 266-75

JANAF Thermochemical Tables, (1971) Nat. Stand. Ref. data Ser. Nat. Bur. Stand. NSRDS-NBS 37

Jemal, M., Thermochemistry and relative stability of apatite phosphates, *Phos. Res. Bull*, 15, (2004), 119-124

Jenkins, H.D.B. and Glasser, L. Standard Absolute Entropy, S°298. Values From Volume or Density. 1. *Inorg. Mat.*, Inorg. Chem., 42, N°26, (2003), 8702-08

Khattech, I. and Jemal, M. Etude de la décomposition thermique de fluorapatites carbonatées, *Thermochim. Acta*, 85, (1985), 119-128

Khattech, I. and Jemal, Décomposition thermique de fluorapatites carbonatées de type B 'inverses', *Thermochim. Acta*, 118, (1987), 267-275

Khattech, I. (1996), *Synthèse, caractérisation et étude thermochimique de phosphates à base de métaux alcalino-terreux*, Thèse de Doctorat, Université de Tunis El Manar, Tunis, Tunisia

Khattech, I. and Jemal, M. Thermochemistry of phosphate products. Part II: Standard enthalpies of formation and mixing of calcium and strontium fluorapatites, *Thermochim. Acta*, 298, (1997), 23-30

LeGeros, R.Z., (1991), Calcium phosphates in oral biology and medecine, Karger , Basel...Sydney.

Ntahomvukiyé I., *Synthèse, caractérisation et étude thermochimique de phospho-apatites au calcium et au plomb*, (1997a) Thèse de Doctorat, Université de Tunis El Manar, Tunis, Tunisia

Ntahomvukiyé, I., Khattech, I. and Jemal, M., Synthèse, caractérisation et thermochimie d'apatites calco- plombeuses fluorées $Ca_{(10-x)}Pb_x(PO_4)_6F_2$; $0 \le x \le 10$, *Ann. Chim. Sci. Mat.*, 22, (1997b) 435-446

Pattengill, M.D. and Sands, D.E., Statistical significance of linear least-square parameters, *J. Chem. Educ*, 56(4), (1979), 244-7

Sands, D.E. Weighting factors in least squares, *J. Chem. Educ.*, 51(7), (1974), 473-4

Sevim, F., Sarac, H., Kocakerim, M.M. and Yartasi, A., Dissolution kinetics of phosphate ore in H2SO4 solutions, *Ind. Eng. Chem. Res.*, 42, (2003), 2052-57

Shakourzadeh, K., Bloise, R. and Baratin, F., Modeling of a wet-process phosphoric acid reactor : Influence of phosphate rock impurities, *Ind. Miner. Tech.*, 9 (1984), 446-52

van der Sluis, S., Meszaros, Y., Marchee, W.G.J., Wesselingh, H.A. and van Rosmalen, G.M., The digestion of phosphate ore in phosphoric acid, *Ind. Eng. Chem. Res.*, 26, (1987), 2501-05

Willson, R.J., Beezer, A.E., Mitchell, J.C. and Loh, W. *J. Phys. Chem.* 99 (1995), 7108

22

The Sintering Behaviour of Fe-Mn-C Powder System, Correlation between Thermodynamics and Sintering Process, Manganese Distribution and Microstructure Composition, Effect of Alloying Mode

Eduard Hryha[1] and Eva Dudrova[2]
[1]Chalmers University of Technology, Gothenburg
[2]Institute of Materials Research, Slovak Academy of Sciences, Košice
[1]Sweden
[2]Slovak Republic

1. Introduction

Among steel-making techniques Powder Metallurgy (PM) concept utilizes unique production cycle, consisting of powder compaction and sintering steps that give high productivity with low energy consumption and high material utilization. Due to the presence of residual porosity, mechanical properties of PM components are inferior in comparison with structural components produced by other technologies. Improvement of mechanical properties at the same level of porosity can be achieved primarily by adding variety of alloying elements. Therefore modern PM technology for production of high-performance PM parts for highly stressed steel components for automotive industry, for example, rely on techniques of utilization of different alloying elements additionally to adjustment of technological process depending on alloying system used. When talking about high-strength low-alloyed structural steels, the most common alloying elements, additionally to carbon, added in order to increase mechanical performance, are chromium, manganese, silicon and some other strong carbide and carbonitride-forming elements (V, Nb, Ti etc.). In comparison with classical steelmaking practice, alloying of PM steels is much more complicated as additionally to influence of alloying elements type and content on microstructure, mechanical properties, hardenability etc., number of additional aspects influencing powder production and further component processing has to be considered. Traditionally, PM high-strength steels are alloyed with Cu, Ni, and Mo. This results in a considerable difference in price of material between conventional and PM steels, used for the same high-load application, as the price of currently employed PM alloying elements like Mo and Ni is dozens of times higher in comparison with that of Cr or Mn. This situation creates a strong economical stimulation to introduce cheaper and more efficient alloying elements to improve the competitiveness of PM structural parts.

So, why the potential of most common for conventional metallurgy alloying elements as Cr, Mn and Si is not utilized in PM? First and basic question that arise is how to introduce these elements in PM - as admixed elemental powder (or master-alloy) or by prealloying of the base steel powder. Chromium prealloyed steels are already successfully introduced on the PM market. However due to the peculiar properties of manganese (oxygen affinity, high vapour pressure, ferrite strengthening etc.) attempts to develop Mn sintered steels are still ongoing. Issue of appropriate alloying mode, that is the starting point of manganese introduction in PM, is the basic question that has to be answered at the beginning and is the basic topic of this chapter.

The easiest way to introduce manganese is by admixing of ferromanganese powder that is cheap and widely available on the market in different grades. This approach was firstly proposes around 30 years ago and have been scrutinized thoroughly from different perspectives (Šalak, 1980; Cias et al., 1999; Šalak et al., 2001; Dudrova et al., 2004; Danninger et al., 2005; Cias&Wronski, 2008, Hryha, 2007). The first thing that has to be considered when dealing with admixed with manganese systems is high affinity of manganese to oxygen, implying possibility of considerable oxidation during component processing due to high activity of manganese in admixed elemental powder. However, the possibility of sintering of admixed with manganese powders was assumed due to so-called "self-cleaning" effect, discovered by Šalak (Šalak, 1980). This effect utilizes unique property of manganese to sublimate at relatively low temperature and created during heating stage manganese vapour protects specimen from further oxidation. Another advantage of admixed manganese systems is manganese homogenization in Fe–Mn powder compacts involving Mn-gaseous phase during the heating stage. Second assumption deals with the alloying by different master-alloys that firstly allowed a successful introduction of high oxygen affinity elements in PM industrial production (Zapf et al., 1975; Schlieper & Thummler, 1979; Hoffmann & Dalal, 1979). First developed master-alloys containing manganese–chromium–molybdenum (MCM) and manganese–vanadium–molybdenum (MVM) had a wide range of mechanical properties depending on alloying content, sintered density and processing conditions. Nevertheless, these master-alloys faced with many problems during application (oxides formation during manufacturing process, high hardness of the particles that lead to intensive wear of compacting tools etc.) and fully disappeared from manufacturing and research areas. Recent development of Fe–Cr–Mn–Mo–C master-alloys was much more successful and show promising properties for their future industrial utilization (Beiss, 2006; Sainz et al., 2006).

High affinity of manganese for oxygen and Mn loss by sublimation can be minimized by lowering the manganese activity that can be done by Mn introduction in pre-alloyed state. However powder alloying by manganese faces some difficulties starting from powder production, handling and following compaction and sintering steps. This is connected with manganese selective oxidation on the powder surface during atomization and further annealing depending on processing conditions during powder production (Hryha et al., 2009-b; Hryha et al., 2010-a). A further negative impact of manganese utilization in pre-alloyed state is the expected lower compressibility of such pre-alloyed powders due to ferrite solid solution strengthening by manganese.

This chapter is focused on the influence of alloying mode, utilizing premix systems with different ferromanganese grades and high-purity electrolytic manganese as well as fully prealloying of water atomized powder. While respecting all the benefits and problems with sintered steels alloyed with manganese some basic directions have been chosen — theoretical evaluations of required sintering atmosphere composition for preventing of manganese alloyed steels from oxidation during every stage of sintering, analyzes of sintering cycle coupled with simultaneous atmosphere monitoring and further analysis of sintered specimens using number of advanced spectroscopy and thermoanalytic techniques. Various phenomena, connected with manganese evaporation and reduction/oxidation behaviour of manganese alloyed sintered steels were theoretically evaluated and tested experimentally applying interrupted sintering experiments, when specimens where sampled at different stages of the sintering cycle for extensive study by HR SEM+EDX, XPS, TG+MS etc. Thermodynamic calculations enabled to determine a required sintering atmosphere composition (maximal permitted partial pressures of active gases $CO/CO_2/H_2O$) for preventing of Mn alloyed steels prepared by different alloying mode from oxidation during every stage of sintering. The results were verified by continual monitoring of $CO/CO_2/H_2O$ profiles in sintering atmosphere and further analysis of sintered specimens.

2. Thermodynamic evaluation of Mn-containing PM steels

2.1 Manganese sublimation

The first effect that has to be considered when talking about manganese containing steels is manganese sublimation. High amount of manganese vapour at elevated temperatures determines number of interrelated processes such as manganese distribution through the pore system, manganese loss and its oxidation through reaction with oxygen and water vapour in the local atmosphere. Intensive manganese sublimation starts already at around 700°C when its partial pressure reaches about 10^{-3} Pa, see Fig.1. Further temperature increasing leads to exponential increasing of manganese partial pressure, that at conventional sintering temperature of 1120°C reaches about 20 Pa.

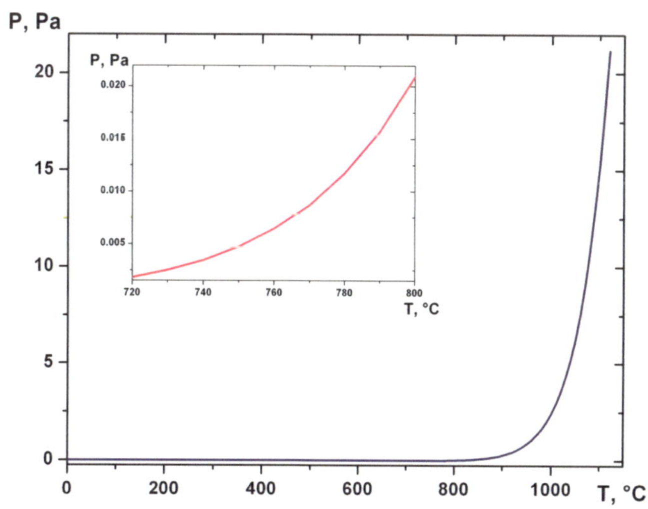

Fig. 1. Equilibrium partial pressure of Mn_{gas} with temperature, ***HSC Chemistry 6.1***

2.2 Oxidation of manganese vapour

It is important to emphasize that at such low temperatures partial pressures of active gases inside pore system has to be incredibly low in order to prevent Mn vapour from oxidation, as it is evident from Ellingham-Richardsson diagram, see Fig.2. Dew-point below -95°C at 700°C is required in typically used in industry $10\%H_2/90\%N_2$ sintering atmosphere to prevent Mn vapour from oxidation that is impossible even in laboratory conditions in powder compact. Phase stability diagram in Mn-O-H and Mn-O-C systems, see Fig.3, indicate that even at typically used in industry sintering temperature of 1120°C in nitrogen/hydrogen atmosphere in presence of admixed graphite oxygen partial pressure below 10^{-21} bar is required to as minimum prevent manganese from oxidation. Diagrams in Fig.3 also indicate that at applied during sintering conditions MnO is the most stable oxide and during all further thermodynamic calculation was considered as only one manganese oxide presented in the system.

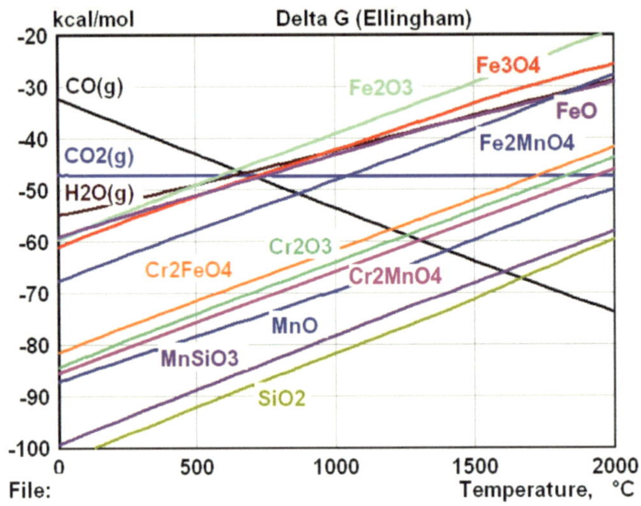

Fig. 2. Ellingham-Richardson diagram indicating stability of oxides

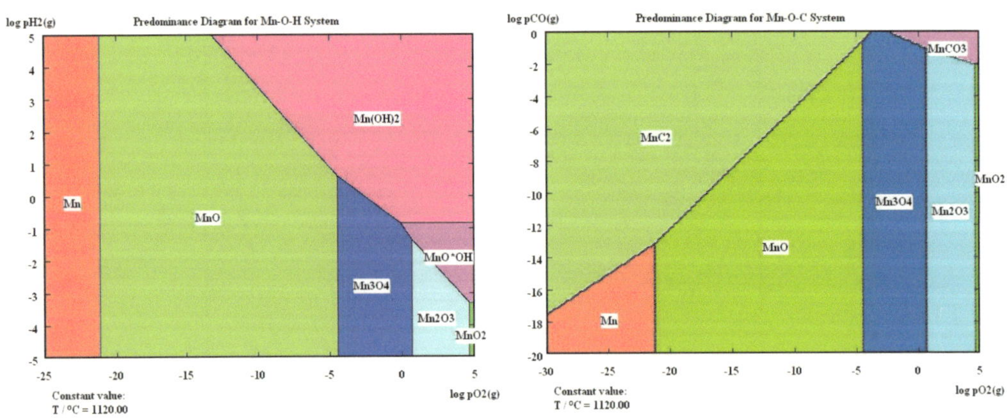

Fig. 3. Mn-O-H (left) and Mn-O-C (right) phase stability diagrams, ***HSC Chemistry 6.1***

Produced at higher temperature manganese vapour reacts with the oxygen in the processing atmosphere according to the following reaction:

$$2Mn_{(gas)}+O_2=2MnO \quad (1)$$

Depending on the oxygen partial pressure in the surrounding manganese source specimen 'microclimate', the reaction between the $Mn_{(gas)}$ and oxygen occurs on different distance. If in the pore system, surrounding manganese source, oxygen partial pressure is higher than equilibrium partial pressure at defined temperature, that is the case during heating stage, manganese vapour will be oxidised close to the manganese carrier particle and manganese oxide will condense on the surface of the surrounding base powder particles. If local microclimate is pure enough, oxidation of manganese vapour take place at larger distances from manganese source, that further purify 'microclimate' conditions and allows transport of manganese through the gas phase in the range of this microclimate, its condensation and further dissolution into base iron powder particles. This is the mechanism of 'self-cleaning' effect described by Šalak (Šalak, 1980), however it is important to note that it leads to desirable 'microclimate' purification preferentially in dry atmospheres and high temperatures, when oxygen partial pressure is slightly higher than equilibrium one. When considering strong oxidising conditions, especially at low temperatures when manganese starts to sublimate and have still rather low partial pressure and oxygen partial pressure at the same time is much higher than equilibrium one, all the available $Mn_{(gas)}$ in the system is oxidised that does not lead to considerable microclimate purification. Extensive manganese oxidation at higher oxygen partial pressure was experimentally observed by *(Šalak, 1980)* that include the formation of a greenish MnO film on the surface of compacts. Another fact that is important to underline is that at the intermediate temperatures (600-900°C), when manganese actually starts to sublimate, local 'microclimate' conditions inside the powder compact are much worse than in the processing atmosphere applied because considerable amount of oxygen is released by the reduction of the surface iron oxide from the base powder. Oxygen, produced by the reduction of surface iron oxide, can be released in different forms depending on sintering technique applied (vacuum or atmosphere sintering, utilizing inert or hydrogen containing atmosphere). When considering vacuum sintering dissociation of iron oxide is one of the dominant mechanism (when carbon is not present or inactive at low temperatures) and oxygen is released. However, most typically manganese PM steels are sintered in hydrogen-containing atmospheres, where due to reduction of iron oxide by hydrogen water vapour is produced. Importance of the reaction of manganese vapour with water vapour was emphasized by (Cias et al., 1999). This reaction results in the formation of H_2 in the microclimate through the reaction:

$$Mn_{(gas)}+H_2O_{(vapour)}=MnO+H_2 \quad (2)$$

and also leads to further atmosphere purification, but, at the same time - considerable manganese vapour oxidation. Nevertheless, there is one more mechanism that leads to more considerable manganese vapour oxidation, especially at inter-mediate temperatures. As it was shown by (Danninger et al., 2002) for a number of iron and alloyed PM steels, and by (Hryha et al., 2010-a) for a manganese prealloyed PM steels, above ~600°C, depending on alloying element type and content, carbothermal reduction of iron surface oxide layer starts, leading to production of carbon monoxide. This carbon monoxide is also oxidising agent for manganese vapour at low temperatures at observed high partial pressures, registered by (Hryha, 2007), according to reaction:

$$Mn_{(gas)}+CO=MnO+C \quad (3)$$

According to this reaction from *HSC Chemistry 6.1* database as minimum to prevent manganese vapour from oxidation, carbon monoxide partial pressure has to be below ~$6.3*10^{-7}$ (bar) at 700°C. However, during atmosphere monitoring of sintering process for manganese containing PM steels (Hryha, 2007) CO-content in the sintering atmosphere was in the range from 10^{-4} to 10^{-3} (bar), clearly indicating oxidation of Mn-vapour. It is important to emphasize that situation inside the pores is even more oxidizing and if sintering is performed with bad processing atmosphere control (low purity/flow), formation of carbon dioxide at inter-mediate temperatures is possible creating even more harmful oxidising conditions. This is the reason why 'self-cleaning' effect does not actually work at inter-mediate temperatures.

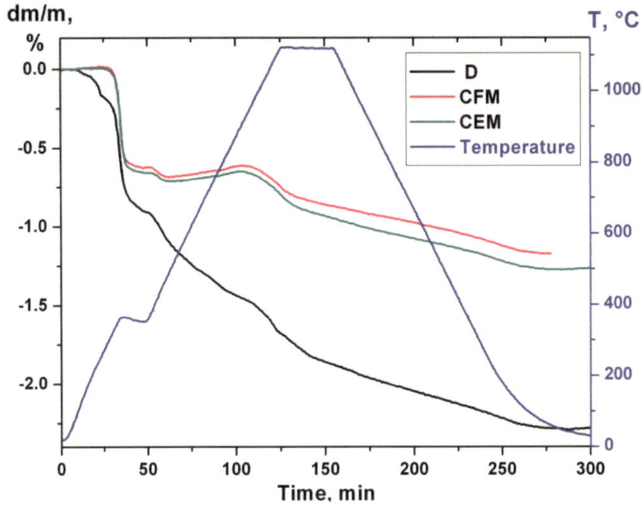

Fig. 4. Thermogravimetry graphs for admixed with electrolytic manganese (***CEM***) and ferromanganese (***CFM)*** materials in comparison with fully prealloyed material (***D)*** of the same composition Fe-0.8Mn-0.5C

Intensive manganese oxidation at medium temperatures is evident even on thermal analysis (Hryha, 2007), especially when comparing with the material based on fully prealloyed powder. The mass-change curves are very similar for both admixed specimens, see Fig.4, but huge difference in comparison with the pre-alloyed material is evident. First high mass loss is observed during heating up to 350°C, including the de-waxing holding, and is connected preferably with lubricant removal and desorption of the physically bonded water. During further heating up to 500°C higher mass loss was detected for prealloyed material - reduction of the iron oxide layer on the powder surface by the hydrogen. The much lower mass-loss in the case of admixed specimens points at their potential oxidation. Temperature increasing up to ~900°C results in increase of the samples mass for admixed specimens - oxidation of manganese carrier particles/manganese vapour as high mass-loss was observed in the case of prealloyed material. After ~920°C high mass-losses were observed for all three materials connected with the carbothermal reduction of the more stable surface oxides and internal oxides for all three materials.

As it was mentioned above, the product of this reaction is fine dispersed MnO that is condensed on the free powder surfaces and hampers the development of interparticle connections. To avoid the formation of oxide networks, Cias (Cias et al., 2003) proposed a method employing a semiclosed container to create an active "microclimate" around and within the Fe-Mn-C compacts. Another alternative is sintering in a hydrogen atmosphere with low enough dew point dictated by the Ellingham-Richardson diagram. As it was emphasized above, in the hydrogen-rich sintering atmosphere it is important to maintain both high atmosphere purity and adequate gas flow to minimize content of water vapour, formed by surface iron oxide reduction. From the Ellingham-Richardson diagram, Fig.2, it is also evident that temperature increasing leads to improved thermodynamic conditions that allows to prevent manganese vapour from intensive oxidation and application of high sintering temperatures (1200-1250°C) allows successful sintering of admixed manganese steels as even reduction of manganese oxides is possible in hydrogen-containing atmospheres in modern sintering furnaces.

Presented in Fig.2 Ellingham-Richardson diagram also emphasize presence in the system and high stability of mixed oxides of spinel type, that are important to take into consideration as iron/steel powders always contain trace amounts of Cr and Si, and formation at high temperatures and presence of these oxides were experimentally confirmed by (Hryha et al., 2010-a). Especially it is important to emphasize presence of Cr-Mn and Mn-Si oxides that due to the lack of experimental information and absence of thermodynamic data were disregarded during previous studies of oxidation/reduction behaviour of Mn and Cr-Mn PM steels. Based on the presented in Fig.2 diagram and the discussion above, the stability of oxides presented in Cr-Mn alloyed PM materials increases in the line:

$$Fe_2O_3 \rightarrow FeO \rightarrow Fe_2MnO_4 \rightarrow Cr_2FeO_4 \rightarrow Cr_2O_3 \rightarrow MnCr_2O_4 \rightarrow MnO \rightarrow MnSiO_3 \rightarrow SiO_2 \quad (4)$$

Detailed study of the thermodynamics of reduction of the presented mixed oxides in the case of powders pre-alloyed with Cr and Mn is presented elsewhere (*Hryha et al., 2010)*.

2.3 Thermodynamic evaluation of sintering requirements

Process conditions for the successful sintering can be calculated based on simple algorithms for reaction energies calculations (Gaskell, 2003). Mitchell (Mitchell & Cias, 2004) performed such calculations for some chromium and manganese containing PM steels. Algorithm and theoretical evaluation of the required atmosphere composition at every sintering stage for manganese contained PM steels in more details are also presented by (Hryha et al., 2010-b).

Sintering atmosphere composition required was evaluated based on standard free-energy calculations using a thermodynamic software package *Outotech HSC Chemistry 6.1*. Calculations are performed for the oxidation/reduction reactions for the oxide that is in direct contact with atmosphere - condensed manganese vapour located on the base powder particles surface. Carbon during calculations was considered as pure non-dissolved graphite admixed to the base powder. According to these considerations, the activities of the surface oxides and carbon can be taken as unity. As the most stable oxide in the system is MnO, see Figs.2 and 3, thermodynamic requirements, calculated for the reduction of this oxide, will meet the thermodynamic requirements for less stable iron and iron-manganese oxides.

The algorithm of the calculations is based on evaluation of maximum pressures of active gases, which can be tolerated in the system for as minimum avoiding oxidation of manganese, at the sintering temperature. The calculations are based on evaluation of free energy for each reaction at the defined sintering temperature based on $\Delta G_T^0 = \Delta H - T\Delta S$. Using relationship $\Delta G_T^0 = -RT\ln K_p$ equilibrium constant for each reaction can be calculated and partial pressures of active gases can be easily evaluated. Four basic chemical reactions were taken into consideration:

$$MnO = Mn + 0.5O_2 \quad (5)$$

$$MnO+C=Mn+CO \tag{6}$$

$$C+CO_2=2CO \tag{7}$$

$$2CO+O_2=2CO_2 \tag{8}$$

$$C+H_2O=CO+H_2 \tag{9}$$

These reactions allow establish equilibrium partial pressures of the most important active gases in the system (oxygen, water vapour, carbon monoxide and carbon dioxide). Based on these calculations it was shown (Hryha et al., 2010-b) that in the case of sintering in nitrogen/10% hydrogen atmosphere at 1120°C for MnO, Mn and C to be in equilibrium with a gaseous atmosphere, this atmosphere must be of composition:

$$\mathbf{P(CO)= 1.674*10^{-2}\ bar}$$

$$\mathbf{P(CO_2)=5.16*10^{-7}\ bar}$$

$$\mathbf{P(O_2)=7.18*10^{-22}\ bar}$$

$$\mathbf{P(H_2O)=6.75*10^{-6}\ (bar)}$$

Water partial pressure $P(H_2O)= 6.75*10^{-6}$ bar corresponds to a dew-point near -63°C. This points on the fact that a dew-point lower than -63°C is required for proper sintering of manganese containing steels in the $90\%N_2/10\%H_2$ atmosphere at 1120°C.

Sintering temperature increasing to 1200°C put less strict requirements on the sintering atmosphere purity:

$$\mathbf{P(CO)=6.09\ *10^{-2}\ bar}$$

$$\mathbf{P(CO_2)=3.13*10^{-6}\ bar}$$

$$\mathbf{P(O_2)=2.79*10^{-20}\ bar}$$

$$\mathbf{P(H_2O)=1.3*10^{-5}\ bar\ (DP\approx -58°C)}$$

easily achievable in a modern sintering furnaces. It is important to note that for inert atmospheres or atmospheres with hydrogen content other than 10% equilibrium pressure of water vapour has to be re-evaluated.

It is important to state that the pre-alloyed powder do not have such strict requirements to the atmosphere purity due to the much lower manganese activity in the prealloyed state. This means that the activity of manganese in the case of the prealloyed powder can be taken equal to its content according to Henry's law (Gaskell, 2003) that will decrease maximum allowed pressures of active gases for about one-two orders. However, it has to be taken into account that presented calculations are still valid when considering necessity of full reduction of manganese particulate oxides on the surface of prealloyed powder, (Hryha et al., 2010-b).

2.4 Liquid phase formation in the Fe-Mn-C system

Šalak (Šalak 1980-a and Šalak 1980-b) experimentally identified one more specific feature of manganese alloying when using high-carbon ferromanganese that is concerned with presence of liquid manganese alloy phase that occurs at certain circumstances at sintering temperatures. Many further detailed observation of fracture surface of Fe-Mn-C PM materials (Hryha, 2007, Dudrova et al., 2010) confirms presence of a transient liquid phase in the areas of prior ferromanganese particle surfaces. Such liquid phase formation was not registered in Fe-Mn-C steel when high-purity electrolytic manages was used as the manganese source (Hryha, 2007). Detailed thermodynamic analysis of low melting point alloys by Gomez-Acebo (Gomez-Acebo et al., 2003) in the case of Fe-Mn-C system using the Thermo-Calc software reveals lowest melting point for the alloy composition Fe-30.3Mn-4.03C at 1077°C. The thermodynamic calculations of the equilibrium isothermal section for the ternary Fe-Mn-C alloy system at 1120°C using Thermo-Calc together with SSOL2 thermodynamic database, Fig. 5, displays a rather extended liquid phase region close to the sintering temperature. Clearly, intensive manganese evaporation during heating together with iron and carbon diffusion into ferromanganese residue can lead to liquid phase formation at around 1120°C for high-carbon and even medium carbon ferromanganese.

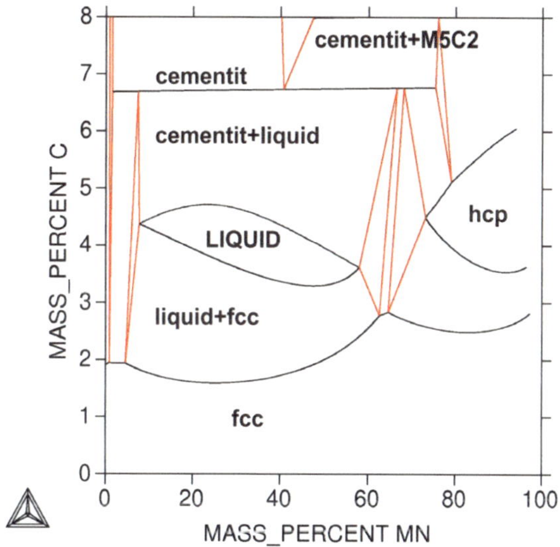

Fig. 5. Equilibrium isothermal section for the Fe-Mn-C alloy system at 1120°C (Thermo-Calc, SSOL2 database)

3. Microstructure development and composition

Tracing of microstructure development during sintering process (heating stage and different dwelling times) proofs itself to be a powerful tool when studying powder systems (Hryha 2007, Dudrova et al., 2010). It allows to trace reduction of surface iron oxide layer, carbon and manganese dissolution and so microstructure development, oxides distribution etc. To be able to follow processes carefully during heating stage, specimens were slowly heated (10 °C·min^{-1}) up to the sintering temperature of 1120/1200°C, and sampled at defined temperatures, as it is shown in Fig.6, and cooled at the cooling rate of 10 or 50 K·min^{-1} to room temperature. Due to the lubricant presence, the specimens were additionally held at ~350°C for de-lubrication. First two sampling temperatures were chosen below the $\alpha \rightarrow \gamma$ transition (onset on dilatometry curve) and next one above transition temperature (offset on dilatometry curve) measured for the studied alloys, (Hryha, 2007).

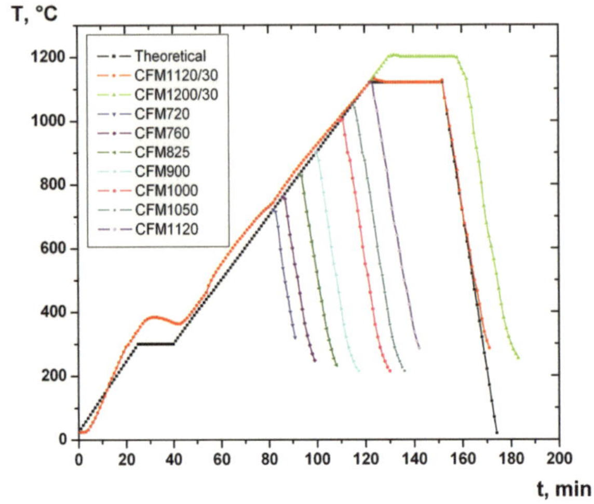

Fig. 6. Thermo-profile during interrupted sintering trials

Microstructure development processes differs considerably for pre-mixed powder systems, where manganese was added as high-purity electrolytic manganese powder and medium carbon ferromanganese powder, compared with fully prealloyed powder (Hryha, 2007; Hryha et al., 2010-b). As an example, Fig.7 presents microstructure development and composition in the case of admixed with ferromanganese powder system, marked as ***CFM***. Heterogeneous ferritic-pearlitic microstructure is characteristic for all sampled at different sintering stages specimens, see Fig. 7. Some coarse pearlite on the grain boundaries near the edges of prior particles extending up to 10 µm was registered after the interrupted sintering at 825°C and its fraction is about 10% at 900°C. Austenitic rims around Mn carrier particles were registered and were shown to have a thickness of up to 1 µm after the 760°C run, ~3 µm after the 825°C run, up to ~4 µm after the 900°C run and about 5 µm for the sampled at 1000°C specimen, indicating rather pronounced Mn dissolution in the iron matrix with increasing temperature. Ferromanganese particles can be observed in material processed up to the sintering temperature. Nevertheless at the beginning of the sinter-holding manganese carriers are shown to occur only as large pores with pronounced "sponge" structure, filled by contaminants. This indicates that manganese is already distributed throughout the surrounding pore system. The fraction of fine pearlite reaches about 20%. The complicated sequence of microstructure constituents around the Mn-residues (austenite that changes to martensite and mixture of upper and lower bainite and pearlite at larger distances from the FeMn particles) was firstly observed at 1120°C and shown to extend up to 10-12 µm from the Mn-carrier particles. After sintering for 30 min at 1120°C, pronounced Mn dissolution in the Fe matrix is evident resulting in the extension of the austenite-martensite-bainite mixture up to 20 µm followed by a more extended pearlite region, while the core of the prior iron particles is still fully ferritic. The pronounced difference between the slow and fast cooled specimens, sintered at 1120°C, is evident in lower portion of pearlite in the slowly cooled one. This is also a result of lower as-sintered carbon content in slowly cooled specimen related to the higher carbon loss due to more oxidising conditions during the slow cooling. The oxidation of the slowly cooled specimen was clearly seen on a non-etched microstructure. The specimens sintered at higher temperature (1200°C) show more homogeneous austenite-martensite mixture in the areas adjoining to residue. The bainitic areas pass into pearlitic rims around ferromanganese residue. Due to the large size of admixed ferromanganese particles even after sintering rather large pores (up to 20 µm) filled with contaminants were observed. The same microstructure heterogeneity as for the admixed with ferromanganese alloy is characteristic for the admixed with electrolytic manganese alloy for all temperatures. The microstructure development includes the same features as the admixed with the ferromanganese system, however more extended dissolution of manganese during heating stage was registered for admixed with electrolytic manganese system, showing austenitic rims nearly twice as thick around electrolytic manganese in comparison with FeMn, see (Hryha et al., 2010-b).

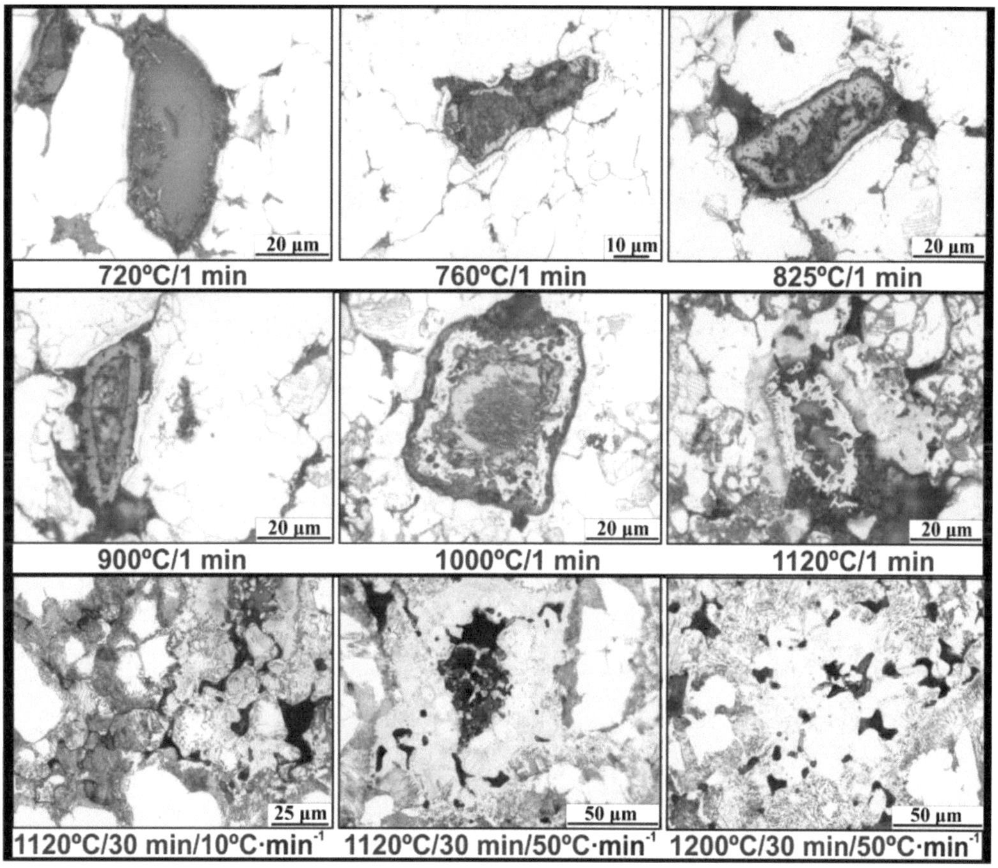

Fig. 7. Microstructure development during sintering of admixed with ferromanganese powder system Fe-0.8Mn-0.5C

The development of the microstructure with increasing temperature and its final state in the case of the pre-alloyed material differs significantly from the admixed systems with the same chemical composition (Fe-0.8Mn-0.5C) and processing conditions (Hryha et al., 2009-a and Hryha et al., 2010-b). Higher purity and presence of some pearlite close to the edges of prior particles was observed even for the first sampled specimen at 740°C, indicating much faster carbon transport in this system in comparison with the admixed one, see Fig.8. After the 810°C run, the pearlite extends up to 10 µm; the comparable amount and distribution of the pearlite for admixed systems was observed only after the 900°C run. After the 900°C-processing of the pre-alloyed material, fine pearlite is evenly distributed in the iron matrix and its fraction is ~ 20%; such pronounced carbon distribution was observed for the admixed system only after processing close to the sintering temperature. Further temperature increasing leads to pearlite portion increasing to 30-35% after the 1000°C run and to about 40% after the 1120°C run that are comparable with the results regarding the as-sintered microstructure of admixed alloy. Then, sintering of the pre-alloyed material at 1120°C for 30 min results in evenly distributed ferrite-pearlite mixture where fine pearlite is prevailing and its portion is up to 55-60%. Higher oxidation of the slow-cooled specimen after sintering at 1120°C was registered as well, with a little bit lower portion of pearlite being coarser than in the case of fast cooled specimen, see Fig.8. Sintering at 1200°C gave microstructure comparable to sintered at 1120°C compact. Distinguishing feature of the prealloyed material is more rounded, smaller and pure pores in comparison with the admixed one.

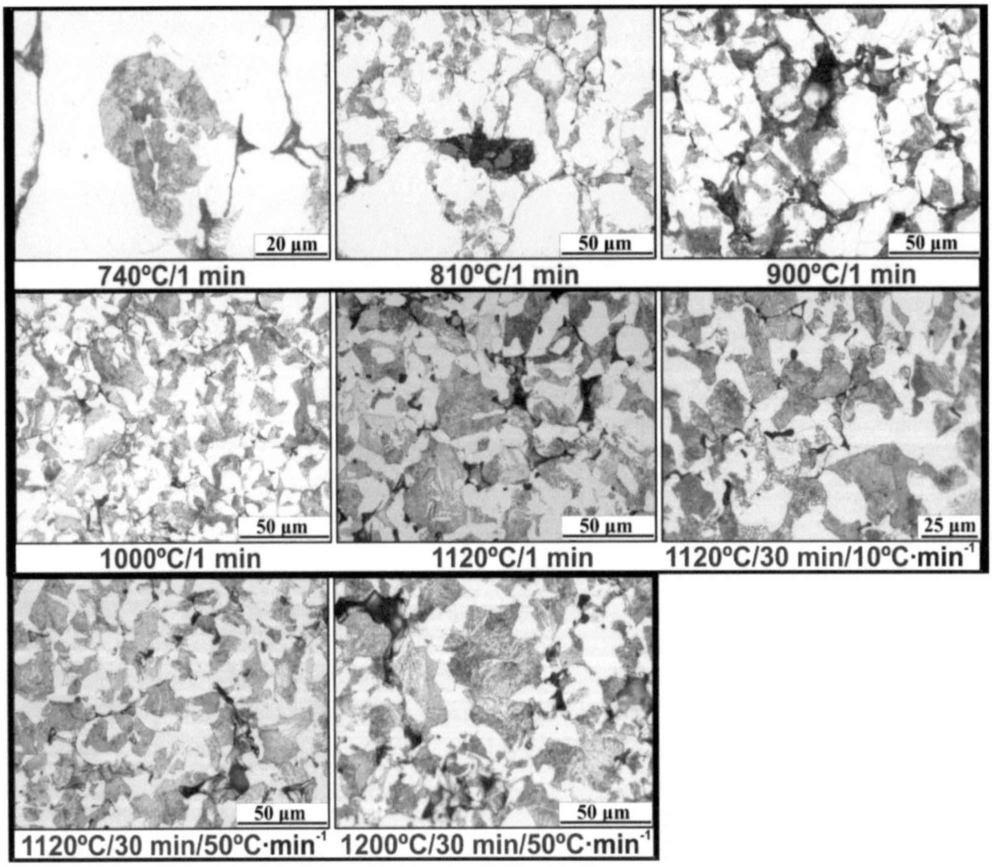

Fig. 8. Microstructure development during sintering of prealloyed powder system Fe-0.8Mn-0.5C

Additionally effect of manganese content on microstructure development was studied on prealloyed powder with higher manganese content of 1.8 wt.% (Hryha et al., 2009-a). The microstructure of this material indicates appearance of some pearlite near the edges of the prior particles even after the 710°C run, see Fig.9. Temperature increasing leads to higher pearlite portion, being about 20% after the 790°C run, and after 900°C run a structure consists of evenly distributed ferrite-pearlite with pearlite fraction about 35%. After the 1000°C run, an almost fully fine-pearlitic microstructure was obtained. This was not observed for any of the materials mentioned. The pearlite had a very fine lamellae structure distinguishable only at high resolution. After reaching of sintering temperature (1120°C), a bainitic microstructure starts to occur and after sintering for 30 min at this temperature the bainite fraction is 10-15 wt.%. The microstructure of the specimens sintered at 1200°C consists of evenly distributed and very fine ferrite-carbide mixture, where the bainitic microstructure is dominant (<50%). Distinguishing features of this alloy are higher pearlite fraction at all temperatures as well as its much finer structure than observed for the prealloyed with 0.8 wt.% Mn material at all temperatures.

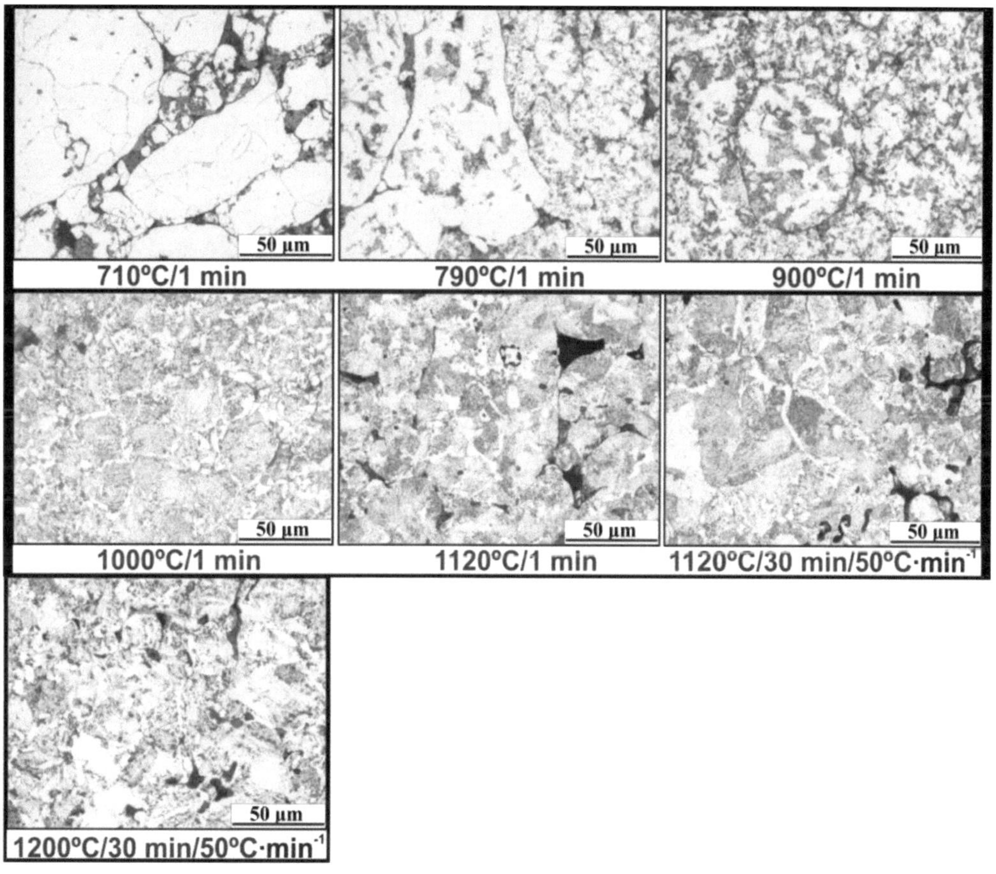

Fig. 9. Microstructure development during sintering of prealloyed powder system Fe-1.8Mn-0.5C

4. Manganese distribution

When comparing microstructure development between the admixed with the ferromanganese and electrolytic manganese systems, detected thickness of the Mn-enriched areas was nearly twice as thick around electrolytic manganese in comparison with the FeMn. Additionally, when comparing measured values during heating stage (increasing from 2-4 μm at 750°C up to

~8 μm at 1000°C) with the calculated depth of Mn solution in the iron matrix in this range of temperatures based on literature data, (Dudrova et al., 2010), see Table 1, it is clearly evident that detected values are as minimum one order of magnitude larger.

The fact that manganese dissolution depth calculated based on the only diffusion mechanism is clearly underestimated was also confirmed by careful measurements of the manganese concentration profiles using energy and wave dispersive X-ray analyses (EDX and WDX, respectively), (Hryha, 2007). Presented in Fig.10 SEM micrograph of ferromanganese residue in the Fe-0.8Mn-0.5C material, sintered at 1120°C for 30 min in nitrogen/hydrogen atmosphere, with marked line of EDX analysis, show that FeMn residue is a complex contaminations, formed by MnS and complex Fe-Mn-Si-O oxides. Line EDX analysis shows gradient in manganese concentration in the areas around FeMn residue. As line EDX analysis gives only relative information about elements distribution, quantitative analysis of Mn-concentration profile was performed by WDX analysis in the area around FeMn residue, see Fig.11. Mn concentration profiles clearly shows manganese enriched area up to ~30 μm from the iron particle edge (Hryha&Dudrova, 2007; Hryha, 2007).

Composition, %Mn/%C	T/t, °C/min	D_{Mn}, $cm^2\ s^{-1}$	Reference*	Mn diffusion depths, μm
Mn in α-Fe, 0/0	770/3	7.0481×10^{-15}	*Nohara&Hirano, 1971*	0.00113
Mn in α-Fe, 20/0	770/3	3.8965×10^{-14}	*Wells&Mehl, 1941*	0.02648
Mn in γ-Fe	1040/3	6.2110×10^{-12}	*Nohara&Hirano, 1971*	0.3344
Mn in γ-Fe, 10/0	1040/3	6.0604×10^{-12}	*Wells&Mehl, 1941*	0.3303
Mn in γ-Fe	1080/3	1.2619×10^{-11}	*Nohara&Hirano, 1971*	0.4766
Mn in γ-Fe, 10/0	1080/3	1.2808×10^{-11}	*Wells&Mehl, 1941*	0.4663
Mn in γ-Fe	1170/3	5.3847×10^{-11}	*Nohara&Hirano, 1971*	0.9845
Mn in γ-Fe, 10/0	1170/3	5.9270×10^{-11}	*Wells&Mehl, 1941*	1.0329
Mn in γ-Fe	1220/3	1.1179×10^{-10}	*Nohara&Hirano, 1971*	1.4185
Mn in γ-Fe, 10/0	1220/3	1.6060×10^{-10}	*Wells&Mehl, 1941*	1.7002
Mn in γ-Fe	1220/30	1.1179×10^{-10}	*Nohara&Hirano, 1971*	4.4858
Mn in γ-Fe, 10/0	1220/30	1.6060×10^{-10}	*Wells&Mehl, 1941*	5.3766

Table 1. Literature Mn diffusion coefficients D_{Mn} and calculated Mn diffusion depths x in iron lattice for temperature range 770–1220°C ; $x=(D_{Mn}\times t)^{1/2}$

Results show the manganese content in the range 2-4 at.% at the distances of about 20-30 μm from the manganese carrier residue. It has to be emphasized that such a large deviation of measured Mn content from the place to place is caused by two main reasons: size of the initial ferromanganese particle (amount of the capable to diffusion manganese) and geometry of the iron particles cut where analysis is performed. There is a high probability of existing pore close under analyzed area, meaning contribution from manganese dissolution from the manganese layer beneath (caused by manganese transport in the adjoining to FeMn particle pores by manganese vapor phase). This also describes such a 'wavy' character of measured manganese profiles, see Fig.11. This means that large number of concentration profiles has to be measured to obtain statistically correct values for evaluation of manganese dissolution depth in Fe-Mn PM steels.

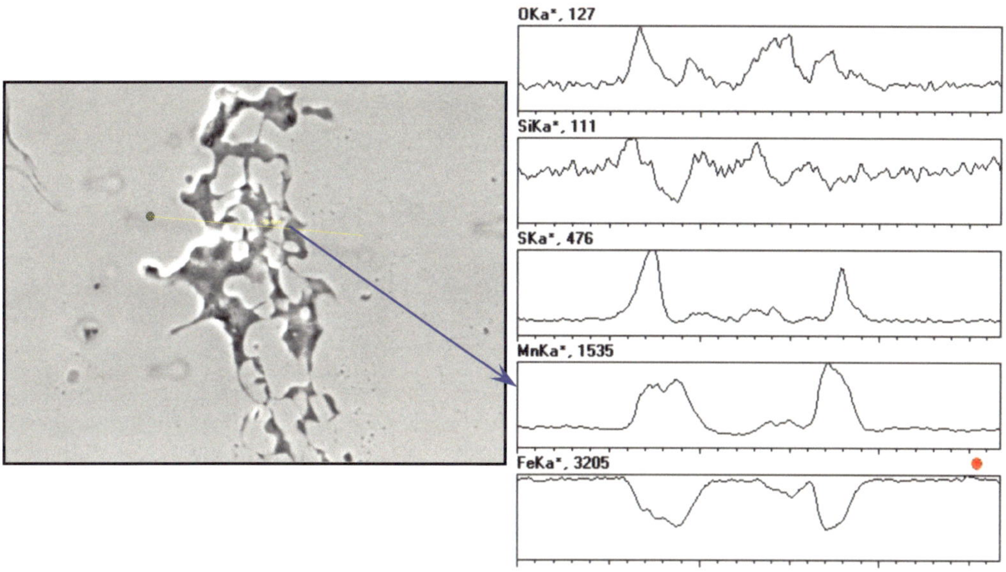

Fig. 10. SEI image of ferromanganese residue for Fe-0.8Mn-0.5C specimen, sintered at 1120°C for 30 min, with marked line of Energy Dispersive X-ray analysis and results of line analysis of O, Si, S, Mn and Fe distribution

Nevertheless, it is no doubt that the observed depth of manganese alloyed areas around manganese carrier particles is much larger than volume diffusion could account for. Even if considering that manganese isoconcentration lines are parallel to the iron particle boundaries, that can point to volume diffusion as the dominant mechanism, its just can be assumed inappropriateness of applying of presented in literature volume diffusion coefficient to the powder Fe-Mn system. However, diffusion profile calculated based on only volume diffusion gives very steep concentration profile (Navara, 1980), contrary to the smooth concentration gradient observed. When discussing manganese dissolution it is important to remember Mn-transport through the gas phase (Mn-vapour). Comparable effect of improved diffusivity of several orders of magnitude in the presence of gas phase was firstly observed by (Hillert&Purdy, 1977), in the case of iron-zinc materials in the atmosphere of zinc. They clearly showed that much higher diffusivity observed is a result of combination of a grain boundary diffusion and chemically induced grain boundary migration (DIGM). Another important conclusion by (Hillert&Purdy, 1977), is that grain boundary diffusion is 2-4 orders of magnitude enhanced by the movement of the grain boundary, that together leads to considerable enhancement of the dissolution depth. The hypothesis about application of alloying mechanism with the presence of the gas phase, as in the case of iron-zinc system, for iron-manganese system, was firstly proposed by Navara (Navara, 1980) and further developed by Šalak (Šalak, 1989).

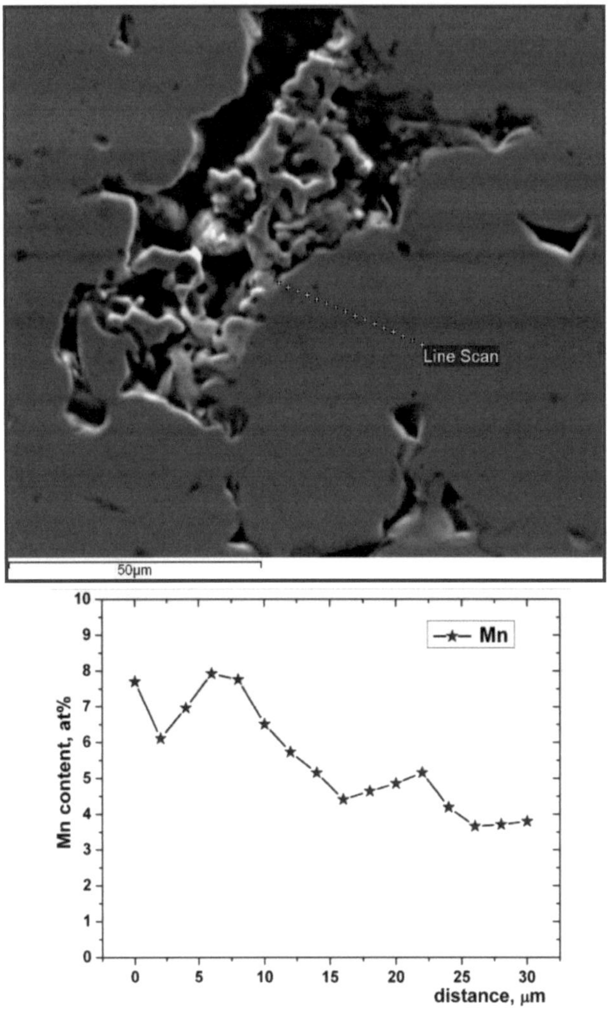

Fig. 11. SEI image of the area around ferromanganese residue for Fe-0.8Mn-0.5C specimen, sintered at 1120 °C for 30 min, with marked points of line Wave-length Dispersive X-ray analysis and results of manganese distribution

More detailed study of the iron-zinc system by Chongmo (Chongmo&Hillert, 1981), indicate formation of fine-grained zinc-rich surface layer on the initial stage of alloying. These fine grains reveal fast growth with increasing heat-treatment time and temperature and disappear rather fast. However, such fine-grained structure is much more difficult to observe by conventional metallography in the Fe-Mn powder system due to the difficulties of specimens preparation (porous and very heterogeneous structure). Nevertheless, such a fine-grain structure around manganese carrier residues was confirmed recently by (Hryha, 2007) by careful analysis of the fracture surface around manganese carrier residue using high-resolution scanning electron microscopy. Example of such fine-grained structure around manganese carrier particle is presented in Fig.12. Metallographic observation of the same material do not allow to clearly distinguish fine-grained structure of manganese-enriched area around ferromanganese residue in neither un-etched nor in etched state, see Fig.13. It is important to note that most probably such fine-grained structure appears in this material at much lower temperature, when the prerequisite for its formation - manganese evaporation/condensation (after ~700°C) become more intensive. However, it is possible to observe only at higher temperature (after ~1000°C) on the fracture surface because there are two main prerequisites of its appearance on the fracture surface: grain boundaries of these fine grains has to be weakened (degraded) and inter-particle necks around manganese carrier particles has to be strong enough (stronger than inter-granular cohesion). Unlike in iron-zinc system, this fine grained structure does not disappear with the temperature increasing and is observed even after sintering at low temperature of 1120°C, see Fig.14. Moreover, after sintering at low temperature, especially at unsatisfactory sintering atmosphere purity (dew point <-45°C), this fine-grained structure is even more pronounced and can be even easily distinguished by metallography, see Fig.14. High-temperature sintering at good conditions allows to avoid fine-grained structure around manganese carriers. Degradation of grain boundaries of these fine grains is very harmful to the mechanical properties of the admixed manganese-containing PM steels, as it is described in next section. Metallographic evidence of fine-grained structure around ferromanganese carrier particles after heating of Fe-3Mn-0.5C material to 1040°C was revealed in the recent work of (Dudrova et al., 2010) by using specially designed etchant.

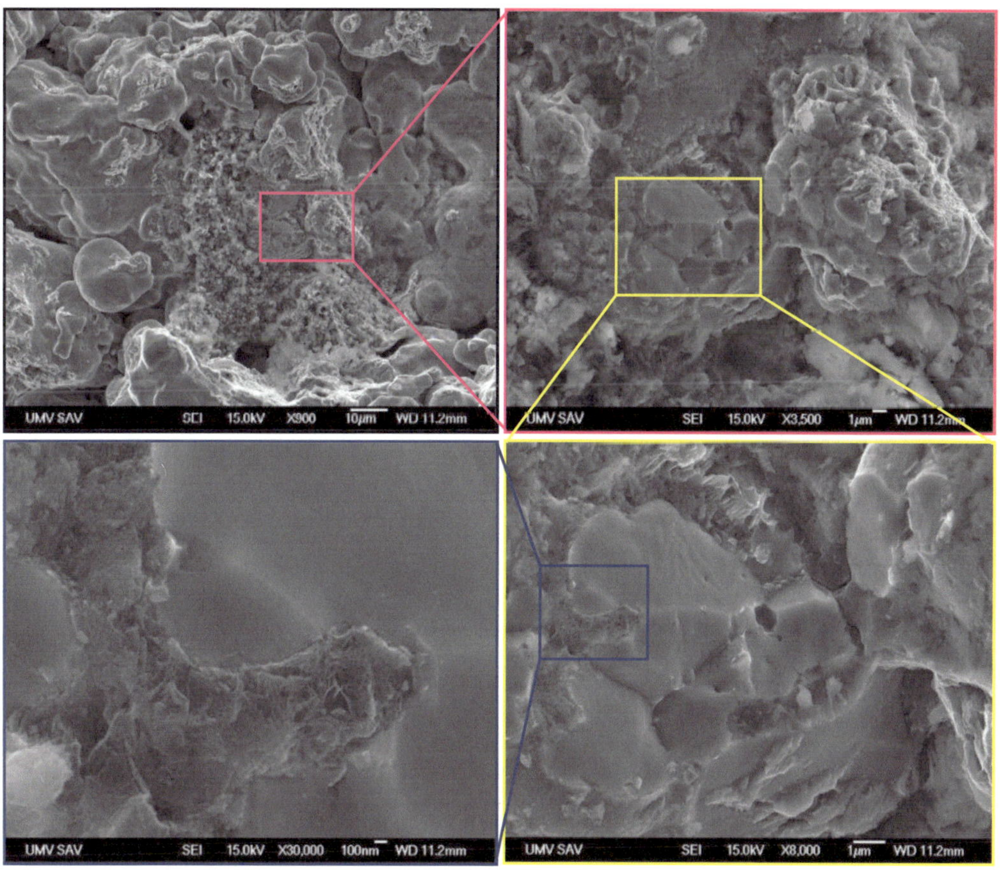

Fig. 12. Fine-grained structure of the manganese-reach surface layer around manganese carrier particle in Fe-0.8Mn-0.5C material, heated up to 1050°C (*Hryha, 2007*)

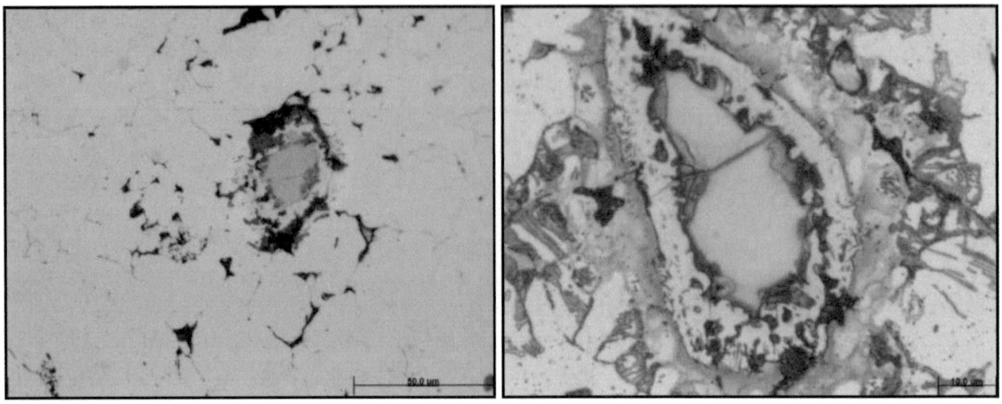

Fig. 13. Microstructure of Fe-0.8Mn-0.5C material, heated up to 1050°C in un-etched (left) and etched (right) state (*Hryha, 2007*)

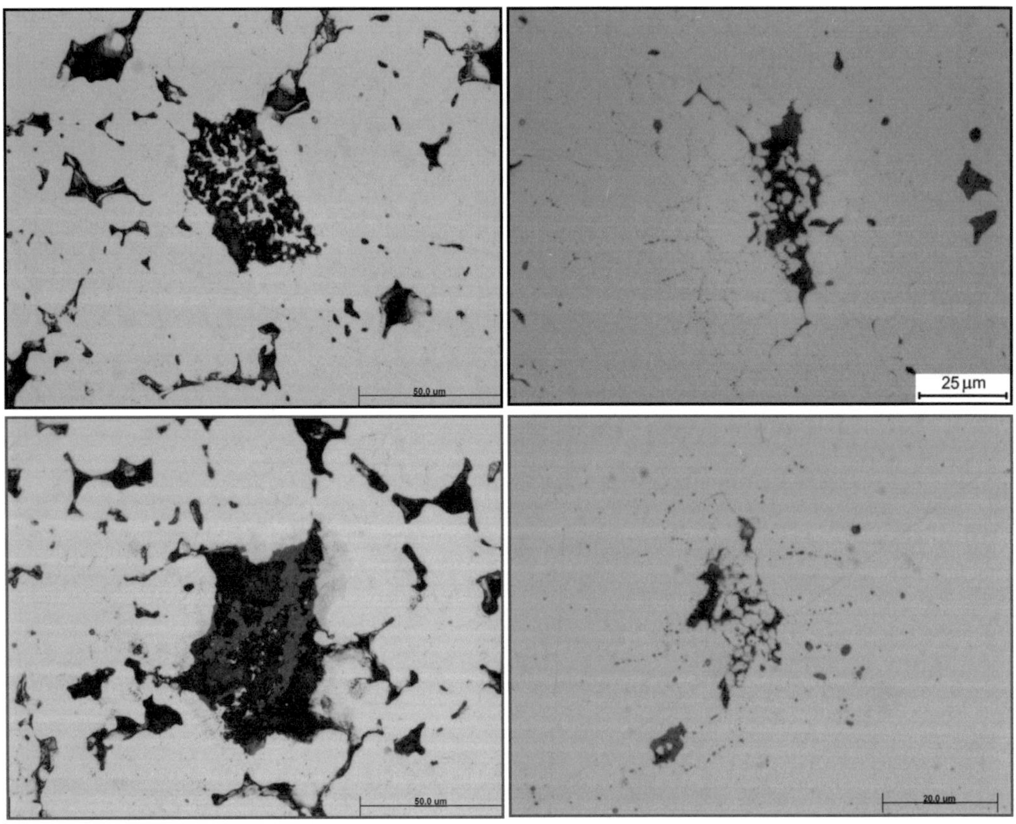

Fig. 14. Microstructure of Fe-0.8Mn-0.5C material, admixed with ferromanganese (upper row) and electrolytic manganese (bottom row), heated up to 1120°C (left) and after sintering at 1120°C for 30 min (right), showing network of 'point' oxides that mark fine grain structure around manganese carrier particles (*Hryha, 2007*)

5. Brittleness of admixed manganese containing PM steels

After about thirty years of an intensive investigation of the admixed manganese containing PM steels, number of inventions concerning adjustment of the sintering process and manganese carrier did not result in wide industrial application of mentioned above materials. The main obstacle for industrial application of admixed manganese PM steels is high brittleness of sintered components. Several hypotheses have been proposed, connected with brittleness caused by microstructure heterogeneity, size of manganese carrier particles, their residue and their oxidation (Dudrova et al., 2004; Dudrova et al., 2005; Danninger et al., 2005; Cias et al., 2003).

5.1 Effect of manganese carrier residues.

Detrimental effect of pores, filled with slag, after manganese source due to its evaporation was emphasized by Dudrova (Dudrova et al., 2004) and Danninger (Danninger et al., 2005). Due to angular shape of such pores with inclusions they act as strong crack-initiation sites. High-temperature sintering leads to rather significant reduction of slag, preferably formed by thermodynamically stable complex oxides, and pore rounding, however it did not result in so considerable mechanical properties improvement as it was expected.

Appearance of electrolytic manganese and ferromanganese residues after heating up to 1120°C, presented in Fig. 15, reveal some differences in the residues appearance and composition. Ferromanganese residue has complex porous shape with the presence of phase on the surface of the pore with smooth surface and numerous cracks, pointing to presence of liquid phase at this temperature. Formation of the liquid phase is thermodynamically probable at this temperature in the wide range of concentration in the Fe-Mn-C system, see Fig.5, taking into account manganese evaporation and diffusion of iron and carbon into manganese carrier particle. Residues of electrolytic manganese particles have a shape of contamination agglomerate, formed by manganese oxide with traces of manganese sulphide. Residues of ferromanganese have more complex chemical composition due to higher amount of silicon oxide in it (Hryha, 2007).

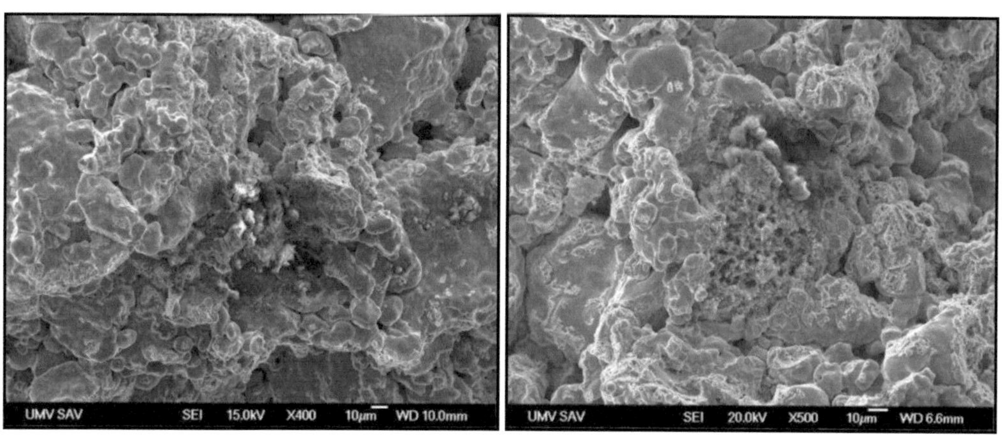

Fig. 15. Appearance of manganese carrier residue after heating to 1120°C in the case of Fe-0.8Mn-0.5C material, admixed with electrolytic manganese (left) and ferromanganese (right) (*Hryha, 2007*)

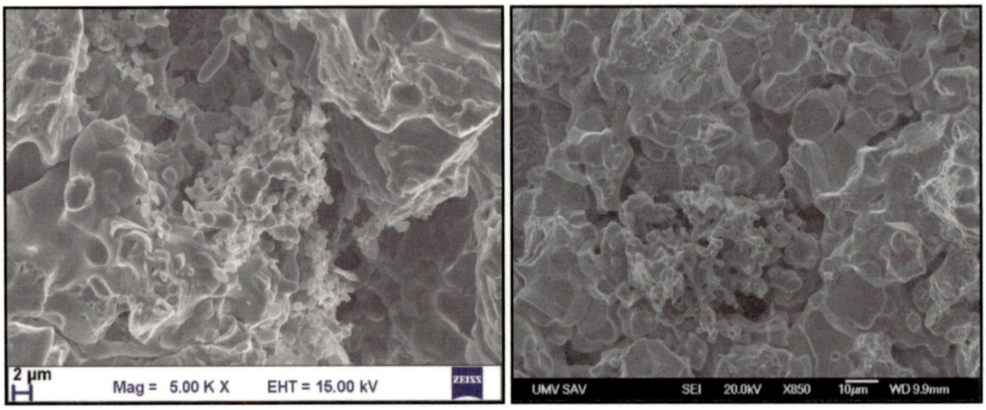

Fig. 16. Appearance of manganese carrier residue after sintering at 1120°C for 30 min in the case of Fe-0.8Mn-0.5C material, admixed with electrolytic manganese (left) and ferromanganese (right)

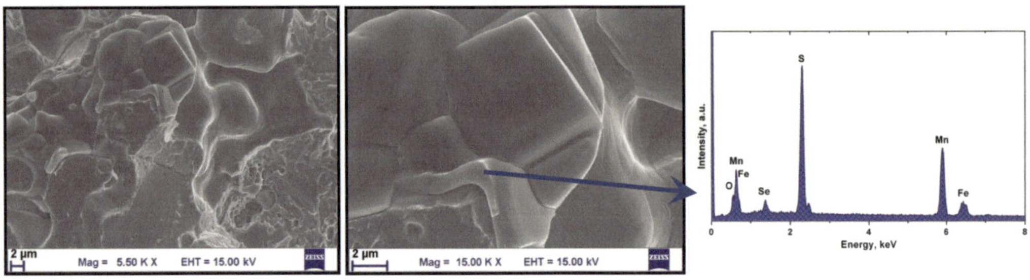

Fig. 17. SEM+EDX analysis of the electrolytic manganese residue after sintering at 1200°C for 30 min on the fracture surface of Fe-0.8Mn-0.5C material

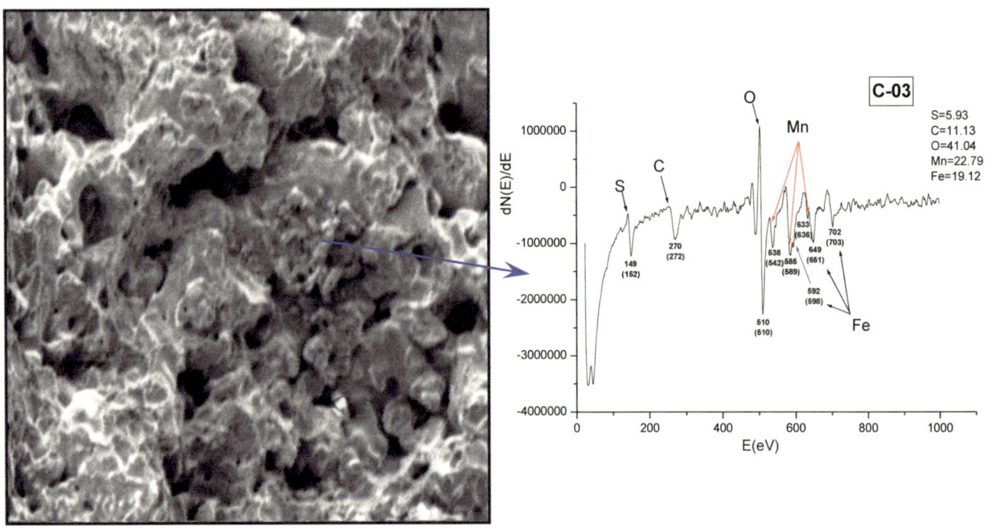

Fig. 18. Auger spectra of the ferromanganese residue after sintering at 1120°C for 30 min on the fracture surface of Fe-0.8Mn-0.5C material

Sintering even at 1120°C for 30 min in good atmosphere purity allows considerable reduction of manganese carrier residues in size and amount of manganese oxide as well as morphology change, see Fig.16. Residue of electrolytic manganese looks more like inclusion agglomerate, whereas ferromanganese residue has specific 'coral-like'-shape. According to the results of SEM+EDX and Auger analyses, see Figs.17 and 18, in both cases residua are composed of manganese oxide with larger amount of manganese sulphide, than it was observed before sinter-holding. However after sintering at high-temperature (1200°C) together with essential decrease in size of the manganese residues the oxygen content decreases as well, while much higher sulphur content is registered. Ferromanganese residue also contain large amount of silicon oxide.

5.2 Inter-granular decohesion brittleness.

Application of high-purity sintering atmospheres combined with high-temperature sintering and utilization of fine manganese carrier powder will considerably minimize detrimental effect of large angular pores on the mechanical properties. However, careful examination of the areas around manganese carrier particles at high magnification reveal more critical defect, presented in the admixed manganese systems. Recent studies of brittleness of admixed with manganese PM steels studied by complex of advanced microscopy and spectroscopy techniques (high-resolution scanning electron microscopy (HR SEM), energy dispersive X-ray analysis (EDX), Auger spectroscopy and X-ray photoelectron spectroscopy (XPS)) clearly indicate that the brittleness of material containing admixed manganese is caused by the weakness of grain boundaries within the base matrix particles around the manganese carrier residuals (Hryha, 2007; Hryha et al., 2008; Hryha et al., 2010). Completely brittle behaviour associated with predominantly inter-granular type of failure caused by the oxide phase at grain boundaries was also assumed in previous works by Dudrova (Dudrova et al., 2004; Dudrova et al., 2005).
Inter-granular type of failure starts to be observed in the admixed materials above ~1000°C, see Fig.12, caused by the presence of very thin manganese oxide layer, formed due to diffusion of condensed manganese from the iron particle surface inside the particles along the grain boundaries and its further oxidation there. However, inter-granular type of failure around manganese carrier residua is much better pronounced after low-temperature sintering, see Fig.19. Large number of fine point inclusions (size <200 nm) can be seen on the inter-granular decohesion facets, that according to SEM+EDX analysis are formed preferably by manganese oxides after sintering at 1120°C, see Fig.20.
These point oxides on the degraded grain boundaries around manganese residuals are evident on the unetched microstructure of admixed materials, see Fig.14, forming on metallographic cut "chains" of point oxides. They develop close to the sintering temperature and are much better pronounced after sintering at 1120°C for 30 min. The networks of point oxides are better pronounced for materials admixed with electrolytic manganese.
Presence of these point inclusions on the grain-boundaries around manganese carrier residues explains why this fine grain structure does not disappear with temperature increasing or after sintering at low temperature, as it can be expected based on behaviour of iron-zinc system with similar dissolution of zinc by DIGM mechanism. Large amount of point oxides hamper grain growth of fine manganese-enriched grains, formed around

manganese carrier source. Sintering at lower atmosphere purity leads to increased amount of inter-granular decohesion facets with higher amount of point oxides on them. This is also reflected on higher microstructure heterogeneity due to the retarded manganese dissolution caused by hampering grain boundary motion by point oxides.

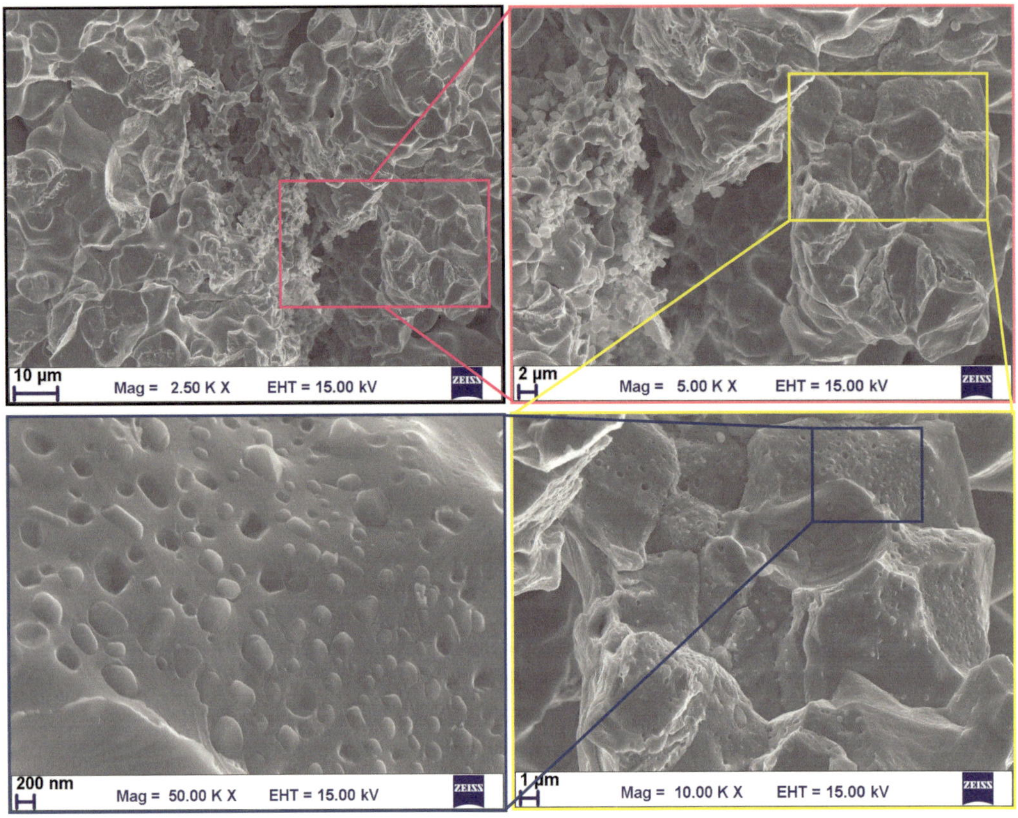

Fig. 19. Inter-granular decohesion facets around electrolytic manganese residue after sintering at 1120°C for 30 min on the fracture surface of Fe-0.8Mn-0.5C material

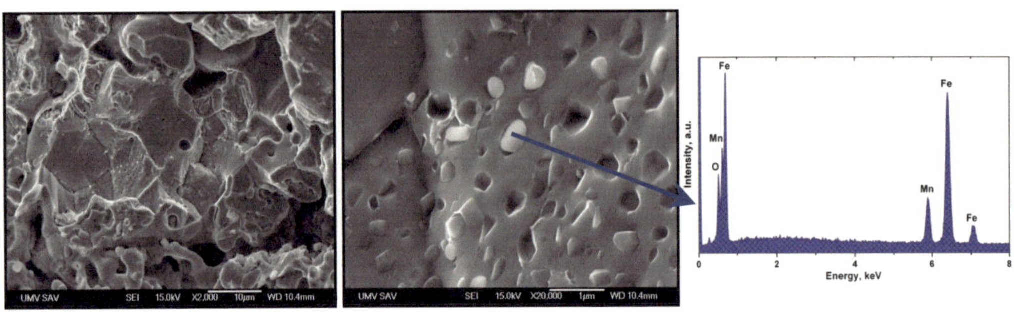

Fig. 20. SEM+EDX analysis of particulate inclusions on the inter-granular decohesion facets around ferromanganese residue after sintering at 1120°C for 30 min on the fracture surface of Fe-0.8Mn-0.5C material

Due to the lower amount of point oxides on the degraded grain boundaries at lower temperatures, such residual inter-granular brittleness is nearly fully removed for material admixed with ferromanganese after high-temperature sintering, see Fig.21, but is still observed for admixed with electrolytic manganese material.

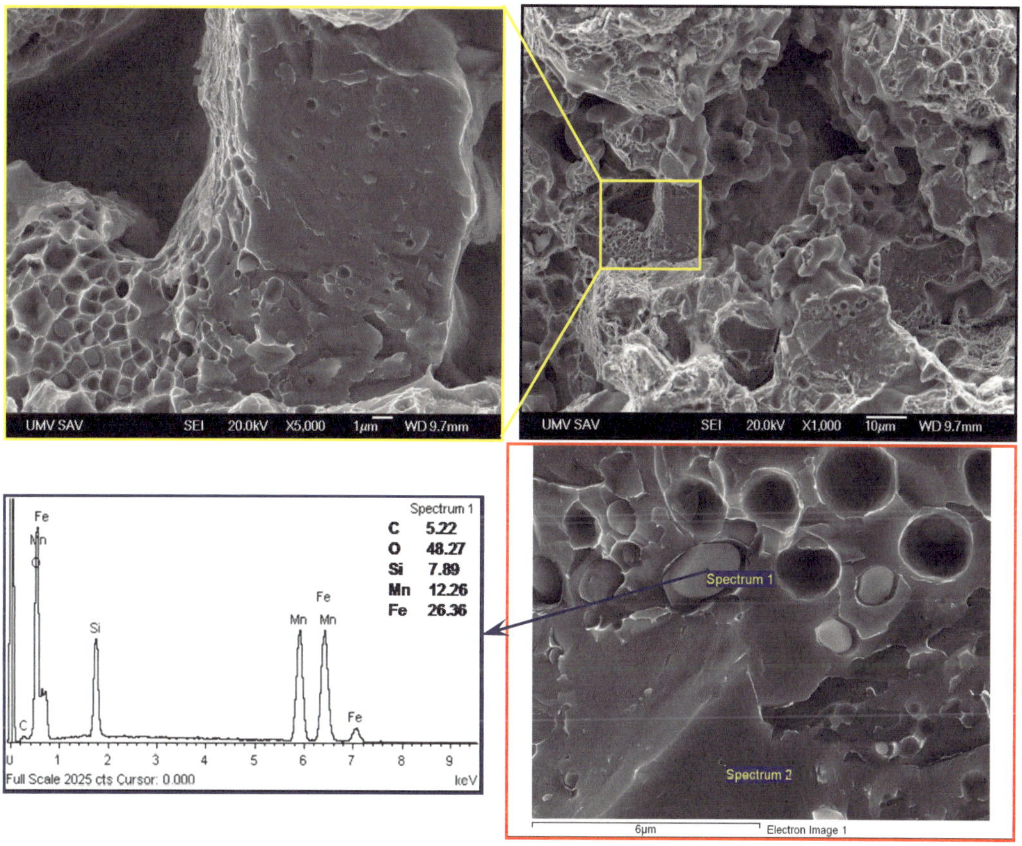

Fig. 21. SEM+EDX analysis of particulate inclusions on the inter-granular decohesion facets around ferromanganese residue after sintering at 1200°C for 30 min on the fracture surface of Fe-0.8Mn-0.5C material

Results of XPS analysis confirm presence of manganese oxides and sulphides on the fracture surface of both admixed specimens even after high-temperature sintering, see Fig.22 (Hryha et al., 2008; Hryha et al., 2010-b). The analyses also indicate that there is no significant difference in surface composition between the two admixed alloys. The XPS analyses show higher amount of oxides/sulphides present on the fracture surface of admixed with ferromanganese material due to larger ferromanganese residues after high-temperature sintering.

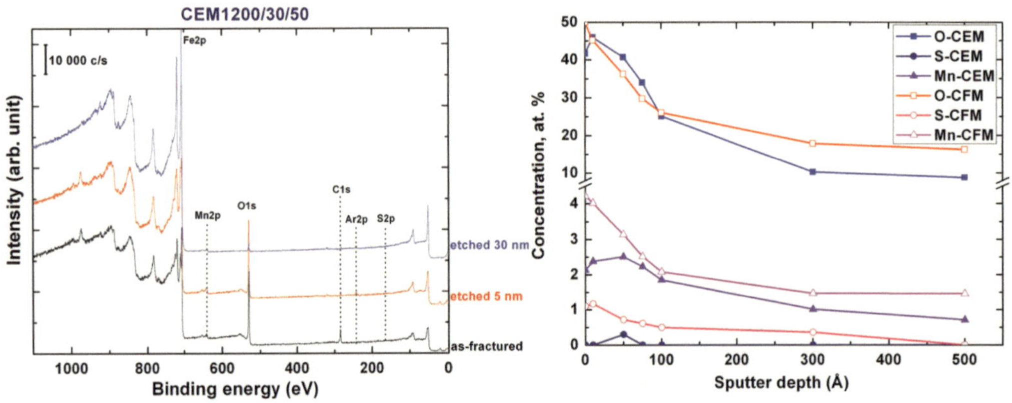

Fig. 22. XPS survey scan of fracture surface of admixed with electrolytic manganese Fe-0.8Mn-0.5C material (left) and comparison of elements concentration profile (right) for admixed with electrolytic (***CEM1200/30***) and ferromanganese (***CFM1200/30***) materials, sintered at1200°C for 30 min

6. Effect of alloying mode

Summarizing the results of number of investigation performed up to now, it is clear that admixed materials are exposed to considerable oxidation during heating stage even when applying high purity sintering atmospheres. Sintering at lower atmosphere purity without hydrogen presence results in even higher oxidation of admixed alloys (Hryha, 2007). Comparison of oxygen content and mass-gain reveals highest oxidation of admixed with electrolytic manganese alloy and lowest oxidation in the case of prealloyed material (Hryha, 2007; Hryha et al., 2010-b). In addition temperature ranges of considerable oxidation of admixed alloys coincides with the onset of intensive manganese sublimation (>720°C), pointing out that these processes are closely related and are connected with the oxidation of manganese carrier particles.

This can be shortly described by a model of evolution of manganese carrier particles, describing manganese sublimation and further oxidation of Mn vapour and its influence on the oxides distribution and development of inter-particle connections in the areas adjoining to manganese carrier, presented in Fig.23. During first stages of heating manganese carrier particle is naturally oxidized and is covered by thick oxide layer (Fig.23 (a)). After 700°C, when manganese starts to sublimate, Mn vapour reacts with oxygen presented in sintering atmosphere, water vapour and carbon oxides (produced by reduction of iron oxide layer). Formed manganese oxide condenses on the surface of the iron particles, adjoining to manganese source. Temperature increasing leads to reduction of iron oxides and starting of the inter-particle necks development and, at the same time, more intensive manganese evaporation and distribution through the pore system, its further oxidation and formation of Mn-oxide layer on the adjoining pores, see Fig.23 (b) - red areas. This stage can be characterized as 'shift' of oxidation - oxygen from the iron oxide from the base powder is transformed to manganese oxide around manganese carrier particles, that coincides with the keeping of the same level of oxygen content during heating for both admixed systems (Hryha et al., 2010-b). The problem with such manganese oxide layer is that it obstructs formation of metal-metal contacts between base powder particles and so development

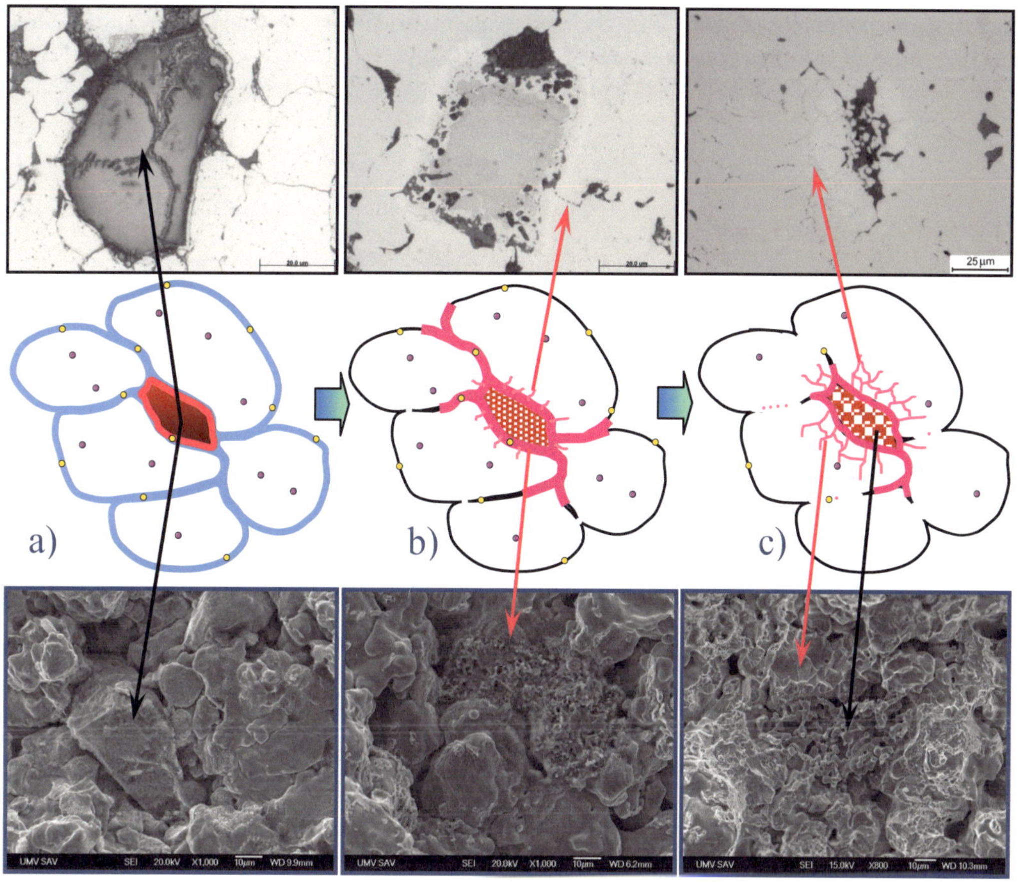

Fig. 23. Model of the evolution of the manganese carrier particles and inter-particle connections around them from 720°C (left), 1000°C (middle) to as-sintered structure (1120°C/30 min)

of inter-particle connections, that are formed only on the larger distance from Mn-carriers, see Fig.23 (b). The extension of such contaminated region is bigger for admixed with electrolytic manganese material and is increasing with temperature - at 1000°C it reaches around 100 µm, depending on surrounding pores structure. Lower total amount of the inter-particle connections is reflected in much lower strength values during heating stage, especially for admixed with electrolytic manganese material, see Fig.24. Temperature increasing leads to improvement in local conditions inside pores that enable reduction of surface manganese oxides and leads to decrease in size of contaminated areas. This enables intensive growth of inter-particle connections around manganese source, see Fig.23 (c), and is reflected on the increasing of the macro-strength of compact between 1000 and 1120°C, see Fig.24. It is important to emphasize that even after sintering at 1120°C properties of admixed with electrolytic manganese material (strength and oxygen content) are inferior in comparison with admixed with ferromanganese material, that now is close to the prealloyed material and only after sintering at 1200°C improvement in properties of admixed with electrolytic manganese material can be seen. This is connected with the slower reduction of manganese oxides, enclosed inside inter-particle necks and formed on the grain-boundaries during manganese dissolution. Presence of manganese carrier particles during the whole heating stage is another confirmation of their intensive oxidation - presence of the thick oxide layer retard intensive manganese evaporation during heating stage.

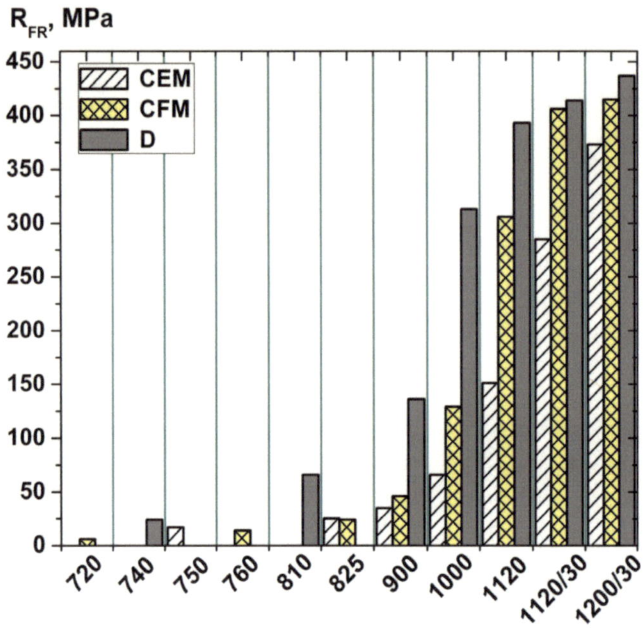

Fig. 24. Rupture strength of interrupted sintered specimens of Fe-0.8Mn-0.5C material, based on admixed with electrolytic manganese (CEM) and ferromanganese (CFM) systems as well as fully prealloyed powder (D), (*Hryha et al., 2010-b*)

7. Summary

Detailed analysis of manganese containing sintered steels based on fully prealloyed powder and admixed systems using high-purity electrolytic manganese and medium-carbon ferromanganese, indicate high brittleness of admixed materials. Lower strength of admixed materials is described based on oxidation of manganese vapor and its further condensation on the surrounding iron particles that suppress inter-particle necks development in the areas around the manganese source and results in extended defect areas. The worst results were obtained for admixed with electrolytic manganese material that is explained by more intensive manganese evaporation and accordingly higher specimen oxidation at low temperatures. Second type of critical defects in admixed systems is connected with the degradation of grain-boundaries around manganese carrier particles due to the presence of complex oxides and manganese sulfide on the inter-granular decohesion facets. The amount and composition of the contaminants after sintering are determined by the type of manganese carrier used and sintering conditions.

Number of presented in literature experimental results indicates that it is no doubt that the observed depth of manganese alloyed areas around manganese carrier particles is much

larger than volume diffusion could account for. Recent author's experimental results clearly show that much higher diffusivity observed can be described only based on complex mechanism - combination of grain boundary diffusion and chemically induced grain boundary migration (DIGM). Alloying mechanism with the presence of the gas phase (manganese vapour) and fine-grained structure of the manganese-rich surface layer on the initial stage of alloying abundantly evidences validity of this hypothesis.
Utilization of hydrogen-containing atmospheres of high-purity and higher heating rate at medium temperatures range (700-1000°C) together with sintering at high temperatures (≥1200°C) are suggested for successful sintering of admixed with manganese PM steels. Lower sensitivity to atmosphere purity, homogeneously distributed microstructure, absence of the large contaminated pores and oxide inclusions indicate definite advantage of pre-alloyed powder system from both economical and technical points of view.

8. Future research

Availability of new thermodynamic data concerning low-melting point systems and application of modern sintering techniques for high-temperature sintering combined with advanced control of sintering atmosphere open space for development of new master-alloys based on Fe-Mn-C systems. During design of such master-alloys stress has to be set on homogenisation and alloying through transient liquid phase and shifting manganese evaporation to as high temperatures as possible otherwise intensive formation of stable manganese oxide is unavoidable. Improved diffusivity of manganese by DIGM mechanism is another strong side of utilization of admixed iron-manganese PM steels. However, recent study (Hryha et al., 2010-a) shows high quality of the manganese prealloyed powder from the surface composition point of view that allow us to expect the appearance of manganese alloyed steel powder on the market in near future.

9. Acknowledgement

The main part of this work was done in the framework of the Höganäs Chair III project. The authors would like to thank Höganäs AB, Sweden, for scientific cooperation and permission to publish these results. Special thanks to Prof. Lars Nyborg, Chalmers University of Technology, Sweden, and Prof. Andrew Wronski, Bradgord University, UK, for the helpful discussions and input to this work.

10. References

Beiss, P. (2006). Alloy Cost Optimization of High Strength Mn-Cr-Mo Steels with Kerosene-Atomized Master Alloy. *Advances in Powder Metallurgy and Particulate Materials-2006,* pp. 727-735, ISSN: 1065-5824, San Diego, CA, June 2006, MPIF.

Chongmo, L. & Hillert, M. (1981). A Metallographic Study of Diffusion-Induced Grain Boundary Migration in the Fe-Zn System. *Acta Metallurgica,* Vol. 29, (1981), pp. 1949-1960, ISSN: 0956-7151.

Cias, A.; Mitchell, S.C.; Watts, A. & Wronski, A.S. (1999). Microstructure and Mechanical Properties of Sintered (2-4)Mn-(0.6-0.8)C Steels. *Powder Metallurgy,* Vol. 42, (1999), pp. 227-233, ISSN (printed): 0032-5899. ISSN (electronic): 1743-2901.

Cias, A.; Mitchell, S.C.; Plich, K., Cias, H.; Sulovski, M. & Wronski, A.S. (2003). Tensile Properties of Fe-3Mn-0.6/0.7C Steels Sintered in Semiclosed Containers in Dry Hydrogen, Nitrogen and Mixtures thereof. *Powder Metallurgy,* Vol. 46, (2003), pp. 165-170, ISSN (printed): 0032-5899. ISSN (electronic): 1743-2901.

Cias, A. & Wronski A.S. (2008). Comparison of Mechanical Poperties of Manganese Steels of the Same Chemical Composition Based on Sponge and Atomised Iron Powders. *Powder Metallurgy Progress,* Vol. 8, No. 2, (2008), pp. 76-82, ISSN: 1335-8987.

Danninger, H.; Gierl, C.; Kremel, S.; Leitner, G.; Jaenicke-Roessler, K. & Yu, Y. (2002). Degassing and Deoxidation Processes During Sintering of Unalloyed and Alloyed PM Steels. *Powder Metallurgy Progress,* Vol. 2, No. 3, (2002), pp. 125-139, ISSN: 1335-8987.

Danninger, H.; Pöttschcher, R.; Bradac, S.; Šalak, A & Seykammer, J. (2005). Comparison of Mn, Cr and Mo alloyed sintered steels prepared from elemental powders. *Powder Metallurgy,* Vol. 48, (2005), pp. 23-32, ISSN (printed): 0032-5899. ISSN (electronic): 1743-2901.

Dudrova, E.; Kabatova, M.; Bidulsky, R. & Wronski, A.S. (2004). Industrial Processing, Microstructures and Mechanical Properties of Fe-(2-4)Mn-0.85Mo-(0.3-0.7)C Sintered Steels. *Powder Metallurgy,* Vol. 47, (2004), pp. 181-190, ISSN (printed): 0032-5899. ISSN (electronic): 1743-2901.

Dudrova, E.; Kabatova, M.; Bures, R.; Bidulsky, R. & Wronski, A.S. (2005). Processing, Microstructure and properties of 2-4%Mn and 0.3/0.7%C sintered steels. *Kovové materiály / Metallic Materials,* Vol. 43, (2005), pp. 404-421, ISSN: 0023-432X.

Dudrova, E.; Kabatova, M.; Mitchell, S.C.; Bidulsky, R. & Wronski, A.S. (2010). Microstructure Evolution in Fe-Mn-C during Sintering. *Powder Metallurgy,* Vol.53, No.3, (2010), pp.244-250, ISSN (printed): 0032-5899. ISSN (electronic): 1743-2901.

Gaskell, D.R. (2003). *Introduction to the Thermodynamics of Materials,* 4-th ed., Taylor&Francis, Publisher, ISBN 1-56032-992-0, New York, 2003, 618 p.

Gomez-Acebo, M.; Sarasola, M. & Castro, F. (2003). Systematic Search of Low Melting Point Alloys in the Fe-Cr-Mn-Mo-C System. *Computer Coupling of Phase Diagrams and Thermochemistry,* Vol. 27, (2003), pp. 325-334, ISSN: 0364-5916.

Hillert, M. & Purdy, G.R. (1978). Chemically Induced Grain Boundary Migration. *Acta Metallurgica,* Vol. 26, (1978), pp. 333-340, ISSN: 0956-7151.

Hoffmann, G. & Dalal, K. (1979). Development and Present Situation of Low Alloyed PM Steels Using MCM and MVM Master Alloys. *Powder Metallurgy International,* Vol. 11, (1979), pp. 177-180, ISSN: 0048-5012.

Hryha, E. & Dudrova, E. (2007). The Sintering Behaviour of Fe-Mn-C Powder System, Correlation between Thermodynamics and Sintering Process, Mn Distribution, Microstructure. *Materials Science Forum,* Vol.534-536, (2007), pp. 761-764, ISSN: 0255-5476.

Hryha, E. (2007). Fundamental Study of Mn Containing PM Steels with Alloying Method of both Premix and Pre-alloy. *PhD Thesis,* (2007), 257 p., IMR SAS, Kosice, Slovakia.

Hryha, E.; Nyborg, L.; Dudrova, E. & Bengtsson, S. (2008). Brittleness of Structural PM Steels Admixed with Manganese Studied by Advanced Electron Microscopy and Spectroscopy. *Powder Metallurgy Progress,* Vol. 8, No. 2, (2008), pp. 109-114, ISSN: 1335-8987.

Hryha, E.; Nyborg, L.; Dudrova, E. & Bengtsson, S. (2009-a). Microstructure Development during Sintering of Manganese Alloyed PM Steels. *Proceedings of International Powder Metallurgy Congress & Exhibition Euro PM2009,* Vol. 1, pp. 17-22, ISBN 9781899072064, Copenhagen, Denmark, October 2009, EPMA, Shrewsbury, UK.

Hryha, E.; Nyborg, L. & Bengtsson, S. (2009-b). Surface Analysis of Prealloyed with Manganese Steel Powder. *Proceedings of International Powder Metallurgy Congress & Exhibition Euro PM2009,* Vol. 2, pp. 169-174, ISBN 9781899072071, Copenhagen, Denmark, October 2009, EPMA, Shrewsbury, UK.

Hryha, E.; Girl. C.; Nyborg, L.; Danninger, D. & Dudrova, E. (2010-a). Surface Composition of the Steel Powders Pre-Alloyed with Manganese. *Applied Surface Science,* Vol. 256, No. 12, (2010), pp. 3946-3961, ISSN: 0169-4332.

Hryha, E.; Nyborg, L. & Dudrova, E. (2010-b). Critical Aspects of Alloying of Sintered Steels with Manganese. *Metallurgical and Materials Transaction A,* Vol. 41A, No.11, (2010), pp. 2880-2897, ISSN: 1073-5623.

Mitchell, S.C. & Cias, A. (2004). Carbothermic Reduction of Oxides During Nitrogen Sintering of Manganese and Chromium Steels. *Powder Metallurgy Progress,* Vol. 4, No. 3, (2004), pp. 132-142, ISSN: 1335-8987.

Navara, E. (1982). Alloying of Sintered Steels with Manganese. *Proceedings of VI-th International Conference on Powder Metallurgy,* Vol. 1, pp. 143-154, Brno, Czech Republic, 1982, EPMA, Shrewsbury, UK.

Nohara, K. & Hirano, K. (1971). Diffusion of Mn in Fe and Fe-Mn Alloys. *Suppl. Trans. Iron Steel Jpn.,* Vol. 11, (1971), pp. 1267-1273, ISSN: .

Sainz, S.; Martinez, V.; Dougan, M.; Baumgaertner, F. & Castro, F. (2006). Sinterability, Hardenability and Mechanical Properties of Mn-Containing PM Steels through the Use of a Specially Designed Fe-Mn-C Master Alloy. *Advances in Powder Metallurgy and Particulate Materials-2006,* pp. 624-637, ISSN: 1065-5824, San Diego, CA, June 2006, MPIF.

Schlieper, G. & Thummler, F. (1979). High Strength Heat-Treatable Sintered Steels Containing Manganese, Chromium, Vanadium and Molybdenum. *Powder Metallurgy International,* Vol. 11, (1979), pp. 172-176, ISSN: 0048-5012.

Šalak, A. (1980-a). Sintered Manganese Steels, Part II: Manganese Evaporation During Sintering. *Powder Metallurgy International,* Vol. 12, (1980), pp. 72-75, ISSN: 0048-5012.

Šalak, A. (1980-b). Manganese Sublimation and Carbon Ferromanganese Liquid Phase Formation During Sintering of Premixed Manganese Steels. *The International Journal of Powder Metallurgy and Powder Technology,* Vol. 16, No.4, (1980), pp. 369-379, ISSN: 0361-3488.

Šalak, A. (1989). Diffusion Induced Grain Boundary Migration by Iron Powder Alloying by Manganese Vapour. *Kovové materiály / Metallic Materials,* Vol. 27, No.2, (1989), pp. 159-170, ISSN: 0023-432X.

Šalak, A; Selecka, M. & Bureš, R. (2001). Manganese in Ferrous Powder Metallurgy. *Powder Metallurgy Progress,* Vol. 1, No. 1, (2001), pp. 41-58, ISSN: 1335-8987.

Wells, C. & Mehl, R.F. (1941). Rate of Diffusion of Manganese in Gamma Iron in Low-carbon and High-carbon Manganese Steels. *Trans. Am. Inst. Min. Met. Pet. Eng.,* Vol. 145, (1941), pp. 315-328, ISSN: .

Zapf, G.; Hoffmann, G. & Dalal, K. (1975). Effect of Additional Alloying Elements on the Properties of Sintered Manganese Steels. *Powder Metallurgy,* Vol. 18, (1975), pp. 214-236, ISSN (printed): 0032-5899. ISSN (electronic): 1743-2901.

23

Molecular-dynamics Calculation of Nanostructures Thermodynamics. Research of Impurities Influence on Results

Igor Golovnev, Elena Golovneva and Vasily Fomin
Siberian Branch of Russian Academy of Sciences,
Khristianovich Institute of Theoretical and Applied Mechanics
Russian Federation

1. Introduction

There is a need to emphasize a number of peculiarities occurring during the study of dynamic phenomena in nanostructures. First, the spatial scales: typical sizes of the nanostructures lie within the ranges from 10 to 1000 Å. Second, the governing parameters have the duration from parts of the molecular vibrations period and atomic vibrations in crystal structures, which is about 10^{-14} s, to the time of a disturbance wave passage over the characteristics scale in the space, i.e. up to 10^{-9} s. Such spatial-temporal scales of the dynamic phenomena make their experimental investigation difficult, at least now. Unique experimental equipment as a scanning tunnel microscope, scanning probe microscope, atomic force microscope created in recent years permits to break fresh ground along this line. Moreover, the experimental investigations are very money- and time-consuming. All these factors generate a need to study the above mentioned fundamental problems theoretically.

According to specialists research, the approaches of quantum mechanics can be used in the above spatial-temporal ranges to calculate electrical, electromagnetic and optical properties of the nanostructures, as well as the molecular dynamics method (MDM) can be utilized to study thermomechanical processes and calculation of respective characteristics and phenomena wherein quantum effects can be ignored.

This method permits, knowing nothing but the potential of interaction between the atoms in the system, which is calculated by the quantum mechanics method, to obtain the complete information about the system – the set of axes and pulses of all atoms at an arbitrary time point. It enables to solve the above mentioned fundamental problems. But the method has even more opportunities. By one or another way of information processing, we can transit from the micro-level to the next meso-level and obtain the system parameters for the concordance with the continuum mechanics macro-characteristics, which is the most important problem in the deformable body mechanics. To do this, it is necessary to research the dependence of the nanosystem characteristics on its parameters and to obtain an asymptotic value when the parameters do not depend on the sizes.

2. Validation of the molecular-dynamic approach to the calculation of the nanostructure state equation

Both the formation of the nanostructures in modern technologies (for example, condensation of nanoclusters from the gas phase, generation of nanofilms in CVD and ALD processes, etc) and their further functional utilization are frequently attended by considerable temperature loads (for example when the current passes through contacts in microelectronic circuits). For this reason, a demand arose to study the thermodynamic properties of nanostructures.

However their spatial and temporal scales make the experimental investigation difficult, at least today, and hence they should be studied theoretically, foremost by the molecular dynamics method.

At the moment, there are a great many works devoted to the study of individual thermodynamic properties of specific materials (see, for example Sekkal et al., 1998; Won Ha Moon & Ho Jung Hwang, 2003; Chen Jan et al., 2007; Berroukche et al., 2008).

Normally, under consideration are structural characteristics, elastic constants, internal energy change regarding the temperature, thermal capacity, thermal expansion factor, the dependence of the compression modulus regarding on the temperature, etc. As a rule, periodical boundary conditions are used. It results in the fact that the surface role is ignored, and volume thermodynamic properties are calculated alone. Few works (see, for example Bolesta et al., 2001; Bolesta et al., 2002; Golovnev et al., 2003; Bolesta et al., 2006; Golovnev et al., 2006 a; Golovnev et al., 2006 b; Alekseev et al., 2008) were devoted to the study of the particular influence of the surface on thermomechanical characteristics of the nanostructures.

But traditionally the thermodynamics is supposed to be developed for a certain system if any thermodynamic potential is found (free energy is the most common parameter in practice).

As it was mentioned above, the nanostructres differ from ordinary materials in that the quantity of surface atoms, their energy and energy of bond with volume atoms are comparable to similar characteristics of the volume atoms, for which, in the case of macrosystems, the phenomenology thermodynamic theory has been developed and respective thermodynamic characteristics have been found. At the same time, the properties of the surface atoms are very much unlike from the volume atoms properties. The most evident way is to construct the thermodynamics of the systems consisting of two sub-systems: the volume and surface ones. In this case, however, the traditional way is unfit because of the following reason. The thermodynamics contains the postulate of additivity: the energy of the thermodynamic system is the sum of its components energies.

The total energy of the nanostructure U_t can be presented as a sum:

$$U_t = U_{sa} + U_V + U_b, \tag{1}$$

where U_V is the energy of volume atoms; U_{sa} is the energy of surface atoms; U_b is the energy of these subsystems binding. To apply the traditional approach, i.e. to use the additivity postulate, the binding energy should be much less than the energies U_V and U_{sa}. This condition is however not satisfied for the case of nano-systems.

All these circumstances caused the investigations of the thermodynamic properties of the nanostructures with the detailed separation of the volume and surface atoms subsystems.

Moreover, the formulation of the equation of nanostructure state is a fundamental challenge of independent importance.

2.1 The structure of solids state equation

The most reasonable approach to the formulation of the equation of an arbitrary system state in thermodynamically-equilibrium state is given by the statistical physics. With the Hamilton function of the system

$$H = \sum_{j=1}^{N} \left(\frac{p_{xj}^2}{2m} + \frac{p_{yj}^2}{2m} + \frac{p_{zj}^2}{2m} \right) + U(\vec{r}_1, ..., \vec{r}_N), \tag{2}$$

the statistical integral can be calculated as

$$Z = \int \exp\left(-\frac{H}{kT} \right) d\Gamma \tag{3}$$

and either free energy

$$F = -kT \ln Z \tag{4}$$

or internal energy

$$E_{in} = kT^2 \frac{\partial \ln Z}{\partial T} \tag{5}$$

can be found.

Having these thermodynamic functions, all thermodynamic characteristics of the systems can be described, and the equation of state can be formulated, too:

$$P = -\left(\frac{\partial F}{\partial V} \right)_T . \tag{6}$$

But this way has been successfully passed only for the calculation of gas thermodynamics. This method was used for solid bodies by Mie and Grüneisen (Mie, 1903; Gruneisen, 1926). When the solid body was considered as a system of classical 3D oscillators, they received the following expression for the free energy:

$$F = U_0(V) + kT \sum_{i=1}^{3N} \ln \frac{h\omega_i}{kT}; \tag{7}$$

Here $U_0(V)$ is the potential energy of atoms interaction at zero temperature, which can depend only on the solid-body volume; T is the temperature; N is the total amount of atoms in the system; ω_i is the atoms oscillation frequency in the crystal lattice. With the aid of Gibbs–Helmholtz theorem

$$E_{in} = F - T\left(\frac{\partial F}{\partial T} \right)_V \tag{8}$$

the expression for the internal energy was found, too:

$$E_{in}(V,T) = E_x(V) + E_T(V,T) \tag{9}$$

here $E_x = U_0(V)$ is the «cold» energy, and $E_T(V,T)$ - the thermal component of the internal energy, which can depend on the system temperature and volume. To the harmonic oscillator approximation, $E_T = 3NkT$.

Using the expression known from the thermodynamics, the authors found also the thermal equation of the solid-body state:

$$P = -\left(\frac{\partial F}{\partial V}\right)_T = -\frac{\partial U_0}{\partial V} - kT\sum_{i=1}^{3N}\frac{1}{\omega_i}\frac{\partial \omega_i}{\partial V}. \tag{10}$$

Evident that the total pressure also consists of the «cold»

$$P_x = -\frac{dU_0(V)}{dV} \tag{11}$$

and thermal components:

$$P_T = -kT\sum_{i=1}^{3N}\frac{1}{\omega_i}\frac{d\omega_i}{dV}. \tag{12}$$

But the dependence of the atoms potential energy and oscillations frequency on the solid body volume has still not been found, neither within the classical nor quantum model framework. For this reason, phenomenology theories based on a certain set of experimental data were developed within the whole XX century. The less phenomenology constants the theory required, and the broader was the parameter space wherein the theory gave reliable results, the better it was deemed to be. In the continuum mechanics, the sum in the last expression is normally replaced by some mean expression

$$\frac{3N}{V}\left\langle\frac{V}{\omega}\frac{\partial \omega}{\partial V}\right\rangle, \tag{13}$$

and the equation of state

$$P - P_x = \gamma\frac{3NkT}{V}, \tag{14}$$

is written through the Grüneisen constant

$$\gamma = -\left\langle\frac{V\,d\omega}{\omega\,dV}\right\rangle, \tag{15}$$

which includes all phenomenology.

In (Stepanyuk, 2009), the direct calculation of the statistical integral with the aid of the molecular dynamics was realized. The authors however did not give any details on the preparation of the initial data for the nanostructure (coordinates and pulses at the zero time), nor on the boundary conditions. Moreover, the accuracy of the numerical model is extremely important for the direct calculation of the statistical integral (averaging on long times). The data on the numerical error for the energy and the method of the system temperature modeling should have been given.

2.2 The stages of the molecular-dynamic modeling

Though the Mie and Grüneisen theory has not enabled to calculate the solids thermodynamics from first principles, but it has given the structure of the thermal and caloric equations of state. This permitted to construct the molecular-dynamic approach to the calculation of solid nanostructures thermodynamics and finally, using the scaling peculiarities, of macroscopic solid bodies.

2.2.1. Since the properties of the nano-strctures depend on the shape and sizes, at the early stage the corresponding atomic structure with an ideal crystal lattice should be chosen; its properties are to be calculated. Below the example of the calculation of the cooper cluster shaped as a sphere of radius 20 Å will be presented.

2.2.2. Since the nanostructure has a very developed surface, its potential energy does not correspond to the potential energy minimum. Hence, at the second stage of the calculation one should find the global minimum of the potential energy of the structure at zero temperature. The found coordinates of atoms will be used as initial data for the third-stage calculation.

2.2.3. The isothermal process of the cluster compression by the controlled external pressure at $T = 0^o K$ must be modeled. This will enable to obtain such «cold» characteristics as the dependence of pressure and internal energy on the volume.

2.2.4. At the fourth stage, the nanostructure undergoes the isochoric heating. Such «cold» characteristics as energy and pressure retain constant. This allows to obtain the dependence of the thermal components of the internal energy and pressure on the volume and temperature (the caloric and thermal equations of state, correspondingly).

2.2.5. The free energy of the nanostructure is calculated from the equation of state with the aid of the Gibbs-Gelmgolc equation.

3. Modeling of the initial state of the nanostructures

3.1 Choice of initial coordinates

So far as the nanostructures properties depend on the shape and sizes, at the first stage the respective atomic systems with calculated thermodynamic properties is chosen. Below is given an example of the calculation of the copper cluster shaped as a sphere with the radius 20 Å. Fig. 1 shows the sample assignment of the initial coordinates of such a structure. First, the coordinates of the atoms in the ideal copper FCC-structure are assigned (the atoms in such a structure are located in the corners of a cube with the edge $a = 3{,}615$ Å and in faces centers). Then those are selected which radii are less than the assigned one (20 Å in our case). To simulate the interaction, we used the EAM Voter potential for copper (Voter & Chen, 1987; Voter, 1993).

3.2 Detection of the global minimum of the system energy

The nanostructure possesses an extremely developed surface and thus the constructed cluster does not lie within the potential energy minimum. To bring the system to the global minimum of the total potential energy, the artificial-viscosity method is used. It enables to find the global minimum of the potential energy of the system at zero temperature. The method consists in the following: the atoms are continuously under the action of the dissipative forces $\vec{F_i} = -\nu \vec{p_i}$. Varying the viscosity factor ν, it is possible to accelerate or to decelerate the process of the system evolution in the state with the potential-energy minimum.

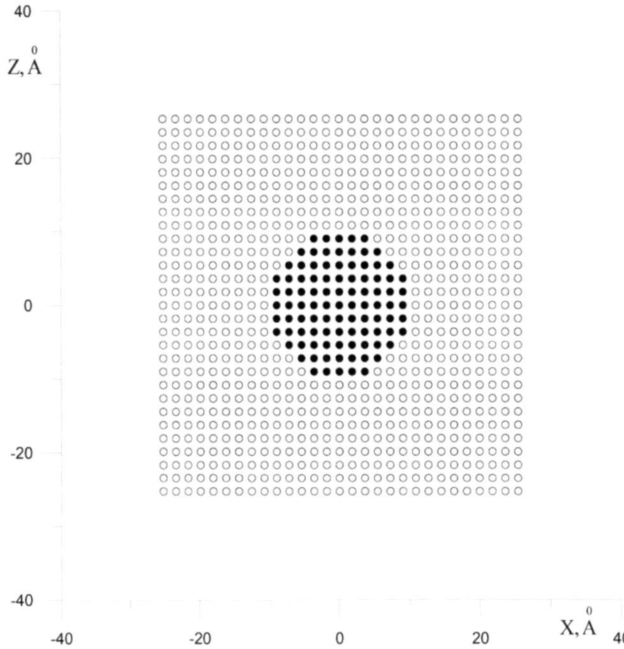

Fig. 1. Positions of the atoms in the XZ plane. Light bobbles – the atoms of cubic cluster; black bubbles – the atoms of the forming spherical cluster

4. Calculation of the «cold» characteristics of the nanostructure

To model the isothermal process of the nanostructure compression by the controlled external pressure at $T = 0^o K$, the following method is used.

The pressure P_0 is assigned on the surface as an external parameter. Evident that the problem of surface definities arises for the nanostructure case. Evaluation of the amount of the closest neighbors is the most common way. For the ideal FCC structure, this amount is 12 (first coordination sphere). Thus, in the nanostructure, those atoms which have the coordination number equal to 12, are referred to the volume atoms. The rest are referred to surface ones.

The forces f_a acting on the surface atoms and directed by the radius from the center (sphere extension) or toward the center (sphere constringency) are calculated by the formula

$$f_a = (S_t P_0) / N_s, \tag{16}$$

where S_t is the surface area, N_s is the total amount of the surface atoms.

To damp the shock-wave processes in the sphere, the methods of linear force amplification f_a from zero to the assigned value and artificial viscosity are used. As a result, at the end of the calculation, the statically-equilibrium state is reached at $T \approx 0^o K$. Then, the finite coordinates of the system are used to calculate the necessary characteristics.
For convenience, the relative variation of the volume is chosen as an independent variable:

$$x = (V - V_{fr}) / V_{fr}. \tag{17}$$

This value is negative for the sphere constringency case, and positive for the sphere extension.
To test the results, the pressure P_c внутри inside the control sphere was additively calculated as follows. The sum of projections of the forces acting on the atoms inside some control sphere is calculated (in the calculations, the radius of this sphere r_c was chosen to be 10 Å) normal to the surface from the side of the atoms with radius above r_c. Then, to calculate P_c, this total force is divided to the area of the control sphere surface. Thus, in the calculations the pressure has the negative sign in the case of compression and positive - in the case of extension. Fig.2 presents the sample dependence of the external and control pressure in the cluster on the time pitches amount.

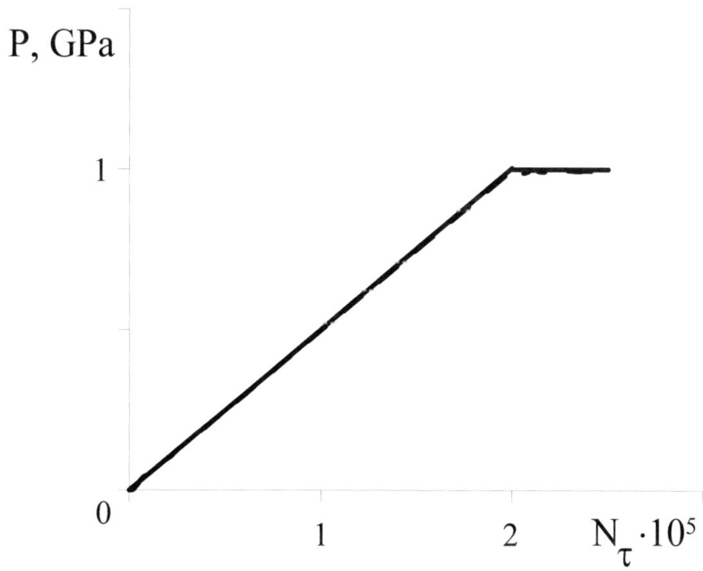

Fig. 2. External pressure (GPa - the solid line) and pressure in the control sphere (dotted line) vs the time pitches amount (10^5)

Fig.3 presents the dependence of the nanoclusters radius on the time pitches amount for the compression case.
Evident, that both the control pressure and radius take the asymptotic values. Final coordinates of the atoms are utilized then for the calculation of the cluster volume and potential energy.

The extension is calculated within the range from 0.1 to 12 GPa, and constringency - from 0.1 GPa to 20 GPa. The sphere state is analyzed in details as a whole unit, both its volume and surface parts. Index α in this case takes the values of t,v,s, respectively.

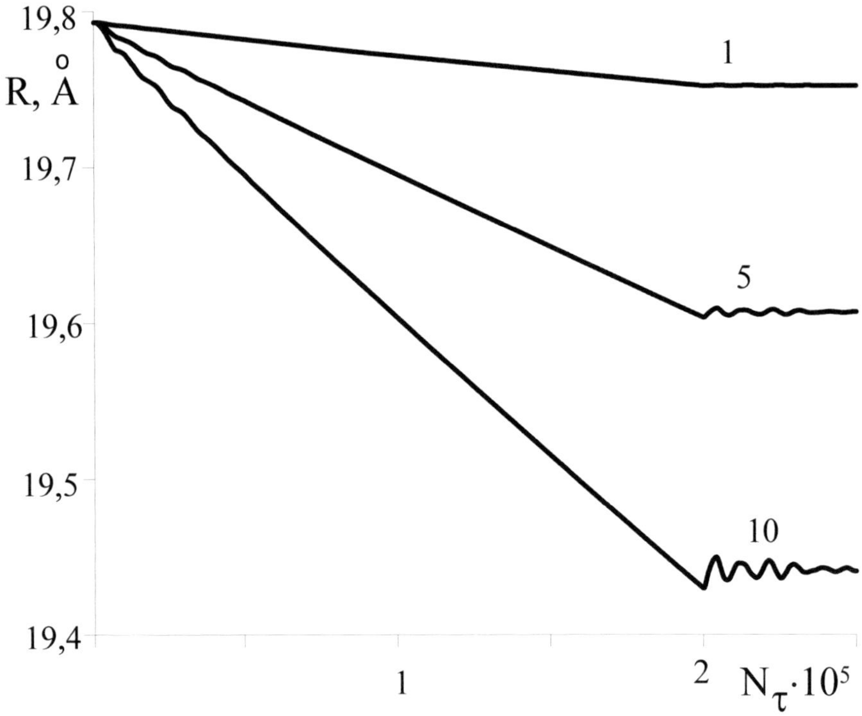

Fig. 3. Cluster radius (Å) vs the time pitches amount (10^5) for various compressing pressures (GPa)

4.1 Calculation of the nanostructure «cold» energy dependence on the volume

It is convenient to perform also the physical analysis of the energy characteristics for the potential energies variations in respect to the initial potential energy, which is the cooled-sphere energy:

$$\Delta U_{\alpha} = U_{\alpha} - U_{\alpha fr}. \qquad (18)$$

The value of the index α is given above.

Fig. 4 shows such a dependence for the total energy of the system on the relative change of the volume.

The dependence is close to quadric, though it differs somehow from it. It results from the fact that for the high values of the external pressure, the deviations of atoms from the equilibrium state are big, and the non-linear part of the interatomic interaction potential has its effect.

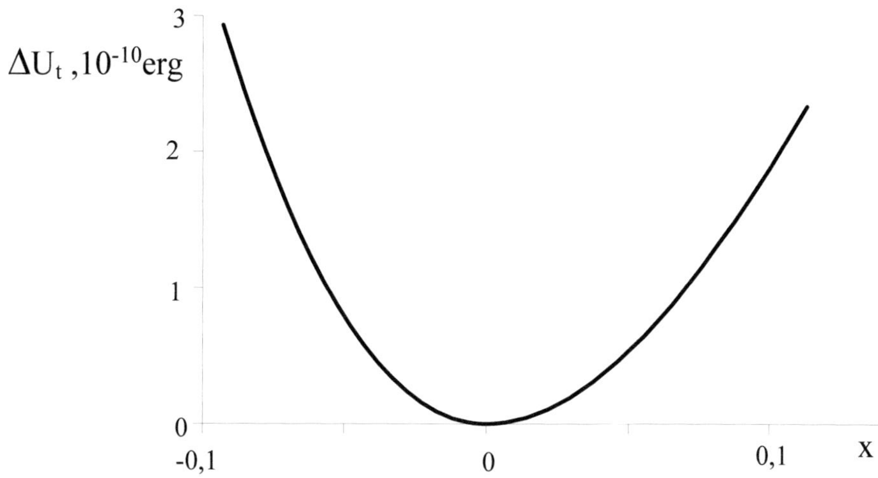

Fig. 4. Variation of the total energy (10^{-10} erg) of the cluster vs the relative change of the volume Title of figure, left justified

Hence, the performed calculations result in the dependence of the «cold» internal energy of the nanocluster on the volume. This characteristic is the most important one for the calculation of the thermodynamic properties of the nanostructures.

4.2 Calculation of the «cold» pressure dependence on the volume

The second most important «cold» characteristic is the dependence of the pressure on the volume at $T \approx 0^o K$. Fig. 5 presents the dependence of the external pressure on the relative variation of the total volume for the pressure range from -20 to 12 GPa. The non-linearity, similarly to the energy case, is caused by the non-linear character of the interatomic interaction potential. For convenience, Fig.6 presents the same graph for the external pressures varying from -1 to 1 GPa. The dependence is also convenient for evaluations.

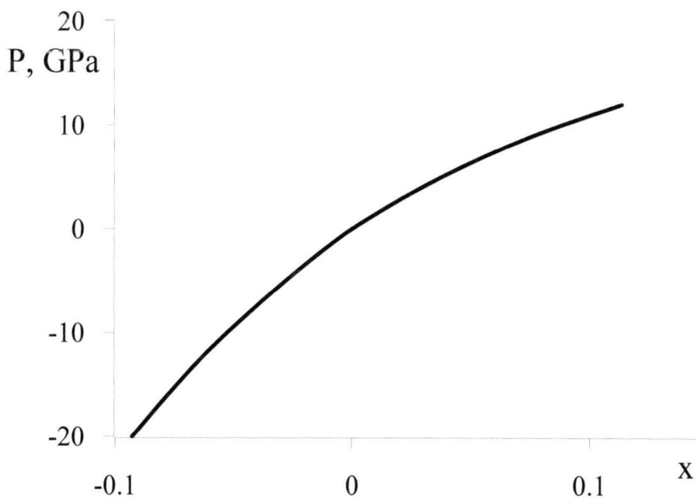

Fig. 5. The cold pressure (GPa) in the cluster vs the relative change of the volume

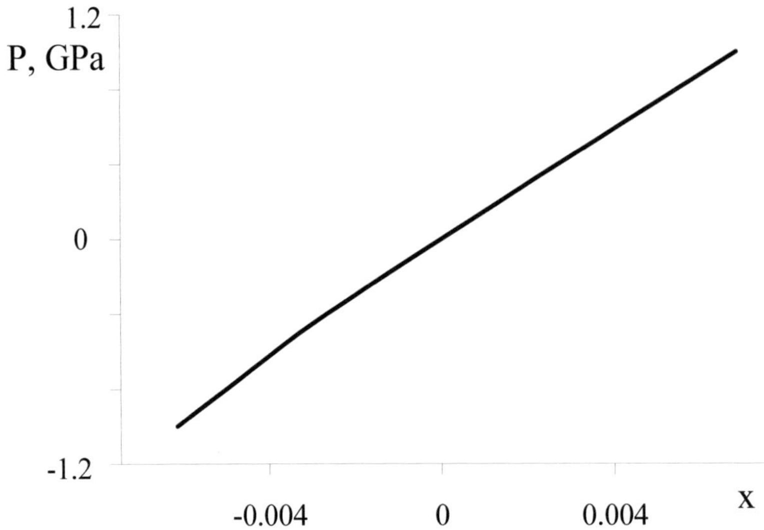

Fig. 6. The cold pressure (GPa) in the cluster vs the small relative change of the volume

Numerically obtained characteristics completely define the "cold" component of the thermal and caloric states of the nanostructure.

5. Molecular-dynamic modeling of the nanostructure heating

When it is necessary to assign the system initial state with T > 0° K, the stochastic forces method is used. Its idea is that atoms are under the action of chaotically directed momenta of constant amplitude. The parameters of the acting random force are: average impact frequency ω along each Cartesian axis and amplitude f_0, which provides the increment of the atom momentum component at one impact $\Delta p = f_0\tau$. Since the atoms coordinates do not change at the impact, the heating speed in this case depends on the kinetic energy increment:

$$\frac{\Delta E}{\Delta t} = \omega \sum_{a=1}^{3}\sum_{i=1}^{N}\left(\frac{\left(p_i^a + \Delta p\right)^2}{2m} - \frac{\left(p_i^a\right)^2}{2m}\right) = \frac{\omega}{2m}\sum_{a=1}^{3}\sum_{i=1}^{N}\left(2p_i^a \Delta p + \left(\Delta p\right)^2\right), \tag{19}$$

Followed by the energy transport from kinetic to the potential component. The first item in parentheses in the right part of (5.1) turns into zero at adding, since the random forces and momenta are not correlated, so finally we have

$$\frac{\Delta E}{\Delta t} = \frac{3\omega N}{2m}\left(\Delta p\right)^2. \tag{20}$$

Hence the heating speed does not depend on the cluster energy and linearly depends on time. This is the major advantage of this heating method in contrast with the commonly used momentum scaling. The random force parameters are chosen according to the requirement of the equilibrium state of the cluster at each time point of the heating process.

6. Molecular-dynamic calculation of the nanostructure state equations

In line with the proposed method of the calculation of the nanostructure state equation, the finite states featuring the arrays of coordinates and moments which are obtained by comprehensive compression and extension at zero temperature, are used then as initial data at the system heating. The heating is carried out for each value of the external pressure and hence for the set of subsystems volumes values. The isochoric process is modeled by means of surface atoms location into the external harmonic potential, which does not allow the atoms to shift toward the area of big radii.

The second peculiarity of the heating process is the fact that over each range of $\Delta T = 25^o K$, the stochastic force is off, and the system relaxes into the thermodynamic-equilibrium state within the time range of $\Delta t = 10^{-12}$ s. All characteristics are averaged by thermal fluctuations on this time range. The temperature is determined by the energy of atoms chaotic motion:

$$\frac{3}{2}NkT_{kin} = E_{kin} = \sum_{i=1}^{N}\frac{\overrightarrow{\Delta p_i}^2}{2m}. \tag{21}$$

Here $\Delta\vec{p}_i = \vec{p}_i - m_i\vec{V}_c$ is the chaotic component of the i -atom momentum. ($\vec{p}_i$ is the i -atom momentum in the laboratory system of coordinates; m_i is its mass; and $\vec{V}_c$ is the velocity of the mass center of the system). The heating was performed within the range from 25 to 1000 K. According to this, below authors use the index of the calculation temperature grid i_T, with this value the corresponding chaotic temperature of the system $T_{kin} = i_T * 25$ (° K) is found easily.

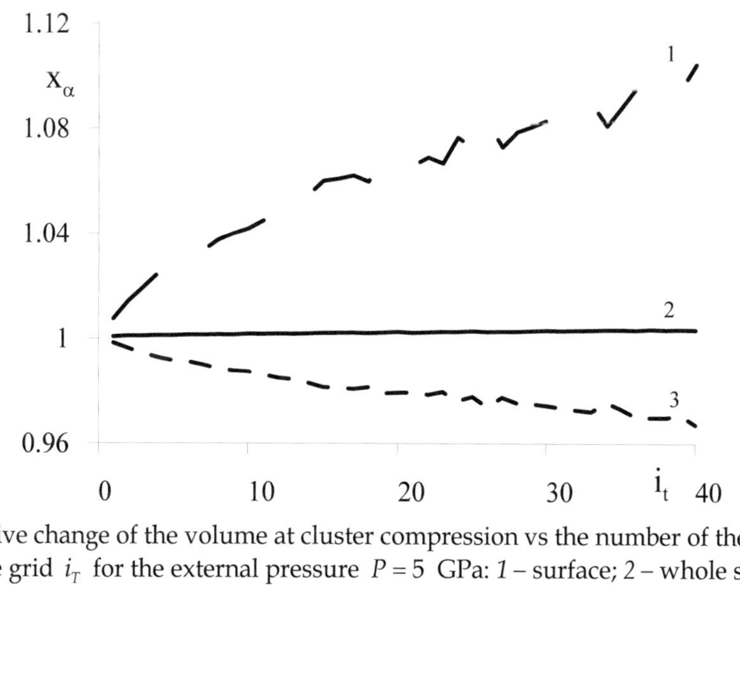

Fig. 7. Relative change of the volume at cluster compression vs the number of the temperature grid i_T for the external pressure $P = 5$ GPa: *1* – surface; 2 – whole system; 3 – volume

Additive restrictions which should be taken into account at the analysis of thermodynamic phenomena in the nano-systems are associated with the classical mechanics application. It is known that at the temperatures below Debye temperature the quantum effects must be regarded. Hence the conclusions obtained below are valid for the temperatures of $T \geq \Theta_D$. For copper $\Theta_D = 315^o$ K (Born & Huang, 1954).

6.1 Isochoric process modeling

To analyze the behavior of the subsystems volumes in the nanocluster, the best example is the ratio of $x_\alpha = V_\alpha / V_{\alpha 0}$ which varies with temperature. In the case of sphere constringency, Fig. 7 shows the dependence of this parameter on index i_T for the initial pressure 5 GPa.

It is evident that the total volume retains constant to a high accuracy and the process can be deemed to be isochoric. On the other hand, the growing x_S means that the surface layer volume increases during the heating, and the volume layer volume decreases, because x_V reduces.

In the case of extension, the similar pattern is observed but up to the external pressures of 1 GPa. For higher extending pressures, such a method of modeling of isochoric processes does not give proper results below about 300° K. At the temperatures above this value, x_t approaches to 1, and the process can be treated as isochoric.

6.2 Calculation of the caloric equation of the nanostructure state

The most fundamental value in thermodynamics is the total internal energy of the system E_{in} , which has the mechanical analog and involves both the potential energy of atoms interaction (including the «cold» one), and the energy of atoms chaotic motion E_{kin} .

The dependences of energy increments $\Delta U = U(T,V) - U_0(V)$ and E_{kin} of the whole system on the temperature grid number are linear for the case of the initial compression of the sphere. Their values are close to each other, which vindicates the presence of the equilibrium thermodynamic state. Similar situation is observed for the sphere extension.

The dependence of the total internal energy on the temperature grid number is given in Fig.8 for the initial constringency, for two values of the cold pressures of 0.5 and 5 GPa. Internal-energy dependence on temperature is also linear, and the graphs coincide within the whole pressure range, both for the compression and extension cases.

It was mentioned above that in the extension case for the pressures above 1 GPa, the dependence of the relation between the finite and initial volumes on the temperature differs dramatically from the case of pressures within the range from -20 to +1 GPa. At the same time it was found that the dependence of the internal energy for the cold pressures of 1 and 5 GPa on the temperature practically coincides, and linear graphs have the same angle of inclination (or nanostructure thermal capacity).

Hence, the internal energy has been found numerically within the wide range of system temperatures and volumes. It should be especially noted that this energy linearly depends on the temperature and satisfies the classical Dulong–Petit law. At the same time, the dependence on the volume is weak and can be neglected for the zero approximation.

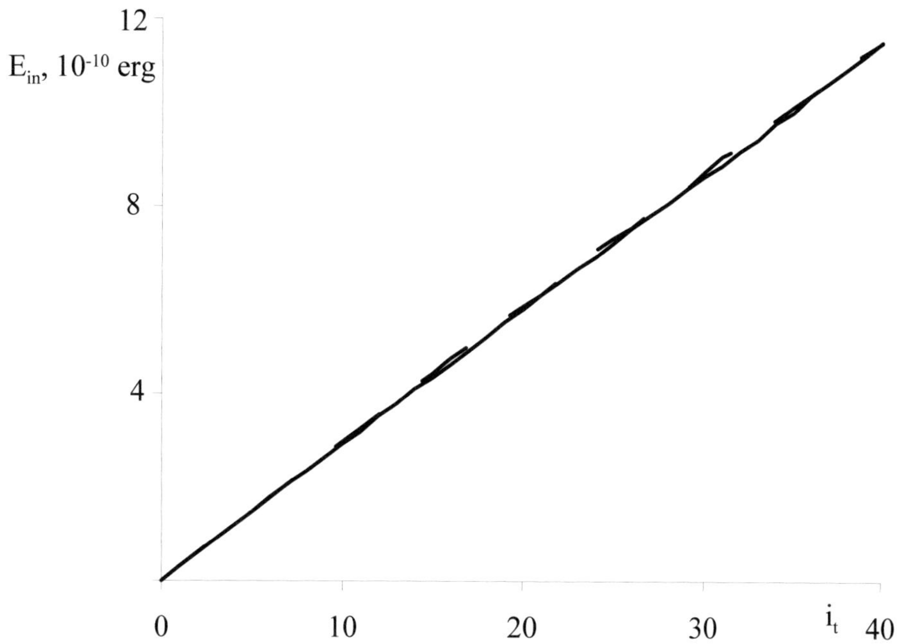

Fig. 8. The increment of the cluster internal energy E_{in} (10^{-10} erg) at the cluster compression vs the temperature grid number i_T for external pressures $P = 0.5$ GPa (solid line) and $P = 5$ GPa (dotted line)

6.3 Calculation of the thermal equation of the nanostructure state

When calculating the thermal equation of state, it is necessary, above all, to choose the model of the pressure calculation in the cluster. To do this, the formula for stress tensor components, obtained within the kinetic theory framework (see, for example, (Zimmermann, 1999; Zimmerman et al., 2009), is used:

$$\sigma_{\alpha\beta} = \frac{1}{V_c}\left[\sum_i \frac{p_i^\alpha p_i^\beta}{m} + \frac{1}{2}\sum_{i \neq j} F_{ij}^\alpha r_{ij}^\beta\right]. \tag{22}$$

Here V_c is the volume of the studied system; p_i^α is the α-component of the i-atom momentum; m is atomic mass; r_{ij}^β is the β-component of the radius-vector between the i- and j-atoms. The summing-up is done by all atoms lying within the studied volume. The first item in the thermodynamically equilibrium state is rearranged in the form of (Allen & Tildesley, 1987):

$$P_{kin} = \frac{NkT}{V}. \tag{23}$$

Evident that the total pressure consists of the kinetic (similar to gases) and potential components:

$$P_T = P_{kin} + P_U, \qquad (24)$$

and the potential component P_U also contains the «cold» pressure. Since the calculation of the kinetic component is not difficult, and the «cold» pressure has been found in the preliminary calculations, it is more convenient to use the expression $\Delta P_U = P_U - P_x$ to analyze the thermal effects.

The obtained data valid for t the coordinates of all atoms enabled to construct the dependence of pressure ΔP_U on the internal energy of a volume unit which is quite commonly used in the continuum mechanics. The obtained parametric dependence on the initial cold pressure is biunique bound with the relative volume variation. Thus, these data give the fundamental thermodynamic dependence $P_T(V, E_{in})$.

It is more usual for thermodynamics to use the thermal equation of state in the variables (P, V, T). Since all variables are calculated numerically, it is rather simple to construct the dependence of the changing thermal pressure ΔP_U on the temperature (Fig.9).

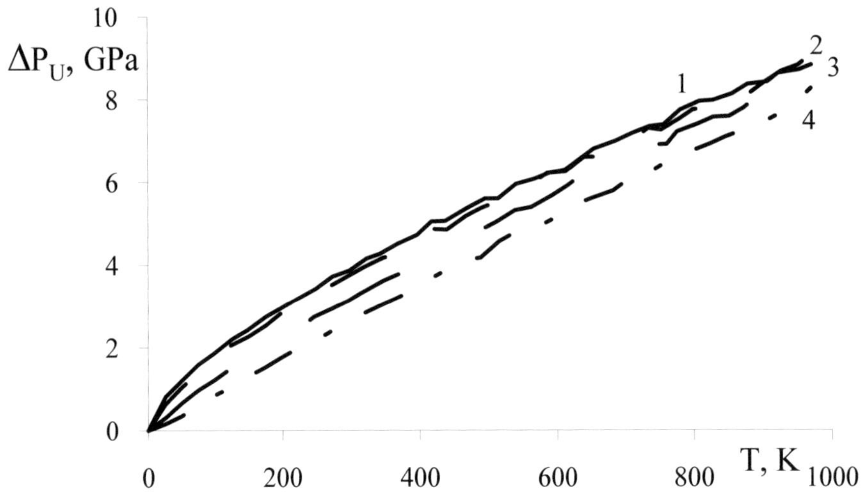

Fig. 9. Thermal pressure in the cluster ΔP_U (GPa) vs the temperature (compression case) for the cold pressure: 1 - 0.1 GPa; 2 - 1 GPa; 3 - 5 GPa; 4 – 10 GPa

Within the range from 0.1 to 1 GPa, the graphs almost coincide, and at higher pressures one can see the parametric dependence on the sphere volume. For the case of sphere extension, in the range of 0.1 - 1 GPa, the data for the cluster extension almost coincide. It should be noted that the dependence is close to linear at the temperature above 300 °K.

6.4 Grüneisen constant calculation

In practice, the expression for the thermal equation of state with the Grüneisen constant is frequently used (see above). Since in the molecular-dynamic calculations the temperature, volume and pressure are found from the first principles, this constant can be found:

$$\gamma = \Delta P / (3NkT / V). \qquad (25)$$

Fig. 10 presents its dependence on the temperature for various values of the constringent cold pressure (or the same - on the value $x = V(P_x)/V_0$).

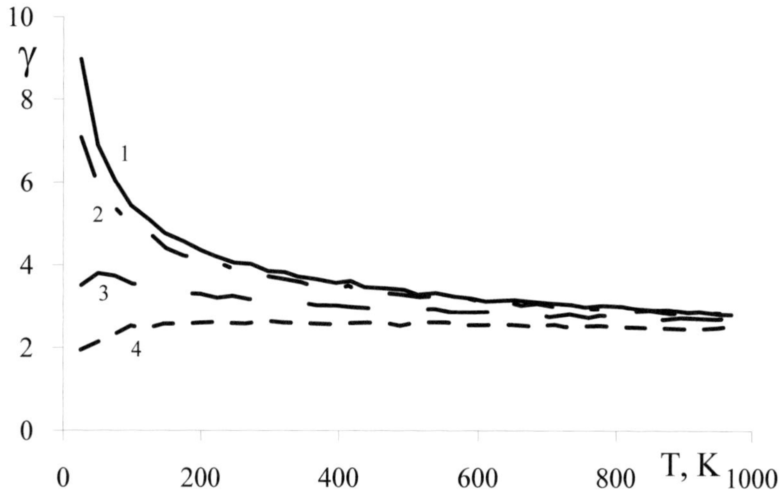

Fig. 10. Grüneisen constant vs the temperature (compression case) for the cold pressure: *1* – 0.1 GPa; *2* – 1 GPa; *3* – 5 GPa; *4* – 10 GPa

Evident that for the temperature of $T \geq 300^o$ K, the slight decrease of γ within the limits of 10 % is evidently observed. At the same time, one can see the strong dependence of this value on the initial cold pressure or on the system volume. For $T = 300^o$ K, the Grüneisen constant values differ almost twice from the case of the pressures of 0.1 and 10 GPa. At the same time, we see the strong dependence of this value on the initial cold pressure or on the system volume. For example, at $T = 300^o$ K, the values of the Grüneisen constant differ almost twice for the pressures of 0.1 and 10 GPa. It should be noted that the experimental magnitude of this value (the case of copper) found in the macroscopic experiments is 1.8. This is approximately twice lower than the similar value for the nanostructures at low pressures.

6.5 Free energy of nanostructures

As it is well known, any known thermodynamic potential gives complete information on the thermodynamics of the whole system. Before the internal energy was found in the variables (T,V) which are the eigen variables of the free energy. This is a typical situation when the internal energy is calculated from *ab initio* by statistical physics methods or, for example, inside of the dynamic method proposed here. In this context, the Gibbs-Helmholtz equation should be used at the following stage:

$$F(T,V) = -T\int \frac{E_{in}(T,V)}{\tilde{T}^2} d\tilde{T} + T\varphi(V), \tag{26}$$

which would enable to transit from the internal energy in the variables (T,V) to the free energy. Here, the internal energy is

$$E_{in}(T,V) = U(T,V) + E_{kin}\,, \tag{27}$$

where $U(T,V)$ is the total energy of the cluster and E_{kin} is the kinetic energy of the chaotic motion of the atoms.
The potential energy can be presented as

$$U(T,V) = U_x(V) + \Delta U(T,V)\,. \tag{28}$$

In return, the potential energy (the cold energy) of the cluster is a sum

$$U_x(V) = U_{0fr}(V_{0fr}) + \Delta U_x(V)\,. \tag{29}$$

of the potential energy of the cooled cluster without external pressure $U_{0fr}(V_{0fr})$ with the initial cluster volume V_{0fr} and positive addition to the energy $\Delta U_x(V)$, which occurs due to the compression by the external pressure to the volume V.
The addition to the potential energy $\Delta U(T, V)$ can depend in the general case on the temperature and volume, similarly to the thermal energy E_{kin}. As was shown before, both the internal energy and $\Delta U(T,V)$ weakly depend on the volume and this dependence can be ignored in the zero approximation, whereas the Dulong-Petit law $\Delta E_{in} = 3NkT$ can be used for the internal energy increment.
Hence, the internal energy of the cluster can be presented as:

$$E_{in} = U_{0fr}(V_{0fr}) + \Delta U_x(V) + 3NkT\,. \tag{30}$$

Substituting this in (26), we have:

$$F(T,V) = -T\int \frac{U_{0fr}(V_{0fr}) + \Delta U_x(V) + 3Nk\tilde{T}}{\tilde{T}^2}\,d\tilde{T} + T\varphi(V)\,. \tag{31}$$

Upon elementary transformations we obtain the expression for the free energy:

$$F(T,V) = U_{0fr}(V_{0fr}) + \Delta U_x(V) - 3NkT\ln\left(\frac{T}{T_0}\right) + T\varphi(V)\,. \tag{32}$$

With the thermodynamics generalities we obtain the expression for the pressure, its numerical values, on the other hand, are found *ab initio*:

$$P = -\left(\frac{\partial F}{\partial V}\right)_T = -\frac{d(\Delta U_x(V))}{dV} - T\frac{d\varphi(V)}{dV} \equiv P_x(V) - T\frac{d\varphi(V)}{dV}\,. \tag{33}$$

To determine the unknown function $\varphi(V)$, the expression $\Delta P_T = P - P_x$ from the molecular-dynamic calculations should be analyzed; The pressure caused by the chaotic motion of the atoms

$$\Delta P_{kin} = T\frac{3Nk}{V}\,, \tag{34}$$

Depends linearly on the temperature and corresponds to the expression

$$P - P_x = -T\frac{d\varphi}{dV}. \tag{35}$$

Evident that the function $\varphi(V)$ can be presented as a sum

$$\varphi(V) = \varphi_{kin}(V) + \varphi_U(V). \tag{36}$$

Of kinetic and potential components. Then

$$-T\frac{d\varphi_{kin}}{dV} = T\frac{3Nk}{V}. \tag{37}$$

Or, when transiting to the relative volume

$$x = \frac{V_{0fr} - V}{V_{0fr}}, \tag{38}$$

we have

$$\varphi_{kin}(V) = -3Nk\ln(1 - x). \tag{39}$$

To find the potential component of this function, the numerical molecular-dynamic calculations should be used. The dependence of ΔP_U on the temperature for various cold pressures (or volumes) is shown in Fig.9. For the temperatures above 300 °K at which these calculations are valid, the pressure dependence is truly approaching to the linearity. In this case, the derivative is

$$\frac{d\Delta P_U}{dT} = \frac{d\varphi_U}{dV}. \tag{39}$$

The treatment of numerical calculations results enabled to construct the value $\frac{d\varphi_U}{dV}$ from the relative change of the volume $x = \frac{V_{0fr} - V}{V_{0fr}}$. (Fig.11.). The polynomial approximant is:

$$\frac{d\varphi_U}{dV} = 0{,}00755251 + 0{,}0324687x - 0{,}416867x^2 \equiv ax^2 + bx + c. \tag{40}$$

As a result, we have

$$\varphi_U(V) = \tilde{a}x^3 + \tilde{b}x^2 + \tilde{c}x + c_0, \tag{41}$$

where

$$\tilde{a} = -\frac{V_{0fr}a}{3};\ \tilde{b} = -\frac{V_{0fr}b}{3};\ \tilde{c} = -\frac{V_{0fr}c}{3}. \tag{42}$$

Finally, the free energy of the nanoclusters is in the form of

$$F(T,V) = U_{0fr}(V_{0fr}) + \Delta U_x(V) - 3NkT\ln\left(\frac{T}{T_0}\right) + \\ + T\left(\tilde{a}x^3 + \tilde{b}x^2 + \tilde{c}x + c_0 - 3Nk\ln(1-x)\right) \tag{43}$$

The obtained expression and numerically found cold energy $\Delta U_x(V)$ together solve completely the challenge of calculation of the thermodynamic properties of the nanostructure.

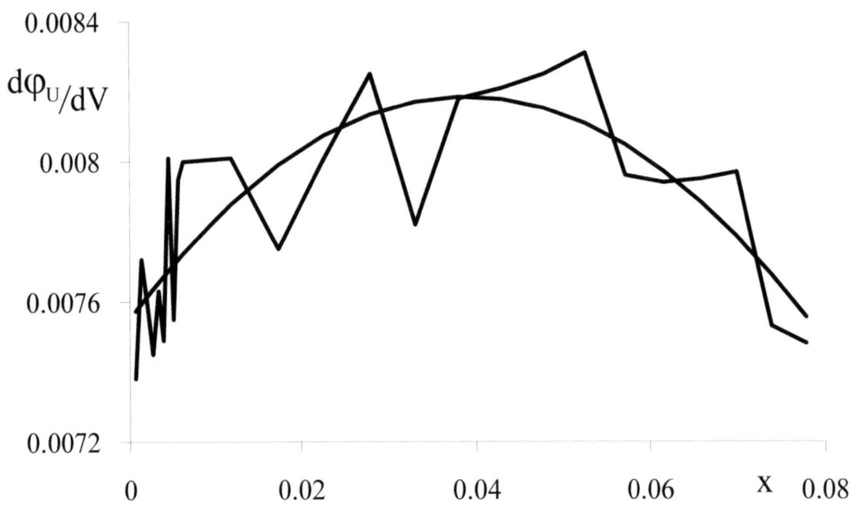

Fig. 11. Dependence of $\frac{d\varphi_U}{dV}$ on the relative change of the volume: fine solid line – exact calculations, heavy line – polynomial approximant

7. Analysis of the influence of admixtures on nanoclusters thermodynamics

The developed approach was applied to the study of nanoclusters thermomechancial properties by the molecular dynamics method. It permitted to find out the criteria of applicability of traditional methods to alloyed nanostructures, to investigate the influence of the stoichiometric coefficient on the thermomechancial properties. Moreover, scaling enables to extend the results of this work to macro-subjects and to obtain similar characteristics for the deformed body mechanics.

7.1 The physical model and basic calculation procedures

A copper sphere with the radius of 20 Å (Golovnev et al., 2006 a) was used to simulate a metal nanoculster with insertions. Then the copper atoms with specific numbers were replaced by silver atoms with the aid of the Monte-Carlo method. Total amount of the replaced atoms corresponded to the percentage of the silver atoms, which was used as the external controlled parameter. The content of the silver atoms in the nanoclusters varied from 1 to 8%.

To take the copper-silver interaction into account, the Voter EAM potential for Cu, Ag, CuAg is used (Chen Jan et al., 2007; Berroukche et al., 2008). To prepare the initial data, the whole system was cooled with the help of an artificial viscosity method (Bolesta et al., 2001), similarly to the case of pure copper, and whereby the system of atoms was at the calculation end in the potential energy minimum. The finite arrays of coordinates and momenta obtained during the cooling were then used as initial data for the simulation of the sphere quasi-statical compression and extension. Fig.12 illustrated the influence of the silver concentration by the cluster radii dependence on time for the cooling process. Evident that as the silver concentration rises, the cluster radius increases. Below we give the results for the cluster compression within the pressure range from 0.1 to 10 GPa.

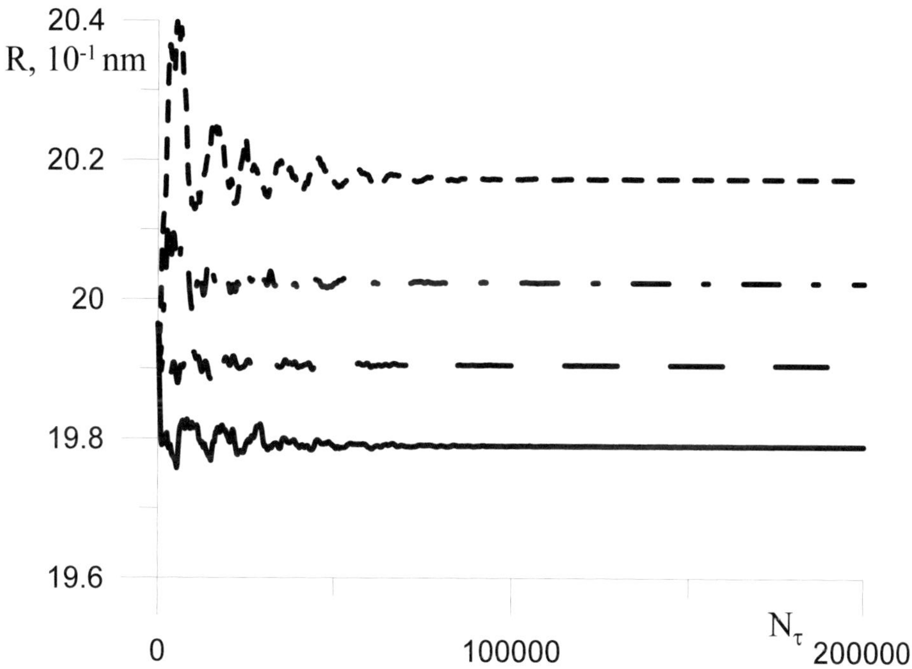

Fig. 12. Radii of nano-spheres varying with time during the cooling. Solid line - pure copper; dashed line - alloy Cu99%Ag1%, chain line - Cu96%Ag4%, dotted- Cu92%Ag8%

7.2 Simulation of the cooled nanosphere compression

Then the sphere was cooled similarly to the pure copper case (see paragraph 4).

The key thermodynamic characteristics of the system are the dependencies of the "cold" pressure P_x and internal energy U_{tx}, which is just the potential energy of the system at $T \approx 0^o K$, on the cluster volume.

In the frame work of the proposed method (Golovneva, 2007), the pressure P_x equals the external pressure, which is the parameter under control. The nondimensional value was used as the characteristics of volume:

$$x_t = 1 - \frac{V}{V_0}. \tag{44}$$

Here V_0 is the volume of cooled cluster without external pressure at the preset silver concentration. The magnitude of the value was taken on the asymptotic after static equilibrium fixing. The calculation results are presented in the Fig.13. The significant influence of the admixtures and abnormal behavior of the dependence (a sharp bend) are seen when the concentration of the silver atoms in the nanoclusters reaches the silver solubility limit in copper (8%).

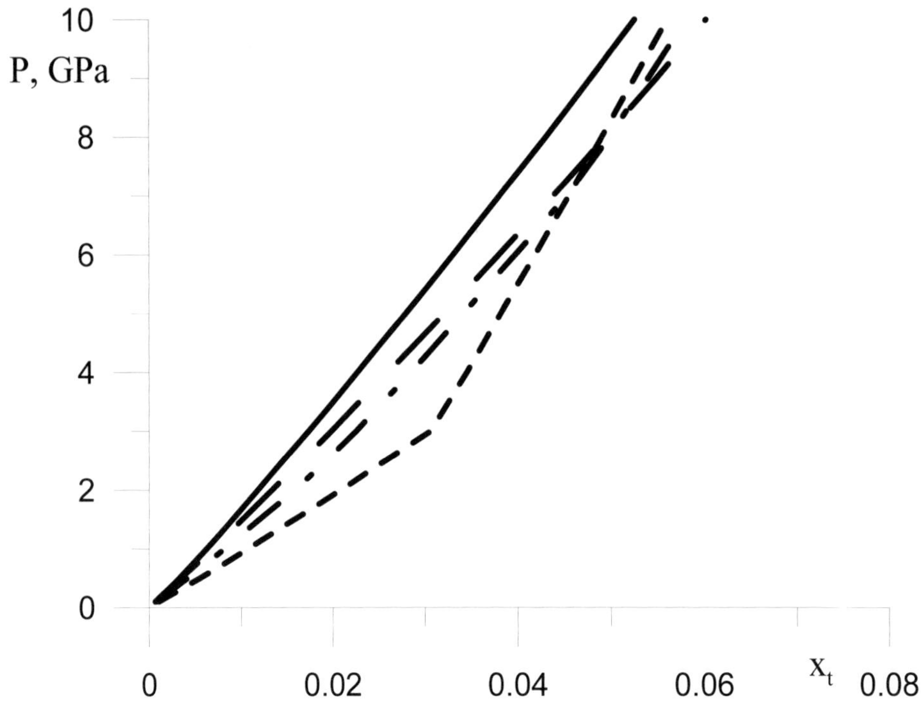

Fig. 13. Pressure vs the relative change of the Cu nanoclusters volume for various concentrations of silver. Solid line - pure copper; dashed line - alloy Cu99%Ag1%, chain line - alloy Cu98%Ag4%, dotted line- Cu92%Ag8%

For greater visualization below we present the dependence of the internal energy variation:

$$\Delta U_{tx} = U_{tx} - U_0 . \tag{45}$$

On the relative volume x_t . Here U_0 is the potential energy of cooled cluster without external pressure at the preset silver concentration. Likewise as in the previous case, the energy value U_{tx} and relative change of volume were taken in the equilibrium state at a certain pressure. The results are shown in Fig.14. Both for pure copper and for small silver concentrations, the dependence approaches to a parabolic one. Increasing concentration of silver causes the reduction of value ΔU_{tx} . However at the limit silver concentration in the alloy CuAg of 8%, the anomalous sharp bend and abrupt energy growth are observed.

To explain this phenomenon, the radial distribution function (RDF) of copper atoms in respect to silver atoms was studied. Fig.15-a presents the RDF with 2%silver for the case of no external pressure (solid line) and for the pressure of 10 GPa (dotted line). Evident that both the peaks size and shape are well preserved, only their centers are slight shifted. At the same time, at 8% silver (Fig.15-b), we see the considerable broadening of the lines and centers shift. In return, it brings many new atoms in the interaction sphere (its radius is 4.96 Å (the fourth peak)), change of the dependence $\Delta U_{tx}(x_t)$, and as consequence, the function $P_x(x_t)$, since $P_x = \frac{dU_{tx}}{dV}$.

7.3 Calculation of caloric and thermal equations of state for the nanoclusters

The nano-sized spherical alloy was heated in the same way as in the pure copper case (see paragraph 6).

The obtained results enable to construct the equation of state in the Mie-Grüneisen form, which is written for the continuum mechanics as:

$$P - P_x = \gamma \frac{3NkT}{V}. \tag{46}$$

or through the internal energy

$$P = \gamma \frac{(E_{in} - U_{tx})}{V}, \tag{47}$$

where E_{in} is the total internal energy of the cluster.

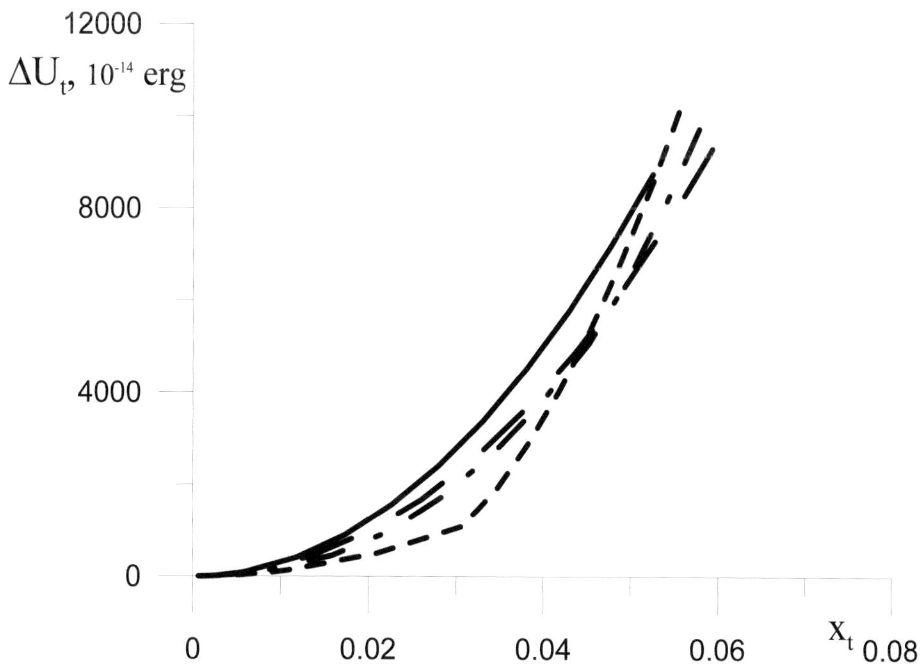

Fig. 14. Relative potential energy vs the relative change of the Cu nanoclusters volume for various concentrations of silver. Solid line - pure copper; dashed line - alloy Cu99%Ag1%, chain line - alloy Cu98%Ag4%, dotted line - Cu92%Ag8%

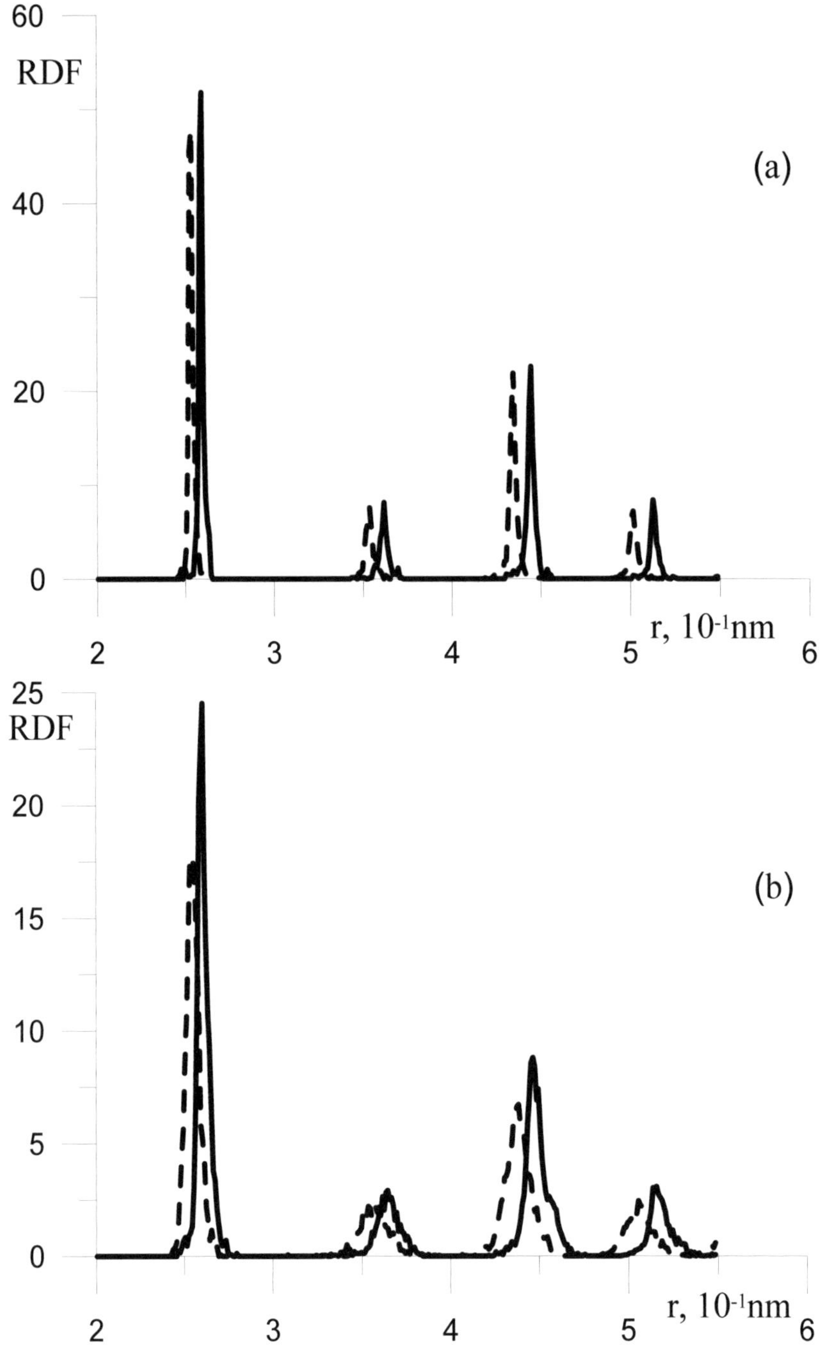

Fig. 15. Radial function of distribution (RDF) of copper atoms in respect to silver atoms regarding the distance at the cold nanocluster compression. Solid line – no external pressure, dotted line – the external prersure is 10 GPa. a) silver atoms concentration 2%; b) silver atoms concentration 8%

It should be noted that in the modeled process, the volume of the total system retains constant during the heating, i.e. the pressure change is caused by the temperature increase alone. As an example, Fig. 16 presents the dependence $\Delta P = P - P_x$ on the temperature for the external pressure of 1 GPa.

The obtained set of such curves for the initial compression pressures from 0.1 to 10 GPa dictates the thermal state of the nanostructure $P(V,T)$.

The caloric equation determining the total internal energy E_{in} which is the temperature and volume function was calculated as a sum of ΔU and kinetic energy of the chaotic motion of atoms. The numerical dependence of the relative total internal energy $\Delta E_{in} = E_{in} - E_{in0}$ on the temperature for various silver concentrations was found. It is important that the inclination angle of nanostructure thermal capacity do not change as the silver concentration varies until it reaches 8%, and coincides with the classical Dulong-Petit law.

The obtained results completely describe the thermodynamic properties of the nanoclusters from pure copper and copper-silver alloy. But frequently it is convenient to use the analytical expression (46), with the complicated dependence of the processes on the volume and temperature being transmitted into the Grüneisen constant. Fig.17 shows such a dependence for the case of the initial external pressure of 1 GPa.

It should be stressed (refer above), that the results were obtained in the context of classical mechanics and can be used for the temperature above the Debye temperature. For comparison, the experimental value of the Grüneisen constant for macroscopic samples of copper is 1.84, for macroscopic samples of silver – 2.46 (22 Grigor'ev & Meilikhov, 1991).

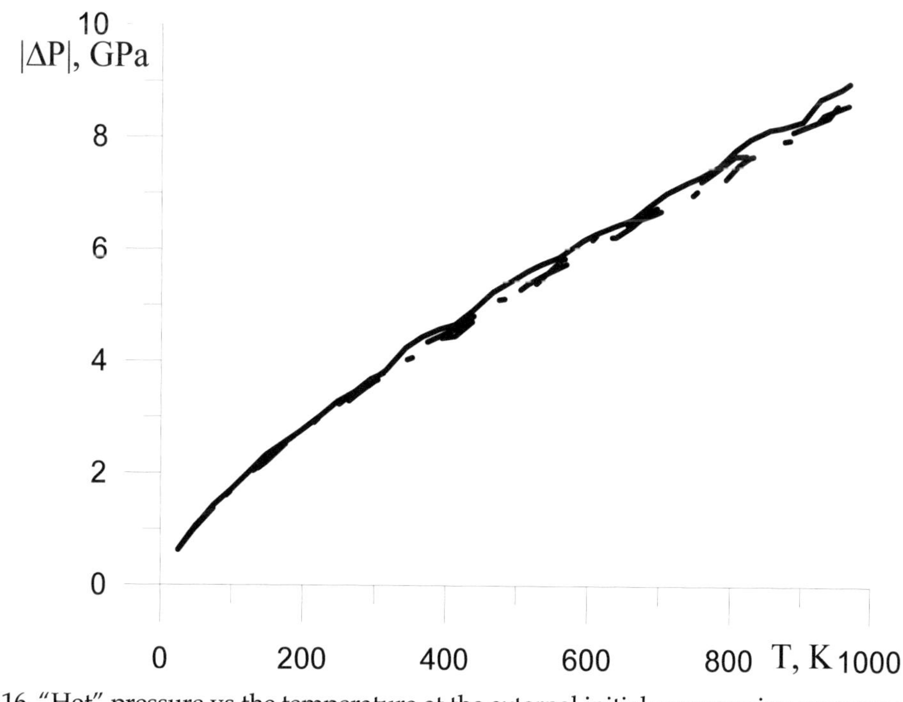

Fig. 16. "Hot" pressure vs the temperature at the external initial compressing pressure of 1 GPa. Solid line – pure copper; dashed line – alloy Cu98%Ag2%, chain line – alloy Cu96%Ag4%, dotted line – Cu92%Ag8%

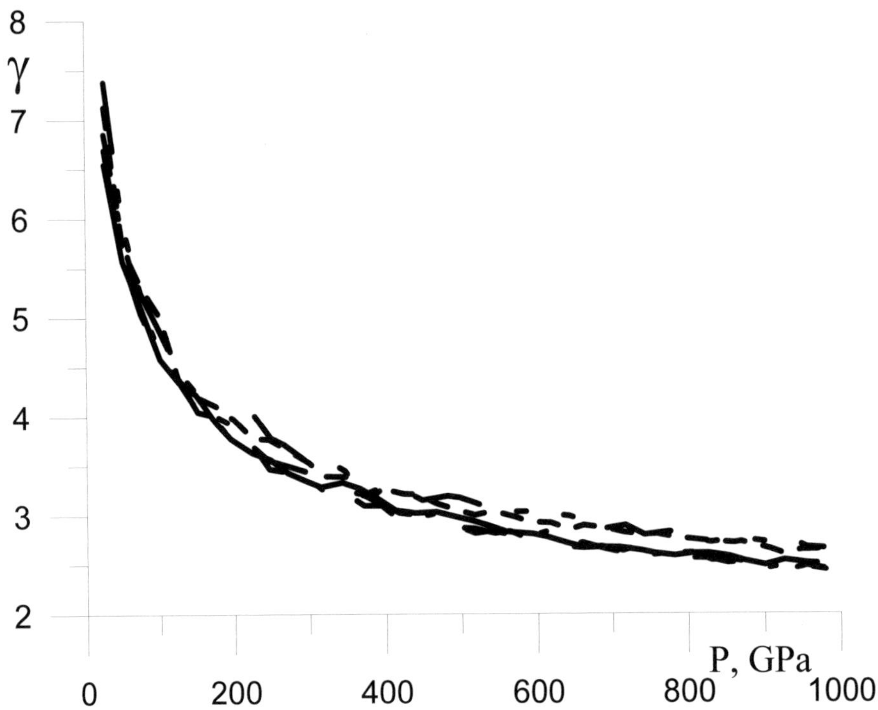

Fig. 17. Grüneisen constant vs the temperature at the external initial compressing pressure of 1 GPa. Solid line - pure copper; dashed line - alloy Cu98%Ag2%, chain line - alloy Cu98%Ag4%, dotted line - Cu92%Ag8%

8. Conclusion

In summary, the procedure of calculation of the thermodynamic properties of nanostructures *ab initio* is proposed and tested for the case of spherical copper clusters.

The study demonstrated that the results for the volume part of the system coincide with the results for macro-subjects. The difference between nanostructure and macrobodies properties results from the significant influence of the surface atoms in the nanostructures.

The proposed method enables to calculate the thermomechanical properties of the macrobodies by means of similar calculations for the microstructures when the influence of the surface atoms can be ignored.

Moreover, the developed method of the calculation of the thermodynamic properties by the molecular-dynamic method *ab initio* was tested on the nano-alloy Cu_xAg_{1-x} shaped as spherical clusters. Furthermore, a program system was produced which permits to calculate the thermomechanical properties of any alloy with known inter-atomic interaction potentialities.

9. Acknowledgments

The work is supported by RFBR №08-01-00585-a "Molecular dynamic study of surfaces metallization" and grant of RF President MK-631.2009.1.

10. References

Alekseev, D.; Saletskii, A. & Stepanyuk, O. (2008). Melting of Copper Nanoclusters on a (100) Copper Surface. *Moscow University Physics Bulletin,* Vol.63, No.2, 137-139, ISSN 0027-1349

Allen, M.; Tildesley, D. (1987). *Computer Simulation of Liquids,* Oxford University Press, 019855645, New York

Berroukche, A.; Soudini, B.; Amara, K. (2008). Molecular dynamics simulation study of structural, elastic and thermodynamic properties of tin below 286K. *International Journal of Nanoelectronics and Materials,* No.1, 41-51, ISSN 1985-5761

Bolesta, A.; Golovnev, I.; Fomin, V. (2001). Contact melting of nickel cluster at collision with rigid wall, *Physical Mesomechanics,* Vol.4, No.1, 5-10, ISSN 1029-9599

Bolesta, A.; Golovnev, I.; Fomin, V. (2002). Molecular dynamic modeling of Al/Ni composition quasi-static tension along the interface, *Physical Mesomechanics,* Vol.5, No.3-4, 101-107, ISSN 1029-9599

Bolesta, A.; Golovnev, I.; Fomin, V. (2006). InGaAs/GaAs nanotubes simulation: Comparison between continual and molecular dynamics approaches. *Computational Materials Science,* No.36, 147-151, ISSN 0927-0256

Born, M., Huang, K. (1954) *Dynamical Theory of Crystal Lattices,* Clarendon Press, ISBN Oxford, England, ISBN 0-19-814763-5

Chen, J.; Chen, D.; Zhang, J. (2007). Molecular dynamics simulation of thermodynamic properties of YAG. *Chines Physics,* Vol.16, No.9, 2779-2785, ISSN 1009-1963

Golovnev, I.; Golovneva, E.; Fomin, V. (2003). Simulation of quasi-static processes in the crystals by molecular dynamics method. *Physical mesomechanics,* Vol.6, No.6, 5-10, ISSN 1029-9599

a Golovnev, I.; Golovneva, E.; Fomin, V. (2006). Molecular-dynamic modeling of mechanical properties of free defect metal nanocrystals. *Computational Materials Science,* No.37, 336-348, ISSN 0927-0256

b Golovnev, I.; Golovneva, E.; Fomin, V. (2006). The influence of a nanocrystal size on the results of molecular-dynamics modelling. *Computational Materials Science,* No.36, 176-179, ISSN 0927-0256

Golovneva, E.; Golovnev, I.; Fomin, V. (2007). The calculation of thermodynamic properties of nanostructure by molecular dynamics method. *Physical Mesomechanics,* Vol.10, No.5, 11-16, ISSN 1029-9599

Grigor'ev, I.; & Meilikhov, E. (1991). *Physical Values. Handbook,* Moscow. Energoatomizdat Publisher House, ISBN 978-5-2830-0842-4

Gruneisen, E. (1926). *Handbuch der Physik,* Berlin: Springer Verlag, Br. 10, 7-59, ISBN (not found)

Mie, G. (1903). Zur kinetischen. Theorie der einatomigen Korper. *Annals of Physics,* No.11, 657, ISSN 978-5-2830-0842-4

Sekkal, W.; Bouhafs, B.; Aourag, H.; Certier, M. (1998). Molecular-dynamics simulation of structural and thermodynamic properties of boron nitride. Journal of Physics: Condensed Matter, No.10, 4975-4984, ISSN 0953-8984

Stepanyuk, O.; Alekseev, D.; Saletskii, A. (2009). Calculation of the Thermodynamic Properties of Copper by Molecular Dynamics Simulation. *Moscow University Physics Bulletin,* Vol.64, No.2, 226-227, ISSN 0027-1349

Voter, A.; Chen, S. (1987). Accurate interatomic potentials for Ni, Al, and Ni_3Al, *Materials Research Society Symposium Proceedings*, Vol. 82., 175-180, ISSN 02729172

Voter, A. (1993). Embedded Atom Method Potentials for Seven FCC Metals: Ni, Pd, Pt, Cu, Ag, Au, and Al. *Los Alamos Unclassified Technical Report LA-UR-93-3901*. (1993). ISSN (does not exist)

Won, H.; Ho, J. (2003). Strustural and thermodynamic properties of GaN: a molecular dynamics simulation. *Physics Letters A*, No.315, 319-324, ISSN 0375-9601

Zimmermann, J. (1999). Continuum and Atomistic Modeling of Dislocation Nucleation at Crystal Surface Ledges. *PhD thesis*, Stanford University, ISSN (does not exist)

Zimmerman, J.; Bammann, D.; Gao, H. (2009). Deformation gradients for continuum mechanical analysis of atomistic simulations. *International Journal of Solids and Structures*, No.46, 238–253, ISSN 0020-7683